Advanced Structured Materials

Volume 108

Common engineering materials reach in many applications their limits and new developments are required to fulfil increasing demands on engineering materials. The performance of materials can be increased by combining different materials to achieve better properties than a single constituent or by shaping the material or constituents in a specific structure. The interaction between material and structure may arise on different length scales, such as micro-, meso- or macroscale, and offers possible applications in quite diverse fields.

This book series addresses the fundamental relationship between materials and their structure on the overall properties (e.g. mechanical, thermal, chemical or magnetic etc) and applications.

The topics of *Advanced Structured Materials* include but are not limited to

- classical fibre-reinforced composites (e.g. glass, carbon or Aramid reinforced plastics)
- metal matrix composites (MMCs)
- micro porous composites
- micro channel materials
- multilayered materials
- cellular materials (e.g., metallic or polymer foams, sponges, hollow sphere structures)
- porous materials
- truss structures
- nanocomposite materials
- biomaterials
- nanoporous metals
- concrete
- coated materials
- smart materials

Advanced Structured Materials is indexed in Google Scholar and Scopus.

More information about this series at http://www.springer.com/series/8611

Bilen Emek Abali · Holm Altenbach ·
Francesco dell'Isola · Victor A. Eremeyev ·
Andreas Öchsner
Editors

New Achievements in Continuum Mechanics and Thermodynamics

A Tribute to Wolfgang H. Müller

Springer

Editors
Bilen Emek Abali
Institute of Mechanics
Technische Universität Berlin
Berlin, Germany

Holm Altenbach
Institut für Mechanik
Otto-von-Guericke-University Magdeburg
Magdeburg, Germany

Francesco dell'Isola
Dipartimento di Ingegneria Strutturale e Geotecnica
Università degli Studi di Roma "La Sapienza"
Roma, Italy

Victor A. Eremeyev
Faculty of Civil and Environmental Engineering
Gdańsk University of Technology
Gdańsk, Poland

Andreas Öchsner
Fakultät Maschinenbau
Esslingen University of Applied Sciences
Esslingen, Germany

ISSN 1869-8433 ISSN 1869-8441 (electronic)
Advanced Structured Materials
ISBN 978-3-030-13306-1 ISBN 978-3-030-13307-8 (eBook)
https://doi.org/10.1007/978-3-030-13307-8

Library of Congress Control Number: 2019934534

This Springer imprint is published by the registered company Springer Nature Switzerland AG
The registered company address is: Gewerbestrasse 11, 6330 Cham, Switzerland

Laudatio

A tale about Wolfgang Müller, the man and the scientist.

Wolfgang Müller came to our thermodynamics group at the Technical University Berlin with the very best possible qualifications. He arrived with an excellent diploma in theoretical physics and he was also an expert in scientific Donaldism, the study of Donald Duck and his fictional family, and associates. In the latter field he is so erudite that he can distinguish the comic figures by who drew them, all the way down from the Good Duck Artist Carl Barks in the 1930s to the present times. Thus Wolfgang Müller discovered that the second law of thermodynamics does not apply to Ducksburg, Entenhausen. Indeed when uncle Donald once shredded a treasure map and threw the shreds into a river, they reassembled miraculously downstream before the eyes of Donald's antagonist Gladstone Gander—in clear violation of the entropy principle.

Recommended by that type of background Wolfgang Müller became a welcome research assistant in our group of thermodynamicists. His assigned task was the quantitative description of *transformation toughening* in ceramics, a phenomenon that occurs in zirconia with alumina inclusions. A crack could conceivably be stopped—or its advance slowed down—when the stress field around the crack tip induced a phase transition in the alumina inclusions. Wolfgang Müller solved that problem most competently, and his work may have contributed to the recognition that the development of a tough, ductile—not brittle—ceramic material is impossible on that basis, although at the time the toughening effect was advertised by the pundits for just that purpose.

It is usually thought—at least professors like to think this—that research assistants are called assistants, because their advisors assist them. Whatever truth there is in this prejudice generally, it was not true in Wolfgang Müller's case. In fact the opposite was true: He assisted *us* and expanded our horizon. Thus he introduced the group to the study of the voluminous memoir of Muskhelishvili and the complex algebra of linear elasticity in plane strain and plane stress, which he used to advantage in completing his doctoral work. The successful solution of the problem gained Wolfgang Müller not only a doctorate "summa cum laude" but also the prestigious Tiburtius award of the Senate of Berlin for the best dissertation of the year. Also he was invited to present his work at the international conference ZIRCONIA in Paris, where his presentation received the Jean Mandel award.

Wolfgang Müller's work ethos was exceptional, nearly autistic. He spent his time working in the institute from early morning to late at night, occasionally interrupted only by a visit to the opera or the cinema, his great passions apart from work. And although his hard work was rewarded by successful research—albeit theoretical and mathematical research—he was not satisfied. He thought that industry was the right place for him to do "real work," and so he moved to Munich to work for Siemens. That was a mistake: In Munich he was put into a dark room to work on CHILL, a telephone exchange program in which Siemens was interested at the time. That was not intellectually very challenging work for a gifted young scientist. Moreover, he got into trouble with the union representatives at Siemens. Because, indeed, after a few months' working at his accustomed rate he had accumulated a vast amount of overtime, and he was asked to slow down. So, in order to keep busy during the enforced "extra-Siemens-activities," he turned his attention to the tax laws. Actually he learned them by heart, as it were.

In that situation we were able to entice him back—for a while—into academia, again as a research assistant. At lunch hour he instructed us on deductible expense items on the annual tax return. For instance, if you paid for postage stamps for career-relevant correspondence, you should make sure to obtain a receipt. We saw him diligently collecting receipts from the post office stamp machines. And rumour has it that he also gathered surplus receipts which the machine spat out when previous users had neglected to procure *their* receipts. That rumour is not true, of course, it was however easy to believe by some because of Wolfgang Müller's background in Donaldism.

Anyway his second stint in our group lasted two years, or so, and he chose to work on thermally induced internal stresses. He built a device to make the stresses visible under polarized light in transparent polymer plates by stress optics, and he interpreted his findings on cracks and crack propagation successfully. Also at that time he was introduced to non-linear continuum mechanics, assisting in building up a course on the subject. That period ended when Wolfgang Müller received a post-doctoral stipend by the Max Kade Foundation to go to the United States for a year.

He chose Santa Barbara where he joined the group of Professor Evans, an eminent scientist in the field of ceramics. We do not know much about that stay, but it seems that, perhaps, Evans was a mite too eminent; he was mostly travelling so that he could rarely be found in his office. Therefore there was not much assistance either way. Wolfgang Müller proceeded north by a few hundred miles and attached himself to George Hermann's group at Stanford. But at that time his quest for "real work" had not yet left him and—after the stipend expired—he joined a private firm called Failure Analysis. In their employ he had to crawl under the monstrous American trucks in order to inspect their transmission shafts for signs of fatigue. This involved bribing truck drivers at service stations so that they would tolerate the inspection; he used chewing gum and chocolate bars for the purpose.

Once again discouraged by "real work" he returned to Germany to a position in Paderborn which gave him the opportunity for habilitation and to become a Privatdozent in the Institute of Mechanics. What qualified him for the position was his earlier work on thermal stresses, and his habilitation thesis was in that field too: Thermal stresses in composites.

After that he became a lecturer at the Heriot-Watt University in Edinburgh and he quickly advanced there from lecturer to reader to professor. That was quite a feat, because in Britain not every lecturer becomes a reader and not every reader becomes a full professor; and certainly not as quickly as Wolfgang Müller did. He taught non-linear continuum mechanics at Heriot-Watt and we flatter ourselves that, perhaps, he learned about the subject thoroughly in Berlin.

Now, success in the academic field very often creates some mild idiosyncrasies in a person; and Wolfgang Müller was not immune to this phenomenon. He brought two cars to Edinburgh, both identical BMWs except that one was red and the other one black. One had a Paderborn licence plate and one—from his time in the US—had a California licence plate. And he used those cars on alternate days to drive from his home to the university. That drew some attention. Also in the hat-less decades of the 20th century he affected a hat; not just any old hat to keep his head dry from the Scottish rain, but a veritable black Borsalino, the broad-rimmed variety. We are told that among the students he was known as the Professor in a Borsalino who changed BMWs on a daily basis.

However, still Wolfgang Müller's wanderlust was not yet satisfied, not quite. He closed the cycle of his travels when he came back to the TU Berlin to accept a chair of continuum mechanics which he now holds. It is true, the cycle was now closed, geographically, on the surface as it were. Below the surface there were internal permanent deformations. Indeed, up to his move back to the TUB one could consider Wolfgang Müller a comet who orbited around our group of thermodynamicists and made his appearance every few years. That constellation underwent a subtle metamorphosis after his latest return to Berlin: In a manner totally inexplicable by Newtonian mechanics the comet morphed into the central star and we—the authors of this laudatio—became the comets. As such both of us were co-authors of two of Wolfgang Müller's books since his return to Berlin.

That situation must have pleased Wolfgang Müller, at least to some extent. To be sure, he put out feeble feelers toward a position on the southern hemisphere, but his heart was not in them and so they came to nothing. What may have contributed to his growing sense of contentment was the fact that the internal dichotomy of Wolfgang Müller's soul was now finally resolved. His quest for a reconciliation of "real work" and intellectual work, which had plagued him throughout his professional life was finally given the proper satisfactory interpretation: The "real work" is now done in the basement of his institute by the nano-indenters, the atomic force microscopes, the computers, and his co-workers who operate these machines. And the thinking is done on the first floor by the boss in a luxurious office behind a door adorned with pages from Tim and Struppi and Donald Duck.

Successful professors travel a lot and so does Wolfgang Müller. More often than not when we try to reach him, we are told that he is away. Favourite destinations are St. Petersburg and Istanbul, with Kiew and Singapoor vying for second place. He assists people there in their research work and is assisted by them in his work. That is ok. No! More than ok, it is essential, because homebound research tends to become sterile rather quickly; and after all, science is a global movement.

Once again, successful professors travel a lot, and sometimes, occasionally, a tiny little bit like members of the IOC, the International Olympic Committee. Thus Wolfgang Müller is a member of a committee for the evaluation of equivalent curricula between universities abroad. This organisation regularly provides outings of mutual recognition and world-wide sightseeing to its members. Thus he travels the world: from the deepest pit of a South African diamond mine all the way up to the highest Andean height of Peru in Machu Pichu.

His life-long ambition, however, is a trip to the moon. Will that be forthcoming? Wolfgang Müller is sixty now. He can look back to half a lifetime as a period of successful learning, teaching and research. And we wish him continued success in these activities—and in his quest for the moon.

Berlin, April 2019 *Ingo Müller & Wolf Weiss*

Preface

Professor Wolfgang H. Müller is one of the leading German scientist in continuum mechanics and constitutive theory contributing to research in several areas including

- Solid body mechanics: in nonlinear buckling in microsystem technologies,
- Experimental mechanics: including nanoindentation, atomic force microscope (AFM), and Raman spectroscopy,
- Generalized mechanics: theory and computation of higher order models incorporating inner substructure,
- Fatigue related damage mechanics: thermomechanical modeling and computation in microelectronics,
- Geomechanics: deformation in self-gravitating planets,
- Polar medium: theory and analytic solution in polar fluids like liquid crystals,
- Diffusion phenomena: modeling of drug delivery in nanopharmaceuticals.

This volume of the Advanced Structured Materials Series is dedicated to his sixtieth birthday; it contains a selection of manuscripts prepared by his friends and colleagues from several countries in many continents such as China, Germany, Japan, Italy, Poland, Russia, Scotland UK, and the USA.

Professor Müller was born on April 13, 1959. After studying physics in Technische Universität Berlin, he finalized his Diploma (equivalent to M.Sc.) in March 1984, his PhD in September 1986, his Habilitation in January 1997. He worked in different countries for example in Siemens (Munich) and in Failure Analysis Associates (San Francisco). He has been working in various research institutions for example at the Paderborn University (Germany), University of California Santa Barbara (USA), Heriot-Watt-University (Edinburgh UK), where he was appointed (full) professor of mechanical engineering in 1999. Since 2001, he has been appointed (full) professor of Continuum Mechanics and Constitutive Theory at the TU Berlin.

W. H. Müller has been honored by several awards like Jean-Mandel Award in CNRS Paris (1985), Joachim Tiburtius Prize (1987), Max-Kade Foundation Postdoctoral Award (1989), Best Paper Awards (1997 Surface Mount International, 2010 EPTC Singapore), and Distinguished Visiting Fellowship of the Royal Academy of Engineering London (2008).

W. H. Müller has been working as the managing editor of the highly-ranked journal in Springer: Continuum Mechanics and Thermodynamics. With an impressive scientific activity,[1] W. H. Müller has contributed to more than 200 manuscripts leading to an h-index greater than 20. Moreover, he has published the following monographs and textbooks:

- Müller, W. H., & Weiss, W. (2016). The State of Deformation in Earthlike Self-Gravitating Objects. SpringerBriefs in Continuum Mechanics. Springer International Publishing.
- Grigorenko, A. Y., Müller, W. H., Grigorenko, Y. M., & Vlaikov, G. G. (2016). Recent Developments in Anisotropic Heterogeneous Shell Theory: General Theory and Applications of Classical Theory (Vol. 1). SpringerBriefs in Continuum Mechanics. Springer International Publishing.
- Grigorenko, A. Y., Müller, W. H., Grigorenko, Y. M., & Vlaikov, G. G. (2016). Recent Developments in Anisotropic Heterogeneous Shell Theory: Applications of Refined and Three-dimensional Theory (Vol. 2). SpringerBriefs in Continuum Mechanics. Springer International Publishing.
- Müller, W. H., & Ferber, F. (2015). Übungsaufgaben zur technischen Mechanik. Carl Hanser Verlag, Munich, Vienna
- Müller, W. H. (2014). An expedition to continuum theory. Solid Mechanics and Its Applications book series vol. 210, Springer, Dordrecht
- Müller, W. H. (2011). Streifzüge durch die Kontinuumstheorie, Springer, Berlin, Heidelberg
- Müller, I., & Müller, W. H. (2009). Fundamentals of thermodynamics and applications: with historical annotations and many citations from Avogadro to Zermelo. Springer, Berlin, Heidelberg
- Müller, W. H., & Ferber, F. (2008). Technische Mechanik für Ingenieure. Carl Hanser Verlag, Munich

Wolfgang Müller dazzles us with his never ending passion about exploring and teaching more and more. We like to learn from him and enjoy advancing in knowledge with him.

Since 2001, he has been advisor or examiner in more than 35 doctoral thesis.[2] Thousands of bachelor and master students around the globe have been experiencing his wonderful lectures in diverse topics. Many universities are inviting him as an instructor for short and long terms; here is a probably incomplete list of cities, where W. H. Müller has been lecturing: Auckland (New Zealand), Edinburgh (Scotland UK), Gdansk (Poland), Kyoto (Japan), Istanbul (Turkey), Melbourne and Sydney (Australia), Paderborn and Berlin (Germany), Singapore, Sydney (Australia), St. Petersburg (Russia), Tbilisi and Kutaisi (Georgia).

In Berlin, since 2001 he has been sponsoring several courses in statics and mechanics of materials, kinematics and dynamics, continuum mechanics, energy methods, finite element method, tensor calculus, continuum physics—always well pre-

[1] According to https://scholar.google.com/citations?user=rGyO6boAAAAJ

[2] www.lkm.tu-berlin.de/menue/dissertationen/

pared, structured, didactic, and simply crystal clear lectures—in front of 1200 freshmen and then within the next hour for 10 master students. His enthusiasm as an instructor is beyond one's imagination, just to give a better feeling, we share one of his lectures continuing on the garden as the lecturing room was occupied in students' strike in July 2006.

As editors, we intend to thank all authors for their crucial contributions as well as all reviewers for their invaluable time and effort. We delightedly acknowledge Dr. Christoph Baumann (Springer Publisher) for support of the book project.

Berlin, Magdeburg, Rome, Gdansk, Esslingen,
April 2019

Bilen Emek Abali
Holm Altenbach
Francesco dell'Isola
Victor Eremeyev
Andreas Öchsner

Contents

List of Contributors

Bilen Emek Abali
Chair of Continuum Mechanics and Constitutive Theory, Institute of Mechanics, Technische Universität Berlin, Einsteinufer 5, 10587 Berlin, Germany,
e-mail: bilenemek@abali.org

Rafael Abreu
University of Münster, Institut für Geophysik, Corrensstr. 24, 48149 Münster, Germany,
e-mail: abreu@uni-muenster.de

Ekaterina Alekseeva
Peter the Great St. Petersburg Polytechnic University, Polytechnicheskaya, 29, St. Petersburg 195294, Russia,
e-mail: alexeeva__ekaterina@mail.ru

Holm Altenbach
Institut für Mechanik, Fakultät für Maschinenbau, Otto-von-Guericke-Universität Magdeburg, Universitätsplatz 2, 39106 Magdeburg, Germany,
e-mail: holm.altenbach@ovgu.de

Faris Alzahrani
NAAM Research Group, Department of Mathematics, King Abdulaziz University, Jeddah 21589, Saudi Arabia,
e-mail: faris.kau@hotmail.com

Ivan Argatov
Department of Materials Science and Applied Mathematics, Malmö University, SE-205 06 Malmö, Sweden,
e-mail: ivan.argatov@gmail.com

Emilio Barchiesi
Università degli Studi di Roma "La Sapienza" Dipartimento di Ingegneria Strutturale e Geotecnica, Via Eudossiana 18, 00184 Rome & International Research Center for the Mathematics and Mechanics of Complex Systems - MeMoCS, Università degli studi dell'Aquila, L'Aquila, Italy,
e-mail: BarchiesiEmilio@gmail.com

Thomas Böhme
Trelleborg Antivibration Solutions Germany, Engineering Off-Highway and Industry, Berliner Str. 17, 16727 Velten (b) Berlin, Germany,
e-mail: thomas.boehme@trelleborg.com

Anatoly M. Bragov
Research Institute for Mechanics, National Research Lobachevsky State University of Nizhny Novgorod, Russian Federation,
e-mail: bragov@mech.unn.ru

Belal M. Dawwas
Chair of Continuum Mechanics and Constitutive Theory, Institute of Mechanics, Technische Universität Berlin, Einsteinufer 5, 10587 Berlin, Germany,
e-mail: belal.m.dawwas@campus.tu-berlin.de

Francesco dell'Isola
Research Institute for Mechanics, National Research Lobachevsky State University of Nizhny Novgorod, Russian Federation,
e-mail: fdellisola@gmail.com

Fabio Di Cosmo
Dipartimento di Ingegneria Civile, Edile, Ambientale e Architettura Università degli Studi dell'Aquila, L'Aquila, Italy & International Research Center M&MoCS, Università degli Studi dell'Aquila. Via Giovanni Gronchi 18 - Zona industriale di Pile, 67100 L'Aquila, Italy,
e-mail: fabiodicosmo@gmail.com

Rafał Drobnicki
Department of Machinery Design and Biomedical Engineering, Warsaw University of Technology, Warsaw, Poland & Dipartimento di Ingegneria e Scienze dell'Informazione e Matematica, Università degli Studi dell'Aquila, Via Vetoio, Coppito, 67100 L'Aquila, Italy,
e-mail: draf1992@gmail.com

Victor A. Eremeyev
Faculty of Civil and Environmental Engineering, Gdańsk University of Technology, ul. Gabriela Narutowicza 11/12, 80-233 Gdańsk, Poland,
e-mail: victor.eremeev@pg.edu.pl
Southern Federal University, Milchakova str. 8a, 344090 & Southern Scientific Center of RASci, Chekhova str. 41, 344006 Rostov on Don, Russia,
e-mail: eremeyev.victor@gmail.com

Vladimir I. Erofeev
Research Institute for Mechanics, National Research Lobachevsky State University & Mechanical Engineering Research Institute of Russian Academy of Sciences, Nizhny Novgorod, Russia,
e-mail: erof.vi@yandex.ru

Simon R. Eugster
Institute for Nonlinear Mechanics, University of Stuttgart, Germany,
e-mail: simon-raphael.eugster@inm.uni-stuttgart.de

Fabian Forsbach
Stability and Failure of Functionally Optimized Structures Group, Institute of Mechanics, Technische Universität Berlin, Einsteinufer 5, 10587 Berlin, Germany,
e-mail: fabianfo@mailbox.tu-berlin.de

Larisa Yu. Frolenkova
Orel State University named after. I.S. Turgenev, 29 Naugorskoe Shosse, 302020 Orel, Russia,
e-mail: Larafrolenkova@yandex.ru

Ksenia Frolova
Institute for Problems in Mechanical Engineering RAS, Bolshoy pr., 61, V.O., St. Petersburg 199178 & Peter the Great St. Petersburg Polytechnic University, Polytechnicheskaya, 29, St. Petersburg 195294, Russia,
e-mail: kspfrolova@gmail.com

Gregor Ganzosch
Chair of Continuum Mechanics and Constitutive Theory, Institute of Mechanics, Technische Universität Berlin, Einsteinufer 5, 10587 Berlin, Germany,
e-mail: ganzosch@tu-berlin.de

Dimitri V. Georgievskii
Mechanical and Mathematical Department, Moscow State University, Moscow 119991, Russia,
e-mail: georgiev@mech.math.msu.su

Ivan Giorgio
Department of Structural and Geotechnical Engineering, Università degli studi di Roma La Sapienza, 18 Via Eudossiana, Rome & International Research Center for the Mathematics and Mechanics of Complex Systems - MeMoCS, Università degli studi dell'Aquila, L'Aquila, Italy,
e-mail: ivan.giorgio@uniroma1.it

Sebastian Glane
Chair of Continuum Mechanics and Constitutive Theory, Institute of Mechanics, Technische Universität Berlin, Einsteinufer 5, 10587 Berlin, Germany,
e-mail: glane@tu-berlin.de

Evgeniya V. Goloveshkina
Institute of Mathematics, Mechanics and Computer Science of Southern Federal University, Milchakova Str. 8a, 344090 Rostov on Don, Russia,
e-mail: evgeniya.goloveshkina@yandex.ru

Vasilii A. Gorohov
Research Institute for Mechanics, National Research Lobachevsky State University of Nizhny Novgorod, Russian Federation,
e-mail: vas-gor@rambler.ru

Lena Goullon
Fraunhofer IZM, Gustav-Meyer-Allee 25, 13355 Berlin, Germany,
e-mail: Lena.Goullon@izm.fraunhofer.de Peter Göransson
KTH Royal Institute of Technology, Marcus Wallenberg Laboratory for Sound and Vibration Research, SE-100 44, Stockholm, Sweden. e-mail: pege@kth.se

Nils Gräbner
Department of Applied Mechanics, Chair of Mechatronics and Machine Dynamics, TU Berlin, 10587 Berlin, Germany,
e-mail: nils.graebner@tu-berlin.de

Mikhail A. Grekov
Saint Petersburg State University, 7/9, Universitetskaya nab., 199034, St. Petersburg, Russia,
e-mail: m.grekov@spbu.ru, magrekov@mail.ru

Elena F. Grekova
Institute of Problems of Mechanical Engineering, Russian Academy of Sciences, Bolshoy pr. V.O., 61, 199178, St. Petersburg, Russia,
e-mail: elgreco@pdmi.ras.ru

Alexander Ya. Grigorenko
S.P. Timoshenko Institute of Mechanics of NAS of Ukraine, Kiev, Ukraine,
e-mail: ayagrigorenko1991@gmail.com

Tasawar Hayat
Department of Mathematics, Quaid-I-Azam University, Islamabad, Pakistan & NAAM Research Group, Department of Mathematics, King Abdulaziz University, Jeddah 21589, Saudi Arabia,
e-mail: pensy_t@yahoo.com

Yongzhong Huo
Institute of Mechanics and Computational Engineering, Department of Aeronautics and Astronautics, Fudan University, Shanghai 200433, China,
e-mail: yzhuo@fudan.edu.cn

Matthias Hutter
Fraunhofer IZM, Gustav-Meyer-Allee 25, 13355 Berlin, Germany,
e-mail: Matthias.Hutter@izm.fraunhofer.de

Leonid A. Igumnov
Research Institute for Mechanics, National Research Lobachevsky State University of Nizhny Novgorod, Russian Federation,
e-mail: Igumnov@mech.unn.ru

Arion Juritza
Chair of Continuum Mechanics and Constitutive Theory, Institute of Mechanics, Technische Universität Berlin, Einsteinufer 5, 10587 Berlin, Germany,
e-mail: juritza@tu-berlin.de

Maria Kashtalyan
Centre for Micro-and Nanomechanics (CEMINACS), School of Engineering, University of Aberdeen, Fraser Noble Building, Aberdeen AB24 3UE, Scotland UK,
e-mail: m.kashtalyan@abdn.ac.uk Dmitri A. Kazakov
Research Institute for Mechanics, National Research Lobachevsky State University of Nizhny Novgorod, Russian Federation,
e-mail: kazakov@mech.unn.ru

Reinhold Kienzler
Bremen Institute for Mechanical Engineering (bime), Department of Production Engineering, University of Bremen, Am Biologischen Garten 2, 28359 Bremen, Germany,
e-mail: rkienzler@uni-bremen.de

Vitaly S. Kirilyuk
S.P. Timoshenko Institute of Mechanics of NASU, Nesterova 3, 03057 Kyiv, Ukraine,
e-mail: kirilyuk_v@ukr.net

Robin J. Knops
The Maxwell Institute of Mathematical Sciences and School of Mathematical and Computing Sciences, Heriot-Watt University, Edinburgh EH14 4AS, Scotland, UK,
e-mail: r.j.knops@hw.ac.uk

Anton Köllner
Technische Universität Berlin, Institute of Mechanics, Stability and Failure of Functionally Optimized Structures Group, Einsteinufer 5, 10587 Berlin, Germany,
e-mail: anton.koellner@tu-berlin.de

Daniil A. Kolesov
Mechanical Engineering Research Institute of Russian Academy of Sciences Nizhny Novgorod, Russia,
e-mail: alandess@yandex.ru

Alexander Yu. Konstantinov
Research Institute for Mechanics, National Research Lobachevsky State University of Nizhny Novgorod, Russian Federation,
e-mail: constantinov.al@yandex.ru

Ulrich Kroll
Institute of Mechanics, Department of Mechanical Engineering, University of Kassel, Mönchebergstraße 7, 34125 Kassel, Germany,
e-mail: ulrich.kroll@uni-kassel.de

Dmitry A. Lamzin
Research Institute for Mechanics, National Research Lobachevsky State University of Nizhny Novgorod, Russian Federation,
e-mail: lamzin.dmitry@yandex.ru,

Marco Laudato
Dipartimento di Ingegneria e Scienze dell'Informazione e Matematica, Università degli Studi dell'Aquila, Via Vetoio, Coppito, 67100 L'Aquila, Italy,
e-mail: marco.laudato@graduate.univaq.it

Olga I. Levchuk
S.P. Timoshenko Institute of Mechanics of NASU, Nesterova 3, 03057 Kyiv, Ukraine,
e-mail: 2013levchuk@gmail.com

Christian Liebold
Chair of Continuum Mechanics and Constitutive Theory, Institute of Mechanics, Technische Universität Berlin, Einsteinufer 5, 10587 Berlin, Germany,
e-mail: christian.liebold@mail.de

Svetlana Yu. Litvinchuk
Research Institute for Mechanics, National Research Lobachevsky State University of Nizhny Novgorod, Russian Federation,
e-mail: litvinchuk@mech.unn.ru

Andrey K. Lomunov
Research Institute for Mechanics, National Research Lobachevsky State University of Nizhny Novgorod, Russian Federation,
e-mail: lomunov@mech.unn.ru

Igor A. Loza
National Transport University, Kiev, Ukraine,
e-mail: lozaigor1956@gmail.com

Alexey O. Malkhanov
Mechanical Engineering Research Institute of Russian Academy of Sciences Nizhny Novgorod, Russia,
e-mail: alexey.malkhanov@gmail.com

Marin Marin
Department of Mathematics and Computer Science, Transilvania University of Brasov, 500093 Brasov, Romania,
e-mail: m.marin@unitbv.ro

Anton Matzenmiller
Institute of Mechanics, Department of Mechanical Engineering, University of Kassel, Mönchebergstraße 7, 34125 Kassel, Germany,
e-mail: amat@uni-kassel.de

Wolfgang Muschik
Institut für Theoretische Physik, Technische Universität Berlin, Hardenbergstr. 36, 10623 Berlin, Germany
e-mail: muschik@physik.tu-berlin.de

Ingo Müller
Technische Universität Berlin, Germany
e-mail: ingo.mueller@alumni.tu-berlin.de

Andreas Öchsner
Faculty of Mechanical Engineering, Esslingen University of Applied Sciences, 73728 Esslingen, Germany,
e-mail: andreas.oechsner@gmail.com

Giuseppe Pezzotti
Ceramic Physics Laboratory, Kyoto Institute of Technology, Sakyo-ku, Matsugasaki, Kyoto 606-8585, Japan, Department of Orthopedic Surgery, Tokyo Medical University, 6-7-1 Nishi-Shinjuku, Shinjuku-ku, 160-0023 Tokyo, Japan, The Center for Advanced Medical Engineering and Informatics, Osaka University, 2-2 Yamadaoka, Suita, Osaka 565-0854, Japan & Department of Immunology, Graduate School of Medical Science, Kyoto Prefectural University of Medicine, Kamigyo-ku, 465 Kajii-cho, Kyoto 602-8566, Japan,
e-mail: pezzotti@kit.ac.jp

Luca Placidi
International Telematic University Uninettuno, Faculty of Engineering, Corso Vittorio Emanuele II, 39, Rome, Italy,
e-mail: luca.placidi@uninettunounirsity.net

Vladimir Polyanskiy
Institute for Problems in Mechanical Engineering RAS, Bolshoy pr., 61, V.O., St. Petersburg 199178 & Peter the Great St. Petersburg Polytechnic University, Polytechnicheskaya, 29, St. Petersburg 195294, Russia,
e-mail: vapol@mail.ru

Valentin L. Popov
Department of System Dynamics and Friction Physics, Institute of Mechanics, Technische Universität Berlin, Straße des 17. Juni 135, 10623 Berlin, Germany,
e-mail: v.popov@tu-berlin.de

Victoria Yu. Presnetsova
Orel State University named after. I.S. Turgenev, 29 Naugorskoe Shosse, 302020 Orel, Russia,
e-mail: alluvian@mail.ru

Vadim M. Presniakov
Orel State University named after. I.S. Turgenev, 29 Naugorskoe Shosse, 302020 Orel, Russia,
e-mail: vadim1024@rambler.ru

Wilhelm Rickert
Chair of Continuum Mechanics and Constitutive Theory, Institute of Mechanics, Technische Universität Berlin, Einsteinufer 5, 10587 Berlin, Germany,
e-mail: rickert@tu-berlin.de

Sergey N. Romashin
Orel State University named after. I.S. Turgenev, 29 Naugorskoe Shosse, 302020 Orel, Russia,
e-mail: sromashin@yandex.ru Giuseppe Rosi
Université Paris-Est, Laboratoire Modélisation et Simulation Multi Echelle, MSME UMR 8208 CNRS, 61 av du Général de Gaulle, 94010 Créteil Cedex, France,
e-mail: giuseppe.rosi@u-pec.fr

Roberta Sburlati
Department of Civil, Chemical and Environmental Engineering (DICCA), Polytechnic School, University of Genova, Via Montallegro 1, 16145 Genova, Italy,
e-mail: roberta.sburlati@unige.it

Dominik Schmid
Chair of Mechatronics and Machine Dynamics, Institute of Mechanics, Technische Universität Berlin, Einsteinufer 5, 10587 Berlin, Germany,
e-mail: dominik.schmid@tu-berlin.de

Patrick Schneider
Institute for Lightweight Construction and Design (KLuB), Technische Universität Darmstadt, Otto-Berndt-Straße 2, 64287 Darmstadt, Germany,
e-mail: patrick.schneider@klub.tu-darmstadt.de

Martin Schneider-Ramelow
Technische Universität Berlin, Fachgebiet Werkstoffe der Hetero-Systemintegration, Gustav-Meyer-Allee 25 13355 Berlin, Germany,
e-mail: Martin.Schneider-Ramelow@izm.fraunhofer.de

Vladimir S. Shorkin
Orel State University named after. I.S. Turgenev, 29 Naugorskoe Shosse, 302020 Orel, Russia,
e-mail: vshorkin@yandex.ru

Emilio Turco
Department of Architecture, Design and Urban planning (DADU), University of Sassari and M&MOCS International Research Center, Italy,
e-mail: emilio.turco@uniss.it

Elena Vilchevskaya
Institute for Problems in Mechanical Engineering RAS, Bolshoy pr., 61, V.O., St. Petersburg 199178 & Peter the Great St. Petersburg Polytechnic University, Polytechnicheskaya, 29, St. Petersburg 195294, Russia,
e-mail: vilchevska@gmail.com

Sorin Vlase
Department of Mechanical Engineering, Transilvania University of Brasov, 500093 Brasov, Romania,
e-mail: svlase@unitbv.ro

Christina Völlmecke
Stability and Failure of Functionally Optimized Structures Group, Institute of Mechanics, Technische Universität Berlin, Einsteinufer 5, 10587 Berlin, Germany,
e-mail: christina.voellmecke@tu-berlin.de

Ivan A. Volkov
Research Institute for Mechanics, National Research Lobachevsky State University of Nizhny Novgorod, Russian Federation,
e-mail: pmptmvgavt@yandex.ru

Utz von Wagner
Chair of Mechatronics and Machine Dynamics, Institute of Mechanics, Technische Universität Berlin, Einsteinufer 5, 10587 Berlin, Germany,
e-mail: utz.vonwagner@tu-berlin.de

Constanze Weber
Fraunhofer IZM, Gustav-Meyer-Allee 25, 13355 Berlin, Germany,
e-mail: Constanze.Weber@izm.fraunhofer.de

Kerstin Weinberg
Universität Siegen, Lehrstuhl für Festkörpermechanik, Department Maschinenbau, Paul-Bonatz-Str. 9-11, 57076 Siegen,
e-mail: kerstin.weinberg@uni-siegen.de

Wolf Weiss
Technische Universität Berlin, Germany
e-mail: wolf.weiss@alumni.tu-berlin.de

Marek Werner
Universität Siegen, Lehrstuhl für Festkörpermechanik, Department Maschinenbau, Paul-Bonatz-Str. 9-11, 57076 Siegen,
e-mail: marek.werner@uni-siegen.de

Svetlana I. Yakushina
Orel State University, 29 Naugorskoe Shosse, 302020 Orel, Russia,
e-mail: jakushina@rambler.ru

Hua Yang
Chair of Continuum Mechanics and Constitutive Theory, Institute of Mechanics, Technische Universität Berlin, Einsteinufer 5, 10587 Berlin, Germany,
e-mail: hua.yang@campus.tu-berlin.de

Sergiy N. Yaremchenko
S.P. Timoshenko Institute of Mechanics of NAS of Ukraine, Kiev, Ukraine,
e-mail: yaremch@gmail.com

Yang Zhang
Institute of Mechanics and Computational Engineering, Department of Aeronautics and Astronautics, Fudan University, Shanghai 200433, China,
e-mail: 14110290008@fudan.edu.cn

Tarek I. Zohdi
Department of Mechanical Engineering, 6195 Etcheverry Hall, University of California, Berkeley, CA, 94720-1740, USA
e-mail: zohdi@berkeley.edu

Leonid M. Zubov
Institute of Mathematics, Mechanics and Computer Science of Southern Federal University, Milchakova Str. 8a, 344090 Rostov on Don, Russia,
e-mail: zubovl@yandex.ru

Chapter 1
Magnetorheological Elastomer's Material Modeling and Parameter Determination by Using the Energy-based Method

Bilen Emek Abali & Hua Yang

Abstract Functionalized materials provide tailored properties to design *smart* structures. For example, by adding polarized particles in a polymer, a composite material is generated, which couples deformation with electromagnetism. This *magnetorheological elastomer* (MRE) is a particle reinforced polymer matrix. Such a composite material deforms under an externally applied magnetic field so materials response is steered without contact. In order to achieve a simulation of an engineering design with MRE, we need an appropriate constitutive (material) equation modeling the deformation behavior accurately under different magnetic fields. We aim at determining the parameters in such a material equation out of experiments by using an inverse analysis. Although the material equation is nonlinear in deformation, its material parameters are mostly in such a way that we acquire a linear regression problem by using the energy-based method. Hence, the obtained parameters are unique and the method is fast allowing us to try out various material models. We present the method for determining the material parameters out of experimental data obtained by a standard rotational rheometer. The proposed material equation with its determined parameters can be used in a computation, for example by the finite element method.

Keywords: Magnetorheological elastomer · Material modeling · Inverse analysis · Rheometer

1.1 Introduction

Magnetorheological elastomers (MRE) are functionalized materials showing a varying stiffness as a consequence of the externally applied magnetic flux. MRE is a

Bilen Emek Abali · Hua Yang
Institute of Mechanics, Technische Universität Berlin, Germany
e-mail: bilenemek@abali.org, hua.yang@campus.tu-berlin.de

B. E. Abali et al. (eds.), *New Achievements in Continuum Mechanics and Thermodynamics*, Advanced Structured Materials 108,
https://doi.org/10.1007/978-3-030-13307-8_1

composite material of magnetic particles such as carbonyl iron particles suspended in a (non-polarized) matrix like polyurethane (PU), rubber, or elastomer, see Gong et al (2005), Park et al (2009), Li and Nakano (2013), Yu et al (2015) and An et al (2017). The externally applied magnetic flux generates a magnetic field within the material causing a motion of particles leading to eigenstress and stiffening of the composite material. We search for a homogenized material model performing this stiffening effect without modeling the iron particles and surface effects between particles and matrix. Effected by the nonlinear materials response of the matrix material, we consider a nonlinear homogenized model with material parameters, where their numerical values depend on the applied magnetic flux.

There are different approaches for defining material equations. We need to consider that the amount of particles in the polymer may change the microstructure completely. In the case of a relatively low amount of iron particles, we can assume that they fail to interact with each other such that the mechanical properties are affected mainly by the matrix. Increased number of particles may lead to agglomerating and even touching particles, thus, the characteristics of the composite material starts changing — we refer to Zohdi and Wriggers (2008) and Zohdi (2012) for a microstructural interpretation and a homogenization procedure in such a case. Herein, we consider a relatively low amount of iron particles embedded in a nonlinear matrix. As a consequence of a magnetic flux, the iron particles are magnetized and change the structural response of the composite material. We use a phenomenological model and ignore the details of the microstructure and its evolution. Hence, because of the dominating matrix material, we expect to obtain a nonlinear material model with parameters varying under different magnetic flux magnitudes. In an experiment, the magnitude of the magnetic flux is controlled and held constant throughout one set of experiments for determining the structural response. The experimental results are indeed very difficult to obtain because of self magnetization as well as due to the soft matter used as the matrix. We refer to Jolly et al (1996) and An et al (2012) for a discussion and results in MREs. In this paper, we use measurement data obtained from Yu et al (2017) and try to find an appropriate material model in the lengthscale of the matrix material. This model is a homogenized model ignoring agglomeration.

This work aims at modeling MRE by a material equation. This material equation can be used in a computation of a structure by using a computational method such as the finite element method (FEM). Elastomer is a soft material performing large elastic deformations. We model this behavior by using a so-called stored energy density, w, depending on the deformation, $\boldsymbol{u}$. The stored energy density is equivalent to the free energy, ψ, for an isothermal deformation, which is a realistic assumption in elasticity. The energy is of particular interest since it is a directly measurable quantity — see Treloar (1975) for stretching measurements of rubber type materials. Energy is expected to depend on the deformation. Thus, we use the deformation gradient, $\boldsymbol{F}$, or the right Cauchy–Green strain tensor, $\boldsymbol{C} = \boldsymbol{F}^{\mathrm{T}}\boldsymbol{F}$, as arguments (inputs) in the scalar function energy. However, the form of this function is a challenging question and is being discussed heavily in the literature. As a consequence of the objectivity, in the case of an isotropic material, the free energy does not depend on the material frame. Therefore, the dependency is usually described by invariants, which are also scalar functions. It is straight-forward to represent a scalar function by other scalar functions and admissible to use any expansion from the thermodynamics point of view. As we want to use this model in FEM computations, there might occur numerical problems when some restrictions are not met. These are called *ellipticity* and for an isotropic material *invertibility*. They assure a smooth deformation gradient, $\boldsymbol{F}$, i.e., the deformation gradient is continuous within the domain (Rosakis, 1990). According to the Hamilton–Cayley theorem, tensor of rank 2 (in three dimensional space) generates a characteristic equation leading to 3 invariants. Thus, the free energy depends on 3 arguments and ellipticity holds in every single argument of the free energy, which is called *quasiconvexity*. The choice of the functional form and the material constants shall not violate the quasiconvexity in order to prevent any numerical problems in a possible FEM computation. Depending on the used model, the analysis of the quasiconvexity might be very challenging.

Hyperelastic behavior is modeled by the free energy depending on the invariants. There are several procedures for obtaining an appropriate energy function, among others see Weber and Anand (1990), Arruda and Boyce (1993), Holzapfel (2000), Attard and Hunt (2004), Itskov and Aksel (2004), Lurie (1990), and Soe et al (2014). There are also prominent material models (satisfying quasiconvexity) in the literature. We can subsume them in three groups:

- motivated by using representation theorems, as in Mooney–Rivlin (Mooney, 1940; Rivlin, 1948), Biderman (Biderman, 1958), Ogden (Ogden, 1972), Haine–Wilson (James et al, 1975), and Yeoh (Yeoh, 1993) models,
- obtained as the best fitting function to experimental data, for example Rivlin–Saunders (Rivlin and Saunders, 1951), Hart–Smith (Hart-Smith, 1966), and Gent (Gent, 1996) models,
- acquired by statistical averaging of a microstructure model, for example neo-Hooke (Flory and Rehner Jr, 1943), Isihara (Isihara et al, 1951), and Arruda–Boyce (Boyce and Arruda, 2000) models.

As material modeling is challenging for hyperelastic materials, additionally, it is even more difficult to find the material model that fits existing experiments, for some rigor discussions we refer to Triantafyllidis and Aifantis (1986), Brigadnov and Dorfmann (2003), Kankanala and Triantafyllidis (2004), Marckmann and Verron (2006), Saxena et al (2014), Spieler et al (2014), Metsch et al (2016), Schubert and Harrison (2016), and Mehnert et al (2017).

Especially for soft polymer materials, it is beneficial to use a rotational rheometer and its oscillatory measurement capabilities. A harmonic shear deformation is applied periodically under different frequencies such that an interval of time scales are covered in one set of experiments. The output of a rheometer is called storage and loss moduli, they depend on the frequencies. Out of this data we aim at determining the parameters in one of the aforementioned nonlinear hyperelastic material models. In order to construct a nonlinear model in rheology, considerable amount of work has been done in Wilhelm (2002), Klein et al (2007), Ewoldt et al (2008), Ewoldt et al (2010), Hyun et al (2011), and Kádár et al (2017). For finding an appropriate material equation in relation to the rheometer measurement, different expansions are suggested and a nonlinear regression problem is formulated by manipulating the rheometer output. These different expansions have mathematical similarities (Argatov et al, 2017). If we want to use one of these models, it is unlikely that the aforementioned polyconvexity is met. Therefore, we want to have a methodology allowing us to use already existing and theoretically sound models based on tensor algebra and determine their material parameters by exploiting oscillatory rheometer experiments.

The energy of a hyperelastic material is a nonlinear function in the invariants of the deformation. However, the energy is mostly linear in the material parameters. This property allows us to advance an approach for determining the material parameters by using a linear regression. This energy-based method is proposed in Abali (2014, Sect. 9) and utilized in Abali et al (2016) and Abali (2018) for an oscillatory rheometer measurement as well as in Yang et al (2018) for a uniaxial tensile test. By using the energy-based method, we generate a linear regression problem such that the solution is unique and computationally fast. Otherwise, we need to implement a nonlinear optimization problem such as in Abramowitch and Woo (2004), Pawlikowski (2014), and Tang et al (2011). Linear regression problem allows us quickly apply various material models and find out the best one. We present exactly this point by using experimental data from Yu et al (2017) and determine the simplest model yet capable of representing experimental values.

1.2 The Energy-based Method for the Inverse Analysis

We use standard continuum mechanics notation and understand a summation over repeated indices. As the temperature and magnetic flux are held constant in one set of oscillatory rheometer measurements, rate of free energy is equal to the rate of stored energy. For the specimen, Ω, in a period of time, τ, the energy reads

$$E = \int_{\tau} \int_{\Omega} \psi^{\bullet} \, \mathrm{d}v \, \mathrm{d}t \ . \tag{1.1}$$

This amount of energy is supplied to the specimen resulting a (rotational) shear deformation. Normally, a standard rheometer fails to output the energy supplied to the system, but it gives out storage and loss moduli, G' and G'', respectively. We refer to Abali (2014, Sect. 9) for the detailed discussion and utilize the following general relations:

$$\begin{aligned} e_1 &= \int_0^T \psi^{\bullet} \, \mathrm{d}t = 2\pi G'' \varepsilon_0^2 \ , \\ e_2 &= \int_0^{T/4} \psi^{\bullet} \, \mathrm{d}t - \frac{1}{4} e_1 = G' \varepsilon_0^2 \end{aligned} \tag{1.2}$$

for an oscillatory experiment at the frequency ν with a period $T = 1/\nu$. The experiment is steered by the given periodic strain,

$$\varepsilon = \varepsilon_0 \sin\left(2\pi\nu t\right) , \tag{1.3}$$

applied as a shear deformation in the so-called deformation gradient, say, on xy-plane,

$$F_{ij} = \begin{pmatrix} 1 & \varepsilon & 0 \\ 0 & 1 & 0 \\ 0 & 0 & 1 \end{pmatrix} . \tag{1.4}$$

We can assume that this shear deformation is constant in the specimen, since the thickness is at least one order smaller than the perimeter of the cylindrical specimen. Hence, the right CAUCHY–GREEN strain tensor, $\boldsymbol{C} = \boldsymbol{F}^{\mathrm{T}}\boldsymbol{F}$, reads

$$C_{ij} = \begin{pmatrix} 1 & 0 & 0 \\ \varepsilon & 1 & 0 \\ 0 & 0 & 1 \end{pmatrix} \begin{pmatrix} 1 & \varepsilon & 0 \\ 0 & 1 & 0 \\ 0 & 0 & 1 \end{pmatrix} = \begin{pmatrix} 1 & \varepsilon & 0 \\ \varepsilon & 1+\varepsilon^2 & 0 \\ 0 & 0 & 1 \end{pmatrix} . \tag{1.5}$$

The rheometer measures the energy and calculates the stored and loss moduli by using the above relations based on the linear rheology. Since we want to go beyond the linear equations, we need to recalculate the energy as in Eq. (1.2) and then use this data for determining the material parameters occurring in ψ. As we calculate the energy within the whole period, G'' quantifies the dissipated energy from the system. For a quarter period, apart the dissipated energy, the rest is stored and seen in connection with G'. Herein we emphasize that we only use G' and G'' for acquiring e_1 and e_2 for the inverse analysis. We intend to handle these energy densities, e_1 and e_2, as the real measurables.

In order to present the energy-based method, we start with the usual assertion and aim at defining the energy density,

$$\psi = \psi(I_1, I_2, I_3) , \tag{1.6}$$

depending on invariants:

$$\begin{aligned} \tilde{I}_1 &= \mathrm{tr}(\boldsymbol{C}) = C_{ii} , \\ \tilde{I}_2 &= \frac{1}{2}\big(\tilde{I}_1^2(\boldsymbol{C}) - \tilde{I}_1(\boldsymbol{C}^2)\big) = \frac{1}{2}\big(C_{ii}C_{jj} - C_{ij}C_{ji}\big) , \\ \tilde{I}_3 &= \frac{1}{3}\big(\tilde{I}_1(\boldsymbol{C}^3) + 3\tilde{I}_1(\boldsymbol{C})\tilde{I}_2(\boldsymbol{C}) - \tilde{I}_1^3(\boldsymbol{C})\big) = \det(\boldsymbol{C}) , \end{aligned} \tag{1.7}$$

of the strain tensor $\boldsymbol{C}$. The definition of invariants follows from the characteristic equation, we refer to (Zyczkowski, 1981, Sect. 5.3), (Betten, 2013, Sect. 12.1), (Altenbach, 2018, Sect. 2.2.5). It is beneficial to construct another set of invariants as follows:

$$I_1 = \mathrm{tr}(\boldsymbol{C}) = C_{ii} , \; I_2 = I_1(\boldsymbol{C}^2) = C_{ij}C_{ji} , \; I_3 = \det(\boldsymbol{C}) , \tag{1.8}$$

The invariants of the right CAUCHY–GREEN strain tensor, $\boldsymbol{C} = \boldsymbol{F}^{\mathrm{T}}\boldsymbol{F}$, are identical to the invariants of the left CAUCHY–GREEN strain tensor, $\boldsymbol{B} = \boldsymbol{F}\boldsymbol{F}^{\mathrm{T}}$, by using the deformation gradient $F_{ij} = \partial(u_i + X_i)/\partial X_j$. With the same functional form, the invariants of $\boldsymbol{B}$ read

$$I_1 = \mathrm{tr}(\boldsymbol{B}) \, , \; I_2 = I_1(\boldsymbol{B}^2) \, , \; I_3 = \det(\boldsymbol{B}) \, . \tag{1.9}$$

The invariants I_1, I_2, I_3 are linear, quadratic, cubic in the strain tensor, respectively. By using the conventional notation, $J = \det(\boldsymbol{F})$, we rewrite the third invariant, $I_3 = J^2$. Since the third invariant is cubic in the strain tensor, $J^{2/3}$ is linear and $J^{4/3}$ is quadratic. Hence, we can construct another set of invariants:

$$\bar{I}_1 = \frac{I_1}{J^{2/3}} \, , \; \bar{I}_2 = \frac{I_2}{J^{4/3}} \, . \tag{1.10}$$

As we understand $\boldsymbol{F}$ as a transformation from the current frame onto the reference frame, by decomposing volumetric and deviatoric deformation and using J as the volumetric contraction, we understand that $\bar{I}_1$ and $\bar{I}_2$ remain constant subject to a volumetric change. Therefore, it is convenient to express

$$\psi(I_1, I_2, I_3) = \hat{\psi}(J) + \bar{\psi}(\bar{I}_1, \bar{I}_2) \, , \tag{1.11}$$

where the volumetric part of the deformation is given by $\hat{\psi}$ and the deviatoric part is dictated by $\bar{\psi}$. In the rheometer experiment, pure deviatoric type of deformation leads to $J = 1$ such that we can determine $\bar{\psi}$ but not $\hat{\psi}$. For obtaining the functional form and the corresponding parameters, a volume expansion test of the same material is necessary. We continue modeling and determining parameters of $\bar{\psi}$. The energy density is constructed in such a way that it vanishes for the case without deformation. This "stress-free" or natural configuration is called the reference frame, usually chosen as the initial frame. The invariants read $\bar{I}_1 = \bar{I}_2 = 3$ in the reference frame. If we consider the shear measurement in the oscillatory rheometer experiment, after using Eq. (1.5) and

$$C_{ij}C_{jk} = \begin{pmatrix} 1+\varepsilon^2 & 2\varepsilon+\varepsilon^3 & 0 \\ 2\varepsilon+\varepsilon^3 & 1+3\varepsilon^2+\varepsilon^4 & 0 \\ 0 & 0 & 1 \end{pmatrix} \, , \tag{1.12}$$

the invariants become

$$\bar{I}_1 = 3 + \varepsilon^2 \, , \; \bar{I}_2 = 3 + 4\varepsilon^2 + \varepsilon^4 \, . \tag{1.13}$$

All experiments are realized by steering the strain, ε, as given in Eq. (1.3). For a specific frequency, a set of experiments are realized by varying the amplitude, ε_0 in Eq. (1.3). Thus, we can calculate the numerical values of invariants in each set of experiments by using the above definitions in Eq. (1.13).

In order to model $\bar{\psi}(\bar{I}_1, \bar{I}_2)$, as an example, we use a MOONEY–RIVLIN type of expansion

$$\bar{\psi} = c_1(\bar{I}_1 - 3) + c_2(\bar{I}_2 - 3) . \tag{1.14}$$

Rate of this energy density is

$$\bar{\psi}^{\cdot} = c_1 \bar{I}_1^{\cdot} + c_2 \bar{I}_2^{\cdot} , \tag{1.15}$$

with

$$\bar{I}_1^{\cdot} = 2\varepsilon\varepsilon^{\cdot} , \ \bar{I}_2^{\cdot} = 8\varepsilon\varepsilon^{\cdot} + 4\varepsilon^3\varepsilon^{\cdot} . \tag{1.16}$$

Rate of invariants can be calculated directly by using Eq. (1.3). Since c_1 and c_2 are constants in time - we remark that the magnetic flux is held constant throughout one set of experiments - we rewrite the energy density,

$$\psi = \int \psi^{\cdot} \,\mathrm{d}t = c_1 \int \bar{I}_1^{\cdot} \,\mathrm{d}t + c_2 \int \bar{I}_2^{\cdot} \,\mathrm{d}t , \tag{1.17}$$

since the material parameters c_1 and c_2 are constants. For a specific frequency ν and amplitude ε_0 we obtain from the latter, the dissipated energy as in Eq. (1.2) for the whole period, T,

$$\begin{gathered} e_1 = A_{11}c_1 + A_{12}c_2 , \\ A_{11} = \int_0^T \bar{I}_1^{\cdot} \,\mathrm{d}t , \ A_{12} = \int_0^T \bar{I}_2^{\cdot} \,\mathrm{d}t , \end{gathered} \tag{1.18}$$

as well as for the quarter period,

$$\begin{gathered} e_2 = A_{21}c_1 + A_{22}c_2 , \\ A_{21} = \int_0^{T/4} \bar{I}_1^{\cdot} \,\mathrm{d}t - \frac{1}{4}A_{11} , \ A_{22} = \int_0^{T/4} \bar{I}_2^{\cdot} \,\mathrm{d}t - \frac{1}{4}A_{12} . \end{gathered} \tag{1.19}$$

We emphasize that the components of $\boldsymbol{A}$ possess strain as given in Eq. (1.3) such that for specific values of the frequency and amplitude, the numerical values of components of $\boldsymbol{A}$ are obtained by a simple integration. By using the measurement data from Eq. (1.2), we obtain the following form:

$$\begin{pmatrix} A_{11} & A_{12} \\ A_{21} & A_{22} \end{pmatrix} \begin{pmatrix} c_1 \\ c_2 \end{pmatrix} = \begin{pmatrix} 2\pi G'' \varepsilon_0^2 \\ G' \varepsilon_0^2 \end{pmatrix} . \tag{1.20}$$

Suppose we have n measurements with different amplitudes, ε_0^1, ε_0^2, ..., ε_0^n, at a specific frequency. Then the matrix $\boldsymbol{A}$ will be $2n \times 3$ as follows

$$\begin{pmatrix} A_{11}|_{\varepsilon_0^1} & A_{12}|_{\varepsilon_0^1} \\ A_{21}|_{\varepsilon_0^1} & A_{22}|_{\varepsilon_0^1} \\ A_{11}|_{\varepsilon_0^2} & A_{12}|_{\varepsilon_0^2} \\ A_{21}|_{\varepsilon_0^2} & A_{22}|_{\varepsilon_0^2} \\ \vdots & \vdots \\ A_{11}|_{\varepsilon_0^n} & A_{12}|_{\varepsilon_0^n} \\ A_{21}|_{\varepsilon_0^n} & A_{22}|_{\varepsilon_0^n} \end{pmatrix} \begin{pmatrix} c_1 \\ c_2 \end{pmatrix} = \begin{pmatrix} 2\pi G''|_{\varepsilon_0^1}(\varepsilon_0^1)^2 \\ G'|_{\varepsilon_0^1}(\varepsilon_0^1)^2 \\ 2\pi G''|_{\varepsilon_0^2}(\varepsilon_0^2)^2 \\ G'|_{\varepsilon_0^2}(\varepsilon_0^2)^2 \\ \vdots \\ 2\pi G''|_{\varepsilon_0^n}(\varepsilon_0^n)^2 \\ G'|_{\varepsilon_0^n}(\varepsilon_0^n)^2 \end{pmatrix} , \tag{1.21}$$

which is tantamount to

$$\boldsymbol{A}\boldsymbol{u} = \boldsymbol{b} \, , \tag{1.22}$$

where the unknowns, $\boldsymbol{u}$, are the material parameters to be determined; we know or calculate the experimental results, $\boldsymbol{b}$, as well as the components in the coefficient matrix, $\boldsymbol{A}$. The unknowns form a linear relation such that we can solve it. However, there are more measurements than unknowns, in other words, $\boldsymbol{A}$ is not quadratic, thus the solution is not unique. In order to find the best material parameters in the sense of minimizing the squared error, we use the so-called *normal equation* in statistics:

$$\boldsymbol{u} = (\boldsymbol{A}^{\mathrm{T}}\boldsymbol{A})^{-1}\boldsymbol{A}^{\mathrm{T}}\boldsymbol{b} \, , \tag{1.23}$$

we refer to Abali et al (2016, Sect. 6.1.1) for its straight-forward derivation, which is well-known in optimization. By reinterpreting rheometer data, we achieved a linear regression problem with a unique solution for the best material parameters fitting the output. We will utilize this method for determining materials parameters.

1.3 Method of Solution and Results

The measurements conducted in Yu et al (2017) are used as the data for constructing $\boldsymbol{b}$ in Eq. (1.23). Concretely, from G' and G'' data, we obtain e_1 and e_2 with the aid of Eq. (1.2) as visualized in Fig. 1.1. Every dot in Fig. 1.1 represents one measurement. Such an amplitude sweep measurement is automatized in modern rheometer softwares. We expect to find a material model capturing the energies for different amplitudes by means of the same material parameters. We construct $\boldsymbol{b}$ as well as the coefficient matrix, $\boldsymbol{A}$, for every strain amplitude ε_0 in the periodic strain as in Eq. (1.3). The coefficient matrix, $\boldsymbol{A}$, is acquired by computing the integrals numerically with SIMPSON's rule. We use open-source packages in Python language by exploiting NumPy (Oliphant, 2007) for computation and MatPlotLIB (Hunter, 2007) for visualization. Building the matrices and solving Eq. (1.23) lasts less than a minute[1].

We start off by trying the MOONEY–RIVLIN type hyperelastic model as in Eq. (1.14) and find out that e_1 cannot be represented as in the experiment. The reason is the

[1] Python 2.7 running in an Ubuntu 16.04 working on a single Intel Core i7-4650U at 1700 MHz processor.

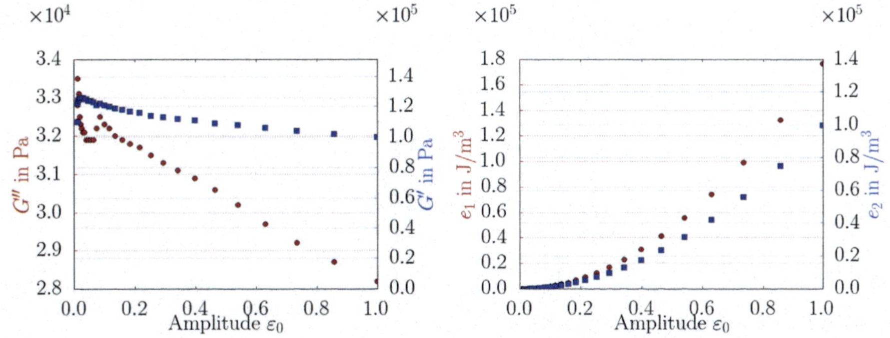

Fig. 1.1 Experimental data for various amplitudes of shear strain, ε_0. Left: storage and loss muduli (no magnetic flux) obtained from Yu et al (2017). Right: data is converted into energies by using Eq. (1.2).

material behavior being viscoelastic. We need a material model incorporating not only the strain tensor, $\boldsymbol{C}$, but also its rate, $\boldsymbol{C}^{\bullet}$, in the energy definition. Analogous to Eq. (1.8), we introduce invariants depending on the rate of strain tensor,

$$
\begin{gathered}
I_1 = \mathrm{tr}(\boldsymbol{C}) = C_{ii}\,,\ I_2 = \mathrm{tr}(\boldsymbol{C}\cdot\boldsymbol{C}) = C_{ij}C_{ji}\,,\ I_3 = \det(\boldsymbol{C})\,,\\
I_4 = \mathrm{tr}(\boldsymbol{C}^{\bullet}) = C^{\bullet}_{ii}\,,\ I_5 = \mathrm{tr}(\boldsymbol{C}^{\bullet}\cdot\boldsymbol{C}^{\bullet}) = C^{\bullet}_{ij}C^{\bullet}_{ji}\,,\ I_6 = \det(\boldsymbol{C}^{\bullet})\,,\\
I_7 = \mathrm{tr}(\boldsymbol{C}\cdot\boldsymbol{C}^{\bullet}) = C_{ij}C^{\bullet}_{ji}\,,
\end{gathered}
\tag{1.24}
$$

for every different possibility of C_{ij}, $C^{\bullet}_{ij}$, and $C_{ij}C^{\bullet}_{jk}$. By using an abusive notation,

$$
\begin{gathered}
J = \sqrt{I_3}\,,\ J^{\bullet} = \sqrt{I_6}\,,\\
\bar{I}_1 = \frac{I_1}{J^{2/3}}\,,\ \bar{I}_2 = \frac{I_2}{J^{4/3}}\,,\ \bar{I}_4 = \frac{I_4}{J^{2/3}}\,,\ \bar{I}_5 = \frac{I_5}{J^{4/3}}\,,\ \bar{I}_7 = \frac{I_7}{J^{4/3}}\,,
\end{gathered}
\tag{1.25}
$$

we obtain

$$
\psi(I_1, I_2, I_3, I_4, I_5, I_6) = \hat{\psi}(J, J^{\bullet}) + \bar{\psi}(\bar{I}_1, \bar{I}_2, \bar{I}_4, \bar{I}_5, \bar{I}_7)\,. \tag{1.26}
$$

Again, only for the deviatoric part, $\bar{\psi}$, we obtain the invariants in the case of shear experiments, $J = 1$, by using Eqs. (1.5), (1.12), as well as

$$
\begin{gathered}
C^{\bullet}_{ij} = \begin{pmatrix} 0 & \varepsilon^{\bullet} & 0 \\ \varepsilon^{\bullet} & 2\varepsilon\varepsilon^{\bullet} & 0 \\ 0 & 0 & 0 \end{pmatrix}\,,\ C^{\bullet}_{ij}C^{\bullet}_{jk} = \begin{pmatrix} (\varepsilon^{\bullet})^2 & 2\varepsilon(\varepsilon^{\bullet})^2 & 0 \\ 2\varepsilon(\varepsilon^{\bullet})^2 & (\varepsilon^{\bullet})^2 + 4\varepsilon^2(\varepsilon^{\bullet})^2 & 0 \\ 0 & 0 & 0 \end{pmatrix}\,,\\
C_{ij}C^{\bullet}_{jk} = \begin{pmatrix} \varepsilon\varepsilon^{\bullet} & \varepsilon^{\bullet} + 2\varepsilon^2\varepsilon^{\bullet} & 0 \\ (1+\varepsilon^2)\varepsilon^{\bullet} & 3\varepsilon\varepsilon^{\bullet} + 2\varepsilon^3\varepsilon^{\bullet} & 0 \\ 0 & 0 & 0 \end{pmatrix}\,,
\end{gathered}
\tag{1.27}
$$

as follows

$$\bar{I}_1 = 3 + \varepsilon^2 \,,\; \bar{I}_2 = 3 + 4\varepsilon^2 + \varepsilon^4 \,,\; \bar{I}_4 = 2\varepsilon\varepsilon^{\boldsymbol{\cdot}} \,, \\ \bar{I}_5 = 2(\varepsilon^{\boldsymbol{\cdot}})^2 + 4\varepsilon^2(\varepsilon^{\boldsymbol{\cdot}})^2 \,,\; \bar{I}_7 = 4\varepsilon\varepsilon^{\boldsymbol{\cdot}} + 2\varepsilon^3\varepsilon^{\boldsymbol{\cdot}} \,. \tag{1.28}$$

Therefore, we acquire the following model

$$\psi = c_1(\bar{I}_1 - 3) + c_2(\bar{I}_2 - 3) + c_3\bar{I}_4 + c_4\bar{I}_5 + c_5\bar{I}_7 \,, \tag{1.29}$$

by assuring $\psi = 0$ in the reference frame. Rate of energy reads

$$\psi^{\boldsymbol{\cdot}} = c_1\bar{I}_1^{\boldsymbol{\cdot}} + c_2\bar{I}_2^{\boldsymbol{\cdot}} + c_3\bar{I}_4^{\boldsymbol{\cdot}} + c_4\bar{I}_5^{\boldsymbol{\cdot}} + c_5\bar{I}_7^{\boldsymbol{\cdot}} \,, \tag{1.30}$$

with the following rate of invariants

$$\bar{I}_1^{\boldsymbol{\cdot}} = 2\varepsilon\varepsilon^{\boldsymbol{\cdot}} \,,\; \bar{I}_2^{\boldsymbol{\cdot}} = 8\varepsilon\varepsilon^{\boldsymbol{\cdot}} + 4\varepsilon^3\varepsilon^{\boldsymbol{\cdot}} \,,\; \bar{I}_4^{\boldsymbol{\cdot}} = 2(\varepsilon^{\boldsymbol{\cdot}})^2 + 2\varepsilon\varepsilon^{\boldsymbol{\cdot\cdot}} \,, \\ \bar{I}_5^{\boldsymbol{\cdot}} = 4\varepsilon^{\boldsymbol{\cdot}}\varepsilon^{\boldsymbol{\cdot\cdot}} + 8\varepsilon(\varepsilon^{\boldsymbol{\cdot}})^3 + 8\varepsilon^2\varepsilon^{\boldsymbol{\cdot}}\varepsilon^{\boldsymbol{\cdot\cdot}} \,,\; \bar{I}_7^{\boldsymbol{\cdot}} = 4(\varepsilon^{\boldsymbol{\cdot}})^2 + 4\varepsilon\varepsilon^{\boldsymbol{\cdot\cdot}} + 6\varepsilon^2(\varepsilon^{\boldsymbol{\cdot}})^2 + 2\varepsilon^3\varepsilon^{\boldsymbol{\cdot\cdot}} \,. \tag{1.31}$$

Since we have experimental values for e_1 and e_2, we need to express them by using the latter material equation. By utilizing Eq. (1.2), we obtain

$$e_1 = A_{11}c_1 + A_{12}c_2 + A_{13}c_3 + A_{14}c_4 + A_{15}c_5 \,, \\ A_{11} = \int_0^T \bar{I}_1^{\boldsymbol{\cdot}}\,\mathrm{d}t \,,\; A_{12} = \int_0^T \bar{I}_2^{\boldsymbol{\cdot}}\,\mathrm{d}t \,,\; A_{13} = \int_0^T \bar{I}_4^{\boldsymbol{\cdot}}\,\mathrm{d}t \,,\; A_{14} = \int_0^T \bar{I}_5^{\boldsymbol{\cdot}}\,\mathrm{d}t \,,\; A_{15} = \int_0^T \bar{I}_7^{\boldsymbol{\cdot}}\,\mathrm{d}t \,, \tag{1.32}$$

for the whole period and analogously, for the quarter period,

$$e_2 = A_{21}c_1 + A_{22}c_2 + A_{23}c_3 + A_{24}c_4 + A_{25}c_5 \,, \\ A_{21} = \int_0^{T/4} \bar{I}_1^{\boldsymbol{\cdot}}\,\mathrm{d}t - \frac{1}{4}A_{11} \,,\; A_{22} = \int_0^{T/4} \bar{I}_2^{\boldsymbol{\cdot}}\,\mathrm{d}t - \frac{1}{4}A_{12} \,,\; A_{23} = \int_0^{T/4} \bar{I}_4^{\boldsymbol{\cdot}}\,\mathrm{d}t - \frac{1}{4}A_{13} \,, \\ A_{24} = \int_0^{T/4} \bar{I}_5^{\boldsymbol{\cdot}}\,\mathrm{d}t - \frac{1}{4}A_{14} \,,\; A_{25} = \int_0^{T/4} \bar{I}_7^{\boldsymbol{\cdot}}\,\mathrm{d}t - \frac{1}{4}A_{15} \,. \tag{1.33}$$

This model in Eq. (1.29) is capable to represent the data adequately, see Fig. 1.2 (left). By using the method herein, we can easily construct different models and test them as well. Since the model in Eq. (1.29) has five parameters, we try to eliminate each of them and simplify the energy equation still representing the experiments fairly well. Analogous to linear YEOH model or incompressible neo-HOOKEan relation, we constructed the following model:

$$\psi = c_1(\bar{I}_1 - 3) + c_3\bar{I}_4 \tag{1.34}$$

that provides a very good agreement to experiments, see Fig. 1.2 (right). Thus, we utilize this YEOH type of model and try to fit all experiments conducted under dif-

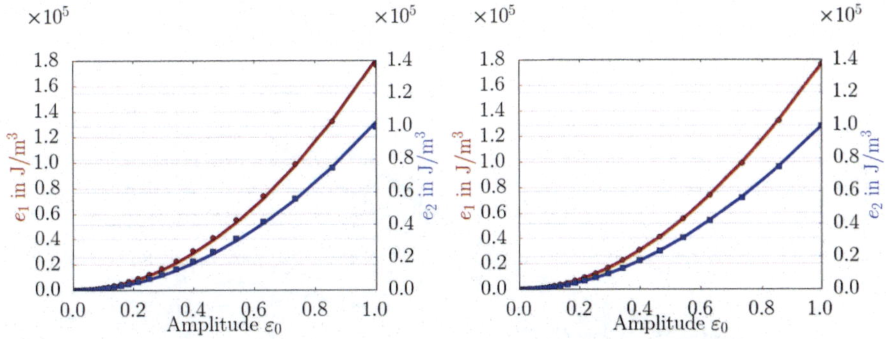

Fig. 1.2 Fit to experimental data. Dots denote experiments, continuous lines express the fit. Left: fit by using Eq. (1.29). Right: fit by using Eq. (1.34).

Table 1.1 Material parameters in the model, $\psi = c_1(\bar{I}_1 - 3) + c_3\bar{I}_4$

B in T	c_1 in Pa	c_3 in Pa s
0.0	147.5×10^3	-34.7×10^9
0.1	198.7×10^3	-47.5×10^9
0.2	300.8×10^3	-71.2×10^9
0.3	422.6×10^3	-98.5×10^9
0.4	538.8×10^3	-121.7×10^9
0.5	642.8×10^3	-141.5×10^9
0.6	726.4×10^3	-155.0×10^9
0.7	778.6×10^3	-161.2×10^9
0.8	797.4×10^3	-160.3×10^9

ferent magnetic flux values from 0 T up to 0.8 T. The accuracy of the result can be depicted in Fig. 1.3. Especially for a magnetic flux up to 0.6 T the model is describing the material response accurately. Within the regime up to 0.6 T, we encourage using this Yeoh type of model in a possible computation. Hence, we compile the results in Table 1.1, between the values a linear interpolation is adequate to implement. Beyond the rough estimate of 0.6 T, the accuracy of the material model decreases. This fact can be comprehended in a better way, when we plot the material parameters in the magnetic flux as presented in Fig. 1.4. Interestingly, the material starts to converge to a state where the increase of the magnetic flux is not affecting the material response. This case might be explained as a consequence of the saturation in polarization. Hence, the characteristics of the homogenized material tends to change as well such that the proposed model fails to represent the behavior ac-

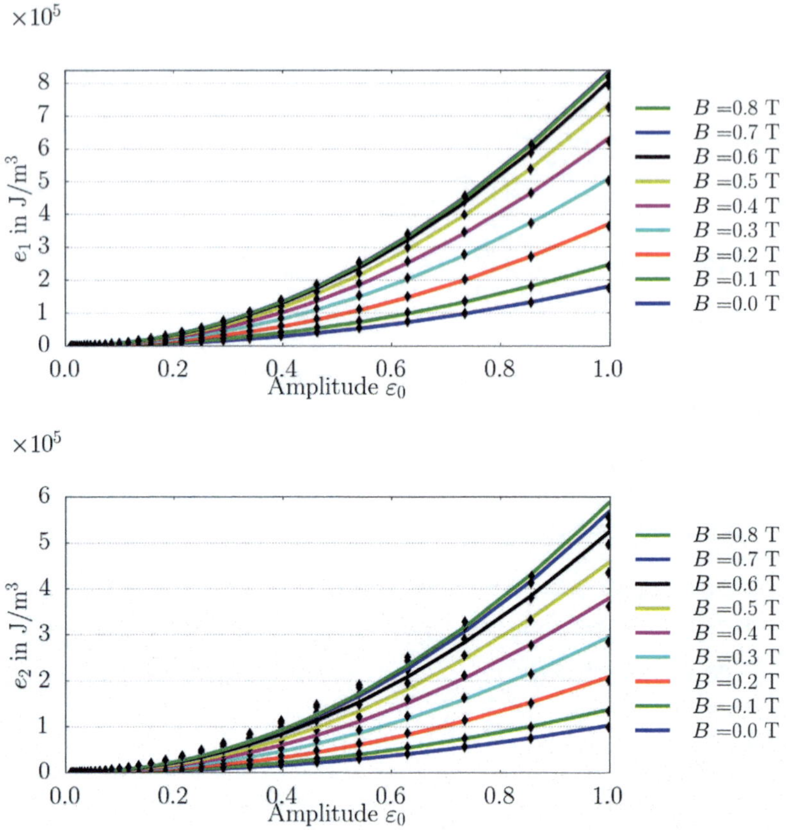

Fig. 1.3 Fit to experimental data under different magnetic fluxes. Dots denote experiments, continuous lines express the fit by using Eq. (1.34).

curately. We suggest to use the model with the parameters as given in Table 1.1 for large deformations up to 100% under magnetic flux up to 0.6 T.

1.4 Conclusion

We have presented a relatively simple approach in order to determine the material parameters of a nonlinear material model out of a rheometer measurement. Magnetorheological elastomer is a composite material of a soft matrix material filled with magnetizable iron particles, thus, the material response changes with respect to the magnetic flux. An oscillatory measurement with a rotational rheometer generates an outcome that is recalculated in order to obtain a linear regression problem. Hence, the inverse analysis let us acquire unique parameters of the hyperelastic material

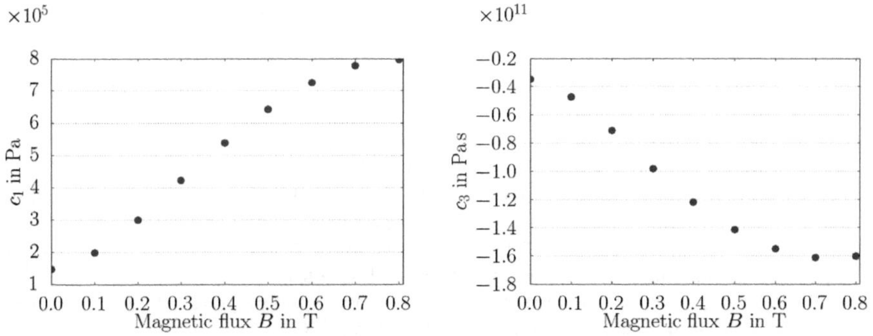

Fig. 1.4 Material parameter change subject to magnetic flux: left: c_1 in Pa, right: c_3 in Pa s taken from Table 1.1.

model. After using different models obtained by the rules of tensor calculus, we have found a simple yet powerful material model, a linear YEOH type of model, capable of accurately representing the material behavior up to 0.6 T. The results are compiled in Table 1.1 for exploiting in a computation, for example by using the finite element model.

References

Abali BE (2014) Thermodynamically compatible modeling, determination of material parameters, and numerical analysis of nonlinear rheological materials. PhD thesis, Technische Universität Berlin, Institute of Mechanics

Abali BE (2018) Inverse analysis of cellulose by using the energy-based method and a rotational rheometer. Applied Science 8:1354

Abali BE, Wu CC, Müller WH (2016) An energy-based method to determine material constants in nonlinear rheology with applications. Continuum Mechanics and Thermodynamics 28(5):1221–1246

Abramowitch SD, Woo SLY (2004) An improved method to analyze the stress relaxation of ligaments following a finite ramp time based on the quasi-linear viscoelastic theory. Journal of biomechanical engineering 126(1):92–97

Altenbach H (2018) Kontinuumsmechanik, 4. Aufl. Springer-Verlag Berlin Heidelberg

An H, Picken SJ, Mendes E (2012) Nonlinear rheological study of magneto responsive soft gels. Polymer 53(19):4164–4170

An JS, Kwon SH, Choi HJ, Jung JH, Kim YG (2017) Modified silane-coated carbonyl iron/natural rubber composite elastomer and its magnetorheological performance. Composite Structures 160:1020–1026

Argatov I, Iantchenko A, Kocherbitov V (2017) How to define the storage and loss moduli for a rheologically nonlinear material? Continuum Mechanics and Thermodynamics 29(1375–1387)

Arruda EM, Boyce MC (1993) A three-dimensional constitutive model for the large stretch behavior of rubber elastic materials. Journal of the Mechanics and Physics of Solids 41(2):389–412

Attard MM, Hunt GW (2004) Hyperelastic constitutive modeling under finite strain. International Journal of Solids and Structures 41(18-19):5327–5350

Betten J (2013) Kontinuumsmechanik: Elastisches und inelastisches Verhalten isotroper und anisotroper Stoffe, vol 2. Springer-Verlag

Biderman V (1958) Calculation of rubber parts (in Russ.). Rascheti na prochnost Moscow

Boyce MC, Arruda EM (2000) Constitutive models of rubber elasticity: a review. Rubber chemistry and technology 73(3):504–523

Brigadnov I, Dorfmann A (2003) Mathematical modeling of magneto-sensitive elastomers. International Journal of Solids and Structures 40(18):4659–4674

Ewoldt R, Winter P, Maxey J, McKinley G (2010) Large amplitude oscillatory shear of pseudoplastic and elastoviscoplastic materials. Rheologica Acta 49(2):191–212

Ewoldt RH, Hosoi A, McKinley GH (2008) New measures for characterizing nonlinear viscoelasticity in large amplitude oscillatory shear. Journal of Rheology 52(6):1427–1458

Flory PJ, Rehner Jr J (1943) Statistical mechanics of cross-linked polymer networks i. rubberlike elasticity. The Journal of Chemical Physics 11(11):512–520

Gent A (1996) A new constitutive relation for rubber. Rubber chemistry and technology 69(1):59–61

Gong X, Zhang X, Zhang P (2005) Fabrication and characterization of isotropic magnetorheological elastomers. Polymer Testing 24(5):669–676

Hart-Smith L (1966) Elasticity parameters for finite deformations of rubber-like materials. Zeitschrift für angewandte Mathematik und Physik ZAMP 17(5):608–626

Holzapfel AG (2000) Nonlinear Solid Mechanics: A Continuum Approach for Engineering. John Wiley & Sons, Inc.

Hunter JD (2007) Matplotlib: A 2d graphics environment. Computing in Science & Engineering 9(3):90–95

Hyun K, Wilhelm M, Klein CO, Cho KS, Nam JG, Ahn KH, Lee SJ, Ewoldt RH, McKinley GH (2011) A review of nonlinear oscillatory shear tests: Analysis and application of large amplitude oscillatory shear (laos). Progress in Polymer Science 36(12):1697–1753

Isihara A, Hashitsume N, Tatibana M (1951) Statistical theory of rubber-like elasticity. iv.(two-dimensional stretching). The Journal of Chemical Physics 19(12):1508–1512

Itskov M, Aksel N (2004) A class of orthotropic and transversely isotropic hyperelastic constitutive models based on a polyconvex strain energy function. International Journal of Solids and Structures 41(14):3833–3848

James AG, Green A, Simpson G (1975) Strain energy functions of rubber. i. characterization of gum vulcanizates. Journal of Applied Polymer Science 19(7):2033–2058

Jolly MR, Carlson JD, Muñoz BC, Bullions TA (1996) The magnetoviscoelastic response of elastomer composites consisting of ferrous particles embedded in a polymer matrix. Journal of Intelligent Material Systems and Structures 7(6):613–622

Kádár R, Abbasi M, Figuli R, Rigdahl M, Wilhelm M (2017) Linear and nonlinear rheology combined with dielectric spectroscopy of hybrid polymer nanocomposites for semiconductive applications. Nanomaterials 7(2):23

Kankanala S, Triantafyllidis N (2004) On finitely strained magnetorheological elastomers. Journal of the Mechanics and Physics of Solids 52(12):2869–2908

Klein C, Spiess H, Calin A, Balan C, Wilhelm M (2007) Separation of the nonlinear oscillatory response into a superposition of linear, strain hardening, strain softening, and wall slip response. Macromolecules 40(12):4250–4259

Li W, Nakano M (2013) Fabrication and characterization of pdms based magnetorheological elastomers. Smart Materials and Structures 22(5):055,035

Lurie AI (1990) Non-linear theory of elasticity. North Holland

Marckmann G, Verron E (2006) Comparison of hyperelastic models for rubber-like materials. Rubber Chemistry and Technology 79(5):835–858

Mehnert M, Hossain M, Steinmann P (2017) Towards a thermo-magneto-mechanical coupling framework for magneto-rheological elastomers. International Journal of Solids and Structures

Metsch P, Kalina KA, Spieler C, Kästner M (2016) A numerical study on magnetostrictive phenomena in magnetorheological elastomers. Computational Materials Science 124:364–374

Mooney M (1940) A theory of large elastic deformation. Journal of Applied Physics 11(9):582–592
Ogden RW (1972) Large deformation isotropic elasticity–on the correlation of theory and experiment for incompressible rubberlike solids. Proc R Soc Lond A 326(1567):565–584
Oliphant TE (2007) Python for scientific computing. Computing in Science & Engineering 9(3):10–20
Park BJ, Kim MS, Choi HJ (2009) Fabrication and magnetorheological property of core/shell structured magnetic composite particle encapsulated with cross-linked poly (methyl methacrylate). Materials Letters 63(24-25):2178–2180
Pawlikowski M (2014) Non-linear approach in visco-hyperelastic constitutive modelling of polyurethane nanocomposite. Mechanics of Time-Dependent Materials 18(1):1–20
Rivlin R (1948) Large elastic deformations of isotropic materials iv. further developments of the general theory. Phil Trans R Soc Lond A 241(835):379–397
Rivlin RS, Saunders D (1951) Large elastic deformations of isotropic materials vii. experiments on the deformation of rubber. Phil Trans R Soc Lond A 243(865):251–288
Rosakis P (1990) Ellipticity and deformations with discontinuous gradients in finite elastostatics. Archive for Rational Mechanics and Analysis 109(1):1–37
Saxena P, Hossain M, Steinmann P (2014) Nonlinear magneto-viscoelasticity of transversally isotropic magneto-active polymers. Proc R Soc A 470(2166):20140,082
Schubert G, Harrison P (2016) Magnetic induction measurements and identification of the permeability of magneto-rheological elastomers using finite element simulations. Journal of Magnetism and Magnetic Materials 404:205–214
Soe SP, Martindale N, Constantinou C, Robinson M (2014) Mechanical characterisation of duraform® flex for fea hyperelastic material modelling. Polymer Testing 34:103–112
Spieler C, Metsch P, Kästner M, Ulbricht V (2014) Microscale modeling of magnetoactive composites undergoing large deformations. Technische Mechanik 34(1):39–50
Tang C, Ng G, Wang Z, Tsui C, Zhang G (2011) Parameter optimization for the visco-hyperelastic constitutive model of tendon using fem. Bio-medical Materials and Engineering 21(1):9–24
Treloar LRG (1975) The Physics of Rubber Elasticity. Oxford University Press, Oxford, USA
Triantafyllidis N, Aifantis EC (1986) A gradient approach to localization of deformation. I. Hyperelastic materials. Journal of Elasticity 16(3):225–237
Weber G, Anand L (1990) Finite deformation constitutive equations and a time integration procedure for isotropic, hyperelastic-viscoplastic solids. Computer Methods in Applied Mechanics and Engineering 79(2):173–202
Wilhelm M (2002) Fourier-transform rheology. Macromolecular Materials and Engineering 287(2):83–105
Yang H, Ganzosch G, Giorgio I, Abali BE (2018) Material characterization and computations of a polymeric metamaterial with a pantographic substructure. Zeitschrift für angewandte Mathematik und Physik (ZAMP) 69(4):105
Yeoh O (1993) Some forms of the strain energy function for rubber. Rubber Chemistry and Technology 66(5):754–771
Yu M, Qi S, Fu J, Yang P, Zhu M (2015) Preparation and characterization of a novel magnetorheological elastomer based on polyurethane/epoxy resin ipns matrix. Smart Materials and Structures 24(4):045,009
Yu M, Qi S, Fu J, Zhu M, Chen D (2017) Understanding the reinforcing behaviors of polyaniline-modified carbonyl iron particles in magnetorheological elastomer based on polyurethane/epoxy resin ipns matrix. Composites Science and Technology 139:36–46
Zohdi TI (2012) Electromagnetic Properties of Multiphase Dielectrics: a Primer on Modeling, Theory and Computation, Lecture Notes in Applied and Computational Mechanics, vol 64. Springer Science & Business Media
Zohdi TI, Wriggers P (2008) An Introduction to Computational Micromechanics. Springer Science & Business Media
Zyczkowski M (1981) Combined loadings in the theory of plasticity. Springer Science & Business Media

Chapter 2
On the Size Effects in Indentation Testing of Elastic Functionally-graded Materials

Ivan Argatov

Abstract The size effect in the small-scale indentation testing is studied for a functionally-graded material (FGM) whose shear elastic modulus varies according to the exponential law. Under the simplifying assumption of zero Poisson's ratio, the asymptotic model of the indentation stiffness for an axisymmetric frictionless indenter is developed in the case when the contact radius is small compared to the inhomogeneity characteristic size. The so-called sample size effect is considered on the example of a simply supported FGM plate indented at the center of its top surface. A certain range of applicability of the first-order asymptotic models has been established by comparison with the approximate analytical solution available in the literature.

Keywords: Indentation stiffness · Functionally graded material · Size effect · Asymptotic model

2.1 Introduction

Indentation techniques represent a simple practical method of nondestructive characterizing mechanical properties of materials, e.g., hardness (Oliver and Pharr, 1992), elastic modulus (Bulychev et al, 1975), plasticity (Müller et al, 2009), fracture toughness (Anstis et al, 1981), adhesion strength (Borodich and Galanov, 2008). For instance, when the thermo-mechanical properties of micromechanical components (e.g., of solder joints, are to be determined realistically from small test volumes) a microindentation technique can be utilized (Villain et al, 2008), as it allows to measure the mechanical properties locally in the material.

Ivan Argatov
Department of Materials Science and Applied Mathematics, Malmö University, SE-205 06 Malmö, Sweden,
e-mail: ivan.argatov@gmail.com

B. E. Abali et al. (eds.), *New Achievements in Continuum Mechanics and Thermodynamics*, Advanced Structured Materials 108,
https://doi.org/10.1007/978-3-030-13307-8_2

An indentation test can be performed with a rigid indenter whose displacement, δ, can be monitored under an externally applied contact load, F, and an appropriate mathematical material model is needed to extract the material parameters from the indentation data. Generally speaking, a complete stress-strain curve should be determined for a full characterization of the elastic-plastic deformation behavior (Müller et al, 2009).

With the development of the Oliver–Pharr method (Oliver and Pharr, 1992), nanoindentation, known as an instrumented indentation test, where the direct inspection of the indent imprint is replaced by an indirect assessment from the force-displacement curve, has emerged as an indispensable technique for evaluation of mechanical properties at micro- and nano-scales (Borodich and Keer, 2004; Argatov, 2010). However, practical application of the nanoindentation method can be accompanied with numerous technological difficulties, especially, if a tested material exhibits a fine microstructure (Albrecht et al, 2005; Gibson, 2014; Argatov and Sabina, 2017) and/or a complicated deformation behavior (Cheng et al, 2000; Koumi et al, 2014). In response to the continuous miniaturization of microelectronic components in modern electronic industry, a number of practically important issues in indentation testing have been resolved by the research group headed by Prof. W.H. Müller (TU Berlin).

In particular, essential for the correct determination of the contact area in the depth-sensing indentation is the precise measurement of the actual indentation depth of the indenter. This practical issue has been addressed by Müller et al (2011). The evaluation of the material properties at elevated temperature reveals the influence of the surface oxidation on the indentation data. To avoid this problem, Müller et al (2009) developed an effective measurement procedure, which is of particular significance for characterizing the solder materials. The effect of crystal grain orientation is another parameter, which influences the accuracy of detailed indentation analysis of local material properties, was studied by Müller et al (2009). For describing the time-dependence of material response shown by low melting solder materials, which under indentation load are susceptible to creep behavior, Müller and Worrack (2012) have developed an enhanced analysis of nanoindentation data based on rheological models. To obtain information on the material's work hardening from experimentally measured load-displacement curves obtained with a blunt probe, the inverse analysis based determination methodology was established by Weinberg et al (2005). A practically important problem arises in application of nanoindentation for determining the mechanical properties of individual phases in heterogeneous materials and, especially, of intermetallic phases in microelectronic structures. It has been shown (Albrecht et al, 2005) that nanoindentation can be effectively used to quantify the growth of intermetallic phases, in particular at the interface of a solder connection.

What interests us in nanoindentation is that in many cases the effect of plastic deformations on the elastic deformation response can be get rid of by considering the indentation unloading (Bulychev et al, 1975), so that by evaluating the incremental indentation stiffness

$$S = \frac{\mathrm{d}F}{\mathrm{d}\delta}, \tag{2.1}$$

one can assess the stiffness property of solid materials, which for an isotropic homogeneous linearly elastic material, is fully characterized by its shear modulus, G, and Poisson's ratio, ν.

In the case of an axisymmetric indenter (e.g., cylinder, sphere, or cone), the evaluation of elastic indentation test is based on the BASh (Bulychev–Alekhin–Shorshorov) relation (Bulychev et al, 1975)

$$S(a) = \frac{4aG}{1-\nu}, \tag{2.2}$$

which reveals the fact that the indentation stiffness is proportional to the radius of contact area, a.

Equations (2.1) and (2.2) show that the ratio $S(a)/\sqrt{A}$, where A is the contact area, should be insensitive to the size of the indenter imprint. However, for a functionally-graded material, when, in addition to its elastic moduli, the material's response to indentation depends on some characteristic size of material inhomogeneity, the ratio $S(a)/a$ (in the axisymmetric setting) will vary with the contact radius a, thereby exhibiting the size effect. This issue was considered using either experimental, analytical or numerical methods (e.g., Suresh et al, 1997; Gouldstone et al, 2007).

Recall (Markworth et al, 1995) that the concept of functionally graded material (FGM) refers to composite materials with spatially variable properties, which is usually achieved by gradual compositional variation of the constituents. The development of indentation methods for a FGM sample requires the solution of contact problems for a semi-infinite elastic medium with a continuous variation of elastic properties (e.g., Giannakopoulos and Suresh, 1997b; Aizikovich et al, 2002; Heß, 2016; Argatov et al, 2018). Moreover, in order to assess the sample size effect, the corresponding contact problem should be formulated for a finite body, and such problems still remain to be solved by analytical methods.

Recently, the three-dimensional elastic deformation of an isotropic functionally graded plate subjected to point loading was solved by Abali et al (2014) in the special case of exponentially graded inhomogeneity using the analytical approach based on the displacement functions method (Plevako, 1971; Kashtalyan, 2004). In the present paper, we make use of the obtained singular solution and employ asymptotic modeling approach (Argatov, 2010) for evaluating the local indentation stiffness of a simply supported FGM plate in the range of small-scale indentation.

2.2 Small-scale Indentation

To fix our ideas, we consider indentation of a FGM sample, which is supposed to be isotropic with a constant Poisson's ratio, ν. To be more precise, we assume that the shear elastic modulus, $G(x_3)$, varies according to an exponential law of the type

$$G(x_3) = G_0 \exp\left(\frac{x_3}{l}\right), \tag{2.3}$$

where G_0 is the value of the shear modulus at the surface, x_3 is a Cartesian coordinate measured along the inner normal to the indented surface $x_3 = 0$, and l is the characteristic size of inhomogeneity.

In the general case of in-depth functionally graded material, the inhomogeneity characteristic size can be introduced as follows:

$$l = \left|\frac{G(0)}{G'(0)}\right|. \tag{2.4}$$

Here, $G'(0)$ is the right-hand derivative of the function $G(x_3)$ at $x_3 = 0$.

Let a be a characteristic size of the contact area, e.g., the contact radius in the case of axisymmetric indenter and a circular contact area. Then, the range of small-scale indentation can be characterized by assuming that $a/l \ll 1$. An approximate solution of the contact problem for a frictionless cylindrical indenter (Fig. 2.1) and an exponentially graded elastic medium (2.3) in the special case $\nu = 0$ was obtained by Giannakopoulos and Suresh (1997b). Based on their results, the indentation stiffness can be evaluated as follows:

$$S(a) \approx 4aG_0\left\{1 - \frac{2}{\pi}\frac{a}{l}\left(C_0 - \frac{C_1}{3}\left(\frac{a}{l}\right)^3 + \frac{2C_1^2}{45\pi}\left(\frac{a}{l}\right)^5\right)\right\}^{-1}. \tag{2.5}$$

Here, $C_0 = 3.7$ and $C_1 = 403.5$.

Observe that the factor $4aG_0$, which stands just before the curly braces in (2.5), corresponds to the isotropic homogeneous case with the surface shear modulus G_0. It is interesting that the relative difference between $S(a)$, as it is given by (2.5), and $4aG_0$ is less than 5% only in a relatively small interval $[0, 0.02)$. At the same time, the 5 percent interval for the first order approximation $S(a) \approx 4aG_0\{1 - (2/\pi)C_0(a/l)\}$ is longer and equal to $[0, 0.16)$. This example shows that due to the size effect, the classical BASh formula (2.2) and the Oliver–Pharr method can be applied only in a limited contact size range.

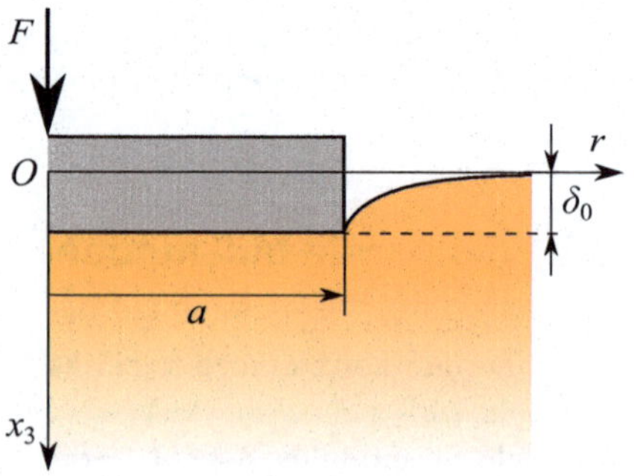

Fig. 2.1 Schematic of the cylindrical flat-ended indentation

2.3 First-order Asymptotic Model for the Indentation Stiffness

Let $G_3(x_1, x_2)$ denote the surface influence function in the Boussinesq problem of acting a unit point force on the surface of an elastic half-space along the normal to the surface and directed inside the half-space. In the case of exponentially graded material (2.3) with zero Poisson's ratio, the following solution holds (Giannakopoulos and Suresh, 1997a):

$$G_3(r) = \frac{1}{2\pi G_0 l} \int_0^\infty \mathcal{L}(u) J_0\Big(\frac{r}{l}u\Big)\, du. \tag{2.6}$$

Here, $r = \sqrt{x_1^2 + x_2^2}$ is a polar radius, $J_0(t)$ is the zeroth order Bessel function of the first kind, and

$$\mathcal{L}(u) = \frac{2u\sqrt{(2u)^2+1}}{\big(1+\sqrt{(2u)^2+1}\big)^2}. \tag{2.7}$$

First of all we observe that

$$\mathcal{L}(u) = 1 + D_1 u^{-1} + O(u^{-2}), \quad u \to \infty, \tag{2.8}$$

with $D_1 = -1$ and, therefore, the integral in (2.6) diverges as $r \to 0$.

To proceed, we recall the known formulas (e.g., Gradshteyn and Ryzhik, 1994, formulas (6.511.1) and (6.532.4))

$$\int_0^\infty J_0(ut)\, du = \frac{1}{t}, \quad \int_0^\infty \frac{uJ_0(ut)}{u^2+1}\, du = K_0(t), \tag{2.9}$$

where $K_0(t)$ is the Macdonald function, which admits the asymptotic expansion

$$K_0(t) = -\ln\frac{t}{2} + \gamma + O(t^2), \quad t \to 0, \tag{2.10}$$

with $\gamma = 0.557\ldots$ being Euler's constant.

Now, by accounting for (2.9) and (2.10), it can be shown that the integral in Eq. (2.6) possesses the asymptotic expansion

$$\int_0^\infty \mathcal{L}(u) J_0\Big(\frac{r}{l}u\Big)\, du = \frac{l}{r} + D_1 \ln\frac{l}{r} - a_0 + O\Big(\frac{r}{l}\Big), \quad r \to 0, \tag{2.11}$$

where $D_1 = -1$ is the asymptotic constant in (2.8) and a_0 is given by

$$a_0 = -D_1(\ln 2 - \gamma) + \int_0^\infty \Big(1 - \mathcal{L}(u) + \frac{D_1 u}{u^2+1}\Big) du. \tag{2.12}$$

Thus, in light of (2.11), we obtain

$$G_3(r) = \frac{1}{2\pi G_0}\Big(\frac{1}{r} + \frac{D_1}{l}\ln\frac{l}{r} - \frac{a_0}{l} + O\Big(\frac{r}{l}\Big)\Big), \quad r \to 0. \tag{2.13}$$

In order to determine the indentation stiffness, we consider the frictionless indentation problem for a flat-ended cylindrical indenter of radius a, which can be formulated in the form of the following integral equation (Vorovich et al, 1974):

$$\int_0^a p(\bar{r})K\Big(\frac{\bar{r}}{l}, \frac{r}{l}\Big)\bar{r}\,d\bar{r} = \theta_0 l\delta_0. \tag{2.14}$$

Here, $p(r)$ is the contact pressure, $\theta_0 = G_0/(1-\nu)$ is an elastic constant, δ_0 is the indenter displacement, and the kernel $K(s,t)$ is given by the integral

$$K(s,t) = \int_0^\infty \mathcal{L}(u)J_0(us)J_0(ut)\,du.$$

It can be shown (e.g., Vorovich et al, 1974) that the solution of Eq. (2.14) is related to the solution $q(r)$ of the integral equation

$$\int_0^a q(\bar{x})\,d\bar{x}\int_0^\infty \mathcal{L}(u)\cos\frac{\bar{x}}{l}u\cos\frac{x}{l}u\,du = \frac{\pi}{2}\theta_0 l\delta_0 \tag{2.15}$$

via the formula

$$p(r) = \frac{2}{\pi}\Bigg(\frac{q(a)}{\sqrt{a^2-r^2}} - \int_r^a \frac{q'(\bar{r})\,d\bar{r}}{\sqrt{\bar{r}^2-r^2}}\Bigg). \tag{2.16}$$

In turn, by introducing the dimensionless variables

$$\varphi(\xi) = \frac{q(\xi a)}{\theta_0 a}, \quad \xi = \frac{x}{a}, \quad \lambda = \frac{l}{a}, \quad f_0 = \frac{\delta_0}{a}, \tag{2.17}$$

the integral equation (2.15) can be transformed to the following form (Vorovich et al, 1974):

$$\varphi(\xi) - \frac{1}{\pi\lambda}\int_{-1}^1 \varphi(\bar{\xi})k\Big(\frac{\xi-\bar{\xi}}{\lambda}\Big)\,d\bar{\xi} = f_0. \tag{2.18}$$

Here we have introduced the notation

$$k(t) = \int_0^\infty \big[1-\mathcal{L}(u)\big]\cos ut\,du. \tag{2.19}$$

Further, for the function $\mathcal{L}(u)$ given by Eq. (2.7) and satisfying the asymptotic expansion (2.8), it can be verified that the following expansion holds (Ajzikovich and Aleksandrov, 1986):

$$k(t) = D_1 \ln |t| - a_{30} + O(t), \quad t \to 0. \tag{2.20}$$

The asymptotic constant a_{30} is given by (cf. Eq. (2.12))

$$a_{30} = \int_0^\infty \left(\mathcal{L}(u) - 1 - \frac{D_1(1 - e^{-u})}{u} \right) du. \tag{2.21}$$

Using the properties of the digamma function, we find

$$a_{30} = -a_0 - D_1 \ln 2, \tag{2.22}$$

where a_0 is the asymptotic constant (2.12).

Finally, substituting the asymptotic approximation (2.20) into Eq. (2.18) and assuming that $\lambda \gg 1$, we readily find the first-order asymptotic approximation of its solution in the form

$$\varphi(\xi) \simeq f_0 \left\{ 1 + \frac{1}{\pi\lambda} \int_{-1}^{1} \left(D_1 \ln \frac{|\xi - \bar{\xi}|}{\lambda} - a_{30} \right) d\bar{\xi} \right\}. \tag{2.23}$$

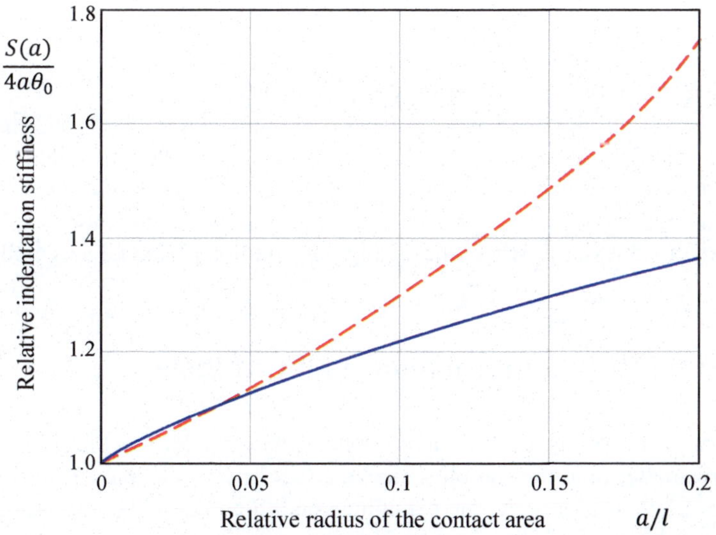

Fig. 2.2 Predictions for the relative indentation stiffness according to the approximate formula (2.5) (dashed line) and the asymptotic formula (2.25) (solid line)

By direct integration, it can be shown that

$$\int_{-1}^{1} \ln \frac{1}{|\xi - \bar{\xi}|} \, d\bar{\xi} = -\ln(1-\xi^2) + 2 + \xi \ln\Big(\frac{1-\xi}{1+\xi}\Big).$$

Thus, the substitution of (2.23) into Eqs. (2.16) and (2.17) yields

$$p(r) \simeq \frac{2}{\pi}\frac{\theta_0\delta_0}{a}\left\{\frac{1}{\sqrt{1-\rho^2}}\left(1 - \frac{1}{\pi\lambda}\left[2D_1\Big(1+\ln\frac{\lambda}{2}\Big) - 2a_{30}\right]\right)\right.$$
$$\left. + \frac{D_1}{\pi\lambda}\int_{\rho}^{1} \ln\Big(\frac{1-\xi}{1+\xi}\Big)\frac{d\xi}{\sqrt{\xi^2-\rho^2}}\right\}, \qquad (2.24)$$

where $\rho = r/a$ is the dimensionless polar radius.

In turn, the substitution of (2.24) into the formula

$$S(a) = \frac{2\pi}{\delta_0}\int_0^a p(r) r \, dr$$

leads to the following first-order asymptotic model for the indentation stiffness (cf. Ajzikovich and Aleksandrov, 1986)

$$S(a) \simeq 4a\theta_0\left\{1 + \frac{1}{\pi\lambda}\left(D_1\Big(3 + 2\ln\frac{\lambda}{2}\Big) + 2a_{30}\right)\right\}^{-1}. \qquad (2.25)$$

Here, λ is the large dimensionless parameter given by (2.17).

Note that the relative difference between the asymptotic solution (2.25) and the approximate solution (2.5) does not exceed 5% in the interval $[0, 0.09)$. It should be emphasized that the expression on the right-hand side of (2.5) has a singularity for a certain value of the ratio a/l. Hence, the accuracy of the approximate solution (2.5) is doubtful for small values of λ, where the asymptotic solution (2.25) fails as well.

2.4 Sample Size Effect in Indentation of a FGM Plate

It is clear that formulas (2.5) and (2.25) can be applied when, in addition, the contact radius a is much smaller than the sample's characteristic size, h. Moreover, in the case of a tested sample of finite size, the clamping conditions should be accounted for as well. To illustrate this issue, we consider the problem of indentation of a square FGM plate, which is assumed to be simply supported at its perimeter. In order to construct an asymptotic model for the indentation stiffness, we need the singular solution of the boundary-value problem of point loading of the plate (see

Fig. 2.3), which was investigated by a combination of analytical and numerical tools by Abali et al (2014).

Let the plate be referred to a Cartesian coordinate system (x_1, x_2, x_3), so that $-b/2 \leq x_1 \leq b/2$, $-b/2 \leq x_2 \leq b/2$, $0 \leq x_3 \leq h$. Let us also introduce the new vertical variable

$$z = h - x_3, \tag{2.26}$$

which has been used in the analysis of Abali et al (2014) and Kashtalyan (2004). According to Plevako's general solution (Plevako, 1971), the vertical component of the displacement vector can be represented in the following form (Kashtalyan, 2004)

$$\mathcal{G}_3(\mathbf{x}) = \frac{1}{G}\sum_{m=1}^{\infty}\sum_{n=1}^{\infty}\left\{-(1-\nu)\left(\frac{1}{l}\frac{\partial^2 L_{mn}}{\partial z^2}(\mathbf{x}) + \frac{\partial^3 L_{mn}}{\partial z^3}(\mathbf{x})\right)\right.$$
$$\left. + \alpha_{mn}^2\left((2-\nu)\frac{\partial L_{mn}}{\partial z}(\mathbf{x}) - \frac{\nu}{l}L_{mn}(\mathbf{x})\right)\right\}. \tag{2.27}$$

Here we have introduced the notation

$$L_{mn}(\mathbf{x}) = \phi_{mn}(z)\sin\frac{\pi m}{b}\left(\frac{b}{2} + x_1\right)\sin\frac{\pi m}{b}\left(\frac{b}{2} + x_2\right), \tag{2.28}$$

$$\phi_{mn}(z) = h^4\left[A_{1mn}f_{1mn}(z) + A_{2mn}f_{2mn}(z) + A_{3mn}f_{3mn}(z) + A_{4mn}f_{4mn}(z)\right],$$

$$f_{1mn}(z) = e^{-z/l}\cosh\frac{\lambda_{mn}z}{h}\cos\frac{\mu_{mn}z}{h},\; f_{2mn}(z) = e^{-z/l}\sinh\frac{\lambda_{mn}z}{h}\cos\frac{\mu_{mn}z}{h},$$

$$f_{3mn}(z) = e^{-z/l}\cosh\frac{\lambda_{mn}z}{h}\sin\frac{\mu_{mn}z}{h},\; f_{4mn}(z) = e^{-z/l}\sinh\frac{\lambda_{mn}z}{h}\sin\frac{\mu_{mn}z}{h}.$$

The constants A_{1mn}, A_{2mn}, A_{3mn}, and A_{4mn} can be found from the boundary conditions on the top and bottom surfaces of the plate, λ_{mn} and μ_{mn} are the roots of the characteristic equations (see Abali et al, 2014; Kashtalyan, 2004, for details), and $\alpha_{mn} = \pi\sqrt{m^2 + n^2}/b$. In the case of unit point loading, we have

$$\sigma_{33}\big|_{x_3=0} = -\frac{4}{b^2}\sum_m\sum_n \cos\frac{\pi m}{b}x_1\cos\frac{\pi m}{b}x_2, \tag{2.29}$$

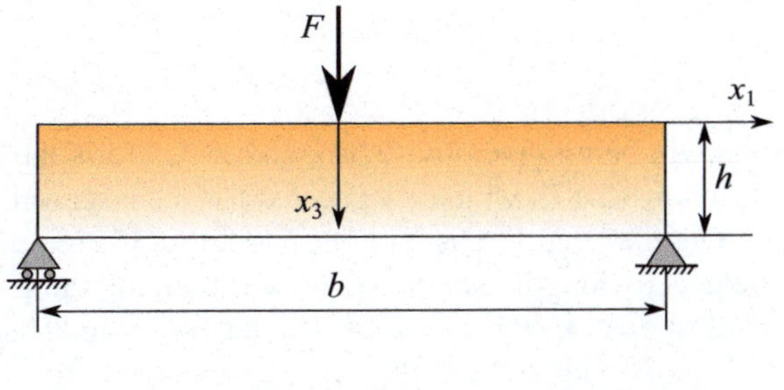

Fig. 2.3 Schematic of the point loading of a square simply supported FGM plate

where $m, n = 1, 3, 5, \ldots$.

Thus, in view of Eqs. (2.27)–(2.29), the surface normal displacement of the FGM plate can be represented in the form

$$\mathcal{G}_3(x_1, x_2, 0) = \frac{1}{2\pi G_0 l} \sum_m \sum_n \mathcal{A}_{mn} \cos\frac{\pi m}{b} x_1 \cos\frac{\pi m}{b} x_2, \tag{2.30}$$

where the dimensionless coefficients $\mathcal{A}_{mn}$ are linear combinations of A_{1mn}, A_{2mn}, A_{3mn}, and A_{4mn} $(m, n = 1, 3, 5, \ldots)$.

For a FGM plate of the type (2.6), it can be shown that the singular function (2.30) satisfies the asymptotic expansion of the type (2.13). In light of this fact, we introduce the notation

$$\mathcal{B}_{mn}^{(1)} = \frac{16l}{b^2} \int_0^{b/2}\int_0^{b/2} \frac{1}{\sqrt{x_1^2 + x_2^2}} \cos\frac{\pi m}{b} x_1 \cos\frac{\pi m}{b} x_2 \, dx_1 dx_2, \tag{2.31}$$

$$\mathcal{B}_{mn}^{(2)} = \frac{16}{b^2} \int_0^{b/2}\int_0^{b/2} \ln\frac{l}{\sqrt{x_1^2 + x_2^2}} \cos\frac{\pi m}{b} x_1 \cos\frac{\pi m}{b} x_2 \, dx_1 dx_2, \tag{2.32}$$

and put

$$\mathcal{A}_0 = -\left(2\pi G_0 l \mathcal{G}_3(x_1, x_2, 0) - \frac{l}{r} + D_1 \ln\frac{l}{r} \right)\bigg|_{r=0}, \tag{2.33}$$

where $D_1 = -1$ is the asymptotic constant from the asymptotic expansion (2.13) for the fundamental solution $G_3(r)$ of the Boussinesq problem for a FGM half-space.

Then, according to (2.30)–(2.33), we find

$$\mathcal{A}_0 = -\sum_m \sum_n \left(\mathcal{A}_{mn} - \mathcal{B}_{mn}^{(1)} - D_1 \mathcal{B}_{mn}^{(2)} \right). \tag{2.34}$$

Finally, by applying the asymptotic modeling approach (Argatov, 2010) it can be shown that the first order model for the indentation stiffness is given by

$$S(a) \approx 4a\theta_0 \left\{ 1 + \frac{1}{\pi\lambda} \left(D_1 \left(3 + 2\ln\frac{\lambda}{2} \right) + 2\mathcal{A}_{30} \right) \right\}^{-1}, \tag{2.35}$$

where, in view of (2.22), we have

$$\mathcal{A}_{30} = -\mathcal{A}_0 - D_1 \ln 2$$

with $\mathcal{A}_0$ being given by (2.34), and $\lambda = l/a$ is the dimensionless parameter introduced by (2.17), which is expected to take large values.

Observe that the asymptotic model (2.35) requires that the characteristic inhomogeneity length l should be smaller than the sample thickness h, so that $a \ll h$, as well as $a \ll b$, where a is the contact radius and b is the plate width.

2.5 Discussion and Conclusion

It should be emphasized that asymptotic models, which are usually derived under certain simplifying assumptions, should not be exploited outside the range of their applicability. With regard to the asymptotic models developed above, we observe that the singular solution (2.6), (2.5) was obtained in the special case where $\nu = 0$. For the more realistic case of non-zero Poisson's ratio, the factor $(2\pi G_0)^{-1}$ in (2.6) should be replaced with $(1-\nu)/(2\pi G_0)$, and, apparently, the asymptotic constant D_1 in (2.8) will be a function of ν.

In the general case of a functionally graded material with constant Poisson's ratio, the first-order asymptotic model (2.25) still can be used, provided the inhomogeneity characteristic size l is defined by formula (2.4).

It should be noted that the 5 percent interval, determined for the asymptotic model (2.25) based on the approximate solution (2.5) is rather small (see Fig. 2.2). This, in particular, implies that further research is needed to understand the strength of the size effect.

References

Abali BE, Völlmecke C, Woodward B, Kashtalyan M, Guz I, Müller WH (2014) Three-dimensional elastic deformation of functionally graded isotropic plates under point loading. Composite Structures 118:367–376

Aizikovich SM, Alexandrov VM, Kalker JJ, Krenev LI, Trubchik IS (2002) Analytical solution of the spherical indentation problem for a half-space with gradients with the depth elastic properties. International Journal of Solids and Structures 39(10):2745–2772

Ajzikovich SM, Aleksandrov VM (1986) Asymptotic solutions to contact problems of elasticity theory for half-space and half-plane inhomogeneous by depth (in Russ.). Izvestiya AN Armjanskoj SSR Mehanika 3:13–28

Albrecht HJ, Hannach T, Häse A, Juritza A, Müller K, Müller WH (2005) NanoIndentation: a suitable tool to determine local mechanical properties in microelectronic packages and materials? Archive of Applied Mechanics 74(11):728–738

Anstis GR, Chantikul P, Lawn BR, Marshall DB (1981) A critical evaluation of indentation techniques for measuring fracture toughness: I. Direct crack measurements. Journal of the American Ceramic Society 64(9):533–538

Argatov I (2010) Frictionless and adhesive nanoindentation: Asymptotic modeling of size effects. Mechanics of Materials 42(8):807–815

Argatov I, Sabina F (2017) A two-phase self-consistent model for the grid indentation testing of composite materials. International Journal of Engineering Science 121:52–59

Argatov I, Heß M, Popov VL (2018) The extension of the method of dimensionality reduction to layered elastic media. ZAMM - Journal of Applied Mathematics and Mechanics / Zeitschrift für Angewandte Mathematik und Mechanik 98(4):622–634

Borodich FM, Galanov BA (2008) Non-direct estimations of adhesive and elastic properties of materials by depth-sensing indentation. Proceedings: Mathematical, Physical and Engineering Sciences (The Royal Society) 464(2098):2759–2776

Borodich FM, Keer LM (2004) Contact problems and depth-sensing nanoindentation for frictionless and frictional boundary conditions. International Journal of Solids and Structures 41(9):2479–2499

Bulychev SI, Alekhin VP, Shorshorov MK, Ternovskij AP, Shnyrev GD (1975) Determination of young modulus by the hardness indentation diagram (in Russ. Zavodskaya Laboratoriya 41(9):1137–1140

Cheng L, Xia X, Yu W, Scriven LE, Gerberich WW (2000) Flat-punch indentation of viscoelastic material. Journal of Polymer Science Part B: Polymer Physics 38(1):10–22

Giannakopoulos AE, Suresh S (1997a) Indentation of solids with gradients in elastic properties: Part I. Point force. International Journal of Solids and Structures 34(19):2357–2392

Giannakopoulos AE, Suresh S (1997b) Indentation of solids with gradients in elastic properties: Part II. Axisymmetric indentors. International Journal of Solids and Structures 34(19):2393–2428

Gibson RF (2014) A review of recent research on nanoindentation of polymer composites and their constituents. Composites Science and Technology 105:51–65

Gouldstone A, Chollacoop N, Dao M, Li J, Minor AM, Shen YL (2007) Indentation across size scales and disciplines: Recent developments in experimentation and modeling. Acta Materialia 55(12):4015–4039

Gradshteyn IS, Ryzhik IM (1994) Tables of Integrals, Series and Products. Academic Press, New York

Heß M (2016) A simple method for solving adhesive and non-adhesive axisymmetric contact problems of elastically graded materials. International Journal of Engineering Science 104:20–33

Kashtalyan M (2004) Three-dimensional elasticity solution for bending of functionally graded rectangular plates. European Journal of Mechanics - A/Solids 23(5):853–864

Koumi KE, Nelias D, Chaise T, Duval A (2014) Modeling of the contact between a rigid indenter and a heterogeneous viscoelastic material. Mechanics of Materials 77:28–42

Markworth AJ, Ramesh KS, Parks WP (1995) Modelling studies applied to functionally graded materials. Journal of Materials Science 30(9):2183–2193

Müller WH, Worrack H (2012) Analysis of nanoindentation experiments by means of rheological models. PAMM 12(1):293–294

Müller WH, Worrack H, Sterthaus J, Villain J, Wilden J, Juritza A (2009) How to extract continuum materials properties for (lead-free) solders from tensile tests and nanoindentation experiments. Microsystem Technologies 15(1):45–55

Müller WH, Worrack H, Sterthaus J, Wilden J (2009) Nanoindentation experiments at elevated temperatures for the determination of mechanical solder material properties. PAMM 9(1):719–720

Müller WH, Worrack H, Zapara M (2011) Analysis of nanoindentation experiments by means of atomic force microscopy. PAMM 11(1):413–414

Oliver WC, Pharr GM (1992) An improved technique for determining hardness and elastic modulus using load and displacement sensing indentation experiments. Journal of Materials Research 7(6):1564–1583

Plevako VP (1971) On the theory of elasticity of inhomogeneous media. Journal of Applied Mathematics and Mechanics 35(5):806–813

Suresh S, Giannakopoulos AE, Alcalá J (1997) Spherical indentation of compositionally graded materials: Theory and experiments. Acta Materialia 45(4):1307–1321

Villain J, Mueller WH, Haese A, Weippert C, Corradi U, Saeed U (2008) Determination of mechanical properties of small test volumes using nanoindentation - a critical review. In: 2008 2nd Electronics System-Integration Technology Conference, pp 155–162

Vorovich II, Aleksandrov VM, Babeshko VA (1974) Non-Classical Mixed Problems in the Theory of Elasticity (in Russ.). Nauka, Moscow

Weinberg K, Sterthaus J, Müller WH (2005) Determining material parameter of solder alloys by nanoindentation. PAMM 5(1):451–452

Chapter 3
The Effect of Mechanical Load-induced Intraosseous Pressure Gradients on Bone Remodeling

Emilio Barchiesi, Ivan Giorgio, Faris Alzahrani & Tasawar Hayat

Abstract It is well established that changes in bone blood and interstitial fluid flows are associated with changes in the bone remodeling process. These flows in bone are a result not only of trans-cortical pressure gradients produced by vascular and hydro-static pressure, but also of mechanical loadings. Mechanical load-induced intraosseous pressure gradients may result in some fluid stimuli effects which, in turn, may enable bone cells to detect external mechanical signals. In this paper, the exploitation of a 2D continuum model based on classical poroelasticity is presented within a variational framework. The investigation is aimed at describing how mechanical actions can affect the remodeling process of a bone tissue. The focus is on the introduction of a physically motivated strain energy contribution aimed to take into account the presence of saturating fluid in the interconnected pores of bone tissue. The interaction with a bio-resorbable organic ceramic material like those used in bone graft implants is also considered in presented model. Numerical results are provided in a relevant exemplary case.

Keywords: Mechanical–biological coupling · Bone functional adaptation · Growth resorption processes · Bone remodeling

Emilio Barchiesi & Ivan Giorgio
Department of Structural and Geotechnical Engineering, Università degli studi di Roma La Sapienza, 18 Via Eudossiana, Rome & International Research Center for the Mathematics and Mechanics of Complex Systems - MeMoCS, Università degli studi dell'Aquila, L'Aquila, Italy,
e-mail: BarchiesiEmilio@gmail.com, ivan.giorgio@uniroma1.it

Faris Alzahrani
NAAM Research Group, Department of Mathematics, King Abdulaziz University, Jeddah 21589, Saudi Arabia,
e-mail: faris.kau@hotmail.com

Tasawar Hayat
Department of Mathematics, Quaid-I-Azam University, Islamabad, Pakistan & NAAM Research Group, Department of Mathematics, King Abdulaziz University, Jeddah 21589, Saudi Arabia,
e-mail: pensy_t@yahoo.com

B. E. Abali et al. (eds.), *New Achievements in Continuum Mechanics and Thermodynamics*, Advanced Structured Materials 108,
https://doi.org/10.1007/978-3-030-13307-8_3

3.1 Introduction

In bone tissue, it is possible to distinguish mainly two kinds of fluids: blood and interstitial fluids like, e.g., bone marrow. Blood carries through the arterial system oxygen and other nutrients, and the blood components depart from this arterial system via smaller channels, i.e. the venous system, to zones containing less oxygen and reduced nutrition (see, e.g., George et al, 2018a; Spingarn et al, 2018). Within the bone, as within other tissues, substances pass from the blood flowing through the arterial walls into the interstitial fluid. The interstitial fluid subsequently carries these substances to the cells within the bone and, at the same time, carries away the waste materials from the cells. Bone tissue would not remain alive without these fluid movements. It is thus clear the reason why it is commonly accepted that vascularization is required for effective bone healing and maintenance. These statements are supported by the fact that changes in bone blood and interstitial fluid flows are associated with changes in bone remodeling and formation (see, e.g., Hillsley and Frangos, 1994). These flows in bone are a result not only of trans-cortical pressure gradients produced by vascular and hydro-static pressure, but are also related to externally applied mechanical loadings. It is observed that flow rates are affected by many factors, like the increase in venous pressure due to hypertension, the fluid shifts occurring in bedrest or microgravity, the increase in vascularization during the injury-healing response, and the mechanical compression/tension and bending/torsion of bone during exercise. Thus, mechanical load-induced intraosseous pressure gradients, like those induced by mechanical loading of bone during exercise, affect the fluid flow rate and, eventually, bone remodeling. The purpose of this article is to present a mathematical model able to describe the role of mechanical actions in the bone osteogenesis process. In this paper we will make use of the classical poroelasticity theory, as formulated by Biot in its 1941's foundational paper (Biot, 1941), suitably complemented with a novel non-local energy contribution purposely introduced to account for fluid compression. In poroelasticity, a field accounting for porosity is usually introduced in addition to the placement function of the solid phase. Poroelasticity is, thus, a so-called micromorphic theory, belonging to the wider class of generalized continua with internal variables or with extra kinematical descriptors. The importance of these continua has been questioned, but, in our opinion, has been proven useful in presence of long range interactions at micro-level, when a macro continuous model is more suitable (see, e.g., Alibert et al, 2003; Eremeyev et al, 2018a; Abali et al, 2017; Pietraszkiewicz and Eremeyev, 2009). The macroscopic theories formulated in the framework of the mechanics of generalized continua is being formulated for 3D and 2D bodies and is increasingly attracting the attention of those researchers interested in non-standard mechanical effects (see, e.g., Altenbach and Eremeyev, 2009; Bertram and Glüge, 2016; Gusev and Lurie, 2017; Camar-Eddine and Seppecher, 2001). The recent literature stresses two aspects of the considered multi-scale mechanical systems: their potentially exotic macroscopic behavior and the corresponding microscopic structure, in which there are eventually active long range interactions. The need for generalized continua, including in this class also higher gradient theories (dell'Isola and

Seppecher, 1997; dell'Isola et al, 2012; dell'Isola and Steigmann, 2015; dell'Isola et al, 2016, 2015b), is unavoidable when one has to describe those mechanical phenomena which involves the activation of deformation modes at microlevel determining the interaction of parts of the micro-structure having high stiffness contrast and bridging distant homogenization cells. These complex deformation patterns cannot be accounted for in standard Cauchy theories: for a series of examples of this circumstance, see e.g., Cuomo et al (2016); dell'Isola et al (2016); Placidi et al (2016) while for theoretical arguments dealing with micro-macro convergence motivating higher gradient theories, see e.g., Abdoul-Anziz and Seppecher (2018); Seppecher et al (2011); Pideri and Seppecher (1997). In the present paper, we deal with a reconstructed bone during its remodeling process. This is surely a multi scale complex system, which involves mechanical, chemical and biological aspects and, therefore the previous modeling concerns are surely appropriate. This approach is present in the literature of biomechanics of bones (see, e.g., Lekszycki and dell'Isola, 2012; Giorgio et al, 2017; Ganghoffer, 2016; Goda et al, 2014, 2012; Ganghoffer, 2012) but has attracted the attention also in view of different possible applications. Generalized continua are indeed considered also for modeling electromechanical systems in biological applications (see, e.g., Steigmann and Agrawal, 2016) memory shape alloys (Shirani et al, 2017) and piezo/flexo-electric materials (see, e.g., Abdalladan et al, 2017; Abd-alla et al, 2017; Pagnini and Piccardo, 2016; Enakoutsa et al, 2017). These studies may have a relevance in the process of bone remodeling, if one of the mechanisms regulating considered bone growth process are regulated by electromagnetically induced biological activity, as it seems to happen when electrical currents are used to favor bone growth. Growing bones are resisting to external load, also in elastic regime: therefore, generalized elastic continua can be of use in the class of biomechanics phenomena which we consider here. In this context, the works (Andreaus et al, 2010; Rosi et al, 2018; Abali et al, 2015; Altenbach and Eremeyev, 2015; Franciosi et al, 2018; Spagnuolo and Andreaus, 2018; Andreaus et al, 2018), which try to capture some aspects of the elastic deformation of reconstructed bones, are relevant. As we hope to have explained already in an exhaustive way, the complexity of behavior of a reconstructed bone does not allow for too drastic simplifications. Surely in the small and larger channels in which interstitial fluids are flowing may activate capillary phenomena: therefore, the analysis of capillary fluids (Auffray et al, 2015; Seppecher, 1993, 2000) and their influence at macro-level may be of relevance. In this context, the results presented in Madeo et al (2013); Sciarra et al (2007) may be of use, as well as the analysis of damage phenomena (see, e.g., Placidi et al, 2018; Rinaldi and Placidi, 2014; Placidi, 2015; Misra and Singh, 2013; Spagnuolo et al, 2017; Goda and Ganghoffer, 2015; Di Nino et al, 2017; Battista et al, 2017b) based on generalized continuum models. Remark that in the last paper a purely discrete model for mechanical phenomena is introduced, based on the postulation involving a generalization of cellular automata. Indeed, among the other feed-back mechanisms which regulate bone remodeling, it has been proven that it is particularly effective that which activates the action of osteoblasts and osteoclast when the microstructure of the bone exhibits some growing damage and micro fracture. One should not, however, believe that the only feed-back control mechanism of growth involves a measure of micro-fracture of bone. It seems that also the dissipation occurring in the interstitial fluids at the level of micro-canalicula and micro-trabecular structure may be effective. As a consequence dissipation phenomena has to be included in the modeling process (e.g. using the ideas presented in Lekszycki et al (2017); Cuomo (2017); Luongo and D'Annibale (2017). If the considered phenomena involve lower scales then the granular structure of the bone and the reconstructing material must be accounted for: in this case the modeling issues addressed in Misra and Poorsolhjouy (2015a); Misra and Singh (2015); Misra and Poorsolhjouy (2015b); Altenbach et al (2010); Eremeyev (2018), may become relevant. Finally, it has to be remarked that the peculiar features exhibited by reconstructed bones and by physiological bones imply some specific corresponding nonstandard characteristics in wave propagation. These peculiarities may be exploited to get informations, using noninvasive methods, about the health and mechanical performance of (possibly reconstructed) bone tissue. In this case wave propagation analysis proposed in Placidi et al (2008); Engelbrecht and Berezovski (2015); Berezovski et al (2018); Abbas et al (2016); Eremeyev et al (2018b) while dynamic analyses performed in Battista et al (2015, 2017a); Ferretti et al (2017) may be of use.

The plan of the work present is the following: first, in Sect. 3.2, we address some basic facts on bone physiology, and will extrapolate the main information needed for the modeling. Then, in Sect. 3.3, we introduce the proposed continuum model. The focus is on the introduction of a physically motivated dedicated strain energy contribution due to the presence of saturating fluid in the interconnected pores of bone tissue and the interaction with a bio-resorbable organic ceramic material like those used in bone graft implants is considered. In Sect. 3.4, we provide numerical results in a relevant exemplary case. Finally, in Sect. 3.5, we give conclusions and perspectives.

3.2 Some Considerations on Bone Physiology

Bone is a 'rigid' living organ that constitutes the vertebrate skeleton; it protects the organs of the body and enables mobility. It has a honeycomb-like matrix internally, which helps to give the bone rigidity and that can be considered porous with respect to the characteristic length that we are concerned with at the macroscopic level. The pores of the solid matrix are filled with interstitial fluid. Inside bone tissue, different types of bone cells act. Osteoblasts and Osteocytes are involved in the formation and mineralization of bone; Osteoclasts are involved in the resorption of bone tissue. Modified (flattened) Osteoblasts become the lining cells that form a protective layer on the bone surface. The mineralized matrix of bone tissue has an organic component of mainly collagen, called ossein, and an inorganic component of bone mineral made up of various salts. Bone tissue is a mineralized tissue of two types, cortical and cancellous bone. Other types of tissue found in bones include bone marrow, endosteum, periosteum, nerves, blood vessels and cartilage. Bone tissue is constantly reshaped by the biological activity which takes place therein. Biology tells us that the basic (feedback) mechanism, relevant for the subsequent modeling, is the one shown in Fig. 3.1. External mechanical actions are sensed by the Osteocytes which, being actors of the mechano-transduction process, transduce the mechanical stimulus into a biological, i.e. bio-chemical, signal, inducing the resorption of bone tissue by the Osteoclasts or its synthesis by the Osteoblasts. Bio-resorbable artificial materials can be roughly divided into polymeric and ceramic materials. Polymers are mainly degraded in the process of hydrolysis, while organic ceramics like, e.g., β-TCP (Tri-Calcium Phosphate) with 5% of HAP (Hydroxyapatite), are instead mostly processed by the Osteoclasts. Both processes always coexist and, depending on the constitution of the graft, one process or the other results to be dominant. In this paper we consider resorption due to Osteoclasts only. Taking into account the presence of a bone graft allows to better understand and predict how mechanical actions can affect the remodeling process of a bone tissue possibly interacting with a bio-resorbable material like those used in bone graft implants (e.g. dental bone grafting), whether the implanted graft is suitable as mechanically supporting element, the capability of the graft to integrate with the bone tissue, so that it can be gradually resorbed and replaced by new natural bone tissue. This last property is conferred by chemical composition, number and size of pores, because they determine the extent to which living cells migration, their expected activities, the development of vascularization for cell survival etc. can occur.

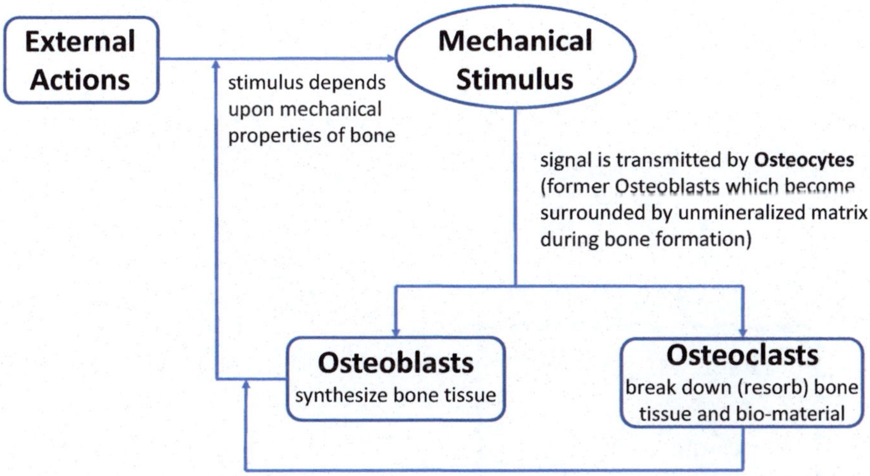

Fig. 3.1 Basic feedback mechanism in a bone. A detailed description of the physiology of bone remodeling would be out of reach.

3.3 Modelling

3.3.1 Kinematics

In this paper we are aimed only at showing the main features of the model and, thus, notwithstanding the fact that the subsequent modeling is suitable also for the study of 3D bodies, it is here sufficient to consider a 2D body. Such a body is made up of a mixture composed by three phases: the binary solid porous matrix of bone (B) and bio-resorbable graft material (M) and the fluid phase (F) that fills the connected pores of the solid matrix. The shape of the body in its undeformed reference configuration is represented by the subset $\mathcal{B}_0 \subset \mathbb{R}^2$. We will not make use of a so-called mixture model in a strict sense, meaning that we are not going to consider as independent kinematical descriptors of the model placement functions for each component of the mixture. The only (sufficiently regular) displacement field $u : \mathbb{R}^2 \supseteq \mathcal{B}_0 \to \mathbb{R}^2$, with $\chi(X) = X + u(X)$ being the corresponding placement function, which we consider as an independent Lagrangian kinematic variable is such that $u(X,t)$ is the displacement of the solid binary mixture in the representative three-phase volume element (see Fig. 3.2) whose barycenter is in X in the reference configuration.

While the relative displacement of the components of the solid matrix can be neglected, the same does not hold for the fluid phase, which in general can move (only, as impermeability of the solid matrix is here assumed) inside the solid matrix. The set $\mathcal{B} \equiv \chi(\mathcal{B}_0) \subset \mathbb{R}^2$ is the current shape of the body. We shall denote the Lagrangian representation $e\left(\chi(X)\right)$ of an Eulerian field $e(x)$, where $x \in \mathcal{B}$ and it

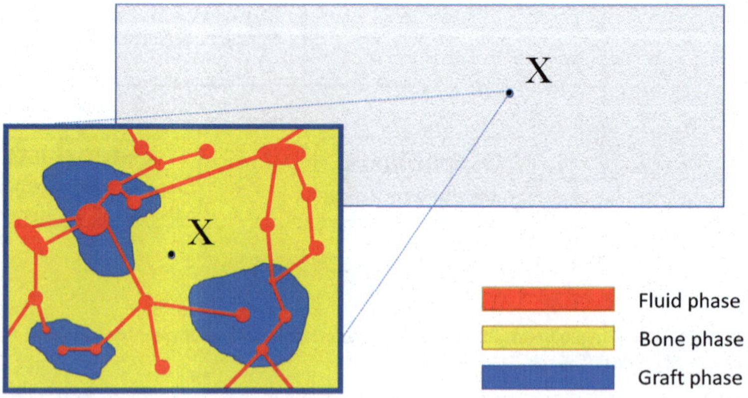

Fig. 3.2 A rectangular 2D body with a zoomed schematic of a representative volume element (RVE).

is such that $x = \chi(X)$, with $e^{\mathcal{L}}(X)$. The (apparent) Lagrangian mass density in the reference configuration $\rho_R(X)$, with $X \in \mathcal{B}_0$, is given by

$$\rho_R(X) = \sum_{i=B,M,F} \rho_{i,R}(X) = \sum_{i=B,M,F} \hat{\rho}_{i,R}(X)\xi_{i,R}(X), \tag{3.1}$$

where $\rho_{i,R}$ is the partial Lagrangian mass of component i in the mixture in the reference configuration, is the $\hat{\rho}_{i,R}$ (true) Lagrangian mass density of component i in a mono-phase mixture in the reference configuration, $\xi_{i,R}$ is the Lagrangian volume fraction of component i in the mixture in the reference configuration which, since the porous solid matrix is saturated with fluid, is also referred to as the porosity in the reference configuration. The (apparent) Eulerian mass density in the current configuration $\rho_C(x)$, $x \in \mathcal{B}$, is given by

$$\rho_C(x) = \sum_{i=B,M,F} \rho_{i,C}(x) = \sum_{i=B,M,F} \hat{\rho}_{i,C}(x)\xi_{i,C}(x), \tag{3.2}$$

with $\rho_{i,C}$ the partial Eulerian mass of component i in the mixture in the current configuration, $\hat{\rho}_{i,C}$ the (true) Eulerian mass density of component i in a mono-phase mixture in the current configuration, $\xi_{i,C}$ the Eulerian volume fraction of component i in the mixture in the current configuration which, since the porous solid matrix is saturated with fluid, is also referred to as the porosity in the current configuration (Wilmanski, 1998). As we mentioned above, in this paper we consider undrained conditions, i.e. impermeability conditions at the boundary. Thus, the following global (as opposed to local) mass conservation constraint holds for the fluid phase

$$\int_{\mathcal{B}_0} \rho^{\mathcal{L}}_{F,C}(X)\, \det\nabla\chi \mathrm{d}X = \int_{\mathcal{B}_0} \rho_{F,C}\left(\chi(X)\right)\, \det\nabla\chi \mathrm{d}X \tag{3.3}$$

$$\int_{\mathcal{B}} \rho_{F,C}(x)\, \mathrm{d}x = M_F = \int_{\mathcal{B}_0} \rho_{F,R}(X)\, \mathrm{d}X, \tag{3.4}$$

with M_F being the total fluid mass in the body. Finally, in the spirit of continuum poroelasticity, we introduce another independent Lagrangian micromorphic field $\vartheta : \mathbb{R}^2 \supseteq \mathcal{B}_0 \to \mathbb{R}$, which is the change of porosity. Following Coussy (2004), we assume that

$$\vartheta(X) := \xi_{F,C}\left(\chi(X)\right) - \xi_{F,R}(X) = \xi^{\mathcal{L}}_{F,C}(X) - \xi_{F,R}(X). \tag{3.5}$$

3.3.2 Elastic Mechanical Energy Stored Within the Body

For fixed $\xi_{B,R}$, $\xi_{M,R}$ (remind that $\xi_{F,R} = 1 - \xi_{B,R} - \xi_{M,R}$), the system is assumed to behave elastically. We consider the quasi-static case, i.e. inertia and micro-inertia forces/energies are negligible, and the total deformation energy of the system is assumed to be

$$\psi = \psi_{\text{STRAIN}} - \psi_{\text{EXT}} \tag{3.6}$$

with

$$\psi_{\text{STRAIN}} = \psi_{\text{POR}} + \psi_{\text{PERIDYN}} \tag{3.7}$$

and

$$\psi_{\text{EXT}} = \int_{\mathcal{B}_0} b^{ext} \cdot u \, \mathrm{d}X + \int_{\partial\mathcal{B}_0} f^{ext} \cdot u \, \mathrm{d}X, \tag{3.8}$$

the quantities b^{ext} and f^{ext} in (3.8) being, respectively, bulk and surface loads (see for more detail on variational formulation e.g. Abali et al, 2017; Eugster and Glocker, 2017). Let us now examine the terms ψ_{POR} in (3.7), which is the poro-elasticity strain energy density contribution in Lagrangian form i.e. the energy stored within the body due to the deformation of the solid matrix and to pores surface tension and related phenomena (Giorgio et al, 2016). In this paper we consider the small strain assumption (i.e. $\nabla u \simeq 0$) and, in what follows, $E = \text{Sym}\nabla u$ is the linearized Green-Saint Venant strain tensor. The purely (no pre-stress) quadratic form in the strain E and micro-strain ϑ is

$$\psi_{\text{POR}}(E,\vartheta) = \int_{\mathcal{B}_0} \Bigg[\underbrace{\frac{Q}{2}\left(\vartheta - \alpha\,\text{tr}(E)\right)^2}_{\text{Biot's contribution}} + \underbrace{\frac{1}{2}\frac{Y(\rho_{B,R},\rho_{M,R})\nu}{(1-2\nu)(1+\nu)}\text{tr}(E)^2 + \frac{1}{2}\frac{Y(\rho_{B,R},\rho_{M,R})}{(1+\nu)}\text{tr}(E^2)}_{\text{isotropic strain energy density of the solid bone+graft mixture}} \Bigg] \mathrm{d}X \tag{3.9}$$

where

$$Y = Y_B \xi_{B,R}^{\beta_B} + Y_M \xi_{M,R}^{\beta_M} \tag{3.10}$$

is the effective bone-graft Young modulus, Y_B is the bone Young modulus, Y_M is the graft Young modulus, β_B and β_M are two constitutive exponents, ν is the effective bone-graft Poisson's ratio (set to be 0.3), $Q > 0$ is the 1st Biot parameter (resistance to change of porosity), and α is the 2nd Biot parameter. Specifically, for the sake of simplicity, we set

$$\alpha = \frac{Y}{H_1\,3(1-2\nu)} \qquad \frac{1}{Q} = \frac{1}{R} - \frac{\alpha}{H_1} \tag{3.11}$$

with H_1 and R positive constants (Biot, 1941). We remark that Biot's contribution includes coupling between u and ϑ an we stress again that, contrarily to what is customarily done in classical continuum poroelasticity, the Biot's contribution does not encode the energy part due to interstitial fluid compression. We now discuss the peridynamic (in the sense given by Piola, see dell'Isola et al (2015a)) contribution in Lagrangian form ψ_{PERIDYN} in (3.7). In Eulerian form, the energy stored within the body due to fluid compression is assumed to be

$$\psi_{\text{PERIDYN}} = \frac{1}{2}\beta \int_{\mathcal{B}} \hat{\rho}^2_{F,C}\, \xi_{F,C}\, \mathrm{d}x, \tag{3.12}$$

with $\beta > 0$ being the fluid resistance to compression and the differential $\mathrm{d}V_F = \xi_{F,C}\,\mathrm{d}x$ indicates that integration is taken with respect to the fluid volume. Assuming $\hat{\rho}_{F,C}$ to be uniform over $\mathcal{B}$, and this is reasonable in a quasi-static framework, we have

$$\psi_{\text{PERIDYN}} = \frac{1}{2}\beta\hat{\rho}^2_{F,C} \int_{\mathcal{B}} \xi_{F,C}\,\mathrm{d}x. \tag{3.13}$$

We now want to transform the integration over the (unknown) deformed shape $\mathcal{B}$ in (3.13) into an integration over the reference shape $\mathcal{B}_0$, i.e. we want to derive ψ_{PERIDYN} in Lagrangian form. To this goal, we perform the change of variable $x = \chi(X)$. We have that $\mathrm{d}x = \det(F)\,\mathrm{d}X + o(\mathrm{d}X)$, with $F = \nabla\chi = I + \nabla u$. Reminding that

$$\det(I + \varepsilon A) = 1 + \varepsilon\,\mathrm{tr}(A) + o(\varepsilon) \tag{3.14}$$

$$\det(I + A) = 1 + \mathrm{tr}(A) + o(A), \tag{3.15}$$

we have $\det(F) = 1 + \mathrm{tr}(\nabla u) + o(\nabla u)$. Since we are working under the small strain hypothesis, observing that $\mathrm{tr}(A) = \mathrm{tr}(\mathrm{Sym}A)$, and neglecting higher order contributions, we have $\det(F) = 1 + \mathrm{tr}(E)$. The volume occupied by the fluid phase in the current configuration is

$$V_F = \int_{\mathcal{B}} \xi_{F,C}\,\mathrm{d}x = \int_{\mathcal{B}_0} \xi^{\mathcal{L}}_{F,C}\,(1 + \mathrm{tr}(E))\,\mathrm{d}X. \tag{3.16}$$

The energy stored within the body due to fluid compression reads thus in Lagrangian form as

$$\psi_{\text{PERIDYN}} = \frac{\beta M_F^2}{2} \frac{1}{\int_{\mathcal{B}_0} \xi^{\mathcal{L}}_{F,C}\,(1 + \mathrm{tr}(E))\,\mathrm{d}X}. \tag{3.17}$$

Of course, we have that

$$\begin{aligned}
\psi_{\text{PERIDYN}} &= \frac{\beta M_F^2}{2} \frac{1}{\int_{\mathcal{B}_0} \xi^{\mathcal{L}}_{F,C}(1 + \mathrm{tr}(E))\,\mathrm{d}X} \frac{\int_{\mathcal{B}} \xi_{F,C}\,\mathrm{d}x}{V_F} = \\
&\frac{1}{V_F}\int_{\mathcal{B}} \frac{\beta M_F^2}{2} \frac{1}{\int_{\mathcal{B}_0} (\xi_{F,R} + \vartheta)(1 + \mathrm{tr}(E))\,\mathrm{d}X}\,\mathrm{d}x = \\
&\frac{1}{\int_{\mathcal{B}_0} (\xi_{F,R} + \vartheta)(1 + \mathrm{tr}(E))\,\mathrm{d}X} \int_{\mathcal{B}_0} \frac{\beta M_F^2}{2} \frac{\xi^{\mathcal{L}}_{F,C}(1 + \mathrm{tr}(E))}{\int_{\mathcal{B}_0} (\xi_{F,R} + \vartheta)(1 + \mathrm{tr}(E))\,\mathrm{d}X}\mathrm{d}X = \\
&\int_{\mathcal{B}_0} \frac{\beta M_F^2}{2} \frac{(\xi_{F,R} + \vartheta)(1 + \mathrm{tr}(E))}{\left[\int_{\mathcal{B}_0} (\xi_{F,R} + \vartheta)(1 + \mathrm{tr}(E))\,\mathrm{d}X\right]^2}\,\mathrm{d}X
\end{aligned} \tag{3.18}$$

where $\xi^{\mathcal{L}}_{F,C} = (\xi_{F,R} + \vartheta)$ has been used. Therefore, the Lagrangian density associated to ψ_{PERIDYN} reads as

$$\frac{\beta M_F^2}{2} \frac{(\xi_{F,R} + \vartheta)\,(1 + \text{tr}(E))}{\left[\int_{\mathcal{B}_0} (\xi_{F,R} + \vartheta)\,(1 + \text{tr}(E))\,\text{d}X\right]^2}. \tag{3.19}$$

We notice that no spatial derivatives of the additional kinematic parameter ϑ appear in the first gradient (with respect to displacement) poroelasticity strain energy ψ_{POR}; hence, the prescription of arbitrary boundary conditions for the porosity field does not yield in general the minimization of the mechanical energy ψ a well-posed problem. Furthermore, we remark that non-locality is given by the peridynamic contribution ψ_{PERIDYN} and not by the dependence of the internal stored energy upon, e.g., higher gradients of the displacement and/or of the change of porosity. Positive definiteness of the isotropic strain energy density of the solid bone-graft mixture is ensured if $\lambda + \mu > 0$ and $\lambda - \mu > 0$, where λ is the effective bone-graft first Lamé parameter and μ is the effective bone-graft shear modulus. Finally, we observe that, following our assumptions, the strain energy is such that remodeling, i.e. a change in the densities of bone tissue and bio-resorbable material in the reference configuration, cannot induce any (local) mechanical anisotropy (for more details see Allena and Cluzel, 2018; Cluzel and Allena, 2018).

3.3.3 Mechanical Stimulus, Bone Remodeling and Graft Resorption

Let $\rho_{OC,R}$ be the Lagrangian density of Osteocytes in the reference configuration. It is assumed to be proportional to the bone density in the reference configuration. Nevertheless, in the literature it is possible to find approaches assuming that the time evolution of the Lagrangian density of Osteocytes in the reference configuration, together with those of Osteoblasts and Osteoclasts, is governed by a distributed cellular population evolution model (Lekszycki and dell'Isola, 2012; George et al, 2018b,c). The Lagrangian mechanical stimulus is defined as

$$S(X,t) = \int_{\mathcal{B}_0} \left[\psi(Y,t)\rho_{OC,R}(Y,t)\, e^{-\frac{(X-Y)^2}{2D^2}}\,\text{d}Y\right] - S_0 \tag{3.20}$$

i.e. it is the 2D Gaussian convolution of the product $\psi\ \rho_{OC,R}$. We remark that the 2D Gaussian $\sim e^{-\frac{(X-Y)^2}{2D^2}}$ has mean X and standard deviation D (variance D^2) and, thus, D is a measure of the circular influence range, since the Gaussian is de facto vanishing at a distance from the mean X greater than three times the standard deviation. We further remark that stimulus is the (spatially) shifted smoothed product $\psi\ \rho_{OC,R}$, which is the Lagrangian strain energy density weighted by the Lagrangian Osteocytes density in the reference configuration; this entails that a non-zero strain energy density in a sufficiently close neighborhood of a point does not necessarily imply a non-zero stimulus in that point, as in that neighborhood there might be no mechanosensors. Finally, it is worth to be noticed that in the literature (Beaupré et al, 1990; Giorgio et al, 2016) it possible to find a slight variation of the definition of stimulus employed herein, by taking into account a 'dead-zone'

$$S(X,t) = \begin{cases} \tilde{S}(X,t) - S_u & \text{if } \tilde{S}(X,t) \geq S_u \\ \tilde{S}(X,t) - S_d & \text{if } \tilde{S}(X,t) \leq S_d \\ 0 & \text{if } S_d < \tilde{S}(X,t) < S_u \end{cases} \tag{3.21}$$

with

$$\tilde{S}(X,t) = \int_{\mathcal{B}_0} \psi(Y,t)\rho_{OC,R}(Y,t)\, e^{-\frac{(X-Y)^2}{2D^2}}\, \mathrm{d}Y. \tag{3.22}$$

The evolution of graft density due to resorption and of bone density due to formation and resorption is described, for each X, by means of a system of ordinary differential equations

$$\begin{cases} \dot{\rho}_{M,R}(X,t) & = A_M(S)\, H(\xi^{\mathcal{L}}_{F,C}) \\ \dot{\rho}_{B,R}(X,t) & = A_B(S)\, H(\xi^{\mathcal{L}}_{F,C}) \\ \rho_{M,R}(X,0) & = \rho_{M0}(X) \\ \rho_{B,R}(X,0) & = \rho_{B0}(X) \end{cases} \tag{3.23}$$

with

$$H(y) = k\, y(1-y), \qquad \text{with } y \in [0,1] \tag{3.24}$$

$$A_M(S) = \begin{cases} 0 & \text{if } S \geq 0 \\ r_M S & \text{if } S < 0, \end{cases} \tag{3.25}$$

$$A_M(S) = \begin{cases} s_B S & \text{if } S \geq 0 \\ r_B S & \text{if } S < 0, \end{cases} \tag{3.26}$$

and k, r_M, r_B, $s_B \geq 0$. We remark that the function $H(\xi^{\mathcal{L}}_{F,R}) = k\xi^{\mathcal{L}}_{F,R}(1 - \xi^{\mathcal{L}}_{F,R})$, a concave symmetric parabola with maximum value $k/4$ attained in $\xi^{\mathcal{L}}_{F,R} = 1/2$ and intersecting the abscissa $H = 0$ in $\xi^{\mathcal{L}}_{F,R} = 0$ and $\xi^{\mathcal{L}}_{F,R} = 1$, accounts for the fact that if the porosity is too low, then living cells will not be able to efficiently resorb the bio-material, neither to form new bone tissue, as the available space will not allow for the activity of a sufficiently large number of actor cells. If the porosity, on the other hand, is too large, then there is not enough solid-phase on which actor cells may deposit, and remodeling will not occur quickly enough. For the sake of simplicity, we set $k = 4$.

3.4 Solution Algorithm and Qualitative Results for Tensile Test

Dimensionless parameters employed for numerical simulations are shown in Tab. 3.1. Specifically, we consider a rectangular sample whose sides are in the ratio 1:3.

The notation $\langle \psi', \delta u \rangle$ denotes the functional derivative of ψ in Eq. (3.6) in the direction δu (displacement variation). For a fixed time instant and corresponding (given) external bulk and surface forces b^{ext}, equation $\langle \psi, \delta u \rangle = 0 \; \forall \delta u$— which corresponds to the so-called weak form— is solved by means of standard finite element techniques included within the weak form package of the commercial software COMSOL Multiphysics^{TM}. From such a computation—$\rho_{B,R}$ and $\rho_{M,R}$ at previous time step are used to retrieve the effective bone-graft Young's modulus—we get the displacement u and change of porosity at time t. Such displacement and $\rho_{OC,R}$— obtained by assuming it to be proportional by a factor K to $\rho_{B,R}$ at the previous time step—are plugged within the Lagrangian mechanical stimulus defined in Eq. (3.20). The so-found Lagrangian mechanical stimulus is inserted with the change of porosity retrieved from the weak form solution in Eq. (3.23) in order to compute $\rho_{B,R}$ and $\rho_{M,R}$ at the current time step. Such values are then used to compute the effective Young modulus to be plugged in the weak form equation at the next time step.

In Fig. 3.3 a block diagram representation of the solution algorithm employed is shown, while a graphical depiction of the test problem is presented in Fig. 3.4. A rectangular slab formed by a rectangular central graft inclusion in bone tissue matrix is subject to a tensile test. In Fig. 3.5, a zoomed detail for the test problem is shown. At the interface between the two phases we consider a spring foundation for both kinematics variables u and ϑ, and a contribution

$$\int_{\partial G} \left[K_u \| u^+ - u^- \|^2 + K_\vartheta \| \vartheta^+ - \vartheta^- \|^2 \right] \mathrm{d}X \tag{3.27}$$

is added to the strain energy. In Fig. 3.6, the evolution in time of minimum graft density (blue), $\min \rho_{M,R}$, and maximum bone tissue density (green), $\max \rho_{B,R}$, over the inclusion zone are shown. Clearly, as time progresses, the graft is resorbed (i.e. the blue curve is monotonously non-increasing) and new bone is formed (i.e. the green curve is monotonously non-decreasing). In Fig. 3.7 the evolution in time of the stimulus S for some relevant time instants is reported. Time increases from left

Table 3.1 Material coefficients in non-dimensional form. Tilde denotes dimensionless quantities.

$\tilde{Y}_B$	$\tilde{Y}_M$	$\tilde{s}_B$	$\tilde{r}_B$	$\tilde{r}_M$
1	1.2	5×10^8	5×10^8	5×10^8
$\tilde{\rho}_{B0} = \tilde{\rho}_{M0}$	$\beta_B = \beta_M$	$\tilde{D}$	K_u	K_ϑ
0.5	2	0.9	0.1	0.1
$\tilde{H}_1$	$\tilde{R}$	$\beta \tilde{M}_F{}^2$	$\tilde{S}_d$	$\tilde{S}_u$
0.8	0.4	1.0×10^{-4}	1.30×10^{-6}	1.30×10^{-6}

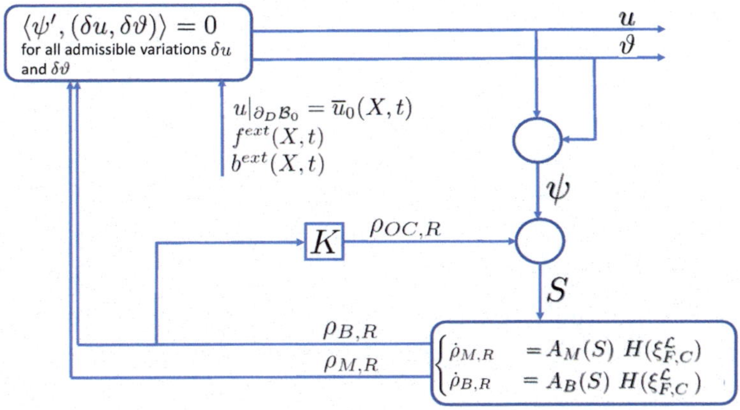

Fig. 3.3 Feedback loop schematic for model solving.

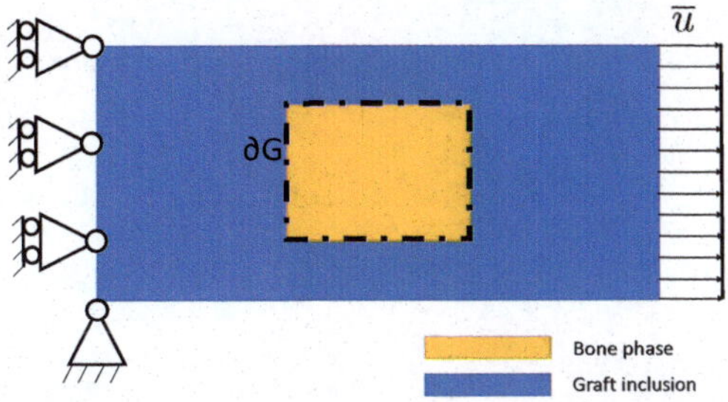

Fig. 3.4 Graphical depiction of the test problem. A rectangular slab formed by a rectangular central graft inclusion in bone tissue matrix is subject to a tensile test.

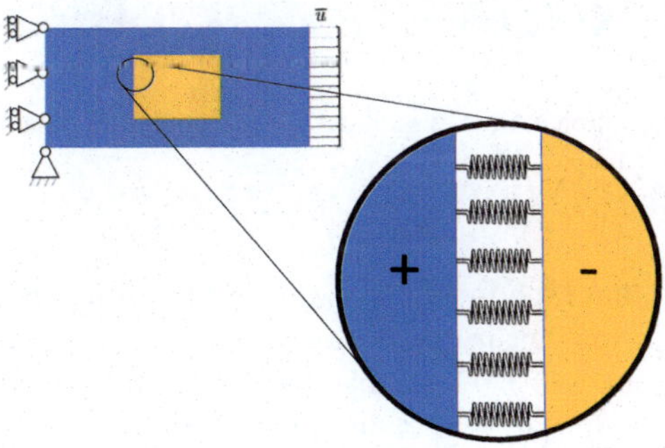

Fig. 3.5 Zoomed detail for the test problem. At the interface between the two phases there is a spring foundation for the variables u and ϑ.

to right and from up to down. As time progresses, the stimulus peaks on the left and on the right of the specimen shift toward the center, and eventually coalesce. The evolution in time of density of Osteocytes $\rho_{OC,R}$ for some relevant time instants is shown in Fig. 3.8. Time increases from left to right and from up to down. Following the feedback behaviour of bone physiology, Osteocytes colonize the graft

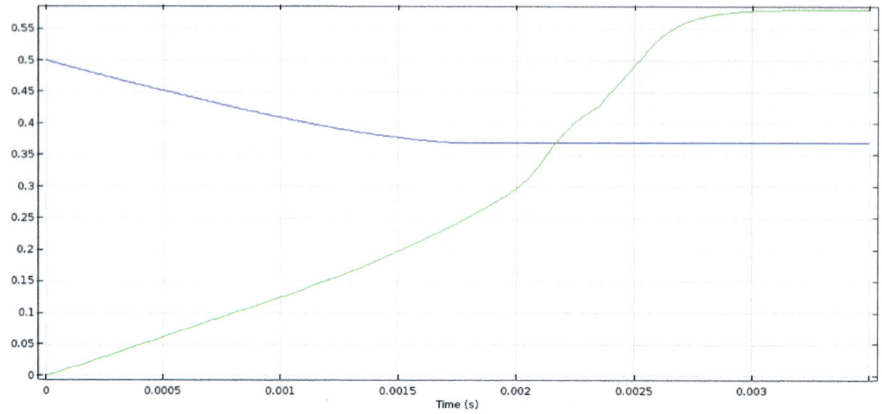

Fig. 3.6 Evolution in time of minimum graft density, $\min \rho_{M,R}$ (blue), and maximum bone tissue density (green), $\max \rho_{B,R}$, over the inclusion zone.

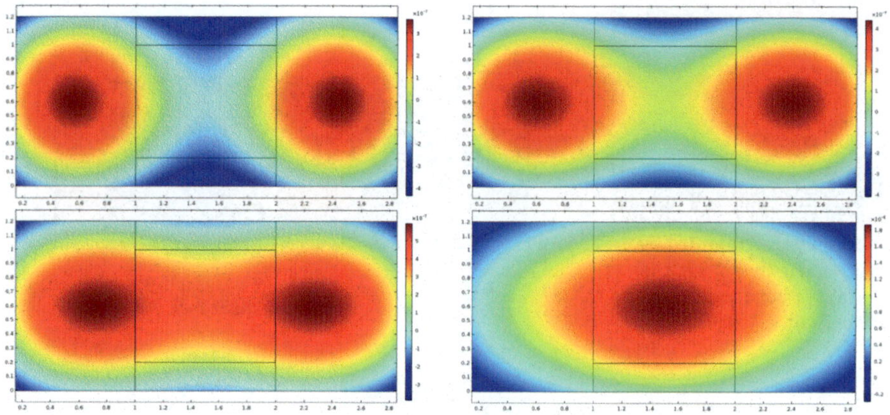

Fig. 3.7 Evolution in time of stimulus for some relevant time instants. Time increases from left to right and from up to down.

gradually, until the inclusion is uniformly saturated by them. The evolution in time of bone tissue density $\rho_{B,R}$ for some relevant time instants is reported as well in Figs. 3.9 and 3.10. Time increases from left to right and from up to down. Finally, evolution in time of the graft density $\rho_{M,R}$ for some relevant time instants is presented in Fig. 3.11. Time increases from left to right and from up to down. Therefore, from Fig. 3.6 it is clear that, as time progresses, the graft is resorbed (i.e. the blue curve is monotonously non-increasing) and new bone is formed (i.e. the green curve is monotonously non-decreasing). Stationary state is reached as both curves are asymptotically approaching limit values.

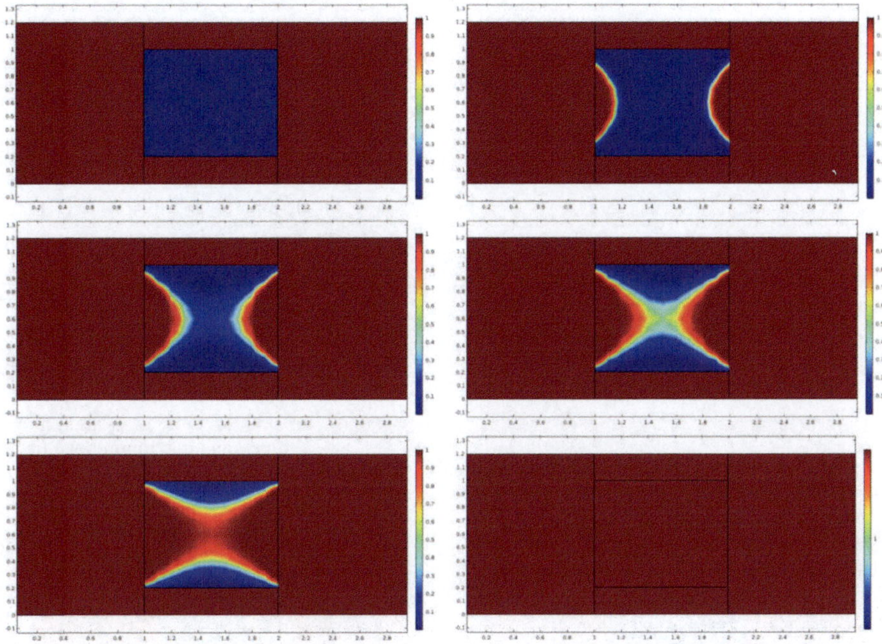

Fig. 3.8 Evolution in time of density of Osteocytes for some relevant time instants. Time increases from left to right and from up to down.

3.5 Conclusion and Outlooks

In this paper, in the framework of poroelasticity, we addressed the study of the interplay between bone remodeling and graft resorption under loading conditions. Aiming at enhancing the modeling proposed in past literature, we considered a physically motivated dedicated strain energy contribution due to the presence of saturating fluid in the interconnected pores, which has some compression resistance. In past literature, the Biot's contribution is not single targeted and includes also, but not only, the effect due to the possible presence of saturating fluid. The classical Biot's strain energy contribution, which is quadratic in the porosity change, is not physically motivated when dealing with the presence of interstitial fluid, and account for the presence of fluid exhibiting resistance to compression must be given in this framework through the 'effective' resistance to the change of porosity. The outlooks of the paper are the following. A parameter estimation from experiments would be useful to allow quantitative (not just qualitative) predictions. Furthermore, the model could be suitably adapted in order to take into account Turner's rules for bone adaptation: (1) remodeling it is driven by dynamic, rather than static, loading; (2) only a short duration of mechanical loading is necessary to initiate an adaptive response; (3) bone cells accommodate to a customary mechanical loading en-

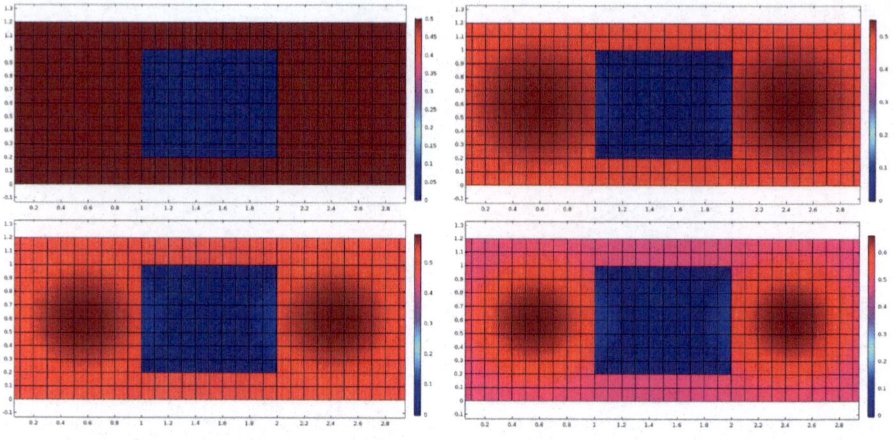

Fig. 3.9 Evolution in time of bone tissue density for some relevant time instants. Time increases from left to right and from up to down.

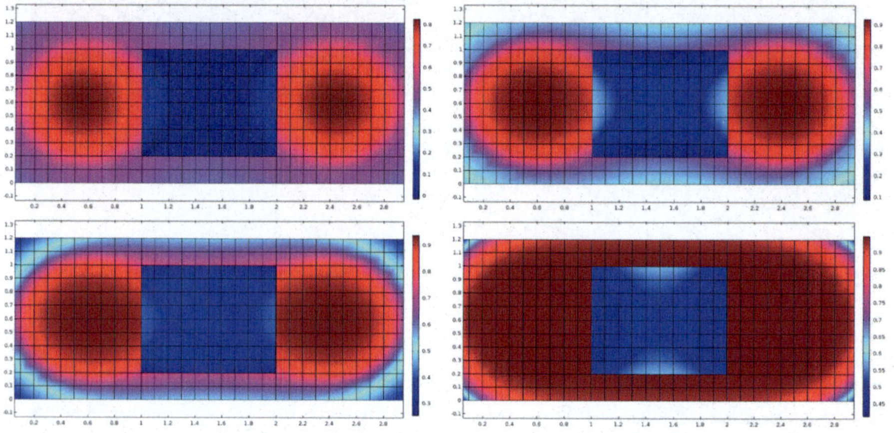

Fig. 3.10 Evolution in time of bone tissue density for some relevant time instants. Time increases from left to right and from up to down.

vironment, making them less responsive to routine loading signals. For example, Turner's rule could be indirectly taken into account by including dissipation. Furthermore, it has to be remarked that, while in this paper a macroscopic continuum model has been formulated directly, a homogenization procedure starting from discrete/continuum descriptions of the phenomena occurring at smaller length scales could give a better insight into the results obtained at the macro-level. In this regard, many procedures, like coarse-graining, hydrodynamical limits (De Masi et al, 2015; De Masi and Olla, 2015; Carinci et al, 2014b,a) for many-particle systems, and computational homogenization (Chatzigeorgiou et al, 2014; Saeb et al, 2016;

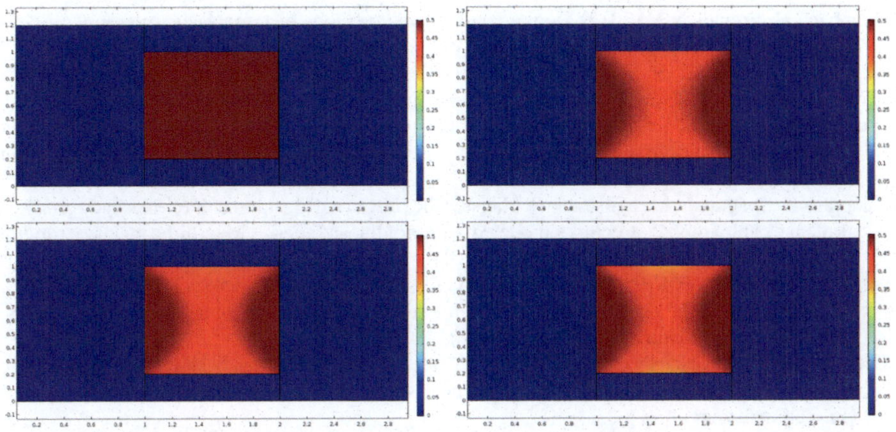

Fig. 3.11 Evolution in time of graft density for some relevant time instants. Time increases from left to right and from up to down.

Javili et al, 2013), are being employed in literature, and they deserve to be better understood.

References

Abali BE, Müller WH, Eremeyev VA (2015) Strain gradient elasticity with geometric nonlinearities and its computational evaluation. Mechanics of Advanced Materials and Modern Processes 1(1):4

Abali BE, Müller WH, dell'Isola F (2017) Theory and computation of higher gradient elasticity theories based on action principles. Archive of Applied Mechanics pp 1–16

Abbas IA, Abdalla AEN, Alzahrani FS, Spagnuolo M (2016) Wave propagation in a generalized thermoelastic plate using eigenvalue approach. Journal of Thermal Stresses 39(11):1367–1377

Abd-alla Aen, Alshaikh F, Del Vescovo D, Spagnuolo M (2017) Plane waves and eigenfrequency study in a transversely isotropic magneto-thermoelastic medium under the effect of a constant angular velocity. Journal of Thermal Stresses 40(9):1079–1092

Abd-alladan AenN, Hamdan AM, Almarashi AA, Battista A (2017) The mathematical modeling for bulk acoustic wave propagation velocities in transversely isotropic piezoelectric materials. Mathematics and Mechanics of Solids 22(4):823–836

Abdoul-Anziz H, Seppecher P (2018) Strain gradient and generalized continua obtained by homogenizing frame lattices. Mathematics and mechanics of complex systems 6(3):213–250

Alibert JJ, Seppecher P, dell'Isola F (2003) Truss modular beams with deformation energy depending on higher displacement gradients. Mathematics and Mechanics of Solids 8(1):51–73

Allena R, Cluzel C (2018) Heterogeneous directions of orthotropy in three-dimensional structures: finite element description based on diffusion equations. Mathematics and Mechanics of Complex Systems 6(4):339–351

Altenbach H, Eremeyev V (2009) On the linear theory of micropolar plates. ZAMM-Journal of Applied Mathematics and Mechanics/Zeitschrift für Angewandte Mathematik und Mechanik 89(4):242–256

Altenbach H, Eremeyev V (2015) On the constitutive equations of viscoelastic micropolar plates and shells of differential type. Mathematics and Mechanics of Complex Systems 3(3):273–283

Altenbach J, Altenbach H, Eremeyev VA (2010) On generalized cosserat-type theories of plates and shells: a short review and bibliography. Archive of Applied Mechanics 80(1):73–92

Andreaus U, Placidi L, Rega G (2010) Numerical simulation of the soft contact dynamics of an impacting bilinear oscillator. Communications in Nonlinear Science and Numerical Simulation 15(9):2603–2616

Andreaus U, Spagnuolo M, Lekszycki T, Eugster SR (2018) A Ritz approach for the static analysis of planar pantographic structures modeled with nonlinear Euler–Bernoulli beams. Continuum Mechanics and Thermodynamics pp 1–21

Auffray N, dell'Isola F, Eremeyev V, Madeo A, Rosi G (2015) Analytical continuum mechanics à la Hamilton–Piola least action principle for second gradient continua and capillary fluids. Mathematics and Mechanics of Solids 20(4):375–417

Battista A, Cardillo C, Del Vescovo D, Rizzi N, Turco E (2015) Frequency shifts induced by large deformations in planar pantographic continua. Nanomechanics Science and Technology: An International Journal 6(2)

Battista A, Del Vescovo D, Rizzi NL, Turco E (2017a) Frequency shifts in natural vibrations in pantographic metamaterials under biaxial tests. Technische Mechanik 37(1):1–17

Battista A, Rosa L, dell'Erba R, Greco L (2017b) Numerical investigation of a particle system compared with first and second gradient continua: Deformation and fracture phenomena. Mathematics and Mechanics of Solids 22(11):2120–2134

Beaupré GS, Orr TE, Carter DR (1990) An approach for time-dependent bone modeling and remodeling—theoretical development. Journal of Orthopaedic Research 8(5):651–661

Berezovski A, Yildizdag ME, Scerrato D (2018) On the wave dispersion in microstructured solids. Continuum Mechanics and Thermodynamics doi: 10.1007/s00161-018-0683-1:1–20

Bertram A, Glüge R (2016) Gradient materials with internal constraints. Mathematics and Mechanics of Complex Systems 4(1):1–15

Biot MA (1941) General theory of three-dimensional consolidation. Journal of applied physics 12(2):155–164

Camar-Eddine M, Seppecher P (2001) Non-local interactions resulting from the homogenization of a linear diffusive medium. Comptes Rendus de l'Academie des Sciences Series I Mathematics 332(5):485–490

Carinci G, De Masi A, Giardinà C, Presutti E (2014a) Hydrodynamic limit in a particle system with topological interactions. Arabian Journal of Mathematics 3(4):381–417

Carinci G, De Masi A, Giardinà C, Presutti E (2014b) Super-hydrodynamic limit in interacting particle systems. Journal of Statistical Physics 155(5):867–887

Chatzigeorgiou G, Javili A, Steinmann P (2014) Unified magnetomechanical homogenization framework with application to magnetorheological elastomers. Mathematics and Mechanics of Solids 19(2):193–211

Cluzel C, Allena R (2018) A general method for the determination of the local orthotropic directions of heterogeneous materials: application to bone structures using μct images. Mathematics and Mechanics of Complex Systems 6(4):353–367

Coussy O (2004) Poromechanics. John Wiley & Sons

Cuomo M (2017) Forms of the dissipation function for a class of viscoplastic models. Mathematics and Mechanics of Complex Systems 5(3):217–237

Cuomo M, dell'Isola F, Greco L, Rizzi N (2016) First versus second gradient energies for planar sheets with two families of inextensible fibres: Investigation on deformation boundary layers, discontinuities and geometrical instabilities. Composites Part B: Engineering

De Masi A, Olla S (2015) Quasi-static hydrodynamic limits. Journal of Statistical Physics 161(5):1037–1058

De Masi A, Galves A, Löcherbach E, Presutti E (2015) Hydrodynamic limit for interacting neurons. Journal of Statistical Physics 158(4):866–902

dell'Isola F, Seppecher P (1997) Edge contact forces and quasi-balanced power. Meccanica 32(1):33–52 dell'Isola F, Steigmann D (2015) A two-dimensional gradient-elasticity theory for woven fabrics. Journal of Elasticity 118(1):113–125

dell'Isola F, Seppecher P, Madeo A (2012) How contact interactions may depend on the shape of Cauchy cuts in Nth gradient continua: approach "à la D'Alembert". Zeitschrift für angewandte Mathematik und Physik 63(6):1119–1141

dell'Isola F, Andreaus U, Placidi L (2015a) At the origins and in the vanguard of peridynamics, non-local and higher-gradient continuum mechanics: An underestimated and still topical contribution of Gabrio Piola. Mathematics and Mechanics of Solids 20(8):887–928

dell'Isola F, Seppecher P, Della Corte A (2015b) The postulations á la D'Alembert and á la Cauchy for higher gradient continuum theories are equivalent: a review of existing results. Proc R Soc A 471(2183):20150,415

dell'Isola F, Madeo A, Seppecher P (2016) Cauchy tetrahedron argument applied to higher contact interactions. Archive for Rational Mechanics and Analysis 219(3):1305–1341

Di Nino S, D'Annibale F, Luongo A (2017) A simple model for damage analysis of a frame-masonry shear-wall system. International Journal of Solids and Structures 129:119–134

Enakoutsa K, DEl Vescovo D, Scerrato D (2017) Combined polarization field gradient and strain field gradient effects in elastic flexoelectric materials. Mathematics and Mechanics of Solids 22(5):938–951

Engelbrecht J, Berezovski A (2015) Reflections on mathematical models of deformation waves in elastic microstructured solids. Mathematics and Mechanics of Complex Systems 3(1):43–82

Eremeyev VA (2018) On the material symmetry group for micromorphic media with applications to granular materials. Mechanics Research Communications 94:8–12

Eremeyev VA, dell'Isola F, Boutin C, Steigmann D (2018a) Linear pantographic sheets: existence and uniqueness of weak solutions. Journal of Elasticity 132(2):175–196

Eremeyev VA, Rosi G, Naili S (2018b) Comparison of anti-plane surface waves in strain-gradient materials and materials with surface stresses. Mathematics and Mechanics of Solids p 1081286518769960

Eugster SR, Glocker C (2017) On the notion of stress in classical continuum mechanics. Mathematics and Mechanics of Complex Systems p 299

Ferretti M, Piccardo G, Luongo A (2017) Weakly nonlinear dynamics of taut strings traveled by a single moving force. Meccanica 52(13):3087–3099

Franciosi P, Spagnuolo M, Salman OU (2018) Mean green operators of deformable fiber networks embedded in a compliant matrix and property estimates. Continuum Mechanics and Thermodynamics doi: 10.1007/s00161-018-0668-0:1–32

Ganghoffer JF (2012) A contribution to the mechanics and thermodynamics of surface growth. application to bone external remodeling. International Journal of Engineering Science 50(1):166–191

Ganghoffer JF (2016) Spatial and material stress tensors in continuum mechanics of growing solid bodies. Mathematics and Mechanics of Complex Systems 3(4):341–363

George D, Allena R, Remond Y (2018a) Cell nutriments and motility for mechanobiological bone remodeling in the context of orthodontic periodontal ligament deformation. Journal of Cellular Immunotherapy

George D, Allena R, Remond Y (2018b) Integrating molecular and cellular kinetics into a coupled continuum mechanobiological stimulus for bone reconstruction. Continuum Mechanics and Thermodynamics pp 1–16

George D, Allena R, Remond Y (2018c) A multiphysics stimulus for continuum mechanics bone remodeling. Mathematics and Mechanics of Complex Systems 6(4):307–319

Giorgio I, Andreaus U, Scerrato D, dell'Isola F (2016) A visco-poroelastic model of functional adaptation in bones reconstructed with bio-resorbable materials. Biomechanics and modeling in mechanobiology 15(5):1325–1343

Giorgio I, Andreaus U, Scerrato D, Braidotti P (2017) Modeling of a non-local stimulus for bone remodeling process under cyclic load: Application to a dental implant using a bioresorbable porous material. Mathematics and Mechanics of Solids 22(9):1790–1805

Goda I, Ganghoffer JF (2015) 3d plastic collapse and brittle fracture surface models of trabecular bone from asymptotic homogenization method. International Journal of Engineering Science 87:58–82

Goda I, Assidi M, Belouettar S, Ganghoffer JF (2012) A micropolar anisotropic constitutive model of cancellous bone from discrete homogenization. Journal of the mechanical behavior of biomedical materials 16:87–108

Goda I, Assidi M, Ganghoffer JF (2014) A 3D elastic micropolar model of vertebral trabecular bone from lattice homogenization of the bone microstructure. Biomech Model Mechanobiol 13:53–83

Gusev AA, Lurie SA (2017) Symmetry conditions in strain gradient elasticity. Mathematics and Mechanics of Solids 22(4):683–691

Hillsley MV, Frangos JA (1994) Bone tissue engineering: the role of interstitial fluid flow. Biotechnology and bioengineering 43(7):573–581

Javili A, Chatzigeorgiou G, Steinmann P (2013) Computational homogenization in magneto-mechanics. International Journal of Solids and Structures 50(25):4197–4216

Lekszycki T, dell'Isola F (2012) A mixture model with evolving mass densities for describing synthesis and resorption phenomena in bones reconstructed with bio-resorbable materials. ZAMM-Zeitschrift für Angewandte Mathematik und Mechanik 92(6):426–444

Lekszycki T, Bucci S, Del Vescovo D, Turco E, Rizzi NL (2017) A comparison between different approaches for modelling media with viscoelastic properties via optimization analyses. ZAMM-Zeitschrift für Angewandte Mathematik und Mechanik 97(5):515–531

Luongo A, D'Annibale F (2017) Nonlinear hysteretic damping effects on the post-critical behaviour of the visco-elastic beck's beam. Mathematics and Mechanics of Solids 22(6):1347–1365

Madeo A, dell'Isola F, Darve F (2013) A continuum model for deformable, second gradient porous media partially saturated with compressible fluids. Journal of the Mechanics and Physics of Solids 61(11):2196–2211

Misra A, Poorsolhjouy P (2015a) Granular micromechanics model for damage and plasticity of cementitious materials based upon thermomechanics. Mathematics and Mechanics of Solids doi: 10.1177/1081286515576821

Misra A, Poorsolhjouy P (2015b) Identification of higher-order elastic constants for grain assemblies based upon granular micromechanics. Mathematics and Mechanics of Complex Systems 3(3):285–308

Misra A, Singh V (2013) Micromechanical model for viscoelastic materials undergoing damage. Continuum Mechanics and Thermodynamics 25(2-4):343–358

Misra A, Singh V (2015) Thermomechanics-based nonlinear rate-dependent coupled damage-plasticity granular micromechanics model. Continuum Mechanics and Thermodynamics 27(4-5):787

Pagnini LC, Piccardo G (2016) The three-hinged arch as an example of piezomechanic passive controlled structure. Continuum Mechanics and Thermodynamics 28(5):1247–1262

Pideri C, Seppecher P (1997) A second gradient material resulting from the homogenization of an heterogeneous linear elastic medium. Continuum Mechanics and Thermodynamics 9(5):241–257

Pietraszkiewicz W, Eremeyev V (2009) On natural strain measures of the non-linear micropolar continuum. International Journal of Solids and Structures 46(3):774–787

Placidi L (2015) A variational approach for a nonlinear 1-dimensional second gradient continuum damage model. Continuum Mechanics and Thermodynamics 27(4-5):623

Placidi L, dell'Isola F, Ianiro N, Sciarra G (2008) Variational formulation of pre-stressed solid–fluid mixture theory, with an application to wave phenomena. European Journal of Mechanics-A/Solids 27(4):582–606

Placidi L, Greco L, Bucci S, Turco E, Rizzi N (2016) A second gradient formulation for a 2d fabric sheet with inextensible fibres. Zeitschrift für angewandte Mathematik und Physik 67(5)(114)

Placidi L, Barchiesi E, Misra A (2018) A strain gradient variational approach to damage: a comparison with damage gradient models and numerical results. Mathematics and Mechanics of Complex Systems 6(2):77–100

Rinaldi A, Placidi L (2014) A microscale second gradient approximation of the damage parameter of quasi-brittle heterogeneous lattices. ZAMM-Journal of Applied Mathematics and Mechanics/Zeitschrift für Angewandte Mathematik und Mechanik 94(10):862–877

Rosi G, Placidi L, Auffray N (2018) On the validity range of strain-gradient elasticity: a mixed static-dynamic identification procedure. European Journal of Mechanics-A/Solids 69:179–191

Saeb S, Steinmann P, Javili A (2016) Aspects of computational homogenization at finite deformations: A unifying review from Reuss' to Voigt's bound. Applied Mechanics Reviews 68(5):050,801

Sciarra G, dell'Isola F, Coussy O (2007) Second gradient poromechanics. International Journal of Solids and Structures 44(20):6607–6629

Seppecher P (1993) Equilibrium of a Cahn-Hilliard fluid on a wall: influence of the wetting properties of the fluid upon the stability of a thin liquid film. European journal of mechanics series B fluids 12:69–69

Seppecher P (2000) Second-gradient theory: application to Cahn-Hilliard fluids. In: Continuum thermomechanics, Springer, pp 379–388

Seppecher P, Alibert JJ, dell'Isola F (2011) Linear elastic trusses leading to continua with exotic mechanical interactions. In: Journal of Physics: Conference Series, IOP Publishing, vol 319, p 012018

Shirani M, Andani MT, Kadkhodaei M, Elahinia M (2017) Effect of loading history on phase transition and martensitic detwinning in shape memory alloys: Limitations of current approaches and development of a 1d constitutive model. Journal of Alloys and Compounds 729:390–406

Spagnuolo M, Andreaus U (2018) A targeted review on large deformations of planar elastic beams: extensibility, distributed loads, buckling and post-buckling. Mathematics and Mechanics of Solids p 1081286517737000

Spagnuolo M, Barcz K, Pfaff A, dell'Isola F, Franciosi P (2017) Qualitative pivot damage analysis in aluminum printed pantographic sheets: numerics and experiments. Mechanics Research Communications

Spingarn C, Wagner D, Remond Y, George D (2018) Theoretical numerical modeling of the oxygen diffusion effects within the periodontal ligament for orthodontic tooth displacement. Journal of Cellular Immunotherapy

Steigmann D, Agrawal A (2016) Electromechanics of polarized lipid bilayers. Mathematics and Mechanics of Complex Systems 4(1):31–54

Wilmanski K (1998) A thermodynamic model of compressible porous materials with the balance equation of porosity. Transport in Porous Media 32(1):21–47

Chapter 4
Mechanical and Thermodynamic Materials Properties Derived by Semi-empirical Atomic Potentials with Special Focus on Ag, Cu, and the Binary Alloy Ag-Cu

Thomas Böhme

Abstract The following contribution deals with the relationship between atomic interaction potentials and macrospcopic materials properties typically required in engineering disciplines such as mechanical engineering or thermodynamics. Special focus is exemplarily turned to the so-called Nearest Neighbor Embedded-Atom-Method, which has proved to reliably calculate various materials properties especially for FCC lattice configurations. An energy expression for binary alloys is derived and linked to the elastic constants as well as to the phase diagram construction. The obtained equations are applied to the binary brazing alloy Ag-Cu, and the results are compared to experimental data. Finally the theory is extended to lattice dynamics/vibrations in order to calculate temperature-depend materials quantities such as the heat capacity.

4.1 Motivation

Today Computational Materials Science represents a well-established discipline within natural and engineering science. Simulation techniques such as Molecular Dynamics (MD) (Hammerschmidt et al, 2005; Xu et al, 2001; Zhang and Ghosh, 2013; Wu et al, 2015) and Monte-Carlo (MC) (Bocchetti and Diep, 2013; Yang et al, 2001; Ramasubramaniam et al, 2008) simulations or Phase Field (PF) approaches (Eastgate et al, 2002; Weinberg et al, 2016; Wang and Li, 2010; Anders et al, 2012; Böhme et al, 2009; Dreyer and Müller, 2000) are widely used in order to predict complex materials behavior such as grain boundary diffusion (Hammerschmidt et al, 2005), hydrogen diffusion (Ramasubramaniam et al, 2008) and embrittlement (Xu et al, 2001), crack propagation (Zhang and Ghosh, 2013; Wu et al, 2015; Eastgate

Thomas Böhme
Trelleborg Antivibration Solutions Germany, Engineering Off-Highway and Industry, Berliner Str. 17, 16727 Velten (b) Berlin, Germany,
e-mail: thomas.boehme@trelleborg.com

B. E. Abali et al. (eds.), *New Achievements in Continuum Mechanics and Thermodynamics*, Advanced Structured Materials 108,
https://doi.org/10.1007/978-3-030-13307-8_4

et al, 2002; Weinberg et al, 2016), plasticity (Wu et al, 2015; Wang and Li, 2010), melting phenomena (Bocchetti and Diep, 2013; Yang et al, 2001), or phase separation and coarsening (Anders et al, 2012; Böhme et al, 2009; Dreyer and Müller, 2000).

Although these approaches are fundamentally different and partially performed on different length scales, they all require materials specific "information" in order to "adjust" the underlying equations for the considered material. Such materials data can be found, on the one hand side, from so-called *ab initio* calculations, which apply quantum-mechanical theories to quantify various atomic interactions and which - in turns - can be used to perform MD and MC simulations.

In contrast to investigations on the atomic scale PF theories are typically used to describe materials behavior on the microscopic scale. Since the seminal works of J. W. CAHN and J. E. HILLIARD (Cahn and Hilliard, 1958; Cahn, 1968) PF approaches are widely used to quantify temporal and spatial changes of microstructures in materials, *e.g.* phase separation and coarsening[1] (Anders et al, 2012; Dreyer and Müller, 2000; Wang and Li, 2010; Böhme, 2008) or solid-liquid phase evolution (Wheeler et al, 1992; Galenko et al, 2009), *cf.* Fig. 4.1. However the governing equations also need material-specific quantities, *e.g.* GIBBS free energy g or mobility M. Here experimental data must be used to provide the required data.

In many cases measurements of materials data are very complex and cost or time-consuming, respectively. Especially heterogeneous materials require large efforts, *e.g.* combinations of simulation and experiment (Yang et al, 2018; Heripre et al, 2007; Constantinides et al, 2006), to identify the macroscopic materials parameter. However, some quantities cannot or, at least, indirectly measured, such as the so-called Higher Gradient Coefficients[2] (HGC), see Dreyer and Müller (2000); Böhme (2008). In order to overcome this shortcoming BÖHME *et al.* (Böhme et al, 2007) has used so-called semi-empirical, atomic potentials in order to calculate the

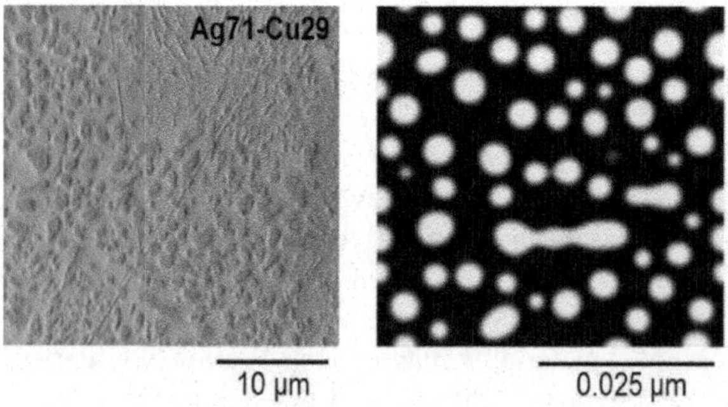

Fig. 4.1 Phase separation and coarsening in eutectic Ag-Cu. The spherical phase represents the Cu-rich (β) phase embedded into the Ag-rich (α) matrix. *Left* : experiment, *Right* : numerical simulation Böhme (2008).

[1] An alternative, mostly equivalent, notation is Spinodal Decomposition and OSTWALD ripening.

[2] Original notation by CAHN and HILLIARD: gradient energy coefficient.

HGCs[3]. Here the Embedded Atom Method (EAM) represents a powerful extension of classical pair potentials to describe the atomic interactions especially in metal lattice configurations. It was developed by M. S. DAW and M. I. BASKES (Daw and Baskes, 1983, 1984), and allows to reasonably reproduce the state of energy of an atomic system without any *ab initio* calculations.

The current work explains, how atomic potentials in general and specifically EAM potentials can be used to derive macroscopic materials properties. Here we start with the considerations in Böhme et al (2007) an calculate stiffness data of Ag, Cu and the corresponding alloy Ag-Cu. Furthermore we illustrate, how the solid lines of the binary phase diagram of Ag-Cu can be obtained. Finally the theory in Böhme et al (2007) is extended and temperature depending materials properties following from lattice vibrations are investigated. It is shown, how the heat capacity can be calculated and the results are compared with experimental findings.

4.2 Lattice Kinematics and Energy

First, let us consider the bulk material (no surfaces) and assume a perfect, periodic lattice[4]. The current positions $\underline{\mathbf{X}}^\alpha, \underline{\mathbf{X}}^\beta, \underline{\mathbf{X}}^\gamma, \ldots$ of all atoms $\alpha, \beta, \gamma, \ldots$ can be written by the reference positions $\underline{\mathbf{X}}_0^\alpha$, $\underline{\mathbf{X}}_0^\beta$, $\underline{\mathbf{X}}_0^\gamma$, ... and the discrete displacements $\underline{\xi}^\alpha$, $\underline{\xi}^\beta$, $\underline{\xi}^\gamma, \ldots$, namely $\underline{\mathbf{X}}^\alpha = \underline{\mathbf{X}}_0^\alpha + \underline{\xi}^\alpha$, $\underline{\mathbf{X}}^\beta = \underline{\mathbf{X}}_0^\beta + \underline{\xi}^\beta, \ldots$ (*cf.* Fig. 4.2). By means of the distance vectors:

$$\underline{\mathbf{R}}_0^{\alpha\beta} = \underline{\mathbf{X}}_0^\beta - \underline{\mathbf{X}}_0^\alpha \quad , \quad \underline{\mathbf{R}}^{\alpha\beta} = \underline{\mathbf{X}}^\beta - \underline{\mathbf{X}}^\alpha = \underline{\mathbf{R}}_0^{\alpha\beta} + \underline{\xi}^\beta - \underline{\xi}^\alpha \tag{4.1}$$

between atom α and β the continuous displacement function $\underline{\mathbf{u}}$ is defined with the definition $\underline{\xi}^\alpha \equiv \underline{\mathbf{u}}(\underline{\mathbf{X}}_0^\alpha) \equiv \underline{\mathbf{u}}(\underline{\mathbf{X}}_0)$ as follows:

$$\underline{\xi}^\beta = \underline{\mathbf{u}}(\underline{\mathbf{X}}_0^\beta) = \underline{\mathbf{u}}(\underline{\mathbf{X}}_0) + \frac{\partial \underline{\mathbf{u}}}{\partial \underline{\mathbf{X}}_0} \cdot \underline{\mathbf{R}}_0^{\alpha\beta} + \ldots , \quad \underline{\mathbf{R}}^{\alpha\beta} = \underline{\mathbf{R}}_0^{\alpha\beta} + \frac{\partial \underline{\mathbf{u}}}{\partial \underline{\mathbf{X}}_0} \cdot \underline{\mathbf{R}}_0^{\alpha\beta} = \underline{\underline{\mathbf{F}}} \cdot \underline{\mathbf{R}}_0^{\alpha\beta} . \tag{4.2}$$

Here the symbol

$$\underline{\underline{\mathbf{F}}} = \underline{\underline{\mathbf{I}}} + \frac{\partial \underline{\mathbf{u}}}{\partial \underline{\mathbf{X}}_0}$$

denotes the deformation gradient well known from the macroscopic continuum mechanics. In order to describe the temperature-independent energy of a lattice the deformed configuration is expanded into a TAYLOR series around the unde-

[3] The HGCs crucially determine the long-time phase evolution within PF simulations and can be directly related to interfacial energy or surface tension and interfacial width of the phase boundary (Ardell, 2012; Ubachs et al, 2004).

[4] The following notation is applied: Vectors are denoted in bold letters with an underline. Matrices and tensors of 2nd order are written in bold letters with double underline; tensors of higher order are noted with blackboard bold letters (double-strike). This, admittedly, slightly redundant notation has proofed for better readability in case of low-resolution copies.

formed lattice state. By neglecting higher order terms the energy of an atom α, $E^\alpha(\underline{\mathbf{R}}^{\alpha 1},\ldots,\underline{\mathbf{R}}^{\alpha N})$, within a deformed lattice consisting of N atoms can be written as:

$$E^\alpha(\underline{\mathbf{R}}^{\alpha 1},\ldots,\underline{\mathbf{R}}^{\alpha N}) = E^\alpha(\underline{\mathbf{R}}_0^{\alpha 1},\ldots,\underline{\mathbf{R}}_0^{\alpha N}) + \sum_{\substack{\beta \\ (\alpha\neq\beta)}} \frac{\partial E^\alpha}{\partial \underline{\mathbf{R}}^{\alpha\beta}}\bigg|_{\underline{\mathbf{R}}_0^{\alpha\beta}} \cdot \left(\underline{\mathbf{R}}^{\alpha\beta} - \underline{\mathbf{R}}_0^{\alpha\beta}\right)$$
$$+ \frac{1}{2}\sum_{\substack{\beta \\ (\alpha\neq\beta)}} \frac{\partial^2 E^\alpha}{\partial \underline{\mathbf{R}}^{\alpha\beta}\partial \underline{\mathbf{R}}^{\alpha\beta}}\bigg|_{\underline{\mathbf{R}}_0^{\alpha\beta}} \cdot\cdot \left(\underline{\mathbf{R}}^{\alpha\beta} - \underline{\mathbf{R}}_0^{\alpha\beta}\right)\left(\underline{\mathbf{R}}^{\alpha\beta} - \underline{\mathbf{R}}_0^{\alpha\beta}\right), \tag{4.3}$$

in which we introduced the double scalar product $\underline{\underline{\mathbf{Y}}}\cdot\cdot\underline{\underline{\mathbf{Y}}} = Y_{ij}Y_{ij}$ for tensors of 2nd order. By considering the scalar product of the atomic distance vector $\underline{\mathbf{R}}^{\alpha\beta}$, *viz.*

$$R^{\alpha\beta\,2} = \underline{\mathbf{R}}^{\alpha\beta}\cdot\underline{\mathbf{R}}^{\alpha\beta} = (\underline{\underline{\mathbf{F}}}\cdot\underline{\mathbf{R}}_0^{\alpha\beta})\cdot(\underline{\underline{\mathbf{F}}}\cdot\underline{\mathbf{R}}_0^{\alpha\beta}) = R_0^{\alpha\beta\,2} + \underline{\mathbf{R}}_0^{\alpha\beta}\cdot(\underline{\underline{\mathbf{C}}} - \underline{\underline{\mathbf{I}}})\cdot\underline{\mathbf{R}}_0^{\alpha\beta}\,, \tag{4.4}$$

Here GREEN's strain tensor

$$\underline{\underline{\mathbf{G}}} = \frac{1}{2}(\underline{\underline{\mathbf{C}}} - \underline{\underline{\mathbf{I}}})$$

with

$$\underline{\underline{\mathbf{C}}} = \underline{\underline{\mathbf{F}}}^{\mathrm{T}}\cdot\underline{\underline{\mathbf{F}}}$$

can be introduced to quantify the deformation. Please note, for small deformations holds $\underline{\underline{\mathbf{G}}} \approx \underline{\underline{\boldsymbol{\varepsilon}}}$ with the so-called linearized strains

$$\nabla\underline{\mathbf{u}} \equiv \frac{\partial\underline{\mathbf{u}}}{\partial\underline{\mathbf{X}}_0} \approx \frac{1}{2}(\nabla\underline{\mathbf{u}} + (\nabla\underline{\mathbf{u}})^{\mathrm{T}}) = \underline{\underline{\boldsymbol{\varepsilon}}},\,.$$

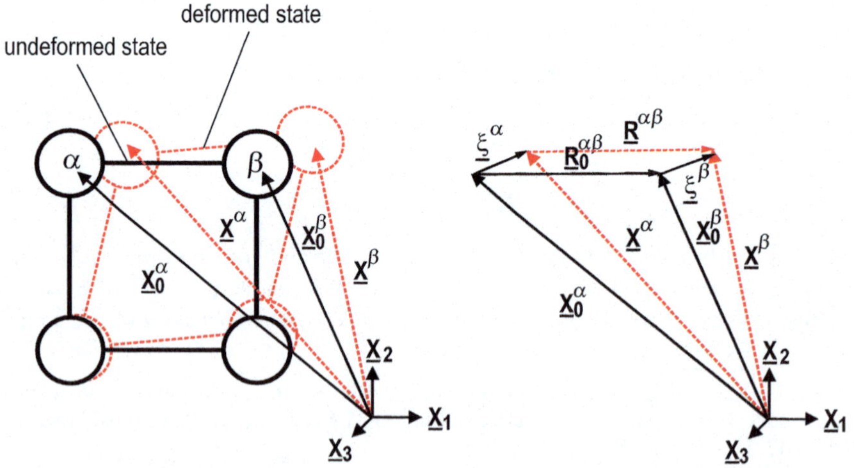

Fig. 4.2 Illustration of the kinematic quantities of the (un-)deformed lattice configuration.

By substituting $\underline{\mathbf{R}}^{\alpha\beta} - \underline{\mathbf{R}}_0^{\alpha\beta}$ by the relation of Eq. (4.2) the energy of Eq. (4.3) can be finally rewritten as follows:

$$E^{\alpha}(R^{\alpha 1^2}, \ldots, R^{\alpha N^2}) = E^{\alpha}(R_0^{\alpha 1^2}, \ldots, R_0^{\alpha N^2}) + 2\underline{\underline{\mathbf{G}}} \cdot\cdot \sum_{\substack{\beta \\ (\alpha \neq \beta)}} E^{\alpha\prime} \underline{\mathbf{R}}_0^{\alpha\beta} \underline{\mathbf{R}}_0^{\alpha\beta}$$
$$+ \frac{4}{2} \underline{\underline{\mathbf{G}}} \cdot\cdot \Big(\sum_{\substack{\beta \\ (\alpha \neq \beta)}} E^{\alpha\prime\prime} \underline{\mathbf{R}}_0^{\alpha\beta} \underline{\mathbf{R}}_0^{\alpha\beta} \underline{\mathbf{R}}_0^{\alpha\beta} \underline{\mathbf{R}}_0^{\alpha\beta} \Big) \cdot\cdot \underline{\underline{\mathbf{G}}} . \qquad (4.5)$$

with $E^{\alpha\prime} = \partial E^{\alpha} / \partial R^{\alpha\beta\,2}|_{R^{\alpha\beta\,2} = R_0^{\alpha\beta\,2}}$. First derivatives of the energy must vanish for equilibrium (minimum of energy). This fact yields the equilibrium condition, which - in turns - allows to calculate the lattice parameter a. The last term of Eq. (4.5) can be linked to the stiffness matrix $\mathbb{C} = [C_{ijkl}]$, which contains the elastic constants of the solid. However, the atomic energy E^{α} in Eq. (4.5) must be formulated in terms of the square of the scalar distances $R^{\alpha\beta}$ between the atoms α, $\beta = 1, \ldots, N$.

4.3 The Embedded Atom Method (EAM)

4.3.1 General Idea of EAM

The Embedded Atom Method was firstly introduced by the works of M. S. DAW and M. I. BASKES (Daw and Baskes, 1983, 1984). Both authors recommended, justified by quantum mechanical arguments, to add a nonlinear function F_{α} to the pairwise interaction term $\phi^{\alpha\beta}$. In particular the nonlinear character of the so-called embedding function remedies *e.g.* artifacts like the CAUCHY paradox $C_{1122} = C_{2323}$ or $C_{12} = C_{44}$ (VOIGT notation). Thus the energy of atom α is written as follows:

$$E^{\alpha} = \frac{1}{2} \sum_{\substack{\beta \\ (\alpha \neq \beta)}} \phi^{\alpha\beta}(R^{\alpha\beta}) + F_{\alpha}(\bar{\rho}_{\alpha}) \quad \text{with} \quad \bar{\rho}_{\alpha} = \sum_{\substack{\beta \\ (\alpha \neq \beta)}} \rho_{\beta}(R^{\alpha\beta}) . \qquad (4.6)$$

The factor $\frac{1}{2}$ is used to avoid double-counting of bonds for $E^{\text{tot}} = \sum E^{\alpha}$. Here the embedding function F_{α} only depends on the electronic density $\bar{\rho}_{\alpha}$ at the position of atom α, whereas $\phi^{\alpha\beta}$ only depends on the scalar distance $R^{\alpha\beta}$ between atom α and β. Furthermore $\bar{\rho}_{\alpha}$ can be interpreted as a constant background electronic density, that "feels" atom α due to the superposition of the atomic charge densities ρ_{β} of its neighbors separated by the distance $R^{\alpha\beta}$. Moreover, F_{α} can be understood as the energy to place an atom α into a homogeneous electron gas with the density $\bar{\rho}_{\alpha}$. Here the embedding function itself only depends on the type of the embedded atom and the argument of F_{α} refers to the medium in which the atom is embedded. Typically the first term of Eq. (4.6) stands for the purely repulsive ion-ion interaction; the second term characterizes the ion-electron interaction, *cf.* Fig. 4.3.

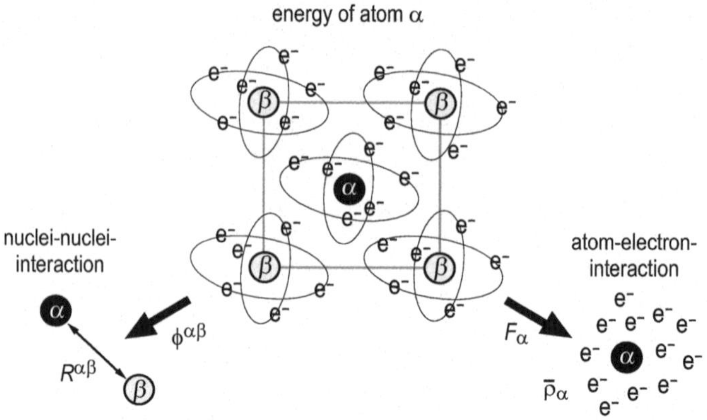

Fig. 4.3 Illustration of the EAM approach; two contributions to the energy E_α.

Especially for metals such decomposition is reasonable because the valence electrons can nearly free move within the metallic ionic lattice. The expression of the atom-specific energy in Eq. (4.6) can be used in Eq. (4.5) for the substitution of E^α, $E^{\alpha\prime}$ and $E^{\alpha\prime\prime}$. However, the different interaction terms, *i.e.* $\phi^{\alpha\beta}$, F_α and ρ_β, must be specified and quantified.

4.3.2 Restriction to Nearest Neighbor Interactions

In the following section we want to restrict ourselves to the so-called analytical EAM introduced by Johnson (1988, 1989). Here only nearest-neighbors-interactions, *i.e.* the atoms only interact with their direct neighbors separated by the nearest neighbor distance $R_0 = a^{(e)}/\sqrt{2}$ or $R = a\sqrt{2}$, are considered. In this notation the symbol a denotes the lattice parameter and the index (e) stands for "equilibrium". By considering the pure substance "A" the following, monotone decreasing form for the atomic electron density[5] and the pairwise interaction term holds:

$$\rho_{\rm A}(R^2) = \rho^{(e)} \exp\left[-\beta\left(\frac{R^2}{R_0^2}-1\right)\right], \quad \phi^{\rm AA}(R^2) = \phi^{(e)} \exp\left[-\gamma\left(\frac{R^2}{R_0^2}-1\right)\right]. \tag{4.7}$$

Here the index "A" at the quantities $\rho^{(e)}$, β, $\phi^{(e)}$, γ and R were omitted by convenience. Furthermore please note that Johnson used in his work the scalar distance R within the above equations, but due to the explanations in Sect. 4.2 the present formulation in terms of R^2 is used by simple substitution, see also Böhme et al (2007).

[5] This form corresponds to spherical s-orbitals and is predestined for isotropic structures, such as FCC lattice configurations.

The embedding function is parametrized by means of the so-called universal equation of state developed by Rose et al (1984), *viz.*

$$F_{\mathrm{A}}(\rho_{\mathrm{A}}) = -E_{\mathrm{sub}}\left[1+\alpha\left(\sqrt{1-\frac{1}{\beta}\ln\frac{\bar{\rho}_{\mathrm{A}}}{\bar{\rho}_{\mathrm{A}}^{(\mathrm{e})}}}-1\right)\right]\exp\left[\alpha\left(1-\sqrt{1-\frac{1}{\beta}\ln\frac{\bar{\rho}_{\mathrm{A}}}{\bar{\rho}_{\mathrm{A}}^{(\mathrm{e})}}}\right)\right] -6\phi^{(\mathrm{e})}\left(\frac{\bar{\rho}_{\mathrm{A}}}{\bar{\rho}_{\mathrm{A}}^{(\mathrm{e})}}\right)^{\frac{\gamma}{\beta}} \tag{4.8}$$

with $\alpha = \sqrt{\kappa\Omega^{(\mathrm{e})}/E_{\mathrm{sub}}}$; ($\Omega^{(\mathrm{e})}$: volume occupied by a single atom). Hence three functions ϕ^{AA}, ρ_{A}, and F_{A} must be specified for the pure substance "A". This can be done by fitting the five parameters $\alpha, \beta, \gamma, \phi^{(\mathrm{e})}, \rho^{(\mathrm{e})}$ to experimental data such as bulk modulus κ, shear modulus G, unrelaxed vacancy formation energy $E_{\mathrm{v}}^{\mathrm{u}}$, and sublimation energy E_{sub} (Böhme et al, 2007). In particular the fact, that experimental data are used to “adjust” the above equation for the material “A” yields to the notation of semi-empirical potentials.

For a binary alloy "A-B" seven functions, namely $\phi^{\mathrm{AA}}, \phi^{\mathrm{BB}}, \phi^{\mathrm{AB}}, \rho_{\mathrm{A}}, \rho_{\mathrm{B}}, F_{\mathrm{A}}, F_{\mathrm{B}}$ must be determined. Here the symbol ϕ^{AB} represents the pairwise interaction between atoms of different type; it is defined by the following equation:

$$\phi^{\mathrm{AB}} = \frac{1}{2}\left(\frac{\rho_{\mathrm{B}}}{\rho_{\mathrm{A}}}\phi^{\mathrm{AA}} + \frac{\rho_{\mathrm{A}}}{\rho_{\mathrm{B}}}\phi^{\mathrm{AA}}\right). \tag{4.9}$$

Thus all functions can be determined by considering pure substances. However, 10 parameters remain for fitting[6]. For the FFC metals Ag and Cu Table 4.1 shows the data, used to fit the EAM parameters. The corresponding EAM energy E^{α} for Ag

Table 4.1 Experimental data and fit parameters of Ag and Cu (for FCC holds $\Omega^{(\mathrm{e})} = {a^{(\mathrm{e})}}^3/4$).

experimental quantity	Ag	Cu
a in Å	4.09	3.61
E_{sub} in eV	2.85	3.54
$E_{\mathrm{v}}^{\mathrm{u}}$ in eV	1.10	1.30
κ in eV/Å^3	0.65	0.86
G in eV/Å^3	0.21	0.34
fitting parameter		
α	5.92	5.08
β	2.98	2.92
γ	4.13	4.00
$\phi^{(\mathrm{e})}$ in eV/Å^3	0.48	0.59
$\rho^{(\mathrm{e})}$ in eV/Å^3	0.17	0.30

[6] According to Böhme et al (2007) the following relations hold: $\phi^{(\mathrm{e})} = E_{\mathrm{sub}}/6$, $\rho^{(\mathrm{e})} = E_{\mathrm{sub}}/\Omega^{(\mathrm{e})}$, $G = \frac{8}{5}\frac{\phi^{(\mathrm{e})}\gamma(\gamma-\beta)}{\Omega^{(\mathrm{e})}}$, $E_{\mathrm{v}}^{\mathrm{u}} = 6\phi^{(\mathrm{e})}\frac{\gamma-\beta}{\beta}$.

Fig. 4.4 Atom-specific energy derived for Ag and Cu from the EAM formalism.

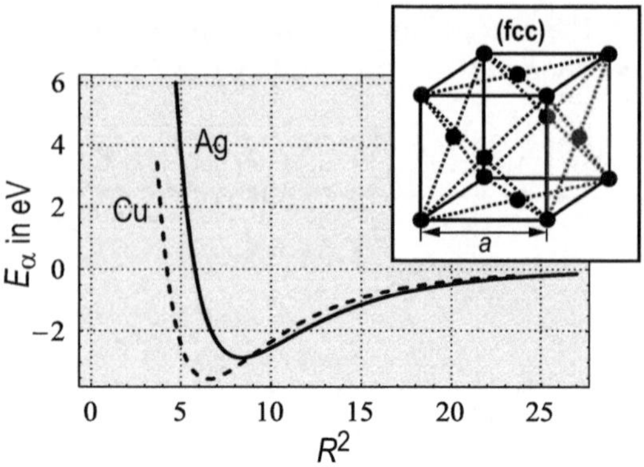

and Cu are illustrated in Fig. 4.4. These functions depending on $R^2_{\alpha\beta}$ can be further exploited to calculate various mechanical and thermodynamic materials data.

4.4 Exploitation of EAM energy expression

4.4.1 Equilibrium Condition and Elastic Constants

4.4.1.1 Pure Metals

The EAM formalism in Eq. (4.6) can be combined with Eq. (4.5). For this reason we use the kinematic relation $R^{\alpha\beta\,2} = R_0^{\alpha\beta\,2} + 2\mathbf{\underline{R}}_0^{\alpha\beta} \cdot \underline{\underline{\mathbf{G}}} \cdot \mathbf{\underline{R}}_0^{\alpha\beta}$ derived in Sect. 4.2 and expand $\phi^{\alpha\beta}(R^{\alpha\beta\,2})$, $\rho_\beta(R^{\alpha\beta\,2})$ as well as $F_\alpha(\sum \rho_\beta(R^{\alpha\beta\,2}))$ around $R_0^{\alpha\beta\,2}$. Thus we obtain for the energy of atom α:

$$E^\alpha = \frac{1}{2}\sum_\beta \phi^{\alpha\beta}(R_0^{\alpha\beta\,2}) + F_\alpha(\bar{\rho}_\alpha^{(\mathrm{e})}) + \underline{\underline{\mathbf{G}}} \cdot\cdot \left[\underline{\underline{\mathbf{A}}}^\alpha + 2F'_\alpha(\bar{\rho}_\alpha^{(\mathrm{e})})\underline{\underline{\mathbf{V}}}^\alpha\right]$$
$$+ \underline{\underline{\mathbf{G}}} \cdot\cdot \left[\mathbb{B}^\alpha + 2F'_\alpha(\bar{\rho}_\alpha^{(\mathrm{e})})\mathbb{W}^\alpha + 2F''_\alpha(\bar{\rho}_\alpha^{(\mathrm{e})})\underline{\underline{\mathbf{V}}}^\alpha\underline{\underline{\mathbf{V}}}^\alpha\right] \cdot\cdot \underline{\underline{\mathbf{G}}} \qquad (4.10)$$

with

$$\underline{\underline{\mathbf{A}}}^\alpha = \sum_\beta \phi'^{\alpha\beta}(R_0^{\alpha\beta\,2})\mathbf{\underline{R}}_0^{\alpha\beta}\mathbf{\underline{R}}_0^{\alpha\beta},\ \mathbb{B}^\alpha = \sum_\beta \phi''^{\alpha\beta}(R_0^{\alpha\beta\,2})\mathbf{\underline{R}}_0^{\alpha\beta}\mathbf{\underline{R}}_0^{\alpha\beta}\mathbf{\underline{R}}_0^{\alpha\beta}\mathbf{\underline{R}}_0^{\alpha\beta},$$
$$\underline{\underline{\mathbf{V}}}^\alpha = \sum_\beta \rho'_\beta(R_0^{\alpha\beta\,2})\mathbf{\underline{R}}_0^{\alpha\beta}\mathbf{\underline{R}}_0^{\alpha\beta},\ \mathbb{W}^\alpha = \sum_\beta \rho''_\beta(R_0^{\alpha\beta\,2})\mathbf{\underline{R}}_0^{\alpha\beta}\mathbf{\underline{R}}_0^{\alpha\beta}\mathbf{\underline{R}}_0^{\alpha\beta}\mathbf{\underline{R}}_0^{\alpha\beta}. \qquad (4.11)$$

The first two terms represent the energy of atom α for an non-deformed lattice. The term within the brackets $[\ldots]$ of the third summand denotes the slope of the energy curves in Figure 4.4. If lattice dynamics is neglected, this expression is equal to zero and represents the *equilibrium condition*, which in turns defines the nearest neighbor distance in equilibrium. The expression within the brackets $\underline{\underline{\mathbf{G}}} \cdot\cdot [\ldots] \cdot\cdot \underline{\underline{\mathbf{G}}}$ of the last term can be related to the macroscopic constitutive equation $E_{\text{elast}}/V = \frac{1}{2}\underline{\underline{\mathcal{E}}} \cdot\cdot \mathbb{C} \cdot\cdot \underline{\underline{\mathcal{E}}}$ with $\underline{\underline{\mathbf{G}}} \approx \underline{\underline{\mathcal{E}}}$ (HOOKE's law). Here $\mathbb{C}$ stands for the *stiffness matrix* and the coefficients $[C_{ijkl}]$ represent the *elastic constants*. Hence it can be summarized:

$$\underline{\underline{\mathbf{A}}}^{\alpha} + 2F'_{\alpha}(\bar{\rho}^{(\mathrm{e})}_{\alpha})\underline{\underline{\mathbf{V}}}^{\alpha} = 0 \text{ (equilibrium condition)},$$

$$\mathbb{C}^{\alpha} = \frac{2}{\Omega^{(\mathrm{e})}}[\mathbb{B}^{\alpha} + 2F'_{\alpha}(\bar{\rho}^{(\mathrm{e})}_{\alpha})\mathbb{W}^{\alpha} + 2F''_{\alpha}(\bar{\rho}^{(\mathrm{e})}_{\alpha})\underline{\underline{\mathbf{V}}}^{\alpha}\underline{\underline{\mathbf{V}}}^{\alpha}] \text{ (stiffness matrix)}. \tag{4.12}$$

For the metals Ag and Cu following values can be calculated by means of Eq. (4.12). Here, for comparison reasons, the literature values (Kittel, 1973; Leibfried, 1955) are additionally noted within the parenthesis.

$$\begin{aligned} C^{\mathrm{Ag}}_{1111} &= 132.6\ (124)\ \mathrm{GPa}\,, & C^{\mathrm{Cu}}_{1111} &= 183.7\ (168)\ \mathrm{GPa},\\ C^{\mathrm{Ag}}_{1122} &= 90.2\ (94)\ \mathrm{GPa}\,, & C^{\mathrm{Cu}}_{1122} &= 115.1\ (121)\ \mathrm{GPa},\\ C^{\mathrm{Ag}}_{2323} &= 42.4\ (46)\ \mathrm{GPa}\,, & C^{\mathrm{Cu}}_{2323} &= 68.7\ (75)\ \mathrm{GPa}. \end{aligned}$$

Please note, for cubic crystals hold: $C_{1111} = C_{2222} = C_{3333}$; $C_{1122} = C_{1133} = C_{2233}$; $C_{2323} = C_{1313} = C_{1212}$ and $C_{ijkl} = C_{klij}$. Consequently there are three non-equivalent elastic constants (Leibfried, 1955).

4.4.1.2 Binary Alloys

Now we turn the attention to the question, how to exploit Eq. (4.10) for binary solid mixtures. For this reason different "types" of atoms must be considered within the above framework. In particular for a binary alloy A-B (with stochastic occupation of lattice sites by A and B) three interactions have to be distinct, *viz*. A $\leftrightarrow$ A, B $\leftrightarrow$ B and A $\leftrightarrow$ B. Following de Fontaine (1975) the discrete concentration is introduced by $\hat{y}_{\alpha} = \delta_{\alpha\mathrm{B}}$; $\alpha = \{1, \ldots, N\}$, where δ_{ij} denotes the KRONECKER symbol. Thus it follows:

$$\phi^{\alpha\beta} = \phi^{\mathrm{AA}} + \left[\hat{y}_{\alpha} + (1 - 2\hat{y}_{\alpha})\hat{y}_{\beta}\right]\phi + (\hat{y}_{\alpha} + \hat{y}_{\beta})\tilde{\phi}\,, \tag{4.13}$$

$$\bar{\rho}^{(\mathrm{e})}_{\alpha} = \sum_{\beta}\left[\hat{y}_{\beta}(\rho_{\mathrm{B}} - \rho_{\mathrm{A}}) + \rho_{\mathrm{A}}\right] \tag{4.14}$$

with

$$\phi = \phi^{\mathrm{AB}} - \frac{1}{2}(\phi^{\mathrm{AA}} + \phi^{\mathrm{BB}}), \quad \tilde{\phi} = \frac{1}{2}(\phi^{\mathrm{BB}} - \phi^{\mathrm{AA}})\,.$$

Here, for example, $\hat{y}_{\alpha}$ and $\hat{y}_{\beta}$ is zero, if A atoms are considered and the relations $\phi^{\alpha\beta} = \phi^{\mathrm{AA}}$ and $\bar{\rho}^{(\mathrm{e})}_{\alpha} = \sum_{\beta}\rho_{\mathrm{A}}$ follow. Now we introduce the continuous concentration y by:

$$\hat{y}_\alpha = y(\underline{\mathbf{X}}_0^\alpha) \equiv y(\underline{\mathbf{X}}_0) \quad , \quad \hat{y}_\beta = y(\underline{\mathbf{X}}_0) + \frac{\partial y}{\partial \underline{\mathbf{X}}_0} \cdot \underline{\mathbf{R}}_0^{\alpha\beta} + \frac{1}{2}\frac{\partial^2 y}{\partial \underline{\mathbf{X}}_0^2} \cdot\cdot \underline{\mathbf{R}}_0^{\alpha\beta}\underline{\mathbf{R}}_0^{\alpha\beta} \, . \tag{4.15}$$

In case of homogeneous mixtures no concentration gradients occur, *i.e.* $\nabla y = \partial y/\partial \underline{\mathbf{X}}_0$ and $\nabla^2 y = \partial y^2/\partial \underline{\mathbf{X}}_0^2$ can be neglected. Then we obtain:

$$\phi^{\alpha\beta} = \phi^{\mathrm{AA}} + 2y(1-y)\phi + 2y\tilde{\phi} + \dots \, , \tag{4.16}$$

$$\bar{\rho}_\alpha^{(\mathrm{e})} = \bar{\rho}_\mathrm{A} + y\bar{\rho}_\Delta + \dots \quad \text{with} \quad \bar{\rho}_\Delta = \sum_\beta (\bar{\rho}_\mathrm{B} - \bar{\rho}_\mathrm{A}) \, . \tag{4.17}$$

For the embedding function F_α in Eq. (4.10) follows by the same procedure:

$$F_\alpha(\bar{\rho}_\alpha^{(\mathrm{e})}) = (1-y)F_\mathrm{A} + yF_\mathrm{B} \, . \tag{4.18}$$

Furthermore all remaining quantities of Eq. (4.10), *viz.* $\underline{\underline{\mathbf{A}}}^\alpha$, $\mathbb{B}^\alpha$, $F'_\alpha \underline{\underline{\mathbf{V}}}^\alpha$, $F''_\alpha \underline{\underline{\mathbf{V}}}^\alpha \underline{\underline{\mathbf{V}}}^\alpha$, and $F'_\alpha \mathbb{W}^\alpha$ can also be treated analogously to Eqs. (4.16-4.18). Following the straight forward calculations in Böhme et al (2007) one finds finally for the atom-specific energy of an arbitrary atom in the binary alloy:

$$E^\alpha(y) = E^\alpha_{\mathrm{undef}}(y) \; + \; \underline{\underline{\mathbf{G}}} \cdot\cdot \underline{\underline{\mathbf{Q}}}(y) \; + \; \frac{\Omega^{(\mathrm{e})}(y)}{2} \underline{\underline{\mathbf{G}}} \cdot\cdot \mathbb{C}(y) \cdot\cdot \underline{\underline{\mathbf{G}}} \tag{4.19}$$

with the following abbreviations

$$\begin{aligned}
E^\alpha_{\mathrm{undef}}(y) &= \frac{1}{2} g^{\mathrm{AA}} + F_\mathrm{A} + y g^{\tilde{\phi}} + y(F_\mathrm{B} - F_\mathrm{A}) + y(1-y) g^\phi , \\
\underline{\underline{\mathbf{Q}}}(y) &= \underline{\underline{\mathbf{A}}}^\mathrm{A} + 2y\underline{\underline{\mathbf{A}}}^{\tilde{\phi}} + 2y(1-y)\underline{\underline{\mathbf{A}}}^\phi + 2\Big(\underline{\underline{\mathbf{V}}}^\mathrm{A} + y\underline{\underline{\mathbf{V}}}^\Delta\Big)\Big(F'_\mathrm{A} + y(F'_\mathrm{B} - F'_\mathrm{A})\Big), \\
\mathbb{C}(y) &= \frac{1}{\Omega^{(\mathrm{e})}(y)} \Bigg[2\mathbb{B}^\mathrm{A} + 4y\mathbb{B}^{\tilde{\phi}} + 2y(1-y)\mathbb{B}^\phi \\
&+4\Big(\mathbb{W}^\mathrm{A} + y\mathbb{W}^\Delta\Big)\Big(F'_\mathrm{A} + y(F'_\mathrm{B} - F'_\mathrm{A})\Big) \\
&+ 4\Big(\underline{\underline{\mathbf{V}}}^\mathrm{A} + y\underline{\underline{\mathbf{V}}}^\Delta\Big)\Big(\underline{\underline{\mathbf{V}}}^\mathrm{A} + y\underline{\underline{\mathbf{V}}}^\Delta\Big)\Big(F''_\mathrm{A} + y(F''_\mathrm{B} - F''_\mathrm{A})\Big)\Bigg]
\end{aligned} \tag{4.20}$$

and $g^{\mathrm{AA}} = \sum_\beta \phi^{\mathrm{AA}}$, $g^\phi = \sum_\beta \phi$, $g^{\tilde{\phi}} = \sum_\beta \tilde{\phi}$. All remaining symbols, in particular $\underline{\underline{\mathbf{A}}}^\mathrm{A}$, $\underline{\underline{\mathbf{A}}}^\phi$, $\underline{\underline{\mathbf{A}}}^{\tilde{\phi}}$, $\mathbb{B}^\phi$, $\mathbb{B}^{\tilde{\phi}}$, $\underline{\underline{\mathbf{V}}}^\Delta$, and $\mathbb{W}^\Delta$ are defined analogous to Eq. (4.11). The indices A, ϕ, $\tilde{\phi}$, and Δ refer to the first argument within the sum, *i.e.* $\phi^{\mathrm{AA}'}$; ϕ' or ϕ''; $\tilde{\phi}'$ or $\tilde{\phi}''$, and $(\rho'_\mathrm{B} - \rho'_\mathrm{A})$ or $(\rho''_\mathrm{B} - \rho''_\mathrm{A})$. The first term of Eq. (4.19), namely $E^\alpha_{\mathrm{undef}}(y)$, denotes the energy of the undeformed lattice.

The second term contains first derivatives of the different energy contributions and must vanish for equilibrium, *i.e.* $\underline{\underline{\mathbf{Q}}}(y) \equiv 0$. This condition represents the defining equation for the equilibrium lattice parameter $a^{(\mathrm{e})}(y)$ as a function of particle concentration y, *cf.* Fig. 4.5. Please note, it is important to clearly distinct between the particle concentration and the mass concentration

$$c(y) = yM_{\mathrm{Cu}}/(yM_{\mathrm{Cu}} + (1-y)M_{\mathrm{Ag}});$$

(M: atomic mass). Obviously a pure linear interpolation (VEGARD's law) between the lattice parameters $a^{(\mathrm{e})}_{\mathrm{Ag}}$ and $a^{(\mathrm{e})}_{\mathrm{Cu}}$ does not hold for values $0 < y < 1$. However, in case of using c as argument the linear dependence follows, *viz.*

$$a^{(\mathrm{e})}(c) = (1-c)a_{\mathrm{Ag}} + ca_{\mathrm{Cu}}.$$

The third term represents the atom-specific elastic energy $E^{\mathrm{elast}} = \frac{1}{2}\underline{\underline{\mathcal{E}}}\cdot\cdot\,\mathbb{C}(y)\cdot\cdot\,\underline{\underline{\mathcal{E}}}$ with $\underline{\underline{\mathcal{E}}} \approx \underline{\underline{\mathbf{G}}}$ and with the (fourth order) stiffness matrix $\mathbb{C}(y)$. By means of Eq. (4.20) the three independent elastic constants can be determined as functions of the particle concentration. Figure 4.6 illustrates the results using $a^{(\mathrm{e})}(y_i)$, with $y_i = 0, 0.1, \ldots, 0.9, 1$. For $y = 0$ (Ag) and $y = 1$ (Cu) the elastic constants of Ag and Cu, illustrated on page 59, result. However, for $0 < y < 1$ the elastic constants do not follow the linear interpolation as indicated in Fig. 4.6.

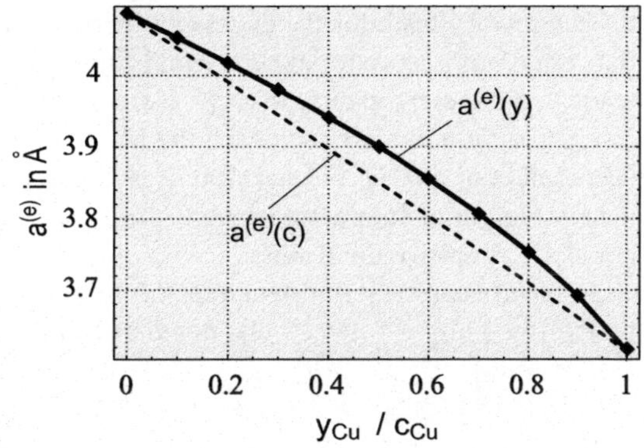

Fig. 4.5 Equilibrium lattice parameters for different particle or mass concentrations, respectively.

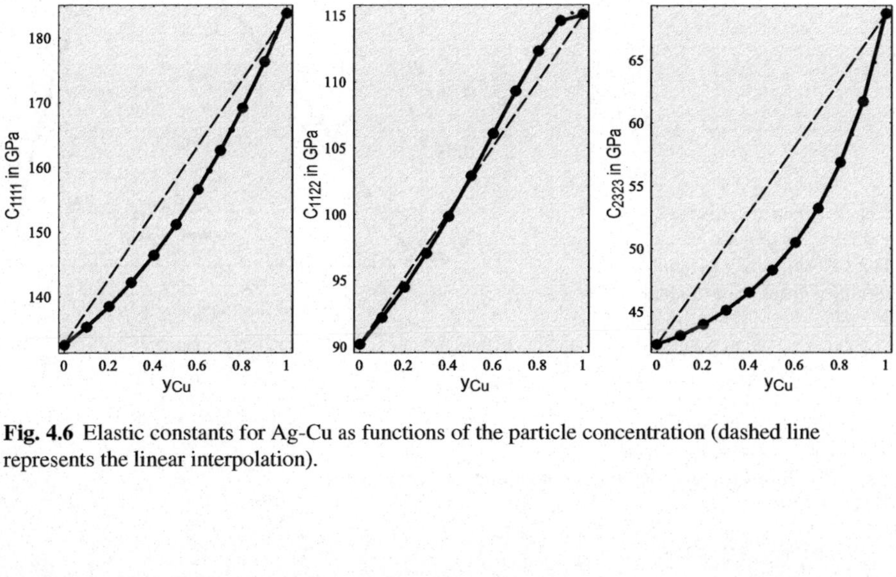

Fig. 4.6 Elastic constants for Ag-Cu as functions of the particle concentration (dashed line represents the linear interpolation).

4.4.2 Excess Enthalpy and Phase Diagram

The excess enthalpy $g^{\rm exc}$ characterizes the deviation of the (solid, binary) mixture from the so-called ideal mixture, in which the enthalpy follows by the "weighted" contributions of the pure substances, *i.e.* $g^{\alpha}_{\rm A-B} = g^{\rm exc} + y_{\rm A} g_{\rm AA} + y_{\rm B} g_{\rm BB}$. Following the arguments in Böhme et al (2007) $g^{\rm exc}$ can be identified with Eq. $(4.20)_1$ as follows:

$$g^{\rm exc} = \Lambda\, y(1-y) \quad \text{with} \quad y = y_{\rm B}\ ,\ y_{\rm A} = 1 - y\ , \tag{4.21}$$

$$\Lambda = \Lambda(y) = g^{\phi}(y) + \underline{\underline{\mathbf{G}}}(y) \cdot\cdot\, \mathbb{B}^{\phi} \cdot\cdot\, \underline{\underline{\mathbf{G}}}\ . \tag{4.22}$$

Note, g^{ϕ} as well as $\mathbb{B}^{\phi}$ directly follow from the interatomic potentials, especially from the pairwise interaction terms, see Eqs. (4.13), (4.20) and the corresponding, subsequent explanations. Any contributions from the nonlinear embedding function are canceled. Moreover, for the determination of both quantities the concentration dependent nearest neighbor distance $R_0(y) = a^{(\rm e)}(y)/\sqrt{2}$ must be used, which is defined by the equilibrium condition $\underline{\underline{\mathbf{Q}}}(y) \equiv 0$, *cf.* Fig. 4.5.

Figure 4.7 illustrates the excess enthalpy of Ag-Cu calculated with EAM potentials and compared with experimental findings.[7] Obviously the theoretical method leads to an overestimation of $g^{\rm exc}$. Here better results are expected with more precise interaction models such each the Modified Embedded Atom Method (MEAM) (Feraoun et al, 2001). Furthermore it is worth-mentioned, that we only incorporate nearest neighbor interactions; here the incorporation of more neighboring atoms could also improve the results.

In macroscopic thermodynamics of mixtures the molare Gibbs free energy of an undeformed binary alloy can be noted as follows (constant pressure p):

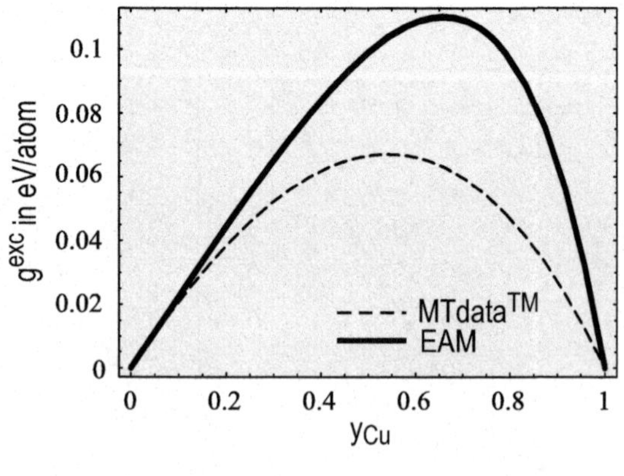

Fig. 4.7 Atomic excess enthalpy for Ag-Cu calculated from the EAM and obtained from experiments, MTdata$^{\rm TM}$ (MTData, 1998).

[7] see also: http://resource.npl.co.uk/mtdata/mtdatasoftware.htm

$$\tilde{g}(y,T) = (1-y_\mathrm{B})\tilde{g}_\mathrm{A}(T) + y_\mathrm{B}\tilde{g}_\mathrm{B}(T) + \\ + N_A k_B T\Big[y_\mathrm{B}\ln y_\mathrm{B} + (1-y_\mathrm{B})\ln(1-y_\mathrm{B})\Big] + \tilde{g}^{\mathrm{exc}}(y,T)\,. \quad (4.23)$$

This equation contains three characteristic parts: **(a)** the first and second term, which stand for the linear contributions of the pure substances; **(b)** the third term, which denotes the entropic part $-T\tilde{s}(y) = -N_A k_B T \sum_{i=1}^{2} y_i \ln y_i$ with $N_A = 6.022 \cdot 10^{23}\ \mathrm{mol}^{-1}$ (AVOGADRO constant) and $k_B = 1.38 \cdot 10^{-23}$ J/K (BOLTZMANN constant). This function depends on y in a convex and symmetric manner, has a minimum at $y = 0.5$ and vanishes for $y = \{0,1\}$; **(c)** the last term, the excess enthalpy (see above), which represents - in case of a miscibility gap - a positive, concave curve. The upper right picture of Fig. 4.8 illustrates the resulting double-well function for Ag-Cu at 1000 K. Here the concave range, $y \approx 0.19 \ldots 0.79$, denotes the so-called unstable area, in which the solid mixture decompose into two different equilibrium concentrations $y_{\alpha,\beta}$, *cf.* Fig. 4.1 and Böhme (2008).

In case of no deformations Eq. (4.23) can be related to $E^{\alpha}_{\mathrm{undef}}$ in Eqs. (4.19), (4.20). By means of $\tilde{g}(y,T) = N_A g(y,T) = N_\mathrm{A}[E^{\alpha} - Ts(y)]$ we get:

$$g(y,T) = (1-y_\mathrm{B})(6\phi^{\mathrm{AA}} + F_\mathrm{A}) + y(6\phi^{\mathrm{BB}} + F_\mathrm{B}) \\ + k_B T\Big[y_\mathrm{B}\ln y_\mathrm{B} + (1-y_\mathrm{B})\ln(1-y_\mathrm{B})\Big] + 12y(1-y)\phi\,. \quad (4.24)$$

From this expression we can calculate the above mentioned equilibrium concentrations for the solid part of the binary phase diagram. Here we apply the so-called *common tangent rule* well-known from macroscopic thermodynamics of mixtures (Böhme et al, 2007). Here the mixture decomposes into two different phase such, that the slope of the energy at equilibrium concentration $y_{(\alpha)}$ is equal to the slope at $y_{(\beta)}$ and is equal to the slope of the connecting line through these points, *cf.* Fig. 4.8 (upper right). From the mathematical point-of-view this rule reads:

$$\left.\frac{\partial g(y,T)}{\partial y}\right|_{y=y_{(\alpha)}} = \left.\frac{\partial g(y,T)}{\partial y}\right|_{y=y_{(\beta)}} = \frac{g(y_{(\beta)},T) - g(y_{(\alpha)},T)}{y_{(\beta)} - y_{(\alpha)}}\,. \quad (4.25)$$

Equations (4.24), (4.25) allow for the determination of equilibrium concentrations at different temperatures; Fig. 4.8 displays the corresponding results. Here the solid lines with spherical markings represent the calculated equilibrium concentrations for different temperatures, whereas the the dashed lines denotes the experimental data adopted from the database MTData^{TM}.

In summary we can say, that there is a good agreement between the experimental and theoretical results. Deviations mainly occur for higher temperatures and - more pronounced - for the (β)-phase. The most obvious reason for the deviations is the discrepancy between the experimental and theoretical excess enthalpy, illustrated in Fig. 4.7. Here the theoretical curve shows a larger asymmetry as well as overestimated values. In particular the asymmetry allows to justify the discrepancy for $y_{(\beta)}$. Better results are expected by using more accurate interaction models, *e.g.* MEAM (Feraoun et al, 2001), and/or by incorporating more neighbors.

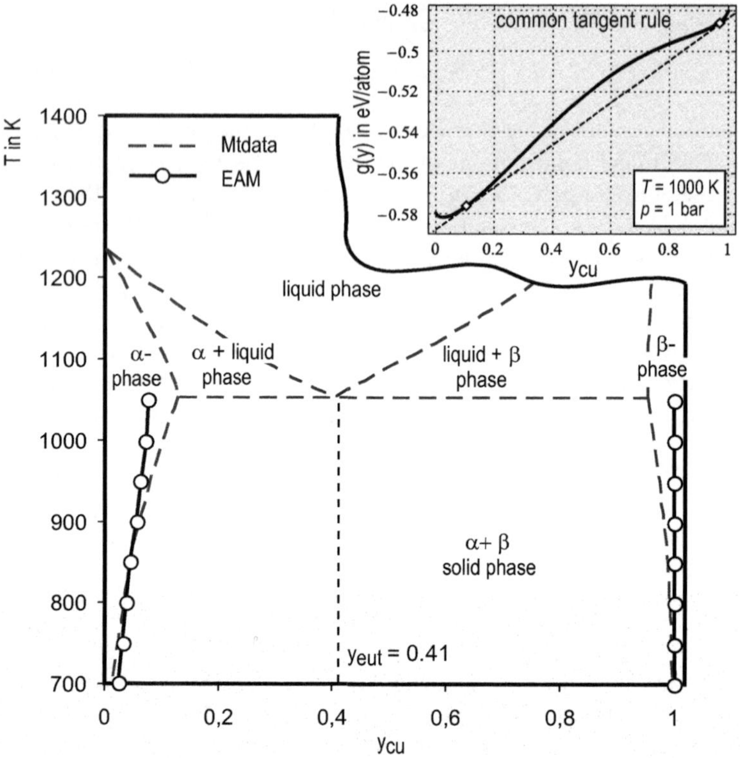

Fig. 4.8 Phase diagram of Ag-Cu; theoretical approach vs. experimental data. *Upper right panel* : Illustration of MAXWELL construction for Ag-Cu at 1000 K.

4.4.3 Temperature-dependent Materials Properties

4.4.3.1 Consideration of Lattice Vibrations

The temperature of a system is related to the (mean) velocity of the atoms. Therefore temperature-dependent materials properties can only be precisely determined on the atomistic scale by incorporating the dynamics of atoms, which - in turns - means lattice vibrations, *i.e.* phonons, in case of metals. In principle atomic vibrations can be modeled by considering a 3D multi-body-system, which consists of mass points (atoms) linked to each other by springs (interatomic forces). Then the resulting equation of motion for atom α can be obtained by using the framework of classical mechanics:

$$m_\alpha \ddot{\underline{\xi}}_\alpha = \underline{\mathbf{F}}_\alpha = -\nabla E^\alpha = -\sum_\beta \frac{\partial^2 E^\alpha}{\partial \underline{\mathbf{R}}^{\alpha\beta} \partial \underline{\mathbf{R}}^{\alpha\beta}}\bigg|_{\underline{\mathbf{R}}_0^{\alpha\beta}} \cdot (\underline{\mathbf{R}}^{\alpha\beta} - \underline{\mathbf{R}}_0^{\alpha\beta}) \,. \tag{4.26}$$

Please note, Eq. (4.26) follows by neglecting higher order terms in Eq. (4.3). Furthermore it holds the identity $\underline{\mathbf{R}}^{\alpha\beta} - \underline{\mathbf{R}}_0^{\alpha\beta} = \underline{\xi}^{\beta} - \underline{\xi}^{\alpha}$. Therefore Eq. (4.26) represents a PDE for $\underline{\xi}^{\alpha}$, which can be solved by the ansatz $\underline{\xi}^{\alpha} = \underline{\mathbf{e}}\ \exp[\ \mathrm{i}(\underline{\mathbf{k}} \cdot \underline{\mathbf{X}}_0^{\alpha} - \omega t)]$ (Leibfried, 1955). Thus we obtain:

$$m_\alpha \omega^2\ \underline{\mathbf{e}} = \sum_\beta \underline{\underline{\mathbf{D}}}^{\alpha\beta}(\underline{\mathbf{R}}_0^{\alpha\beta}) \cdot \underline{\mathbf{e}} \left(1 - \mathrm{e}^{\mathrm{i}\underline{\mathbf{k}}\cdot\underline{\mathbf{R}}_0^{\alpha\beta}}\right) \tag{4.27}$$

with

$$\underline{\underline{\mathbf{D}}}^{\alpha\beta}(\underline{\mathbf{R}}_0^{\alpha\beta}) = \left.\frac{\partial^2 E^\alpha}{\partial \underline{\mathbf{R}}^{\alpha\beta} \partial \underline{\mathbf{R}}^{\alpha\beta}}\right|_{\underline{\mathbf{R}}_0^{\alpha\beta}}$$

Here $\underline{\underline{\mathbf{D}}}^{\alpha\beta}$ stands for the stiffness matrix and $\underline{\mathbf{e}}$ represents the normalized vector parallel to the corresponding displacement. Furthermore $\underline{\mathbf{k}}$ defines the wave vector with the wave length $\lambda = 2\pi/|\underline{\mathbf{k}}|$ and $\omega = 2\pi\nu$ denotes the angular velocity with the corresponding frequency ν.

It is worth-mentioning that the derived EAM potentials within this work depend on $R^{\alpha\beta\,2}$; consequently the chain rule must be applied for derivatives of E^α, *viz.*

$$\partial^2 E^\alpha/(\partial \underline{\mathbf{R}}^{\alpha\beta})^2 = (\partial^2 E^\alpha/\partial x^2)(\partial x/\partial \underline{\mathbf{R}}^{\alpha\beta})^2 + (\partial E^\alpha/\partial x)(\partial^2 x/\partial \underline{\mathbf{R}}^{\alpha\beta\,2})$$

Equation $(4.4.3.1)_1$ defines an eigenvalue problem, which can be solved by

$$\det[\underline{\underline{\mathring{\mathbf{D}}}}^{\alpha\beta} - m_\alpha \omega^2 \underline{\underline{\mathbf{I}}}] = 0$$

with the following relation:

$$\underline{\underline{\mathring{\mathbf{D}}}}^{\alpha\beta}(\underline{\mathbf{k}}) = \sum_\beta \underline{\underline{\mathbf{D}}}^{\alpha\beta}(\underline{\mathbf{R}}_0^{\alpha\beta}) \left(1 - e^{\mathrm{i}\underline{\mathbf{k}}\cdot\underline{\mathbf{R}}_0^{\alpha\beta}}\right). \tag{4.28}$$

The three eigenvalues $\mathring{D}_{\mathrm{I/II/III}}(\underline{\mathbf{k}}) = 4\pi^2 m_\alpha \nu_{\mathrm{I/II/III}}(\underline{\mathbf{k}})$ allows to determine the eigenfrequencies $\nu_{\mathrm{I/II/III}}(\underline{\mathbf{k}})$ of the lattice. Here three orthonormal eigenvectors $\underline{\mathbf{e}}^{\mathrm{I/II/III}}$ exist, which characterize the oscillation direction of atoms (wave polarization). In particular we have one longitudinal wave (L with $\underline{\mathbf{e}}^k \perp \underline{\mathbf{k}}$) and two transversal waves (T_1 and T_2 with $\underline{\mathbf{e}}^k || \underline{\mathbf{k}}$).

For finite temperature crystal vibrations show a wide range of wave vectors and frequencies, as illustrated in Fig. 4.9 for Ag und Cu. Here the calculated so-called lattice phonon dispersion is displayed for the three selected symmetry directions of the FCC-structure, namely $\xi[100]$, $\xi[011]$, and $\xi[111]$ with $\xi \in [0, 2\pi/a^{(\mathrm{e})}]$ or $\xi \in [0, \pi/a^{(\mathrm{e})}]$. Although we have only used nearest neighbor interactions during the calculations the results are comparable to experimental results as presented *e.g.* in Svensson et al (1967); Bian et al (2008).

Please note, $\underline{\underline{\mathbf{D}}}^{\alpha\beta}$ in Eqs. $(4.4.3.1)_2$ and (4.28) can be calculated for the pure substances from Eq. (4.10) as well as for alloys by means of Eq. (4.19), *cf.* Fig. 4.10. Here we exclusively considered the undeformed terms, i.e. $\underline{\underline{\mathbf{G}}} \equiv 0$. Furthermore, in case of alloys, $\underline{\underline{\mathbf{D}}}^{\alpha\beta} = \underline{\underline{\mathbf{D}}}^{\alpha\beta}(\underline{\mathbf{R}}_0^{\alpha\beta}, y_\beta)$ additionally depends on the particle concentration. Moreover, the atomic distance $\underline{\mathbf{R}}_0^{\alpha\beta} = \underline{\mathbf{R}}_0^{\alpha\beta}(y_\beta)$ itself also depends on y_β and must be determined in advance by the equilibrium condition $\underline{\underline{\mathbf{Q}}}(y) \equiv 0$, *cf.* Eq. (4.20)$_2$.

4.4.3.2 Kinetic Energy and Heat Capacity

The calculated eigenfrequencies of the lattice system allow to quantify the resulting kinetic energy E^α_{kin}. For this reason we follow the quantum mechanical and statistical arguments in Leibfried (1955) and note:

$$E^\alpha_{\text{kin}}(T) = \frac{1}{N}\sum_{i=1}^{3N}\sum_{\underline{\mathbf{k}}} \frac{h\,\nu_i(\underline{\mathbf{k}})}{2} + \frac{1}{N}\sum_{i=1}^{3N}\sum_{\underline{\mathbf{k}}} \frac{h\,\nu_i(\underline{\mathbf{k}})}{\exp\left[\frac{h\,\nu_i(\underline{\mathbf{k}})}{k_B T}\right] - 1}\,, \tag{4.29}$$

with PLANCK's constant $h = 6.626 \cdot 10^{-34}$ Js. The most sensible point for a realistic reproduction of the kinetic energy denotes the summation of the eigenfrequencies $\nu_1, \ldots, \nu_{3N}$ of the N atoms within the lattice system and the wave vectors $\underline{\mathbf{k}}$. The question of which and how many frequencies ν_i and wave vectors $\underline{\mathbf{k}}$ are used to quantify $E^{\text{tot}}_{\text{kin}}$ may strongly determine the accuracy of all subsequently derived quantities. For instances, Bian et al (2008) used a weighted sum of 256 different wave vectors, which requires considerable computational capacities since the eigenvalue-

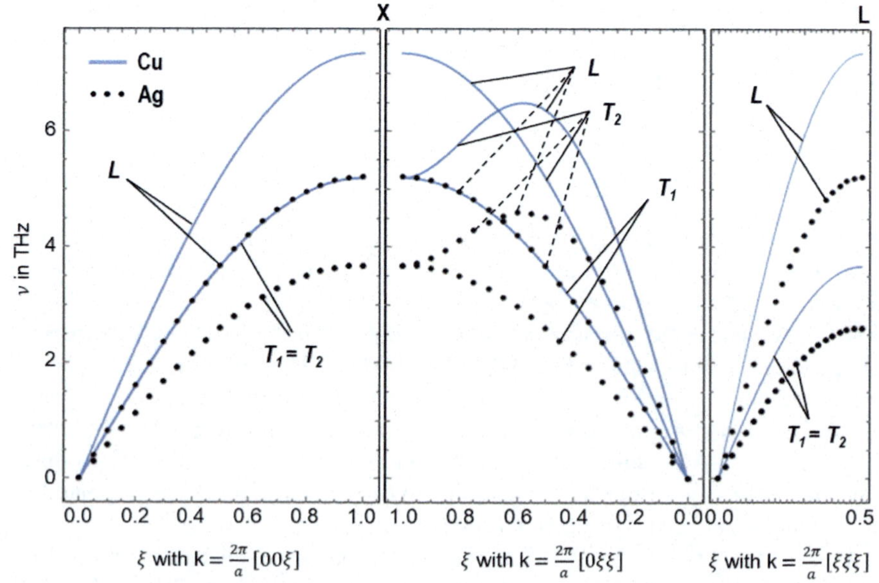

Fig. 4.9 EAM-calculation of the phonon dispersion of Cu and Ag for the three FCC symmetry directions [001], [011] and [111].

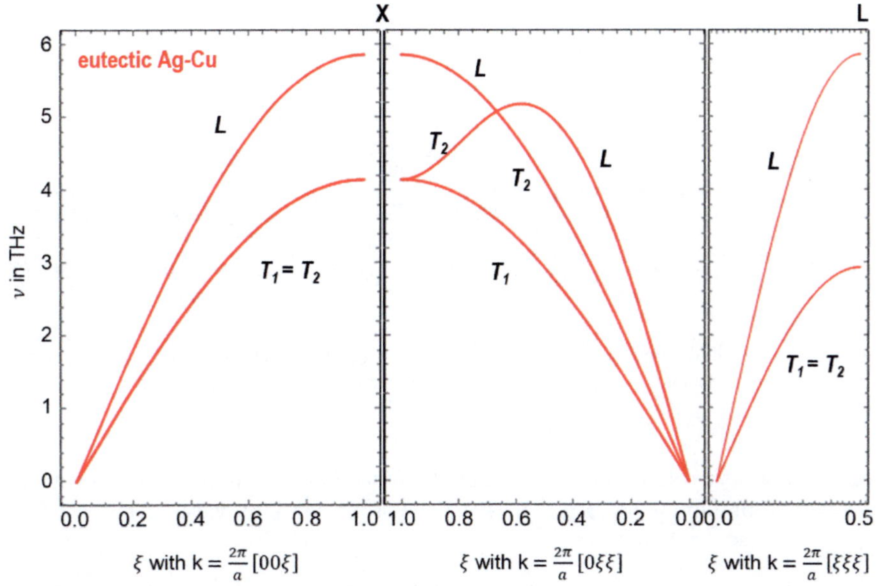

Fig. 4.10 EAM-calculation of the phonon dispersion of the eutectic binary alloy Ag-Cu for [001], [011] and [111] directions.

problem of Eq. (4.28) must be solved for each choice of $\underline{\mathbf{k}}$. Here we exclusively investigated a weighted sum of the eigenfrequencies of the three elemental symmetry directions [001], [011], and [111].

Equation (4.29) results from considering the $6N$-dimensional phase space, well-known in statistical mechanics, and by adding the energy-contribution of each oscillator to the partition function Z. The total energy of atom α as well as the particle-specific heat capacity (at constant volume, *i.e.* $a = a^{(\mathrm{e})}$) can now be written as:

$$E^{\alpha}_{\mathrm{tot}}(T,y) = E^{\alpha}_{(\mathrm{EAM})}(y) + E^{\alpha}_{\mathrm{kin}}(T), \qquad c_{\mathrm{v}}(T,y) = \frac{\partial E^{\alpha}_{\mathrm{tot}}(T,y)}{\partial T} . \tag{4.30}$$

By means of the results of Fig. 4.9 and 4.10 $c_{\mathrm{v}}(T,y)$ can be calculated for different temperatures and - in case of alloys - for different concentrations. Figure 4.11 illustrates the results for the pure metals as well as for the eutectic mixture. Here, for compsarison reasons, we also displayed the heat capacity according to the EINSTEIN model, *viz.*

$$c_{\mathrm{v,E}} = 3N_{\mathrm{A}}k_{\mathrm{B}}\left(\frac{T_{\mathrm{E}}}{T}\right)^2 \frac{\exp(T_{\mathrm{E}}/T)}{[\exp(T_{\mathrm{E}}/T)-1]^2} \quad \text{with} \quad \nu_{\mathrm{E}} = \frac{T_{\mathrm{E}}k_{\mathrm{B}}}{h} . \tag{4.31}$$

Obviously the EAM values corresponds to theoretical curves in a very good manner. Here experimental data were used for T_{E} and ν_{E} in order to fit the curves for Ag

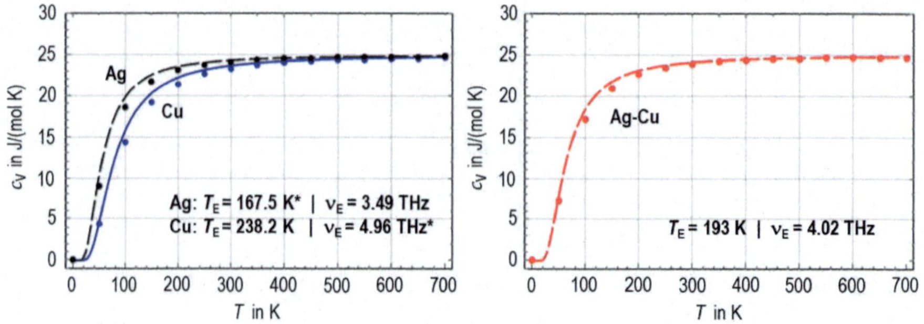

Fig. 4.11 Temperature dependent heat capacity for Ag, Cu, and eutectic Ag-Cu including EINSTEIN temperature and frequency (values with ⋆ represent experimental data used as input).

and Cu (Fornasini et al, 2004; Dubiel et al, 2009). For the mixture no experimental data were found in the literature; here we adjust the curve by varying T_{E}.

4.5 Summary and Conclusions

In the preceding sections it was demonstrated, how atomic interactions can be linked to macroscopic materials behavior. Starting with various basic considerations of lattice kinematics the general idea of EAM was illustrated and an energy expression for an atom in pure metals (*i.e.* one atom type within the lattice) and in binary alloys (*i.e.* two atom types stochastically distributed within the lattice) was developed. Furthermore JOHNSON's analytical nearest neighbor approach was considered and applied to the FCC metals Cu and Ag as well as to the binary alloy Ag-Cu. Moreover, lattice vibrations were analyzed in order to additionally incorporate temperature-dependent, kinetic energy contributions.

Without any doubt, the crucial element denotes the energy expressions for the single atom α. The equations for E^{α} are continuously used for any further exploitations and allow to derive the macroscopic materials properties, *e.g.* the elastic constants, equilibrium lattice parameter, excess enthalpy or specific heat capacity. This procedure, namely using calculations on the atomic scale and ending up with macroscopic materials properties, represents a multiscale approach - a widely used slogan with the scientific environment in the recent time.

However, various improvements and open tasks remain. The world of metals does not only consists of FCC lattices. Consequently the presented theory should be extended to further lattice configurations, such as BCC (Fe, Li, Cr, ...), HCP (Zn, Mg, Ti, ...) or BCT (β-Sn). Here the so-called Modified Embedded-Atom-Method (MEAM), *e.g.* Feraoun et al (2001), were developed, which allow to consider directional-specific bonding. Finally, leaving the restriction to nearest neighbor interactions as well as the incorporation of more eigenfrequencies to the kinetic energy could (further) improve the calculated quantities, such as excess enthalpy or specific heat capacity.

References

Anders D, Hesch C, Weinberg K (2012) Computational modeling of phase separation and coarsening in solder alloys. International Journal of Solids and Structures 49(13):1557–1572

Ardell AJ (2012) Gradient energy, interfacial energy and interface width. Scripta Materialia 66(7):423–426

Bian Q, Bose S, Shukla R (2008) Vibrational and thermodynamic properties of metals from a model embedded-atom potential. Journal of Physics and Chemistry of Solids 69(1):168–181

Bocchetti V, Diep HT (2013) Monte carlo simulation of melting and lattice relaxation of the (111) surface of silver. Surface Science 614:46–52

Böhme T (2008) Investigations of microstructural changes in lead-free solder alloys by means of phase field theories. Cuvillier Verlag

Böhme T, Dreyer W, Müller WH (2007) Determination of stiffness and higher gradient coefficients by means of the embedded-atom method. Continuum Mechanics and Thermodynamics 18(7-8):411–441

Böhme T, Müller WH, Weinberg K (2009) Numerical modeling of diffusion induced phase transformations in mechanically stressed lead-free alloys. Computational Materials Science 45(3):837–844

Cahn JW (1968) Spinodal decomposition. Trans Metall Soc AIME 242:166–180

Cahn JW, Hilliard JE (1958) Free energy of a nonuniform system. I. Interfacial free energy. J Chem Phys 28(2):258–267

Constantinides G, Chandran KR, Ulm FJ, Van Vliet K (2006) Grid indentation analysis of composite microstructure and mechanics: Principles and validation. Materials Science and Engineering: A 430(1-2):189–202

Daw MS, Baskes MI (1983) Semiempirical, quantum mechanical calculation of hydrogen embrittlement in metals. Physical review letters 50(17):1285–1288

Daw MS, Baskes MI (1984) Embedded-atom method: Derivation and application to impurities, surfaces, and other defects in metals. Physical Review B 29(12):6443–6453

Dreyer W, Müller WH (2000) A study of the coarsening in tin/lead solders. International Journal of solids and structures 37(28):3841–3871

Dubiel M, Haug J, Kruth H, Hofmeister H, Seifert W (2009) Temperature dependence of EXAFS cumulants of Ag nanoparticles in glass. Journal of Physics: Conference Series 190:012,123

Eastgate L, Sethna J, Rauscher M, Cretegny T, Chen CS, Myers C (2002) Fracture in mode I using a conserved phase-field model. Physical review E 65(3):036,117

Feraoun H, Aourag H, Grosdidier T, Klein D, Coddet C (2001) Development of modified embedded atom potentials for the cu–ag system. Superlattices and microstructures 30(5):261–271

de Fontaine D (1975) Clustering effects in solid solutions. In: Changes of State, Springer, pp 129–178

Fornasini P, a Beccara S, Dalba G, Grisenti R, Sanson A, Vaccari M, Rocca F (2004) Extended x-ray-absorption fine-structure measurements of copper: Local dynamics, anharmonicity, and thermal expansion. Physical Review B 70(17):174,301/1–12

Galenko P, Reutzel S, Herlach D, Fries S, Steinbach I, Apel M (2009) Dendritic solidification in undercooled Ni–Zr–Al melts: experiments and modeling. Acta Materialia 57(20):6166–6175

Hammerschmidt T, Kersch A, Vogl P (2005) Embedded atom simulations of titanium systems with grain boundaries. Physical Review B 71(20):205,409

Heripre E, Dexet M, Crepin J, Gélebart L, Roos A, Bornert M, Caldemaison D (2007) Coupling between experimental measurements and polycrystal finite element calculations for micromechanical study of metallic materials. International Journal of Plasticity 23(9):1512–1539

Johnson R (1988) Analytic nearest-neighbor model for fcc metals. Physical Review B 37(8):3924–3931

Johnson R (1989) Alloy models with the embedded-atom method. Physical Review B 39(17):12,554–12,559

Kittel C (1973) Einführung in die Festkörperphysik (3. edition)

Leibfried G (1955) Mechanische und thermische Eigenschaften von Kristallen (ed. by s. flügge)

MTData N (1998) Databank for materials thermochemistry. National Physical Laboratory, Queens Road, Teddington, Middlesex, TW11 0LW

Ramasubramaniam A, Itakura M, Ortiz M, Carter E (2008) Effect of atomic scale plasticity on hydrogen diffusion in iron: Quantum mechanically informed and on-the-fly kinetic monte carlo simulations. Journal of Materials Research 23(10):2757–2773

Rose JH, Smith JR, Guinea F, Ferrante J (1984) Universal features of the equation of state of metals. Phys Rev B 29:2963–2969

Svensson E, Brockhouse B, Rowe J (1967) Crystal dynamics of copper. Physical Review 155(3):619–632

Ubachs R, Schreurs P, Geers M (2004) A nonlocal diffuse interface model for microstructure evolution of tin–lead solder. Journal of the Mechanics and Physics of Solids 52(8):1763–1792

Wang Y, Li J (2010) Phase field modeling of defects and deformation. Acta Materialia 58(4):1212–1235

Weinberg K, Dally T, Schuß S, Werner M, Bilgen C (2016) Modeling and numerical simulation of crack growth and damage with a phase field approach. GAMM-Mitteilungen 39(1):55–77

Wheeler AA, Boettinger WJ, McFadden GB (1992) Phase-field model for isothermal phase transitions in binary alloys. Physical Review A 45(10):7424–7440

Wu Z, Francis M, Curtin W (2015) Magnesium interatomic potential for simulating plasticity and fracture phenomena. Modelling and Simulation in Materials Science and Engineering 23(1):015,004

Xu X, Wen M, Fukuyama S, Yokogawa K (2001) Simulation of hydrogen embrittlement at crack tip in nickel single crystal by embedded atom method. Materials Transactions 42(11):2283–2289

Yang H, Ganzosch G, Giorgio I, Abali BE (2018) Material characterization and computations of a polymeric metamaterial with a pantographic substructure. Zeitschrift für angewandte Mathematik und Physik 69(4):105

Yang Y, Zhou X, Johnson R, Wadley H (2001) Monte carlo simulation of hyperthermal physical vapor deposition. Acta materialia 49(16):3321–3332

Zhang J, Ghosh S (2013) Molecular dynamics based study and characterization of deformation mechanisms near a crack in a crystalline material. Journal of the Mechanics and Physics of Solids 61(8):1670–1690

Chapter 5
Mechanical Response Change in Fine Grain Concrete Under High Strain and Stress Rates

Francesco dell'Isola, Anatoly M. Bragov, Leonid A. Igumnov, Bilen Emek Abali, Andrey K. Lomunov, Dmitry A. Lamzin & Alexander Yu. Konstantinov

Abstract Experimental results on assessing the effects of strain and stress rates on the behavior of fine-grain concretes are presented. Specimens of fine-grain and fiber-reinforced concretes were dynamically tested using the Kolsky method and its modification, the "Brazilian test". As a result of the experiments, values of the Dynamic Increase Factor (DIF) were determined for both the materials studied. Their curves as a function of strain and stress rates were constructed. The experimental data is compared with the theoretically obtained values of DIF as a function of strain rate available in the literature

Keywords: Experiments · Fine grain concrete · Reinforced concrete · Kolsky method · Dynamic increase factor (DIF)

5.1 Introduction

Buildings and structures are constantly under various mechanical loadings. Hopefully, all of these loading conditions were foreseen at the stage of construction. Especially impact loading is more critical in the case of safety. Some extreme examples are earthquakes, explosions, collisions of vehicles with parts of structures, etc. Adequate and comprehensive understanding of how structures behave under impact

Francesco dell'Isola · Anatoly M. Bragov · Leonid A. Igumnov · Andrey K. Lomunov · Dmitry A. Lamzin · Alexander Yu. Konstantinov
Research Institute for Mechanics, National Research Lobachevsky State University of Nizhny Novgorod, Russian Federation,
e-mail: fdellisola@gmail.com, bragov@mech.unn.ru, Igumnov@mech.unn.ru, lomunov@mech.unn.ru, lamzin.dmitry@yandex.ru, constantinov.al@yandex.ru

Bilen Emek Abali
Institute of Mechanics, Technische Universität Berlin, Germany,
e-mail: bilenemek@abali.org

B. E. Abali et al. (eds.), *New Achievements in Continuum Mechanics and Thermodynamics*, Advanced Structured Materials 108,
https://doi.org/10.1007/978-3-030-13307-8_5

loading requires knowledge of mechanical properties of structural materials under strain rates. An experimental approach leads to material models and parameters; they are necessary for simulations of applications by utilizing computations with the aid of the finite element method as in Abali (2017). Determining the properties of materials under impact loading is challenging such that procedures and outcome are reported heavily in the literature (Bischoff and Perry, 1991; Malvar and Crawford, 1998; Wang et al, 2011; Thomas and Sorensen, 2017; Scerrato et al, 2016). Modeling concrete by an appropriate constitutive relation is challenging as well. For example in Scerrato et al (2014), Mazars et al (2015) and Giorgio and Scerrato (2017), different constitutive equations are presented in order to represent materials behavior of concrete. Subject to impact loading, the response changes such that the strain rate dependency plays an important role in the material modeling as studied in Chiaia et al (2015), Xiao et al (2015) and Kezmane et al (2017). There are different techniques to define a model, see Contrafatto et al (2016) and Contrafatto et al (2017). Even in the case of multiphysics as in Altenbach et al (2011) and Misra and Poorsolhjouy (2015) there are suggested material models for brittle materials. In order to define the parameters in any of the introduced material models, we need experimental studies.

This work presents the results of studies on evaluating the effect of strain and stress increase rates on the mechanical behavior of fine-grain concrete. Specimens out of fine-grain concrete and fiber-reinforced concrete were dynamically tested by using the so-called Kolsky method and its modified version known as "Brazilian test." Experimental outcome is deformation diagrams for different stress and strain rates. This data is used to obtain the strength of the underlying material subject to various stress and strain rates. Strength increases with an increasing rate, for the brittle materials this increase is characterized by the so-called *dynamic increase Factor* (DIF). Diagrams of DIF as a function of strain or stress increase rates were constructed. The experimentally obtained data was compared with the theoretical diagrams of DIF as described in Wang et al (2011).

5.2 Specimen Preparation

Specimens of blended fine-grain concrete with class B25 axial compression strength were tested. The composition of the concrete mixture corresponded to Russian standard GOST 27006-86. The concrete mixture included grade 400 cement, sand with fineness modulus 2, as well as Muraplast FK-63 plasticizer and Reostab stabilizer. The component ratio used was C:W:F=1:0.6:3, where C denotes cement, W is water, and F stands for fine filling. Cylindrical specimen with 20 mm diameter and 10 mm axial height were prepared for dynamic compression tests, and additionally, diameter of 20 mm and height of 20 mm were constructed for quasi-static tests (up to a strain rate of $0.001\,\mathrm{s}^{-1}$). The mass density is measured as $2000\,\mathrm{kg/m}^3$. The quasi-static strength in compression was equal to 12.1 MPa, and in tension was 1.31 MPa.

The above values of the quasi-static strength were used as reference values for the further computations.

Apart from the tests on the aforementioned concrete B25, we conducted experiments with fine-grain fiber-reinforced concrete CARDIFRC developed and produced in the University of Cardiff (Great Britain) as studied in Benson and Karihaloo (2005a), Benson et al (2005), Benson and Karihaloo (2005b) and Nicolaides et al (2010). In this composite material, as fibers, zinc-coated steel wires were employed. For quasi-static tests—again with the constant strain rate of $0.001\,\mathrm{s}^{-1}$—we used cylindrical specimens of 20 mm diameter and 20 mm height. The quasi-static strength in compression read 94.4 MPa and 23.11 MPa in tension. In the case of dynamic tests, we chose cylinder specimens of 15 mm diameter with 10 mm height on RSG-20 stand (with 20 mm diameter of measuring bars); cylinder specimens of 60 mm diameter with 30 mm height on RSG-60 stand (with 60 mm diameter of measuring bars). The experiments using the two stands with measuring bars of different diameters allowed us to cover a wide range of stress increase rates. Effected by high stresses (locally exceeding the tensile strength), spallation is observed in several tests.

5.3 Method of Impact Loading

One of the most widely used approaches of dynamic testing is the so-called Kolsky method using a Split Hopkinson Bar as introduced in Kolsky (1949). Various modifications of this approach have been discussed in the literature, among others in Bragov and Lomunov (1995), Bragov et al (1994), Bragov et al (1996), Bragov et al (2001), Bragov et al (2008), Bragov et al (2017) and Rodriguez et al (1994). These amendments allow one to determine mechanical properties of various solid materials subjected to high strain rates. In the tests on fine-grain concretes, the classical Kolsky methodology for determining strength under uniaxial compression was used, as well as its modification, the "Brazilian" or spallation test as in Rodriguez et al (1994), which is tantamount to the tensile test configuration as shown in Fig. 5.1.

In compression tests via the Kolsky approach, the loading pulse is assumed to be much longer than the traveling time of the pressure along the specimen. This assumption is indeed accurate since the specimen height is in mm length scale and the speed of sound in solid materials is in the order of km/s. Hence, multiple reflections of the waves from both ends of the specimen result in uniform stress and strain states during the experiment. Therefore, even in the case of high strain rates, we can handle the experimental results in the same way as in quasi-static tests. Propagation of waves along measuring bars is assumed dispersion-free. Otherwise, special corrections must be used as in Bragov et al (2011).

We briefly sum up the expressions used for computing stresses, strains, and strain rates in the specimen as follows:

$$
\begin{aligned}
\sigma_s(t) &= \frac{EA}{A_0}\varepsilon^T(t)\,, \\
\dot{\varepsilon_s}(t) &= -\frac{2C}{L_0}\varepsilon^R(t)\,, \\
\varepsilon_s(t) &= -\frac{2C}{L_0}\int_0^t \varepsilon^R(\bar{t})\,\mathrm{d}\bar{t}\,,
\end{aligned}
\tag{5.1}
$$

where E, A, c denote elastic modulus, cross-sectional area, speed of sound in the Hopkinson bars, respectively; A_0 is the initial cross-sectional area of the specimen as well as L_0 is the initial length of the specimen. The measured quantities, reflected strain, ε^R, and transmitted strain, ε^T, are recorded at the cross-sections of the Hopkinson bars. Concretely, the reflected strain is measured at the loading bar and the transmitted strain is received at the supporting bar. Thus, the material response depends on the amplitude and form of the reflected and transmitted strain pulses.

The Brazilian or spallation test is a modified version of the Kolsky method and it is used for determining tensile strength of brittle materials, see Rodriguez et al (1994). In a conventional compression test of a cylindrical specimen, loading is applied along the longitudinal axis of the specimen. In spallation experiments, a cylindrical specimen is rotated by 90° relative to the transversal axis, and a load is applied along the diametric plane of the specimen, i.e., along the generatrix of the cylinder, see Fig. 5.1. In this case, tensile stresses are induced normal to the loading direction, and the tensile stress, σ_t, is calculated by

$$
\sigma_t(t) = \frac{2EA}{\pi L_0 D_0}\varepsilon^T(t)\,, \tag{5.2}
$$

where D_0 is the diameter of the specimen such that $\pi D_0 L_0$ denotes the curved surface area of the cylindrical specimen.

Two different stands are employed in the experiments:

- RSG-20 composed of 20 mm diameter measuring bars,

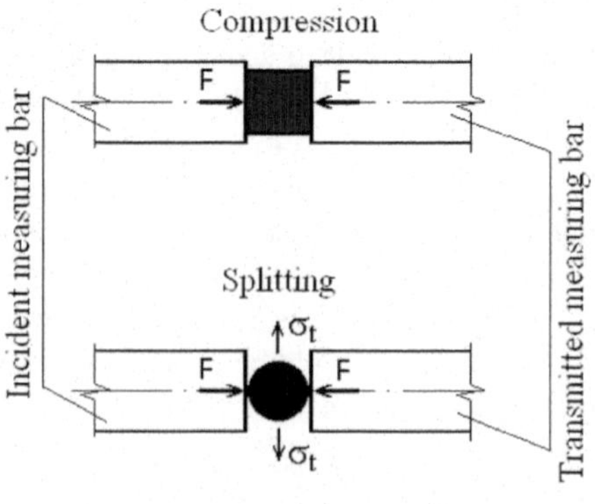

Fig. 5.1 Drawings of the specimen positioning in a conventional Kolsky method (upper) and a spallation test (lower).

- RSG-60 with bars of 60 mm diameter.

The stands were equipped with a pneumatic loading unit (a gas gun with a controlling system), sensors for recording the strains, ε^{R}, ε^{T}, at the two measuring (cylindrical) bars named after Hopkinson. The initial condition is determined by using strain gauges cemented to the curved surface of the measuring bars. Strain gauge signals were transmitted to a digital oscilloscope, the data is post-processed in oscilloscope's own software package.

5.4 Results and Discussion

Results of dynamical tests of the brittle material has been thoroughly analyzed in order to examine the main assumption in Kolsky's method: homogeneous stress within the specimen. Otherwise, the dynamic strength of the material fails to be calculated accurately. In Li and Meng (2003), Zhang et al (2009) and Li et al (2009), the effect of lateral inertia confinement is described in a Split Hopkinson Bar test, especially, in the case of brittle materials. According to the given recommendations, the ratio of the length and the diameter of specimens must be within the range of 0.3-1.0, and the contact surfaces of the measuring bars and the specimen should be lubricated to decrease the effect of friction forces when the specimen expands. These recommendations were followed that provided the equilibrium deformation state in the specimens. This is corroborated by the synchronized strain pulses in the measuring bars, as shown in Fig. 5.2.

Dynamic deformation diagrams showing stress versus strain are obtained out of the latter measurements of transmitted and reflected strains with the aid of Eq. (5.1). The material response for various loading rates is depicted on Fig. 5.3. The deformation diagrams of the concrete and the fiber-reinforced concrete are similar; they indicate a nonlinear stress-strain relation. For each experiment, one deformation di-

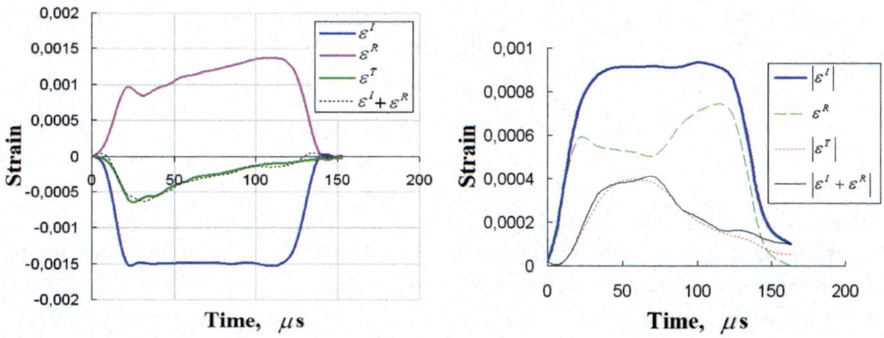

Fig. 5.2 Measured strain pulses at the measuring bars during the compression tests: on fine-grain concrete (left) and fiber-reinforced concrete (right).

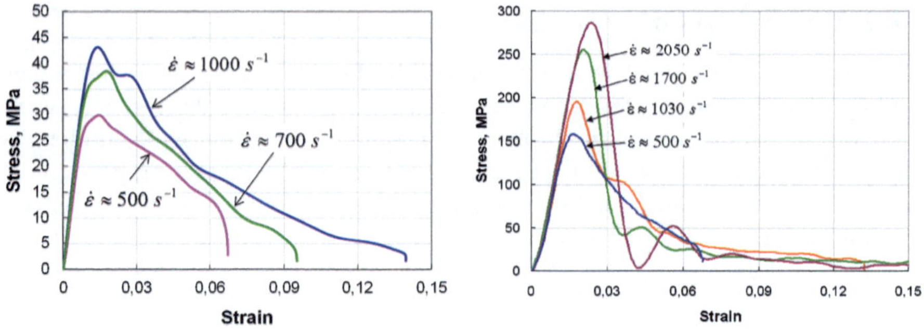

Fig. 5.3 Stress-strain plots obtained from the compression tests: on fine-grain concrete (left) and fiber-reinforced concrete (right).

agram was used to determine strength of the material. The strength of the material let us evaluate DIF. Then a diagram of DIF as a function of strain rate was generated with the following approach. As also noted in Wang et al (2011), the formulas for DIF recommended by Comité Euro-International du Béton adequately describe the experimental results for the high-strength concrete; but they overestimate the strength for the fiber-reinforced concrete. The transition strain rate—where the inaccuracy starts growing—is approximately $30\,\mathrm{s}^{-1}$. In order to obtain DIF from the experimental results, we use a modified version of the formula for DIF of concretes, with a parameter $i = 0$ for the plain concrete and $i = 1$ for the fiber-reinforced concrete, as follows:

$$\mathrm{DIF} = \frac{\sigma_\mathrm{d}}{\sigma_\mathrm{s}} = \begin{cases} \left(\dfrac{\varepsilon^\bullet_\mathrm{d}}{\varepsilon^\bullet_\mathrm{s}}\right)^{1.026\alpha_\mathrm{s}} & \text{for } \varepsilon^\bullet \le (30+23i)\,, \\ \left(\eta\dfrac{\varepsilon^\bullet_\mathrm{d}}{\varepsilon^\bullet_\mathrm{s}}\right)^{k} & \text{for } \varepsilon^\bullet > (30+23i)\,, \end{cases} \tag{5.3}$$

where the fit parameters read

$$\begin{gathered} \alpha_\mathrm{s} = \frac{1}{5+9\frac{\sigma_\mathrm{s}}{\sigma_{\mathrm{c}0}}}\,,\ \sigma_{\mathrm{c}0} = 10\,\mathrm{MPa}\,,\ \eta = \gamma_\mathrm{s}(1-0.3392i)\,, \\ \ln\gamma_\mathrm{s} = 6.156\alpha_\mathrm{s} - 2\,,\ k = \frac{1+0.05i}{3}\,,\ \varepsilon^\bullet_\mathrm{s} = 30\times10^{-6}\,\mathrm{s}^{-1}\,. \end{gathered} \tag{5.4}$$

The diagrams comparing the experimental data with the theoretically obtained curves of DIF are presented in Fig. 5.4. Obviously, the experimental results fail to be represented accurately by the relation as in Eq. (5.3). We can set the following assumptions in question. As also indicated in Rodriguez et al (1994), the applicability of the Brazilian test needs to be analyzed for determining tensile strength of brittle materials. The authors indicate that the Brazilian test can be used for determining tensile strength of brittle materials under the conditions:

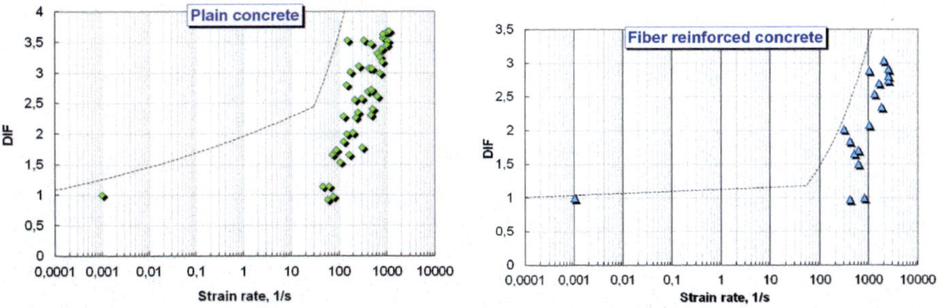

Fig. 5.4 Experimental results for DIF as a function of strain rate: for fine-grain concrete with green diamond markers (left) and for fiber-reinforced concrete with blue triangles (right). The dashed line corresponds to the relation as given in Eq. (5.3).

- Material deforms elastically.
- Deformation occurs in equilibrium.
- Compression creates a deformation solely on the diametral plane of the cylinder.

The measured strain pulses of the measuring bars are presented in Fig. 5.5. Since they are synchronized, the deformation is in equilibrium during the process of spallation. Cracks as well as total disintegration of specimens occurred on the diametral plane. Therefore, we use a slightly different approach to compare relations from Eq. (5.3) experimental results. In each of the spallation experiment, time histories of tensile stress were constructed. Then this relation was used for determining strength of the material, the value of DIF and stress rate. To determine stress rate, the following algorithm was used. A nearly linear part of the time history of tensile stress was chosen, and the experimental curve was approximated by a linear function. The

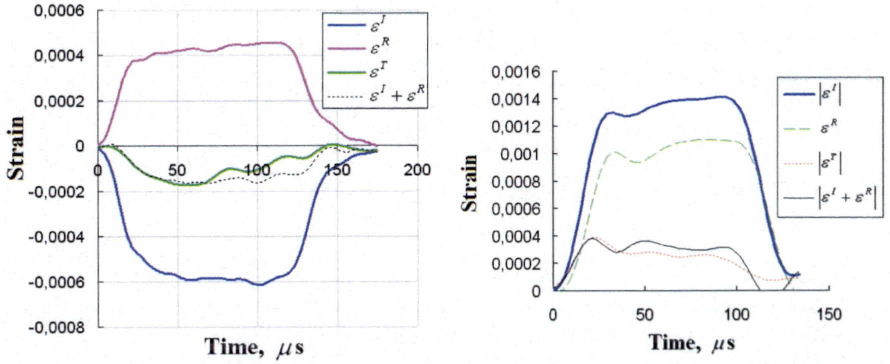

Fig. 5.5 Experimental results for DIF as a function of strain rate: for fine-grain concrete with green diamond markers (left) and for fiber-reinforced concrete with blue triangles (right). The dashed line corresponds to the relation as given in Eq. (5.3).

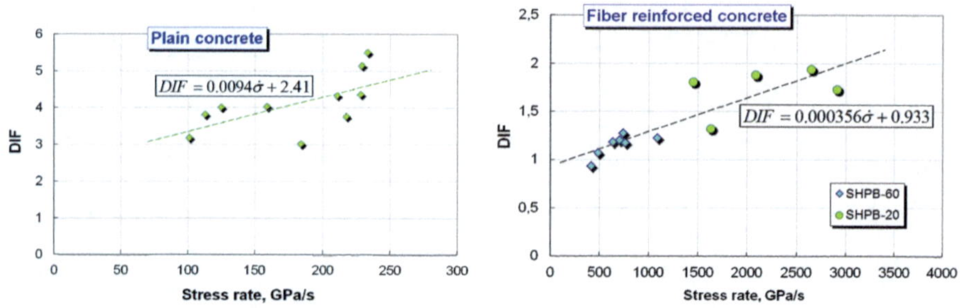

Fig. 5.6 DIF as a function of stress rate: for fine-grain concrete with green diamond markers (left) and for fiber-reinforced concrete with blue diamonds from Hopkinson bar RSG-20 experiment as well as green circles from RSG-60 experiment (right).

slope ratio of the approximating straight line was taken to be the average stress rate. In this way, we have generated the diagrams of DIF as a function of stress rate obtained from spallation test data as shown in Fig. 5.6.

5.5 Conclusion

The Kolsky method and its modification—the Brazilian test—have been used in uniaxial compression and tension spallation tests with specimens of fine-grain and fiber-reinforced concretes. The obtained experimental results demonstrate the effect of strain and stress rates on the strength of the tested materials. The increase of strength observed both in compression and tension has been characterized by DIF. It is noted that the theoretical relations between DIF and strain rate available in the literature do not describe the experimental data adequately.

Acknowledgements This work was supported by a grant from the Government of the Russian Federation (contract No. 14.Y26.31.0031).

References

Abali BE (2017) Computational Reality, Solving Nonlinear and Coupled Problems in Continuum Mechanics, Advanced Structured Materials, vol 55. Springer Nature, Singapore

Altenbach H, Eremeyev VA, Kutschke A, Naumenko K (2011) Conservation laws and prediction methods for stress concentration fields. Acta Mechanica 218(3-4):349–355

Benson SDP, Karihaloo BL (2005a) CARDIFRC®-development and mechanical properties. Part I: Development and workability. Magazine of Concrete Research 57(6):347–352

Benson SDP, Karihaloo BL (2005b) CARDIFRC®-development and mechanical properties. Part III: Uniaxial tensile response and other mechanical properties. Magazine of Concrete Research 57(8):433–443

Benson SDP, Nicolaides D, Karihaloo BL (2005) CARDIFRC®-development and mechanical properties. Part II: Fibre distribution. Magazine of Concrete Research 57(7):421–432

Bischoff PH, Perry SH (1991) Compressive behaviour of concrete at high strain rates. Materials and structures 24(6):425–450

Bragov AM, Lomunov AK (1995) Methodological aspects of studying dynamic material properties using the Kolsky method. International Journal of Impact Engineering 16(2):321–330

Bragov AM, Grushevsky GM, Lomunov AK (1994) Use of the Kolsky method for studying shear resistance of soils. Dymat Journal 3(1):253–259

Bragov AM, Grushevsky GM, Lomunov AK (1996) Use of the Kolsky method for confined tests of soft soils. Experimental Mechanics 36(3):237–242

Bragov AM, Lomunov AK, Sergeichev IV (2001) Modification of the Kolsky method for studying properties of low-density materials under high-velocity cyclic strain. Journal of Applied Mechanics and Technical Physics 42(6):1090–1094

Bragov AM, Konstantinov AY, Lomunov AK (2008) Determining dynamic friction using a modified Kolsky method. Technical Physics Letters 34(5):439

Bragov AM, Konstantinov AY, Medvedkina MV (2011) Dispersiya voln v razreznyh sterzhnyah Gopkinsona pri dinamicheskih ispytaniyah hrupkih materialov (in Russ.). Vestnik Nizhegorodskogo Universiteta im NI Lobachevskogo 6:158–162

Bragov AM, Lomunov AK, Konstantinov AY, Lamzin DA (2017) A modified Kolsky method for determining the shear strength of brittle materials. Technical Physics Letters 43(1):130–132

Chiaia B, Kumpyak O, Placidi L, Maksimov V (2015) Experimental analysis and modeling of two-way reinforced concrete slabs over different kinds of yielding supports under short-term dynamic loading. Engineering Structures 96:88–99

Contrafatto L, Cuomo M, Gazzo S (2016) A concrete homogenisation technique at meso-scale level accounting for damaging behaviour of cement paste and aggregates. Computers & Structures 173:1–18

Contrafatto L, Cuomo M, Greco L (2017) Meso-scale simulation of concrete multiaxial behaviour. European Journal of Environmental and Civil Engineering 21(7-8):896–911

Giorgio I, Scerrato D (2017) Multi-scale concrete model with rate-dependent internal friction. European Journal of Environmental and Civil Engineering 21(7-8):821–839

Kezmane A, Chiaia B, Kumpyak O, Maksimov V, Placidi L (2017) 3d modelling of reinforced concrete slab with yielding supports subject to impact load. European Journal of Environmental and Civil Engineering 21(7-8):988–1025

Kolsky H (1949) An investigation of the mechanical properties of materials at very high rates of loading. Proceedings of the physical society Section B 62(11):676

Li Q, Meng H (2003) About the dynamic strength enhancement of concrete-like materials in a split Hopkinson pressure bar test. International Journal of Solids and Structures 40(2):343–360

Li QM, Lu YB, Meng H (2009) Further investigation on the dynamic compressive strength enhancement of concrete-like materials based on split hopkinson pressure bar tests. part ii: numerical simulations. International Journal of Impact Engineering 36(12):1335–1345

Malvar LJ, Crawford JE (1998) Dynamic increase factors for concrete. Tech. rep., Naval Facilities Engineering Service Center Port hueneme CA

Mazars J, Hamon F, Grange S (2015) A new 3d damage model for concrete under monotonic, cyclic and dynamic loadings. Materials and Structures 48(11):3779–3793

Misra A, Poorsolhjouy P (2015) Granular micromechanics model for damage and plasticity of cementitious materials based upon thermomechanics. Mathematics and Mechanics of Solids pp 1–26

Nicolaides D, Kanellopoulos A, Karihaloo BL (2010) Fatigue life and self-induced volumetric changes of cardifrc. Magazine of Concrete Research 62(9):679–683

Rodriguez J, Navarro C, Sanchez-Galvez V (1994) Splitting tests: an alternative to determine the dynamic tensile strength of ceramic materials. Le Journal de Physique IV 4(C8):C8–101

Scerrato D, Giorgio I, Madeo A, Limam A, Darve F (2014) A simple non-linear model for internal friction in modified concrete. International Journal of Engineering Science 80:136–152

Scerrato D, Giorgio I, Della Corte A, Madeo A, Dowling NE, Darve F (2016) Towards the design of an enriched concrete with enhanced dissipation performances. Cement and Concrete Research 84:48–61

Thomas R, Sorensen AD (2017) Review of strain rate effects for uhpc in tension. Construction and Building Materials 153:846–856

Wang SS, Zhang MH, Quek ST (2011) Compressive behavior of plain and fiber-reinforced high-strength concrete subjected to high strain rate loading. In: Applied Mechanics and Materials, Trans Tech Publ, vol 82, pp 57–62

Xiao J, Li L, Shen L, Poon CS (2015) Compressive behaviour of recycled aggregate concrete under impact loading. Cement and Concrete Research 71:46–55

Zhang M, Wu HJ, Li QM, Huang FL (2009) Further investigation on the dynamic compressive strength enhancement of concrete-like materials based on split Hopkinson pressure bar tests. part i: Experiments. International Journal of Impact Engineering 36(12):1327–1334

Chapter 6
Estimating Fatigue Related Damage in Alloys under Block-type Non-symmetrical Low-cycle Loading

Francesco dell'Isola, Ivan A. Volkov, Leonid A. Igumnov, Simon R. Eugster, Svetlana Yu. Litvinchuk, Dmitri A. Kazakov, Vasilii A. Gorohov & Bilen Emek Abali

Abstract Processes of plastic deformation and damage accumulation in polycrystalline structural alloys are investigated under block-type, nonstationary, non-symmetric cyclic loading. In the framework of damage mechanics, a mathematical model is proposed that effectively describes elastoplastic deformation and fatigue related damage accumulation processes under low-cycle loading. This model can be subsumed under three main parts: the relations defining elastoplastic behavior of the material; the equations describing damage accumulation kinetics; the strength criterion of the damaged material. For validating the model, we perform a numerical analysis and a comparison with the data from full-scale experiments. We demonstrate that the proposed model qualitatively and quantitatively describes the main effects of plastic deformation and damage accumulation processes in structural alloys under complex loading scenarios. Moreover, fatigue related lifetime of the structure is accurately captured by this model as well.

Keywords: Low-cycle fatigue · Plasticity · Non-symmetric loading · Damage mechanics · Numerical analysis · Experiments

---

Francesco dell'Isola · Ivan A. Volkov · Leonid A. Igumnov · Svetlana Yu. Litvinchuk · Dmitri A. Kazakov · Vasilii A. Gorohov
Research Institute for Mechanics, National Research Lobachevsky State University of Nizhny Novgorod, Russian Federation,
e-mail: fdellisola@gmail.com, pmptmvgavt@yandex.ru, igumnov@mech.unn.ru, litvinchuk@mech.unn.ru, kazakov@mech.unn.ru, vas-gor@rambler.ru

Simon R. Eugster
Institute for Nonlinear Mechanics, University of Stuttgart, Germany,
e-mail: simon-raphael.eugster@inm.uni-stuttgart.de

Bilen Emek Abali
Institute of Mechanics, Technische Universität Berlin, Germany,
e-mail: bilenemek@abali.org

B. E. Abali et al. (eds.), *New Achievements in Continuum Mechanics and Thermodynamics*, Advanced Structured Materials 108,
https://doi.org/10.1007/978-3-030-13307-8_6

6.1 Introduction

Materials response under cyclic loading is an important parameter for estimating the reliability and service life of a structure. In n experiment, service life is measured as the number of cycles until failure under repeated loading cycles. This measure is used to compare various designs. In order to estimate the number of cycles until failure, we aim to develop an accurate numerical analysis, for example by using the finite element method (FEM) as demonstrated in Abali (2017b). For a successful computation, the characteristics of the inelastic deformation in hazardous zones requires the formulation of governing equations of thermal plasticity accounting realistic properties of materials as proposed in Mitenkov et al (2007).

Currently, special attention is given to experimental study of cyclic deformation processes since an experiment is of utmost importance to validate any model describing this complex phenomenon. A repetitive loading pattern causes a stationary cyclic deformation meaning that the same amount of energy dissipates in each cycle. Interestingly, it has been detected that preceded by a transition stage the energy dissipation in each cycle starts deviating from the mean value. We may say that there occurs a *cyclic* hardening, softening or relaxation in the response. We need parameters defining a plastic hysteresis loop.

In the process of non-symmetric cyclic deformation of the material, unilateral accumulation of plastic deformation may be observed. Under cyclic loading with initial anisotropy of the stress amplitude at half-cycles of tension and compression, relaxation is observed in average stresses up to zero in a finite number of loading cycles. Under a combined effect of mechanical and thermal loading, the process of cyclic variation generates a multiaxial and non-proportional response leading to additional effects in materials response. The results of an experimental investigation of such processes show that the behavior of structural materials under cyclic proportional loading substantially differ in the case of monotonous deformation processes—modeling a *cyclic* hardening has to be reconsidered. Analogously, multiaxial non-proportional cyclic processes substantially deviate from proportional cyclic processes as in Lemba (1978); Makdauel (1985); Ohasi et al (1985); Tanaka et al (1985a,b); Hassan et al (2008); Huang et al (2014); Jiang and Zhang (2008); Taleb et al (2014).

Governing equations are often developed on the basis of monotonous loading processes. They fail to model the specific features of cyclic deformation under both proportional and non-proportional loading phenomena. Hence, it is challenging to expect a good estimation of fatigue related damage by using these equations. We suggest to model materials response by evaluating service life characteristics of materials with the aid of experimental studies under proportional as well as non-proportional loading, see Bodner and Lindholm (1976); Lemaitre (1985); Chaboche (1989); Bondar and Danshin (2008); Volkov and Korotkikh (2008); Mitenkov et al (2015). Classical methods for predicting fatigue life of materials utilize semi-empirical formulas based on the assumption that the energy dissipation remains constant through the service life. These methods not only require large bulks of experimental information, they also hold only for a small class of loading regimes within the limits of available basic experimental data as in Collins (1984).

A novel scientific approach is proposed to overcome the aforementioned difficulties, see Murakami (1983); Volkov and Igumnov (2017). This so-called mechanics of damaged materials (MDM) studies the processes of growth of microdefects, mechanical behavior of damaged materials by describing the effect of distributed microdefects, using certain mechanical parameters, and the formation of macroscopic cracks (processes of damage accumulation), trying to combine the viewpoints of materials science and continuum mechanics. The current practice of using this approach for various mechanisms of exhausting the service life allows us to state that such an approach is effective enough for the practical purposes of evaluating service life characteristics of materials and can accurately evaluate the process of exhausting the service life of structural elements and parts of load-carrying structures.

Ample studies consider novel developments in plasticity. Governing equations in coupled examples has been studied in several works, for example see Papadopoulos and Lu (1998); Miehe et al (2009); Soyarslan and Tekkaya (2010); Altenbach and Eremeyev (2014b); Abali (2017a). Especially models involving porosity as well as viscoelasticity is challenging, among others, see Altenbach and Eremeyev (2014a); Misra and Poorsolhjouy (2015); Placidi (2016). Complex phenomena and their numerical implementation in 3D depends on the success of the accurate modeling of the material behavior as presented in Papadopoulos and Lu (2001); Schröder et al (2002); Montáns and Bathe (2005); Mazière and Forest (2015); Eremeyev et al (2016); Giorgio et al (2016).

The present paper proposes a mathematical model of MDM describing processes of complex plastic deformation and damage accumulation in structural materials (metals and their alloys) under monotonous and cyclic proportional and non-proportional thermal-mechanical loading regimes. To assess the reliability and the scope of applicability of the developed defining relations of MDM, the processes of plastic deformation and fatigue damage accumulation in stainless steels (X10CrNiTi18-10, X10CrNiTi18-9) under block-type, nonstationary, non-symmetric, low-cycle loading were numerically analyzed. The obtained numerical results are compared with the data from full-scale experiments.

6.2 Defining Relations of Mechanics of Damaged Media

A model as in Volkov and Korotkikh (2008); Volkov and Igumnov (2017) describing the damage in a body consists of the following:

- relations defining the elastoplastic behavior of the material, accounting for its dependence on the failure process;
- evolutionary equations describing damage accumulation kinetics;
- a strength criterion of the damaged material.

6.2.1 Defining Relations in Plasticity

The definition of relations modeling plasticity is based on the following assumptions:

- Components of strain tensors e_{ij} and strain rates $e^{\bullet}_{ij}$ include elastic, ${}^{\mathrm{e}}e_{ij}$, ${}^{\mathrm{e}}e^{\bullet}_{ij}$, and plastic strains, ${}^{\mathrm{p}}e_{ij}$, ${}^{\mathrm{p}}e^{\bullet}_{ij}$, in other words, the reversible and irreversible components.
- For various temperatures, the initial yield surface is described by a Mises type surface. The evolution of the yield surface is described by the variation of its radius ${}^{\mathrm{p}}C$ and the motion of its center, so-called back stress, ρ_{ij}.
- The volume change is reversible, in other words, plastic deformation is deviatoric.
- Continuum body is isotropic at the reference frame; but anisotropy may be caused by plasticity.

For the elastic regime, the spherical and deviatoric parts of the stress and elastic strain tensors,

$$\sigma_{ij} = \sigma\delta_{ij} + \sigma_{|ij|} \, , \ {}^{\mathrm{e}}e_{ij} = e\delta_{ij} + e_{|ij|} \, , \tag{6.1}$$

are modeled by HOOKE's law with the DUHAMEL–NEUMANN extension as follows:

$$\sigma = 3K\Big({}^{\mathrm{e}}e - \alpha(T - T_{\text{ref.}})\Big) \, , \ \sigma_{|ij|} = 2G\,{}^{\mathrm{e}}e_{|ij|} \, , \tag{6.2}$$

with the material parameters, K, α, G, depending on the temperature. Usually, the initial temperature of the simulation is assumed to be the reference temperature, $T_{\text{ref.}}$, in order to achieve the natural state initially. As a consequence, their rates read

$$\sigma^{\bullet} = 3K\Big({}^{\mathrm{e}}e^{\bullet} - \alpha^{\bullet}T - \alpha T^{\bullet}\Big) + \frac{K^{\bullet}}{K}\sigma \, , \ \sigma^{\bullet}_{|ij|} = 2G\,{}^{\mathrm{e}}e^{\bullet}_{|ij|} + \frac{G^{\bullet}}{G}\sigma_{|ij|} \, . \tag{6.3}$$

In classical or so-called PRANDTL–REUSS plasticity, a yield surface is introduced,

$$F_S = S_{ij}S_{ij} - {}^{\mathrm{p}}C \, , \ S_{ij} = \sigma_{|ij|} - \rho_{ij} \, , \tag{6.4}$$

by using an isotropic hardening with the center, ${}^{\mathrm{p}}C$, and a kinematic hardening with its motion, ρ_{ij}. For modeling complex cyclic deformation modes in the stress space, we introduce the following yield surface:

$$F_\rho = \rho_{ij}\rho_{ij} - \rho^2_{\max} \, , \tag{6.5}$$

providing a "memory" with the aid of the maximal modulus, $\rho_{\max}$. We consider such a temperature range, where annealing effects fail to be significant. Then the isotropic hardening, modeling the evolution of ${}^{\mathrm{p}}C$, consists of three parts: effected by monotonous, cyclic, and temperature related phenomena. We follow Volkov and Korotkikh (2008); Mitenkov et al (2015) and implement the following evolution law:

$$
\begin{gathered}
{}^{\mathrm{p}}C^{\bullet} = \Big(q_\chi H(F\rho) + a(Q_{\mathrm{S}} - {}^{\mathrm{p}}C)\Gamma(F_\rho)\Big)\chi^{\bullet} + q_3 T^{\bullet} , \\
{}^{\mathrm{p}}C = {}^{\mathrm{p}}C^0 + \int_0^t {}^{\mathrm{p}}C^{\bullet}\,\mathrm{d}t , \ \chi^{\bullet} = \sqrt{\frac{2}{3}\,{}^{\mathrm{p}}e^{\bullet}_{ij}\,{}^{\mathrm{p}}e^{\bullet}_{ij}} , \ \chi_m = \int_0^t \chi^{\bullet} H(F_\rho)\,\mathrm{d}t , \ \chi = \int_0^t \chi^{\bullet}\,\mathrm{d}t , \\
q_\chi = \frac{q_2 A\psi_1 + (1-A)q_1}{A\psi_1 + (1-A)} , \ Q_{\mathrm{S}} = \frac{Q_2 A\psi_2 + (1-A)Q_1}{A\psi_2 + (1-A)} , \ 0 \le \psi \le 1 , \ i = 1,2 , \\
A = 1 - \cos^2\theta , \ \cos\theta = n^e_{ij} n^s_{ij} , \ n^e_{ij} = \frac{e^{\bullet}_{|ij|}}{\sqrt{e^{\bullet}_{|ij|}e^{\bullet}_{|ij|}}} , \ n^s_{ij} = \frac{S_{ij}}{\sqrt{S_{ij}S_{ij}}} , \\
\Gamma(F_\rho) = 1 - H(F_\rho) , \ H(F_\rho) = \begin{cases} 1 , & F_\rho = 0 \wedge \rho_{ij}\rho^{\bullet}_{ij} \ge 0 \\ 0 , & F_\rho < 0 \vee \rho_{ij}\rho^{\bullet}_{ij} < 0 \end{cases} ,
\end{gathered}
\tag{6.6}
$$

where q_1, q_2, q_3 denote moduli of *monotonous* isotropic hardening; Q_1, Q_2 indicate moduli of *cyclic* isotropic hardening; a is a constant defining evolution of the hysteresis loop of cyclic deformation; Q_{S} is the corresponding yield surface radius for the known ρ_{max} and T; ${}^{\mathrm{p}}C^0$ is the initial value of the yield surface radius. Evolution of the back-stress is obtained by introducing it as an internal variable and postulated to have the following form:

$$
\rho^{\bullet}_{ij} = f(\chi_m)\Big(g_1\,{}^{\mathrm{p}}e^{\bullet}_{ij} - g_2\rho_{ij}\chi^{\bullet}\Big) + g_T\rho_{ij}\langle T^{\bullet}\rangle + \tilde{\rho}^{\bullet}_{ij} , \ \rho_{ij} = \int_0^t \rho^{\bullet}_{ij}\,\mathrm{d}t , \tag{6.7}
$$

with

$$
\begin{gathered}
f(\chi_m) = 1 + k_1\big(1 - \exp(-k_2\chi_m)\big) , \\
\tilde{\rho}^{\bullet}_{ij} = g_3\,{}^{\mathrm{p}}e^{\bullet}_{ij}H(F_\rho) - g_4\tilde{\rho}_{ij}\chi^{\bullet}\Gamma(F_\rho)\langle\cos(\gamma)\rangle , \\
\langle\cos(\gamma)\rangle = \frac{\rho^{\bullet}_{ij}\tilde{\rho}_{ij}}{\sqrt{\rho^{\bullet}_{kl}\rho^{\bullet}_{kl}}\sqrt{\tilde{\rho}_{mn}\tilde{\rho}_{mn}}} ,
\end{gathered}
\tag{6.8}
$$

where g_1, g_2, g_3, g_4, g_T, as well as k_1, k_2 are material parameters to be determined experimentally.

Equations $(6.7)_{1,2}$ describe the anisotropic part of deformation hardening. Equation $(6.7)_3$ models the evolution of ρ_{ij} by including the effect of the temperature rate as well. Equation $(6.7)_4$ indicates an anisotropic hardening due to the unilaterally accumulated plastic deformation. Weighting factor $f(\chi_m)$ allows to describe the evolution of ρ_{ij} under cyclic deformation regimes (Korotkikh, 1985). For a non-symmetric cyclic loading, $\tilde{\rho}^{\bullet}_{ij}$ models the cyclic plastic hysteresis loop differently for various loading strengths. In the case of vanishing $g_T = g_3 = g_4 = k_1 = 0$, Eqs. (6.7) reduce to the special case of ARMSTRONG–FREDERIK–KADASHEVICH equations as follows:

$$
\rho^{\bullet}_{ij} = g_1\,{}^{\mathrm{p}}e^{\bullet}_{ij} - g_2\rho_{ij}\chi^{\bullet} . \tag{6.9}
$$

In order to include a *memory* effect for the yield surface, it is necessary to generate an evolution equation for ρ_{max} as well,

$$
\rho^{\bullet}_{\mathrm{max}} = \frac{\rho_{ij}\rho^{\bullet}_{ij}H(F_\rho)}{\sqrt{\rho_{mn}\rho_{mn}}} - g_2\rho_{\mathrm{max}}\chi^{\bullet} - g_T\rho_{\mathrm{max}}\langle T^{\bullet}\rangle . \tag{6.10}
$$

The plastic strain rate tensor components fulfill the orthogonality restriction at the yield surface

$$ {}^{\mathrm{p}}e^{\bullet}_{ij} = \lambda S_{ij} \,, \tag{6.11} $$

where λ is proportionality coefficient determined from the condition that a new yield surface passes through the end of the stress deviator vector at the end of the loading stage.

As damage is modeled by a field function, ω, we can easily introduce the effect of damage on the material properties by introducing effective stresses as in Volkov and Igumnov (2017),

$$ \begin{aligned} \tilde{\sigma}_{|ij|} &= F_1(\omega)\sigma_{|ij|} = \frac{G}{\tilde{G}}\sigma_{|ij|} = \frac{\sigma_{|ij|}}{(1-\omega)\Big(1-\frac{6K+12G}{9K+8G}\Big)} \,, \\ \tilde{\sigma} &= F_2(\omega)\sigma = \frac{K}{\tilde{K}}\sigma = \frac{\sigma}{(1-\omega)\frac{4G}{4G+3K\omega}} \,, \end{aligned} \tag{6.12} $$

where $\tilde{G}$, $\tilde{K}$ are effective elastic moduli determined by the MACKENZIE formulas (Mackenzie, 1950). Analogously, we propose to obtain

$$ \tilde{\rho}_{ij} = F_1(\omega)\rho_{ij} \,. \tag{6.13} $$

6.2.2 Evolutionary equations of fatigue damage accumulation

Rate of damage is modeled by an evolution equation, for low-cycle fatigue (LCF), we define this relation by the following model as in Bodner and Lindholm (1976); Lemaitre (1985); Volkov and Korotkikh (2008); Volkov and Igumnov (2017):

$$ \omega^{\bullet} = f_1(\beta) f_2(\omega) f_3(W) f_4(W^{\bullet}) \,, \tag{6.14} $$

where the function $f_1(\beta)$ denotes the effects of volume change with $\beta = \sigma/\sigma_{\mathrm{u}}$, the function $f_2(\omega)$ incorporates the degree of the present damage accumulated over the time, the function $f_3(W)$ models the effect of the dissipated work (energy) on damage for creating a fracture, the function $f_4(W^{\bullet})$ embodies the effects of the rate of damage energy. We model these functions in the following way:

$$ \begin{aligned} f_1(\beta) &= \exp(\beta)\,,\; f_2(\omega) = \begin{cases} 0\,, & W \le W_{\mathrm{a}} \\ \omega^{1/3}(1-\omega)^{2/3} & W > W_{\mathrm{a}} \wedge \omega \le \frac{1}{3} \\ \frac{16^{1/3}}{9}\omega^{-1/3}(1-\omega)^{-2/3} & W > W_{\mathrm{a}} \wedge \omega > \frac{1}{3} \end{cases} \\ f_3(W) &= \frac{W - W_{\mathrm{a}}}{W_{\mathrm{f}}}\,,\; f_4(W^{\bullet}) = \frac{W^{\bullet}}{W_{\mathrm{f}}}\,, \end{aligned} \tag{6.15} $$

where $\beta = \sigma/\sigma_{\mathrm{u}}$ gives the voluminosity of stressed state, W_{a} is the corresponding energy used for damage at the end of the stage of nucleation of microcracks under low cycle fatigue, and W_f is the energy used for macroscopic crack formation process. The duration of the microcracks nucleation phase will be related with the value of parameter W_a. When microcracks reach the dimensions in the length scale of the mean distance between microcracks, the process of merging (agglomeration) starts. We circumvent introducing a detailed micromechanical model of this agglomeration and model this phenomenon via kinetic equation by introducing term $f_2(\omega)$ in such a way that relation $\omega^\cdot = f_1(\omega)$ considers for the avalanche-like increase of the damage when damage reaches the value of $\omega = 1/3$.

6.2.3 Strength Criterion of the Damaged Material

We implement a simple approach and terminate accumulation of microcracks in the case of damage approaching its critical value

$$\omega = \omega_\text{f} \,, \tag{6.16}$$

where this critical value has to be smaller than 1, otherwise numerical instabilities occur. For engineering alloys $\omega_\text{f} = 0.3$ and for pure materials $\omega_\text{f} = 0.7$ (Lemaitre, 1985).

6.3 Numerical Results

Specimens of stainless steel 12X18H10T were experimentally studied under a uniaxial tension-compression test at the ambient temperature in the Laboratory for Testing Physical-Mechanical Properties of Structural Materials, Research Institute for Mechanics, Nizhniy Novgorod Lobachevski State University. The testing procedure consisted of five subsequent tests including monotonous and cyclic loading as in Korotkikh (1985):

I. 20 cycles of symmetric cyclic loading with a deformation amplitude of $e_{11} = 0.08\%$
II. Monotonous tension up to the deformation of $e_{11} = 5\%$
III. 200 cycles of non-symmetric cyclic loading with the deformation amplitude of $\Delta e_{11} = 1.2\%$ and the mean deformation of $e_{11}^{(m)} = 4.4\%$ (during this test, plastic hysteresis loop occurs)
IV. Monotonous tension up to deformation $e_{11} = 1\%$
V. Non-symmetric cyclic loading with the deformation amplitude of $\Delta e_{11} = 1.2\%$ and the mean deformation of $e_{11}^{(m)} = 9.4\%$ up to failure (the number of cycles to failure is $N_f = 2800$, again, plastic hysteresis loop takes place)

We compile in Tables 6.1-6.3 the properties of the presented MDM model for steel X10CrNiTi18-10 determined from the results of experiments in Volkov and

Table 6.1 Obtained parameters of the MDM model for X10CrNiTi18-10

K	G	PC^0 in MPa	g_1 in MPa	g_2	g_3 in MPa	g_4	k_1	k_2	a
165 277	76 282	203	20 850	297	660	3	0.48	0.2	5

Table 6.2 Modulus of cyclic hardening $Q_S(\rho_{max})$ in MPa for X10CrNiTi18-10

Q_S in MPa	203	210	232	232	232	232	232
ρ_{max} in MPa	0	30	60	90	100	110	120

Table 6.3 Modulus of monotonous hardening q_χ in MPa for X10CrNiTi18-10

q_χ in MPa	-17 000	-4 634	-811	371	737	849	897	900	900	900	900	900	900
χ	0	0.002	0.004	0.006	0.008	0.01	0.015	0.02	0.03	0.04	0.05	0.09	0.15

Korotkikh (2008); Mitenkov et al (2015); Volkov and Igumnov (2017) and used for the following computations.

In order to present the results, we use dots for experiments and a continuous line for the computation in the following figures for the aforementioned 5 subsequent loading scenarios I-V demonstrated in Figs. 6.1-6.3

We emphasize the quantitative and qualitative agreement between the experimental data and its computation. Moreover, another set of experiments were conducted for stainless steel X10CrBiTi18-9 under non-stationary, non-symmetric cyclic loading as follows:

VI. The specimen is compressed up to $e_{11} = 0.01$ and then in pulled in tensile up to $e_{11} = 0.05$

VII. Non-symmetric cyclic loading is applied with a strain interval of 1% up to the failure, $N_f = 850$. Plastic hysteresis loop occurs and after 500th loading cycle, the loop becomes nearly symmetric.

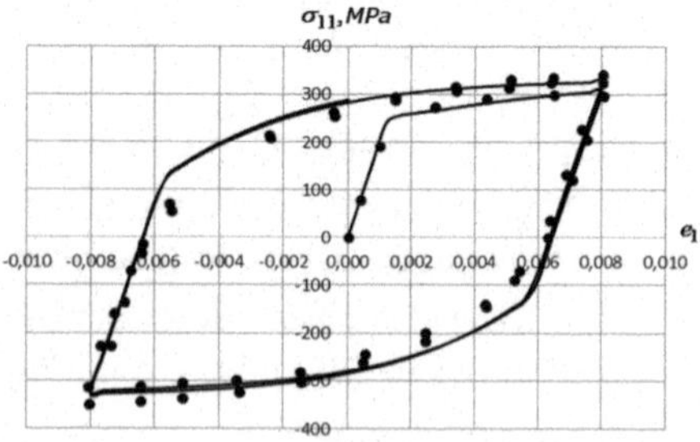

Fig. 6.1 Comparison of the numerical results with the proposed model (continuous line) with the experimental data (dots). One single symmetric loading cycle as in I.

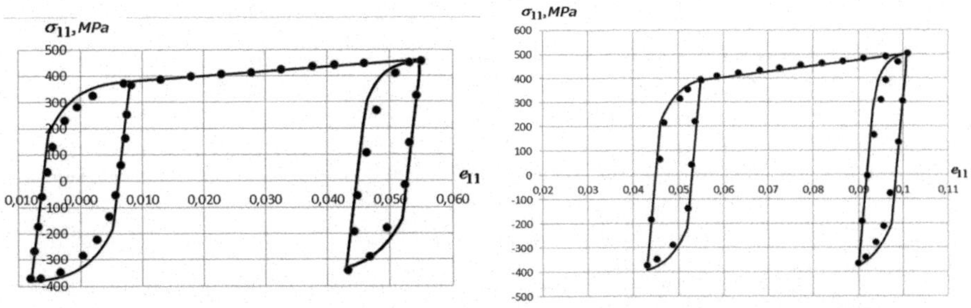

Fig. 6.2 Comparison of the numerical results with the proposed model (continuous line) with the experimental data (dots). Left: 20th cycle of symmetric loading in II followed by a monotonous tensile test until $e_{11} = 0.05$ and a subsequent non-symmetric cyclic loading. Right: 200th cycle of III is followed by one cycle of V.

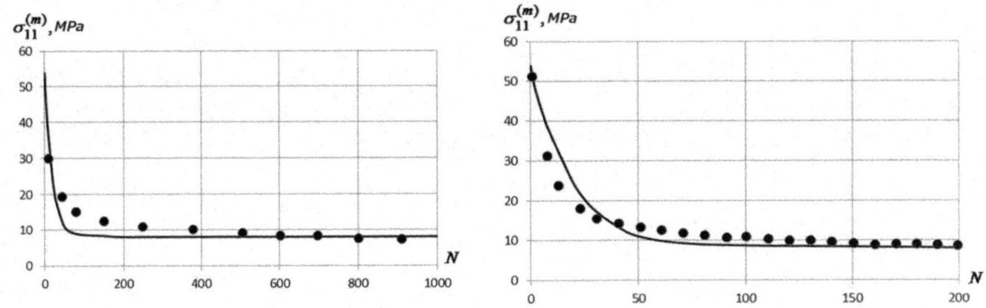

Fig. 6.3 Comparison of the numerical results with the proposed model (continuous line) with the experimental data (dots). The dissipated energy density given as the maximum stress amplitude by III (left) and V (right) loading scenarios.

Table 6.4 Obtained parameters of the MDM model for X10CrNiTi18-10

K	G	$^{p}C'^{0}$ in MPa	g_1 in MPa	g_2	g_3 in MPa	g_4	k_1	k_2	a	W_a	W_f
165 277	76 282	190	24 090	286	800	2	0.415	0.2	5	0	800

Table 6.5 Modulus of cyclic hardening $Q_S(\rho_{max})$ in MPa for X10CrNiTi18-9

Q_S in MPa	190	205	210	215	220	225	225
ρ_{max} in MPa	0	20	40	60	80	100	120

For calculations, we used the parameters as compiled in Tables 6.4-6.6 for the steel X10CrNiTi18-9 obtained from the results of experiment in Volkov and Korotkikh (2008); Mitenkov et al (2015). Analogously, the experiments VI-VII are utilized to validate the accuracy of the proposed model as demonstrated in Fig. 6.4. Two different models for the evolution equation show significant discrepancies in Fig. 6.4

Table 6.6 Modulus of monotonous hardening q_χ in MPa for X10CrNiTi18-9

q_χ in MPa	-5 000	-4 471	-4 188	-3 859	-2 460	-182	888	1 531	1 274	913	913	913
χ	0	0.002	0.004	0.006	0.008	0.01	0.015	0.02	0.03	0.04	0.05	0.06

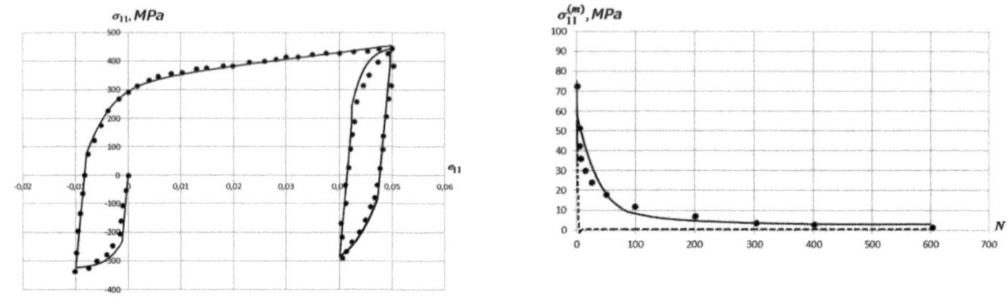

Fig. 6.4 Comparison of the numerical results with the proposed model (continuous line) with the experimental data (dots) for X10CrNiTi18-9. Left: at the 500th cycle during the experiment VII. Right: mean stress for various cycles during the experiment VII, model in Eq. (6.7)$_1$ as the continuous line and model in Eq. (6.9) as the dash line.

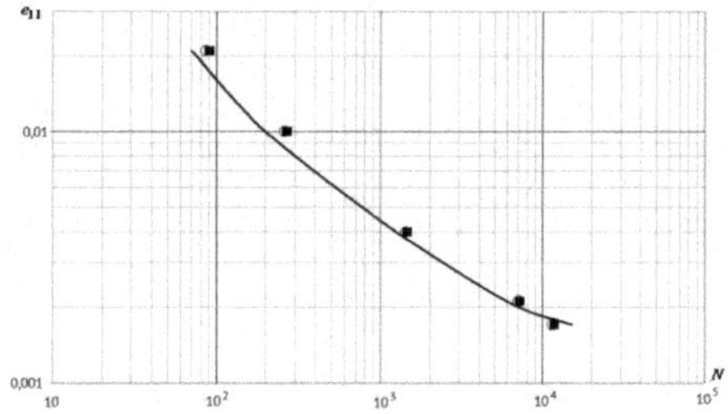

Fig. 6.5 Comparison of the fatigue curves obtained by the two proposed models (dots and squares) with the experimental data (continuous line). Circles denote the thermal plasticity model with th evolution equation as in Eq. (6.7)$_1$ and squares indicate the model with Eq. (6.9).

(right). By using the model in Eq. (6.9), we observe that cyclic hysteresis loop is immediately stated in the first loading cycle such that the physically-important relations of plasticity fail to be described accurately. By employing the model in Eq. (6.7), material behavior during the process is perfectly modeled. In Fig. 6.5, a fatigue curve is shown for the same material, X10CrNiTi18-9, experimented as in VII. The experimental data is quantitatively well represented by the numerical results acquired by the models in Eqs. (6.7), (6.9). In general, the comparison of the numerical and experimental results demonstrate the strength of the proposed MDM model. It describes the processes of fatigue life of polycrystalline structural alloys under non-symmetric low-cycle loading.

6.4 Conclusion

A novel model has been proposed for modeling fatigue life of polycrystalline structural alloys under non-symmetric cyclic loading. The model is verified by utilizing experiments of block-type, transient, non-symmetric, low-cycle loading for two stainless steels (X10CrNiTi18-10, X10CrNiTi18-9). We have demonstrated that even different conditions are perfectly captured by this model, all material parameters are compiled in Tables 6.1-6.6. Two different versions have been implemented and their differences have been discussed. Fatigue life estimation is equally accurate in both versions for the case of non-symmetric low-cycle loading. We recommend to use the simpler version given in Eq. (6.9).

Acknowledgements This work was supported by a grant from the Government of the Russian Federation (contract No. 14.Y26.31.0031).

References

Abali BE (2017a) Computational Reality, Solving Nonlinear and Coupled Problems in Continuum Mechanics, Advanced Structured Materials, vol 55. Springer Nature, Singapore

Abali BE (2017b) Computational study for reliability improvement of a circuit board. Mechanics of Advanced Materials and Modern Processes 3(1):1–11

Altenbach H, Eremeyev V (2014a) Strain rate tensors and constitutive equations of inelastic micropolar materials. International Journal of Plasticity 63:3–17

Altenbach H, Eremeyev VA (2014b) Basic equations of continuum mechanics. In: Plasticity of Pressure-Sensitive Materials, Springer, pp 1–47

Bodner SR, Lindholm US (1976) Kriteriy prirasheniya povrezhdeniya dlya zavisyashego ot vremeni razrusheniya materialov (in Russian). Trudy Amer Ob-va inzh-meh Ser D Teoret Osnovy inzh Raschetov 100(2):51–58

Bondar VS, Danshin VV (2008) Plastichnost, proportsyonalnye i neproportsyonalalnye nagruzheniya (in Russian). M: Fizmatlit p 176

Chaboche JL (1989) Constitutive equations for cyclic plasticity and cyclic viscoplasticity. International journal of plasticity 5(3):247–302

Collins J (1984) Povrezhdeniye materialov v konstruktziyah. Analiz. Predskazaniye. Predotvrasheniye (in Russian). M: Mir

Eremeyev VA, Skrzat A, Stachowicz F (2016) On FEM evaluation of stress concentration in micropolar elastic materials. Nanoscience and Technology: An International Journal 7(4)

Giorgio I, Andreaus U, Scerrato D, dell'Isola F (2016) A visco-poroelastic model of functional adaptation in bones reconstructed with bio-resorbable materials. Biomechanics and modeling in mechanobiology 15(5):1325–1343

Hassan T, Taleb L, Krishna S (2008) Influence of non-proportional loading on ratcheting responses and simulations by two recent cyclic plasticity models. International Journal of Plasticity 24(10):1863–1889

Huang ZY, Chaboche JL, Wang QY, Wagner D, Bathias C (2014) Effect of dynamic strain aging on isotropic hardening in low cycle fatigue for carbon manganese steel. Materials Science and Engineering: A 589:34–40

Jiang Y, Zhang J (2008) Benchmark experiments and characteristic cyclic plasticity deformation. International Journal of Plasticity 24(9):1481–1515

Korotkikh YG (1985) Opisaniye protsessov nakopleniya povrezhdeniy materiala pri neizotermicheskom vyazkoplasticheskom deformirovanii (in Russian). Problemy prochnosti 1:18–23
Lemaitre J (1985) Kontinualnaya model povrezhdeniya, ispolzuemaya dlya rascheta razrusheniya plastichnykh materialov (in Russian). Trudy Amer Ob-va inzh-meh Ser D Teoret Osnovy inzh Raschetov 107(1):90–98
Lemba (1978) Sisebottom plastichnost pri cyklicheskim nagruzhenii po neproportsionalnym traektoriyam (in Russian). Teoreticheskiye osnovy inzhenernykh raschetov 100(1):108–126
Mackenzie J (1950) The elastic constants of a solid containing spherical holes. Proceedings of the Physical Society Section B 63(1):2
Makdauel (1985) Eksperimentalnoye izuchenie struktury opredelyayushih uravneniy dlya neproportsionalnoy cyklicheskoy plastichnosti (in Russian). Teoreticheskiye osnovy inzhenernykh raschetov 107(4):98–111
Mazière M, Forest S (2015) Strain gradient plasticity modeling and finite element simulation of lüders band formation and propagation. Continuum Mechanics and Thermodynamics 27(1-2):83–104
Miehe C, Göktepe S, Diez JM (2009) Finite viscoplasticity of amorphous glassy polymers in the logarithmic strain space. International Journal of Solids and Structures 46(1):181–202
Misra A, Poorsolhjouy P (2015) Granular micromechanics model for damage and plasticity of cementitious materials based upon thermomechanics. Mathematics and Mechanics of Solids p 1081286515576821
Mitenkov AM, Kaydalov VB, Korotkikh YG (2007) Metody obosnovaniya resursa yaeu (in Russian). Mashinostroyeniye p 445
Mitenkov FM, Volkov IA, Igumnov LA (2015) Prikladnaya teoriya plastichnosti (in Russian). Fizmatlit p 284
Montáns FJ, Bathe KJ (2005) Computational issues in large strain elasto-plasticity: an algorithm for mixed hardening and plastic spin. International Journal for Numerical Methods in Engineering 63(2):159–196
Murakami S (1983) Sushnost mehaniki povrezhdennoy sredy i eyo prilozheniye k teorii anizotropnykh povrerzhdeniy pri polzuchesti (in Russian). TOIR 2:44–50
Ohasi, Kavai, Kaito (1985) Neuprugoye povedeniye stali 316 pri mnogoosnykh neproportsionalnykh zyklicheskikh nagruzheniyakh pri povyshennoy temperature (in Russian). Teoreticheskiye osnovy inzhenernykh raschetov 107(2):6–15
Papadopoulos P, Lu J (1998) A general framework for the numerical solution of problems in finite elasto-plasticity. Computer Methods in Applied Mechanics and Engineering 159(1-2):1–18
Papadopoulos P, Lu J (2001) On the formulation and numerical solution of problems in anisotropic finite plasticity. Computer Methods in Applied Mechanics and Engineering 190(37-38):4889–4910
Placidi L (2016) A variational approach for a nonlinear one-dimensional damage-elasto-plastic second-gradient continuum model. Continuum Mechanics and Thermodynamics 28(1-2):119–137
Schröder J, Gruttmann F, Löblein J (2002) A simple orthotropic finite elasto–plasticity model based on generalized stress–strain measures. Computational Mechanics 30(1):48–64
Soyarslan C, Tekkaya A (2010) A damage coupled orthotropic finite plasticity model for sheet metal forming: Cdm approach. Computational Materials Science 48(1):150–165
Taleb L, Cailletaud G, Saï K (2014) Experimental and numerical analysis about the cyclic behavior of the 304L and 316L stainless steels at 350°C. International Journal of Plasticity 61:32–48
Tanaka E, Murakami S, Ōoka M (1985a) Effects of plastic strain amplitudes on non-proportional cyclic plasticity. Acta Mechanica 57(3-4):167–182
Tanaka E, Murakami S, Ōoka M (1985b) Effects of strain path shapes on non-proportional cyclic plasticity. Journal of the Mechanics and Physics of Solids 33(6):559–575
Volkov IA, Igumnov LA (2017) Vvedenie v kontinualnuyu mehaniku povrezhdennoy sredy (in Russian). M: Fizmatlit p 304
Volkov IA, Korotkikh YG (2008) Uravneniya sostoyaniya vyazkouprugoplasticheskikh sred s povrezhdeniyami (in Russian). Fizmatlit p 424

Chapter 7
On Non-holonomic Boundary Conditions within the Nonlinear Cosserat Continuum

Victor A. Eremeyev

Abstract Within the framework of the nonlinear micropolar elastic continuum we discuss non-holonomic kinematic boundary conditions. By non-holonomic boundary conditions we mean linear relations between virtual displacements and virtual rotations given on the boundary. Such boundary conditions can be used for modelling of complex material interactions in the vicinity of the boundaries and interfaces.

7.1 Introduction

The model of micropolar medium known also as Cosserat continuum was proposed by Cosserat brothers, see Cosserat and Cosserat (1909) and the contributions by Nowacki (1986) for infinitesimal deformations and by Eringen and Kafadar (1976); Eringen (1999); Eremeyev et al (2013); Altenbach and Eremeyev (2013); Eremeyev and Altenbach (2017) for finite deformations, where the further references can be found. The Cosserat model found various applications to description of such microstructured media as foams, granular media, composites, magnetic fluids, and thin-walled structures. Within the micropolar continuum the fields of translations and rotations are used as kinematical descriptors. In addition to stress tensor the couple stress tensor is also introduced in the theory which describes the rotational (moment-type) interactions in the medium.

Considering initial boundary-value problems of the micropolar mechanics one usually assumes kinematic or/and static boundary conditions expressed through

Victor A. Eremeyev
Faculty of Civil and Environmental Engineering, Gdańsk University of Technology, ul. Gabriela Narutowicza 11/12 80-233 Gdańsk, Poland,
e-mail: victor.eremeev@pg.edu.pl
Southern Federal University, Milchakova str. 8a, 344090 Rostov on Don, Russia,
Southern Scientific Center of RASci, Chekhova str. 41, 344006 Rostov on Don, Russia,
e-mail: eremeyev.victor@gmail.com

B. E. Abali et al. (eds.), *New Achievements in Continuum Mechanics and Thermodynamics*, Advanced Structured Materials 108,
https://doi.org/10.1007/978-3-030-13307-8_7

translations and rotations or/and stress and couple stress vectors, respectively. These conditions play a role of principal and natural boundary conditions which follow from the stationarity of the corresponding functionals or from the principle of virtual work, see, e.g., Nowacki (1986); Eringen (1999); Pietraszkiewicz and Eremeyev (2009). Here we consider more general case of boundary conditions (BCs) when the latter cannot be derived from any functional, in general. For example, such type of boundary conditions one has in the case of nonconservative loading (Bolotin, 1963) or when some relations between linear and angular velocities are prescribed on a micropolar fluid surface (Migoun and Prokhorenko, 1984; Łukaszewicz, 1999).

The paper is organized as follows. First, in Section 7.2 we briefly recall basic equations of the micropolar continuum undergoing finite deformations. Considering the principle of virtual work in Section 7.3 we discuss the weak formulations of boundary conditions. In Section 7.4 we introduce non-holonomic boundary relations expressed as linear relations between virtual displacements and virtual rotations. Finally, we present few examples of non-holonomic boundary conditions.

7.2 Constitutive Relations

The deformation of a micropolar medium is described through kinematically independent fields of translations and rotations. So the kinematics of a micropolar continuum is described through the following vectorial fields:

$$\mathbf{x} = \mathbf{x}(\mathbf{X}), \tag{7.1}$$

$$\mathbf{d}_k = \mathbf{d}_k(\mathbf{X}), \quad k = 1 \dots 3, \tag{7.2}$$

where $\mathbf{x}$ and $\mathbf{X}$ are positions vectors defined in current and reference placements, whereas $\mathbf{d}_k$ are unit orthogonal vectors called directors, see (Eringen and Kafadar, 1976; Eringen, 1999; Eremeyev et al, 2013) for details. Instead of (7.2) one can use the microrotation tensor defined as follows

$$\mathbf{Q} = \mathbf{D}_k \otimes \mathbf{d}_k, \tag{7.3}$$

where $\mathbf{D}_k$ are directors in a reference placement, $\otimes$ stands for the tensor (diadic) product, and Einstein's summation rule is utilized. Note that without loss of generality $\mathbf{Q}$ can be defined as a proper orthogonal tensor. To this end one have to chose the same orientation of triples $\mathbf{D}_k$ and $\mathbf{d}_k$. In what follows we use the direct (coordinate-free) tensor calculus presented in Lurie (1990); Simmonds (1994); Eremeyev et al (2018).

For an hyperelastic material there exists a strain energy density $\mathcal{W}$. We assume that $\mathcal{W}$ depends on $\mathbf{x}$, $\mathbf{Q}$ and their gradients

$$\mathcal{W} = \mathcal{W}(\mathbf{x}, \mathbf{F}, \mathbf{Q}, \nabla\mathbf{Q}), \tag{7.4}$$

where $\mathbf{F} = \nabla \mathbf{x}$ is the deformation gradient, ∇ is the gradient operator defined in the reference placement. For example, in the Cartesian coordinates X_k we have

$$\nabla = \mathbf{i}_k \frac{\partial}{\partial X_k},$$

where $\mathbf{i}_k$ are the Cartesian base vectors, $\mathbf{i}_k \cdot \mathbf{i}_m = \delta_{km}$, so $\mathbf{X} = X_k \mathbf{i}_k$, "$\cdot$" denotes scalar product, and δ_{mn} is the Kronecker symbol.

The principle of the material frame-indifference (Truesdell and Noll, 2004) says that $\mathcal{W}$ is invariant under changes

$$\mathbf{x} \to \mathbf{O} \cdot \mathbf{x} + \mathbf{a}, \quad \mathbf{d}_k \to \mathbf{O} \cdot \mathbf{d}_k \tag{7.5}$$

for any constant proper orthogonal tensor $\mathbf{O}$ and any constant vector $\mathbf{a}$. From (7.3) and (7.5) it follows that $\mathbf{Q}$ and $\nabla \mathbf{Q}$ change as follows

$$\mathbf{Q} \to \mathbf{D}_k \otimes (\mathbf{O} \cdot \mathbf{d}_k) = \mathbf{D}_k \otimes \mathbf{d}_k \cdot \mathbf{O}^T = \mathbf{Q} \cdot \mathbf{O}^T, \quad \nabla \mathbf{Q} \to \nabla \mathbf{Q} \cdot \mathbf{O}^T. \tag{7.6}$$

As a result of the invariance we get

$$\begin{aligned} \mathcal{W} =& \mathcal{W}(\mathbf{F}, \mathbf{Q}, \nabla \mathbf{Q}) \\ =& \mathcal{W}(\mathbf{F} \cdot \mathbf{O}^T, \mathbf{Q} \cdot \mathbf{O}^T, \nabla \mathbf{Q} \cdot \mathbf{O}^T), \end{aligned} \tag{7.7}$$

where the superscript T stands for the transpose tensors. Choosing $\mathbf{O} = \mathbf{Q}$ we have

$$\mathcal{W} = \mathcal{W}(\mathbf{F} \cdot \mathbf{Q}^T, \mathbf{I}, \nabla \mathbf{Q} \cdot \mathbf{Q}^T). \tag{7.8}$$

Hereinafter $\mathbf{I}$ is the unit tensor. This choice is possible as $\mathbf{O}$ can be any proper orthogonal tensor, so it can also coincide with $\mathbf{Q}$ given in any point. On the other hand Eq. (7.8) verifies the principle of material frame-indifference.

Note that $\frac{\partial \mathbf{Q}}{\partial X_k} \cdot \mathbf{Q}^T$ is a skew tensor. So it can be represented as

$$\frac{\partial \mathbf{Q}}{\partial X_k} \cdot \mathbf{Q}^T = -\mathbf{k}_k \times \mathbf{I} \tag{7.9}$$

through an axial vector $\mathbf{k}_k$. Here $\times$ stands for the cross product. Thus, the third-order tensor $\nabla \mathbf{Q} \cdot \mathbf{O}^T$ has the form

$$\nabla \mathbf{Q} \cdot \mathbf{O}^T = -\mathbf{K} \times \mathbf{I}, \quad \mathbf{K} = \mathbf{i}_k \otimes \mathbf{k}_k. \tag{7.10}$$

Note that in (7.10) we introduce the cross-product between two second-order tensors. For diads it was introduced by Gibbs, see (Wilson, 1901, p. 281), as follows

$$(\mathbf{a} \otimes \mathbf{b}) \times (\mathbf{c} \otimes \mathbf{d}) = \mathbf{a} \otimes (\mathbf{b} \times \mathbf{c}) \otimes \mathbf{d}$$

and can be easily extended for tensors of any order, see Eremeyev et al (2018).

Using Gibbsian cross operation $(\ldots)_\times$ introduced again by Gibbs (Wilson, 1901, p. 275), we get the formula

$$\mathbf{K} = \frac{1}{2}\mathbf{i}_k \otimes \left(\frac{\partial \mathbf{Q}}{\partial X_k} \cdot \mathbf{Q}^T\right)_\times. \tag{7.11}$$

Hereinafter for any second-order tensor $\mathbf{T} = T_{mn}\mathbf{i}_m \otimes \mathbf{i}_n$ the notation $\mathbf{T}_\times$ denotes the vectorial invariant of $\mathbf{T}$ defined as follows $\mathbf{T}_\times = T_{mn}\mathbf{i}_m \times \mathbf{i}_n$.

As a result, the strain energy density depends on two natural strain measures $\mathbf{E} = \mathbf{F} \cdot \mathbf{Q}^T$ and $\mathbf{K}$, see, e.g., Pietraszkiewicz and Eremeyev (2009),

$$\mathcal{W} = \mathcal{W}(\mathbf{E}, \mathbf{K}). \tag{7.12}$$

Various examples of the micropolar constitutive equations can be found in the literature, see, e.g., Eringen (1999); Eremeyev and Pietraszkiewicz (2012, 2016).

7.3 Principle of Virtual Work

In order to formulate the virtual work principle we consider the first variation of the energy functional

$$\mathcal{E} = \int_V \mathcal{W} \mathrm{d}V,$$

where V is the micropolar body volume. Calculating $\delta\mathcal{E}$ we can find the consistent form of the work $\delta\mathcal{A}$ of external loads. We introduce first the variations of translations

$$\mathbf{u} = \delta\mathbf{x}. \tag{7.13}$$

In order to introduce the variation of rotations we consider $\delta\mathbf{d}_k$. As $\mathbf{d}_k$ are unit vectors, that is $\mathbf{d}_m \cdot \mathbf{d}_n = \delta_{mn}$, we have that $\delta\mathbf{d}_m \cdot \mathbf{d}_n + \mathbf{d}_m \cdot \delta\mathbf{d}_n = 0$, and

$$\delta\mathbf{d}_1 \cdot \mathbf{d}_1 = 0, \quad \delta\mathbf{d}_2 \cdot \mathbf{d}_2 = 0, \quad \delta\mathbf{d}_3 \cdot \mathbf{d}_3 = 0.$$

As a result, $\delta\mathbf{d}_k$, $k = 1, 2, 3$, can be represented through the same vector $\boldsymbol{\psi}$

$$\delta\mathbf{d}_k = \boldsymbol{\psi} \times \mathbf{d}_k. \tag{7.14}$$

From (7.14) it follows that

$$\delta\mathbf{Q} = -\mathbf{Q} \times \boldsymbol{\psi}.$$

Note that unlike $\mathbf{u}$, vector $\boldsymbol{\psi}$ does not coincide with a variation of any vector, in general.

Calculating $\delta\mathcal{W}$ we get

$$\delta\mathcal{W} = \frac{\partial\mathcal{W}}{\partial\mathbf{E}} : \delta\mathbf{E} + \frac{\partial\mathcal{W}}{\partial\mathbf{K}} : \delta\mathbf{K},$$

where ":" stands for the scalar product in the space of second-order tensors, for example, $\mathbf{T} : \mathbf{E} = \mathrm{tr}\,(\mathbf{T} \cdot \mathbf{E}^T)$, tr is the trace operator, and

$$\begin{aligned}\delta \mathbf{E} =& \delta(\mathbf{F}\cdot\mathbf{Q}^T) = \delta\mathbf{F}\cdot\mathbf{Q}^T + \mathbf{F}\cdot\delta\mathbf{Q}^T\\ =& \nabla\mathbf{u}\cdot\mathbf{Q}^T + \mathbf{F}\cdot(\boldsymbol{\psi}\times\mathbf{I})\cdot\mathbf{Q}^T, \qquad (7.15)\\ \delta\mathbf{K} =& \nabla\boldsymbol{\psi}\cdot\mathbf{Q}^T. \qquad (7.16)\end{aligned}$$

For the derivation here we used the relations

$$\delta\mathbf{Q} = \mathbf{D}_k\otimes\delta\mathbf{d}_k = \mathbf{D}_k\otimes\mathbf{d}_k\times\boldsymbol{\psi} = -\mathbf{Q}\times\boldsymbol{\psi}, \quad (\mathbf{Q}\times\boldsymbol{\psi})^T = -\boldsymbol{\psi}\times\mathbf{Q}^T,$$

see Eremeyev and Zubov (1994); Eremeyev et al (2013) for details. So we have

$$\delta W = \left(\frac{\partial W}{\partial\mathbf{E}}\cdot\mathbf{Q}\right):\nabla\mathbf{u} + \left(\frac{\partial W}{\partial\mathbf{K}}\cdot\mathbf{Q}\right):\nabla\boldsymbol{\psi} + \left(\frac{\partial W}{\partial\mathbf{E}}\cdot\mathbf{Q}\right):(\mathbf{F}\times\boldsymbol{\psi}). \quad (7.17)$$

Introducing the first Piola–Kirchhoff stress $\mathbf{T}$ and couple stress $\mathbf{M}$ tensors by the formulae

$$\mathbf{T} = \frac{\partial W}{\partial\mathbf{E}}\cdot\mathbf{Q}, \quad \mathbf{M} = \frac{\partial W}{\partial\mathbf{K}}\cdot\mathbf{Q}, \quad (7.18)$$

we transform δW into the compact form

$$\delta W = \mathbf{T}:\nabla\mathbf{u} + \mathbf{M}:\nabla\boldsymbol{\psi} + \mathbf{T}:(\mathbf{F}\times\boldsymbol{\psi}).$$

Calculating the first variation of the energy functional with the use of the integration by parts we get

$$\begin{aligned}\delta\mathcal{E} =& \int_V \delta W\,\mathrm{d}V\\ =& -\int_V \left[(\nabla\cdot\mathbf{T})\cdot\mathbf{u} + (\nabla\cdot\mathbf{M} + (\mathbf{F}^T\cdot\mathbf{T})_\times)\cdot\boldsymbol{\psi}\right]\mathrm{d}V\\ &+ \int_{\partial V}(\mathbf{n}\cdot\mathbf{T}\cdot\mathbf{u} + \mathbf{n}\cdot\mathbf{M}\cdot\boldsymbol{\psi})\,\mathrm{d}S. \qquad (7.19)\end{aligned}$$

Here $\mathbf{n}$ is the vector of outer unit normal to the boundary ∂V. The form of $\delta\mathcal{E}$ dictates the possible consistent expression of the external loadings work

$$\mathcal{A} = \int_V(\mathbf{f}\cdot\mathbf{u} + \mathbf{m}\cdot\boldsymbol{\psi})\,\mathrm{d}V + \int_{\partial V}(\boldsymbol{\phi}\cdot\mathbf{u} + \boldsymbol{\mu}\cdot\boldsymbol{\psi})\,\mathrm{d}S. \quad (7.20)$$

In (7.20) $\mathbf{f}$ and $\boldsymbol{\phi}$ are external forces given in the volume and on its boundary, respectively, whereas $\mathbf{m}$ and $\boldsymbol{\mu}$ are external volumetric and surface couples (moments).

Finally, the virtual work principle takes the following form

$$\begin{aligned}\delta\mathcal{E} - \delta\mathcal{A} =& \int_V\left[-(\nabla\cdot\mathbf{T} - \mathbf{f})\cdot\mathbf{u} - (\nabla\cdot\mathbf{M} + (\mathbf{F}^T\cdot\mathbf{T})_\times - \mathbf{m})\cdot\boldsymbol{\psi}\right]\mathrm{d}V\\ &+ \int_{\partial V}\left[(\mathbf{n}\cdot\mathbf{T} - \boldsymbol{\phi})\cdot\mathbf{u} + (\mathbf{n}\cdot\mathbf{M} - \boldsymbol{\mu})\cdot\boldsymbol{\psi}\right]\mathrm{d}S = 0. \qquad (7.21)\end{aligned}$$

Another method of derivation of (7.21) is presented in Pietraszkiewicz and Eremeyev (2009). Considering admissible variations, from (7.21) it follows the equilibrium equations and the corresponding natural boundary conditions.

For example, when the translations and rotations are both fixed on ∂V we have that

$$\mathbf{u} = \mathbf{0}, \quad \boldsymbol{\psi} = \mathbf{0} \quad \mathbf{X} \in \partial V \tag{7.22}$$

and the surface integral in (7.21) vanishes. So (7.22) play a role of incremental kinematic boundary conditions in the micropolar elasticity. Obviously, there is a straightforward correspondence between (7.22) and standard kinematic relations

$$\mathbf{x} = \mathbf{x}_0, \quad \mathbf{Q} = \mathbf{Q}_0 \quad \mathbf{X} \in \partial V,$$

where $\mathbf{x}_0$ and $\mathbf{Q}_0$ are given vector and tensor-valued functions, $\mathbf{Q}_0^T = \mathbf{Q}_0^{-1}$.

If $\mathbf{u}$ and $\boldsymbol{\psi}$ do not vanish on ∂V from (7.21) we have

$$\int_{\partial V} [(\mathbf{n} \cdot \mathbf{T} - \boldsymbol{\phi}) \cdot \mathbf{u} + (\mathbf{n} \cdot \mathbf{M} - \boldsymbol{\mu}) \cdot \boldsymbol{\psi}] \, \mathrm{d}S = 0. \tag{7.23}$$

Equation (7.23) constitutes a weak form of a natural boundary conditions. In particular, if $\mathbf{u}$ and $\boldsymbol{\psi}$ are arbitrary, Eq. (7.23) results in the natural static boundary conditions

$$\mathbf{n} \cdot \mathbf{T} = \boldsymbol{\phi}, \quad \mathbf{n} \cdot \mathbf{M} = \boldsymbol{\mu}. \tag{7.24}$$

In what follows we consider a case intermediate between (7.22) and (7.24). In other word we will consider kinematic constraints that are relations between $\mathbf{u}$ and $\boldsymbol{\psi}$ given on ∂V or its part.

7.4 Non-holonomic Kinematic Boundary Conditions

In the analytical mechanics are known various incremental constraints on generalized variables. These constraints can be holonomic or non-holonomic, see, e.g., Lurie (2001). First, we formulate an incremental boundary condition as a linear relations between $\mathbf{u}$ and $\boldsymbol{\psi}$

$$\mathbf{L}_1 \cdot \mathbf{u} + \mathbf{L}_2 \cdot \boldsymbol{\psi} = \mathbf{0}, \tag{7.25}$$

where second-order tensors $\mathbf{L}_1$ and $\mathbf{L}_2$ depend on $\mathbf{x}$, $\mathbf{Q}$, and their spatial gradients, in general. Let us note that (7.25) does not correspond to any constrain written in terms of $\mathbf{x}$ and $\mathbf{Q}$, in general. So we call (7.25) *non-holonomic boundary conditions*. Such incremental constraints are known in the analytical mechanics, see, e.g., Lurie (2001). Such constraints can be applied using Lagrange multiplier technique or through the direct solving of (7.25) with respect to one on the variables. The conservatives conditions for micropolar solids and rigid bodies including action of external moments were discussed by Eremeyev and Zubov (1994); Zelenina and Zubov (2000).

For example, assuming that $\mathbf{L}_2$ is invertible from (7.25) we get that

$$\boldsymbol{\psi} = -\mathbf{L}_2^{-1} \cdot \mathbf{L}_1 \cdot \mathbf{u}.$$

Substituting this into (7.23) we have

$$\mathbf{n} \cdot \mathbf{T} - (\mathbf{n} \cdot \mathbf{M} - \boldsymbol{\mu}) \cdot \mathbf{L}_2^{-1} \cdot \mathbf{L}_1 = \boldsymbol{\phi}. \tag{7.26}$$

Obviously, the using of the Lagrange multipliers technique gives the same result. Indeed, introducing a Lagrange multiplier $\boldsymbol{\lambda}$ we add to (7.23) the expression

$$\int_{\partial V} \boldsymbol{\lambda} \cdot (\mathbf{L}_1 \cdot \mathbf{u} + \mathbf{L}_2 \cdot \boldsymbol{\psi}) \, \mathrm{d}S = 0.$$

So we get

$$\int_{\partial V} [(\mathbf{n} \cdot \mathbf{T} - \boldsymbol{\phi}) \cdot \mathbf{u} + (\mathbf{n} \cdot \mathbf{M} - \boldsymbol{\mu}) \cdot \boldsymbol{\psi} + \boldsymbol{\lambda} \cdot (\mathbf{L}_1 \cdot \mathbf{u} + \mathbf{L}_2 \cdot \boldsymbol{\psi})] \, \mathrm{d}S = 0. \tag{7.27}$$

From (7.27) it follows that

$$\mathbf{n} \cdot \mathbf{T} - \boldsymbol{\phi} + \boldsymbol{\lambda} \cdot \mathbf{L}_1 = \mathbf{0}, \tag{7.28}$$

$$\mathbf{n} \cdot \mathbf{M} - \boldsymbol{\mu} + \boldsymbol{\lambda} \cdot \mathbf{L}_2 = \mathbf{0}. \tag{7.29}$$

Assuming again that $\mathbf{L}_2$ is invertible we exclude $\boldsymbol{\lambda}$ from (7.29)

$$\boldsymbol{\lambda} = -(\mathbf{n} \cdot \mathbf{M} - \boldsymbol{\mu}) \cdot \mathbf{L}_2^{-1}. \tag{7.30}$$

Finally, substituting $\boldsymbol{\lambda}$ into (7.28) we get (7.26).

Let us consider particular cases of (7.25). Obviously, Eqs. (7.22) present the trivial case of (7.25). Indeed, (7.22) follows from (7.25) with $\mathbf{L}_1 = \mathbf{L}_2 = \mathbf{0}$. Another case is sliding with free rotations at the boundary, $\mathbf{n} \cdot \mathbf{u} = 0$, $\boldsymbol{\psi}$ has arbitrary values. This case corresponds to $\mathbf{L}_1 = \mathbf{n} \otimes \mathbf{n}$, $\mathbf{L}_2 = \mathbf{0}$. Eq. (7.23) results in the following static boundary conditions

$$(\mathbf{n} \cdot \mathbf{T} - \boldsymbol{\varphi}) \cdot \mathbf{A} = \mathbf{0}, \quad \mathbf{n} \cdot \mathbf{M} = \boldsymbol{\mu},$$

where $\mathbf{A} = \mathbf{I} - \mathbf{n} \otimes \mathbf{n}$.

Another assumption leading to (7.25) can be motivated as follows. Let us assume that the material particles behave as rolling stones in the vicinity of the body boundary. Assuming the same relation between $\mathbf{u}$ and $\boldsymbol{\psi}$ as for linear and angular velocities of a rigid body rolling on a surface we get

$$\mathbf{u} = r\mathbf{n} \times \boldsymbol{\psi}. \tag{7.31}$$

Here r plays a role of a characteristic length of a micropolar medium. For example, it is the distance between the mass center of a material particle and its boundary. Eq. (7.31) means that $\mathbf{L}_1 = \mathbf{I}$ and $\mathbf{L}_2 = r\mathbf{n} \times \mathbf{I}$. Note that here $\mathbf{L}_2$ is a singular tensor whereas $\mathbf{L}_1$ is invertible. This constraint results in the following natural boundary condition

$$\mathbf{n}\cdot\mathbf{M} - r\mathbf{n}\cdot\mathbf{T}\times\mathbf{n} = \boldsymbol{\mu} - r\boldsymbol{\phi}\times\mathbf{n}. \tag{7.32}$$

Let us consider more general case of non-holonomic surface constraints. We use the following linear relation

$$\mathbf{L}_1\cdot\mathbf{u} + \mathbf{L}_2\cdot\boldsymbol{\psi} + \mathbf{L}_3 : \nabla\mathbf{u} + \mathbf{L}_4 : \nabla\boldsymbol{\psi} = \mathbf{0}, \quad \mathbf{X}\in\partial V. \tag{7.33}$$

Here $\mathbf{L}_3$ and $\mathbf{L}_4$ are third-order tensors depending on $\mathbf{x}$, $\mathbf{Q}$ and their gradients. Eq. (7.33) is a system of first-order partial differential equations which first integral can be found through the characteristic technique, see, e.g., Arnold (2004). Instead we again use the Lagrange multiplier approach. Now instead of (7.23) we have

$$\int_{\partial V} [(\mathbf{n}\cdot\mathbf{T} - \boldsymbol{\phi})\cdot\mathbf{u} + (\mathbf{n}\cdot\mathbf{M} - \boldsymbol{\mu})\cdot\boldsymbol{\psi} \\ +\boldsymbol{\lambda}\cdot(\mathbf{L}_1\cdot\mathbf{u} + \mathbf{L}_2\cdot\boldsymbol{\psi} + \mathbf{L}_3 : \nabla\mathbf{u} + \mathbf{L}_4 : \nabla\boldsymbol{\psi})]\,\mathrm{d}S = 0. \tag{7.34}$$

In order to transform (7.34) using the integration by parts we represent ∇ as a sum of the surface gradient and normal derivative

$$\nabla = \nabla_s + \mathbf{n}\frac{\partial}{\partial n},$$

where $\frac{\partial}{\partial n}$ is the derivative with respect to the coordinate normal to ∂V. Using the surface divergence theorem (Eremeyev et al, 2018) we apply the following integration by parts formula

$$\int_A \mathbf{Y} : \nabla_s\mathbf{y}\,\mathrm{d}S = \int_{\partial A} \boldsymbol{\nu}\cdot\mathbf{Y}\cdot\mathbf{y}\,\mathrm{d}s - \int_A [(\nabla_s\cdot\mathbf{Y})\cdot\mathbf{y} + 2H\mathbf{n}\cdot\mathbf{Y}\cdot\mathbf{y}]\,\mathrm{d}S \tag{7.35}$$

for any fields $\mathbf{Y}$ and $\mathbf{y}$. Here $2H = -\nabla_s\cdot\mathbf{n}$ is the mean curvature of a surface A with the contour ∂A, and $\boldsymbol{\nu}$ is the normal to ∂A such that $\boldsymbol{\nu}\cdot\mathbf{n} = 0$, see Fig 7.1.

With (7.35) we have

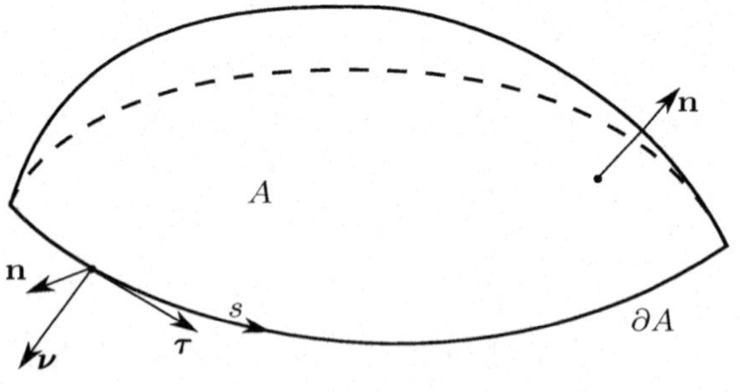

Fig. 7.1 For the surface divergence theorem: surface A with contour ∂A. The unit vectors $\mathbf{n}$, $\boldsymbol{\nu}$, and $\boldsymbol{\tau}$ are defined along ∂A. Here $\mathbf{n}$ is the unit vector normal to A, $\boldsymbol{\tau}$ is the unit vector tangent to ∂A, whereas $\boldsymbol{\nu}$ is the unit vector lying in the tangent plane to A and normal to ∂A.

$$\int_{\partial V} \boldsymbol{\lambda} \cdot \mathbf{L}_3 : \nabla \mathbf{u} \, \mathrm{d}S = \int_{\partial V} \boldsymbol{\lambda} \cdot \mathbf{L}_3 : \left(\mathbf{n} \frac{\partial \mathbf{u}}{\partial n} \right) \mathrm{d}S$$
$$- \int_{\partial V} [\nabla \cdot (\boldsymbol{\lambda}_s \cdot \mathbf{L}_3) + 2H \mathbf{n} \cdot (\boldsymbol{\lambda} \cdot \mathbf{L}_3)] \cdot \mathbf{u} \, \mathrm{d}S, \quad (7.36)$$

$$\int_{\partial V} \boldsymbol{\lambda} \cdot \mathbf{L}_4 : \nabla \boldsymbol{\psi} \, \mathrm{d}S = \int_{\partial V} \boldsymbol{\lambda} \cdot \mathbf{L}_4 : \left(\mathbf{n} \frac{\partial \boldsymbol{\psi}}{\partial n} \right) \mathrm{d}S$$
$$- \int_{\partial V} [\nabla_s \cdot (\boldsymbol{\lambda} \cdot \mathbf{L}_4) + 2H \mathbf{n} \cdot (\boldsymbol{\lambda} \cdot \mathbf{L}_4)] \cdot \boldsymbol{\psi} \, \mathrm{d}S. \quad (7.37)$$

Here we assumed that $\partial\partial V = \emptyset$.

With (7.36) and (7.37) Eq. (7.34) results in

$$(\boldsymbol{\lambda} \otimes \mathbf{n}) : \mathbf{L}_3 = \mathbf{0}, \quad (7.38)$$
$$(\boldsymbol{\lambda} \otimes \mathbf{n}) : \mathbf{L}_4 = \mathbf{0}, \quad (7.39)$$
$$\mathbf{n} \cdot \mathbf{T} - \boldsymbol{\phi} + \boldsymbol{\lambda} \cdot \mathbf{L}_1 = \nabla_s \cdot (\boldsymbol{\lambda} \cdot \mathbf{L}_3) + 2H \mathbf{n} \cdot (\boldsymbol{\lambda} \cdot \mathbf{L}_3), \quad (7.40)$$
$$\mathbf{n} \cdot \mathbf{M} - \boldsymbol{\mu} + \boldsymbol{\lambda} \cdot \mathbf{L}_2 = \nabla_s \cdot (\boldsymbol{\lambda} \cdot \mathbf{L}_4) + 2H \mathbf{n} \cdot (\boldsymbol{\lambda} \cdot \mathbf{L}_4). \quad (7.41)$$

From (7.38) and (7.39) it follows that the curvature dependent terms in (7.40) and (7.41) are vanishing. So we get

$$\mathbf{n} \cdot \mathbf{T} - \boldsymbol{\phi} + \boldsymbol{\lambda} \cdot \mathbf{L}_1 = \nabla_s \cdot (\boldsymbol{\lambda} \cdot \mathbf{L}_3), \quad (7.42)$$
$$\mathbf{n} \cdot \mathbf{M} - \boldsymbol{\mu} + \boldsymbol{\lambda} \cdot \mathbf{L}_2 = \nabla_s \cdot (\boldsymbol{\lambda} \cdot \mathbf{L}_4). \quad (7.43)$$

For simplicity let us restrict ourselves by the case of $\mathbf{L}_3$ and $\mathbf{L}_4$ which have the following property

$$(\mathbf{a} \otimes \mathbf{n}) : \mathbf{L}_3 = \mathbf{0}, \quad (\mathbf{a} \otimes \mathbf{n}) : \mathbf{L}_4 = \mathbf{0} \quad \forall \mathbf{a}. \quad (7.44)$$

Then (7.38) and (7.39) vanish identically. The properties (7.44) means that the non-holonomic kinematic boundary constraint (7.34) takes the form

$$\mathbf{L}_1 \cdot \mathbf{u} + \mathbf{L}_2 \cdot \boldsymbol{\psi} + \mathbf{L}_3 : \nabla_s \mathbf{u} + \mathbf{L}_4 : \nabla_s \boldsymbol{\psi} = \mathbf{0}, \quad \mathbf{X} \in \partial V. \quad (7.45)$$

As an example of (7.33) or (7.45) let us recall the boundary conditions used in the micropolar hydrodynamics, see Migoun and Prokhorenko (1984); Łukaszewicz (1999), where the following relations between angular $\boldsymbol{\omega}$ and linear $\mathbf{v}$ velocity was discussed

$$\boldsymbol{\omega} = \frac{\alpha}{2} \nabla \times \mathbf{v}.$$

Here α is a material parameter, $0 \le \alpha \le 1$. Note that the constraint $\boldsymbol{\omega} = \frac{1}{2} \nabla \times \mathbf{v}$ is used in the theory of couple stresses (Cosserat continuum with constrained rotations or Cosserat pseudocontinuum), see Nowacki (1986).

Assuming similar relation between virtual rotations and translations we get

$$\boldsymbol{\psi} = \frac{\alpha}{2} \nabla \times \mathbf{u}, \quad (7.46)$$

which is a particular case of (7.33). Indeed, (7.46) follows from (7.33) with

$$\mathbf{L}_1 = \mathbf{0}, \quad \mathbf{L}_2 = \mathbf{I}, \quad \mathbf{L}_3 = \tfrac{1}{2}\alpha\mathbf{I}\times\mathbf{I}, \quad \mathbf{L}_4 = \mathbf{0}.$$

Here we used the following identity:

$$(\mathbf{I}\times\mathbf{I}) : \nabla\mathbf{u} = -\nabla\times\mathbf{u}.$$

From the physical point of view (7.46) means that the micro-rotations depends on macro-rotations on the boundary. In other words with (7.46) we model interactions between the medium and its boundary.

Thus, (7.38) and (7.39) result in one constraint

$$\begin{aligned}(\boldsymbol{\lambda}\otimes\mathbf{n}) : \mathbf{L}_3 =&\mathbf{n}\cdot(\boldsymbol{\lambda}\cdot\mathbf{L}_3) = \frac{\alpha}{2}\mathbf{n}\cdot(\boldsymbol{\lambda}\cdot\mathbf{I}\times\mathbf{I}) = \frac{\alpha}{2}\mathbf{n}\cdot(\boldsymbol{\lambda}\times\mathbf{I}) = \frac{\alpha}{2}\mathbf{n}\cdot(\mathbf{I}\times\boldsymbol{\lambda})\\ =&\frac{\alpha}{2}\mathbf{n}\times\boldsymbol{\lambda} = \mathbf{0},\end{aligned}$$

which means that $\boldsymbol{\lambda}$ is normal to ∂V: $\boldsymbol{\lambda} = \Lambda\mathbf{n}$.

Eq. (7.41) transforms into

$$\mathbf{n}\cdot\mathbf{M} - \boldsymbol{\mu} + \boldsymbol{\lambda} = \mathbf{0},$$

so one easily finds $\boldsymbol{\lambda}$ from it

$$\boldsymbol{\lambda} = \Lambda\mathbf{n}, \quad \Lambda = -(\mathbf{n}\cdot\mathbf{M} - \boldsymbol{\mu})\cdot\mathbf{n}. \tag{7.47}$$

With (7.47) and identities

$$\nabla_s\cdot(\boldsymbol{\lambda}\cdot\mathbf{L}_3) = \frac{\alpha}{2}\nabla_s\cdot(\boldsymbol{\lambda}\cdot\mathbf{I}\times\mathbf{I}) = \frac{\alpha}{2}\nabla_s\times\boldsymbol{\lambda},$$

$$\nabla_s\times(\Lambda\mathbf{n}) = \nabla_s\Lambda\times\mathbf{n},$$

we exclude $\boldsymbol{\lambda}$ from Eq. (7.40), which takes the following form

$$\mathbf{n}\cdot\mathbf{T} + \frac{\alpha}{2}\nabla_s\left[(\mathbf{n}\cdot\mathbf{M} - \boldsymbol{\mu})\cdot\mathbf{n}\right]\times\mathbf{n} = \boldsymbol{\phi}, \tag{7.48}$$

which plays a role of the natural boundary condition complementary to (7.46).

Let us note that (7.48) can be also derived without using of Lagrange multiplier technique. To this end one have to transform the variational equation

$$\int_{\partial V}\left[(\mathbf{n}\cdot\mathbf{T} - \boldsymbol{\phi})\cdot\mathbf{u} + \frac{\alpha}{2}(\mathbf{n}\cdot\mathbf{M} - \boldsymbol{\mu})\cdot(\nabla\times\mathbf{u})\right]\,\mathrm{d}S = 0.$$

with integration by parts.

7.5 Conclusions

Within the nonlinear micropolar elasticity we introduced the non-holonomic kinematic boundary conditions. These conditions are formulated as linear relations between virtual translations and rotations. In other words we presented new incremental kinematic boundary conditions. The corresponding natural static boundary conditions are also derived. As for the derivation we used the principle of virtual work, the discussed results extend the class of possible boundary conditions also for inelastic micropolar materials such as considered by Altenbach and Eremeyev (2014). Let us note that, though the boundary condition for the translation field and its natural static counterpart is physically clear, for microrotation there is no general agreement on the vorticity of complex materials on the boundary and on the type of the corresponding boundary condition for the field of microrotation.

It is worth to underline that after Sedov (1965) and Germain (1973a,b) this variational approach became a powerful tool for modelling of media with microstructure, see also discussion by dell'Isola et al (2017); Eugster and dell'Isola (2017, 2018a,b). So in a similar way non-holonomic boundary conditions can be introduced for other generalized media, such as strain gradient elasticity. For the virtual work and the least action principles in strain gradient solids and fluids we refer to Auffray et al (2015); Abali et al (2015, 2017); Eremeyev and Altenbach (2014); Eremeyev (2016) and the reference therein. In particular, such boundary equations could be useful for modelling of the behaviour of complex fluids in the vicinity of a free surface and/or interface.

Acknowledgements The author acknowledges financial support from the Russian Science Foundation under the grant "*Methods of microstructural nonlinear analysis, wave dynamics and mechanics of composites for research and design of modern metamaterials and elements of structures made on its base*" (No 15-19-10008-P).

References

Abali BE, Müller WH, Eremeyev VA (2015) Strain gradient elasticity with geometric nonlinearities and its computational evaluation. Mechanics of Advanced Materials and Modern Processes 1(1):4

Abali BE, Müller WH, dell'Isola F (2017) Theory and computation of higher gradient elasticity theories based on action principles. Archive of Applied Mechanics 87(9):1495–1510

Altenbach H, Eremeyev VA (2013) Cosserat media. In: Altenbach H, Eremeyev VA (eds) Generalized Continua from the Theory to Engineering Applications, CISM Courses and Lectures, vol 541, Springer, Wien, pp 65–130

Altenbach H, Eremeyev VA (2014) Strain rate tensors and constitutive equations of inelastic micropolar materials. International Journal of Plasticity 63:3–17

Arnold VI (2004) Lectures on Partial Differential Equations. Springer, Berlin

Auffray N, dell'Isola F, Eremeyev VA, Madeo A, Rosi G (2015) Analytical continuum mechanics à la Hamilton–Piola least action principle for second gradient continua and capillary fluids. Mathematics and Mechanics of Solids 20(4):375–417

Bolotin VV (1963) Nonconservative Problems of Theory of Elastic Stability. Pergamon Press, Oxford

Cosserat E, Cosserat F (1909) Théorie des corps déformables. Herman et Fils, Paris

dell'Isola F, Corte AD, Giorgio I (2017) Higher-gradient continua: The legacy of Piola, Mindlin, Sedov and Toupin and some future research perspectives. Mathematics and Mechanics of Solids 22(4):852–872

Eremeyev VA (2016) On equilibrium of a second-gradient fluid near edges and corner points. In: Advanced Methods of Continuum Mechanics for Materials and Structures, Springer, pp 547–556

Eremeyev VA, Altenbach H (2014) Equilibrium of a second-gradient fluid and an elastic solid with surface stresses. Meccanica 49(11):2635–2643

Eremeyev VA, Altenbach H (2017) Basics of mechanics of micropolar shells. In: Altenbach H, Eremeyev V (eds) Shell-like Structures, CISM Courses and Lectures, vol 572, Springer, Wien, pp 63–111

Eremeyev VA, Pietraszkiewicz W (2012) Material symmetry group of the non-linear polar-elastic continuum. International Journal of Solids and Structures 49(14):1993–2005

Eremeyev VA, Pietraszkiewicz W (2016) Material symmetry group and constitutive equations of micropolar anisotropic elastic solids. Mathematics and Mechanics of Solids 21(2):210–221

Eremeyev VA, Zubov LM (1994) On the stability of elastic bodies with couple stresses (in Russ.). Mechanics of Solids 3:181–190

Eremeyev VA, Lebedev LP, Altenbach H (2013) Foundations of Micropolar Mechanics. SpringerBriefs in Applied Sciences and Technologies, Springer, Heidelberg et al.

Eremeyev VA, Cloud MJ, Lebedev LP (2018) Applications of Tensor Analysis in Continuum Mechanics. World Scientific, New Jersey

Eringen AC (1999) Microcontinuum Field Theory. I. Foundations and Solids. Springer, New York

Eringen AC, Kafadar CB (1976) Polar field theories. In: Eringen AC (ed) Continuum Physics. Vol. IV, vol IV, Academic Press, New York, pp 1–75

Eugster SR, dell'Isola F (2017) Exegesis of the introduction and sect. I from "Fundamentals of the Mechanics of Continua"** by E. Hellinger. ZAMM 97(4):477–506

Eugster SR, dell'Isola F (2018a) Exegesis of Sect. II and III.A from "Fundamentals of the Mechanics of Continua" by E. Hellinger. ZAMM 98(1):31–68

Eugster SR, dell'Isola F (2018b) Exegesis of Sect. III.B from "Fundamentals of the Mechanics of Continua" by E. Hellinger. ZAMM 98(1):69–105

Germain P (1973a) La méthode des puissances virtuelles en mécanique des milieux continus. première partie: théorie du second gradient. J Mécanique 12:236–274

Germain P (1973b) The method of virtual power in continuum mechanics. part 2: Microstructure. SIAM Journal on Applied Mathematics 25(3):556–575

Łukaszewicz G (1999) Micropolar Fluids: Theory and Applications. Birkhäuser, Boston

Lurie AI (1990) Nonlinear Theory of Elasticity. North-Holland, Amsterdam

Lurie AI (2001) Analytical Mechanics. Springer, Berlin

Migoun NP, Prokhorenko PP (1984) Hydrodynamics and Heattransfer in Gradient Flows of Microstructured Fluids (in Russian). Nauka i Technika, Minsk

Nowacki W (1986) Theory of Asymmetric Elasticity. Pergamon-Press, Oxford et al.

Pietraszkiewicz W, Eremeyev VA (2009) On natural strain measures of the non-linear micropolar continuum. International Journal of Solids and Structures 46(3–4):774–787

Sedov LI (1965) Mathematical methods for constructing new models of continuous media. Russian Mathematical Surveys 20(5):123

Simmonds JG (1994) A Brief on Tensor Analysis, 2nd edn. Springer, New Yourk

Truesdell C, Noll W (2004) The Non-linear Field Theories of Mechanics, 3rd edn. Springer, Berlin

Wilson EB (1901) Vector Analysis, Founded upon the Lectures of G. W. Gibbs. Yale University Press, New Haven

Zelenina AA, Zubov LM (2000) A criterion for forces applied to a rigid body to be potential. Mechanics of Solids 35(3):158–160

Chapter 8
Nonlinear Localized Waves of Deformation in the Class of Metamaterials as Set as the Mass-in-mass Chain

Vladimir I. Erofeev, Daniil A. Kolesov & Alexey O. Malkhanov

Abstract A well-known mathematical model representing a chain of oscillators consisting of elastic elements and masses, each containing an internal oscillator and describing the class of acoustic metamaterials "mass-in-mass", is generalized by taking into account the nonlinearity of the external and (or) internal elastic elements. As a result of analysis of the long-wavelength approximation of the obtained system, it is shown that spatially localized nonlinear deformation waves (solitons) can be formed in a metamaterial, under dynamic influence on it. The dependencies connecting the parameters of a localized wave are determined: amplitude, velocity and width with inertial and elastic characteristics of the metamaterial.

Keywords: Mathematical modeling · Nonlinear waves · Metamaterial · Mass-in-mass chain · One-dimensional system

8.1 Introduction

The development of modern technologies is impossible without the creation of new promising materials with unusual properties. For example, defect-free carbon nanotubes are two orders of magnitude stronger and four times lighter than steel. Currently, a new class of substances with a complexly organized internal structure (microstructure) and possessing unique physicomechanical properties is called meta-

Vladimir I. Erofeev
Research Institute for Mechanics, National Research Lobachevsky State University & Mechanical Engineering Research Institute of Russian Academy of Sciences, Nizhny Novgorod, Russia,
e-mail: erof.vi@yandex.ru

Daniil A. Kolesov · Alexey O. Malkhanov
Mechanical Engineering Research Institute of Russian Academy of Sciences Nizhny Novgorod, Russia,
e-mail: alandess@yandex.ru, alexey.malkhanov@gmail.com

B. E. Abali et al. (eds.), *New Achievements in Continuum Mechanics and Thermodynamics*, Advanced Structured Materials 108,
https://doi.org/10.1007/978-3-030-13307-8_8

materials. They first appeared in the field of optics and photonics (Cummer et al, 2016; Zhu and Zhang, 2018), but now they are increasingly found in other areas. For example, acoustic metamaterials (Zhang et al, 2009; Burov et al, 2011; Norris and Haberman, 2012; Deymier, 2013; Craster and Guenneau, 2013; di Cosmo and Laudato, 2018; Abali et al, 2017; Giorgio et al, 2017; Ming-Hui et al, 2009; Madeo et al, 2016; dell'Isola et al, 2015, 2016; El Sherbiny and Placidi, 2018) are widely used, in particular as sound and vibration absorbers as in Bobrovnitskii (2014, 2015); Bobrovnitskii et al (2016); Fedotovskii (2015, 2018); Bobrovnitskii and Tomilina (2018). Another example of materials with unusual properties are fullerites—solid structures formed based on fullerenes (Sidorov et al, 2005). Super- and ultra-hard fullerites are characterized by uniquely high velocities of longitudinal elastic waves and a wide diapason of these values ranging from 11 km/s to 26 km/s, depending on their structure, determined by the conditions of synthesis (Blank et al, 1998). The value of 26 km/s measured in one of the fullerite phases is a record—it is almost 20% more than the speed of longitudinal waves in graphite along atomic layers equal to 21.6 km/s (until recently this value was the highest for all known substances) and 40% more than the corresponding speed in diamond (18.6 km/s). The speeds of transverse waves in solid fullerite phases are also high (their values range from 7 km/s to 9.7 km/s), but they are still smaller than in diamond (11.6–12.8 km/s) which remain the highest among currently known substances.

Acoustic (or mechanical) metamaterials, being, in fact, not materials, but cellular periodic structures, in the long-wavelength range behave like continuous materials. The study of the features of dispersion, dissipation, and the appearance of nonlinearity of acoustic waves in metamaterials is of high interest (Altenbach et al, 2010; Dreyer et al, 2005; Agranovich et al, 2004; Berezovski et al, 2016; Engelbrecht et al, 2007; Madeo et al, 2015).

Guided by a mathematical analogy between acoustic and electromagnetic waves, many researchers have tried to construct continuous models of mechanical metamaterials. However, great success on this path was not achieved, since the mechanical analogs of actually existing materials with negative dielectric constant are deformable solids with negative mass, density or negative modulus of elasticity (Li and Chan, 2004; Fang et al, 2006; Ding et al, 2007; Cheng et al, 2008; Chan et al, 2006). And such materials do not exist in the reality.

It is obvious that an adequate description of the physicomechanical properties of metamaterials within the framework of the classical theory of elasticity is impossible. Recently, generalized micropolar theories of the Cosserat continuum type (Huang et al, 2009) have become widespread for modeling structurally inhomogeneous materials. However, these theories include a large number of material constants that require experimental determination and whose relationship with the structure of the material is not clear. This disadvantage is devoid of an alternative direction—structural modeling as in Altenbach et al (2011); Pavlov and Potapov (2008). In Pavlov (2010), a one-dimensional chain was considered containing identical masses m_1 connected by elastic elements (springs), having the same rigidity k_1, at the same time each mass inside itself contains another mass m_2 and one more elastic element—a spring with rigidity k_2 (see Fig. 8.1). Such a model, called the mass-in-mass chain, does not give the mentioned absurd results.

8.2 Mathematical Model

We generalize the model in Pavlov (2010) by taking into account the quadratic nonlinearity of the external and internal elastic elements. The potential energy of the unit cell of the mass-in-mass chain is written as:

$$W^{(j)} = \frac{1}{2}\left[k_1\left(u_1^{(j+1)} - u_1^{(j)}\right)^2 + k_2\left(u_2^{(j)} - u_1^{(j)}\right)^2 \right. \\ \left. + h_1\left(u_1^{(j+1)} - u_1^{(j)}\right)^3 + h_2\left(u_2^{(j)} - u_1^{(j)}\right)^2\right], \tag{8.1}$$

and its kinetic energy in the form:

$$T^{(j)} = \frac{1}{2}\left[m_1\left(\ddot{u}_1^{(j)}\right)^2 + m_2\left(\ddot{u}_2^{(j)}\right)^2\right]. \tag{8.2}$$

Let us suppose that $u_1(x)$ and $u_2(x)$ are continuous functions, which describe the displacements of all masses m_1 and m_2, respectively. Taking into account the expansion of displacements in a Taylor series up to the second term, we obtain

$$u_1^{(j+1)} = u_1(x+L) = u_1(x) + \frac{\partial u_1}{\partial x}L = u_1^{(j)} + \frac{\partial u_1}{\partial x}L. \tag{8.3}$$

The technique of expansion displacements in (8.3) was effectively applied by Kunin (1982) in the transformation of multimass discrete systems into a quasicontinuum.

The densities of the potential and kinetic energies for the equivalent continuum, obtained from (8.1) and (8.2), can be written in the form:

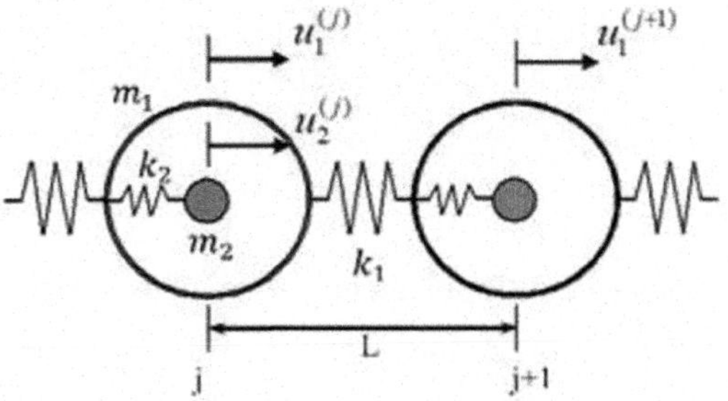

Fig. 8.1 Infinite mass-in-mass lattice structure

$$W=\frac{1}{2L}\left[k_1\left(\frac{\partial u_1}{\partial x}L\right)^2+k_2(u_2-u_1)^2+h_1\left(\frac{\partial u_1}{\partial x}L\right)^3+h_2(u_2-u_1)^2\right], \quad (8.4)$$

$$T=\frac{1}{2}\left[m_1(\ddot{u}_1)^2+m_2(\ddot{u}_2)^2\right]. \quad (8.5)$$

Let us construct from (8.4) and (8.5) the Lagrange function

$$\mathcal{L}=T-W=\mathcal{L}(\dot{u}_1,\dot{u}_2,u_{1x},u_1,u_2)$$

and take into account equations well known from analytical mechanics

$$\begin{cases}\dfrac{\partial}{\partial t}\left(\dfrac{d\mathcal{L}}{d\dot{u}_1}\right)+\dfrac{\partial}{\partial x}\left(\dfrac{\partial\mathcal{L}}{\partial u_{1x}}\right)-\dfrac{\partial\mathcal{L}}{\partial u_1}=0\\ \dfrac{\partial}{\partial t}\left(\dfrac{d\mathcal{L}}{d\dot{u}_2}\right)-\dfrac{\partial\mathcal{L}}{\partial u_2}=0\end{cases} \quad (8.6)$$

to get the system of equations in in displacements:

$$\begin{aligned}&\frac{m_1}{L}\ddot{u}_1-k_1L\frac{\partial^2u_1}{\partial x^2}-3h_1L^3\frac{\partial u_1}{\partial x}\frac{\partial^2u_1}{\partial x^2}-\frac{k_2}{L}(u_2-u_1)-\frac{3h_2}{2L}(u_2-u_1)^2=0,\\ &\frac{m_1}{L}\ddot{u}_2-\frac{k_2}{L}(u_2-u_1)-\frac{3h_2}{2L}(u_2-u_1)^2=0.\end{aligned} \quad (8.7)$$

Further consider a particular case of system (8.6), where $h_1\neq 0,\ h_2=0$, i.e.:

$$\begin{aligned}\frac{m_1}{L}\ddot{u}_1-k_1L\frac{\partial^2u_1}{\partial x^2}-3h_1L^3\frac{\partial u_1}{\partial x}\frac{\partial^2u_1}{\partial x^2}-\frac{k_2}{L}(u_2-u_1)=0,\\ \frac{m_2}{L}\ddot{u}_2+\frac{k_2}{L}(u_2-u_1)=0\end{aligned} \quad (8.8)$$

The system (8.8) can be rewritten in the form of single equation:

$$\begin{aligned}&\frac{\partial^2u_2}{\partial t^2}-\frac{k_1L^2}{m_1+m_2}\frac{\partial^2u_2}{\partial x^2}+\frac{m_1m_2}{k_2(m_1+m_2)}\frac{\partial^4u_2}{\partial t^4}-\frac{k_1L^2m_2}{k_2(m_1+m_2)}\frac{\partial^4u_2}{\partial x^2\partial t^2}\\ &-\frac{3h_1L^4}{m_1+m_2}\left(\frac{\partial u_2}{\partial x}\frac{\partial^2u_2}{\partial x^2}+\frac{m_2}{k_2}\frac{\partial^2u_2}{\partial x^2}\frac{\partial^3u_2}{\partial t^2\partial x}+\frac{m_2}{k_2}\frac{\partial u_2}{\partial x}\frac{\partial^4u_2}{\partial t^2\partial x^2}\right.\\ &\left.+\left(\frac{m_2}{k_2}\right)^2\frac{\partial^3u_2}{\partial t^2\partial x}\frac{\partial^4u_2}{\partial t^2\partial x^2}\right)=0.\end{aligned} \quad (8.9)$$

Let us introduce dimensionless variables—time, coordinate, and displacement:

$$\tau=\frac{t}{T},\quad y=\frac{x}{X},\quad u_2=u_0u. \quad (8.10)$$

The transformed equation (8.8) with the new variables (8.10) takes the form:

$$\frac{\partial^2 u}{\partial \tau^2} - \frac{k_1 L^2}{m_1 + m_2}\frac{T^2}{X^2}\frac{\partial^2 u}{\partial y^2} + \frac{m_1 m_2}{k_2(m_1+m_2)}\frac{1}{T^2}\frac{\partial^4 u}{\partial \tau^4}$$
$$-\frac{k_1 L^2 m_2}{k_2(m_1+m_2)}\frac{1}{X^2}\frac{\partial^4 u}{\partial y^2 \partial \tau^2} - \frac{3h_1 L^4}{m_1+m_2}\frac{T^2 u_0}{X^3}\left(\frac{\partial u}{\partial y}\frac{\partial^2 u}{\partial y^2} + \frac{m_2}{k_2}\frac{1}{T^2}\frac{\partial^2 u}{\partial y^2}\frac{\partial^3 u}{\partial \tau^2 \partial y}\right.$$
$$\left. + \frac{m_2}{k_2}\frac{1}{T^2}\frac{\partial u}{\partial y}\frac{\partial^4 u}{\partial \tau^2 \partial y^2} + + \left(\frac{m_2}{k_2}\frac{1}{T^2}\right)^2 \frac{\partial^3 u}{\partial \tau^2 \partial y}\frac{\partial^4 u}{\partial \tau^2 \partial y^2}\right) = 0. \tag{8.11}$$

We require that all the coefficients (8.11) are finite or small. We choose them so that among the nonlinear terms we can distinguish only one, the main item.

All the subsequent arguments are valid if two conditions are satisfied:

$$\frac{k_1 L^2}{m_1+m_2}\frac{T^2}{X^2} = 1 \text{ and } \frac{m_1 m_2}{k_2(m_1+m_2)}\frac{1}{T^2} = \varepsilon, \ \varepsilon \ll 1. \tag{8.12}$$

When these conditions are fulfilled, in equation (8.11) some of the terms can be discarded, since they have a larger order of smallness and do not have a significant effect on dynamic processes. Thus, equation (8.11) takes the form:

$$\frac{\partial^2 u}{\partial \tau^2} - \frac{\partial^2 u}{\partial y^2} + \varepsilon\frac{\partial^2}{\partial \tau^2}\left[\frac{\partial^2 u}{\partial \tau^2} - \alpha\frac{\partial^2 u}{\partial y^2}\right] = \delta\frac{\partial u}{\partial y}\frac{\partial^2 u}{\partial y^2}, \tag{8.13}$$

where

$$\frac{m_1 m_2}{k_2(m_1+m_2)}\frac{1}{T^2} = \varepsilon \ll 1,$$
$$\frac{3h_1 L u_0 \sqrt{\varepsilon}\alpha}{k_1\sqrt{\frac{k_1}{k_2}\frac{m_2}{m_1}}} = \delta \ll 1, \ \alpha = \frac{m_1+m_2}{m_1} > 1. \tag{8.14}$$

Returning to the original dimensional variables in equation (8.13), we obtain the simplified equation (8.8) in the form:

$$\frac{\partial^2 u_2}{\partial t^2} - \frac{k_1 L^2}{m_1+m_2}\frac{\partial^2 u_2}{\partial x^2} + \frac{m_1 m_2}{k_2(m_1+m_2)}\frac{\partial^4 u_2}{\partial t^4} - \frac{k_1 L^2 m_2}{k_2(m_1+m_2)}\frac{\partial^4 u_2}{\partial x^2 \partial t^2}$$
$$-\frac{3h_1 L^4}{m_1+m_2}\frac{\partial u_2}{\partial x}\frac{\partial^2 u_2}{\partial x^2} = 0. \tag{8.15}$$

8.3 Nonlinear Stationary Waves

We will seek the solution to this equation in the class of traveling stationary waves: $u_2 = u_2(\xi)$, $\xi = x - Vt$, V – velocity of the stationary wave (unknown beforehand). With respect to deformation $\frac{du_2}{d\xi} = U$ the nonlinear partial differential equation (8.15) reduces to the anharmonic oscillator equation:

$$\frac{d^2U}{d\xi^2} + aU + bU^2 = 0,$$
$$a = \frac{k_2\left(m_1+m_2\right)\left(V^2-c_1^2\right)}{m_1 m_2\left(V^2-c_2^2\right)},\ b = \frac{3h_1 L^4 k_2}{2V^2 m_1 m_2\left(V^2-c_2^2\right)}, \tag{8.16}$$
$$c_1^2 = \frac{k_1 L^2}{m_1+m_2},\ c_2^2 = \frac{k_1 L^2}{m_1}.$$

Let us note that $c_2 > c_1$. Depending on the value of the velocity V, there are qualitatively different wave patterns, since equation (8.16) has different solutions in Erofeev et al (2002). Only the solutions which at infinity do not give a constant component for the strain wave U have physical meaning.

If the velocity of the stationary wave satisfies inequality: $c_2 > c_1 > V$, then equation (8.16) has a periodic solution expressed in terms of an elliptic sine:

$$U\left(\xi\right) = \frac{A}{3s^2}\left(1+s^2-\sqrt{1-s^2+s^4}\right) - A \cdot sn^2\left(Q\xi, s\right), \tag{8.17}$$

where $A = -\frac{3a}{2b}\frac{s^2}{\sqrt{1-s^2+s^4}}$ – amplitude of the stationary wave, s– elliptic function module, $Q = \sqrt{\frac{k_1}{4V^2 m_2\sqrt{1-s^2+s^4}}}$ – nonlinear analog of the wave number.

It can be seen that $Q \sim \sqrt{k_1}$, $Q \sim \frac{1}{\sqrt{m_1}}$, $Q \sim \frac{1}{V}$ when the other variables are fixed.

In the Fig. 8.2 the dependence $Q \sim \sqrt{k_1}$ is depicted: curve 1—the qualitative form of this dependence at a fixed velocity and internal mass of the element; curve 2—the trend of the behavior of the graph of the dependence with increasing mass and fixed value of velocity; curve 3—increase in velocity by the same order that the mass was increased in the previous case; curve 4—increase of both parameters. From the analysis of curves 2 and 3 and from the assumption that the mass, in comparison with the velocity, is a much more static parameter, we can conclude that the most significant effect on this dependence is exerted by speed.

In the Fig. 8.3 the dependence $Q \sim \frac{1}{\sqrt{m_1}}$ is shown: curve 1—a qualitative representation of this relationship at a fixed velocity and stiffness of the external spring of

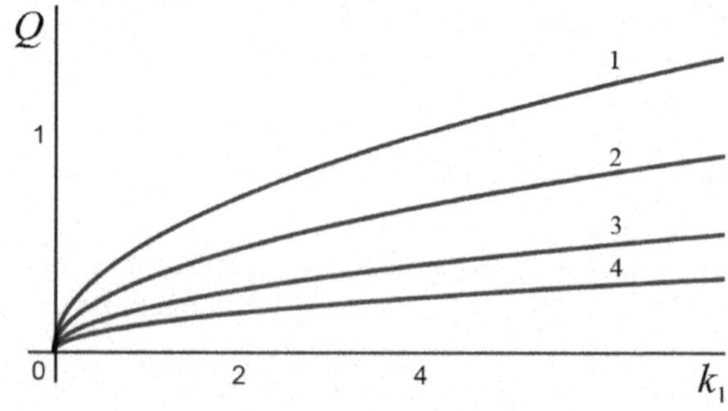

Fig. 8.2 The dependence of the wave number on the rigidity of the external elastic element of the system

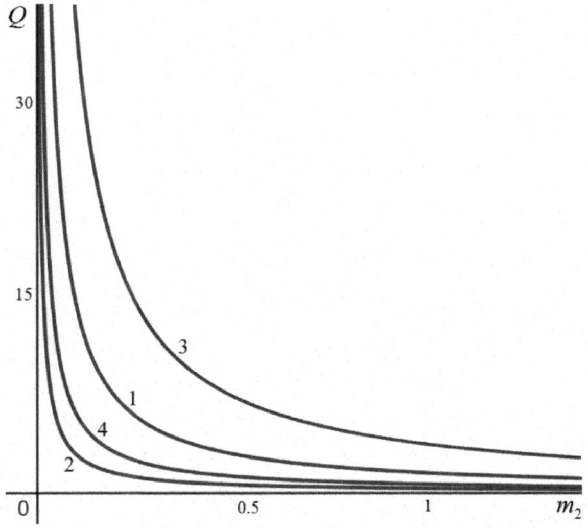

Fig. 8.3 The dependence of the wave number on the internal mass of the system element

the system element; curve 2—the trend of behavior of the graph of the dependence with increasing velocity and constant rigidity; curve 3—increase in rigidity by the same order, which was increased velocity in the previous case; line 4—increase of both parameters. From the analysis of the form of the curves 2 and 3 it can be seen that an increase in the parameters leads to a shift in the dependence curve in different directions. From the analysis of the curve 4 it follows that shows that the change in speed is "stronger."

In the Fig. 8.4 the dependence $Q \sim 1/V$ is depicted: curve 1—qualitative form of the relationship with fixed internal mass and stiffness of the external spring of the system element; curve 2—trend of behavior of the graph of the dependence with increasing mass and constant rigidity; curve 3—increase in rigidity by the same order that the mass was increased in the previous case. When both parameters are increased by the same order, the function graph coincides with curve 1. From the analysis of the curves it can be concluded that these parameters have the same influence on the indicated dependence.

The qualitative form of the periodic wave is shown in Fig. 8.5, where, through $K(s)$ the elliptic integral of the first kind is denoted.

If the velocity of the stationary wave satisfies the inequality $c_2 > V > c_1$, then equation (8.16) has an aperiodic solution, expressed in terms of the hyperbolic cosine:

$$U(\xi) = A_c \cosh^{-2}\left(\frac{\xi}{\Delta}\right). \tag{8.18}$$

This relation describes a solitary stationary wave (soliton) of deformation. Here $A_c = -\frac{3a}{2b}$ – the amplitude of soliton, $\Delta = \frac{2}{\sqrt{-a}}$ – the width of soliton. The analysis of the latter shows that $\Delta \sim V$, $\Delta \sim \frac{1}{\sqrt{k_2}}$, $\Delta \sim \sqrt{m_2}$.

In the Fig. 8.6 the dependence $\Delta \sim V$ is shown: curve 1 is a qualitative represen-

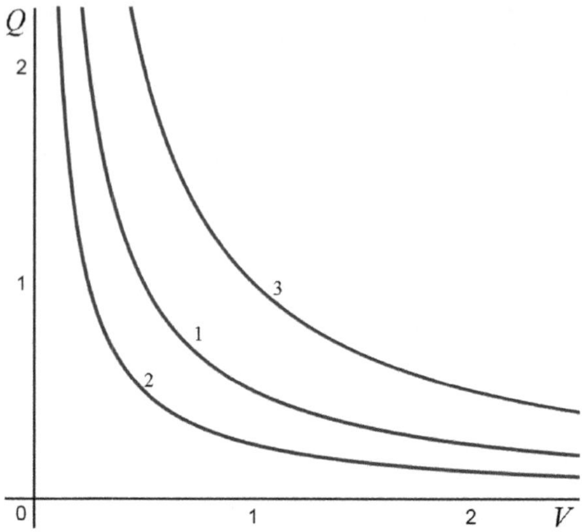

Fig. 8.4 The dependence of the wave number of a stationary wave on its speed

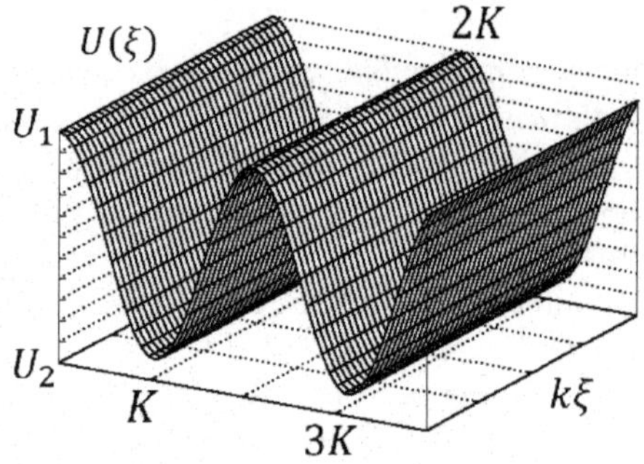

Fig. 8.5 Qualitative representation of the periodic wave

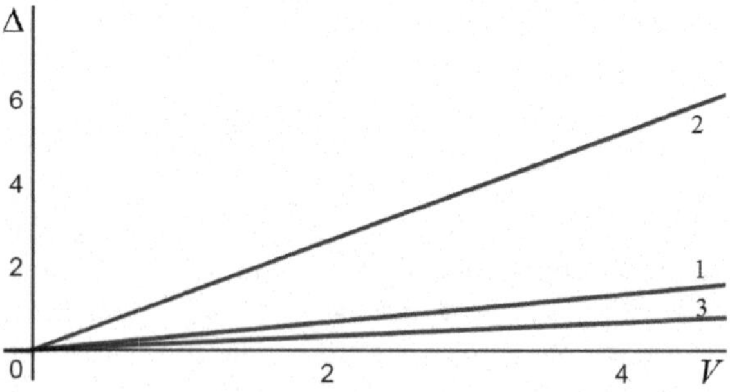

Fig. 8.6 The dependence of the soliton width on its speed

tation of this relationship for fixed internal spring stiffness and internal mass of the system element; curve 2—the trend of behavior of the graph of the dependence with increasing of the mass and constant rigidity; curve 3—increase in rigidity by the same order that the mass was increased in the previous case. When both parameters are increased by the same order, the function graph coincides with curve 1. From the analysis of the curves it can be concluded that these parameters have the same effect on the indicated dependence.

In the Fig. 8.7 the dependence $\Delta \sim 1/\sqrt{k_2}$ is depicted: curve 1—the qualitative form of this dependence at a fixed velocity and internal mass of the element of the system; curve 2—the trend of the behavior of the graph of the dependence with increasing velocity and constant mass; curve 3—increase in mass by the same order, which was increased velocity in the previous case; line 4—increase of both parameters. From the analysis of the curves 2 and 3 it can be seen that an increase in speed leads to a stronger shift up of the graph. Consequently, the change in velocity "stronger" affects the width of the soliton.

In the Fig. 8.8 the dependence $\Delta \sim \sqrt{m_2}$ is shown: curve 1—a qualitative representation of this relationship at a fixed velocity and stiffness of the internal spring of the element; curve 2—the trend of behavior of the graph of the dependence with increasing velocity and constant rigidity; curve 3—increase in rigidity by the same order, which was increased velocity in the previous case; curve 4—increase of both

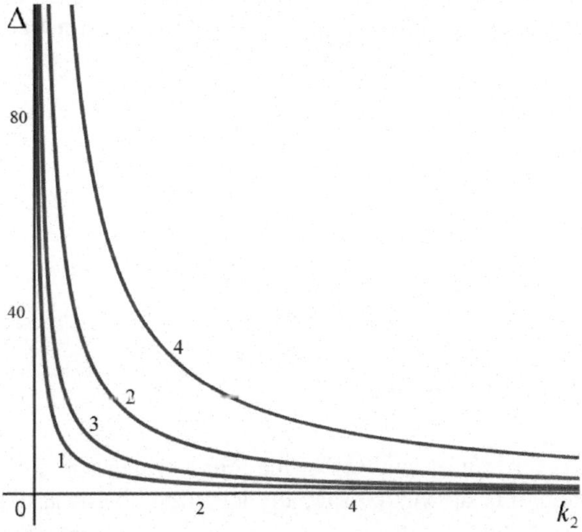

Fig. 8.7 The dependence of the soliton width on the rigidity of the internal elastic element of the system

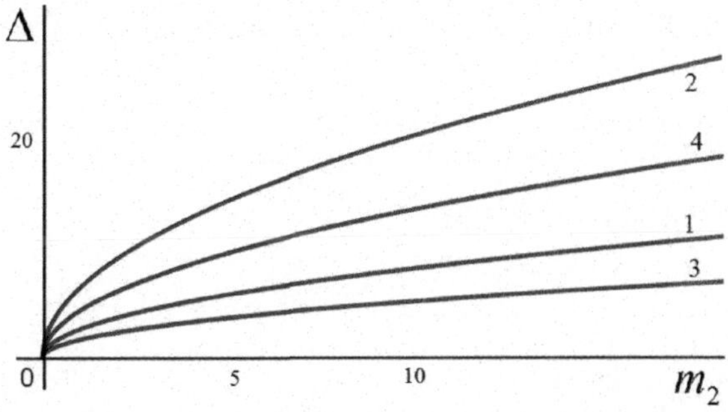

Fig. 8.8 The dependence of the soliton width on the internal mass of a system element

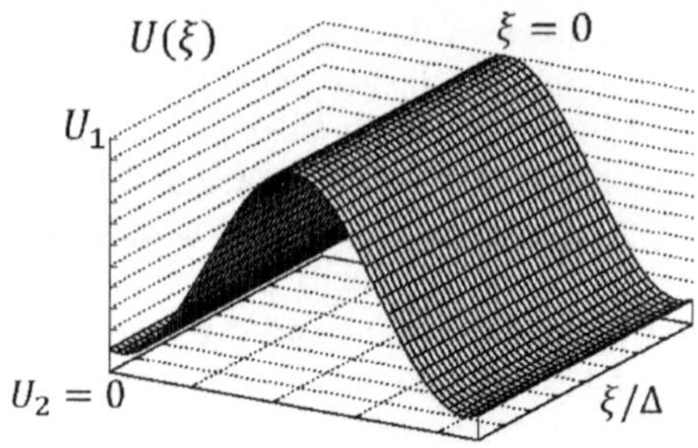

Fig. 8.9 The qualitative representation of a soliton

parameters. From the analysis of the form of curves 2 and 3 and from the assumption that the rigidity is a much more static parameter in comparison with the velocity, it can be concluded that the most significant effect on this dependence is exerted by speed.

The qualitative form of the soliton of deformations is shown in Fig. 8.9.

If the velocity of the stationary wave satisfies the inequality: $V > c_2 > c_1$, then the equation (8.16) has a periodic solution expressed in terms of an elliptic sine:

$$U(\xi) = \frac{A}{3s^2}\left(\sqrt{1-s^2+s^4} - 1 - s^2\right) + A \cdot sn^2(Q\xi, s), \tag{8.19}$$

where

$$A = -\frac{3a}{2b}\frac{s^2}{\sqrt{1-s^2+s^4}},\ Q^2 = \frac{a}{4\sqrt{1-s^2+s^4}}. \tag{8.20}$$

The dependencies shown in Figs. 8.2, 8.3, 8.4 remain valid for this case.

8.4 Conclusions

As a result of analysis of the long-wavelength approximation of the obtained system, it is shown that spatially localized nonlinear deformation waves (solitons) can be formed in a metamaterial, under dynamic influence on it. The dependencies connecting the parameters of a localized wave are determined: amplitude, velocity and width with inertial and elastic characteristics of the metamaterial.

Acknowledgements This work was supported by a grant from the Government of the Russian Federation (contract No 14.Y26.31.0031).

References

Abali BE, Müller WH, dell'Isola F (2017) Theory and computation of higher gradient elasticity theories based on action principles. Archive of Applied Mechanics 87(9):1495–1510

Agranovich VM, Shen YR, Baughman RH, Zakhidov AA (2004) Linear and nonlinear wave propagation in negative refraction metamaterials. Phys Rev B 69:165,112

Altenbach H, Eremeyev VA, Lebedev LP, Rendón LA (2010) Acceleration waves and ellipticity in thermoelastic micropolar media. Archive of Applied Mechanics 80(3):217–227

Altenbach H, Maugin GA, Erofeev VI (eds) (2011) Mechanics of Generalized Continua. Springer-Verlag, Berlin, Heidelberg, 350 p.

Berezovski A, Giorgio I, Corte AD (2016) Interfaces in micromorphic materials: wave transmission and reflection with numerical simulations. Mathematics and Mechanics of Solids 21(1):37–51

Blank VD, Levin VM, Prokhorov VM, Buga SG, Dybitskii GA, Serebryanaya NR (1998) Elastic properties of ultrahard fullerites. Journal of Experimental and Theoretical Physics 87(4):741–746

Bobrovnitskii YI (2014) Effective parameters and energy of acoustic metamaterials and media. Acoustical Physics 60 (2):134–141

Bobrovnitskii YI (2015) Models and general wave properties of two-dimensional acoustic metamaterials and media. Acoustical Physics 61 (3):255–264

Bobrovnitskii YI, Tomilina TM (2018) Sound absorption and metamaterials: a rewiew. Acoustical Physics 64(5):519–526

Bobrovnitskii YI, Tomilina TM, Laktionova MM (2016) A discrete model of damped acoustic metamaterials. Acoustical Physics 62 (1):1–7

Burov VA, Voloshinov VB, Dmitriev KV, Polikarpova NV (2011) Acoustic waves in metamaterials, crystals, and anomalously refracting structures. Physics Uspekhi 54 (11):1165–1170

Chan CT, Li J, Fung KH (2006) On extending the concept of double negativity to acoustic waves. JZUS A 7:24–28

Cheng Y, Xu JY, Liu XJ (2008) One-dimensional structured ultrasonic metamaterials with simultaneously negative dynamic density and modulus. Phys Rev B 77:(045,134)

di Cosmo F, Laudato MSM (2018) Acoustic metamaterials based on local resonances: homogenization, optimization and applications, Springer, pp 247–274

Craster RV, Guenneau SE (2013) Acoustic metamaterials: negative refraction, imaging, lensing and cloaking. Springer, Dordrecht

Cummer SA, Christensen J, Alú A (2016) Controlling sound with acoustic metamaterials. Nat Rev Mater 1:16,001

dell'Isola F, Lekszycki T, Pawlikowski M, Grygoruk R, Greco L (2015) Designing a light fabric metamaterial being highly macroscopically tough under directional extension: first experimental evidence. Zeitschrift fur angewandte Mathematik und Physik ZAMP 66 (6):3473–3498

dell'Isola F, Giorgio I, Pawlikowski M, Rizzi NL (2016) Large deformations of planar extensible beams and pantographic lattices: heuristic homogenization, experimental and numerical examples of equilibrium. Proceedings of the Royal Society of London A 472 (2185):

Deymier PAE (2013) Acoustic metamaterials and phononic crystals. Springer-Verlag, Berlin

Ding Y, Liu Z, Qiu C, Shi J (2007) Metamaterial with simultaneously negative bulk modulus and mass density. Phys Rev Lett 99:(093,904)

Dreyer W, Herrmann M, Mielke A (2005) Micro-macro transition in the atomic chain via Whitham's modulation equation. Nonlinearity 19(2):471–500

El Sherbiny MG, Placidi L (2018) Discrete and continuous aspects of some metamaterial elastic structures with band gaps. Archive of Applied Mechanics pp 1–18

Engelbrecht J, Berezovski A, Salupere A (2007) Nonlinear deformation waves in solids and dispersion. Wave Motion 44 (6):493–500

Erofeev VI, Kazhaev VV, Semerikova NP (2002) Waves in rods. Dispersion. Dissipation. Nonlinearity. Fizmatlit, Moscow (in Russian)

Fang N, Xi D, Xu J, Ambati M, Srituravanich W, Sun C, Zhang X (2006) Ultrasonic metamaterials with negative modulus. Nat Mater 5:452–456

Fedotovskii VS (2015) Transverse waves in a dispersive metamaterial with spherical inclusions. Acoustical Physics 61 (3):281–286

Fedotovskii VS (2018) A porous medium as an acoustic metamaterial with negative inertial and elastic properties. Acoustical Physics 64(5):548–554

Giorgio I, Della Corte A, dell'Isola F (2017) Dynamics of 1d nonlinear pantographic continua. Nonlinear Dynamics 88(1):21–31

Huang HH, Sun CT, Huang GL (2009) On the negative effective mass density in acoustic metamaterials. Int J Eng Sci 47:610–617

Kunin IA (1982) Elastic Media with Microstructure I & II. Springer-Verlag, Berlin, Heidelberg, New York

Li J, Chan CT (2004) Double-negative acoustic metamaterial. Phys Rev E 70:(055,602)

Madeo A, Neff P, Ghiba ID, Placidi L, Rosi G (2015) Wave propagation in relaxed micromorphic continua: modeling metamaterials with frequency-gaps. Continuum Mechnics and Thermodynamics 27(4–5):551–570

Madeo A, Barbagallo G, d'Agostino MV, Placidi L, Neff P (2016) First evidence of non-locality in real band-gap metamaterials: determining parameters in the relaxed micromorphic model. Proceedings of the Royal Society of London 472(2190):20160,169

Ming-Hui L, Liang F, Yan-Feng C (2009) Phononic crystals and acoustic metamaterials. Materials today 12(12):34–42

Norris AN, Haberman MR (2012) Introduction to the special issue on acoustic metamaterials. The Journal of the Acoustical Society of America 132(4):2783–2783

Pavlov IS (2010) Acoustic identification of the anisotropic nanocrystalline medium with non-dense packing of particles. Acoustical Physics 56(6):924–934

Pavlov IS, Potapov AI (2008) Structural models in mechanics of nanocrystalline media. Doklady Physics 53(7):408–412

Sidorov LN, Yurovskaya MA, Borschevsky AY, Trushkov IV, Ioffe IN (2005) Fullerenes: Tutorial. Publishing Exam, Moscow, 688 p. (in Russian)

Zhang X, Yin L, Fang N (2009) Focusing ultrasound with an acoustic metamaterial network. Phys Rev Lett 102 (19):194301:1–4

Zhu S, Zhang X (2018) Metamaterials: artificial materials beyond nature. National Science Review 5(2):131

Chapter 9
Modelling of a Hydrogen Saturated Layer Within the Micropolar Approach

Ksenia Frolova, Elena Vilchevskaya, Vladimir Polyanskiy & Ekaterina Alekseeva

Abstract This paper is concerned with modeling the strongly inhomogeneous hydrogen distribution over a sample by means of the micropolar continuum approach. The presence of micro-cracks covering the lateral surface of the sample is modeled by means of a distributed couple stress prescribed as a boundary condition. The applied couple stress produces a longitudinal displacement in return, which quickly fades away from the surface. The tensile displacement increases the intergranular space in the vicinity of the sample boundary and initiates hydrogen absorption from the environment. A comparison between widths of the surface layer that were experimentally determined and the ones that were analytically obtained allows estimating a value of one of the non-classical elastic parameters.

9.1 Introduction

Modern materials operate under extreme loads and in corrosive environments. The combined effect leads to stress corrosion cracking in metals (Jones, 2017), as well as to hydrogen-induced embrittlement (Koyama et al, 2012; Kyoung et al, 2009; Zhang et al, 2015). Experiments based on mass spectrometry and electron microscopy show that water vapor in the air is a source of hydrogen and that, in turn, the hydrogen leads to a decrease of strength, crack resistance and endurance lim-

Ksenia Frolova · Elena Vilchevskaya · Vladimir Polyanskiy
Institute for Problems in Mechanical Engineering RAS, Bolshoy pr., 61, V.O., St. Petersburg 199178 & Peter the Great St. Petersburg Polytechnic University, Polytechnicheskaya, 29, St. Petersburg 195294, Russia,
e-mail: kspfrolova@gmail.com, vilchevska@gmail.com, vapol@mail.ru

Ekaterina Alekseeva
Peter the Great St. Petersburg Polytechnic University, Polytechnicheskaya, 29, St. Petersburg 195294, Russia,
e-mail: alexeeva__ekaterina@mail.ru

B. E. Abali et al. (eds.), *New Achievements in Continuum Mechanics and Thermodynamics*, Advanced Structured Materials 108,
https://doi.org/10.1007/978-3-030-13307-8_9

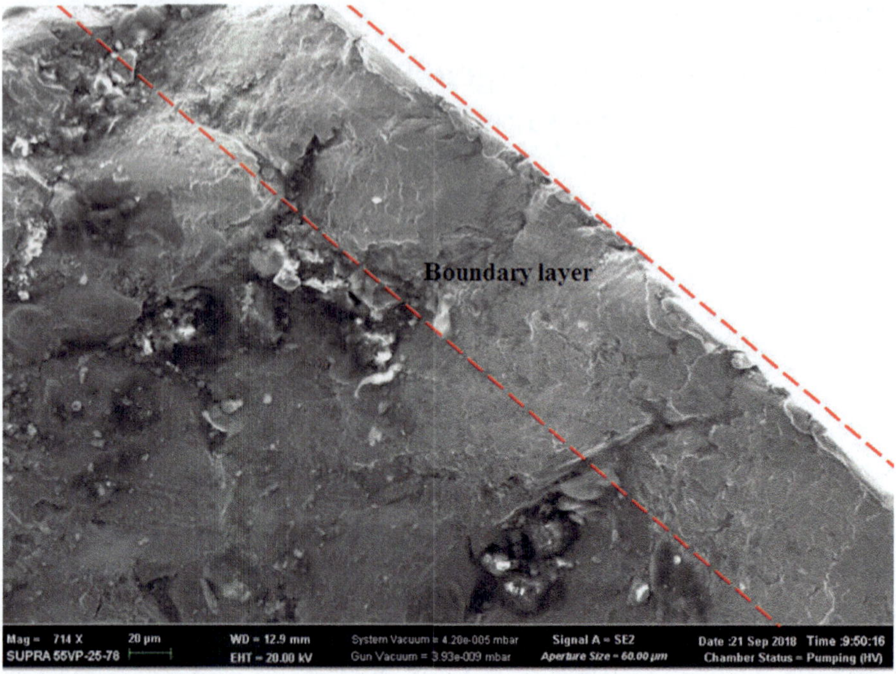

Fig. 9.1 Fractography of the sample after hydrogenation and fatigue test

its (Khrustalev et al, 1989). These effects are observed in samples saturated with hydrogen made from different types of materials, namely nickel, titanium, copper, aluminum alloys, and steel.

A number of articles were devoted to the investigation of hydrogen accumulation inside metals placed in an aggressive environment (Hadam and Zakroczymski, 2009; Martinsson and Sandström, 2012; Omura et al, 2016; Wu et al, 2015; Yagodzinskyy et al, 2011). Direct measurements of hydrogen concentrations, as well as mathematical modeling of hydrogen accumulation, show that the hydrogen is unevenly distributed and that there is a significant excess of its concentration within a thin boundary layer of the metal (Martinsson and Sandström, 2012; Wu et al, 2015). A highly heterogeneous distribution of hydrogen within the samples is also observed when fatigue test are performed with non-hydrogen-charged samples placed in air environment (Belyaev et al, 2017a,b). In Fig. 9.1 the fractography of a sample made of alloy 718 after hydrogen charging and fatigue test is shown. In this case, the hydrogen concentration in a thin boundary layer with an approximate thickness of about 100 µm exceeds 10–100 times the volumetric concentration.

It is known that the lateral surfaces of samples are covered with fractal oriented dislocations that, in turn, lead to micro-crack formation as in Betekhtin et al (2009); Kramer et al (2005); Steffens et al (1987). Detailed examinations of the surface layer in Steffens et al (1987) show that its grains or parts of a uniform mono-crystal rotate and create empty spaces near the boundary. These additional empty spaces initiate the seepage of hydrogen from the external environment into the material. Thus, the non-homogeneous distribution of hydrogen over the sample can be explained by the existence of a lengthwise displacement appearing due to grain rotations in the vicinity of the lateral surface.

In other words, the theoretical interpretation of this phenomena should be connected with the material inner structure, in particular, with introduction of the intrinsic length scale parameter characterizing the width of the surface layer with a high hydrogen concentration. The dependence of mechanical response on the structure size could not be explained by the classical continuum mechanics since no length scale enters the constitutive equations. However, different generalized continuum theories have been successful in addressing a size effect problem. Examples for generalized continuum theories are the strain-gradient-theory as in Aifantis (1984); dell'Isola and Seppecher (1997); dell'Isola et al (2012), the micropolar theory as in Eringen and Kafadar (1976); Eremeyev et al (2012), and the surface-theory as in Altenbach et al (2010); Eremeyev and Pietraszkiewicz (2014), just to mention the most popular ones. A modern comparison of the different modelling options, which must be driven by phenomenology, can be found in dell'Isola et al (2017). In the present paper we are going to use a micropolar medium theory. As the first attempt to start the description and see how that approach works we will restrict ourself to isotropic linear continuum. A detailed discussion about other possible constitutive choices can be found in Eremeyev and Pietraszkiewicz (2016, 2012). Also it should be noted that although direct consideration of the cracking process is beyond the scope of the paper, the initial presence of the surface dislocations and microcracks is modeled by means of a distributed couple stress on the lateral surface of the sample. As a result, it will be shown that the prescription of couple stresses on the boundary leads to the appearance of additional tensile displacements increasing the intergranular space within the thin surface layer and therefore initiating the highly heterogeneous distribution of hydrogen.

Considering more complicated models usually involves introducing new parameters. There are four additional isotropic elastic constants in micropolar theory. Three of them provide a sensitivity to the rotation gradient and the remaining one quantifies the degree of coupling between macro and micro-rotation. Both theoretical and experimental investigations have been undertaken in order to determine these material coefficients (Adomeit, 1968; Askar, 1972; Askar and Cakmak, 1968; Ellis and Smith, 1967; Gauthier and Jahsman, 1975; Lake, 1983, 1995; Liebold and Müller, 2015, 2016; Perkins and Thompson, 1973; Schijve, 1966; Yang and Lakes, 1981). However, the accomplished work is still far from being complete and these coefficients are obtained only for a few different materials, namely, graphite, human bone, and foams. A comparison between the thickness of the high hydrogen concentration layer observed in experiments and the width of the layer with the analytically obtained tensile displacements allows for an estimate of the value of one of the non-classical elastic constants. This will be demonstrated in the last section for the aluminum alloy D16.

9.2 Basic Equations of Micropolar Media

Let us consider a geometrically linear micropolar continuum. Its deformations are described by the standard displacement field, $\boldsymbol{u}$, and an independent microrotation vector, $\boldsymbol{\theta}$. Then the stretch tensor $\boldsymbol{e}$ and wryness tensor $\boldsymbol{\kappa}$ can be introduced in the following way[1] (Eremeyev et al, 2012):

$$\boldsymbol{e} = \nabla \otimes \boldsymbol{u} + \boldsymbol{I} \times \boldsymbol{\theta}, \tag{9.1}$$

$$\boldsymbol{\kappa} = \nabla \otimes \boldsymbol{\theta}, \tag{9.2}$$

where ∇ is the gradient operator, and $\mathbf{I}$ is the unit tensor.

Note that the microrotation vector $\boldsymbol{\theta}$ is kinematically distinct from the "macrorotation," $\boldsymbol{w}$, determined by the antisymmetric part of the displacement gradient:

$$\boldsymbol{w} = \frac{1}{2} \nabla \times \boldsymbol{u}. \tag{9.3}$$

The vector $\boldsymbol{\theta}$ refers to the rotation of body particles, whereas $\boldsymbol{w}$ refers to the rotation associated with translational motion of nearby body particles. Thus, the decomposition of the stretch tensor into symmetric and antisymmetric parts yields

$$\boldsymbol{e} = \boldsymbol{\epsilon} + \boldsymbol{I} \times (\boldsymbol{\theta} - \boldsymbol{w}), \tag{9.4}$$

where $\boldsymbol{\epsilon} = 1/2(\nabla \otimes \boldsymbol{u} + \boldsymbol{u} \otimes \nabla)$ is the strain tensor of classical linear elasticity.

The constitutive equations for the stress tensor $\boldsymbol{T}$ and the couple stress tensor $\boldsymbol{M}$ for linear isotropic Cosserat elasticity are as follows (Eringen and Kafadar, 1976):

$$\boldsymbol{T} = \lambda \left(\nabla \cdot \boldsymbol{u} \right) \boldsymbol{I} + 2\mu \boldsymbol{\epsilon} + \gamma \boldsymbol{I} \times (\boldsymbol{\theta} - \boldsymbol{w}), \tag{9.5}$$

$$\boldsymbol{M} = \beta_1 \left(\nabla \cdot \boldsymbol{\theta} \right) \boldsymbol{I} + \beta_2 \boldsymbol{\kappa}^{\top} + \beta_3 \boldsymbol{\kappa}, \tag{9.6}$$

where λ, μ, γ and β_i $(i = 1, 2, 3)$ are independent elastic moduli. λ and μ are the classical elastic moduli, and γ is a modulus quantifying the degree of coupling between macro and micro rotation fields. The elastic moduli β_2 and β_3 allow to introduce material lengths reflecting effects of the couple stress, for example, $l_t = \sqrt{(\beta_2 + \beta_3)/2\mu}$ for torsion, or $l_b = \sqrt{\beta_3/4\mu}$ for bending (see, for example, Gauthier and Jahsman (1975)). When these length parameters vanish, the solutions obtained from couple stress theory reduce to those of classical elasticity theory. In perfect crystals and amorphous materials l_b and l_t are probably submicroscopic; but might be of the order of the averaged radius of roots of surface cracks (Mindlin, 1964).

The displacement and microrotation vectors can be found from the equilibrium equations, that in case of absence of body forces and body couples are

---

[1] A cross product between a second-rank tensor $\boldsymbol{A} = \boldsymbol{a}_k \boldsymbol{b}_k$ and vector $\boldsymbol{c}$ is realized in the following way: $\boldsymbol{A} \times \boldsymbol{c} = \boldsymbol{a}\boldsymbol{b} \times \boldsymbol{c} = \boldsymbol{a} \left(\boldsymbol{b} \times \boldsymbol{c} \right)$, i.e. the second vector in the tensor dyad is attached with the vector $\boldsymbol{c}$.

$$\nabla \cdot \boldsymbol{T} = 0. \tag{9.7}$$

$$\nabla \cdot \boldsymbol{M} + \boldsymbol{T}_{\times} = 0, \tag{9.8}$$

where $(\boldsymbol{a} \otimes \boldsymbol{b})_{\times} = \boldsymbol{a} \times \boldsymbol{b}$.

Within the framework of this paper we shall consider a case in which $\gamma \to \infty$. This corresponds to a case in which the antisymmetric part of the stretch tensor goes to zero or, in other words, to an assumption that the macrorotation and microrotation vectors are equal. It follows that it is the medium with constrained rotation, and the microrotation is determined by the displacement:

$$\boldsymbol{\theta} = \frac{1}{2} \nabla \times \boldsymbol{u}. \tag{9.9}$$

This means that $\nabla \cdot \boldsymbol{\theta} = 0$ and therefore parameter β_1 does not play any role in this model. Then Eqn.(9.6) simplifies to

$$\boldsymbol{M} = \frac{1}{2} \left(\beta_3 \nabla \otimes (\nabla \times \boldsymbol{u}) + \beta_2 (\nabla \times \boldsymbol{u}) \otimes \nabla \right). \tag{9.10}$$

According to Eqn. (9.8) together with Eqns. (9.5), (9.10), (9.4), and (9.3) we can obtain the following equality:

$$\boldsymbol{T}_{\times} = -2\gamma(\boldsymbol{\theta} - \boldsymbol{w}) = -\beta_3 \triangle \boldsymbol{w}, \tag{9.11}$$

where $\triangle = \nabla \cdot \nabla$ is the Laplace operator.

Consequently, the antisymmetric part of the stress tensor can be rewritten as follows:

$$\boldsymbol{T}^A = \frac{\beta_3}{4} \triangle \left(\boldsymbol{u} \otimes \nabla - \nabla \otimes \boldsymbol{u} \right). \tag{9.12}$$

As a result we obtain the following equation for the displacement field:

$$(\lambda + \mu) \nabla (\nabla \cdot \boldsymbol{u}) + \mu \triangle \boldsymbol{u} + \frac{\beta_3}{4} \nabla \cdot \triangle \left(\boldsymbol{u} \otimes \nabla - \nabla \otimes \boldsymbol{u} \right) = 0. \tag{9.13}$$

9.3 Axially-symmetrical Problem

In this paper we model the behavior of a cylindrical sample made of metal in an aggressive environment. In this context let us consider a boundary-value problem in cylindrical coordinates (r, φ, z) dealing with a solid cylinder of radius r_0 and length L subjected to a distributed couple stress, $-M_0 \boldsymbol{e}_{\varphi}$, on its lateral surface. Whereas a zero displacement field satisfies the equilibrium conditions and guarantees a traction-free lateral surface, the applied couple stress on the lateral surface produces an additional displacement which is fading away quickly from the surface.

We assume that the problem is axially symmetric and look for a solution with the following ansatz:

$$\boldsymbol{u} = \boldsymbol{u}(r,z) = u_r(r,z)\,\boldsymbol{e}_r + u_z(r,z)\,\boldsymbol{e}_z, \tag{9.14}$$

where $\boldsymbol{e}_r$ is the radial unit vector, and $\boldsymbol{e}_z$ is the unit vector along the cylinder axis.

From Eqns. (9.9) and (9.14) it follows that the microrotation vector has only one component, namely

$$\boldsymbol{\theta} = \theta_\varphi \boldsymbol{e}_\varphi = \frac{1}{2}\left(\frac{\partial u_r}{\partial z} - \frac{\partial u_z}{\partial r}\right)\boldsymbol{e}_\varphi. \tag{9.15}$$

In order to obtain the solution for the boundary layer in closed form, we have to take the following inequalities into account:

$$\begin{aligned} \frac{\partial^2 u_r}{\partial z^2} &\ll \frac{\partial^2 u_r}{\partial r^2}, \\ \frac{\partial^2 u_z}{\partial z^2} &\ll \frac{\partial^2 u_z}{\partial r^2}, \\ \frac{\partial u_z}{\partial z} &\ll \frac{\partial^2 u_z}{\partial r^2}. \end{aligned} \tag{9.16}$$

Consequently, Eqn. (9.13) reduces to

$$\begin{aligned} &\left[\lambda\left(\frac{\partial^2 u_r}{\partial r^2} + \frac{\partial^2 u_z}{\partial r \partial z}\right) + 2\mu\left(\frac{\partial^2 u_r}{\partial r^2} + \frac{1}{2}\frac{\partial^2 u_z}{\partial z \partial r}\right) - \frac{\beta_3}{2}\frac{\partial^3 \theta_\varphi}{\partial z \partial r^2}\right]\boldsymbol{e}_r + \\ &\qquad \left[\lambda\frac{\partial^2 u_r}{\partial z \partial r} + \mu\left(\frac{\partial^2 u_z}{\partial r^2} + \frac{\partial^2 u_r}{\partial r \partial z}\right) + \frac{\beta_3}{2}\frac{\partial^3 \theta_\varphi}{\partial r^3}\right]\boldsymbol{e}_z = 0. \end{aligned} \tag{9.17}$$

Here the terms of higher order of smallness are discarded.

By introducing non-dimensional parameters

$$\begin{aligned} &x = 1 - \frac{r}{r_0}, \quad \tilde{z} = \frac{z}{L}, \quad \xi = \frac{r_0}{L}, \quad u_x = \frac{u_r}{r_0}, \quad u_{\tilde{z}} = \frac{u_z}{L}, \\ &\qquad\qquad \tilde{\lambda} = \frac{\lambda}{\mu}, \qquad \delta = \sqrt{\frac{\beta_3}{4\mu r_0^2}} = \frac{l_b}{r_0}. \end{aligned} \tag{9.18}$$

we can rewrite Eqn. (9.17) by the following system of equations:

$$\begin{aligned} \left(\tilde{\lambda}+2\right)\frac{\partial^2 u_x}{\partial x^2} - \left(\tilde{\lambda}+1\right)\frac{\partial^2 u_{\tilde{z}}}{\partial x \partial \tilde{z}} - \delta^2 \frac{\partial^4 u_{\tilde{z}}}{\partial x^3 \partial \tilde{z}} &= 0, \\ -\left(\tilde{\lambda}+1\right)\frac{\partial^2 u_x}{\partial x \partial \tilde{z}} + \frac{1}{\xi^2}\frac{\partial^2 u_{\tilde{z}}}{\partial x^2} - \delta^2\left(\frac{1}{\xi^2}\frac{\partial^4 u_{\tilde{z}}}{\partial x^4} + \frac{\partial^4 u_x}{\partial x^3 \partial \tilde{z}}\right) &= 0. \end{aligned} \tag{9.19}$$

By integrating the first equation, we obtain

$$\frac{\partial u_x}{\partial x} = \frac{\tilde{\lambda}+1}{\tilde{\lambda}+2}\frac{\partial u_{\tilde{z}}}{\partial \tilde{z}} + \frac{\delta^2}{\tilde{\lambda}+2}\frac{\partial^3 u_{\tilde{z}}}{\partial x^2 \partial \tilde{z}}. \tag{9.20}$$

Then the second equation in (9.19) yields

$$u_{\widetilde{z}} - \delta^2 \frac{\partial^2 u_{\widetilde{z}}}{\partial x^2} = 0. \tag{9.21}$$

Since δ is of small value and the displacement has a finite value, we arrive at the following solution for $u_{\widetilde{z}}$:

$$u_{\widetilde{z}} = A\,(\widetilde{z}) \exp\Big(-\frac{x}{\delta}\Big). \tag{9.22}$$

The solution for the dimensionless radial displacement reads

$$u_x = -A'\,(\widetilde{z})\,\delta \exp\Big(-\frac{x}{\delta}\Big). \tag{9.23}$$

By applying non-dimensional parameters, Eqn. (9.15) for the microrotations can be finally rewritten in the following way:

$$\theta_\varphi = \frac{1}{2\xi\delta} A(\widetilde{z}) \exp\Big(-\frac{x}{\delta}\Big). \tag{9.24}$$

As a result the stress tensor has the form

$$\widetilde{\boldsymbol{T}} = \frac{1}{\mu}\boldsymbol{T} = 2\exp\Big(-\frac{x}{\delta}\Big)\Big(A'\,(\widetilde{z})\,(\boldsymbol{e}_z \otimes \boldsymbol{e}_z - \boldsymbol{e}_r \otimes \boldsymbol{e}_r) + \frac{1}{\xi\delta} A\,(\widetilde{z})\,\boldsymbol{e}_z \otimes \boldsymbol{e}_r\Big). \tag{9.25}$$

Obviously, the smallness of the first two terms is of higher order than the last one and formally they can be disregarded. However, in order to be sure that the lateral surface is traction free we will treat $A(z)$ as a constant. Thus the couple stress is given by

$$\widetilde{\boldsymbol{M}} = \frac{1}{\mu L}\boldsymbol{M} = 2A\exp\Big(-\frac{x}{\delta}\Big)\Big(\frac{\beta_2}{\beta_3}\boldsymbol{e}_\varphi \otimes \boldsymbol{e}_r - \boldsymbol{e}_r \otimes \boldsymbol{e}_\varphi\Big). \tag{9.26}$$

From (9.26) we obtain the equation for the unknown constant:

$$\boldsymbol{e}_r \cdot \boldsymbol{M}\,|_{x=0} = -M_0 \boldsymbol{e}_\varphi, \qquad \rightarrow \qquad A = \frac{M_0}{2\mu L}. \tag{9.27}$$

So far we have considered a problem without external loads on the faces at the end. However, either axial tension

$$\boldsymbol{F} = F\boldsymbol{e}_z = \int_S T_{zz}\boldsymbol{e}_z \,\mathrm{d}S$$

or torsion

$$\boldsymbol{M}_t = M_t\boldsymbol{e}_z = \int_S (\boldsymbol{r} \times (\boldsymbol{e}_z \cdot \boldsymbol{T}) + M_{zz}\boldsymbol{e}_z)\,\mathrm{d}S$$

can be added. Here S is the cross-sectional area of the cylinder. The particular solutions of these Saint-Venant's problems are given by Gauthier and Jahsman (1975) and the total solutions for the medium with constrained rotation are

$$\boldsymbol{u} = -\frac{\nu F r}{ES}\boldsymbol{e}_r + \left(\frac{Fz}{ES} + \frac{M_0}{2\mu}\exp\left(-\frac{r_0 - r}{l_b}\right)\right)\boldsymbol{e}_z, \qquad (9.28)$$

$$\boldsymbol{T} = \frac{F}{S}\boldsymbol{e}_z\boldsymbol{e}_z + \frac{M_0}{l_b}\exp\left(-\frac{r_0 - r}{l_b}\right)\boldsymbol{e}_z\boldsymbol{e}_r$$

for tension and

$$\boldsymbol{u} = C_1 r z \boldsymbol{e}_\varphi + \frac{M_0}{2\mu}\exp\left(-\frac{r_0 - r}{l_b}\right)\boldsymbol{e}_z,$$

$$\boldsymbol{T} = C_1 \mu r\left(\boldsymbol{e}_z \otimes \boldsymbol{e}_\varphi + \boldsymbol{e}_\varphi \otimes \boldsymbol{e}_z\right) + \frac{M_0}{l_b}\exp\left(-\frac{r_0 - r}{l_b}\right)\boldsymbol{e}_z \otimes \boldsymbol{e}_r,$$

$$\boldsymbol{M} = C_1(\beta_2 + \beta_3)\left(\boldsymbol{e}_z \otimes \boldsymbol{e}_z - \frac{1}{2}\left(\boldsymbol{e}_r \otimes \boldsymbol{e}_r + \boldsymbol{e}_\varphi \otimes \boldsymbol{e}_\varphi\right)\right) + \qquad (9.29)$$

$$+ M_0 \exp\left(-\frac{r_0 - r}{l_b}\right)\left(\frac{\beta_2}{\beta_3}\boldsymbol{e}_\varphi \otimes \boldsymbol{e}_r - \boldsymbol{e}_r \otimes \boldsymbol{e}_\varphi\right),$$

$$C_1 = \frac{M_t}{S}\left(\beta_2 + \beta_3 + \mu\frac{r_0^2}{2}\right)^{-1}$$

for torsion. Here $\nu = \lambda/(2(\lambda + \mu))$ and $E = \mu(3\lambda + 2\mu)/(\lambda + \mu)$ are Poisson's ratio and Young's modulus, respectively.

In both cases, the additional exponential terms play a role only within a thin surface layer and serves as a correction terms in order to satisfy the boundary condition on the lateral surface. Note that the torsion solution provided in Gauthier and Jahsman (1975) is traction-free on the lateral surface. However, in the case of the medium with constrained rotation it has a couple stress on the boundary:

$$M_{rr} = -\frac{C_1}{2}\left(\beta_2 + \beta_3\right). \qquad (9.30)$$

9.4 Results

As one can see from the previous section, considering a micropolar media allows us to obtain an additional displacement along the cylinder axis in the vicinity of the lateral surface. This tensile displacement increases the intergranular space and can initiate hydrogen absorption from the environment. In order to estimate the parameters of the model let us consider a cylindrical sample made of the aluminum alloy D16, because this material is frequently used in experiments in Andronov et al (2017). The following values will be used: $r_0 = 0.0045$ m, $L = 0.035$ m, $\mu = 27000$ MPa.

In order to estimate the small parameter δ we correlate the width of the layer with the additional displacements within the thickness of the boundary layer containing a higher concentration of hydrogen, x_*. Since u_z is exponentially decreasing from of the surface we define its limit value by $\chi u_z|_{x=0}$, where χ is a small parameter. Thus, the equation for δ can be written in the following way:

$$\delta = -\frac{x_*}{Ln(\chi)}. \tag{9.31}$$

When taking $x_* = 100\ \mu\text{m}/r_0 \approx 0.02$ and $\chi = 0.01$ we obtain $\delta \approx 0.005$. Recall that δ is related to the non-classic elastic modulus β_3 through Eqn. (9.18) as follows:

$$\beta_3 = 4\mu r_0^2 \delta^2. \tag{9.32}$$

Therefore, the elastic modulus can be estimated as: $\beta_3 \approx 5 \cdot 10^{-5}$ MPa m^2.

The dependence of the displacement along the cylinder axis for this value of β_3 is shown in Fig. 9.2. The curve is plotted for $M_0 = -189$ kPa corresponding to $A = 0.0001$.

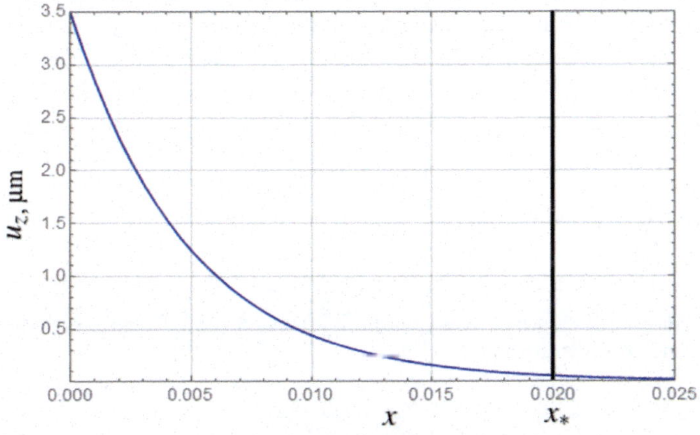

Fig. 9.2 Dependence of the tensile displacement on the relative distance from the lateral surface

9.5 Conclusions and Outlook

In this paper two main tasks were accomplished:

- First, an attempt was made to model the strongly inhomogeneous distribution of hydrogen observed in experiments within a micropolar continuum approach. The stress-strain state in the sample can be obtained within the theory of classical elasticity while the situation near the body surface could be depicted more accurately by a solution, which is quickly fading away from the border. This solution gives additional tensile displacements in the vicinity of the lateral surface by means of which the absorption of hydrogen from the environment can be initiated. Note that we did not consider the hydrogen diffusion but concentrated on possible reasons of the hydrogen seepage.

- Second, a new method for estimating the additional elastic constants of a Cosserat medium was proposed. It is based on the comparison between the experimentally measured width of the surface layer containing the high hydrogen concentration and the analytically obtained characteristic length from the fading solution.

More research in this field is planned for the future in order to study the boundary layer emergence within Cosserat elasticity without restriction. At this stage a medium with constrained rotation was considered. The adopted simplification allowed us to obtain the analytical solution and to demonstrate feasibility of the suggested approach. However, the strong coupling between the macro and microrotations does not describe the rotation of grains in a proper manner and should be softened.

Moreover, it the displacement was supposed to be axially symmetric. This is a "classical" assumption for a cylindrical sample, but it limits the types of boundary condition to be prescribed at the lateral surface, namely, the distributed couple stress caused by microcracks. The microcracks distribution might be non-homogeneously over the sample surface and therefore leads to a dependence of the couple stress on the azimuth angle φ. Evidently a numerical approach will be required in that case.

In order to estimate the value of the non-classical elastic parameters, further efforts should be devoted to providing experiments with different materials and specimens sizes.

Acknowledgements Support of this work by a grant from Russian Science Foundation by RSF grant no. 18-19-00160 is gratefully acknowledged.

References

Adomeit G (1968) Mechanics of Generalized Continua (Edited by E. Kroner), chapter Determination of elastic constants of a structured material. IUTAM Symposia, Springer-Verlag Berlin Heidelberg

Aifantis EC (1984) On the microstructural origin of certain inelastic models. Journal of Engineering Materials and technology 106(4):326–330

Altenbach J, Altenbach H, Eremeyev VA (2010) On generalized Cosserat-type theories of plates and shells: a short review and bibliography. Archive of Applied Mechanics 80(1):73–92

Andronov DY, Arseniev DG, Polyanskiy AM, Polyanskiy VA, Yakovlev YA (2017) Application of multichannel diffusion model to analysis of hydrogen measurements in solid. International Journal of Hydrogen Energy 42(1):699–710

Askar A (1972) Molecular crystals and the polar theories of the continua experimental values of material coefficients for kno3. International Journal of Engineering Science 10(3):293–300

Askar A, Cakmak AS (1968) A structural model of a micropolar continuum. International Journal of Engineering Science 6(10):583–589

Belyaev AK, Grishchenkȯ AI, Polyanskiy VA, Semenov AS, Tretyakov DA, Shtukin LV, Arseniev DG, A YY (2017a) Acoustic anisotropy and dissolved hydrogen as an indicator of waves of plastic deformation. In: Days on Diffraction, 2017, pp 39–44

Belyaev AK, Polyanskiy VA, Yakovlev YA, Mansyrev DE, Polyanskiy AM (2017b) Surface effect of the waves of plastic deformation and hydrogen distribution in metals. In: Days on Diffraction, 2017, pp 45–50

Betekhtin VI, Gilyarov VL, Kadomtsev AG, Korsukov VE, Korsukova MM, Obidov BA (2009) Fractalization of the surface relief of an amorphous alloy as an indication of rupture. Bulletin of the Russian Academy of Sciences: Physics 73(10):1419

dell'Isola F, Seppecher P (1997) Edge contact forces and quasi-balanced power. Meccanica 32(1):33–52

dell'Isola F, Seppecher P, Madeo A (2012) How contact interactions may depend on the shape of Cauchy cuts in Nth gradient continua: approach "à la D'Alembert". Zeitschrift für angewandte Mathematik und Physik 63(6):1119–1141

dell'Isola F, Corte AD, Giorgio I (2017) Higher-gradient continua: The legacy of Piola, Mindlin, Sedov and Toupin and some future research perspectives. Mathematics and Mechanics of Solids 22(4):852–872

Ellis RW, Smith CW (1967) A thin-plate analysis and experimental evaluation of couple-stress effects. Experimental Mechanics 7(9):372–380

Eremeyev VA, Pietraszkiewicz W (2012) Material symmetry group of the non-linear polar-elastic continuum. International Journal of Solids and Structures 49(14):1993–2005

Eremeyev VA, Pietraszkiewicz W (2014) Refined theories of plates and shells. ZAMM-Journal of Applied Mathematics and Mechanics/Zeitschrift für Angewandte Mathematik und Mechanik 94(1-2):5–6

Eremeyev VA, Pietraszkiewicz W (2016) Material symmetry group and constitutive equations of micropolar anisotropic elastic solids. Mathematics and Mechanics of Solids 21(2):210–221

Eremeyev VA, Lebedev LP, Altenbach H (2012) Foundations of micropolar mechanics. Springer Science & Business Media

Eringen AC, Kafadar CB (1976) Polar field theories. In: Continuum Physics, Volume 4, Elsevier, pp 1–73

Gauthier RD, Jahsman WE (1975) A quest for micropolar elastic constants. Journal of Applied Mechanics 42:369–374

Hadam U, Zakroczymski T (2009) Absorption of hydrogen in tensile strained iron and high-carbon steel studied by electrochemical permeation and desorption techniques. International Journal of Hydrogen Energy 34(5):2449–2459

Jones RH (2017) Stress-Corrosion Cracking, Materials performance and evaluation. ASM international

Khrustalev YA, Simakov YS, Glazunov MP, Gubin VV (1989) Formation of hydrogen under the metal friction (in russian). Russian Journal of Physical Chemistry A 63(5):1355–1357

Koyama M, Akiyama E, Tsuzaki K (2012) Hydrogen-induced delayed fracture of a fe–22mn–0.6 c steel pre-strained at different strain rates. Scripta Materialia 66(11):947–950

Kramer DE, Savage MF, Levine LE (2005) Afm observations of slip band development in al single crystals. Acta Materialia 53(17):4655–4664

Kyoung HS, Ji SK, Young SC, Kyung-Tae P, Young-Kook L, Chong SL (2009) Hydrogen delayed fracture properties and internal hydrogen behavior of a fe–18mn–1.5 al–0.6 c twip steel. ISIJ International 49(12):1952–1959

Lake R (1995) Experimental methods for study of Cosserat elastic solids and other generalized elastic continua. Continuum models for materials with microstructure pp 1–25

Lake RS (1983) Size effects and micromechanics of a porous solid. Journal of Materials Science 18(9):2572–2580

Liebold C, Müller WH (2015) Are microcontinuum field theories of elasticity amenable to experiments? a review of some recent results. In: Differential Geometry and Continuum Mechanics, Springer, pp 255–278

Liebold C, Müller WH (2016) Applications of higher-order continua to size effects in bending: Theory and recent experimental results. In: Generalized Continua as Models for Classical and Advanced Materials, Springer, pp 237–260

Martinsson A, Sandström R (2012) Hydrogen depth profile in phosphorus-doped, oxygen-free copper after cathodic charging. Journal of Materials Science 47(19):6768–6776

Mindlin RD (1964) Micro-structure in linear elasticity. Archive for Rational Mechanics and Analysis 16(1):51–78

Omura T, Nakamur J, Hirata H, Jotoku K, Ueyama M, Osuki T, Terunuma M (2016) Effect of surface hydrogen concentration on hydrogen embrittlement properties of stainless steels and ni based alloys. ISIJ International 56(3):405–412

Perkins RW, Thompson D (1973) Experimental evidence of a couple-stress effect. AIAA Journal 11(7):1053–1055

Schijve J (1966) Note on couple stresses. Journal of the Mechanics and Physics of Solids 14(2):113–120

Steffens T, Schwink C, Korner A, Karnthaler HP (1987) Transmission electron microscopy study of the stacking-fault energy and dislocation structure in cumn alloys. Philosophical Magazine A 56(2):161–173

Wu R, Ahlström J, Magnusson H, Frisk K, Martinsson A (2015) Charging, degassing and distribution of hydrogen in cast iron. Swerea KIMAB pp 1–41

Yagodzinskyy Y, Todoshchenko O, Papula S, Hänninen H (2011) Hydrogen solubility and diffusion in austenitic stainless steels studied with thermal desorption spectroscopy. steel research international 82(1):20–25

Yang JFC, Lakes RS (1981) Transient study of couple stress effects in compact bone: torsion. Journal of Biomechanical Engineering 103(4):275–279

Zhang S, Huang Y, Sun B, Liao Q, Lu H, Jian B, Mohrbacher H, Zhang W, Guo A, Zhang Y (2015) Effect of nb on hydrogen-induced delayed fracture in high strength hot stamping steels. Materials Science and Engineering: A 626:136–143

Chapter 10
Types of Physical Nonlinearity in the Theory of Constitutive Relations and the Generalized Poynting Effect

Dimitri V. Georgievskii

Abstract The certain class of constitutive relations are considered that connect the symmetric stress tensor and the symmetric strain tensor by means of isotropic potential tensor nonlinear functions in three-dimensional space. The various definitions of tensor nonlinearity are given as well as their equivalence is shown. From the perspective of mathematical theory about the tensor nonlinear functions, an interpretation of the Poynting effect is given, which is well known in experimental mechanics. It is demonstrated that such an effect is not necessarily the consequence of tensor nonlinearity in constitutive relations; instead, it is effected by the quadratic dependence on invariants in certain material functions. Therefore, in the physically linear case for a *small* strain, this dependence is absent. Concerning this "order of smallness," the Poynting effect is investigated and a possibility is discussed for simulating such an effect by means of the tensor linear constitutive relations.

Keywords: stress, strain, constitutive relation, material function, invariant, scalar potential, establishing experiment, the Poynting effect, tensor nonlinearity

10.1 Various Definitions of Tensor Nonlinearity and Their Equivalence

In the theory of constitutive relations for isotropic media, the considerable place belongs to the scleronomous models for which a connection of the strain tensor, $\tilde{\varepsilon}$, and the stress tensor, $\tilde{\sigma}$, are given in three-dimensional space by means of the isotropic tensor nonlinear function,

$$\tilde{\varepsilon} = B_0 \tilde{\mathrm{I}} + B_1 \tilde{\sigma} + B_2 \tilde{\sigma}^2 , \tag{10.1}$$

Dimitri V. Georgievskii
Mechanical and Mathematical Department, Moscow State University, Moscow 119991, Russia, e-mail: georgiev@mech.math.msu.su

B. E. Abali et al. (eds.), *New Achievements in Continuum Mechanics and Thermodynamics*, Advanced Structured Materials 108,
https://doi.org/10.1007/978-3-030-13307-8_10

where $\tilde{\text{I}}$ is the identity tensor of the second rank, B_0, B_1 and B_2 are the material functions of three independence invariants:

$$I_{\sigma 1} = \text{tr}\tilde{\sigma}, \quad I_{\sigma 2} = \sqrt{\text{tr}(\tilde{\sigma}^2)}, \quad I_{\sigma 3} = \sqrt[3]{\text{tr}(\tilde{\sigma}^3)}\,, \tag{10.2}$$

of the tensor $\tilde{\sigma}$. An extensive literature (see, for example, both the classic and recent works Rivlin, 1953; Rivlin and Ericksen, 1955; Il'yushin, 1963; Altenbach et al, 1995; Abali et al, 2013; Devendiran et al, 2017; Kulvait et al, 2017) is devoted to the problems of generality in continuum mechanics of the representation (10.1) and the inverse one

$$\tilde{\sigma} = A_0\tilde{\text{I}} + A_1\tilde{\varepsilon} + A_2\tilde{\varepsilon}^2\,, \tag{10.3}$$

where A_0, A_1 and A_2 are the material functions of the invariants:

$$I_{\varepsilon 1} = \text{tr}\tilde{\varepsilon}, \quad I_{\varepsilon 2} = \sqrt{\text{tr}(\tilde{\varepsilon}^2)}, \quad I_{\varepsilon 3} = \sqrt[2]{\text{tr}(\tilde{\varepsilon}^3)}\,, \tag{10.4}$$

where they can be expressed in B_0, B_1 and B_2.

If the medium has the scalar potential $w(I_{\sigma 1}, I_{\sigma 2}, I_{\sigma 3})$ such that $\tilde{\varepsilon} = \partial w/\partial\tilde{\sigma}$ then the following three potentiality conditions are fulfilled,

$$\frac{\partial B_0}{\partial I_{\sigma 2}} = I_{\sigma 2}\frac{\partial B_1}{\partial I_{\sigma 1}}, \quad \frac{\partial B_0}{\partial I_{\sigma 3}} = I_{\sigma 3}^2\frac{\partial B_2}{\partial I_{\sigma 1}}, \quad I_{\sigma 2}\frac{\partial B_1}{\partial I_{\sigma 3}} = I_{\sigma 3}^2\frac{\partial B_2}{\partial I_{\sigma 2}}\,, \tag{10.5}$$

that relate the material functions B_0, B_1 and B_2. The set of conditions (10.5) may be considered as the system of differential equations with respect to B_0, B_1 and B_2, which has the first integrals in the certain cases (Georgievskii, 2016b).

Tensor nonlinearity of the function (10.1) is stipulated by presence of the last term in the right part. If $B_2 \equiv 0$ then this function—just as the corresponding class of materials—is called quasilinear, i. e. linear in the tensorial sense; but possibly nonlinear in scalar sense. Among the latter the case of physical linearity corresponds that B_0 linearly depends on $I_{\sigma 1}$ and does not depend on $I_{\sigma 2}$ and $I_{\sigma 3}$, as B_1 is constant.

In this way, a difference from identical zero of the material function B_2 in (10.1) represents the natural definition of tensor nonlinearity. This definition is equivalent to the fact that the angle between deviators $\tilde{s} = \tilde{\sigma} - I_{\sigma 1}\tilde{\text{I}}/3$ and $\tilde{e} = \tilde{\varepsilon} - I_{\varepsilon 1}\tilde{\text{I}}/3$ is not equal to zero identically. Let us prove this.

We assume that $\tilde{\sigma}$ and $\tilde{\varepsilon}$ are not spherical tensors (according to (10.1) and (10.3) they are either spherical or nonspherical, simultaneously) such that $\tilde{s}$ and $\tilde{e}$ are not identically zero tensors of the second rank and the angle $\alpha = (\tilde{s}; \tilde{e})$ is defined. We calculate $\cos\alpha$:

$$\cos\alpha = \frac{\tilde{s} : \tilde{e}}{\sqrt{\tilde{s} : \tilde{s}}\,\sqrt{\tilde{e} : \tilde{e}}} \equiv \frac{\tilde{s} : \tilde{e}}{I_{s2} I_{e2}} \tag{10.6}$$

$$\tilde{s} : \tilde{s} = J, \quad \tilde{s} : \tilde{e} = JB_1 + KB_2, \quad \tilde{e} : \tilde{e} = JB_1^2 + 2KB_1B_2 + LB_2^2 \tag{10.7}$$

where J, K and L are the invariants of stress state depending on $I_{\sigma 1}$, $I_{\sigma 2}$ and $I_{\sigma 3}$ (10.2):

$$J = I_{\sigma 2}^2 - \frac{1}{3} I_{\sigma 1}^2, \quad K = I_{\sigma 3}^3 - \frac{1}{3} I_{\sigma 1} I_{\sigma 2}^2, \quad L = \frac{4}{3} I_{\sigma 1} I_{\sigma 3}^3 - I_{\sigma 1}^2 I_{\sigma 2}^2 + \frac{1}{6}(I_{\sigma 1}^4 + I_{\sigma 2}^4) \tag{10.8}$$

Since $\tilde{s} : \tilde{s} > 0$ and $\tilde{e} : \tilde{e} > 0$ then

$$J > 0, \quad JL - K^2 > 0 \tag{10.9}$$

Using the Hamilton – Cayley theorem, after calculations we write

$$\cos\alpha = \left(1 + \frac{(JL - K^2)B_2^2}{(JB_1 + KB_2)^2}\right)^{-1/2} \tag{10.10}$$

It should be noted that the material function B_0 is not present in the expression for α.

As is obvious from (10.10) that if $B_2 = 0$ and $B_1 > 0$ then $\tilde{s}$ and $\tilde{e}$ are co-directed, i. e. the unit directing tensors $\tilde{s}^0 = \tilde{s}/I_{s2}$ and $\tilde{e}^0 = \tilde{e}/I_{e2}$ are the same. The statement is also truly in reverse (here it is necessary to use both the inequalities (10.9). An equivalence of two definitions has been established.

A relative smallness of the tensor nonlinearity effect usually observable in experiments with deformable solids may be treated as a smallness of the angle α. The relation (10.10) results in the connection in linear approximation of the low values α and the dimensionless material function B_2:

$$\alpha = \sqrt{JL - K^2}\,\frac{B_2}{JB_1} + O(B_2^2) \tag{10.11}$$

If $\alpha \ll 1$ then tensor nonlinear effects of material behavior are said to have the second order of smallness. It is implied that the first order is inherent in the parameters of stress-strain state caused by presence in (10.1) of the material function B_1.

10.2 Establishing experiments to find the material functions B_0, B_1 and B_2

Let us pay attention to the establishing experiments to find the function B_0, B_1 and B_2 at any point $(I_{\sigma 1}, I_{\sigma_2}, I_{\sigma_3})$ in the domain of their definition (Georgievskii (2016a)). For this purpose it is proposed to use long hollow cylindrical specimens suitable to implement any combination of the following realizable stress states (the cylindrical coordinates r, θ and z associated with the specimen under consideration are used)

- uniaxial tension, $\sigma_{zz} = a = \text{const}$;
- torsion, $\sigma_{r\theta} = b = \text{const}$;
- longitudinal shear, $\sigma_{rz} = c = \text{const}$;
- uniform compression, $\sigma_{rr} = \sigma_{\theta\theta} = \sigma_{zz} = -d = \text{const}$.

In the above four cases, the other components of the stress tensor are assumed to be equal to zero. It is also assumed that in a certain range of the loads a, b, c and d the stress-strain relation is described by (10.1) and (10.3) with a sufficient accuracy.

Similar establishing experiments are proposed in Georgievskii et al (2012) for the case when $\tilde{\sigma}$ and $\tilde{\varepsilon}$ are deviators ($I_{\varepsilon 1} = 0$, $I_{\sigma 1} = 0$). For the material functions $\{A_0, A_1, A_2\}$ and $\{B_0, B_1, B_2\}$ the following additional relations are valid: $A_0 = -A_2 I_{\varepsilon 2}^2/3$ and $B_0 = -B_2 I_{\sigma 2}^2/3$. In the establishing experiments, hence, it is required to find two functions B_1 and B_2 dependent on $I_{\sigma 2}$ and $I_{\sigma 3}$. By virtue of incompressibility, in Georgievskii et al (2012) the tensor $\tilde{\varepsilon}$ is considered as a strain-rate tensor in a tensor nonlinear non-Newtonian viscous fluid. It is necessary to mention here the work Placidi et al (2015) devoted to the Gedanken experiments for the determination of two-dimensional linear second gradient elasticity coefficients as well as the work Placidi et al (2017) dealing with identification of two-dimensional pantographic structures.

For our original problem, we have

$$\begin{aligned}
\sigma_{rr} &= \sigma_{\theta\theta} = -d, \quad \sigma_{zz} = a - d, \quad \sigma_{r\theta} = b, \quad \sigma_{rz} = c, \quad \sigma_{\theta z} = 0,\\
(\sigma^2)_{rr} &= b^2 + c^2 + d^2, \quad (\sigma^2)_{\theta\theta} = b^2 + d^2, \quad (\sigma^2)_{zz} = c^2 + (a-d)^2,\\
(\sigma^2)_{r\theta} &= -2bd, \quad (\sigma^2)_{rz} = c(a - 2d), \quad (\sigma^2)_{\theta z} = bc,\\
I_{\sigma 1} &= a - 3d,\\
I_{\sigma 2}^2 &= a^2 + 2b^2 + 2c^2 + 3d^2 - 2ad,\\
I_{\sigma 3}^3 &= a(a^2 + 3c^2 + 3d^2) - 3d(a^2 + 2b^2 + 2c^2 + d^2)
\end{aligned} \tag{10.12}$$

Considering d as some parameter, from (10.12) we express a, b and c in terms of the invariants (10.2)

$$\begin{aligned}
a &= I_{\sigma 1} + 3d,\\
b^2 &= \frac{1}{2}(I_{\sigma 2}^2 - I_{\sigma 1}^2 - 4I_{\sigma 1}d - 6d^2) - c^2,\\
c^2 &= \frac{1}{3(I_{\sigma 1} + 3d)}(I_{\sigma 3}^3 - I_{\sigma 1}^3 - 9I_{\sigma 1}^2 d - 24 I_{\sigma 1} d^2 - 24d^3 + 3I_{\sigma 2}^2 d)
\end{aligned} \tag{10.13}$$

Using (10.1) and (10.3) we determine the strain components ε_{zz}, $\varepsilon_{r\theta}$ and ε_{rz}:

$$\begin{aligned}
\varepsilon_{zz} &= B_0 + B_1(a - d) + B_2[c^2 + (a-d)^2],\\
\varepsilon_{r\theta} &= B_1 b - 2B_2 bd,\\
\varepsilon_{rz} &= B_1 c + B_2 c(a - 2d)
\end{aligned} \tag{10.14}$$

The relations expressed by (10.14) can be considered as the system of equations to obtain B_0, B_1 and B_2 using the strain components ε_{zz}, $\varepsilon_{r\theta}$ and ε_{rz} measured experimentally. This system has the solution

$$\begin{aligned}
B_0 &= \varepsilon_{zz} + (d^2 - a^2 - c^2)\frac{\varepsilon_{rz}}{ac} + (c^2 - d^2 + ad)\frac{\varepsilon_{r\theta}}{ab},\\
B_1 &= 2d\frac{\varepsilon_{rz}}{ac} + (a - 2d)\frac{\varepsilon_{r\theta}}{ab},\\
B_2 &= \frac{\varepsilon_{rz}}{ac} - \frac{\varepsilon_{r\theta}}{ab}
\end{aligned} \tag{10.15}$$

which is unique if

$$a \neq 0, \quad b \neq 0, \quad c \neq 0 \tag{10.16}$$

Thus, in order to specify the material functions, B_0, B_1 and B_2, the experiments should follow the following steps:

1. The choice of the invariants $I_{\sigma 1}$, I_{σ_2} and I_{σ_3} as the arguments of the functions B_0, B_1 and B_2.
2. The calculation of a, b and c on the basis of (10.13) with a fixed value of d as a parameter.
3. The perfomance of experiments using a hollow cylindrical specimen with the stresses a, b, c, d and the measurements of the components ε_{zz}, $\varepsilon_{r\theta}$ and ε_{rz}.
4. The calculation of B_0, B_1 and B_2 on the basis of (10.15) at the chosen at step 1 point $(I_{\sigma 1}, I_{\sigma_2}, I_{\sigma_3})$ of three-dimensional space of arguments.

Although the parameter d is not mentioned in (10.16) and the formulas expressed by (10.13) and (10.15) are also valid for $d = 0$ the above discussion shows its importance. Changing this parameter, in the space $(I_{\sigma 1}, I_{\sigma_2}, I_{\sigma_3})$ we can enlarge the domain where the quantities a, b and c exist and where the denominator of (10.13) is not equal to zero.

10.3 The Generalized Poynting Effect

Returning to the notion "an order of smallness of tensor nonlinearity effects" discussed in Sect. 10.1, we should set the question which order of smallness is inherent in the Poynting effect. During the last century it attracted an attention in experimental mechanics of solids (Green, 1954; Lurie, 2005; Chen and Chen, 1991; dell'Isola et al, 1998; Akinola, 1999; Gavrilyachenko and Karyakin, 2000; Goldstein et al, 2015; Misra et al, 2018). Let us at once talk about the so-called generalized Poynting effect defined in the following way.

The stress tensor is supposed to have in some orthogonal coordinate system the only identically nonzero component $\sigma_{\alpha\beta} = \sigma_0(\mathbf{x})$, $\alpha \neq \beta$. The stress state of this type is characterized by the following invariants (10.2), (10.8)

$$I_{\sigma 1} = I_{\sigma 3} = 0, \quad I_{\sigma 2} = \sqrt{2}\,|\sigma_0|, \quad J = 2\sigma_0^2, \quad K = 0, \quad L = 2\sigma_0^4/3\,. \tag{10.17}$$

The domain of definition of the material functions B_0, B_1 and B_2 represents the positive real axis in three-dimensional invariant space. According to Eq. (10.1) the tensor $\tilde{\varepsilon}$ has the following nonzero components:

$$\begin{aligned} \varepsilon_{\alpha\alpha} &= \varepsilon_{\beta\beta} = B_0(0, I_{\sigma 2}, 0) + \sigma_0^2 B_2(0, I_{\sigma 2}, 0), \\ \varepsilon_{\gamma\gamma} &= B_0(0, I_{\sigma 2}, 0), \\ \varepsilon_{\alpha\beta} &= \sigma_0 B_1(0, I_{\sigma 2}, 0)\,. \end{aligned} \tag{10.18}$$

Difference from zero of the components $\varepsilon_{\alpha\alpha}$ and $\varepsilon_{\beta\beta}$ just makes up the essence of the generalized Poynting effect. We should not dwell here on the equations of equilibrium as well as on the compatibility equations which the stresses (10.17) and the strains (10.18) must comply with. These equations define more exactly a choice of the material functions B_0, B_1, B_2 and determine the stress $\sigma_0(\mathbf{x})$.

The particular case of the cylindrical coordinate system ($\alpha \to \theta, \beta \to z, \gamma \to r$) corresponds to the classical Poynting effect observable in experiments with twisted specimens and thoroughly described in the literature. To simulate this, in Goldstein et al (2015) they use initial or deformation anisotropy, in other works they choose various physical nonlinear models of isotropic solids. The formulae (10.18) show that one can attach the constitutive relations (10.1) to the second group.

The material function B_2 being the indication of tensor nonlinearity (as follows from Sect. 10.1 is contained only in the components $\varepsilon_{\alpha\alpha}$ and $\varepsilon_{\beta\beta}$ in (10.18). This fact means that

a) the effect of tensor nonlinearity in the stress-strain state (10.17), (10.18) appears only in difference of the component $\varepsilon_{\gamma\gamma}$ from two remaining diagonal components $\varepsilon_{\alpha\alpha}$ and $\varepsilon_{\beta\beta}$ which are equal to each other;
b) difference of $\varepsilon_{\beta\beta}$ (or $\varepsilon_{\alpha\alpha}$) from zero can be a consequence both tensor nonlinearity and dependence of the function B_0 on the quadratic invariant $I_{\sigma 2}$; this dependence may be realized among them in tensor linear materials when

$$B_0 = B_0(I_{\sigma 1}, I_{\sigma 2}), \quad B_1 = B_1(I_{\sigma 1}, I_{\sigma 2}), \quad B_2 \equiv 0 \tag{10.19}$$

c) the order of smallness of the component $\varepsilon_{\beta\beta}$ (or $\varepsilon_{\alpha\alpha}$), i. e. the generalized Poynting effect, determines by simultaneous smallness of the angle α which by virtue of (10.11) equal to

$$\alpha = \frac{|\sigma_0|}{\sqrt{3}} \frac{B_2}{B_1} + O(B_2^2) \tag{10.20}$$

and smallness of values of the dimensionless function $B_0(0, I_{\sigma 2}, 0)$ along the axis $I_{\sigma 2} > 0$.

Below we describe briefly a possibility of simulation of the generalized Poynting effect using the tensor linear constitutive relations (10.1) with the material functions (10.19). The second and the third potentiality conditions (10.5) are fulfilled identically while the first condition (10.5) connects the functions B_0 and B_1 as follows:

$$B_0 = -\frac{\nu}{E} I_{\sigma 1} + \frac{b_0}{E^2} I_{\sigma 2}^2, \quad B_1 = \frac{1+\nu}{E} + \frac{2b_0}{E^2} I_{\sigma 1} \tag{10.21}$$

$$\tilde{\varepsilon} = \frac{1}{E}\Big[\Big(-\nu I_{\sigma 1} + \frac{b_0}{E} I_{\sigma 2}^2\Big)\tilde{\mathrm{I}} + \Big(1 + \nu + \frac{2b_0}{E} I_{\sigma 1}\Big)\tilde{\sigma}\Big] \tag{10.22}$$

$$w(I_{\sigma 1}, I_{\sigma 2}) = -\frac{\nu}{2E} I_{\sigma 1}^2 + \frac{1+\nu}{2E} I_{\sigma 2}^2 + \frac{b_0}{E^2} I_{\sigma 1} I_{\sigma 2}^2 \tag{10.23}$$

Here E and ν are the material constants known as Young's modulus and Poisson's ratio, respectively; b_0 is the dimensionless material constant which characterizes a scalar nonlinearity of the constitutive relations (10.22). The potential (10.23) includes three constant and when $b_0 = 0$ it turns into the ordinary in linear elasticity potential of stress with respect to strains. Some other variants of a choice of the potential in conformity to the Poynting effect estimate, contain in Gavrilyachenko and Karyakin (2000).

By substituting (10.21) to (10.18) we receive

$$\varepsilon_{\alpha\alpha} = \varepsilon_{\beta\beta} = \varepsilon_{\gamma\gamma} = \frac{2b_0}{E^2}\sigma_0^2, \quad \varepsilon_{\alpha\beta} = \frac{1+\nu}{E}\sigma_0 \tag{10.24}$$

The value $\varepsilon_{\beta\beta}$ possesses more high order of smallness in comparison with $\varepsilon_{\alpha\beta}$ if $\varepsilon_{\beta\beta}/\varepsilon_{\alpha\beta} \ll 1$, i.e. $b_0\sigma_0 \ll E$. It is just the condition that the generalized Poynting effect within tensor linear connection of stresses and strains represents a phenomenon of the second order. In case of the classical Poynting effect ($\alpha \to \theta$, $\beta \to z$, $\gamma \to r$) the formulae (10.24) show that the relative extension ε_{zz} is in proportion to square of the strain $\varepsilon_{\theta z}$, which is conversely in proportion to the angle of twisting. This feature of the Poynting effect is often exploited in the literature.

It is necessary to mention here a so-called inverse Poynting effect as in Goldstein et al (2015) consisting in twisting of a specimen by action of one-dimensional stretching loading. Some off-diagonal components of the strain tensor are not equal to zero. It is obvious that this phenomenon can not be described by the relations (10.1) even by arbitrary form of tensor nonlinearity. However, one can simulate it using anisotropic models of continuum.

References

Abali BE, Müller WH, Georgievskii DV (2013) A discrete-mechanical approach for computation of three-dimensional flows. ZAMM - Journal of Applied Mathematics and Mechanics / Zeitschrift für Angewandte Mathematik und Mechanik 93(12):868–881

Akinola A (1999) An energy function for transversely-isotropic elastic material and the Poynting effect. Korean J Comput Appl Math 6(3):639 – 649

Altenbach H, Altenbach J, Zolochevsky A (1995) Erweiterte Deformationsmodelle und Versagenskriterien der Werkstoffmechanik. Deutscher Verlag für Grundstoffindustrie, Stuttgart

Chen M, Chen Z (1991) Second-order effect of an elastic circular shaft during torsion. Appl Math Mech 12(9):769 – 776

dell'Isola F, Ruta GC, Batra RC (1998) Generalized poynting effects in predeformed prismatic bars. Journal of Elasticity 50(2):181 – 196

Devendiran VK, Sandeep RK, Kannan K, Rajagopal KR (2017) A thermodynamically consistent constitutive equation for describing the response exhibited by several alloys and the study of a meaningful physical problem. International Journal of Solids and Structures 108(1):1 – 10

Gavrilyachenko TV, Karyakin MI (2000) Specific features of the nonlinearly elastic behavior of cylindrical compressible bodies in torsion. Journal of Applied Mechanics and Technical Physics 41(2):377–381

Georgievskii DV (2016a) Establishing experiments in tensor nonlinear theories of continuum mechanics. Moscow University Mechanics Bulletin 71(2):49–50

Georgievskii DV (2016b) Potentiality of isotropic nonlinear tensor functions relating two deviators. Mechanics of Solids 51(5):619–622

Georgievskii DV, Müller WH, Abali BE (2012) Establishing experiments to find material functions in tensor nonlinear constitutive relations. Bulletin of the Russian Academy of Sciences: Physics 76(12):1374–1377

Goldstein RV, Gorodtsov VA, Lisovenko DS (2015) Linear Poynting's effect at torsion and extension of curvilinearly anisotropic tubes. Doklady Physics 60(9):396–399

Green AE (1954) A note on second-order effects in the torsion of incompressible cylinders. Mathematical Proceedings of the Cambridge Philosophical Society 50(3):488–490

Il'yushin AA (1963) Plasticity. Foundations of the General Mathematical Theory (in Russ.). USSR Acad. Sci. Publ., Moscow

Kulvait V, Málek J, Rajagopal KR (2017) Modeling gum metal and other newly developed titanium alloys within a new class of constitutive relations for elastic bodies. Arch Mech 69(3):223 – 241

Lurie AI (2005) Theory of Elasticity. Foundations of Engineering Mechanics, Springer, Berlin, Heidelberg

Misra A, Lekszycki T, Giorgio I, Ganzosch G, Müller WH, dell'Isola F (2018) Pantographic metamaterials show atypical poynting effect reversal. Mechanics Reserch Communications 89:6–10

Placidi L, Andreaus U, Della Corte A, Lekszycki T (2015) Gedanken experiments for the determination of two-dimensional linear second gradient elasticity coefficients. ZAMP - Journal of Applied Mathematics and Physics / Zeitschrift für Angewandte Mathematik und Physik 66(6):3699–3725

Placidi L, Barchiesi E, Della Corte A (2017) Identification of two-dimensional pantographic structures with a linear d4 orthotropic second gradient elastic model accounting for external bulk double forces. Mathematical Modelling in Solid Mechanics Ser Advanced Structured Materials 69:211–232

Rivlin RS (1953) The solution of problems in second order elasticity theory. Journal of Rational Mechanics and Analysis 2:53–81

Rivlin RS, Ericksen JL (1955) Stress-deformation relations for isotropic materials. Journal of Rational Mechanics and Analysis 4:323–425

Chapter 11
Eigenstresses in a Nonlinearly Elastic Sphere with Distributed Dislocations

Evgeniya V. Goloveshkina & Leonid M. Zubov

Abstract The problem of the eigenstresses due to distributed edge and screw dislocations in a hollow nonlinearly elastic sphere is considered. The dislocation density is given by an arbitrary spherically symmetric tensor field. For a general isotropic elastic material, the problem is reduced to a one-dimensional nonlinear boundary value problem. By replacing the unknown functions, the boundary value problem with nonlinear boundary conditions is transformed to a problem with linear ones. Numerical solutions are constructed for specific models of compressible and incompressible materials. The analysis of the influence of dislocations on a stress state of an elastic sphere at large deformations is carried out.

Keywords: Nonlinear elasticity · Dislocation density · Eigenstresses · Large deformations · Spherical symmetry · Rotation tensor

11.1 Introduction

A microstructure of a solid body largely determines the deformation, strength and other properties. Therefore, a study of the microstructure and its defects is necessary for analyzing the mechanical behavior of many crystalline bodies. There are many studies on this subject which emphasize such defects as dislocations (Bilby et al, 1955; Kondo, 1952; Kröner, 1960; Zubov, 1997; Derezin and Zubov, 2011, 1999). Dislocation models are applicable to the description of such phenomena as crystal growth, fatigue, failure, plastic flow, inelasticity, and also other defects of crystalline and nanostructured materials (Clayton, 2011; Clayton et al, 2006; Gutkin and Ovid'ko, 2004; Maugin, 2012; Zhbanova and Zubov, 2016). When there is a lot

Evgeniya V. Goloveshkina · Leonid M. Zubov
Institute of Mathematics, Mechanics and Computer Science of Southern Federal University, Milchakova Str. 8a, 344090 Rostov on Don, Russia,
e-mail: evgeniya.goloveshkina@yandex.ru, zubovl@yandex.ru

B. E. Abali et al. (eds.), *New Achievements in Continuum Mechanics and Thermodynamics*, Advanced Structured Materials 108,
https://doi.org/10.1007/978-3-030-13307-8_11

of dislocations in a bounded volume, it makes sense to go to their continuous distribution. In this case, the continuum dislocation theory is used. In the present paper, in the context of the continuum dislocation theory, the nonlinear eigenstress problem for a hollow elastic sphere is solved.

The general formulation of the nonlinear equilibrium problem for an elastic isotropic sphere with an arbitrary spherically symmetric dislocation distribution was given in Zubov (2014). In a number of special cases, exact spherically symmetric solutions of the nonlinear dislocation theory were found (Zubov, 2014; Zhbanova and Zubov, 2016; Goloveshkina and Zubov, 2018). In Zhbanova and Zubov (2016) within the framework of the harmonic (semi-linear) material model, the exact solution was found for any function characterizing the density of edge dislocations. In particular, the case of dislocations concentrated on a spherical surface inside a body was investigated. It was established that this surface was a surface of discontinuity of strains and stresses. In addition to the eigenstress problem, the problem for a hollow sphere under loading by external or internal hydrostatic pressure was solved in Zhbanova and Zubov (2016). In Zubov (2014) an analytical solution of nonlinear elasticity for a hollow sphere made of incompressible material with distributed screw dislocations of radial direction was obtained. In Goloveshkina and Zubov (2018), for a special distribution of screw and edge dislocations, a solution universal in the class of isotropic incompressible elastic bodies was found. With the help of the solution obtained, the eigenstresses in a solid elastic sphere and in an infinite space with a spherical cavity were determined. The interaction of dislocations with an external hydrostatic loading was also investigated. The dislocation distribution determining the spherically symmetric quasi-solid state of an elastic body characterized by zero stresses and a nonuniform elementary volumes rotation field was found.

In this paper, we investigate the general case of a spherically symmetric dislocation distribution. In this case, the exact solution can not be obtained analytically. Therefore, the nonlinear boundary value problem is solved numerically. In the eigenstress problem for an elastic sphere, we use a special technique that allows one to transform a boundary value problem with nonlinear boundary conditions into a problem with linear ones. This makes the numerical solving the boundary value problem for a nonlinear differential equation remarkably easy to perform. A numerical analysis is carried out for the semi-linear material model and the incompressible Bartenev–Khazanovich material model also known as the Varga model. The solution obtained describes the effect of distributed screw and edge dislocations on large spherically symmetric deformations of an elastic sphere.

11.2 Input Relations

We define the dislocation density as a second-rank tensor field α such that the total Burgers vector of dislocations crossing an arbitrary surface coincides with the flux of the tensor α through this surface (Nye, 1953; Vakulenko, 1991). The dislocation density tensor field must satisfy the solenoidality condition

$$\mathrm{div}\boldsymbol{\alpha} = 0\,. \tag{11.1}$$

Hereinafter, the divergence, rotor, and gradient operators (Lurie, 1990; Lebedev et al, 2010) are written in coordinates of the reference configuration. We introduce the deformation gradient (Lurie, 1990; Lebedev et al, 2010)

$$\mathbf{F} = \mathrm{grad}\mathbf{R}, \tag{11.2}$$

where $\mathbf{R} = X_k \mathbf{i}_k$ is the radius vector of a point of the elastic medium in the deformed configuration, X_k ($k = 1, 2, 3$) are Cartesian coordinates of the body in the final state, $\mathbf{i}_k$ are the fixed coordinate base vectors.

In the presence of dislocations in the body, the vector field $\mathbf{R}$ does not exist and the geometric relations (11.2) are replaced by the tensor incompatibility equation with respect to $\mathbf{F}$:

$$\mathrm{rot}\mathbf{F} = \boldsymbol{\alpha}, \tag{11.3}$$

and the tensor $\mathbf{F}$ is called the distortion tensor.

In the absence of mass forces, the equilibrium equations for an elastic medium (Lurie, 1990; Ogden, 1997) have the form

$$\mathrm{div}\mathbf{D} = 0, \tag{11.4}$$

where $\mathbf{D}$ is the asymmetric Piola stress tensor associated with the distortion tensor $\mathbf{F}$ by the constitutive equations of an elastic material (Lurie, 1990; Truesdell, 1977; Ogden, 1997)

$$\mathbf{D}(\mathbf{F}) = \mathrm{d}W(\mathbf{G})/\mathrm{d}\mathbf{F}\,, \qquad \mathbf{G} = \mathbf{F}\cdot\mathbf{F}^{\mathrm{T}}\,. \tag{11.5}$$

Here, W is the specific energy, $\mathbf{G}$ is the metric tensor (the Cauchy strain measure).

In the finite strain theory, along with the Piola stress tensor $\mathbf{D}$ we use the symmetric Cauchy tensor (Lurie, 1990; Ogden, 1997; Truesdell, 1977)

$$\mathbf{T} = (\mathrm{det}\mathbf{F})^{-1}\mathbf{F}^{\mathrm{T}}\cdot\mathbf{D} \tag{11.6}$$

and the symmetric Kirchhoff stress tensor also called the second Piola–Kirchhoff stress tensor

$$\mathbf{P} = \mathbf{D}\cdot\mathbf{F}^{-1}\,. \tag{11.7}$$

11.3 Spherically Symmetric State

We introduce the spherical coordinates r, φ, θ:

$$x_1 = r\cos\varphi\cos\theta, \qquad x_2 = r\sin\varphi\cos\theta, \qquad x_3 = r\sin\theta,$$

where x_s $(s = 1, 2, 3)$ are the Cartesian coordinates of a sphere in the reference state. Then $\mathbf{e}_r$, $\mathbf{e}_\varphi$, and $\mathbf{e}_\theta$ are the unit vectors tangent to the coordinate lines, forming the basis.

The spherically symmetric dislocation distribution (Zubov, 2014) is represented by the dislocation density tensor

$$\boldsymbol{\alpha} = \alpha_1(r)\mathbf{g} + \alpha_2(r)\mathbf{d} + \alpha_3(r)\mathbf{e}_r \otimes \mathbf{e}_r \,, \tag{11.8}$$
$$\mathbf{g} = \mathbf{e}_\varphi \otimes \mathbf{e}_\varphi + \mathbf{e}_\theta \otimes \mathbf{e}_\theta \,, \quad \mathbf{d} = \mathbf{e}_\varphi \otimes \mathbf{e}_\theta - \mathbf{e}_\theta \otimes \mathbf{e}_\varphi \,.$$

The first and last terms describe the distribution of screw dislocations, while the second one describes the distribution of edge dislocations. Note that the spherical symmetry of the tensor field (11.8) means that at all points of the sphere on a spherical surface $r =$ const, the components of the tensor field in the considered basis are equal. At the same time, the tensor itself is invariant under rotations about the vector $\mathbf{e}_r$, that is, for an arbitrary function $\chi(r)$, the following equality holds

$$\mathbf{Q} \cdot \boldsymbol{\alpha} \cdot \mathbf{Q}^{\mathrm{T}} = \boldsymbol{\alpha}, \quad \mathbf{Q} = \cos\chi(r)\mathbf{g} + \sin\chi(r)\mathbf{d} + \mathbf{e}_r \otimes \mathbf{e}_r \,.$$

By virtue of (11.8), the solenoidality condition (11.1) implies the equation determining the relation between the components α_1 and α_3 of the dislocation density tensor:

$$\alpha_1 = \alpha_3 + \frac{1}{2} r\alpha_3', \tag{11.9}$$

where $'$ denotes the derivative with respect to the radial coordinate. In the following, the scalar dislocation densities α_2 and α_3 are assumed to be the given functions of the radial coordinate r.

According to (Zubov, 2014), for an isotropic material the distortion tensor as well as the stress tensor are found in the form analogous to the dislocation density tensor:

$$\mathbf{F} = F_1(r)\mathbf{g} + F_2(r)\mathbf{d} + F_3(r)\mathbf{e}_r \otimes \mathbf{e}_r, \tag{11.10}$$
$$\mathbf{D} = D_1(r)\mathbf{g} + D_2(r)\mathbf{d} + D_3(r)\mathbf{e}_r \otimes \mathbf{e}_r \,. \tag{11.11}$$

Taking into account (11.8) and (11.10), the incompatibility equation (11.3) is reduced to three scalar equations

$$(rF_2)' = r\alpha_1, \quad F_2 = \frac{r\alpha_3}{2}, \quad F_3 = (rF_1)' + r\alpha_2, \tag{11.12}$$

and the equilibrium equations (11.4) due to (11.11) reduce to a single equation

$$\frac{\mathrm{d}D_3}{\mathrm{d}r} + \frac{2(D_3 - D_1)}{r} = 0 \,. \tag{11.13}$$

If the prescribed hydrostatic pressure q_0 acts on the outer surface of the sphere $r = r_0$, and the pressure q_1 acts on the inner surface $r = r_1$, then the boundary conditions for the equation (11.13) will be

$$D_3(r_i) = -q_i \left[F_1^2(r_i) + \frac{1}{4}\alpha_3^2 r_i^2\right], \quad i = 0, 1\,.$$

In the eigenstress problem, i. e. in the absence of external loads, the boundary conditions become

$$D_3(r_1) = 0, \quad D_3(r_0) = 0\,. \tag{11.14}$$

We compute the determinant of the distortion tensor:

$$\det\mathbf{F} = F_3\left(F_1^2 + \frac{1}{4}r^2\alpha_3^2\right)\,. \tag{11.15}$$

For physically realizable deformation, it is necessary that $\det\mathbf{F} > 0$. Therefore, from (11.15) it follows that $F_3 > 0$.

The polar decomposition of the distortion tensor has the form $\mathbf{F} = \mathbf{U}\cdot\mathbf{A}$, where the positive definite stretch tensor $\mathbf{U}$ and the proper orthogonal rotation tensor $\mathbf{A}$ in compliance with (11.10) are determined by the formulas

$$\mathbf{U} = \mathbf{G}^{1/2} = \sqrt{F_1^2 + \frac{1}{4}r^2\alpha_3^2}\,\mathbf{g} + F_3\mathbf{e}_r\otimes\mathbf{e}_r, \tag{11.16}$$

$$\mathbf{A} = \mathbf{U}^{-1}\cdot\mathbf{F} = \cos\psi(r)\mathbf{g} + \sin\psi(r)\mathbf{d} + \mathbf{e}_r\otimes\mathbf{e}_r\,. \tag{11.17}$$

Here,

$$\cos\psi = \frac{F_1}{\sqrt{F_1^2 + \frac{1}{4}r^2\alpha_3^2}}, \qquad \sin\psi = \frac{r\alpha_3}{2\sqrt{F_1^2 + \frac{1}{4}r^2\alpha_3^2}}\,. \tag{11.18}$$

From the representation (11.17) one can see that the orthogonal tensor $\mathbf{A}$ describes a rotation through an angle ψ around the vector $\mathbf{e}_r$.

Given (11.10), we find the inverse distortion tensor $\mathbf{F}^{-1}$ and the metric tensor $\mathbf{G}$:

$$\mathbf{F}^{-1} = \left(F_1^2 + \frac{1}{4}r^2\alpha_3^2\right)^{-1}\left(F_1\mathbf{g} - \frac{1}{2}r\alpha_3F_2\mathbf{d}\right) + F_3^{-1}\mathbf{e}_r\otimes\mathbf{e}_r, \tag{11.19}$$

$$\mathbf{G} = \mathbf{F}\cdot\mathbf{F}^{\mathrm{T}} = \left(F_1^2 + \frac{r^2\alpha_3^2}{4}\right)\mathbf{g} + F_3^2\mathbf{e}_r\otimes\mathbf{e}_r\,. \tag{11.20}$$

The invariants of the tensor $\mathbf{G}$ for spherically symmetric deformation are expressed as follows:

$$\begin{aligned} I_1 &= \mathrm{tr}\mathbf{G} = 2\left(F_1^2 + \frac{1}{4}r^2\alpha_3^2\right) + F_3^2, \\ I_2 &= \frac{1}{2}\left(\mathrm{tr}^2\mathbf{G} - \mathrm{tr}\mathbf{G}^2\right) = \left(F_1^2 + \frac{1}{4}r^2\alpha_3^2\right)^2 + 2F_3^2\left(F_1^2 + \frac{1}{4}r^2\alpha_3^2\right), \\ I_3 &= \det\mathbf{G} = F_3^2\left(F_1^2 + \frac{1}{4}r^2\alpha_3^2\right)^2\,. \end{aligned} \tag{11.21}$$

For an arbitrary isotropic elastic material, the constitutive equations are representable in the form (Lurie, 1990; Truesdell, 1977; Ogden, 1997)

$$\mathbf{D} = (\tau_1 + I_1\tau_2)\mathbf{F} - \tau_2\mathbf{G}\cdot\mathbf{F} + I_3\tau_3\mathbf{F}^{-\mathrm{T}}, \quad \tau_k = 2\frac{\partial W(I_1, I_2, I_3)}{\partial I_k}, \quad k = 1,2,3. \tag{11.22}$$

Here, τ_k are the material response functions dependent on the strain measure invariants.

Substituting (11.10) with account of (11.12) as well as (11.19) and (11.20) into the constitutive equations (11.22), we obtain the following representations of the Piola stress tensor components:

$$\begin{aligned}
D_1 &= (\tau_1 + I_1\tau_2)F_1 - \tau_2 F_1\left(F_1^2 + \frac{1}{4}r^2\alpha_3^2\right) + I_3\tau_3 F_1\left(F_1^2 + \frac{1}{4}r^2\alpha_3^2\right)^{-1}, \\
D_2 &= (\tau_1 + I_1\tau_2)\frac{r\alpha_3}{2} - \frac{1}{2}r\alpha_3\tau_2\left(F_1^2 + \frac{1}{4}r^2\alpha_3^2\right) + \frac{1}{2}r\alpha_3 I_3\tau_3\left(F_1^2 + \frac{1}{4}r^2\alpha_3^2\right)^{-1}, \\
D_3 &= (\tau_1 + I_1\tau_2)F_3 - \tau_2 F_3^3 + I_3\tau_3 F_3^{-1}.
\end{aligned} \tag{11.23}$$

Since the stresses (11.23) are defined through the invariants (11.21) dependent on F_1 and F_3, then taking into account the relationship between the tensor $\mathbf{F}$ components (11.12), the stresses are expressed in terms of one distortion component F_1. Thus, for any isotropic material, the boundary value problem consists of the equilibrium equation (11.13), which is a second-order nonlinear ordinary differential equation with respect to the function $F_1(r)$, and the nonlinear boundary conditions (11.14).

As an example, we write this equation explicitly for a semi-linear (harmonic) material having the following constitutive equations (Lurie, 1990; Ogden, 1997; John, 1960):

$$\mathbf{D} = \frac{2\mu}{1-2\nu}\left(\nu \mathrm{tr}\mathbf{U} - 1 - \nu\right)\mathbf{A} + 2\mu\mathbf{F}, \tag{11.24}$$

where μ and ν are the material constants. In the small strain region, the semi-linear material follows Hooke's law with the shear modulus μ and the Poisson's ratio ν. The differential equation with respect to the distortion $F_1(r)$ for the material is written as follows:

$$\begin{aligned}
F_1'' &= \frac{2(2-3\nu)}{(\nu-1)r}F_1' + \frac{(\alpha_2 + r\alpha_2')(1-\nu) + 2\alpha_2(1-2\nu)}{(\nu-1)r} \\
&+ \frac{2\nu}{(\nu-1)r}\left(F_1^2 + \frac{r^2\alpha_3^2}{4}\right)^{-1/2}\left(F_1F_1' + \frac{r\alpha_3(\alpha_3 + r\alpha_3')}{4}\right) \\
&+ \frac{2}{(\nu-1)r^2}\left[2\nu\sqrt{F_1^2 + \frac{r^2\alpha_3^2}{4}} + \nu(F_1 + rF_1' + r\alpha_2) - 1 - \nu\right] \\
&\times\left[1 - F_1\left(F_1^2 + \frac{r^2\alpha_3^2}{4}\right)^{-1/2}\right].
\end{aligned} \tag{11.25}$$

The constitutive equations in terms of the Kirchhoff stress tensor for any, including an anisotropic, elastic body have the form:

$$\mathbf{P} = 2\mathrm{d}W(\mathbf{G})/\mathrm{d}\mathbf{G}\,. \tag{11.26}$$

The semi-linear material belongs to models of an elastic medium, the specific energy of which is given as a function of the stretch tensor $\mathbf{U}$ and not the metric tensor $\mathbf{G}$. In this case, the symmetric Biot stress tensor is convenient to use:

$$\mathbf{S} = \mathrm{d}W/\mathrm{d}\mathbf{U}\,. \tag{11.27}$$

From (11.26) and (11.27), we obtain the formulas connecting the Biot stress tensor with the Kirchhoff and Piola stress tensors

$$\mathbf{S} = \frac{1}{2}(\mathbf{P}\cdot\mathbf{U} + \mathbf{U}\cdot\mathbf{P}) = \frac{1}{2}\left(\mathbf{D}\cdot\mathbf{A}^{\mathrm{T}} + \mathbf{A}\cdot\mathbf{D}^{\mathrm{T}}\right). \tag{11.28}$$

If the material is isotropic, then the specific energy depends on three invariants of the stretch tensor, i.e. $W = W(J_1, J_2, J_3)$, where $J_1 = \mathrm{tr}\mathbf{U}$, $J_2 = \frac{1}{2}\left(\mathrm{tr}^2\mathbf{U} - \mathrm{tr}\mathbf{U}^2\right)$, $J_3 = \det\mathbf{U}$. Consequently, the tensor $\mathbf{S}$ can be rewritten as

$$\mathbf{S} = \left(\frac{\partial W}{\partial J_1} + J_1\frac{\partial W}{\partial J_2}\right)\mathbf{I} - \frac{\partial W}{\partial J_2}\mathbf{U} + J_3\frac{\partial W}{\partial J_3}\mathbf{U}^{-1}\,, \tag{11.29}$$

where $\mathbf{I}$ is the unit tensor.

In an isotropic body, the tensors $\mathbf{P}$ and $\mathbf{U}$ are coaxial and therefore commute: $\mathbf{P}\cdot\mathbf{U} = \mathbf{U}\cdot\mathbf{P}$. Then, $\mathbf{S} = \mathbf{P}\cdot\mathbf{U} = \mathbf{U}\cdot\mathbf{P}$ and $\mathbf{S} = \mathbf{D}\cdot\mathbf{A}^{\mathrm{T}}$.

Constitutive equations in terms of the Piola tensor for the material with the specific energy $W = W(J_1, J_2, J_3)$ due to (11.29) will be

$$\mathbf{D} - (\eta_1 + J_1\eta_2)\,\mathbf{A} - \eta_2\mathbf{F} + J_3\eta_3\mathbf{F}^{-\mathrm{T}}, \quad \eta_k - \frac{\partial W}{\partial J_k}\,. \tag{11.30}$$

In the spherically symmetric problem, the invariants J_1, J_2, J_3 are expressed in F_1, F_2, F_3 using formulas

$$J_1 = 2\sqrt{F_1^2 + \frac{1}{4}r^2\alpha_3^2} + F_3,$$
$$J_2 = F_1^2 + \frac{r^2\alpha_3^2}{4} + 2F_3\sqrt{F_1^2 + \frac{1}{4}r^2\alpha_3^2},$$
$$J_3 = F_3\left(F_1^2 + \frac{1}{4}r^2\alpha_3^2\right).$$

Taking into account (11.10), (11.19), and (11.17), the components of the Piola tensor are written in the form

$$D_1 = (\eta_1 + J_1\eta_2)\left(F_1^2 + \frac{1}{4}r^2\alpha_3^2\right)^{-1/2} F_1 - \eta_2 F_1 + J_3\eta_3\left(F_1^2 + \frac{1}{4}r^2\alpha_3^2\right)^{-1} F_1,$$

$$D_2 = \frac{1}{2}(\eta_1 + J_1\eta_2)\left(F_1^2 + \frac{1}{4}r^2\alpha_3^2\right)^{-1/2} r\alpha_3 - \eta_2 F_2 + \frac{J_3\eta_3 r\alpha_3}{2}\left(F_1^2 + \frac{1}{4}r^2\alpha_3^2\right)^{-1}, \qquad (11.31)$$

$$D_3 = \eta_1 + J_1\eta_2 - \eta_2 F_3 + J_3\eta_3 F_3^{-1}.$$

In the case of an incompressible material, the condition det$\mathbf{F} = I_3 = J_3 = 1$ is satisfied. Therefore, for the material, the constitutive equations (11.22) are modified as follows:

$$\mathbf{D} = (\tau_1 + I_1\tau_2)\mathbf{F} - \tau_2\mathbf{G}\cdot\mathbf{F} - p\mathbf{F}^{-\mathrm{T}}, \qquad (11.32)$$

where p is a pressure in an incompressible body not expressed in terms of strain. Given the incompressibility property, the constitutive equations (11.29) and (11.30) are reduced to

$$\mathbf{S} = (\eta_1 + J_1\eta_2)\,\mathbf{I} - \eta_2\mathbf{U} - p\mathbf{U}^{-1},$$
$$\mathbf{D} = (\eta_1 + J_1\eta_2)\,\mathbf{A} - \eta_2\mathbf{F} - p\mathbf{F}^{-\mathrm{T}}.$$

11.4 Transformation of the Boundary Value Problem

Since the boundary conditions (11.14) of the eigenstress problem are represented by the stress constraints, then, taking into account their expressions (11.31) in terms of the distortion, we finally obtain boundary conditions on the distortion. They represent a nonlinear relation with respect to the function F_1 and its derivative F_1'. To obtain a boundary value problem with linear boundary conditions, it is necessary to replace the unknown function. Instead of the equation with respect to $F_1(r)$, we derive a system of equations with respect to $D_2(r)$ and $D_3(r)$. For this, we need to solve the problem of inversion of the Piola stress tensor as a function of the distortion tensor: $\mathbf{D} = h(\mathbf{F})$, that is, find the tensor function H, inverse to the function h: $\mathbf{F} = H(\mathbf{D})$. The way to solve the problem for an isotropic material is indicated in Zubov (1976) and consists of the following. First, a more simple problem of inversion of the dependence $\mathbf{S} = l(\mathbf{U})$ between the symmetric tensors is solved, i. e. the function L such that $\mathbf{U} = L(\mathbf{S})$ is founded. Further we have

$$\mathbf{F} = \mathbf{U}\cdot\mathbf{A} = L(\mathbf{S})\cdot\mathbf{A} = L\left(\mathbf{D}\cdot\mathbf{A}^{\mathrm{T}}\right)\cdot\mathbf{A}.$$

The problem of constructing the function $\mathbf{F} = H(\mathbf{D})$ will be solved if we express the rotation tensor $\mathbf{A}$ in terms of the Piola stress tensor $\mathbf{D}$. This can be done by solving the equation with respect to $\mathbf{A}$, expressing the symmetry property of the Biot stress tensor

$$\mathbf{D}\cdot\mathbf{A}^{\mathrm{T}} = \mathbf{A}\cdot\mathbf{D}^{\mathrm{T}}. \qquad (11.33)$$

In the spherically symmetric problem considered here, the last equation in view of (11.11) and (11.17) is equivalent to one scalar relation

$$D_1 \sin\psi = D_2 \cos\psi\,. \tag{11.34}$$

This equation has two solutions:

$$\cos\psi = \sqrt{\frac{D_1^2}{D_1^2 + D_2^2}}, \quad \sin\psi = \frac{D_2}{D_1}\sqrt{\frac{D_1^2}{D_1^2 + D_2^2}} \tag{11.35}$$

and

$$\cos\psi = -\sqrt{\frac{D_1^2}{D_1^2 + D_2^2}}, \quad \sin\psi = -\frac{D_2}{D_1}\sqrt{\frac{D_1^2}{D_1^2 + D_2^2}}\,. \tag{11.36}$$

If we assume that $-\pi \le \psi \le \pi$, then the first solution is described by the inequalities

$$-\frac{\pi}{2} \le \psi \le \frac{\pi}{2},$$

and the second by the inequalities

$$-\pi \le \psi \le -\frac{\pi}{2}, \quad \frac{\pi}{2} \le \psi \le \pi\,.$$

From the formula (11.18), it is clear that the first solution corresponds to the positive F_1, and the second to the negative. As shown in Zhbanova and Zubov (2016), F_1 is negative when the eversion deformation of a sphere occurs (Zubov and Moiseyenko, 1983), and positive in case of spherically symmetric deformation of a sphere without eversion. Consequently, the second solution corresponds to the eigenstress problem for the everted sphere with distributed dislocations. We note that in the absence of dislocations, in a sphere without eversion the stresses are identically equal to zero, while in an everted sphere the stresses are not zero due to eversion.

11.5 Problem for Semi-linear Material

Given the constitutive equations of the semi-linear material (11.24), the tensor $\mathbf{S}$ is written as

$$\mathbf{S} = \frac{2\mu\nu}{1-2\nu}\mathbf{I}\mathrm{tr}(\mathbf{U}-\mathbf{I}) + 2\mu(\mathbf{U}-\mathbf{I}). \tag{11.37}$$

We invert the expression (11.37):

$$\mathbf{U} = \mathbf{I} + \frac{1}{2\mu}\left(\mathbf{S} - \frac{\nu}{1+\nu}\mathbf{I}\mathrm{tr}\mathbf{S}\right)\,. \tag{11.38}$$

Let us find the distortion tensor $\mathbf{F}$ by the formula

$$\mathbf{F} = \mathbf{U} \cdot \mathbf{A}\,. \tag{11.39}$$

Taking into account (11.38) we have

$$\mathbf{F} = \mathbf{A} + \frac{1}{2\mu}\left[\mathbf{D} - \frac{\nu}{1+\nu}\mathbf{A}\mathrm{tr}\left(\mathbf{D}\cdot\mathbf{A}^{\mathrm{T}}\right)\right]. \tag{11.40}$$

Based on (11.11), (11.17), and (11.40), we finally obtain the tensor $\mathbf{F}$, expressed in terms of the Piola stresses:

$$\mathbf{F} = \frac{1}{2\mu}\left[(D_1 + K\cos\psi)\,\mathbf{g} + (D_2 + K\sin\psi)\,\mathbf{d} + (D_3 + K)\mathbf{e}_r \otimes \mathbf{e}_r\right], \tag{11.41}$$
$$K = \frac{2\mu(1+\nu) - \nu\left[2\left(D_1\cos\psi + D_2\sin\psi\right) + D_3\right]}{1+\nu},$$

where $\cos\psi$ and $\sin\psi$ are computed from the formulas (11.35) or (11.36).

The tensor $\mathbf{D}$ components are calculated numerically from the system of equations (11.9), (11.12), (11.13), and (11.41) with the boundary conditions (11.14). Knowing the Piola stresses, we find the distortion by the formulas (11.41).

11.6 Problem for Incompressible Material

We consider the problem for incompressible material using the Bartenev–Khazanovich model (Lurie, 1990) as an example. The equations of state of the material have the form:

$$\mathbf{D} = 2\mu\mathbf{A} - p\mathbf{F}^{-\mathrm{T}},$$

where p is the internal pressure in an incompressible body. Then the tensor $\mathbf{S}$ by virtue of (11.39) is written as

$$\mathbf{S} = 2\mu\mathbf{I} - p\mathbf{U}^{-1}\,. \tag{11.42}$$

Then, from the incompressibility condition $\det\mathbf{U} = 1$ we find

$$p = \sqrt[3]{\det\left(2\mu\mathbf{I} - \mathbf{S}\right)}\,. \tag{11.43}$$

Let us invert (11.42) by expressing $\mathbf{U}$ and substituting (11.43):

$$\mathbf{U} = \sqrt[3]{\det\left(2\mu\mathbf{I} - \mathbf{S}\right)}\left(2\mu\mathbf{I} - \mathbf{S}\right)^{-1}. \tag{11.44}$$

According to (11.17) and (11.44) we derive the tensor (11.39) in the form

$$\begin{aligned}
\mathbf{F} &= \frac{\sqrt[3]{2\mu - D_3}(B\cos\psi - C\sin\psi)}{A^{2/3}}\mathbf{g} \\
&+ \frac{\sqrt[3]{2\mu - D_3}(B\sin\psi + C\cos\psi)}{A^{2/3}}\mathbf{d} + \frac{\sqrt[3]{A}}{(2\mu - D_3)^{2/3}}\mathbf{e}_r \otimes \mathbf{e}_r, \qquad (11.45) \\
A &= 4\mu^2 + D_1^2 + D_2^2 - 4\mu(D_1\cos\psi + D_2\sin\psi), \\
B &= 2\mu - D_1\cos\psi - D_2\sin\psi, \quad C = D_2\cos\psi - D_1\sin\psi,
\end{aligned}$$

with $\cos\psi$ and $\sin\psi$ computed from (11.35) or (11.36).

From the system of equations (11.9), (11.12), (11.13), and (11.45) with the boundary conditions (11.14), we find the numerical solution of the problem by calculating the stresses and strains.

11.7 Numerical Results

For numerical calculations within the framework of the semi-linear material, we choose the dislocation distribution

$$\alpha_1 = \frac{\gamma_0}{r}, \quad \alpha_2 = \frac{\beta_0}{r}, \quad \alpha_3 = \frac{2\gamma_0}{r},$$

and for the incompressible material

$$\alpha_3 = \frac{2\gamma_0}{r^2}, \quad \alpha_1 = \alpha_2 = 0,$$

where β_0 and γ_0 are some constants. The outer radius of the sphere is considered to be equal to one ($r_0 = 1$), which is equivalent to introducing a dimensionless radial coordinate. The following numerical results correspond to the value $r_1 = 0.5$. Similarly, assuming $\mu = 1$, we deal with the dimensionless stresses. For the dimensionless constant ν we take the value $\nu = 0.3$.

According to (11.34), the problem has two solutions since $\cos\psi$ can be positive (11.35) or negative (11.36). For the incompressible material, the numerical results are displayed in the case $\cos\psi > 0$ (Figs. 11.11–11.16), and for the semi-linear material, in the case $\cos\psi > 0$ (Figs. 11.1–11.5) as well as $\cos\psi < 0$ (Figs. 11.6–11.10).

It is established that for both material models, the stresses D_1 and D_2 in absolute value are maximal on the inner surface of the sphere, and D_3 on the surface close to the inner surface. For the incompressible material, for $\cos\psi > 0$ the maximum stress D_1 is an order of magnitude higher than the maximum stresses D_2 and D_3.

Different curves in each figure illustrate the influence of the dislocation intensity on a stress-strain state. Thus, for the semi-linear material, the stresses D_1 and D_3 decrease and become more uniformly distributed over the thickness of the sphere (Figs. 11.1, 11.3, 11.6, and 11.8), and the stress D_2 increases and its distribution becomes less uniform (Figs. 11.2 and 11.7). In addition, there are spherical sur-

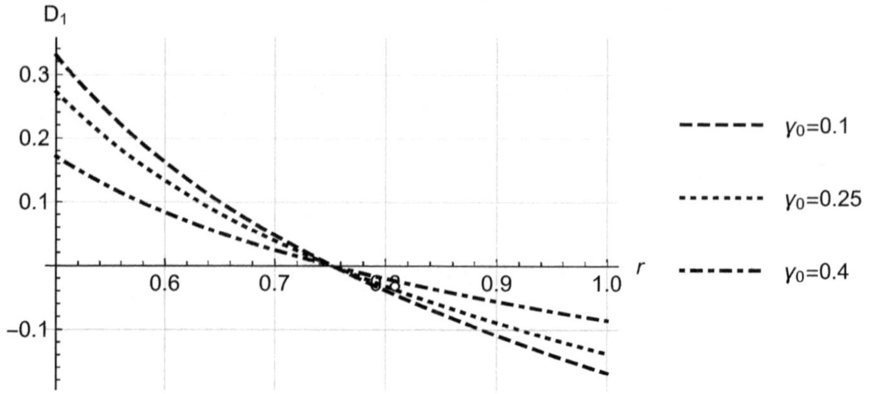

Fig. 11.1 Semi-linear material, $\cos\psi > 0$, $\beta_0 = 0.2$, stress D_1

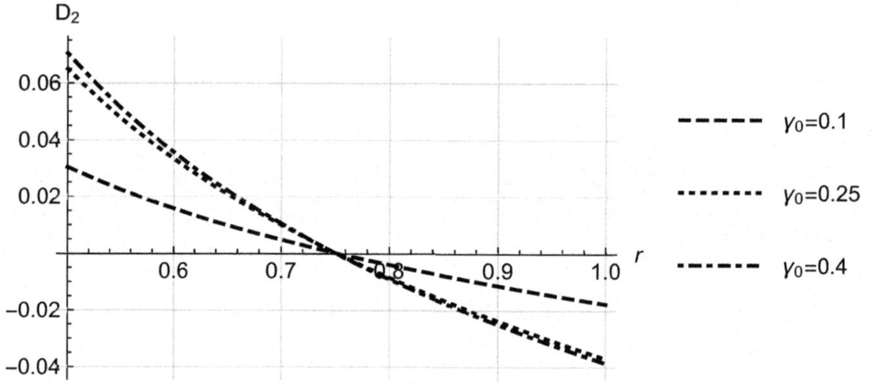

Fig. 11.2 Semi-linear material, $\cos\psi > 0$, $\beta_0 = 0.2$, stress D_2

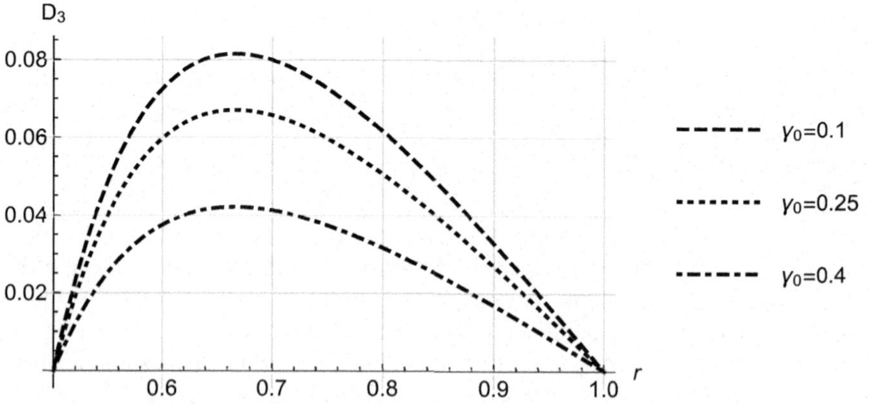

Fig. 11.3 Semi-linear material, $\cos\psi > 0$, $\beta_0 = 0.2$, stress D_3

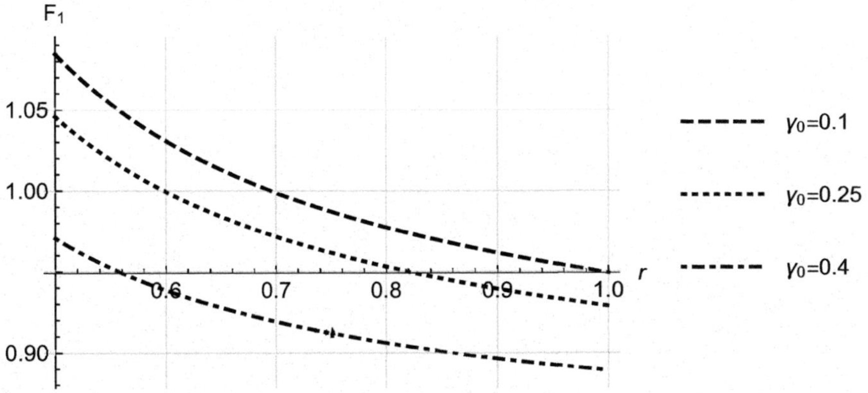

Fig. 11.4 Semi-linear material, $\cos\psi > 0$, $\beta_0 = 0.2$, distortion F_1

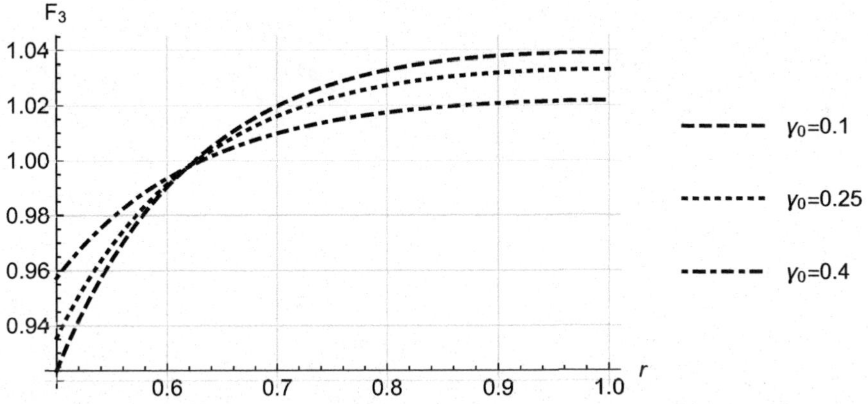

Fig. 11.5 Semi-linear material, $\cos\psi > 0$, $\beta_0 = 0.2$, distortion F_3

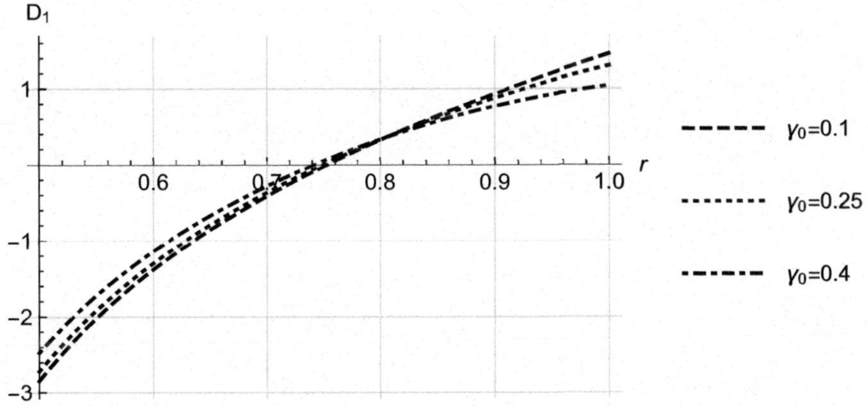

Fig. 11.6 Semi-linear material, $\cos\psi < 0$, $\beta_0 = 0.2$, stress D_1

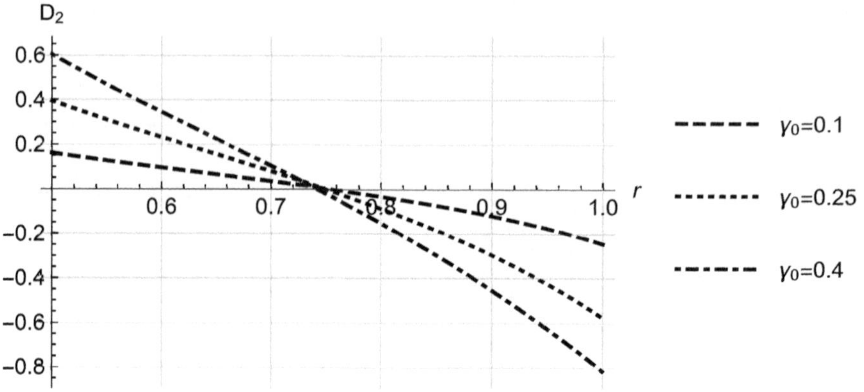

Fig. 11.7 Semi-linear material, $\cos\psi < 0$, $\beta_0 = 0.2$, stress D_2

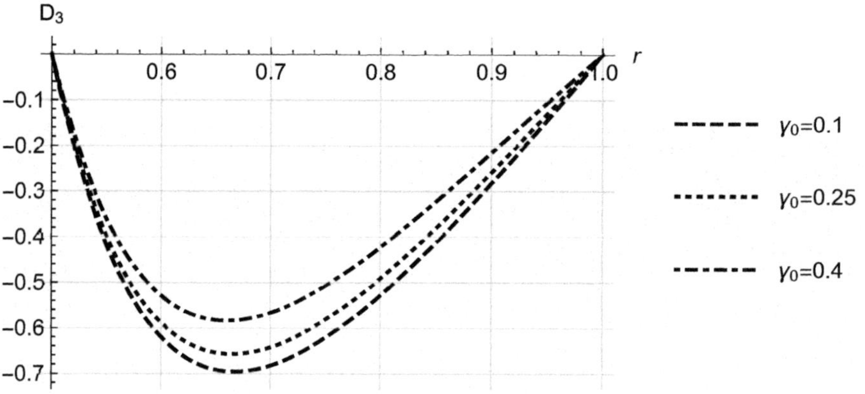

Fig. 11.8 Semi-linear material, $\cos\psi < 0$, $\beta_0 = 0.2$, stress D_3

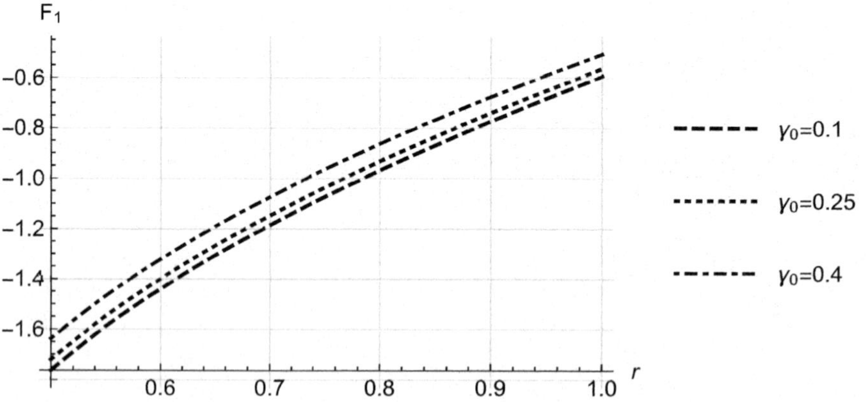

Fig. 11.9 Semi-linear material, $\cos\psi < 0$, $\beta_0 = 0.2$, distortion F_1

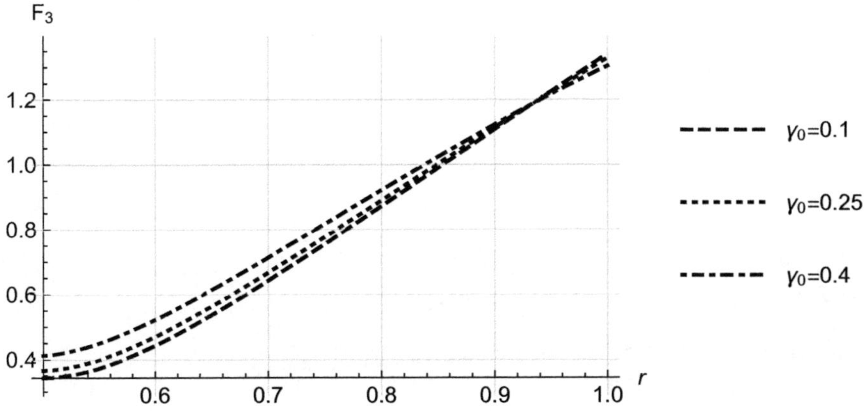

Fig. 11.10 Semi-linear material, $\cos\psi < 0$, $\beta_0 = 0.2$, distortion F_3

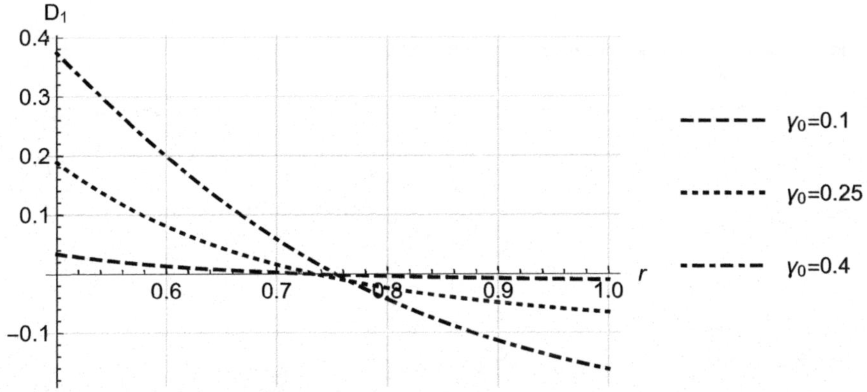

Fig. 11.11 Incompressible material, $\gamma_0 = 0.1$, stress D_1

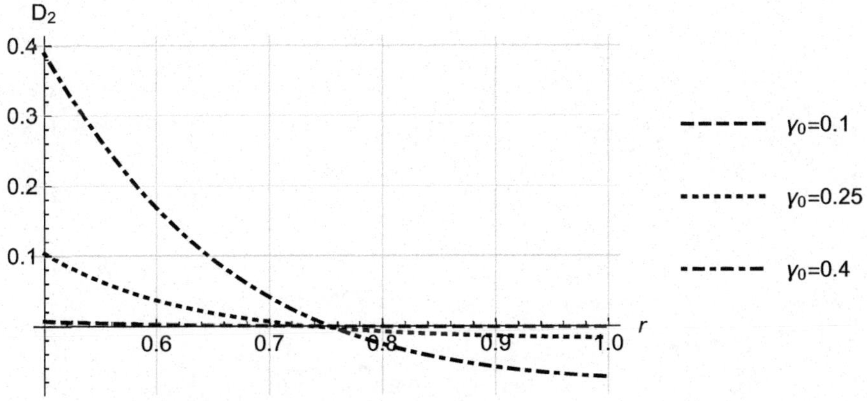

Fig. 11.12 Incompressible material, $\gamma_0 = 0.1$, stress D_2

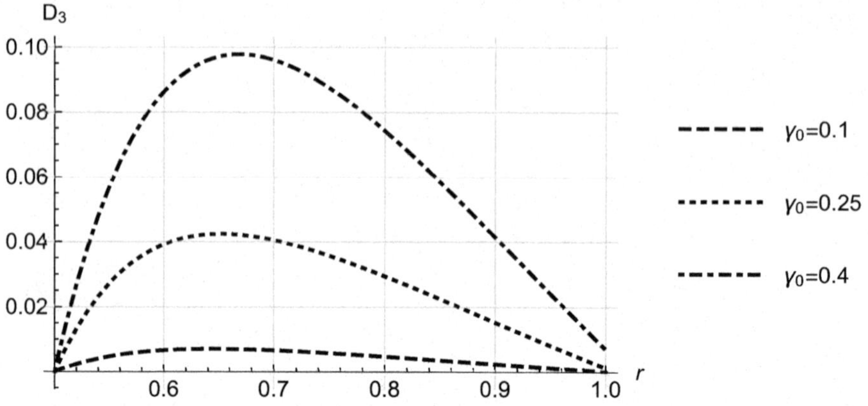

Fig. 11.13 Incompressible material, $\cos\psi > 0$, stress D_3

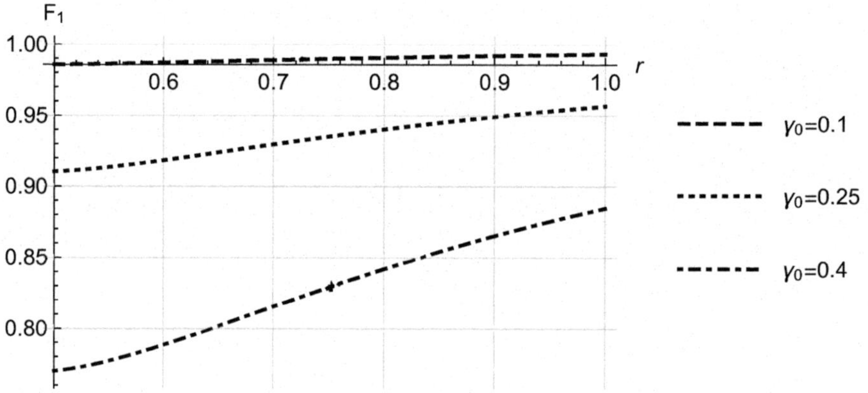

Fig. 11.14 Incompressible material, $\cos\psi > 0$, distortion F_1

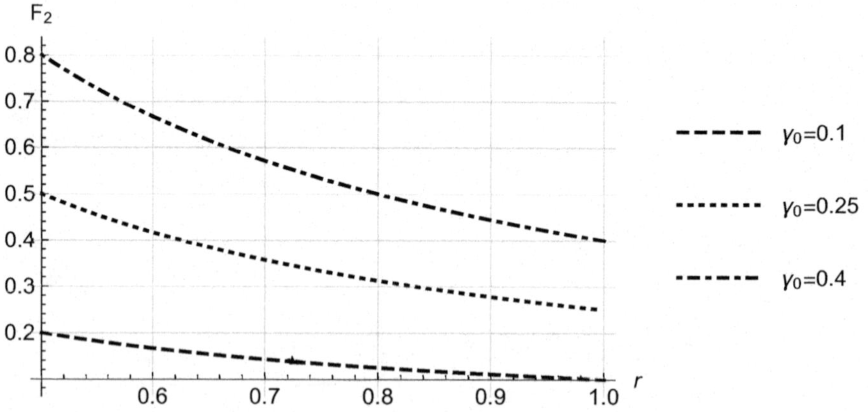

Fig. 11.15 Incompressible material, $\cos\psi > 0$, distortion F_2

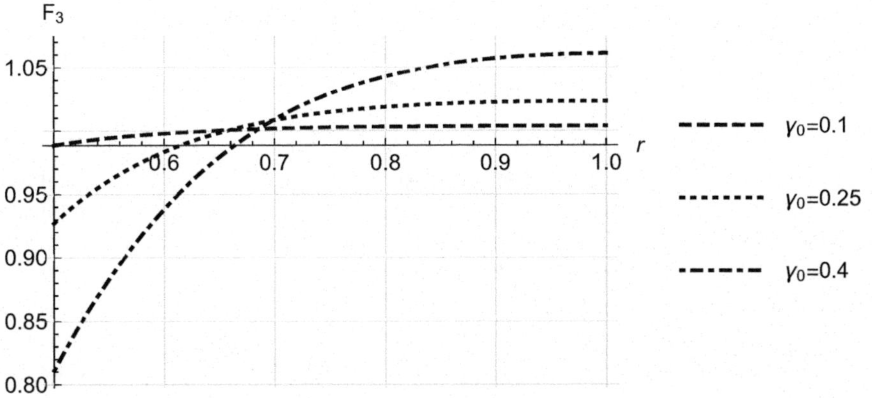

Fig. 11.16 Incompressible material, $\cos\psi > 0$, distortion F_3

faces on which the stress D_1 or D_2 does not depend on the dislocation density. For $\cos\psi < 0$, this kind of surface is located in the middle between the inner and outer surfaces of the sphere for D_2 (Fig. 11.7) and near the outer surface for D_1 (Fig. 11.6). In the case $\cos\psi > 0$, for D_1 and D_2 these surfaces coincide and are located in the middle (Figs. 11.1 and 11.2). Moreover, on said surfaces these stresses are zero. For the distortion F_3 there also exists a surface $r = r_*$ on which $F_3(r_*)$ does not depend on the dislocation density, with F_3 increasing before the surface $r = r_*$ and decreasing after it as moving from the inner surface of the sphere to the outer one (Figs. 11.5 and 11.10). In the sphere without eversion, the surface considered is near the inner surface, and in the everted sphere near the outside one. In both cases, the distortion F_2 due to (11.12) and (11.7) is a constant value. The distortion F_1 decreases in absolute value with increasing dislocation density (Figs. 11.4 and 11.9).

For the incompressible material, the stresses D_1 and D_2, which are approximately equal, also do not depend on the dislocation density on a certain spherical surface. Moreover, on this surface they vanish (Figs. 11.11 and 11.12). At different dislocation densities, on another spherical surface closer to the inner surface of the sphere, the distortion F_3 is about the same. With increasing the dislocations, when moving away from the inner surface of the sphere, F_3 decreases before the considered surface and increases after it (Fig. 11.16). The higher the dislocation density the higher the Piola stresses and the less uniformly the distribution of latter. With that the distortion F_1 decreases (Fig. 11.14), and F_2 increases (Fig. 11.15).

11.8 Conclusion

In the present paper, we have considered the problem of the nonlinear continuum dislocation theory for an elastic hollow sphere for an arbitrary spherically symmetric distribution of screw and edge dislocations. The system of solving equations consists of the equilibrium equations, the incompatibility equations, and the constitutive equations of the elastic medium. Using the properties of spherically symmetric tensor fields, for a general isotropic material we have reduced the original problem to a nonlinear boundary value problem for an ordinary second-order differential equation with respect to one component of the distortion tensor. This equation is obtained in two cases: the specific energy of the material is a function of the metric tensor invariants and a function of the stretch tensor invariants. The boundary conditions for a one-dimensional boundary value problem with respect to the distortion are nonlinear. To simplify the numerical solution of this problem, we have transformed it to a boundary value problem with the linear boundary conditions. The unknown functions of the radial coordinate in the transformed problem are the components of the Piola stress tensor.

We have established that the eigenstress problem for a hollow sphere always has two spherically symmetric solutions, one of which describes the equilibrium of an everted hollow sphere with dislocations.

For two specific models of an elastic medium: the compressible semi-linear material and the incompressible Bartenev–Khazanovich material, we have constructed a numerical solution of a one-dimensional boundary value problem, describing the eigenstresses due to given densities of screw and edge dislocations. Based on the obtained numerical results, we have analyzed the effect of the intensity of the dislocation distribution and its behavior on a stress state of the sphere at large deformations.

Acknowledgements The authors acknowledge support by the Russian Science Foundation (18-11-00069).

References

Bilby BA, Bullough R, Smith E (1955) Continuous distributions of dislocations: a new application of the methods of non-Riemannian geometry. Proceedings of the Royal Society of London A: Mathematical, Physical and Engineering Sciences A231:263–273

Clayton JD (2011) Nonlinear Mechanics of Crystals. Springer, Dordrecht

Clayton JD, McDowell DL, Bammann DJ (2006) Modeling dislocations and disclinations with finite micropolar elastoplasticity. Int J Plast 22(2):210–256

Derezin SV, Zubov LM (1999) Equations of a nonlinear elastic medium with continuously distributed dislocations and disclinations. Doklady Physics **44**(6):391–394

Derezin SV, Zubov LM (2011) Disclinations in nonlinear elasticity. Ztsch Angew Math und Mech 91:433–442

Goloveshkina E, Zubov LM (2018) Universal spherically symmetric solution of nonlinear dislocation theory for incompressible isotropic elastic medium. Archive of Applied Mechanics Https://doi.org/10.1007/s00419-018-1403-9

Gutkin MY, Ovid'ko IA (2004) Plastic Deformation in Nanocrystalline Materials. Springer, Berlin

John F (1960) Plane strain problems for a perfectly elastic material of harmonic type. Commun Pure Appl Math XIII:239–296

Kondo K (1952) On the geometrical and physical foundations in the theory of yielding. In: Proc. 2nd Jap. Nat. Congress of Appl. Mechanics, Tokyo, pp 41–47

Kröner E (1960) Allgemeine Kontinuumstheorie der Versetzungen und Eigenspannungen. Arch Ration Mech Anal 4:273–334

Lebedev LP, Cloud MJ, Eremeyev VA (2010) Tensor Analysis with Applications in Mechanics. World Scientific, New Jersey

Lurie AI (1990) Nonlinear Theory of Elasticity. North-Holland, Amsterdam

Maugin GA (2012) Defects, dislocations and the general theory of material inhomogeneity. In: Sansour C, Skatulla S (eds) Generalized continua and dislocation theory. CISM courses and lectures, vol 537, Springer, Vienna, pp 1–83

Nye JF (1953) Some geometrical relations in dislocated crystals. Acta Metall 1(2):153–162

Ogden RW (1997) Non-linear Elastic Deformations. Dover, New York

Truesdell C (1977) A First Course in Rational Continuum Mechanics. Academic Press, New York

Vakulenko AA (1991) The relationship of micro- and macroproperties in elastic-plastic media (in Russian). Itogi Nauki Tekh, Ser: Mekh Deform Tverd Tela 22(3):3–54

Zhbanova EV, Zubov LM (2016) The influence of distributed dislocations on large deformations of an elastic sphere. In: Naumenko K, Aßmus M (eds) Advanced Methods of Continuum Mechanics for Materials and Structures, Advanced Structured Materials, vol 60, Springer, Singapore, pp 61–76

Zubov LM (1976) On the representation of the displacement gradient of an isotropic elastic body through the Piola stress tensor. PMM 40(6):1070–1077

Zubov LM (1997) Nonlinear Theory of Dislocations and Disclinations in Elastic Bodies. Springer, Berlin

Zubov LM (2014) Spherically symmetric solutions in the nonlinear theory of dislocations. Doklady Physics 59(9):419–422

Zubov LM, Moiseyenko SI (1983) Stability of equilibrium of an elastic sphere turned inside out. Izv Akad Nauk SSSR Mekh Tverd Tela 5:148–155

Chapter 12
Fundamental Solution for the Generalized Plane Stress of a Nanoplate

Mikhail A. Grekov

Abstract The fundamental solution for the generalized plane-stress problem of an infinite, isotropic elastic plate subjected to a point force is presented taking into account surface stresses in the plate faces. Constitutive equations is derived using the stress-strain relations for the bulk material and Gurtin–Murdoch's linearized surface elasticity equations for the surfaces of the plate supposing that the residual surface stress is negligibly small compared with the surface elasticity parameters. The complex relations (Green functions) for the stresses and displacements in the explicit form are evaluated using Goursat–Kolosov complex potentials and Muskhelishvili representations. It is shown that in the case of the generalized plane stress, the fundamental solution depends on the thickness of the plate that is the size effect intrinsic to the nanoobjects.

Keywords: Generalized plane stress · Surface stress · Green functions

12.1 Introduction

Fundamental solutions of linear partial differential equations are a basis for the theory of boundary integral equations (e.g. Crouch and Starfield, 1983; Grekov, 2001; Linkov, 2002)). In the case of the linear theory of elasticity, the simplest solution is the Kelvin–Somigliana one for the point force in an infinite elastic medium (Love, 2013; Lurie, 2015). Other fundamental solutions of elasticity problems are more complicated and depend on the shape of boundary as in the case of deformation of functionally graded isotropic plates under point loading (Abali et al, 2014). Only a few fundamental solutions for a bounded solid have been obtained in an explicit form, as, for example, the solutions for an external (Boussinesq, Cerruti) or internal

Mikhail A. Grekov
Saint Petersburg State University, 7/9, Universitetskaya nab., 199034, St. Petersburg, Russia,
e-mail: m.grekov@spbu.ru, magrekov@mail.ru

B. E. Abali et al. (eds.), *New Achievements in Continuum Mechanics and Thermodynamics*, Advanced Structured Materials 108,
https://doi.org/10.1007/978-3-030-13307-8_12

(Mindlin) point force in a half-space with a planar boundary, the similar solutions (Michell, Flaman) for a half-plane (see references in Lurie, 2015) and the solutions for a point force in an elastic plane with circular inhomogeneity (Grekov, 2010). The value of such analytical solutions for point forces is that they are used directly for constructing the boundary integral equations (Grekov, 2001; Savelyeva and Pronina, 2015) and analysis of different problems with the boundary element method (Crouch and Starfield, 1983).

It is another matter if the properties and behavior of nanomaterials and nanoobjects having at least one dimension in the range of 1–100 nm, such as nanosized beams, plates and shells, wires and films (e.g. Cammarata, 1994; Miller and Shenoy, 2000; Digrevile et al, 2005; Eremeyev et al, 2009; Eremeyev and Morozov, 2010), and nanostructures with nanosized inhomogeneities, as inclusions, voids, cracks, etc. (e.g. Gutkin et al, 2013; Jammes et al, 2009; Mogilevskaya et al, 2008; Povstenko, 1993; Shodja et al, 2012; Tian and Rajapakse, 2007; Fu et al, 2010) are studied. Classical approaches developed for solving different problems of solids at the macrolevel can not be directly applied to the nanostructured materials because of surface stress effects. In particular, the surface stresses are responsible for the size-effect, that means the properties of a nanosstructure depend highly on its size (e.g. Altenbach et al, 2010; Duan et al, 2009; Eremeyev et al, 2009; Eremeyev and Morozov, 2010; Goldstein et al, 2010; Grekov and Kostyrko, 2016; Grekov and Yazovskaya, 2014; Miller and Shenoy, 2000; Shenoy, 2005; Shodja et al, 2012; Tian and Rajapakse, 2007). At the same time, the classical elasticity has been successfully extended to the nanoscale by implementation of the theory of elasticity allowing for the surface stresses the notation of which for the solids was introduced by Gibbs (1906). The most popular and extensively used model of surface elasticity for elastic solids was proposed by Gurtin and Murdoch (1975, 1978). The Gurtin–Murdoch model has been recently applied to many problems of nanomechanics. (see, for example, Duan et al, 2009; Eremeyev, 2016; Wang et al, 2011). In this model, the surface of the material is represented as a two-dimensional membrane type layer of negligible thickness, which does not resist a flexure. For the problems that comprise wrinkling, bending deformation or large curvature change, the total surface elasticity energy should be used allowing for both stretch and flexure stiffness of a surface or interface (Chhapadia et al, 2011; Gao et al, 2017). The corresponding surface elasticity models have been developed by various ways (e.g. related to coating (or reinforcement) or surface-substrate interactions, see Šilhavý, 2013; Steigmann and Ogden, 1997, among others).

In order to study the effect of surface stresses, numerous boundary value problems have been solved for elastic solids with nano-inhomogeneities, based on the Gurtin–Murdoch theory, particularly in the case of the plane problem (e.g. Bauer et al, 2014; Bochkarev and Grekov, 2014, 2015, 2017; Grekov and Yazovskaya, 2014; Gutkin et al, 2013; Jammes et al, 2009; Mogilevskaya et al, 2008; Shodja et al, 2012; Tian and Rajapakse, 2007). Only a few that works consider the plane stress (more exactly, the generalized plane stress) which relates to a thin plate. The stress field in an infinite plane with the nanometer sized circular hole was derived in Grekov and Yazovskaya (2014) for the case of plane stress, taking into account surface stresses only in the surface of the hole. So, this solution can be only applied to a thick macro-plate with nanohole when surface stresses existing in the faces of the plate can be disregarded. The same conclusion is valid also with regard to the results of investigations of the local instability of a plate with a circular nanohole under uniaxial tension, addressed in Bauer et al (2014); Bochkarev and Grekov (2014, 2015), because the authors of these works use the solution presented in Grekov and Yazovskaya (2014). Unlike the solutions obtained in Bauer et al (2014); Bochkarev and Grekov (2014, 2015), the same problem of instability has been correctly solved for the plate of a nanometer thickness in Bochkarev and Grekov (2017) by taking into account surface stresses in plate faces. Within the framework of approach (Bochkarev and Grekov, 2017), the modified Kirsch problem for the case of the generalized plane stress of a nanoplate was constructed in terms of the effective Young module and Poisson ratio similar to those introduced by Altenbach and Eremeyev (2011, 2017); Altenbach et al (2009, 2010) in the theory of plates and shells with surface stresses.

One can find a few papers where fundamental solutions (Green functions) are derived allowing for surface stresses. Based on different versions of Gurtin–Murdoch model, these solutions were obtained either under conditions of the plane strain (e.g. Intarit et al, 2010; Grekov et al, 2017; Grekov and Sergeeva, 2018; Gutkin et al, 2013; Mogilevskaya et al, 2011; Shodja et al, 2012) or for the half-space (He and Lim, 2006; Koguchi, 2008). All these solutions are expressed in terms of very bulky series; they are unsuitable for applications in the methods of the boundary integral equations.

In the present paper, the fundamental solution of the problem for an elastic infinite plate being under the generalized plane stress is constructed incorporating surface stresses in the faces of the plate. First, the constitutive relations are derived in terms of elastic parameters of bulk materials and surface materials and the thickness of the plate, based on the Hooke's law and simplified Gurtin–Murdoch's surface elasticity equations. Then, the explicit complex formulas for the stresses and displacements (Green functions) arising due to the action of a point force in the plate are evaluated using Goursat–Kolosov complex potentials and Muskhelishvili representations.

12.2 Problem formulation

We consider the infinite elastic plate $D = \{(x_1, x_2, x_3) : (x_1, x_2) \in \mathbf{R}^2, |x_3| \leq h/2\}$. The plate is subjected to the action of the forces $\mathbf{P}$ uniformly distributed at the section $[-h/2, +h/2]$ of the x_3-axis, $\mathbf{P} = (P_1, P_2, 0)$ in the Cartesian coordinates x_j $(j = 1, 2, 3)$. The strains and stresses arising in the bulk material and in the plate faces due to the applied forces satisfy the following constitutive equations, respectively:

$$\boldsymbol{\Sigma} = 2\mu \boldsymbol{E} + \lambda \boldsymbol{I} \operatorname{tr} \boldsymbol{E}, \quad |x_3| < h/2, \tag{12.1}$$

$$\boldsymbol{\Sigma}^{\mathrm{s}}_{\pm} = 2\mu^{\mathrm{s}}\boldsymbol{E}^{\mathrm{s}}_{\pm} + \lambda^{\mathrm{s}}\boldsymbol{A}\,\mathrm{tr}\,\boldsymbol{E}^{\mathrm{s}}_{\pm}, \quad x_3 = \pm h/2. \tag{12.2}$$

Equation (12.1) is the Hooke's law and Eq. (12.2) is the simplified Gurtin–Murdoch surface elasticity model as in Gurtin and Murdoch (1975, 1978) in which the residual surface stress (surface tension) is neglected. In Eqs. (12.1), (12.2), $\boldsymbol{\Sigma}$ and $\boldsymbol{E}$ are the three-dimensional stress and strain tensors, $\boldsymbol{I}$ and $\boldsymbol{A} = \boldsymbol{I} - \mathbf{e}_3 \otimes \mathbf{e}_3$ ($\mathbf{e}_3$ is the unit vector of the x_3-axis) are the three- and two-dimensional unit tensors, respectively, λ and μ are Lamé's coefficients of the bulk material whereas λ^{s} and μ^{s} are the similar elastic characteristics of the surface, $\boldsymbol{\Sigma}^{\mathrm{s}}_{\pm}$ and $\boldsymbol{E}^{\mathrm{s}}_{\pm}$ are the two-dimensional surface stress and strain tensors at the plate faces ($\boldsymbol{E}^{\mathrm{s}}_{+} = \boldsymbol{E}^{\mathrm{s}}_{-}$). Following Gurtin and Murdoch (1975); Povstenko (1993), one can express the inseparability condition in terms of the strains

$$\boldsymbol{A} \cdot \boldsymbol{E}|_{x_3=\pm h/2} = \boldsymbol{E}^{\mathrm{s}}_{\pm}. \tag{12.3}$$

Supposing that normal component σ_{33} of the stress tensor $\sum$ equals zero anywhere in the plate and passing to the averages of all quantities (displacements, strains and stresses) over the thickness of the plate by mean of the standard procedure (Love, 2013; Muskhelishvilli, 1977), the constitutive equation for the generalized plane stress is brought to the following relation:

$$\widetilde{\boldsymbol{\Sigma}} = 2\mu\widetilde{\boldsymbol{E}} + \frac{2\lambda\mu}{\lambda + 2\mu}\boldsymbol{A}\,\mathrm{tr}\,\widetilde{\boldsymbol{E}} + \frac{2}{h}\left(2\mu^{\mathrm{s}}\widetilde{\boldsymbol{E}} + \lambda^{\mathrm{s}}\boldsymbol{A}\,\mathrm{tr}\widetilde{\boldsymbol{E}}\right). \tag{12.4}$$

Here $\widetilde{\boldsymbol{\Sigma}}$ and $\widetilde{\boldsymbol{E}}$ are the two-dimensional tensors of average stresses $\widetilde{\sigma}_{ij}$ and strains $\widetilde{\varepsilon}_{ij}$ $(i, j = 1, 2)$, respectively. Equation (12.4) can be rewritten in terms of components of these tensors as follows

$$\begin{aligned} \widetilde{\sigma}_{11} &= (\lambda^* + 2\mu^*)\widetilde{\varepsilon}_{11} + \lambda^*\widetilde{\varepsilon}_{22}, \\ \widetilde{\sigma}_{22} &= \lambda^*\widetilde{\varepsilon}_{11} + (\lambda^* + 2\mu^*)\widetilde{\varepsilon}_{22}, \\ \widetilde{\sigma}_{12} &= 2\mu^*\widetilde{\varepsilon}_{12}, \end{aligned} \tag{12.5}$$

where effective elastic modules λ^*, μ^* equal

$$\lambda^* = \frac{2\lambda\mu}{\lambda + 2\mu} + \frac{2\lambda^{\mathrm{s}}}{h}, \quad \mu^* = \mu + \frac{2\mu^{\mathrm{s}}}{h}. \tag{12.6}$$

Constitutive equations (12.5) coincide with those addressed in Grekov and Yazovskaya (2014) if surface stresses in the plate faces are not taken into account.

Equations (12.5), corresponding equilibrium equation for the stresses $\widetilde{\sigma}_{ij}$ and continuity equation for the strains $\widetilde{\varepsilon}_{ij}$ (Bochkarev and Grekov, 2017) lead to the 2-D problem of elasticity for an infinite isotropic plane with the point force $\mathbf{P} = (P_1, P_2)$ acting in the origin of coordinates x_1, x_2. The solution of this problem at the macrolevel, called by fundamental solution, both for the plane strain and plane stress is described, for example, in Love (2013); Muskhelishvilli (1977). It is worth noting that the statement of the generalized plane stress problem presented here differs from the classical one (Love, 2013; Muskhelishvilli, 1977) only in the coefficients (12.6) of Eqs. (12.5). So, it allows using Goursat–Kolosov complex potentials and Muskhelishvili technique to derive the fundamental solution of the plane stress problem incorporating surface stresses in the plate faces.

12.3 Green Functions

Complex potentials which correspond to the point force **P** acting in the point (0,0) of an infinite elastic plane are defined as (Muskhelishvilli, 1977)

$$\varPhi(z) = -\frac{H}{z}, \quad \varPsi(z) = \frac{æ^* \overline{H}}{z}, \tag{12.7}$$

where

$$z = x_1 + ix_2, \; i = \sqrt{-1}, \quad H = \frac{P}{2\pi(æ^* + 1)}, \quad P = P_1 + iP_2, \; æ^* = \frac{5\lambda^* + 6\mu^*}{3\lambda^* + 2\mu^*}.$$

According to Muskhelishvilli (1977), the stresses and displacements in the elastic plane are related to the complex potentials $\varPhi, \varPsi$ by the equalities

$$\begin{aligned} \widetilde{\sigma}_{nn} + i\widetilde{\sigma}_{nt} &= \varPhi(z) + \overline{\varPhi(z)} + \left(z\overline{\varPhi'(z)} + \overline{\varPsi(z)}\right) e^{-2i\alpha}, \\ 2\mu^* \frac{d\widetilde{u}}{dz} &= æ^* \varPhi(z) - \overline{\varPhi(z)} - \left(z\overline{\varPhi'(z)} + \overline{\varPsi(z)}\right) e^{-2i\alpha}. \end{aligned} \tag{12.8}$$

In Eq. (12.8), $\widetilde{u} = \widetilde{u}_1 + i\widetilde{u}_2$, $\widetilde{u}_1, \widetilde{u}_2$ are the displacements along corresponding axes of coordinates x_1, x_2; $\widetilde{\sigma}_{nn}$, $\widetilde{\sigma}_{nt}$ are the stress tensor components in the local Cartesian coordinates n, t with the angle α between the t and x_1 axes. A bar over a symbol denotes complex conjugation, and a prime denotes the derivative with respect to the argument.

Substituting Eq. (12.7) in Eq. (12.8) yields the following complex relations for stresses and displacements that are the Green functions or fundamental (singular) solution

$$\begin{aligned} \widetilde{\sigma}_{11} - i\widetilde{\sigma}_{12} &= -2\mathrm{Re}\frac{H}{z} + \frac{1}{\overline{z}^2}\left(\overline{H}z + æ^* H\overline{z}\right), \\ \widetilde{\sigma}_{11} + \widetilde{\sigma}_{22} &= -4\mathrm{Re}\frac{H}{z}, \end{aligned} \tag{12.9}$$

$$2\mu^* \widetilde{u}(z) = -æ^* H \ln\frac{z}{\overline{z}} + \overline{H}\frac{z}{\overline{z}}. \tag{12.10}$$

Equations (12.9) are obtained from Eq. (12.8) taking $\alpha = 0$ and $\alpha = \pi/2$, and Eq. (12.10) is evaluated by integrating the second equation (12.8) under $\alpha = 0$. It should be noted that complex displacement u is defined in Eq. (12.10) with accuracy of translation and rotation of a plane as a rigid one.

As it is seen from Eq. (12.6), the effective elastic constants λ^*, μ^* and, as a consequence, displacements and stresses depend on the thickness h of the plate that is the size effect. The less h, the more significant this effect related with the existing surface stresses in the plate faces. If the surface effect is neglected, Eqs. (12.9) and (12.10) are reduced to the classical solution given, for example, by Love (2013) in terms of the real functions for the case when a concentrated force acts in the x_1-direction.

12.4 Summary and Conclusions

The problem on infinite elastic plate subjected to the point forces uniformly distributed along the line perpendicular to the plate faces has been reduced to the generalized plane stress problem, incorporating surface stresses in the faces. The constitutive equations of the generalized plane stress of the plate have been derived in terms of effective elastic modules which depend on the elastic properties of the plate material, surface elastic properties and the plate thickness. The complex relations (Green functions) for the stresses and displacements have been evaluated using Goursat–Kolosov complex potentials and Muskhelishvili representations. It has been shown that in the case of the generalized plane stress, the fundamental solution depends on the thickness of the plate that is the size effect intrinsic to the nanoobjects. The explicit formulae of the complex fundamental solution derived in the paper can be directly used (similarly as in the classical case) for constructing the boundary integral equations and analysis of different problems with the boundary element method in order to study the elastic fields in the nanoplates containing voids, cracks, inclusions etc.

Acknowledgements The work was supported by the Russian Foundation for Basic Research (grant 18-01-00468).

References

Abali BE, Völlmecke C, Woodward B, Kashtalyan M, Guz I, Müller WH (2014) Three-dimensional elastic deformation of functionally graded isotropic plates under point loading. Compos Struct 118:367–376

Altenbach H, Eremeyev VA (2011) On the shell theory on the nanoscale with surface stresses. Int J Eng Sci 49:1294–1301

Altenbach H, Eremeyev VA (2017) On the elastic plates and shell theory with residual surface stresses. Procedia IUTAM 21:25–32

Altenbach H, Eremeyev VA, Morozov NF (2009) Linear theory of shells taking into account surface stresses. Doklady Phys 54:531–535

Altenbach H, Eremeyev VA, Morozov NF (2010) On equations of the linear theory of shells with surface stresses taken into account. Mech Solids 45:331–342

Bauer SM, Kashtanova SV, Morozov NF, Semenov BN (2014) Stability of a nanoscale-thickness plate weakened by a circular hole. Doklady Phys 59:416–418

Bochkarev AO, Grekov MA (2014) Local instability of a plate with circular nanohole under uniaxial tension. Doklady Phys 59:330–334

Bochkarev AO, Grekov MA (2015) The influence of the surface stress on the local buckling of a plate with a circular nanohole. In: 2015 Int. Conf. on "Stability and Control Processes" in Memory of V. I. Zubov (SCP), IEEE, pp 367–370

Bochkarev AO, Grekov MA (2017) Influence of surface stresses on the nanoplate stiffness and stability in the Kirsch problem. Phyzicheskaya Mezomechanika 20:62–76

Cammarata RC (1994) Surface and interface stresses effects in thin films. Progress Surface Sci 46:1–38

Chhapadia P, Mohammadi P, Sharma P (2011) Curvature-dependent surface energy and implications for nanostructures. J Mech Phys Solids 59:2103–2115

Crouch SI, Starfield AM (1983) Boundary Element Method in Solid Mechanics. George Allen and Unwin, London

Digrevile R, Qu JM, Cherkaoui M (2005) Surface free energy and its effect on elastic behavior of nano-sized particles, wires and films. J Mech Phys Solids 53:1827–1854

Duan HL, Wang J, Karihaloo BL (2009) Theory of elasticity at the nanoscale. Adv Appl Mech 42:1–68

Eremeyev VA (2016) On effective properties of materials at the nano-and microscales considering surface effects. Acta Mech 227:29–42

Eremeyev VA, Morozov NF (2010) The effective stiffness of a nanoporous rod. Doklady Phys 55:279–282

Eremeyev VA, Altenbach H, Morozov NF (2009) The influence of surface tension on the effective stiffness of nanosize plates. Doklady Phys 54:98–100

Fu XL, Wang GF, Feng XQ (2010) Surface effects on mode-1 crack tip fields: A numerical study. Eng Fract Mech 77:1048–1057

Gao X, Huang Z, Fang D (2017) Curvature-dependent interfacial energy and its effects on the elastic properties of nanomaterials. Int J Solids Struct 113–114:100–107

Gibbs JW (1906) The Scientific Papers of J. Willard Gibbs, vol 1. Longmans-Green, London

Goldstein RV, Gorodtsov VA, Ustinov KV (2010) Effect of residual stress and surface elasticity on deformation of nanometer spherical inclusions in an elastic matrix. Phys Mesomech 13:318–328

Grekov MA (2001) Singular Plane Problems in Elasticity. St. Petersburg State University, St. Petersburg, Russia

Grekov MA (2010) Joint deformation of a circular inclusion and a matrix. Vestnik St Petersburg University: Mathematics 43:114–121

Grekov MA, Kostyrko SA (2016) Surface effects in an elastic solid with nanosized surface asperities. Int J Solids Struct 96:153–161

Grekov MA, Sergeeva TS (2018) Periodic Green functions for two-component medium with interface stresses at the planar interface. In: AIP Conf. Proc, AIP, vol 1959, p 070017

Grekov MA, Yazovskaya AA (2014) The effect of surface elasticity and residual surface stress in an elastic body with an elliptic nanohole. J Appl Math Mech 78:172–180

Grekov MA, Sergeeva TS, Pronina YG, Sedova OS (2017) A periodic set of edge dislocations in an elastic solid with a planar boundary incorporating surface effects. Eng Frac Mech 186:423–435

Gurtin ME, Murdoch AI (1975) A continuum theory of elastic material surfaces. Arch Rat Mech Anal 57:291–323

Gurtin ME, Murdoch AI (1978) Surface stress in solids. Int J Solids Struct 14:431–440

Gutkin MY, Enzevaee C, Shodja HM (2013) Interface effects behavior of an edge dislocation in core-shell nanowire embedded to an infinite matrix. Int J Solids Struct 50:1177–1186

He LH, Lim CW (2006) Surface Green function for a soft elastic half-space: influence of surface stress. Int J Solids Struct 43:132–143

Intarit P, Senjuntichai T, Rajapakse RKND (2010) Dislocation and internal loading in semi-infinite elastic medium with surface stress. Eng Fract Mech 77:3592–3603

Jammes M, Mogilevskaya SG, Crouch SI (2009) Multiple circular nano-inhomogeneities and/or nano-pores in one of two joined isotropic elastic half-planes. Eng Anal Bound Elem 33:233–248

Koguchi H (2008) Surface Green function with surface stresses and surface elasticity using stroh's formalism. Trans ASME J Appl Mech 75:061,014

Linkov AM (2002) Boundary Integral Equations in Elastisity Theory. Kluwer Acad. Pub., The Netherlands

Love AEH (2013) A Treatise on the Mathematical Theory of Elasticity. Cambridge University Press, Cambridge

Lurie AI (2015) Theory of Elasticity. Springer, Berlin, Heidelberg, New York

Miller RE, Shenoy VB (2000) Size-dependent elastic properties of nanosized structural elements. Nanotechnology 11:139–147

Mogilevskaya SG, Crouch SI, Stolarski HK (2008) Multiple interacting circular nano-inhomogeneities with surface/interface effects. J Mech Phys Solids 56:2298–2327

Mogilevskaya SG, Pyatigorets AV, Crouch SI (2011) Green function for the problem of a plane containing a circular hole with surface effects. Trans ASME J Appl Mech 78:021,008

Muskhelishvilli NI (1977) Some Basic Problems of the Mathematical Theory of Elasticity. Springer, The Netherlands

Povstenko YZ (1993) Theoretical investigation of phenomena caused by heterogeneous surface tention in solids. J Mech Phys Solids 41:1499–1514

Savelyeva M, Pronina Y (2015) Integral equations for the mixed boundary value problem of a notched elastic half-plane. In: 2015 Int. Conf. on "Stability and Control Processes" in Memory of V. I. Zubov (SCP), IEEE, pp 432–435

Shenoy VB (2005) Atomistic calculations of elastic properties of metallic fcc crystal surfaces. Phys Rev B 71:94–104

Shodja HM, Ahmadzadeh-Bakhshayesh H, Gutkin MY (2012) Size-dependent interaction of an edge dislocation with an elliptical nano-inhomogeneity incorporating interface effects. Int J Solids Struct 49:759–770

Steigmann DJ, Ogden RW (1997) Plane deformations of elastic solids with intrinsic boundary elasticity. Proc R Soc A 453:853–877

Tian L, Rajapakse RKND (2007) Elastic field of an isotropic matrix with nanoscale elliptical inhomogeneity. Int J Solids Struct 44:7988–8005

Šilhavý M (2013) A direct approach to nonlinear shells with application to surface-substrate interactions. Math Mech Compl Syst 1:211–232

Wang J, Huang Z, Duan H, Yu S, Feng X, Wang G, Zhang W, Wang T (2011) Surface stress effect in mechanics of nanostructured materials. Acta Mech Solida Sin 24:52–82

Chapter 13
Isotropic Linear Viscoelastic Reduced Cosserat Medium: an Acoustic Metamaterial and a First Step to Model Geomedium

Elena F. Grekova & Rafael Abreu

Abstract The reduced Cosserat medium is a continuum whose body points possess rotational degrees of freedom, and there is a reaction to the rotation of a body point relatively to the background of centres of mass, but no stresses are caused by the gradient of micro-rotation. This theory is useful for modelling rocks and soils containing heterogeneities, a geomedium with blocky structure, certain composites with inclusions as well as seismic metamaterials. In this work we consider the influence of viscosity in the linear isotropic reduced Cosserat medium on the propagation of shear waves. We find that viscosity may change drastically the wave propagation. In some cases, the material behaves as a double negative acoustic metamaterial for shear waves, i.e. there is a decreasing part of the dispersion curve for a certain band of frequencies. We also observe that the attenuation in such a continuum does not necessarily increases with frequency, as it happens in the classical viscoelastic medium. It may have one maximum at a certain frequency, or have maximum and minimum. Similar phenomena are observed in the range of seismic frequencies for a geomedium (Sato et al, 2012). The theory considered in this work is only the first step to model the geomedium since it does not take into account previous stress state and existing couplings between pressure and shear-rotational waves due to rock anisotropy caused by gravity.

Keywords: Reduced Cosserat medium · Viscoelasticity · Acoustic metamaterials · Rotational seismology

Elena F. Grekova
Institute of Problems of Mechanical Engineering, Russian Academy of Sciences, Bolshoy pr. V.O., 61, 199178, St. Petersburg, Russia,
e-mail: elgreco@pdmi.ras.ru

Rafael Abreu
University of Münster, Institut für Geophysik, Corrensstr. 24, 48149 Münster, Germany,
e-mail: abreu@uni-muenster.de

B. E. Abali et al. (eds.), *New Achievements in Continuum Mechanics and Thermodynamics*, Advanced Structured Materials 108,
https://doi.org/10.1007/978-3-030-13307-8_13

13.1 Introduction

Many works are devoted to the mechanics of media with enriched kinematics, in particular, to the theory of Cosserat media (polar, or micropolar continua). The pioneering work by Cosserat and Cosserat (1909) introduced the theory for an elastic 3D medium with rotational degrees of freedom. Later on, Cosserats' theory was developed by Kafadar and Eringen (1971). Altenbach and Zhilin (1988) suggested a method to obtain constitutive equations of the 2D elastic polar medium that can be applied also for the 3D case. Among books on elastic Cosserat continua one can mention Eremeyev et al (2013); Eringen and Maugin (2012); Erofeyev (2003). Various works developing Cosserats' ideas can be found in Maugin and Metrikine (2010). The role of microinertia in enriched continuum mechanics is discussed in Madeo et al (2017). Waves in 1D nonlinear Cosserat continuum are investigated in Porubov et al (2009), and in 2D hexagonal lattice in Erofeev et al (2018). Depending on the different type of Cosserat continua and restrictions on the strain energy, we have to use different strain measures to obtain specific type of constitutive equations in Eremeyev and Pietraszkiewicz (2012, 2016); Pietraszkiewicz and Eremeyev (2009). Some typical features of granular materials and suspensions can be described in terms of micromorphic media which are not sensible to certain types of deformations (Eremeyev, 2018). In order to describe the deformation phenomena in an elastic medium possessing a microstructure, we need micropolar models (Müller and Vilchevskaya, 2017).

The reduced Cosserat medium is a medium whose particles have independent translational and rotational degrees of freedom, but the medium does not react to the gradient of rotation of point bodies, no internal moments work on the gradient of angular velocity. Let us discuss a possible motivation for such a model. In granular materials rotations of particles and agglomerates can be important in many processes, in particular, in restructurisation of the medium and onset of instability (Gilabert et al, 2007; Merkel et al, 2010, 2011; Misra and Poorsolhjouy, 2015; Turco, 2018; Vardoulakis, 1989). However, we do not see there an interaction that reduces the gradient of rotation of grains or aggregates. Indeed, neighbouring particles often tend to rotate in the opposite directions rather than reduce their relative rotation. In these media there is no ordered structure of rotations, contrary to the case of magnetic materials, where the exchange interaction reduces the gradient of spin directions. However, there is an average resistance to the rotation of a body point relatively to the whole background (continuum of centres of mass). As an approximation, we can consider a model where stresses depend on this kind of strain, as well as on the purely translational deformations (i.e. on the gradient of translational displacement $\nabla \boldsymbol{u}$ and rotation tensor $\boldsymbol{P}$), but do not depend on the gradient of micro-rotation $\nabla \boldsymbol{P}$ (Fig. 13.1). In the elastic case, we have instead of the strain energy $U(\nabla \boldsymbol{u}, \boldsymbol{P}, \nabla \boldsymbol{P})$ for the full Cosserat medium a function of the two first arguments $U(\nabla \boldsymbol{u}, \boldsymbol{P})$. Schwartz et al (1984) have suggested to consider such a medium in its isotropic linear variant to describe granular materials. This is a special case of Cosserat medium with a specific constraint.

Fig. 13.1 Reduced Cosserat continuum: $\boldsymbol{P}$ is the microrotation tensor (turn-tensor), $\boldsymbol{u}$ is the body point displacement. The strain energy U is not positively defined with respect to $\nabla \boldsymbol{P}$, and we neglect on average dependence U on $\nabla \boldsymbol{P}$ due to the non-ordered structure of particles' rotations.

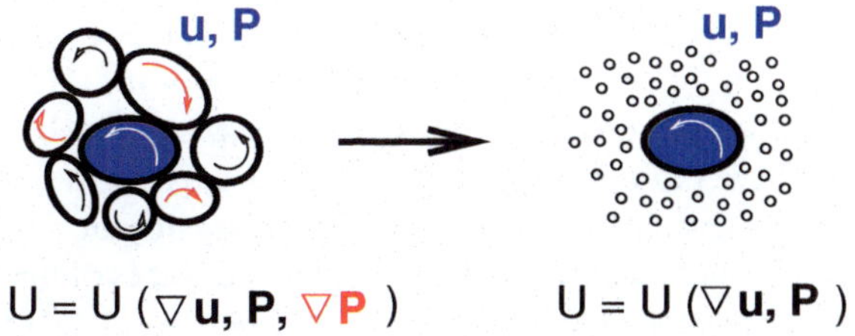

Waves in the elastic reduced Cosserat continuum were investigated in Grekova et al (2009); Kulesh et al (2009). In the isotropic case, the pressure wave (P-wave) is the same as in the classical medium, but the shear-rotational wave (S-wave) has a band gap limited by the boundary and cut-off frequencies. Thus, the reduced Cosserat medium is a single negative acoustic metamaterial, and a strong dispersion is observed in the vicinity of these frequencies (Fig. 13.2). In Misra and Poorsolhjouy (2016) we can find a more sophisticated model describing granular materials. Based on microstructural considerations, the authors introduce a micromorphic theory, where each point is characterised by macro-deformation (describing the strain of the continuum of centroids of the grains) and micro-deformation (related to the strain of each grain). Contrary to the reduced Cosserat continuum, in Misra and Poorsolhjouy (2016) the authors suppose that the rotation tensor of each grain can be expanded into Taylor series in material space, and there is an interaction reducing the relative rotation of grains, which leads to non-zero couple stresses. However, since the model is richer than the full Cosserat continuum, it also allows the existence of band gaps. Curiously, qualitatively the dispersion curves of this continuum are similar to those for transverse waves in the elastic reduced Cosserat continuum with an axisymmetric tensor of inertia (Grekova, 2018b).

Some cases of anisotropy of different nature (a coupling between pressure and shear-rotational waves, non-spherical tensor of inertia, particular cases of the prestressed state) were considered in Grekova (2016, 2017, 2018b). If there is a coupling between volumetric and shear–rotational strain in the strain energy (which

Fig. 13.2 Dispersion curves for the shear-rotational waves in the isotropic reduced Cosserat continuum (ω is the frequency, k is the wave number). Velocities at low and high frequencies $c_s = \sqrt{\mu/\rho}$, $c_{s\alpha} = \sqrt{(\mu+\alpha)/\rho}$, respectively.

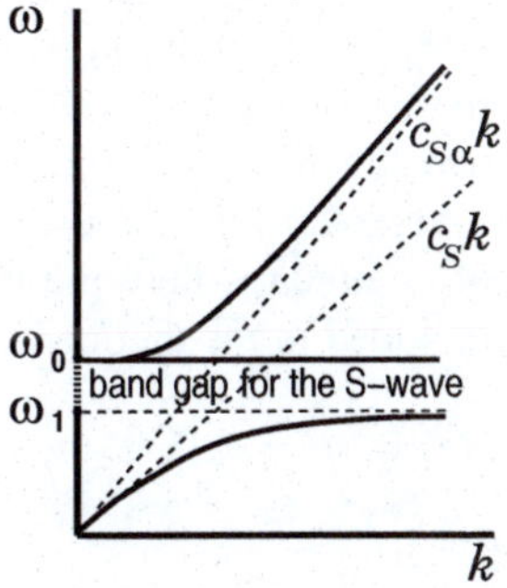

only may exist for anisotropic materials), then it results in existence of band gaps for almost all plane waves in the bulk, with exception for some specific direction of wave propagation and parameter domains. However, if such a coupling is not present in the strain energy, but the inertia tensor is anisotropic, for instance, axisymmetric, then the pressure wave looks as the classical one, but the shear waves are dispersive and have more band gaps than in the case of spherical tensor of inertia. In classical anisotropic elastic media plane waves are non-dispersive. However, in architectured anisotropic materials, where rotational degrees of microstructure are important, they are dispersive (see, for instance, dell'Isola et al, 2015, 2018).

The frequency band with high attenuation draws our attention to the so called single negative acoustic metamaterials (media having a band gap due to existence of internal degrees of freedom). Some heterogeneous media behave as metamaterials due to the local resonances caused by the existence of their inner structure (see, for instance, Chesnais et al, 2012; Hans et al, 2014). Indeed, such an effect cannot be described only in terms of dissipation since the attenuation would then increase with frequency. There may exist various explanations of this phenomenon, for example, relaxation mechanisms in rocks Liu et al (1976). The idea that a time-dispersive material can be interpreted in terms of an "extended" material with hidden internal degrees of freedom is suggested, for instance, in Figotin and Schenker (2007). Complex reduced elastic media ("bearing continuum" enhanced by a distributed non-interconnected "dynamic absorber"), in particular, those with rotational degrees of freedom, are acoustic metamaterials with respect to certain types of waves in Grekova (2019). In this work we will consider the influence of viscosity for a particular kind of continuum, reduced Cosserat medium.

It is widely known that the geomedium consists of blocks and has heterogeneities of various sizes. Depending on the frequencies of dynamic processes, these heterogeneities or blocks may perform rotations that, even being small, may change the character of the wave propagation (Sadovskii and Sadovskaya, 2015), especially in the domain near the partial frequency of a block or heterogeneity in the continuum. The new branch of science, rotational seismology (http://rotational-seismology.org) deals with rotations in the geomedium. These rotations can be taken into account in terms of Cosserat media, where a heterogeneity with surrounding medium or a block play role of a "particle" (Twiss, 2009; Twiss and Marrett, 2010; Twiss et al, 1991, 1993; Twiss and Unruh, 2007; Unruh et al, 2003, 1996).

In this work we consider the simplest variant of the reduced Cosserat medium with dissipation: linear isotropic reduced Cosserat medium with spherical tensor of inertia of particles and dissipation in the rotational and/or translational term and investigate how the attenuation of the shear wave depends on the frequency. It is a first step towards to the modelling of a real geomedium using enriched theories of continua, where the loss of elastic energy is due to a clear physical mechanism included in the theory.

13.2 Reduced Linear Isotropic Cosserat Viscoelastic Model

13.2.1 General Equations

Let $\boldsymbol{\tau}$ be the Cauchy stress tensor, $\boldsymbol{u}$ the vector of translational displacement of the particle, $\boldsymbol{\theta}$ the infinitesimal vector of its rotation (vector of turn, or vector of micro-rotation), $\boldsymbol{P} = \boldsymbol{E} + \boldsymbol{\theta} \times \boldsymbol{E}$ the tensor of infinitesimal micro-rotation (tensor of infinitesimal turn), ρ the mass density, $I\boldsymbol{E}$ the density of the tensor of inertia (we consider it spherical for sake of simplicity), $\boldsymbol{E}$ the identity tensor. We denote by $(\cdot)^{\cdot}$ the time derivative. For infinitesimal rotations the angular velocity equals $\dot{\boldsymbol{\theta}}$.

For any second rank tensor $\boldsymbol{\Lambda} = \lambda_{mn}\boldsymbol{e}_m\boldsymbol{e}_n$ we denote $\boldsymbol{\Lambda}^\top = \lambda_{nm}\boldsymbol{e}_m\boldsymbol{e}_n$, symmetric part of $\boldsymbol{\Lambda}$ as $\boldsymbol{\Lambda}^S = (\boldsymbol{\Lambda}+\boldsymbol{\Lambda}^\top)/2$, its antisymmetric part as $\boldsymbol{\Lambda}^A = (\boldsymbol{\Lambda}-\boldsymbol{\Lambda}^\top)/2$, and its vectorial invariant (accompanying vector of $\boldsymbol{\Lambda}^A$) as $\boldsymbol{\Lambda}_\times = \lambda_{mn}\boldsymbol{e}_m \times \boldsymbol{e}_n$. Note that $\boldsymbol{\Lambda}^A = -\frac{1}{2}\boldsymbol{\Lambda}_\times \times \boldsymbol{E}$.

The balance of force in the linear reduced Cosserat medium in the absence of external loads takes form (Grekova et al, 2009)

$$\nabla \cdot \boldsymbol{\tau} = \rho\ddot{\boldsymbol{u}}, \tag{13.1}$$

and the balance of torque is

$$\boldsymbol{\tau}_\times = I\ddot{\boldsymbol{\theta}}. \tag{13.2}$$

For the full Cosserat continuum, the balance of torque includes also the term $\nabla \cdot \boldsymbol{\mu}$, where $\boldsymbol{\mu}$ is the couple stress. However, we limit ourselves to the consideration of the *reduced* Cosserat medium, in which no work is performed on the gradient of the angular velocity, therefore $\boldsymbol{\mu} \equiv \boldsymbol{0}$. Indeed, for isothermic processes the law of balance of energy for the full linear Cosserat medium is

$$\rho\dot{U} = \boldsymbol{\tau}^\top \cdot\cdot \nabla\dot{\boldsymbol{u}} - \boldsymbol{\tau}_\times \cdot \dot{\boldsymbol{\theta}} + \boldsymbol{\mu}^\top \cdot\cdot \nabla\dot{\boldsymbol{\theta}} = \boldsymbol{\tau}^S \cdot\cdot \nabla\dot{\boldsymbol{u}}^S - \boldsymbol{\tau}_\times \cdot (\dot{\boldsymbol{\theta}} - \nabla \times \boldsymbol{u}/2) + \boldsymbol{\mu}^\top \cdot\cdot \nabla\dot{\boldsymbol{\theta}}, \tag{13.3}$$

where U is the strain energy. Therefore, if we consider a particular case (reduced Cosserat medium), where internal moments do not work on $\nabla\dot{\boldsymbol{\theta}}$, we obtain $\boldsymbol{\mu} = \boldsymbol{0}$.

In the linear elastic isotropic case, the constitutive equations satisfying the law of balance of energy and principle of material objectivity (Noll, 1958) are (Grekova et al, 2009)

$$\boldsymbol{\tau}^e = \lambda\nabla \cdot \boldsymbol{u}\boldsymbol{E} + 2\mu(\nabla\boldsymbol{u})^S + 2\alpha(\nabla\boldsymbol{u} + \boldsymbol{\theta} \times \boldsymbol{E})^A, \tag{13.4}$$

where λ and μ are the conventional Lamé parameters and α is the Cosserat couple modulus. In order to introduce viscosity in the model, we have to satisfy the same fundamental balance laws, as well as material objectivity and the second principle of thermodynamics. The latter for the linear theory in the adiabatic or isothermic case is equivalent to the requirement

$$\boldsymbol{\tau}^\top \cdot\cdot (\nabla\boldsymbol{u} + \boldsymbol{\theta} \times \boldsymbol{E})^{\cdot} \geqslant 0. \tag{13.5}$$

This is a particular case of linear micropolar viscoelastic model suggested by Eringen (1967).

We can write $\boldsymbol{\tau} = \boldsymbol{\tau}^e + \boldsymbol{\tau}^d$, where $\boldsymbol{\tau}^e$ refers to the pure elastic part of the stress, and the dissipative part of the stress $\boldsymbol{\tau}^d$ can be expressed in the form

$$\boldsymbol{\tau}^d = \lambda\kappa\nabla\cdot\dot{\boldsymbol{u}}\boldsymbol{E} + \mu\nu(\nabla\dot{\boldsymbol{u}}^s) + \alpha\beta(\nabla\boldsymbol{u}^A + \boldsymbol{\theta}\times\boldsymbol{E})^{\cdot}, \tag{13.6}$$

where coefficients κ, ν, β characterize the dissipation in the medium.

The equations of motion in displacements take the form

$$\begin{aligned}&(\lambda+2\mu)\nabla\nabla\cdot\boldsymbol{u} - \mu\nabla\times(\nabla\times\boldsymbol{u}) + 2\alpha\nabla\times(\boldsymbol{\theta}-\nabla\times\boldsymbol{u}/2)\\ &+(\lambda\kappa+2\mu\nu)\nabla\nabla\cdot\dot{\boldsymbol{u}} - \mu\nu\nabla\times(\nabla\times\dot{\boldsymbol{u}}) + 2\alpha\beta\nabla\times(\boldsymbol{\theta}-\nabla\times\boldsymbol{u}/2)^{\cdot} = \rho\ddot{\boldsymbol{u}},\\ &-4\alpha(\boldsymbol{\theta}-\nabla\times\boldsymbol{u}/2) - 4\alpha\beta(\boldsymbol{\theta}-\nabla\times\boldsymbol{u}/2)^{\cdot} = I\ddot{\boldsymbol{\theta}}.\end{aligned} \tag{13.7}$$

There exist different ways to estimate material constants using the information on the microstructure of the material. One of them is to use Piola's Ansatz, as it is done for the second gradient continuum model in dell'Isola et al (2016), and to obtain the moduli solving several model problems. Note that a modification of such a method for the reduced Cosserat continuum would require always dynamical test problems in the absence of body torque, for instance, interaction of waves with discontinuities, considered in dell'Isola et al (2012) to determine elastic moduli for second gradient theories. Indeed, for zero body torque in statics $\boldsymbol{\tau}_\times = \mathbf{0}$ due to the absence of couple stresses. In this work we do not consider any homogeneisation procedure.

13.2.2 Constrained Reduced Cosserat Medium

Note that if the micro-rotation $\boldsymbol{\theta}$ is identically equal to the macro-rotation $\nabla\times\boldsymbol{u}/2$ (constrained medium), we cannot write the constitutive equations for $\boldsymbol{\tau}_\times$ in the reduced Cosserat medium, since it works on $\boldsymbol{\theta} - \nabla\times\boldsymbol{u}/2$ (see (13.3)). In this case the antisymmetric part of $\boldsymbol{\tau}$ will be determined from the dynamic equations. Indeed, for the constrained medium the balance of torque (13.2) results in

$$\boldsymbol{\tau}_\times = I\ddot{\boldsymbol{\theta}} = I\nabla\times\ddot{\mathbf{u}}/2, \tag{13.8}$$

and the balance of force looks as

$$\begin{aligned}\rho\ddot{\boldsymbol{u}} = \nabla\cdot\boldsymbol{\tau}^S + \nabla\cdot\boldsymbol{\tau}^A &= \nabla\cdot\boldsymbol{\tau}^S + \nabla\cdot(-\boldsymbol{\tau}_\times\times\boldsymbol{E}/2)\\ &= \nabla\cdot\boldsymbol{\tau}^S - \nabla\times(I\nabla\times\ddot{\mathbf{u}})/4.\end{aligned} \tag{13.9}$$

If I does not depend on the point of space, we obtain

$$\nabla\cdot\boldsymbol{\tau}^S = \rho\ddot{\boldsymbol{u}} + I(\nabla\nabla\cdot\ddot{\mathbf{u}} - \triangle\ddot{\mathbf{u}})/4 \tag{13.10}$$

Here $\boldsymbol{\tau}^S$ is the same as in the classical theory. Note that due to the absence of couple stresses we have a non-classical inertial term, but the theory is not the second gradient theory as it would be in the case of the full constrained Cosserat continuum in Bleustein (1967).

13.3 Dispersional Relations and Attenuation Factor

13.3.1 General Equations

Equation (13.7) in the frequency domain can be obtained by substituting the elastic constants for linear functions: λ for $\lambda(1+i\kappa\omega)$, μ for $\mu(1+i\nu\omega)$ and α for $\alpha(1+i\beta\omega)$. This approach was suggested for the description of frequency independent dissipation in viscoelastic materials by Sorokin (1960).

Therefore, in the reduced Cosserat medium, instead of the dispersion relation for the longitudinal wave

$$\omega^2 = \frac{\lambda + 2\mu}{\rho} k^2 \quad \text{(P-wave in the elastic medium)} \tag{13.11}$$

we will have

$$\omega^2 = \frac{\lambda + 2\mu + i(\lambda\kappa + 2\mu\nu)\omega}{\rho} k^2 \quad \text{(P-wave in the viscoelastic medium),} \tag{13.12}$$

and instead of the dispersion relation for the shear wave

$$k^2 = \frac{\omega^2}{(\mu+\alpha)/\rho} \frac{\omega^2 - 4\alpha/I}{\omega^2 - 4\mu\alpha/((\mu+\alpha)I)} \quad \text{(S-wave in the elastic medium)} \tag{13.13}$$

we will have the dispersion relation for the S-wave in the viscoelastic linear reduced Cosserat medium

$$k^2 = \frac{\rho\omega^2}{\mu(1+i\nu\omega) + \alpha(1+i\beta\omega)} \frac{\omega^2 - 4\alpha(1+i\beta\omega)/I}{\omega^2 - \dfrac{4\mu(1+i\nu\omega)\alpha(1+i\beta\omega)}{(\mu(1+i\nu\omega)+\alpha(1+i\beta\omega))I}}. \tag{13.14}$$

The dissipation for various plane waves with the wave number k is characterized by the inverse quality factor (or attenuation factor) (Liu et al, 1976)

$$Q^{-1} \stackrel{\text{def}}{=} \frac{|\operatorname{Im} k|}{|\operatorname{Re} k|}. \tag{13.15}$$

In the frame of this theory, we observe that the P-wave has the same dispersion and attenuation as in the classical viscoelastic medium, namely,

$$k = \frac{\omega}{c_p{}^4\sqrt{1+B^2\omega^2}} \exp(-\frac{i}{2}\arctan(B\omega)), \tag{13.16}$$

$$B = \frac{\lambda\kappa + 2\mu\nu}{\lambda + 2\mu}, \quad c_p^2 = \frac{\lambda + 2\mu}{\rho}, \tag{13.17}$$

$$Q_P^{-1} = \frac{|\operatorname{Im} k|}{|\operatorname{Re} k|} = \frac{B\omega}{1+\sqrt{1+B^2\omega^2}}. \tag{13.18}$$

We can observe in equation (13.18), that like in the classical theory, Q_P^{-1} increases with ω. Therefore this variant of the reduced Cosserat viscoelastic medium cannot explain experimental fact that for seismic waves $Q_P^{-1}(\omega)$ has a decreasing part.

13.3.2 Detailed Analysis of Shear Waves Dispersion and Dissipation

On one hand, wave dispersion and dissipation predicted by the micropolar model have been useful for explaining physical experiments in metamaterials where the classical linear theory of elasticity fails to give realistic predictions (Frenzel et al, 2017). On the other hand, existence of internal rotational degrees of freedom has been observed in experiments describing materials present in the earth lower most mantle (geomedium) (Kawai and Tsuchiya, 2015). It thus becomes important to investigate in detail the viscoelastic behaviour of the micropolar model for applications in the earth sciences. The frequency dependence attenuation Q^{-1} in the Earth is often approximated by a power law

$$\tilde{q} \propto \tilde{q}_0\omega^\alpha \qquad \text{with} \quad \tilde{q} \equiv \frac{1}{Q}, \tag{13.19}$$

where α is a model dependent, usually smaller than 0.5 (Anderson and Hart, 1978). This, however, does not fit quantitatively experimental data (Sato et al, 2012). The character of $Q^{-1}(\omega)$, according to Sato et al (2012), varies depending on the Earth region, and often it has a maximum. We can suppose that different types of dependencies can be met in geophysical experiments in different regions.

To explain (13.19) in terms of the micropolar model, we shall investigate in detail its dispersion relation (13.14). After some algebra we obtain

$$\begin{aligned} k^2 = {} & \frac{\omega^2}{c_{s\alpha d}^2((\omega^2-\omega_{1d}^2)^2 + T^2\omega^2(\omega^2-\omega_i^2)^2)} \\ & \Big[(\omega^2-\omega_0^2)(\omega^2-\omega_{1d}^2) - T\beta\omega_0^2\omega^2(\omega^2-\omega_i^2) \\ & - i\omega(\beta\omega_0^2(\omega^2-\omega_{1d}^2) + T(\omega^2-\omega_i^2)(\omega^2-\omega_0^2))\Big] \end{aligned} \tag{13.20}$$

where we have introduced the following notation

$$c_s^2 = \frac{\mu}{\rho}, \quad c_{s\alpha}^2 = \frac{\mu + \alpha}{\rho}, \quad c_{s\alpha d}^2 = c_{s\alpha}^2 + c_s^2 \omega_0^2 \nu \beta, \tag{13.21}$$

$$\omega_0^2 = \frac{4\alpha}{I}, \quad \omega_{1d}^2 = \frac{c_s^2}{c_{s\alpha d}^2}\omega_0^2, \quad \omega_i^2 = \frac{c_s^2(\nu + \beta)\rho}{\mu\nu + \alpha\beta}\omega_0^2, \quad T = \frac{\mu\nu + \alpha\beta}{\rho c_{s\alpha d}^2}. \tag{13.22}$$

Note that

$$q \stackrel{\text{def}}{=} \frac{\operatorname{Im} k^2}{\operatorname{Re} k^2} = -\frac{\omega(\beta\omega_0^2(\omega^2 - \omega_{1d}^2) + T(\omega^2 - \omega_i^2)(\omega^2 - \omega_0^2))}{(\omega^2 - \omega_0^2)(\omega^2 - \omega_{1d}^2) - T\beta\omega_0^2\omega^2(\omega^2 - \omega_i^2)}. \tag{13.23}$$

The seismic attenuation factor Q^{-1} equals

$$Q^{-1} \stackrel{\text{def}}{=} \frac{|\operatorname{Im} k|}{|\operatorname{Re} k|} = \frac{|q|}{1 + (1 + q^2)^{-1/2}}. \tag{13.24}$$

Let us introduce parameters and quantities

$$b = \omega_0 \beta, \quad n = \omega_0 \nu, \quad \Omega = \omega/\omega_0, \tag{13.25}$$

$$\Omega_i^2 = \frac{\omega_i^2}{\omega_0^2} = \frac{\nu + \beta}{\nu + \beta\alpha/\mu} = \frac{n + b}{n + b\alpha/\mu}, \quad \Omega_{1d}^2 = \frac{\omega_{1d}^2}{\omega_0^2} = \frac{1}{1 + \alpha/\mu + nb}, \tag{13.26}$$

$$\tau = \omega_0 T = \frac{\omega_0(\nu + \beta\alpha/\mu)}{1 + \frac{\alpha}{\mu} + \nu\beta\omega_0^2} = \frac{\Omega_{1d}^2}{\Omega_i^2}(n + b). \tag{13.27}$$

Then the dispersion relation (13.20) can be written as

$$k^2 = \frac{\omega_0^2 \Omega^2}{c_{s\alpha d}^2((\Omega^2 - \Omega_{1d}^2)^2 + \tau^2\Omega^2(\Omega^2 - \Omega_i^2)^2)} \Big((\Omega^2 - 1)(\Omega^2 - \Omega_{1d}^2) - \tau b \Omega^2(\Omega^2 - \Omega_i^2) - i\Omega(b(\Omega^2 - \Omega_{1d}^2) + \tau(\Omega^2 - \Omega_i^2)(\Omega^2 - 1))\Big). \tag{13.28}$$

In the following we consider $c_{s\alpha d} = 1$ in all the numerical examples since it only scales the dispersion relation.

13.3.2.1 Low Frequencies

At small Ω relation (13.28) takes the form (up to the higher order terms)

$$k^2 = \frac{{\omega_0}^2}{c_{s\alpha d}^2} \frac{\Omega^2}{\Omega_{1d}^2}(1 - in\Omega) = \frac{\omega^2}{c_s^2}(1 - in\frac{\omega}{\omega_0}), \tag{13.29}$$

$$k = \frac{\omega}{c_s}(1 - i\nu\omega/2), \quad Q^{-1} \to -0 \text{ as } \Omega \to 0. \tag{13.30}$$

13.3.2.2 High Frequencies

As $\Omega \to \infty$, if $\tau \neq 0$ (i.e. viscosity is present), relation (13.28) takes the form up to the higher order terms

$$k^2 = \frac{{\omega_0}^2}{c_{s\alpha d}^2}\Big(\frac{1-\tau b}{\tau^2} - i\frac{\Omega}{\tau}\Big), \tag{13.31}$$

so we have

$$k \approx \frac{\sqrt{\omega}}{c_{s\alpha d}\sqrt{T}} e^{-i\pi/4}, \qquad Q^{-1} \to 1. \tag{13.32}$$

At large Ω attenuation decreases if $\tau b > 1$ and increases vice versa (see Figs. 13.3 and 13.4).

13.3.2.3 Attenuation Factor

Relation (13.23) can be rewritten as

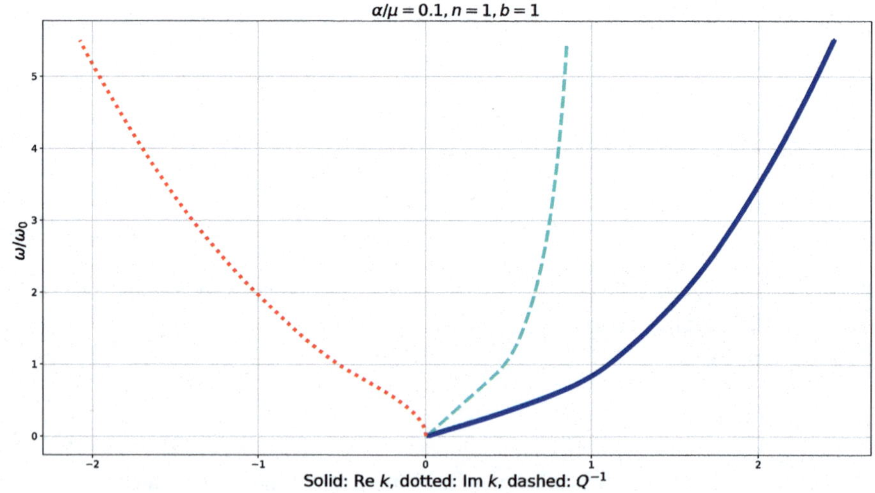

Fig. 13.3 Dispersional characteristics for the shear–rotational wave, $\tau b < 1$

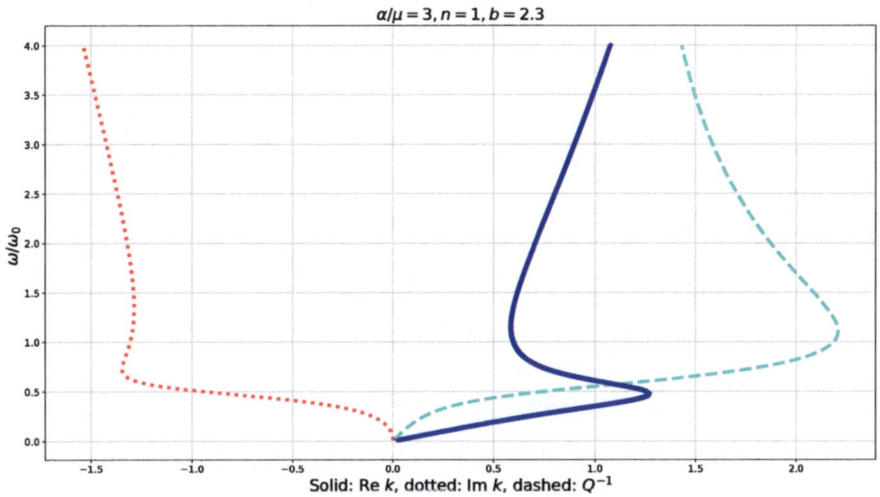

Fig. 13.4 Dispersional characteristics for the shear–rotational wave, $\tau b > 1$

$$\tan(2\arctan Q^{-1}) = \frac{\mathrm{Im}\,k^2}{\mathrm{Re}\,k^2} = -\frac{\Omega(b(\Omega^2-\Omega_{1d}^2)+\tau(\Omega^2-\Omega_i^2)(\Omega^2-1))}{(\Omega^2-1)(\Omega^2-\Omega_{1d}^2)-\tau b\Omega^2(\Omega^2-\Omega_i^2)} \tag{13.33}$$

We observe that Q^{-1} is determined by three parameters: $\alpha/\mu, b, n$. Equation (13.23) does not determine it completely since the sign of $\mathrm{Re}\,k^2$ is important. For instance, if the ratio described by (13.23) has a small absolute value, two cases are possible: small attenuation Q^{-1} in the case $\mathrm{Re}\,k^2 > 0$ and large attenuation if $\mathrm{Re}\,k^2 < 0$ ($|\,\mathrm{Im}\,k| >> |\,\mathrm{Re}\,k|$, phase close to $\pi/2$).

Behaviour of Q^{-1} may change its character depending on the medium parameters. In the numerical examples we tried to represent various types of dependence $Q^{-1}(\omega)$. For the example presented in Fig. 13.3 with $\alpha/\mu = 0.1, n = 1, b = 1$ dependence $Q^{-1}(\omega)$ can be more or less approximated by the power law (13.19), suggested in Anderson and Hart (1978), as follows (see Fig. 13.5):

$$Q^{-1} = 0.51(\omega/\omega_0)^{0.33}. \tag{13.34}$$

Power laws can relatively well approximate micropolar $Q^{-1}(\omega)$ for some domains of parameters and frequencies, but we do not present here similar examples. In the power law model function $Q^{-1}(\omega)$ has no maximum, which is observed in some geophysical experiments Sato et al (2012) and is present in other numerical examples for the viscoelastic reduced Cosserat model (see further).

At $\Omega = 1$ we obtain

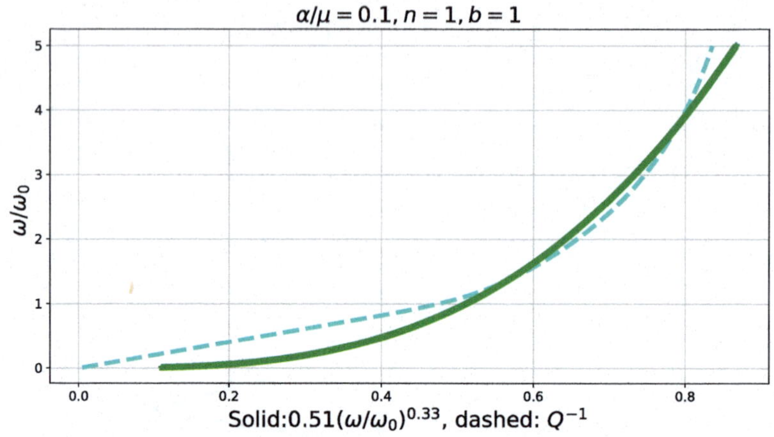

Fig. 13.5 Attenuation factor and its fit to the power law for the shear–rotational wave, $\tau b < 1$

$$\tan(2\arctan Q^{-1}) = \frac{\mathrm{Im}\, k^2}{\mathrm{Re}\, k^2} = \frac{1-\Omega_{1d}^2}{\tau(1-\Omega_i^2)} \tag{13.35}$$

Equation (13.28) yields in

$$\mathrm{sgn}\,\mathrm{Re}\, k^2 = \mathrm{sgn}((1-\tau b)\Omega^4 - \Omega^2(1+\Omega_{1d}^2 - \tau b\Omega_i^2) + \Omega_{1d}^2). \tag{13.36}$$

13.3.2.4 Translational Viscosity ($b = 0$)

Consider particular case $b = 0$, i.e. when there exist only the classical viscosity determined by ν. In this case $\Omega_i = 1$, $c_{s\alpha d} = c_{s\alpha}$ and the dispersion relation (13.28) looks as follows

$$k^2 = \frac{\omega_0^2\Omega^2(\Omega^2-1)}{c_{s\alpha}^2((\Omega^2-\Omega_{1d}^2)^2+\tau^2\Omega^2(\Omega^2-1)^2)}(\Omega^2 - \Omega_{1d}^2 - i\Omega\tau(\Omega^2-1)) \tag{13.37}$$

We observe that at $\Omega = \Omega_{1d}$ we have $\mathrm{Re}\, k^2 = 0$, $\mathrm{Im}\, k^2 < 0$. As $\Omega \to 1$ $k \to 0$, and $\mathrm{Im}\, k^2 \to 0$, $\mathrm{sgn}\,\mathrm{Re}\, k^2 = \mathrm{sgn}(\Omega - 1)$. Therefore, as $\omega \to \omega_0 - 0$ ($\Omega \to 1-0$) we have $\mathrm{Re}\, k/\,\mathrm{Im}\, k \to 0$ and as $\omega \to \omega_0 + 0$ we obtain $\mathrm{Im}\, k/\,\mathrm{Re}\, k \to 0$. This is in accordance to the fact that at zero viscosity we have a band gap below ω_0, and the "optical" branch starts at the cut-off frequency ω_0 as a parabola in the first approximation in the vicinity of ω_0.

Note that eq. (13.37) can also be written in the following form

$$k^2 = -\frac{{\omega_0}^2}{c_{s\alpha}^2}\frac{\Omega F}{1+\tau^2F^2}(1+i\tau F), \tag{13.38}$$

where

$$F(\Omega) = -\Omega \frac{\Omega^2 - 1}{\Omega^2 - \Omega_{1d}^2}. \tag{13.39}$$

It is straightforward to see that in the domain $\Omega > 0$, the function $F(\Omega)$ is positive if and only if $\Omega \in (\Omega_{1d}; 1)$, and it decreases in this domain. As $\Omega \to 1$ we have $F \to 0$, and as $\Omega \to \Omega_{1d} \pm 0$ we have $F \to \pm\infty$.

Using the form of the dispersion relation (13.38), it is straightforward to check that $\Omega(\mathrm{Re}\, k)$ decreases at least as $\Omega \to 1 - 0$ and as $\Omega \to \Omega_{1d} + 0$, which corresponds to the double negative acoustic metamaterial behaviour, $Q^{-1} \to \infty$ as $\Omega \to 1 - 0$, $Q^{-1} \to 0$ as $\Omega \to 1 + 0$, $Q^{-1} \to 1$ as $\Omega \to \Omega_{1d}$. We skip technical details.

Numerical examples show a zone below ω_0 (including ω_{1d}), where $\omega(\mathrm{Re}\, k)$ decreases, and simulations confirm that the attenuation in this zone is very high (Figs. 13.6, 13.7). The part of the wave which passes has a group velocity which is opposite to the phase velocity: in this zone the medium is a double negative acoustic metamaterial with respect to the shear wave.

When the ratio between the Cosserat couple and shear moduli α/μ is small, the zone with high attenuation, where $\omega(\mathrm{Re}\, k)$ decreases, is very narrow (see Fig. 13.7). Then, if we miss in our measurements this zone of frequencies, where the medium behaves as a double negative acoustic metamaterial, we could see that Q^{-1} passes through the maximum, decreases, and then starts to increase again slowly, but not notice the narrow zone of very high attenuation.

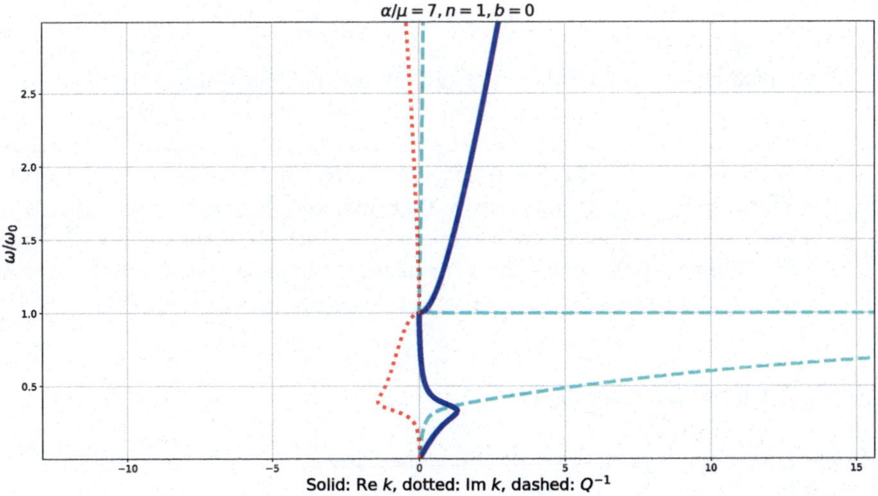

Fig. 13.6 Dispersional characteristics for the case of viscosity with respect to the translational strains

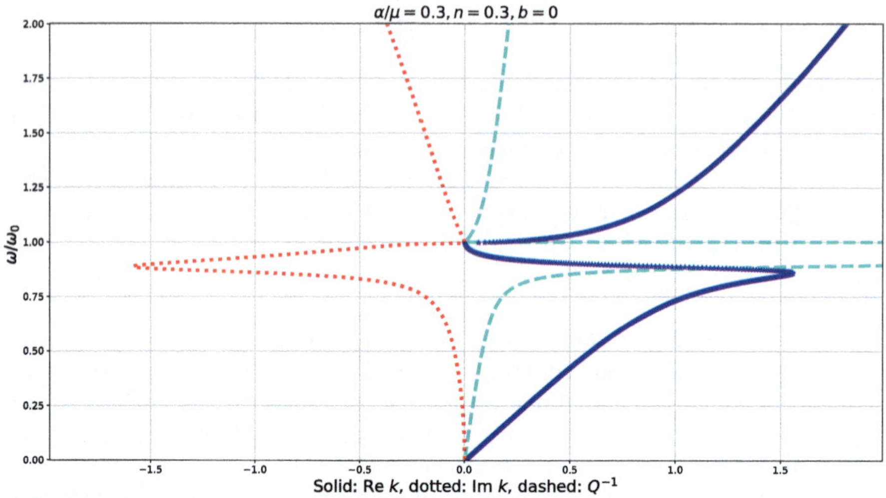

Fig. 13.7 Dispersional characteristics for the case of viscosity with respect to the translational strains, α/μ is small.

Summarising results of this subsection, we can claim that for the case of viscosity with respect to translational strains ($n \neq 0, b = 0$)

1. There is a cut-off frequency ω_0 both for the upper branch of $\mathrm{Re}\, k, \mathrm{Im}\, k$. This is the same frequency as in the purely elastic case. It does not depend on the dissipation coefficient.
2. Attenuation is very low above ω_0 and infinite just below it.
3. There is a frequency band (or frequency bands) starting just below ω_0 and including ω_{1d}, where $\omega(\mathrm{Re}\, k)$ decreases. This means that the medium behaves there as a double negative acoustic metamaterial with respect to the shear wave.
4. In numerical examples we observe that this zone is unique. It is situated just below ω_0 and continues somewhat below ω_{1d}, the attenuation factor Q^{-1} there is very high, and has monotonely increasing with ω parts below and above this zone, starting from 0 in both intervals. It becomes narrow if α/μ is small.

13.3.2.5 Rotational viscosity ($n = 0$)

In this case, parameters can be simplified as follows:

$$\Omega_i^2 = \frac{\mu}{\alpha}, \quad \Omega_{1d}^2 = \frac{\mu}{\mu+\alpha}, \quad \tau = b\,\frac{\alpha}{\mu+\alpha} = b(1-\Omega_{1d}^2) = \frac{b}{1+\Omega_i^2}. \tag{13.40}$$

If $\alpha << \mu$, we have $\Omega_i \to \infty, \Omega_{1d} \to 1$. If $\alpha >> \mu$, we have $\Omega_i \to 0, \Omega_{1d} \to 0$. In general case we may somewhat simplify the dispersion relation; since

$$b(\Omega^2 - \Omega_{1d}^2) + \tau(\Omega^2 - \Omega_i^2)(\Omega^2 - 1) = \Omega^4 b \frac{\alpha}{\mu + \alpha} = \tau\Omega^4, \tag{13.41}$$

equation (13.28) takes form

$$k^2 = \frac{\omega_0^2 \Omega^2}{c_{s\alpha d}^2((\Omega^2 - \Omega_{1d}^2)^2 + \tau^2\Omega^2(\Omega^2 - \frac{\Omega_{1d}^2}{1-\Omega_{1d}^2})^2)}\cdot$$
$$\left((\Omega^2 - 1)(\Omega^2 - \Omega_{1d}^2) - \frac{\tau^2}{1 - \Omega_{1d}^2}\Omega^2(\Omega^2 - \frac{\Omega_{1d}^2}{1 - \Omega_{1d}^2}) - i\tau\Omega^5\right). \tag{13.42}$$

Thus in this case

$$q = \frac{\operatorname{Im} k^2}{\operatorname{Re} k^2} = -\frac{\tau\Omega^5}{(\Omega^2 - 1)(\Omega^2 - \Omega_{1d}^2) - \frac{\tau^2}{1-\Omega_{1d}^2}\Omega^2(\Omega^2 - \frac{\Omega_{1d}^2}{1-\Omega_{1d}^2})} \tag{13.43}$$

Denote $\alpha/\mu = \varepsilon$. Then we can rewrite this expression as follows

$$q = -\frac{\varepsilon b\Omega^5}{(\Omega^2 - 1)(\Omega^2(1 + \varepsilon) - 1) + b^2\Omega^2(1 - \varepsilon\Omega^2)} \tag{13.44}$$

As $\Omega \to 0$, we obtain

$$q \approx -\frac{\varepsilon b\Omega^5}{1 + (b^2 - 2 - \varepsilon)\Omega^2} \approx -\varepsilon b\Omega^5(1 + (2 + \varepsilon - b^2)\Omega^2). \tag{13.45}$$

We have

$$q|_{\Omega \to 0} \to -0, \tag{13.46}$$

$$q|_{\Omega = \Omega_{1d} = 1/\sqrt{1+\varepsilon}} = -\frac{\varepsilon}{b\sqrt{1 + \varepsilon}}. \tag{13.47}$$

At these values of Ω for any medium parameters $\operatorname{Re} k^2 > 0, \operatorname{Im} k^2 < 0$, therefore $\operatorname{Re} k > 0, \operatorname{Im} k < 0$.

For the following values of Ω we have $\operatorname{Re} k^2 > 0, \operatorname{Im} k^2 < 0$, when $\mu > \alpha$, but if $\mu < \alpha$, then $\operatorname{Re} k^2 < 0$:

$$q|_{\Omega=1} = -\frac{\varepsilon}{b(1 - \varepsilon)}, \tag{13.48}$$

$$q|_{\Omega=\Omega_i=1/\sqrt{\varepsilon}} = -\frac{b\sqrt{\varepsilon}}{1 - \varepsilon}. \tag{13.49}$$

Consider separately the case when $\alpha << \mu$, **i.e.** $\varepsilon = o(1)$. Then (13.44) can be simplified. Far from 0 and from 1 up to the higher order terms it is

$$q \approx -\frac{\varepsilon b\Omega^5}{(\Omega^2 - 1)^2 + b^2\Omega^2}, \qquad Q^{-1} \approx |q|/2. \tag{13.50}$$

As $\Omega \to 0$ we have

$$q \approx -\varepsilon b \Omega^5 (1 + (2 - b^2)\Omega^2), \qquad Q^{-1} \approx |q|/2 \to 0 \quad \text{as } \Omega \to 0. \tag{13.51}$$

As $\Omega \to 1$, we obtain for finite b

$$q \approx -\frac{\varepsilon}{b}, \qquad Q^{-1} \approx \frac{\varepsilon}{2b}. \tag{13.52}$$

An example of dispersion curves and the graph of $Q^{-1}(\Omega)$ is given in Fig. 13.8. We observe that if b is small, Q^{-1} has a maximum close to $\omega = \omega_0$, then it decreases and then starts to increase again (slowly in the considered numerical example). This is similar to the behaviour that is observed in a geomedium Sato et al (2012) for shear waves (see equation (13.19)). However, we are still far from trying to fit exactly the experimental data, since our model is very simplified and does not take into account the prestressed state in geomedium, as well as the coupling between shear–rotational and pressure waves.

In Figure 13.8 we observe only the beginning of the dispersion curve. In a larger scale it looks as in Fig. 13.9: at large ω, as it was shown in the beginning of the section, $k \propto \omega^2$.

For $\alpha > \mu$ dispersional behaviour is shown in Fig. 13.10. We observe there a zone where the medium is a double negative acoustic metamaterial, close to this zone the attenuation is high.

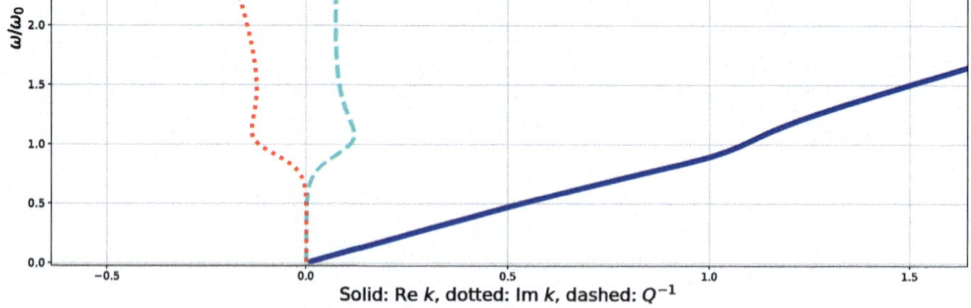

Fig. 13.8 Dispersional characteristics for the case of viscosity with respect to the rotational strains for small α/μ (low frequencies).

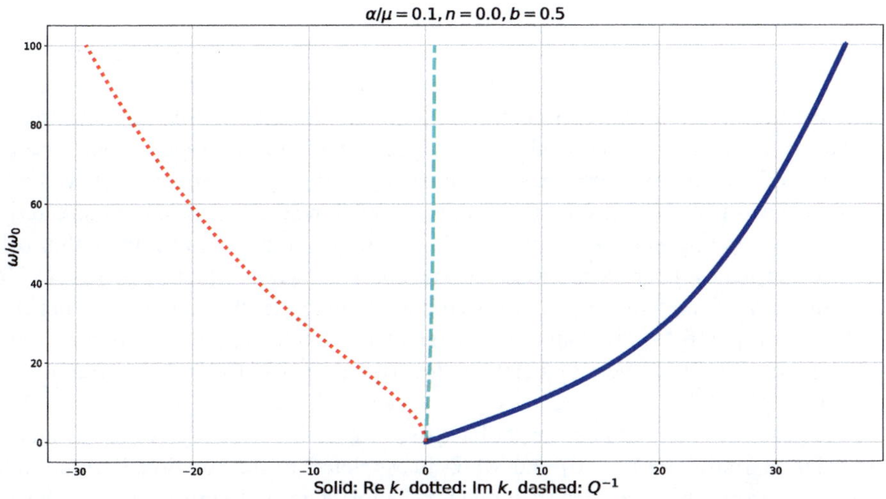

Fig. 13.9 Dispersional characteristics for the case of viscosity with respect to the rotational strains

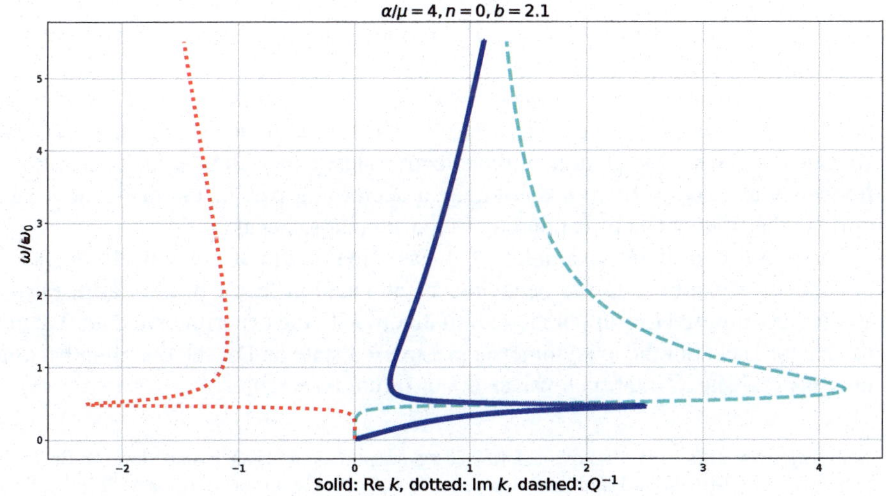

Fig. 13.10 Dispersional characteristics (viscosity with respect to the rotational strains, $\alpha > \mu$).

13.4 Conclusion

We have considered linear isotropic viscoelastic reduced Cosserat medium in the absence of external loads and considered plane waves in this continuum. We observe that the introduction of the viscosity in the model leads to interesting phenomena: band gap present in the reduced elastic Cosserat medium, disappears, but for a big domain of parameters there exist a decreasing part of the dispersion curve $\omega(k)$. In this zone of frequencies the medium is a double negative acoustic metamaterial, energy flux is opposed to the phase velocity. In all numerical examples this zone is situated near a peak of attenuation. We observe these frequency zones also for certain shear–rotational waves in the gyrocontinuum under the action of the negative follower torque (for some domains of parameters) (Grekova, 2018a). We considered asymptotics for the dispersion curves at low (relation (13.30)) and high frequencies (relation (13.32)).

We also investigated two particular cases, when the viscosity is only related to translational strains ($b = 0$) and when it is related to the rotational strains only ($n = 0$). In the first case, the attenuation factor is zero at zero frequency, then it increases rapidly, tends to infinity and decreases abruptly to zero, starting to increase again with frequency. The band gap, present for the medium with the same parameters, but zero viscosity, disappears. In the considered numerical examples the viscosity corresponding only to the translational strains transforms the single negative acoustic metamaterial (elastic reduced Cosserat medium) into the double negative acoustic metamaterial, i.e. the former band gap becomes a zone of high attenuation where $\omega(k)$ decreases.

For rotational viscosity we have a rich variety of behaviour depending on the parameters (especially on α/μ and τb). For instance, at small α/μ in some cases we can observe a rapid change of the effective elastic moduli near the characteristic frequency ω_0, and at $\alpha > \mu$ there exist a decreasing part of the dispersion curve close to the frequency corresponding to the attenuation peak.

In several considered examples Q^{-1} has a maximum at a certain frequency. It is qualitatively similar to the behaviour of attenuation factor in the Earth. Further studies are required to compare it quantitatively with the experimental data. We plan to take into account the axisymmetric prestressed state and/or the coupling between pressure and shear–rotational waves in our future research.

Acknowledgements E.F. Grekova acknowledges support of Russian Foundation for Basic Research (grant 17-01-00230) and partial support of Spanish Government Agency "Ministerio de Economia y Competitividad" (project N FIS2014-54539-P) and of Andalusian Government (group FQM-253). R. Abreu acknowledges support from the MISS ("Minderung der Störwirkung von Windenergieanlagen auf seismologische Stationen") Project.

References

Altenbach H, Zhilin P (1988) A general theory of elastic simple shells. Uspekhi Mekhaniki 11(4):107–148

Anderson D, Hart R (1978) Attenuation models of the earth. Physics of the Earth and Planetary Interiors 16(4):289–306

Bleustein J (1967) A note on the boundary conditions of toupin's strain-gradient theory. International Journal of Solids and Structures 3(6):1053–1057

Chesnais C, Boutin C, Hans S (2012) Effects of the local resonance on the wave propagation in periodic frame structures: Generalized Newtonian mechanics. The Journal of the Acoustical Society of America 132(4):2873–2886

Cosserat E, Cosserat F (1909) Théorie des corps déformables. A. Hermann et fils, in French

dell'Isola F, Madeo A, Placidi L (2012) Linear plane wave propagation and normal transmission and reflection at discontinuity surfaces in second gradient 3d continua. ZAMM-Journal of Applied Mathematics and Mechanics/Zeitschrift für Angewandte Mathematik und Mechanik 92(1):52–71

dell'Isola F, Giorgio I, Andreaus U (2015) Elastic pantographic 2d lattices: a numerical analysis on the static response and wave propagation. Proceedings of the Estonian Academy of Sciences 64(3):219

dell'Isola F, Giorgio I, Pawlikowski M, Rizzi N (2016) Large deformations of planar extensible beams and pantographic lattices: heuristic homogenization, experimental and numerical examples of equilibrium. Proc R Soc A 472(2185):20150,790

dell'Isola F, Seppecher P, Alibert JJ, Lekszycki T, Grygoruk R, Pawlikowski M, Steigmann D, Giorgio I, Andreaus U, Turco E, et al (2018) Pantographic metamaterials: an example of mathematically driven design and of its technological challenges. Continuum Mechanics and Thermodynamics pp 1–34

Eremeyev V (2018) On the material symmetry group for micromorphic media with applications to granular materials. Mechanics Research Communications 94:8–12

Eremeyev V, Pietraszkiewicz W (2012) Material symmetry group of the non-linear polar-elastic continuum. International Journal of Solids and Structures 49(14):1993–2005

Eremeyev V, Pietraszkiewicz W (2016) Material symmetry group and constitutive equations of micropolar anisotropic elastic solids. Mathematics and Mechanics of Solids 21(2):210–221

Eremeyev V, Lebedev L, Altenbach H (2013) Foundations of Micropolar Mechanics. Springer

Eringen A (1967) Linear theory of micropolar viscoelasticity. International Journal of Engineering Science 5(2):191–204

Eringen AC, Maugin GA (2012) Electrodynamics of continua I: foundations and solid media. Springer Science & Business Media

Erofeev V, Pavlov I, Porubov A, Vasiliev A (2018) Dispersion properties of a closed-packed lattice consisting of round particles. In: Generalized Models and Non-classical Approaches in Complex Materials 2, Springer, pp 101–117

Erofeyev V (2003) Wave processes in solids with microstructure, vol 8. World Scientific

Figotin A, Schenker J (2007) Hamiltonian structure for dispersive and dissipative dynamical systems. Journal of Statistical Physics 128(4):969–1056

Frenzel T, Kadic M, Wegener M (2017) Three-dimensional mechanical metamaterials with a twist. Science 358(6366):1072–1074

Gilabert F, Roux JN, Castellanos A (2007) Computer simulation of model cohesive powders: influence of assembling procedure and contact laws on low consolidation states. Physical review E 75(1):011,303

Grekova E (2016) Plane waves in the linear elastic reduced cosserat medium with a finite axially symmetric coupling between volumetric and rotational strains. Mathematics and Mechanics of Solids 21(1):73–93

Grekova E (2017) Waves in the reduced elastic cosserat medium with transversal anisotropy of the coupling between linear rotational and translational deformations: Linearization near natural and axisymmetric prestressed state. special directions. In: Days on Diffraction (DD), 2017, IEEE, pp 147–153

Grekova E (2018a) Harmonic waves in the simplest reduced kelvin's and gyrostatic media under an external body follower torque. In: 2018 Days on Diffraction (DD), IEEE, pp 142–148

Grekova E (2018b) Waves in elastic reduced cosserat medium with anisotropy in the term coupling rotational and translational strains or in the dynamic term. In: Advances in Mechanics of Microstructured Media and Structures, Springer, pp 143–156

Grekova E, Kulesh M, Herman G (2009) Waves in linear elastic media with microrotations, part 2: Isotropic reduced cosserat model. Bulletin of the Seismological Society of America 99(2B):1423–1428

Grekova EF (2019) Reduced enhanced elastic continua as acoustic metamaterials. In: Altenbach H, Belyaev A, Eremeyev VA, Krivtsov A, Porubov AV (eds) Dynamical Processes in Generalized Continua and Structures, Advanced Structured Materials, vol 103, Springer

Hans S, Boutin C, Chesnais C (2014) Atypical dynamic behavior of periodic frame structures with local resonance. The Journal of the Acoustical Society of America 136(4):2077–2077

Kafadar C, Eringen A (1971) Micropolar media — I: the classical theory. International Journal of Engineering Science 9(3):271–305

Kawai K, Tsuchiya T (2015) Small shear modulus of cubic casio3 perovskite. Geophysical Research Letters 42(8):2718–2726

Kulesh M, Grekova E, Shardakov I (2009) The problem of surface wave propagation in a reduced cosserat medium. Acoustical Physics 55(2):218–226

Liu HP, Anderson D, Kanamori H (1976) Velocity dispersion due to anelasticity; implications for seismology and mantle composition. Geophysical Journal of the Royal Astronomical Society 47(1):41–58

Madeo A, Neff P, Aifantis E, Barbagallo G, d'Agostino M (2017) On the role of micro-inertia in enriched continuum mechanics. Proc R Soc A 473(2198):20160,722

Maugin G, Metrikine A (2010) Mechanics of generalized continua. One Hundred Years After the Cosserats. Chapter II. Cosserat media (Rigidly Rotating Microstructure). Springer

Merkel A, Tournat V, Gusev V (2010) Dispersion of elastic waves in three-dimensional noncohesive granular phononic crystals: properties of rotational modes. Physical Review E 82(3):031,305

Merkel A, Tournat V, Gusev V (2011) Experimental evidence of rotational elastic waves in granular phononic crystals. Physical Review Letters 107(22):225,502

Misra A, Poorsolhjouy P (2015) Micro-macro scale instability in 2D regular granular assemblies. Continuum Mechanics and Thermodynamics 27((1-2)):63–82

Misra A, Poorsolhjouy P (2016) Granular micromechanics based micromorphic model predicts frequency band gaps. Continuum Mechanics and Thermodynamics 28(1-2):215–234

Müller W, Vilchevskaya E (2017) Micropolar theory from the viewpoint of mesoscopic and mixture theories. Physical Mesomechanics 20(3):263–279

Noll W (1958) A mathematical theory of the mechanical behavior of continuous media. Archive for rational Mechanics and Analysis 2(1):197–226

Pietraszkiewicz W, Eremeyev V (2009) On natural strain measures of the non-linear micropolar continuum. International Journal of Solids and Structures 46(3-4):774–787

Porubov A, Aero E, Maugin G (2009) Two approaches to study essentially nonlinear and dispersive properties of the internal structure of materials. Physical Review E 79(4):046,608

Sadovskii V, Sadovskaya O (2015) Modeling of elastic waves in a blocky medium based on equations of the cosserat continuum. Wave motion 52:138–150

Sato H, Fehler M, Maeda T (2012) Seismic wave propagation and scattering in the heterogeneous earth, vol 496. Springer

Schwartz L, Johnson D, Feng S (1984) Vibrational modes in granular materials. Physical review letters 52(10):831

Sorokin E (1960) On the internal friction theory of the oscillations of elastic systems. Gosstroyizdat, Moscow In Russian

Turco E (2018) In-plane shear loading of granular membranes modeled as a Lagrangian assembly of rotating elastic particles. Mechanics Research Communications 92:61–66

Twiss R (2009) An asymmetric micropolar moment tensor derived from a discrete-block model for a rotating granular substructure. Bulletin of the Seismological Society of America 99(2B):1103–1131

Twiss R, Marrett R (2010) Determining brittle extension and shear strain using fault-length and displacement systematics: Part i: Theory. Journal of Structural Geology 32(12):1960–1977

Twiss R, Unruh J (2007) Structure, deformation, and strength of the loma prieta fault, northern california, usa, as inferred from the 1989–1990 loma prieta aftershock sequence. GSA Bulletin 119(9-10):1079–1106

Twiss R, Protzman G, Hurst S (1991) Theory of slickenline patterns based on the velocity gradient tensor and microrotation. Tectonophysics 186(3-4):215–239

Twiss R, Souter B, Unruh J (1993) The effect of block rotations on the global seismic moment tensor and the patterns of seismic p and t axes. Journal of Geophysical Research: Solid Earth 98(B1):645–674

Unruh J, Twiss R, Hauksson E (1996) Seismogenic deformation field in the mojave block and implications for tectonics of the eastern california shear zone. Journal of Geophysical Research: Solid Earth 101(B4):8335–8361

Unruh J, Humphrey J, Barron A (2003) Transtensional model for the sierra nevada frontal fault system, eastern california. Geology 31(4):327–330

Vardoulakis I (1989) Shear-banding and liquefaction in granular materials on the basis of a cosserat continuum theory. Ingenieur-Archiv 59(2):106–113

Chapter 14
Numerical Analysis of Free Vibrations of Piezoelectric Cylinders

Alexander Ya. Grigorenko, Igor A. Loza & Sergiy N. Yaremchenko

Abstract Various approaches to solving the linear electroelasticity problems of finite-length cylinders with the discrete-continuous (spline-collocation) and discrete (finite-element) methods are considered. An axisymmetric problem on free vibrations of hollow finite-length cylinders made of piezoelectric materials is solved within the framework of the 3D electroelasticity theory. The lateral surfaces of the cylinders are free of external actions and covered by short-circuited electrodes. The cylinder material is radially polarized. The cylinders with a clamped end are analyzed numerically using two approaches. Practical agreement of the results obtained testifies that the solution is accurate.

Keywords: 3D electroelasticity · Piezoceramic material · Finite element and spline collocation methods · Cylinder of finite length

14.1 Introduction

The wide application of cylindrical elements as functional members of modern acoustoelectronics has motivated significant interest to defining their dynamic characteristics. Piezoelectric solids can be used as actuators in structural control and as sub-elements of more complex smart materials (Giorgio et al, 2015). The dynamic characteristics of the object in a low-frequency range, where the wave length is considerably greater then the size of a structure member (either its thickness or radius), can be sufficiently accurately determined using the applied shell theories (Adelman

Alexander Ya. Grigorenko · Sergiy N. Yaremchenko
S.P. Timoshenko Institute of Mechanics of NAS of Ukraine, Kiev, Ukraine,
e-mail: ayagrigorenko1991@gmail.com, yaremch@gmail.com

Igor A. Loza
National Transport University, Kiev, Ukraine,
e-mail: lozaigor1956@gmail.com

B. E. Abali et al. (eds.), *New Achievements in Continuum Mechanics and Thermodynamics*, Advanced Structured Materials 108,
https://doi.org/10.1007/978-3-030-13307-8_14

et al, 1975; Drumheller and Kalnins, 1970; Hughes, 2012; Piefort, 2001; Shul'ga and Grigorieva, 2009b; Ulitko, 1979). However, in the case of thick-walled members that operate in a high-frequency range, the employment of the applied theories is under a question. For this reason, the problem stated should be considered in three-dimensional formulation.

There are only single studies devoted to investigation of vibrations of cylinders within the framework of the three-dimensional electroelasticity theory. The significant anisotropy of piezoceramic materials in combination with the necessity to allow for a coupled electric field complicates the problem solving in mathematical sense. It should be noted that the problem can be solved analytically only in the case when the physical-mechanical properties of the material manifest certain symmetry, while loading means, type of boundary conditions, and other features are known. It is the case either of the axial polarization of piezoceramic, when the loading function on the lateral surfaces can be expanded into series with respect to cylindrical functions, while the cylinder ends are hinged or the torsional vibrations of a piezoceramic cylinder with circular polarization. These problems can be solved by the means of special functions (Ishihara et al, 2017; Maurini et al, 2006; Paul, 1966; Paul and Nelson, 1996; Paul and Natarajan, 1994b,a; Paul and Venkatesan, 1986). In the case of a more general problem, the method of homogeneous solutions and eigensolution method must be employed as in Grinchenko (1978); Haskins and Walsh (1957). However, in order for these methods to be implemented, it is necessary to have overall information about all roots (real, imaginary, and complex) and the dispersion equation of the problem on propagation of electroelastic waves in an infinite-length cylinder. Defining of these roots meets considerable difficulties. Besides, they should be combined with each other. This process is not simple and what is more important, it can be hardly implemented in the form of a ready algorithm of constructing a computer-aided program. Because of this, numerical methods such ones as finite-element, finite-difference, variational-difference, spline-collocation, and others (Allik and Hughes, 1970; Grigorenko et al, 2012, 2014; Grigorenko and Loza, 2012; Kagawa and Yamabuchi, 1976; Karnaukhov et al, 2001; Ishihara et al, 2018; Reddy, 1981; Shul'ga and Grigorieva, 2009b,a; Tiersten, 1967; Tzou and Zhong, 1994) are basic ones in solving similar problems. Note that the last method demonstrates such advantage as possibility to study dynamic processes in structural members made of functionally gradient materials, whose properties vary continuously along one coordinate or several ones. Moreover, this method makes it possible to satisfy exactly boundary conditions and those on interfaces. As is known, solving of this problem with the finite-element method is either impossible or fraught with large computer losses. Further discussion on inclusion of piezoelectric effect in 3D mechanical theories can be found in Abali and Reich (2017); Maurini et al (2006); Nasedkin and Eremeyev (2013).

14.2 Basic Relations

Vibrations of piezoelectric bodies, as a mechanical process, are described by the continuum equations of linearly elastic deformable solids and, from the viewpoint of electric processes in a piezoelectric continuum, by the equations of electrodynamics. Both groups are interrelated, solved simultaneously and, taken together, form the equations of electroelasticity, which are also called acoustoelectric equations. In particular this includes

- the mechanical equations of motion

$$\mathrm{div}\boldsymbol{\sigma} = \rho \frac{\partial^2 \boldsymbol{u}}{\partial t^2}, \tag{14.1}$$

  where $\boldsymbol{\sigma}$ is the Cauchy stress tensor, ρ is the density of the material, and $\boldsymbol{u}$ is the vector of displacements, and
- Maxwell's equations

$$\mathrm{rot}\boldsymbol{E} = -\frac{1}{c}\frac{\partial \boldsymbol{B}}{\partial t},\ \mathrm{rot}\boldsymbol{H} = -\frac{1}{c}\frac{\partial \boldsymbol{D}}{\partial t},\ \mathrm{div}\boldsymbol{B} = 0,\ \mathrm{div}\boldsymbol{D} = 0, \tag{14.2}$$

  where $\boldsymbol{B}$ is the magnetic flux vector, $\boldsymbol{E}$ is the electric field vector, $\boldsymbol{D}$ is the electric displacement field vector, $\boldsymbol{H}$ is the magnetic field intensity vector, and c is the velocity of light.

Actually, only the first two Maxwell equations are equations of electrodynamics in the sense that they show an explicit time-dependent relation between the electric and magnetic fields: From the first equation it follows that the rate of change of the magnetic field must be rather large. Otherwise the multiplier $\frac{1}{c}$ will decrease it practically to zero. An analogous feature is observed in the second equation for the electric field. In that follows, we consider acoustoelectric vibrations, i.e., Maxwell's equations can be simplified in context with the fact that the acoustic vibrations are much slower than the electromagnetic ones. In reality the acoustic waves propagate at the velocity of sound, and the electromagnetic waves propagate at the velocity of light. If the vibrations are slow, an electromagnetic field splits into an electric field and a magnetic field, which are mutually uncoupled. We are only interested in the electric field because, as will be shown below, it is precisely this field than enters into the constitutive relations for a piezoelectric medium. Thus, in our case, Maxwell's equations take the form:

$$\mathrm{rot}\boldsymbol{E} = 0,\ \mathrm{div}\boldsymbol{D} = 0, \tag{14.3}$$

From the first equation it follows that the electric field has a potential and, hence, the following expression holds:

$$\boldsymbol{E} = -\mathrm{grad}\varphi, \tag{14.4}$$

where φ is the electrostatic potential. These equations are also known as quasistatic approximations of Maxwell's equations or equations of forced electrostatics of dielectrics. Thus, finally, the system of equations that describes vibrations of the piezoelectric continuum has the form:

$$\mathrm{div}\boldsymbol{\sigma} = \rho\frac{\partial^2 \boldsymbol{u}}{\partial t^2}, \ \mathrm{div}\boldsymbol{D} = 0, \ \boldsymbol{E} = -\mathrm{grad}\varphi. \tag{14.5}$$

The relations describing axisymmetric vibrations of a cylinder

$$\begin{aligned}\frac{\partial\sigma_{rr}}{\partial r} + \frac{\sigma_{rr}+\sigma_{\theta\theta}}{r} + \frac{\partial\sigma_{rz}}{\partial z} + \rho\omega^2 u_r = 0,\\ \frac{\partial\sigma_{rz}}{\partial r} + \frac{\sigma_{rz}}{r} + \frac{\partial\sigma_{zz}}{\partial z} + \rho\omega^2 u_z = 0.\end{aligned} \tag{14.6}$$

can be obtained from the motion equations. In (14.6) r, θ, z are cylindrical coordinates σ_{rr}, $\sigma_{\theta\theta}$, σ_{zz}, σ_{rz} are the components of the stress tensor, u_r and u_z are the components of a displacement vector, ω is frequency.

The expressions for the strain tensor expressed in the terms of displacements (Cauchy relations) are

$$\varepsilon_{rr} = \frac{\partial u_r}{\partial r}, \quad \varepsilon_{\theta\theta} = \frac{u_r}{r}, \quad \varepsilon_{zz} = \frac{\partial u_z}{\partial z}, \quad 2\varepsilon_{rz} = \frac{\partial u_r}{\partial z} + \frac{\partial u_z}{\partial r}. \tag{14.7}$$

Electrostatic equations take the form

$$\frac{\partial D_r}{\partial r} + \frac{D_r}{r} + \frac{\partial D_z}{\partial z} = 0, \quad E_r = -\frac{\partial\varphi}{\partial r}; E_z = -\frac{\partial\varphi}{\partial z}, \tag{14.8}$$

where D_r and D_z are the components of the vector of electric-flux density, E_r, E_z are the components of the vector of electric-field strength.

Constitutive relations for a radially polarized piezoceramic material are as follows:

$$\begin{aligned}\sigma_{rr} &= c_{33}\varepsilon_{rr} + c_{13}(\varepsilon_{\theta\theta}+\varepsilon_{zz}) - e_{33}E_r,\\ \sigma_{\theta\theta} &= c_{13}\varepsilon_{rr} + c_{11}\varepsilon_{\theta\theta} + c_{12}\varepsilon_{zz} - e_{13}E_r,\\ \sigma_{zz} &= c_{13}\varepsilon_{rr} + c_{12}\varepsilon_{\theta\theta} + c_{11}\varepsilon_{zz} - e_{13}E_r, \quad \sigma_{rz} = 2c_{55}\varepsilon_{rz} - e_{15}E_z,\\ D_r &= e_{33}\varepsilon_{rr} + e_{13}(\varepsilon_{\theta\theta}+\varepsilon_{zz}) + \varepsilon_{33}E_r, \quad D_z = 2e_{15}\varepsilon_{rz} + \varepsilon_{11}E_z.\end{aligned} \tag{14.9}$$

where c_{ij} are the components of the tensor of elastic modules, e_{ij} are the components of the tensor of piezomodules, ε_{ij} are the components of the tensor of the material dielectric permittivity.

In solving the axisymmetric electroelasticity problem by the finite-element method, we will use the Hamilton variational principle. In this case, the original functional is expressed by

$$J = \pi \int_0^L \int_{r_0}^{r_1} \Big(\sigma_{rr}\varepsilon_{rr} + \sigma_{\theta\theta}\varepsilon_{\theta\theta} + \sigma_{zz}\varepsilon_{zz} + 2\sigma_{rz}\varepsilon_{rz}$$
$$- E_r D_r - E_z D_z - \rho\omega^2 (u_r^2 + u_z^2) \Big) r dr dz. \tag{14.10}$$

where L is the cylinder length, r_0 and r_1 are internal and external radiuses.

Suppose that the cylinder ends $z = 0$ and $z = L$ are clamped, i.e. $u_r = u_z = 0$, and are covered by thin shortened electrodes, $\varphi = 0$. The lateral surfaces are free of external forces, so that $\sigma_{rr} = \sigma_{rz} = 0$, and potential difference $\varphi = V_0 z$ is applied to them. In order to avoid passing of the frequencies corresponding to asymmetric over z modes, function φ is given as variable over z.

14.3 Spline-collocation Method

Adopting the displacements u_r, u_z and electrostatic potential φ appearing in (14.6)–(14.9) as unknowns, we arrive at the governing system of equations

$$\frac{\partial^2 \varphi}{\partial r^2} = -\frac{c_{33}\,\varepsilon_{11} + e_{15}e_{33}}{\Delta}\frac{\partial^2 \varphi}{\partial z^2} + \frac{e_{13}e_{33} - \Delta}{r\Delta}\frac{\partial \varphi}{\partial r} + \frac{e_{33}\left(c_{11} - \rho\,\omega^2 r^2\right) u_r}{r^2 \Delta}$$
$$+ \frac{c_{33}e_{15} - c_{55}e_{33}}{\Delta}\frac{\partial^2 u_r}{\partial z^2} + \frac{c_{33}e_{13}}{r\Delta}\frac{\partial u_r}{\partial r} + \frac{e_{33}(c_{12} - c_{13}) + c_{33}e_{13}}{r\Delta}\frac{\partial u_z}{\partial z}$$
$$+ \frac{c_{33}(e_{13} + e_{15}) - e_{33}(c_{13} + c_{55})}{\Delta}\frac{\partial^2 u_z}{\partial z \partial r},$$

$$\frac{\partial^2 u_r}{\partial r^2} = \frac{e_{33}\,\varepsilon_{11} - e_{15}\varepsilon_{33}}{\Delta}\frac{\partial^2 \varphi}{\partial z^2} + \frac{\varepsilon_{33}\,e_{13}}{r\Delta}\frac{\partial \varphi}{\partial r} + \frac{\varepsilon_{33}\left(c_{11} - \rho\omega^2 r^2\right) u_r}{r^2 \Delta}$$
$$- \frac{\varepsilon_{33}\,c_{55} + e_{15}\,e_{33}}{\Delta}\frac{\partial^2 u_r}{\partial z^2} - \frac{e_{13}\,e_{33} + \Delta}{r\Delta}\frac{\partial u_r}{\partial r} + \frac{\varepsilon_{33}(c_{12} - c_{13}) - e_{13}\,e_{33}}{r\Delta}\frac{\partial u_z}{\partial z}$$
$$- \frac{\varepsilon_{33}(c_{13} + c_{55}) + e_{33}(e_{13} + e_{15})}{\Delta}\frac{\partial^2 u_z}{\partial z \partial r},$$

$$\frac{\partial^2 u_z}{\partial r^2} = -\frac{e_{15}}{c_{55}\,r}\frac{\partial \varphi}{\partial z} - \frac{e_{15} + e_{13}}{c_{55}}\frac{\partial^2 \varphi}{\partial z \partial r} - \left(1 + \frac{c_{12}}{c_{55}}\right)\frac{1}{r}\frac{\partial u_r}{\partial z}$$
$$- \left(1 + \frac{c_{13}}{c_{55}}\right)\frac{\partial^2 u_r}{\partial z \partial r} - \frac{\rho\omega^2 u_z}{c_{55}} - \frac{c_{11}}{c_{55}}\frac{\partial^2 u_z}{\partial z^2} - \frac{1}{r}\frac{\partial u_z}{\partial r},$$

where $\Delta = c_{33}\varepsilon_{33} + e_{33}^2$. The boundary conditions should be expressed through governing functions as well.

Let the solution of the boundary-value problem be

$$\varphi = \sum_{i=0}^{N} \varphi_i(r)\phi_{1i}(z);\ u_r = \sum_{i=0}^{N} u_{ri}(r)\phi_{2i}(z);\ u_z = \sum_{i=0}^{N} u_{zi}(r)\phi_{3i}(z), \tag{14.11}$$

where φ_i, u_{ri}, u_{zi} are unknown functions, while ϕ_{ji} are linear combinations of B-splines of the third-degree that satisfy boundary conditions at the ends. Let us substitute expressions (14.11) into (14.11) and boundary conditions and require for the equations obtained at $N+1$ collocation points to be satisfied. As a result, we arrive at one-dimensional boundary-value problem for the frequency ω, to solve which we will use the discrete-orthogonalization method. Next, applying the step-by-step search method, i.e., by varying ω with some step, we define eigenfrequencies.

14.4 Finite-element Method

Considering (14.7) – (14.9) equality (14.10) becomes

$$J = \pi \int_0^L \int_{r_0}^{r_1} \Big\{ c_{33} \left(\frac{\partial u_r}{\partial r}\right)^2 + c_{11} \left(\frac{\partial u_z}{\partial z}\right)^2 + c_{55} \left(\frac{\partial u_r}{\partial z}\right)^2 + c_{55} \left(\frac{\partial u_z}{\partial r}\right)^2$$
$$- \varepsilon_{33} \left(\frac{\partial \varphi}{\partial r}\right)^2 - \varepsilon_{11} \left(\frac{\partial \varphi}{\partial z}\right)^2 + c_{11} \left(\frac{u_r}{r}\right)^2 + 2c_{13}\frac{u_r}{r}\frac{\partial u_r}{\partial r} + 2c_{12}\frac{u_r}{r}\frac{\partial u_z}{\partial z}$$
$$+ 2e_{13}\frac{u_r}{r}\frac{\partial \varphi}{\partial r} + 2c_{13}\frac{\partial u_r}{\partial r}\frac{\partial u_z}{\partial z} + 2e_{33}\frac{\partial u_r}{\partial r}\frac{\partial \varphi}{\partial r} + 2e_{13}\frac{\partial u_z}{\partial z}\frac{\partial \varphi}{\partial r}$$
$$+ 2e_{15}\frac{\partial u_r}{\partial z}\frac{\partial \varphi}{\partial z} + 2e_{15}\frac{\partial u_z}{\partial r}\frac{\partial \varphi}{\partial z} + 2c_{55}\frac{\partial u_r}{\partial z}\frac{\partial u_z}{\partial r} - \rho\omega^2 \left(u_r^2 + u_z^2\right) \Big\} r dr dz \tag{14.12}$$

To solve the problem, we will employ the fourth-node rectangular finite elements [12]. Let the solution be

$$\varphi = \sum_{i=1}^{4} \varphi_i N_i; u_r = \sum_{i=1}^{4} u_{ri} N_i; u_z = \sum_{i=1}^{4} u_{zi} N_i, \tag{14.13}$$

where φ_i, u_r, and u_z are the values of the unknown functions at the nodes of the finite element, N_i are the shape functions.

Substituting (14.13) into (14.12) with the condition $\delta J = 0$ we get the following system of equations for the unit area S_k

$$2\pi \iint_{S_k} \Big\{ c_{33} r \left(u_{ri} N_{ri}\right) N_{rj} + c_{55} r \left(u_{ri} N_{zi}\right) N_{zj} + \Big(\frac{c_{11}}{r} - r\rho\omega^2\Big)(u_{ri} N_i) N_j$$
$$+ c_{13}((u_{ri} N_{ri}) N_j + (u_{ri} N_i) N_{rj}) + c_{12}(u_{zi} N_{zi}) N_j + c_{55} r (u_{zi} N_{ri}) N_{zj}$$
$$+ c_{13} r (u_{zi} N_{zi}) N_{rj} + e_{33} r (\varphi_i N_{ri}) N_{rj} + e_{15} r (\varphi_i N_{zi}) N_{zj}$$
$$+ e_{13}(\varphi_i N_{ri}) N_j \Big\} dr dz = 0,$$

$$2\pi \iint\limits_{S_k} \Big\{ c_{12}(u_{ri}N_i)N_{zj} + c_{13}r(u_{ri}N_{ri})N_{zj} + c_{55}r(u_{ri}N_{zi})N_{rj}$$
$$+c_{11}r\left(u_{zi}N_{zi}\right)N_{zj} + c_{55}r\left(u_{zi}N_{ri}\right)N_{rj} - r\rho\omega^2(u_{zi}N_i)N_j$$
$$+e_{13}r(\varphi_iN_{ri})N_{zj} + e_{15}r(\varphi_iN_{zi})N_{rj}\Big\}drdz = 0,$$

$$2\pi \iint\limits_{S_k} \Big\{ 2e_{13}(u_{ri}N_i)N_{rj} + e_{33}r(u_{ri}N_{ri})N_{rj} + e_{15}r(u_{ri}N_{zi})N_{zj}$$
$$+e_{13}r(u_{zi}N_{zi})N_{rj} + e_{15}r(u_{zi}N_{ri})N_{zj} - \varepsilon_{33}r\left(\varphi_iN_{ri}\right)N_{rj} \quad (14.14)$$
$$-\varepsilon_{11}r\left(\varphi_iN_{zi}\right)N_{zj}\Big\}drdz = 0.$$

In (14.14) it is assumed that the summation is carried out over the paired indexes, i.e.,

$$u_iN_i = \sum_{i=1}^{4} u_iN_i,$$

for example. Here the following notion is used:

$$N_{ri} = \frac{\partial N_i}{\partial r}, \qquad N_{zi} = \frac{\partial N_i}{\partial z}.$$

To perform integration in (14.14) we have involved the Gauss quadrature method. The global system obtained by assembling, with the boundary conditions on the lateral surfaces of the cylinder, is inhomogeneous. The linear systems of algebraic equations obtained have been solved with the Gauss method. Thus, as in the case of the spline-collocation method, we will determine the eigenfrequencies of the cylinder vibration using the step-by-step solving of the system of linear algebraic equations for the certain value of ω.

14.5 The Results Obtained

Let us compare the results obtained for the cylinder of PZT-4 piezceramic with different methods. The cylinder has the following input parameters: $r_0 = 3$ cm, $r_1 = 5$ cm, $L = 10$ cm, $\rho = 7.5{\cdot}10^3$kg/m^3, $c_{11} = 13.9{\cdot}10^{10}$N/m^2, $c_{12} = 7.43{\cdot}10^{10}$N/m^2, $c_{13} = 7.78 \cdot 10^{10}$N/m^2, $c_{33} = 11.5 \cdot 10^{10}$N/m^2, $c_{55} = 2.56 \cdot 10^{10}$N/m^2, $e_{13} = -5.2$ C/m^2, $e_{15} = 12.7$ C/m^2, $e_{33} = 15.1$ C/m^2, $\varepsilon_{11} = 730\varepsilon_0$; $\varepsilon_{33} = 635\varepsilon_0$, $V_0 = 1$ V. ε_0 is vacuum permittivity. Such dimensionless quantities were used in the calculations: $\tilde{r} = r/h$, $\tilde{z} = z/h$, $\tilde{u}_r = u_r/h$, $\tilde{u}_z = u_z/h$, $\tilde{\varphi} = \varphi\sqrt{\varepsilon_0}/(h\sqrt{\lambda})$, $\Omega = \omega h\sqrt{\rho/\lambda}$, $\tilde{c}_{ij} = c_{ij}/\lambda$, $\tilde{e}_{ij} = e_{ij}/\sqrt{\varepsilon_0\lambda}$, $\tilde{\varepsilon}_{ij} = \varepsilon_{ij}/\varepsilon_0$, where $\lambda = 10^{10}$N/m^2, $h = (r_1 - r_0)/2$.

Fig. 14.1 shows how the first four frequencies are determined using as an example finding of the resonance of displacements u. The displacement distributions obtained with the finite-element method are shown by solid line, while those obtained with the spline-collocation method are shown by dotted line. In this case,

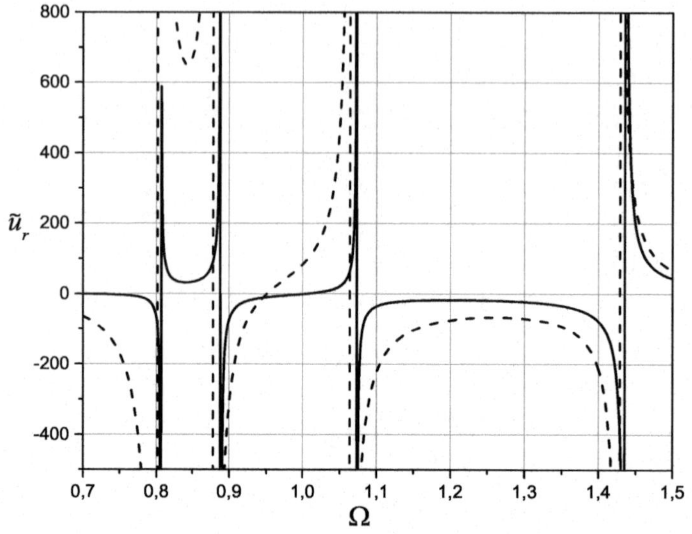

Fig. 14.1 Frequency-dependent distributions of the displacement $\tilde{u}_r$ (continuous line: FEM, dashed line: SCM)

in employing the spline-collocation method, we have chosen 20 collocation points ($N = 19$) and 2000 of 0.1×0.1 square elements.

Six frequencies at different levels of approximation, i.e., for 16, 20, and 24 collocation points in the spline-collocation method and K being equal to 500, 2000, and 8000 in the case of the finite-element method, are collected in Table 1. The linear dimensions of the element at $K = 500$ are doubled, while at $K = 8000$ they decrease in comparison with $K = 2000$.

As is seen from the plots and Table 14.1, the results obtained with both methods agree with high accuracy. Values of the frequencies obtained with spline-collocation method at $N = 19$ and 23 differ insignificantly as in the case of the finite-element method at $K = 2000$ and 8000. The difference in the results obtained by these methods does not exceed 0.8%.

14.6 Conclusions

Using the spline-collocation method with discrete orthogonalization in combination with the finite-element method, we have studied vibrations of a radially polarized piezoceramic cylinder. The values of the frequencies determined with different methods are compared. As it is shown, they are agreed with high accuracy.

Table 14.1 Values of the frequencies obtained with different methods

Ω	SCM			FEM		
	$N=15$	$N=19$	$N=23$	$K=500$	$K=2000$	$K=8000$
Ω_1	0.8	0.802	0.803	0.807	0.806	0.806
Ω_2	0.875	0.879	0.881	0.891	0.889	0.888
Ω_3	1.062	1.065	1.066	1.076	1.074	1.073
Ω_4	1.431	1.428	1.428	1.441	1.436	1.434
Ω_5	1.841	1.846	1.849	1.866	1.862	1.861
Ω_6	1.999	1.99	1.988	2.002	1.992	1.989

References

Abali BE, Reich FA (2017) Thermodynamically consistent derivation and computation of electro-thermo-mechanical systems for solid bodies. Computer Methods in Applied Mechanics and Engineering 319:567–595

Adelman N, Stavsky Y, Segal E (1975) Axisymmetric vibrations of radially polarized piezoelectric ceramic cylinders. Journal of Sound and Vibration 38(2):245–254

Allik H, Hughes TJ (1970) Finite element method for piezoelectric vibration. International Journal for Numerical Methods in Engineering 2(2):151–157

Drumheller DS, Kalnins A (1970) Dynamic shell theory for ferroelectric ceramics. The Journal of the Acoustical Society of America 47(5B):1343–1353

Giorgio I, Galantucci L, Della Corte A, Del Vescovo D (2015) Piezo-electromechanical smart materials with distributed arrays of piezoelectric transducers: current and upcoming applications. International Journal of Applied Electromagnetics and Mechanics 47(4):1051–1084

Grigorenko AY, Loza IA (2012) Solution of the problem of nonaxisymmetric free vibrations of piezoceramic hollow cylinders with axial polarization. Journal of Mathematical Sciences 184(1):69–77

Grigorenko AY, Müller WH, Wille R, Loza IA (2012) Nonaxisymmetric vibrations of radially polarized hollow cylinders made of functionally gradient piezoelectric materials. Continuum Mechanics and Thermodynamics 24(4-6):515–524

Grigorenko AY, Müller WH, Wille R, Loza IA (2014) Nonaxisymmetric electroelastic vibrations of a hollow sphere made of functionally gradient piezoelectric material. Continuum Mechanics and Thermodynamics 26(6):771–781

Grinchenko VT (1978) Equilibrium and Steady-State Vibrations of Finite Elastic Bodies (in Russian). Naukova Dumka, Kiev

Haskins JF, Walsh JL (1957) Vibrations of ferroelectric cylindrical shells with transverse isotropy. i. radially polarized case. The Journal of the Acoustical Society of America 29(6):729–734

Hughes TJ (2012) The Finite Element Method: Linear Static and Dynamic Finite Element Analysis. Courier Corporation

Ishihara M, Ootao Y, Kameo Y, Saito T (2017) Thermoelectroelastic response of a piezoelectric cylinder with d^∞ symmetry under axisymmetric mechanical and thermal loading. Mechanical Engineering Journal 4(5):16–00,609

Ishihara M, Ootao Y, Kameo Y (2018) Analytical technique for thermoelectroelastic field in piezoelectric bodies with d^∞ symmetry in cylindrical coordinates. Journal of Thermal Stresses 41(1):17–36

Kagawa Y, Yamabuchi T (1976) Finite element approach for a piezoelectric circular rod. IEEE Transactions on Sonics and Ultrasonics 23(6):379–385

Karnaukhov VG, Kirichok IF, Kozlov VI (2001) Electromechanical vibrations and dissipative heating of viscoelastic thin-walled piezoelements. International Applied Mechanics 37(2):182–212

Maurini C, Pouget J, dell'Isola F (2006) Extension of the Euler–Bernoulli model of piezoelectric laminates to include 3D effects via a mixed approach. Computers & Structures 84(22-23):1438–1458

Nasedkin AV, Eremeyev VA (2013) Spectral properties of piezoelectric bodies with surface effects. In: Surface Effects in Solid Mechanics, Springer, pp 105–121

Paul HS (1966) Vibrations of circular cylindrical shells of piezoelectric silver iodide crystals. The Journal of the Acoustical Society of America 40(5):1077–1080

Paul HS, Natarajan K (1994a) Axisymmetric free vibrations of piezolectric finite cylindrical bone. The Journal of the Acoustical Society of America 96(1):213–220

Paul HS, Natarajan K (1994b) Flexural vibration in a finite piezoelectric circular cylinder of crystal class 6 mm. International Journal of Engineering Science 32(8):1303–1314

Paul HS, Nelson VK (1996) Flexural vibration of piezoelectric composite hollow cylinder. The Journal of the Acoustical Society of America 99(1):309–313

Paul HS, Venkatesan M (1986) Axisymmetric vibration of a piezoelectric solid cylinder guided by a thin film. The Journal of the Acoustical Society of America 80(4):1091–1096

Piefort V (2001) Finite element modelling of piezoelectric active structures. PhD thesis, Ph. D. thesis. Bruxelles, Belgium: Université Libre de Bruxelles, Department ...

Reddy JN (1981) Finite-element modeling of layered, anisotropic composite plates and shells: A review of recent research

Shul'ga NA, Grigorieva LO (2009a) Radial elaso-electric nonstationary oscillations of piezoceramic cylinder under mechanical loading. International Applied Mechanics 45:466–471

Shul'ga NA, Grigorieva LO (2009b) Radial electromechanical non-stationary vibrations of a hollow piezoceramic cylinder under elastic excitation. International Applied Mechanics 45:230–235

Tiersten HF (1967) Hamilton's principle for linear piezoelectric media. Proceedings of the IEEE 55(8):1523–1524

Tzou HS, Zhong JP (1994) A linear theory of piezoelastic shell vibrations. Journal of Sound and Vibration 175(1):77–88

Ulitko AF (1979) Method of Eigen-Vector Functions in Spatial Elasticity Problems (in Russian). Naukova Dumka, Kyiv

Chapter 15
Qualitative Investigations of Experiments Performed on 3D-FDM-printed Pantographic Structures Made out of PLA

Arion Juritza, Hua Yang & Gregor Ganzosch

Abstract Additive manufacturing methods, commonly known as 3D printing, enable the design and manufacturing of complex and sophisticated material fabrics with a special substructure resulting in extraordinary macroscopic deformation behavior. Such a man-made structure is also referred to as a metamaterial. So called pantographic structures, which can be described as metamaterials with a substructure that is composed of two orthogonal arrays of beams connected by internal cylinders, were manufactured using fused deposition modeling technique. In order to further understand the peculiarity of its deformation behaviors, a plane sheet was also printed to be used as a comparison. Different types of experiments were performed and evaluated qualitatively by the means of digital image correlation being able to localize the initial area of out-of-plane movements in shearing tests for both specimen. Results of quasi-static standard tension and shearing tests indicate a resilient material behavior during high elastic deformations resulting in a high resistance against total failure of the structure. Furthermore, cyclic long-term tests show a viscoelastic deformation behavior of the thermoplastic material. PSs show linear as well as non-linear elastic deformation response in all experiments except the cyclic tension test.

Keywords: Experiment · Metamaterial · Pantographic structure · Digital image correlation · Elasticity

Arion Juritza · Hua Yang · Gregor Ganzosch
Institute of Mechanics, Chair of Continuum Mechanics and Materials Theory, Technische Universität Berlin, Germany,
e-mail: juritza@tu-berlin.de, hua.yang@campus.tu-berlin.de, ganzosch@tu-berlin.de

B. E. Abali et al. (eds.), *New Achievements in Continuum Mechanics and Thermodynamics*, Advanced Structured Materials 108,
https://doi.org/10.1007/978-3-030-13307-8_15

15.1 Introduction

The development of 3D-printing technology enables the fabrication of structures with high degrees of complexity. Such complex structures are known as metamaterials whose effective properties are carefully designed by tailoring the so-called microscopic constituents (Gibson, 2005; Gibson and Ashby, 1999; Barchiesi et al, 2018b). Due to extraordinary properties like light weight and high stiffness, metamaterials have gained considerable attention in the recent past. One of the common methods to construct metamaterials is to repeat its substructures engineered to achieve special desired properties. For example, structures with negative Poisson's ratio, large non-linear elastic deformations because of temperature gradients, instabilities, or buckling of metamaterials are exploited to trigger those desired behaviors. It should be noted that the mechanical performance of metamaterials depends not only on the global structure, but also on the morphology of its substructure.

In order to design and manufacture metamaterials for specific engineering applications, a possibility to predict their performance can be achieved with the aid of finite element method involving detailed mesh of substructure (Yang et al, 2018; Tekoğlu and Onck, 2008; Chen and Fleck, 2002). This is a cumbersome process and computationally very expensive due to the large numbers of unknowns. Because of these reasons homogenization of such discrete structures towards an equivalent Cauchy continuum has been an active research field for many years (Ghosh et al, 1996; Bensoussan et al, 2011; Noor, 1988). The effective properties of the homogenized cellular structure can be predicted with a good agreement with experimental data (Sun and Vaidya, 1996).

Nevertheless, the classical homogenization encounters limitations in many situations (Li, 2011; Askes and Aifantis, 2011). For example, if the variation of macroscopic stress or strain fields is rapid over the length scale of the whole domain, or if the size of the domain is in a comparable order of scale with respect to the considered specimen. In these cases, the macroscopic stress or strain fields can no longer be regarded as uniform in the scale of the domain, and their gradients cannot be neglected in homogenization procedure. In other words, notably size effect can not be captured by standard homogenization with classical elasticity theory; thus classical first gradient Cauchy theories need to be improved by introducing either additional degrees of freedom like the Cosserat medium (Cosserat and Cosserat, 1896, 1909), micropolar parameters (Eringen and Suhubi, 1964; Eringen, 1968; Müller and Vilchevskaya, 2017, 2018) or additional higher order gradients (Mindlin and Eshel, 1968; Toupin, 1962; Abali et al, 2017, 2015; dell'Isola et al, 2009; Placidi et al, 2018b,a). The micropolar continuum presents several limitations to the study of localization of deformations when the gradients of stresses and strains are not excessive; a second order gradient continuum is more suitable for this purposes (Kumar and McDowell, 2004; Rahali et al, 2015; Barchiesi et al, 2018a). A pantographic structure (PS) is a metamaterial which is composed of two orthogonal arrays of beams connected by internal cylinders called pivots (Placidi et al, 2016c), see Fig. 15.1. Interestingly, PSs are reported to investigate the higher gradient theory and thus it is regarded as a kind of second gradient continuum. A large supply

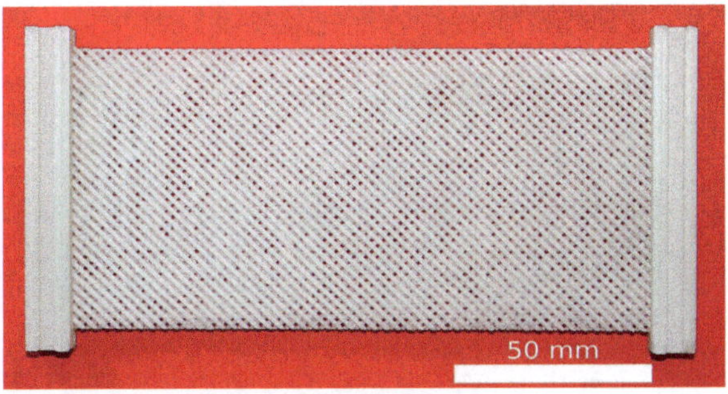

Fig. 15.1 In-house manufactured specimen made out of polylactic acid (PLA): Metamaterial with pantographic substructure (Sample **A**).

of papers (Placidi et al, 2016b) has been published with regarding to the necessities and ways on modeling the PS as a second order gradient continuum (Boutin et al, 2017). However, new material parameters are introduced into the constitutive laws and experiments have to be designed to calibrate and determine these parameters (Placidi et al, 2016a). In this paper results of different types of experiments applied to in-house manufactured metamaterials with pantographic substructure (sample **A**, Fig. 15.1) will be discussed qualitatively. In order to interpret the peculiar deformation behaviors of PSs, those experimental results will be compared to a plane sheet structure without a visible macroscopic substructure (sample **B**, Fig. 15.3b). The metamaterial shows a high resilient and linear as well as non-linear elastic material behavior.

15.2 Materials and Methods

Experiments performed at Technische Universität Berlin, Institute of Continuum Mechanics and Material Theories, are divided into quasi-static and cyclic standard extension tests as well as shearing tests respectively. Experimental setups are presented in Fig. 15.2).

PSs, consisting of rectangular beams (with a quadratic cross-section of $A = a \times a = 1\,\text{mm} \times 1\,\text{mm}$) and cylindrical joints (with a diameter of $d_\text{p} = 1\text{mm}$ and a pivot height of $h_\text{p} = 1\,\text{mm}$), were 3D-printed using Fused Deposition Modeling (FDM) procedure at Technische Universität Berlin, Institute of Continuum Mechanics and Material Theories. 3D-CAD-models were converted into gCode, which was used as input for the in-house located FDM-printer (Ultimaker 3 Extended, Ultimaker B.V., Geldermalsen, Netherlands) controlled by the device own software Cura. Polylactic acid (PLA - Ultimaker B.V., Geldermalsen, Netherlands) was used as raw material for all investigated structures. To increase the printing quality of the specimen, water-soluble polyvinyl acetate (PVA - Ultimaker B.V., Geldermalsen, Netherlands) was used additionally as support-structure during the manufacturing process and solved in water for 24 hours afterwards.

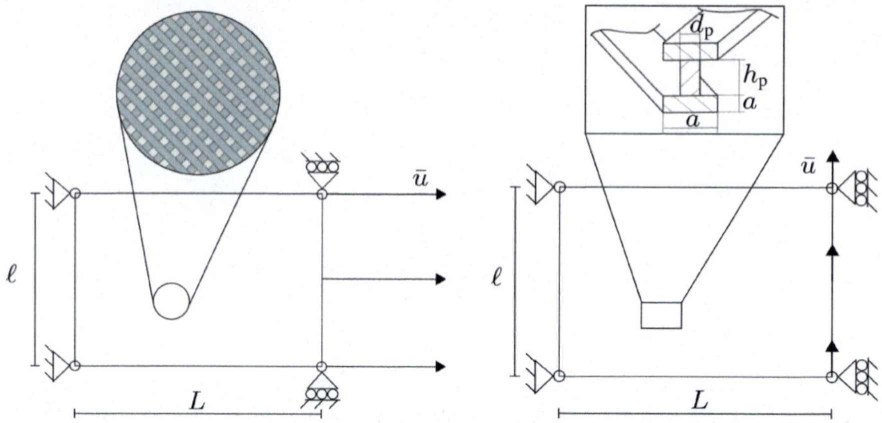

Fig. 15.2 Schematic setups of different experiments performed on sample **A**: setup of tension test on the left and setup of shearing test on the right.

Since the microscopic substructures influence the macroscopic deformation behaviors significantly (Rahali et al, 2015; Barchiesi et al, 2018a), two variations of specimens with different geometries but equal material densities were investigated. Figures 15.3a and 15.3b show both types of specimens: rectangular metamaterial consisting of pantographic substructure (sample **A**) and rectangular plane structure (sample)**B**). Details of the dimensions can be found in Tab. 15.1.

Table 15.1 Outer dimensions (L = width, l = height, t = depth) of sample **A** and sample **B** as well as inner dimensions of the substructure of sample **A** corresponding to the schematic in Fig. 15.2 (a = quadratic width and height of beam, d_p = diameter of pivot, h_p = height of pivot).

Sample	L [mm]	l [mm]	t [mm]	a [mm]	d_p [mm]	h_p [mm]
A	140	70	3	1	1	1
B	140	70	1	–	–	–

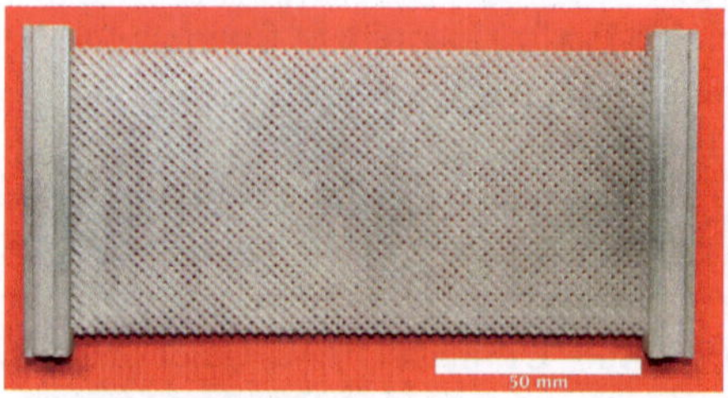

(a) Speckled sample **A**.

(b) Speckled sample **B**.

Fig. 15.3 Investigated specimens after speckling procedure was applied (undeformed state).

With the help of the non-invasive measurement method digital image correlation (DIC) it is possible to measure and calculate displacements and strains on a surface. Therefore, a camera records the surface during deformation with the help of so-called facets. Each facet consists of a rectangle in which pixels of different contrasts are recorded during the deformation. This pixel-data is binarized and displacement vectors of each pixel are calculated in real time during the deformation-process. To achieve better results, a speckled surface is needed. Because of lack of contrast all specimen had to be speckled. With the help of an airbrush-compressor-system a speckle pattern of black ink was sprayed on top of the prime. A homogeneous pattern was achieved by exploiting the ink-inertia. Figure 15.3a and Fig. 15.3b show the full speckled specimen before the experiments were performed. DIC evaluation was performed by means GOM Correlate software (GOM GmbH, Braunschweig, Germany).

Figure 15.4 shows the deformation of sample **A** during a tension test. A MTS Tytron 250 testing-device (MTS Systems Corporation, Eden Prairie, USA) controlled by the software Stationsmanager V 3.14 was used for all experiments. Reaction-force was measured by a device-own load cell, which is able to record axial forces in a range of $N = \pm 250$ N with an accuracy of 0.2 percent. The displacement was imposed horizontally in case of tension load on the right side of the specimen with a loading rate of $v = 15$ mm/min (displacement-controlled). In shear experiments displacement was imposed on the upper side of the specimen horizontally. Displacement was measured and monitored by a device-own encoder unit. Almost frictionless movement was achieved by using an air-film-bearing. External vibration was avoided by arranging the system horizontally on a massive concrete-substructure. Pictures of the surface during deformation were taken (1 picture/2 seconds) by means of a commercial Canon EOS 600D camera with a resolution of about 4272×2848 pixels by arranging the camera orthogonal to the plane surface of the specimen. Each picture was synchronized with the recorded force-displacement data in real time.

Figure 15.4 shows the failure of the structure during plastic deformation of sample **A** in a tension test and Fig. 15.5 shows the deformation of sample **A** leading to total failure in a shearing test. Both cases will be described and discussed in Sect. 15.3.1 for quasi-static tests and in Sect. 15.3.2 for cyclic tests.

15.3 Results and Discussion

Quasi-static tensile and shearing tests as well as cyclic tensile and shearing tests were performed on sample **A** as well as on sample **B** respectively. Reaction-force and its applied displacement were recorded and translated into stress-strain diagrams to enable a qualitative comparison of both specimen.

15.3.1 Quasi-static experiments

Deformation of sample **A** in an extension test is shown in Fig. 15.4. Corresponding stress-strain diagram of same experiment is plotted in Fig. 15.6a. Almost linear elastic material behavior can be recognized up to about 4 percent of elongation (Fig. 15.4a). Then non-linear behavior and plastic deformation occurs leading to permanent necking of the global structure, which is in good agreement with the DIC evaluation (Fig. 15.4b). First rupture of a pivot occurs in the lower right region of Fig. 15.4c next to the mounting reinforcement at an elongation of about 6.2 percent. Notice, that at this point the structure resists total failure and is able to absorb even higher loads after the first failure (Fig. 15.4c). Because of the complex geom-

(a) Before deformation: mounting on the left is fixed while axial-displacement is induced by the mounting on right hand side.

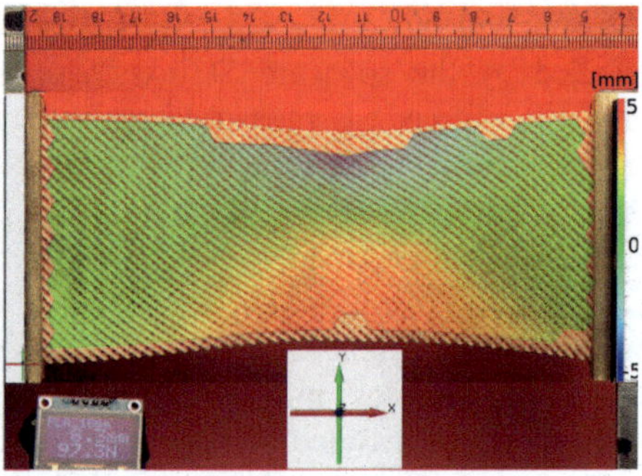

(b) Before first failure: major constriction/necking of the structure can be recognized with the help of DIC (about $u_{yy} = \pm 5$ mm).

(c) After first failure: local broken pivots on the lower right corner of the sample do not lead to total failure of the whole structure (rupture not visible).

(d) After maximal deformation: structure still resists total failure and shows high resilient material behavior. Local rupture of a beam is visible in the lower right corner.

Fig. 15.4 Axial deformation of sample **A** during quasi-static tensile test.

(a) Before deformation: mounting on the bottom is fixed while shear-displacement is induced by the upper mounting.

(b) Before first failure: symetric out-of-plane movement of the structure can be recognized.

(c) After first failure: local broken beams on the lower left corner of the sample lead to failure of the whole structure (rupture not visible).

(d) Before total failure: structure still resists total failure and shows high resilient material behavior. Local rupture can be recognized on the left bottom next to the mounting.

(e) After total failure: structure divided into two broken pieces.

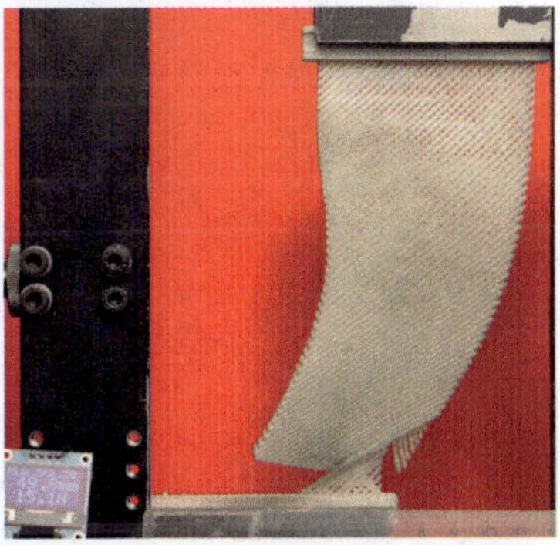

(f) After maximal deformation: broken pieces touching each other explain the measured forces.

Fig. 15.5 Shear deformation of sample **A** during quasi-static shearing test.

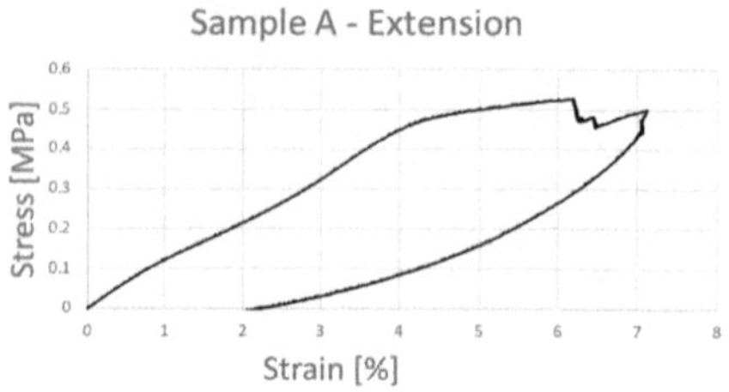

(a) Stress-strain diagram of extension test.

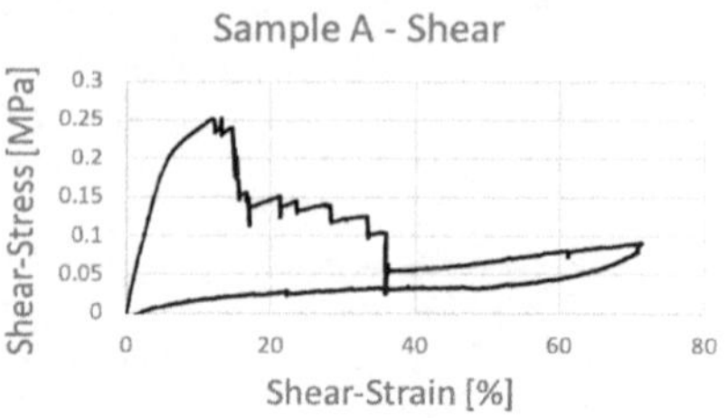

(b) Stress-strain diagram of shearing test.

Fig. 15.6 Stress-strain diagrams of sample **A** during quasi-static experiments.

etry, beams and pivots reorganize themselves indicating a smart resilient material behavior (Cennamo et al, 2014; Ganzosch et al, 2017) of the whole structure.

The same resistance deformation behavior to outer loads was observed in the shear-test of sample **A**. Deformation of sample **A** is shown in Fig. 15.5. Corresponding stress-strain diagram of same experiment is plotted in Fig. 15.6b. Almost linear elastic material behavior can be recognized up to about 5.6 percent of elongation (Fig. 15.5a). Symmetric out-of-plane movement starts to show up inside the elastic range (Fig. 15.5b). Because of plastic deformation a first rupture of a beam in the lower left region of Fig. 15.5c next to the mounting reinforcement appears at an elongation of about 12.08 percent and corresponding load of 251 kPa. This sample is capable to resist considerable external shear loads without leading to complete failure of the whole structure. Because of the complex geometry, beams and pivots reorganize themselves resulting in an higher resistance to outer load, so that even higher loads can be carried after failure (at 13.25 percent of elongation at a maximal load of 252 kPa). At 15.72 percent of elongation the local rupture results in a global failure of the whole structure (Fig. 15.5d) still resisting total failure of the whole structure. After 36.25 percent of elongation the whole structure fails (Fig. 15.5e). Friction of broken beams among each other are the reason for the small increase of the stress-strain regime above 36.25 percent of elongation (Fig. 15.5f).

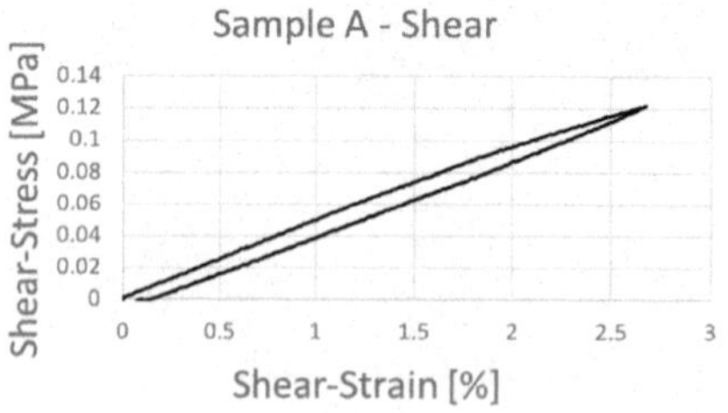

(a) Stress-strain diagram of shearing test of sample **A**.

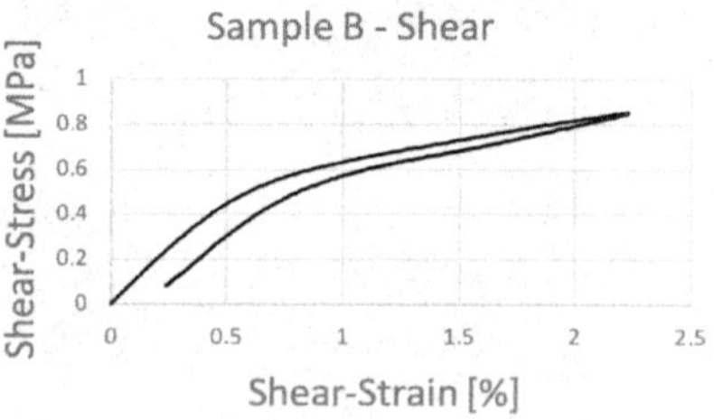

(b) Stress-strain diagram of shearing test of sample **B**.

Fig. 15.7 Stress-strain diagrams of sample **A** and sample **B** during quasi-static experiments.

Results of quasi-static shear experiments performed on both samples are plotted in Fig. 15.7. Sample **A** shows an almost linear elastic material behavior. Small differences between increasing slope and decreasing slope can be neglected because of small friction effects in the mounting device. In contrast sample **B** shows a much more complex material behavior. After transition of linear to non-linear elasticity, small plasticity effects need also to be taken into account for explaining the dissipative deformation behavior. Discrepancy between the start value at zero-stress and the end value of the shear-strain indicates plastic deformation. In both experiments the specimens experience an out-of-plane movement, which was measured for small values of about ±0.2 percent of buckling in Fig. 15.8 with the help of a

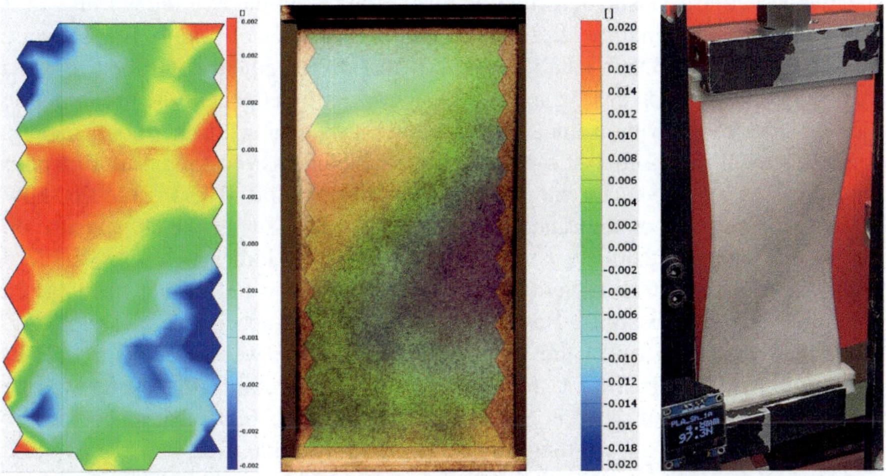

(a) Symetric out-of-plane movement of sample **A**: positive buckling in red area and negative buckling in blue area (about ±0.2 percent).

(b) Overlay of symetric out-of-plane movement of sample **B** and plane structure: positive buckling in red area and negative buckling in blue area (about ±0.2 percent).

(c) Sideview of out-of-plane movement of sample **B** during shear test.

Fig. 15.8 Out-of-plane movement of sample **A** and sample **B** in a quasi-static shear experiment.

so called constriction-function in the evaluation process of 2D-DIC software GOM Correlate. With this technique it is possible to measure buckling effects in the very early state qualitatively, but not quantitatively, because of lack of information in the out-of-plane direction. Results generated with this technique should be threaten very carefully because of the bad standard error deviation. This is also the reason for the negative buckling marked as blue are in the upper left corner of Fig. 15.8a and Fig. 15.8b. Nevertheless, results concerning out-of-plane movements of the initial area of PS are also in good agreement with Barchiesi et al (2018a); Ganzosch et al (2018).

Furthermore, quasi-static extension tests of sample **B** were performed and evaluated also, but because of limitations due to the loading-device only ordinary deformation behavior in the elastic range up to a maximum load of 250 Newton could be observed without providing any new results.

15.3.2 Cyclic Long-term Experiments

In Fig. 15.9 stress-strain diagrams of both samples under shear-load are plotted for the case of twenty repeated load cycles up to about 2.5 percent of elongation, respectively. A linear elastic deformation hysteresis of sample **A** is shown Fig. 15.9a and a non-linear but mainly elastic deformation hysteresis of sample **B** is shown in Fig. 15.9a. In Fig. 15.9a a small decrease of the displacement-controlled maximal load value in each loop can be neglected due to friction effects of the mounting device in order to describe a linear elastic material behavior. Furthermore, material damage as well as viscoelastic effects can be excluded even though out-of-plane movements are involved for sample **A**. In contrast, sample **B** in Fig. 15.9b shows a higher discrepancy of start and end value of the shear-stress, which was also observed in the quasi-static shear test in Fig. 15.7b. This indicates a viscoelastic or even plastic deformation behavior and should be investigated further in long-term relaxation tests for example. Based on the fact that all specimens had to stay 24 hours in water to get rid of the support structure which was manufactured in the prototyping process (see Sect. 15.2), increases the possibility that water was absorbed into the bulk of the structure resulting in the probability that viscoelasticity is involved. As well known, thermoplastics like polyamide used for selective laser sintering (Yang et al, 2018) or FDM show similar viscoelastic material behavior.

Figure 15.10 shows the stress-strain diagram of sample **A** of an extension test in which twenty displacement-controlled load circles up to about 3.3 percent of elongation have been applied. Obviously, a significant decrease of maximal load during

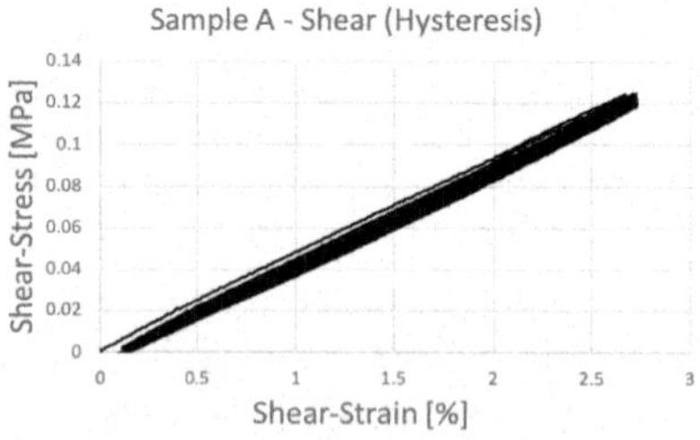

(a) Stress-strain diagram of shearing test of sample **A** with 20 iterations of load protocol.

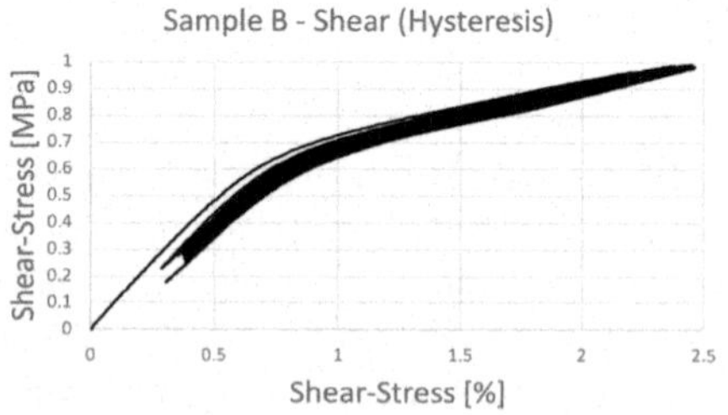

(b) Stress-strain diagram of shearing test of sample **B** with 20 iterations of load protocol.

Fig. 15.9 Stress-strain diagrams of sample **A** and sample **B** during cyclic shear-load.

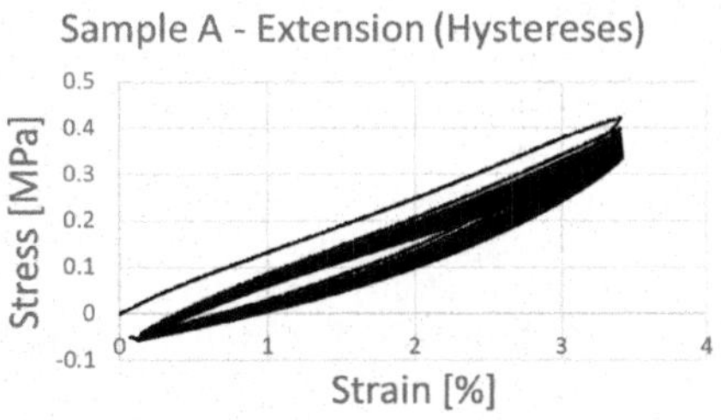

Fig. 15.10 Stress-strain diagram of tension test of sample **A** with 20 iterations of load in a cyclic experiment.

the hysteresis can be recognized. A constant slope would indicate an almost linear viscoelastic deformation behavior, but viscoelasticity is not the only reason for the irreversible dissipation. Negative stress values of the decreasing regimes indicate a plastic deformation. On the other hand, these negative stress-values could also be based on the preload which is induced into the structure because of insufficient mounting conditions such as slipping of the specimen.

15.4 Conclusion

In this paper two types of specimens with different substructures were designed and manufactured using FDM technique. Different types of experiments (quasi-static and cyclic extension as well as shearing tests, respectively) have been applied to those specimens and investigated qualitatively with the means of 2D-DIC. We were able to localize the initial area of out-of-plane movements in shearing tests for both specimen with the help of 2D-DIC.

Furthermore, PSs show linear as well as non-linear elastic deformation response in all experiments except the cyclic tension test, in which viscoelasticity was observed. PSs (sample **A**) enable a high elastic deformation range up to about 4 percent. In contrast elastic deformation range of the plane sheet structures (sample **B**) is only at about 0.5 percent. Furthermore, PS are able to take even higher loads after a rupture occurs resisting total failure of the structure resulting in a high resilient deformation behavior. This behavior makes PS very attractive for applications in which very large deformations are inescapable like in the aircraft or maritime industries.

References

Abali BE, Müller WH, Eremeyev VA (2015) Strain gradient elasticity with geometric nonlinearities and its computational evaluation. Mechanics of Advanced Materials and Modern Processes 1(4)

Abali BE, Müller WH, dell'Isola F (2017) Theory and computation of higher gradient elasticity theories based on action principles. Archive of Applied Mechanics 87(9):1495–1510, URL

http://dx.doi.org/10.1007/s00419-017-1266-5
Askes H, Aifantis EC (2011) Gradient elasticity in statics and dynamics: an overview of formulations, length scale identification procedures, finite element implementations and new results. International Journal of Solids and Structures 48(13):1962–1990
Barchiesi E, Ganzosch G, Liebold C, Placidi L, Grygoruk R, Müller WH (2018a) Out-of-plane buckling of pantographic fabrics in displacement-controlled shear tests: experimental results and model validation. Continuum Mechanics and Thermodynamics pp 1–13
Barchiesi E, Spagnuolo M, Placidi L (2018b) Mechanical metamaterials: a state of the art. Mathematics and Mechanics of Solids p 1081286517735695
Bensoussan A, Lions JL, Papanicolaou G (2011) Asymptotic analysis for periodic structures, vol 374. American Mathematical Soc.
Boutin C, dell'Isola F, Giorgio I, Placidi L (2017) Linear pantographic sheets: Asymptotic micro-macro models identification. Mathematics and Mechanics of Complex Systems 5(2):127–162
Cennamo C, Chiaia B, Biagi VD, Placidi L (2014) Monitoring and compartmentalized structures. ZAMM - Journal of Applied Mathematics and Mechanics / Zeitschrift für Angewandte Mathematik und Mechanik 95(6):638–648
Chen C, Fleck N (2002) Size effects in the constrained deformation of metallic foams. Journal of the Mechanics and Physics of Solids 50(5):955–977
Cosserat E, Cosserat F (1896) Sur la théorie de l'élasticité. premier mémoire. In: Annales de la Faculté des sciences de Toulouse: Mathématiques, Gauthier-Villars et Fils, Imprimeurs-Libraires, vol 10 (3-4), pp I1–I116
Cosserat E, Cosserat F (1909) Théorie des corps déformables. A. Hermann et fils,
dell'Isola F, Sciarra G, Vidoli S (2009) Generalized hooke's law for isotropic second gradient materials. In: Proceedings of the Royal Society of London A: Mathematical, Physical and Engineering Sciences, The Royal Society, pp rspa–2008
Eringen AC (1968) Mechanics of micromorphic continua. In: Mechanics of Generalized Continua, Springer, pp 18–35
Eringen AC, Suhubi ES (1964) Nonlinear theory of simple micro-elastic solids—i. International Journal of Engineering Science 2(2):189–203
Ganzosch G, dell'Isola F, Turco E, Lekszycki T, Müller W (2017) Shearing tests applied to pantographic structures. Acta Polytechnica CTU Proceedings 7(1-6)
Ganzosch G, Hoschke K, Lekszycki T, Giorgio I, Turco E, Müller WH (2018) 3d-measurements of 3d-deformations of pantographic structures. Technische Mechanik 38(3):233–245, DOI 10.24352/UB.OVGU-2018-031
Ghosh S, Lee K, Moorthy S (1996) Two scale analysis of heterogeneous elastic-plastic materials with asymptotic homogenization and voronoi cell finite element model. Computer methods in applied mechanics and engineering 132(1-2):63–116
Gibson LJ (2005) Biomechanics of cellular solids. Journal of biomechanics 38(3):377–399
Gibson LJ, Ashby MF (1999) Cellular solids: structure and properties. Cambridge university press
Kumar RS, McDowell DL (2004) Generalized continuum modeling of 2-d periodic cellular solids. International Journal of solids and structures 41(26):7399–7422
Li J (2011) A micromechanics-based strain gradient damage model for fracture prediction of brittle materials–part i: Homogenization methodology and constitutive relations. International Journal of Solids and Structures 48(24):3336–3345
Mindlin RD, Eshel N (1968) On first strain-gradient theories in linear elasticity. International Journal of Solids and Structures 4(1):109–124
Müller WH, Vilchevskaya EN (2017) Micropolar theory from the viewpoint of mesoscopic and mixture theories. Physical Mesomechanics 20(3):263–279
Müller WH, Vilchevskaya EN (2018) Micropolar theory with production of rotational inertia: A rational mechanics approach. In: Generalized Models and Non-classical Approaches in Complex Materials 1, Springer, pp 581–606
Noor AK (1988) Continuum modeling for repetitive lattice structures. Applied Mechanics Reviews 41(7):285–296

Placidi L, Andreaus U, Giorgio I (2016a) Identification of two-dimensional pantographic structure via a linear d4 orthotropic second gradient elastic model. Journal of Engineering Mathematics 103(1):1–21

Placidi L, Barchiesi E, Turco E, Rizzi NL (2016b) A review on 2D models for the description of pantographic fabrics. Zeitschrift für angewandte Mathematik und Physik 67(5):121

Placidi L, Greco L, Bucci S, Turco E, Rizzi NL (2016c) A second gradient formulation for a 2d fabric sheet with inextensible fibres. Zeitschrift für angewandte Mathematik und Physik 67(5)

Placidi L, Barchiesi E, Misra A (2018a) A strain gradient variational approach to damage: a comparison with damage gradient models and numerical results. Mathematics and Mechanics of Complex Systems 6(2):77–100

Placidi L, Misra A, Barchiesi E (2018b) Two-dimensional strain gradient damage modeling: a variational approach. Zeitschrift für angewandte Mathematik und Physik 69(3):56

Rahali Y, Giorgio I, Ganghoffer J, Dell'Isola F (2015) Homogenization à la piola produces second gradient continuum models for linear pantographic lattices. International Journal of Engineering Science 97:148–172

Sun C, Vaidya R (1996) Prediction of composite properties from a representative volume element. Composites Science and Technology 56(2):171–179

Tekoğlu C, Onck PR (2008) Size effects in two-dimensional voronoi foams: a comparison between generalized continua and discrete models. Journal of the Mechanics and Physics of Solids 56(12):3541–3564

Toupin RA (1962) Elastic materials with couple-stresses. Archive for Rational Mechanics and Analysis 11(1):385–414

Yang H, Ganzosch G, Giorgio I, Abali BE (2018) Material characterization and computations of a polymeric metamaterial with a pantographic substructure. Zeitschrift für angewandte Mathematik und Physik 69(4):105

Chapter 16
Calculation of Stress Intensity Factors for an Arbitrary Oriented Penny-shaped Crack Under Inner Pressure in an Orthotropic Electroelastic Material

Vitaly S. Kirilyuk, Olga I. Levchuk & Holm Altenbach

Abstract The electroelasticity problem for an arbitrarily oriented disc-like crack under internal pressure in an orthotropic electroelastic material was considered. Generalizing of the Willis approach for an elastic material, using the Fourier transform of the Green's function for an infinite anisotropic electroelastic space, the problem of electroelasticity is reduced to finding unknowns of the jumps of displacements and electric potential through the surface of a circular crack. Quadrature Gauss formulas were used to calculate one-dimensional integrals. Testing the approach in the particular case of the problem for which an exact solution is known confirms the effectiveness of the used approach. The distribution of stress intensity factors (SIF) along the boundary of a disc-shaped crack (under internal pressure) in a piezoelectric orthotropic material under various orientations of a crack was studied. A significant effect of the crack orientation on the SIF distributions was established.

Keywords: Electroelasticity problem · Orthotropic piezoelectric material · Penny-shape crack · Arbitrary orientation · Inner pressure · Stress intensity factors

16.1 Introduction

The wide use of piezoelectric materials, which have considerable brittleness, when creating energy converters and various sensors for measuring instruments, stimulates the interest in studying and analyzing the stress state and concentration of

Vitaly S. Kirilyuk and Olga I. Levchuk
S.P. Timoshenko Institute of Mechanics of NASU, Nesterova 3, 03057 Kyiv, Ukraine,
e-mail: kirilyuk_v@ukr.net, 2013levchuk@gmail.com

Holm Altenbach
Institut für Mechanik, Fakultät für Maschinenbau, Otto-von-Guericke-Universität Magdeburg, Universitätsplatz 2, 39106 Magdeburg, Germany,
e-mail: holm.altenbach@ovgu.de

B. E. Abali et al. (eds.), *New Achievements in Continuum Mechanics and Thermodynamics*, Advanced Structured Materials 108,
https://doi.org/10.1007/978-3-030-13307-8_16

force and electric fields in electroelastic bodies near defects such as cavities, inclusions, cracks. However, the solution of three-dimensional problems of electroelasticity is a rather complex problem, since the system of equations of electroelasticity is related to force and electric fields (Grinchenko et al, 1989; Shulga and Karlash, 2008). Therefore, up to now, two-dimensional problems of electroelasicity for a piezoelectric material with stress concentrators had been studied with greater completeness (Dai et al, 2006; Kaloerov, 2007; Wang et al, 2015; Xu et al, 2015; Zhang et al, 2017; Zhang and Gao, 2004; Zhao et al, 2016). In Podil'chuk (1998b); Wang and Zheng (1995) two similar approaches to the construction of general solutions of the three-dimensional coupled system of equations of static electroelasticity for transversely isotropic electroelastic bodies were proposed. With the help of these approaches, it was possible to obtain solutions of a number of electroelastic problems with stress concentrators in the form of cavities, inclusions, cracks with special orientation (Chen and Lim, 2005; Chiang and Weng, 2005; Kirilyuk, 2005, 2006; Lin et al, 2003; Podil'chuk, 1998a; Shang et al, 2003; Wang et al, 2015). When applying the approaches (Podil'chuk, 1998b; Wang and Zheng, 1995) it was assumed that the axis of symmetry of the transversely isotropic electroelastic material coincides with the axis of rotation of the stress concentrator, and for plane cracks it was believed that they are located in a plane that is perpendicular to the axis of symmetry of the material. For other orientations of stress concentrators in the material, these approaches are not effective. Some space problems of electroelasticity for transversely isotropic piezoelectric bodies (with special orientation of the stress concentrator in the material) by means of other methods (FEM and BEM) were considered in Sladek et al (2018); Ricoeur and Kuna (2009); Wippler and Kuna (2007); Wippler et al (2004); Zhao et al (2018).

Note that, for circular cracks in elastic bodies, several important results of investigations of the stress state and stress intensity factors are contained in Lekhnitskii (1981); Willis (1968), and for electroelastic transversely isotropic bodies (with special restrictions on the orientation of the circular crack), interesting studies are performed in Chiang and Weng (2005); Dunn and Taya (1994); Kirilyuk (2005, 2006); Lin et al (2003); Podil'chuk (1998a); Sladek et al (2018); Zhao et al (2018). The influence of the orientation of a circular crack in a transversely isotropic electroelastic material on the distribution of stress intensity factors was considered (Kirilyuk, 2008), and the stress state in an orthotropic electroelastic material with an elliptical crack located in one of the principal planes of symmetry under homogeneous loading was investigated in Kirilyuk and Levchuk (2017), respectively. The solutions of two problems of electroelasticity for orthotropic piezoelectric laminates with interfacial imperfections were obtained in Chen et al (2004); Zhou et al (2010).

In this paper, the distribution of stress intensity factors for an arbitrarily oriented circular crack in an orthotropic electroelastic material is studied for the first time. The investigations are based on the generalization of the approach (Willis, 1968, for an anisotropic elastic medium with an elliptical crack) to the case of an electroelastic orthotropic material. The problem is reduced to the search for unknown jumps of displacements and electric potential through the surface of a circular crack. In solving the problem, we also used the Fourier transform of the Green's function for an anisotropic electroelastic space and Gauss quadrature formulas for calculating of contour integrals. For a particular case of the considered problem when it has an exact solution, an approbation of the used approach has been carried out and its effectiveness has been confirmed. Investigations of stress intensity factors along the boundary of a circular crack (for different orientations in the material) were performed, and the influence of the crack orientation on their distribution was revealed.

16.2 Basic Equations and Statement of the Problem

Let the orthotropic electroelastic space contains a plane circular crack. Let us assume that one of the axes of material symmetry coincides with the axis z, the angle between which and the normal to the plane of the crack is equal α. We assume that the electroelastic space is under the action of homogeneous force and electric loads. The presence of a crack in the material, as a concentrator, leads to the appearance of perturbations of electric and stressed states.

The complete system of static equations of electroelasticity has the following form:

- Equilibrium equations in the absence of volume forces

$$\sigma_{ij,j} = 0; \tag{16.1}$$

- Equations of electrostatics

$$D_{i,j} = 0, \qquad E_i = -\Psi_{,i}; \tag{16.2}$$

- Cauchy relations

$$\varepsilon_{ij} = \frac{1}{2}(u_{i,j} + u_{j,i});$$

- Equations of state

$$\sigma_{ij} = C_{ijmn}\varepsilon_{mn} + e_{nij}\Psi_{,n}, \qquad D_i = e_{imn}\varepsilon_{mn} - k_{in}\Psi_{,n}; \tag{16.3}$$

where $\sigma_{ij}, \varepsilon_{ij}, u_i, D_i, E_i, \Psi$ - components of stresses, deformations, displacements, electrical displacements (electric induction), electric field intensity and electric potential, respectively. In addition, let us introduce the notations of following tensors: $C_{ijmn}, e_{imn}, k_{ij}$ - tensor of elastic modules, tensor of piezoelectric modules, dielectric permittivity tensor. For piezoelectrical bodies that are orthotropic in their properties, the elastic characteristics of the material are described by nine independent constants $c_{11}, c_{22}, c_{33}, c_{12}, c_{13}, c_{23}, c_{44}, c_{55}, c_{66}$, the piezomodules – by five constants $e_{15}, e_{24}, e_{31}, e_{32}, e_{33}$, the dielectric permittivity – by three independent constants k_{11}, k_{22}, k_{33}. The components of the indicated tensors are related to the corresponding independent constants as follows

$$
\begin{gathered}
C_{1111} = c_{11}, \qquad C_{2222} = c_{22}, \qquad C_{3333} = c_{33}, \\
C_{1122} = C_{2211} = c_{12}, \qquad C_{1133} = C_{3311} = c_{13}, \qquad C_{2233} = C_{3322} = c_{23}, \\
C_{2323} = C_{2332} = C_{3232} = C_{3223} = c_{44}, \\
C_{3131} = C_{3113} = C_{1331} = C_{1313} = c_{55}, \\
C_{1212} = C_{1221} = C_{2121} = C_{2112} = c_{66}, \\
e_{113} = e_{131} = e_{15}, \qquad e_{223} = e_{232} = e_{24}, \\
e_{311} = e_{31}, \qquad e_{322} = e_{32}, \qquad e_{333} = e_{33}
\end{gathered} \tag{16.4}
$$

The other components of mentioned two tensors are equal to zero. Also, three components of the dielectric permittivity are nonzero.

When solving the problem, it is convenient to introduce a new coordinate system in which the direction of one of the axes coincides with the direction of the normal to the plane of the crack. Suppose that the original coordinate system $0x^1x^2x^3$ is associated with a new (local) system in such a way that it is obtained from the initial coordinate system by rotating about an axis $0x$ by an angle α. Then tensors of elastic moduli, piezomodulaes and dielectric constants $C^{(\alpha)}_{ijkl}$, $e^{(\alpha)}_{ijk}$, $k^{(\alpha)}_{ij}$, in the new coordinate system, we obtain according to transformations of tensors of corresponding orders, where α_{ij} is the transformation matrix of the next form

$$
\alpha_{ij} = \begin{bmatrix} 1 & 0 & 0 \\ 0 & \cos\alpha & -\sin\alpha \\ 0 & \sin\alpha & \cos\alpha \end{bmatrix} \tag{16.5}
$$

An arbitrary orientation of the crack in the material can be obtained by sequential rotations at angles α, β, γ around the axes of the old coordinate system $0x, 0y, 0z$, respectively. Then the transformation matrix T_{ij} has the form

$$
T_{ij} = \begin{bmatrix} \cos\beta\cos\gamma & -\cos\beta\sin\gamma & \sin\beta \\ \sin\alpha\sin\beta\cos\gamma + \cos\alpha\sin\gamma & -\sin\alpha\sin\beta\sin\gamma + \cos\alpha\cos\gamma & -\sin\alpha\cos\beta \\ -\cos\alpha\sin\beta\cos\gamma + \sin\alpha\sin\gamma & \cos\alpha\sin\beta\sin\gamma + \sin\alpha\cos\gamma & \cos\alpha\cos\beta \end{bmatrix}
$$

Matrix T_{ij} is the result of the multiplication of three matrices, like (16.5), reflecting the right of rotation around each of the coordinate axes. New tensors of elastic moduli, piezomodules and dielectric constants $C^{(\alpha,\beta,\gamma)}_{ijkl}$, $e^{(\alpha,\beta,\gamma)}_{ijk}$, $k^{(\alpha,\beta,\gamma)}_{ij}$ we obtain using the rules for transforming of tensors of corresponding orders

$$
C^{(\alpha,\beta,\gamma)}_{ijkl} = C_{mnpq}T_{im}T_{jn}T_{kp}T_{lq}, \; e^{(\alpha,\beta,\gamma)}_{ijk} = e_{mnp}T_{im}T_{jn}T_{kp}, \; k^{(\alpha,\beta,\gamma)}_{ij} = k_{mn}T_{im}T_{jn},
$$

where summation is carried out over repeated indexes.

We note that further in the text of the paper we used the usual tensor notation of expressions, that is, by summation over the indexes repeated in the expressions. To describe the electroelastic state, we use more universal notations (Dunn and Taya, 1994), based on which we write the elastic displacements and the electric potential

$$
U_M = \begin{cases} u_m, & M = 1,2,3, \\ \Psi, & M = 4, \end{cases} \tag{16.6}
$$

elastic deformations or electric potential

$$Z_{Mn} = \begin{cases} \varepsilon_{mn}, \ M = 1, 2, 3, \\ \Psi_{,n}, \ \ M = 4, \end{cases} \tag{16.7}$$

stress or electrical displacements

$$\Sigma_{iJ} = \begin{cases} \sigma_{ij}, \ J = 1, 2, 3, \\ D_i, \ M = 4, \end{cases} \tag{16.8}$$

electroelastic modules

$$E^{(\alpha,\beta,\gamma)}_{iJMn} = \begin{cases} C^{(\alpha,\beta,\gamma)}_{ijmn}, & J, M = 1, 2, 3, \\ e^{(\alpha,\beta,\gamma)}_{nij}, & J = 1, 2, 3, M = 4, \\ e^{(\alpha,\beta,\gamma)}_{imn}, & J = 4, M = 1, 2, 3, \\ -k^{(\alpha,\beta,\gamma)}_{in}, & J, M = 4 \end{cases} \tag{16.9}$$

Using the notation (16.6)-(16.9), the equation of state (16.3) can be written in the form

$$\Sigma_{iJ} = E^{(\alpha,\beta,\gamma)}_{iJMn} Z_{Mn}$$

Under uniform force and electrical loads, we obtain the following boundary conditions

$$\tau^{\pm}_{13}\Big|_S = f^{(\alpha,\beta,\gamma)}_1, \tau^{\pm}_{23}\Big|_S = f^{(\alpha,\beta,\gamma)}_2, \sigma^{\pm}_{33}\Big|_S = -P^{(\alpha,\beta,\gamma)}_0, D^{\pm}_3\Big|_S = -D^{(\alpha,\beta,\gamma)}_0,$$
$$U_M(\boldsymbol{x}) \to 0 \quad \text{when} \quad |\boldsymbol{x}| \to \infty \tag{16.10}$$

where S - the double-sided surface of the crack, is referred to a new coordinate system (it is obtained by rotation at angles α, β, γ around the axes of the old system), and the load is also written in the new coordinate system.

16.3 Solution Method

We consider the Green's function $G_{IJ}(\boldsymbol{x}-\boldsymbol{x}')$ for an infinite electroelastic anisotropic space. It satisfies the following equations

$$E^{(\alpha,\beta,\gamma)}_{kJMn} G_{JM,kn} + \delta_{JM}\delta(\boldsymbol{x} - \boldsymbol{x}') = 0, \tag{16.11}$$

where $\delta(\boldsymbol{x} - \boldsymbol{x}')$ - is the Dirac delta function; δ_{JM} - the symbol of Kronecker. A comma after an index means differentiation with respect to the corresponding variable. We use the integral expression of the fundamental solution

$$G_{JM}(\boldsymbol{x}-\boldsymbol{x}') = \frac{1}{(2\pi)^3} \iiint\limits_{-\infty}^{\infty} A_{JM}(\boldsymbol{\xi}) D^{-1} \mathrm{e}^{\imath \boldsymbol{\xi}(\boldsymbol{x}-\boldsymbol{x}')} \mathrm{d}\xi_1 \mathrm{d}\xi_2 \mathrm{d}\xi_3, \tag{16.12}$$

where $\imath = \sqrt{-1}$ and $A_{JM}(\boldsymbol{\xi})$ - are the corresponding algebraic complements of the elements of the matrix

$$\{K_{JM}(\boldsymbol{\xi})\} = \{E^{(\alpha,\beta,\gamma)}_{iJMn} \xi_i \xi_n\} \tag{16.13}$$

$D(\boldsymbol{\xi})$ - is its determinant, which is a polynomial of the eighth order.

Using the further transformations of expressions (16.10)-(16.13), we represent the perturbed electric and stress states, generalizing the purely elastic case, by means of unknown jumps of displacements and electric potential through a two-sided crack surface

$$U_I(\boldsymbol{x}) = \frac{1}{4\pi^2} \sum_{N=1}^{4} \iint\limits_{-\infty}^{\infty} \frac{E^{(\alpha,\beta,\gamma)}_{lJM3} \xi^N_l A_{IJ}(\boldsymbol{\xi}^N)}{\dfrac{\partial D(\boldsymbol{\xi}^N)}{\partial \xi_3}} \iint\limits_{S} b_M(\boldsymbol{x}') \mathrm{e}^{\imath \boldsymbol{\xi}^N(\boldsymbol{x}-\boldsymbol{x}')} \mathrm{d}\xi_1 \mathrm{d}\xi_2 \mathrm{d}x'_1 \mathrm{d}x'_2, \tag{16.14}$$

where for a circular crack an unknown vector $\boldsymbol{b}(\boldsymbol{x})$ has the form

$$\boldsymbol{b}(\boldsymbol{x}) = \boldsymbol{b}^{(0)} \sqrt{1 - \frac{x_1^2}{a^2} - \frac{x_2^2}{a^2}} \tag{16.15}$$

a - radius of a circular crack; $\boldsymbol{b}^{(0)}$ is a constant fourth-order vector whose components are complex numbers in the general case. We note that the summation in Eq. (16.14) using expression (16.15) is carried out for the ξ_3^M - roots of the equation $D(\xi) = 0$ with negative imaginary part, and the vector $\boldsymbol{\xi}^M$ has the form $\boldsymbol{\xi}^M = (\xi_1, \xi_2, \xi_3^M(\xi_1, \xi_2))$. Components of stresses and electrical induction are determined by the following expressions

$$\Sigma_{iJ}(\boldsymbol{x}) = E^{(\alpha,\beta,\gamma)}_{iJKl} U_{K,l}$$

$$= \frac{-\imath}{4\pi^2} \sum_{N=1}^{4} \iiiint\limits_{-\infty\ S}^{\infty} \frac{E^{(\alpha,\beta,\gamma)}_{iJKl} E^{(\alpha,\beta,\gamma)}_{pQM3} \xi^N_p \xi^N_l A_{KQ}(\boldsymbol{\xi}^N)}{\dfrac{\partial D(\boldsymbol{\xi}^N)}{\partial \xi_3}} b_M(\boldsymbol{x}') \mathrm{e}^{\imath \boldsymbol{\xi}^M(\boldsymbol{x}-\boldsymbol{x}')} \mathrm{d}\xi_1 \mathrm{d}\xi_2 \mathrm{d}x_1 \mathrm{d}x_2$$

Using transformations analogous to the elastic case (Willis, 1968) we obtain

$$\Sigma_{iJ}(\boldsymbol{x}) = \frac{-\imath}{4} \int\limits_0^{2\pi} \sum_{N=1}^{4} F^{(\alpha,\beta,\gamma)}_{iJM} \left(\frac{\eta_1}{a}, \frac{\eta_2}{a}, \xi_3^N\left(\frac{\eta_1}{a}, \frac{\eta_2}{a}\right)\right) b_M^{(0,0)} \mathrm{d}\varphi, \tag{16.16}$$

where the function $F^{(\alpha,\beta,\gamma)}_{iJM}(\xi_1, \xi_2, \xi_3)$ is determined by the formula

$$F_{iJM}^{(\alpha,\beta,\gamma)}(\xi_1,\xi_2,\xi_3) = \frac{E_{iJKl}^{(\alpha,\beta,\gamma)}E_{pQM3}^{(\alpha,\beta,\gamma)}\xi_p^N\xi_l^N A_{KQ}(\boldsymbol{\xi}^N)}{\dfrac{\partial D(\boldsymbol{\xi}^N)}{\partial\xi_3}} \tag{16.17}$$

Further, the calculation of the one-dimensional integral in Eq. (16.16) (using the notation (16.17)) is carried out by means of the Gauss quadrature formula.

After further analysis of the asymptotic expressions for stresses and electric induction in the crack plane and, using the definition of the stress intensity coefficients K_I, K_{II}, K_{III} and electric induction K_D, we obtain

$$k_{iJ} = \imath\sqrt{\pi a}\frac{a}{\sqrt[4]{x_1^2+x_2^2}}\sum_{N=1}^{4} F_{iJM}^{(\alpha,\beta,\gamma)}\left[\frac{x_1}{a^2},\frac{x_2}{a^2},\xi_3^N\left(\frac{x_1}{a^2},\frac{x_2}{a^2}\right)\right]b_M^{(0,0)},$$
$$K_I = k_{33},\ \ K_{II} = k_{31}n_1 + k_{32}n_2,\ \ K_{III} = k_{31}(-n_2) + k_{32}n_1,\ \ K_{IV} = K_D = k_{34} \tag{16.18}$$

For a circular crack, the components of the normal vector have the form

$$n_1 = \frac{x_1}{\sqrt{x_1^2+x_2^2}}, \qquad n_2 = \frac{x_2}{\sqrt{x_1^2+x_2^2}}$$

Using the method of Gauss quadratures to calculate one-dimensional integrals and satisfying the boundary conditions on the crack surface, we find unknown values of displacement jumps and electric induction, and then calculate the stress intensity and electric induction coefficients from Eqs. (16.18).

To approbate the approach used, we first consider the problem of a circular crack in an electroelastic transversely isotropic material located in the plane of its isotropy, at a constant pressure P_0 on the surface of the crack, as well as in shear forces $\sigma_{13}^0 \neq 0$. The normal component of the electric induction was considered zero when considering the problem. In this case, for an impenetrable crack, the stress intensity factors K_I according to the results of Kirilyuk (2005) are independent of the properties of the material and coincide with their expression for a purely elastic isotropic material (for the same shape of the crack and the same symmetrical load), and the value $K_D = 0$. At the same time, the stress intensity factors, under shear, depends on the elastic and electrical properties of the material (Kirilyuk, 2006). According to the studies of Kirilyuk (2006), in order to find the SIF K_{II}, K_{III}, in an electroelastic transversally isotropic material with a flat crack in the isotropy plane of the material, it is sufficient to take their expressions for an isotropic elastic material under the same shear loads, and instead of the Poisson ratio ν in the corresponding SIF expressions it is necessary to substitute the value ν_{PIEZO}, which depends on the properties of the electro-elastic material (Kirilyuk, 2006). Since the formula for finding this value is rather complex and the procedure for obtaining it is described in detail in Kirilyuk (2006), we give the value ν_{PIEZO} for some piezoelectrical materials in Table 16.1 according to Kirilyuk (2006). The output data regarding electrical and elastic properties correspond to the characteristics given in Grinchenko et al (1989); Kirilyuk (2005, 2006). The second row of the table contains values founded only for the elastic properties of a transversely isotropic material (without considering its electrical properties).

According to the results of Kirilyuk (2005, 2006), for an electroelastic space containing a penny-shaped crack, with internal pressure P_0 at the crack surface and shear σ_{13}^0 in the piezoelectric material, we obtain the following SIF expressions along the crack front

$$K_I = \frac{2}{\sqrt{\pi}}(P_0\sqrt{a}), K_{II} = \frac{4}{(2-\nu_{\text{PIEZO}})\sqrt{\pi}}(\sigma_{13}^0\sqrt{a})\cos\phi, K_{III} = \frac{-4(1-\nu_{\text{PIEZO}})}{(2-\nu_{\text{PIEZO}})\sqrt{\pi}}(\sigma_{13}^0\sqrt{a})\sin\phi \tag{16.19}$$

To validate the approach used, SIF calculations were performed, and for the case of the location of a circular crack in the isotropy plane of an electroelastic transversely isotropic material by means of analytical expressions (16.19) and Eqs. (16.18). For piezoelectric materials from Table 16.1, the results of the investigations by means of two approaches coincide up to 7 significant digits. In calculating one-dimensional integrals, the Gauss quadrature formula with 24 nodes was used in expressions (16.16). Further testing of the approach and the developed on this base computer program for the electroelastic orthotropic material with a circular crack in particular cases (an arbitrarily oriented penny - shaped crack in elastic orthotropic and electroelastic transversal-isotropic materials, a circular crack in the orthotropic plane of the electroelastic orthotropic material) confirmed the coincidence of the obtained results and data in (Kirilyuk, 2008; Kirilyuk and Levchuk, 2017).

16.4 Analysis of the Results of Numerical Investigations

Consider the following materials:

1. piezoelectric orthotropic material $Ba_2NaNb_5O_{15}$, whose electroelastic properties (total 17 independent parameters) are given in Shulga and Karlash (2008, pp. 37 and 39);
2. electroelastic piezoceramic material PZT-4 (10 independent electroelastic parameters, the values of which are given in Grinchenko et al, 1989);
3. elastic orthotropic material whose elastic properties (9 independent parameters) correspond to the values of orthogonally reinforced 2:1 fiberglass and are given in Lekhnitskii (1981, p. 66).

We assume that the circular crack can be arbitrarily located in the electro elastic material and is under internal pressure, and its orientation is determined by the

Table 16.1 Special values for expressions of SIF

Values	PZT-4	PXE-5	PZT-5	PZT-7A	BaTiO3	PZT-5H
ν_{PIEZO}	0,48513	0,48815	0,51190	0,47324	0,34369	0,37867
ν_{ELAST}	0,35034	0,34591	0,36965	0,35239	0,29768	0,30074
ν_{CONTROL}	0,35034	0,34591	0,36965	0,35239	0,29768	0,30074

rotation angle α around the axis $0x$ and, respectively, by the coordinate transformation matrix (16.5). In Figs. 16.1-16.3 reflects the change in the SIF K_{I} along the front of a circular cracks in piezoelectric orthotropic material $Ba_2NaNb_5O_{15}$, electroelastic piezoceramic material PZT-4 and orthogonally reinforced with 2:1 fiberglass (elastic orthotropic material), respectively, for different cases of orientation of the circular crack. Curves 1, 2, 3 and 4 in these figures correspond to values $\alpha = 0, \alpha = \pi/6, \alpha = \pi/3$ and $\alpha = \pi/2$ for uniform pressure P_0 on a two-sided crack surface. It can be seen that the orientation of the crack in the material significantly affects the value and character of the distribution of stress intensity factors along the crack boundary. In some cases, the effect of orientation on the SIF K_{I} value exceeded 23 %.

Figures 16.4 and 16.5 show the appearance of nonzero values of the stress intensity factors K_{II}, and K_{III} for a symmetrical load, which is the internal pressure, due to the orientation of the crack in orthotropic material (2:1 orthogonally reinforced fiberglass), when the crack is located not in the plane of symmetry of the material. Curves 1 and 2 in Figs. 16.4 and 16.5 correspond to the values of the angles $\alpha = \pi/6$ and $\alpha = \pi/3$. Note that for materials $Ba_2NaNb_5O_{15}$ and PZT-4, when the crack is located not in the plane of symmetry of the material and under the pressure of its surface, nonzero values of SIF K_{II} and K_{III} appear also, but they are some smaller in value than for orthogonally reinforced 2:1 fiberglass.

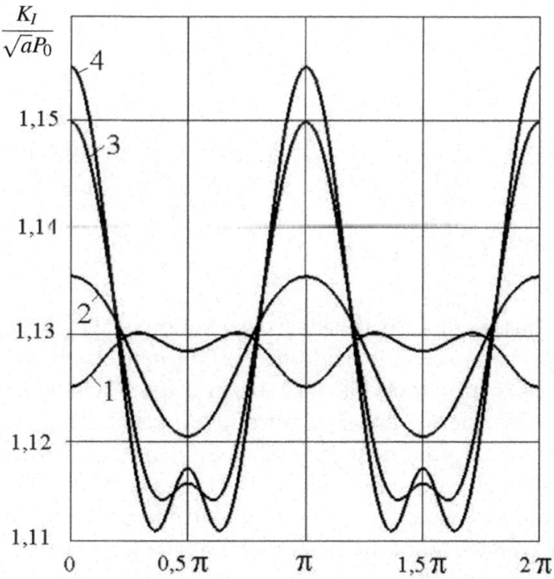

Fig. 16.1 Distribution of SIF K_I in the electroelastic orthotropic material

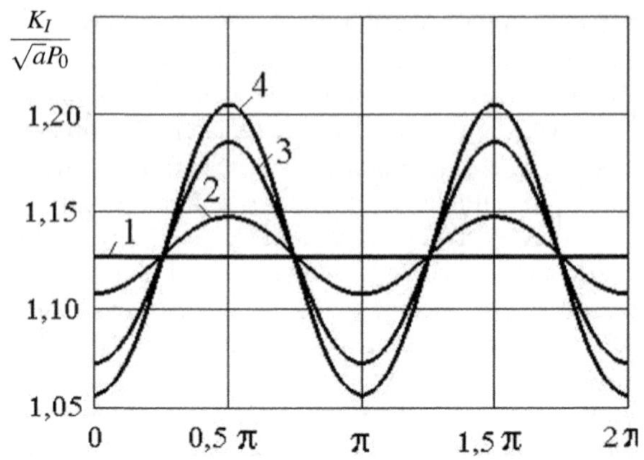

Fig. 16.2 Changes of SIF K_I in the transversely isotropic piezoelectric material

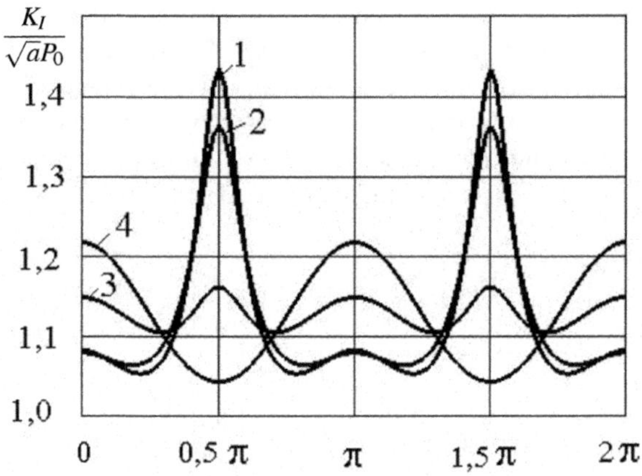

Fig. 16.3 Distribution of SIF K_I in the elastic orthotropic material

16.5 Conclusion

Thus, in this paper, using the solution of the electroelasticity problem for coupled forces and electric fields in an orthotropic electroelastic material, stress intensity factors in an orthotropic piezoelectric space with an arbitrarily oriented circular crack under internal pressure were investigated. The distribution of stress intensity factors along the boundary of a circular crack under internal pressure is studied for different cases of its orientation in the material. A significant influence of the orientation of the circular crack (under internal pressure) in the orthotropic material on the character of the distribution of stress intensity factors along the crack boundary was established.

Acknowledgements Olga Levchuk gratefully acknowledge the financial support by the German Academic Exchange Service (DAAD) via grant 91697662.

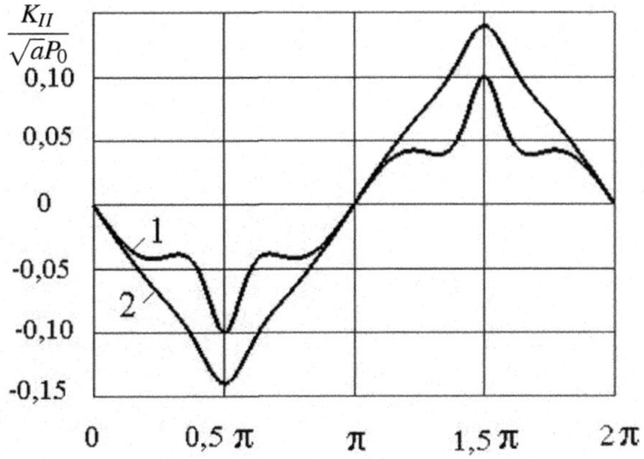

Fig. 16.4 The appearance of nonzero values of SIF K_{II} for the crack under pressure due to its orientation in the material

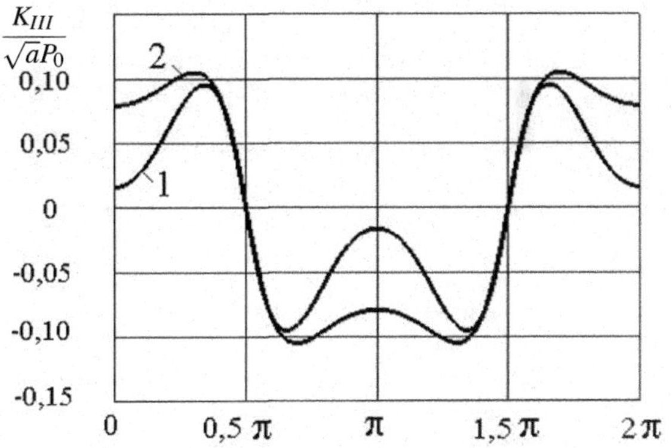

Fig. 16.5 Non-zero distribution of SIF K_{III} for the crack under pressure due to its orientation in the material

References

Chen WQ, Lim CW (2005) 3D point force solution for a permeable penny-shaped crack embedded in an infinite transversely isotropic piezoelectric medium. International Journal of Fracture 31(3):231–246

Chen WQ, Cai JB, Ye GR, Wang YF (2004) Exact three-dimensional solutions of laminated orthotropic piezoelectric rectangular plates featuring interlaminar bonding imperfections modeled by a general spring layer. International Journal of Solids and Structures 41(18):5247–5263

Chiang CR, Weng GJ (2005) The nature of stress and electric-displacement concentrations around a strongly oblate cavity in a transversely isotropic piezoelectric material. International Journal of Fracture 134(3):319–337

Dai L, Guo W, Wang X (2006) Stress concentration at an elliptic hole in transversely isotropic piezoelectric solids. International Journal of Solids and Structures 43(6):1818–1831

Dunn ML, Taya M (1994) Electroelastic field concentrations in and around inhomogeneities in piezoelectric solids. Journal of Applied Mechanics 61(4):474–475

Grinchenko VT, Ulitko AF, Shulga NA (1989) Electroelasticity (in Russ.), Mechanics of Coupled Fields in the Elements of Constructions, vol 5. Naukova Dumka, Kiev

Kaloerov SA (2007) Determining the intensity factors for stresses, electric-flux density, and electric-field strength in multiply connected electroelastic anisotropic media. International Applied Mechanics 43(6):631–637

Kirilyuk VS (2005) On the stress state of a piezoceramic body with a flat crack under symmetric loads. International Applied Mechanics 41(11):1263–1271

Kirilyuk VS (2006) Stress state of a piezoelectric ceramic body with a plane crack under antisymmetric loads. International Applied Mechanics 42(2):152–161

Kirilyuk VS (2008) Elastic state of a transversely isotropic piezoelectric body with an arbitrarily oriented elliptic crack. International Applied Mechanics 44(2):150–157

Kirilyuk VS, Levchuk OI (2017) Stress state of an orthotropic piezoelectric material with an elliptic crack. International Applied Mechanics 53(3):305–312

Lekhnitskii SG (1981) Theory of Elasticity of an Anisotropic Body. Mir, Moscow

Lin S, Narita F, Shindo Y (2003) Electroelastic analysis of a penny-shaped crack in a piezoelectric ceramic under mode I loading. Mechanics Research Communications 30(4):371–386

Podil'chuk YN (1998a) Electroelastic equilibrium of transversally isotropic, piezoceramic media containing cavities, inclusions, and cracks. International Applied Mechanics 34(10):1023–1034

Podil'chuk YN (1998b) Representation of the general solution of statics equations of the electroelasticity of a transversally isotropic piezoceramic body in terms of harmonic functions. International Applied Mechanics 34(7):623–628

Ricoeur A, Kuna M (2009) Electrostatic tractions at dielectric interfaces and their implication for crack boundary conditions. Mechanics Research Communications 36(3):330–335

Shang F, Kuna M, Kitamura T (2003) Theoretical investigation of an elliptical crack in thermopiezoelectric material. Part I: Analitical development. Theoretical and Applied Fracture Mechanics 40(3):247–253

Shulga NA, Karlash VL (2008) Resonant Electromechanical Oscillations of Piezoelectric Plates (in Ukrain.). Naukova Dumka, Kiev

Sladek J, Sladek V, Wünsche M, Zhang C (2018) Effects of electric field and strain gradients on cracks in piezoelectric solids. European Journal of Mechanics - A/Solids 71:187–198

Wang YJ, Gao CF, Song H (2015) The anti-plane solution for the edge cracks originating from an arbitrary hole in a piezoelectric material. Mechanics Research Communications 65:17–23

Wang Z, Zheng B (1995) The general solution of three-dimensional problems in piezoelectric media. International Journal of Solids and Structures 32(1):105–115

Willis JR (1968) The stress field around an elliptical crack in an anisotropic elastic medium. International Journal of Engineering Science 6(5):253–263

Wippler K, Kuna M (2007) Crack analyses in three-dimensional piezoelectric structures by the BEM. Computational Materials Science 39(1):261–266

Wippler K, Ricoeur A, Kuna M (2004) Towards the computation of electrically permeable cracks in piezoelectrics. Engineering Fracture Mechanics 71(18):2567–2587

Xu CH, Zhou ZH, Xu XS, Leung AYT (2015) Electroelastic singularities and intensity factors for an interface crack in piezoelectric-elastic bimaterials. Applied Mathematical Modelling 39(9):2721–2739

Zhang QY, Fan CY, Xu GT, Zhao MH (2017) Iterative boundary element method for crack analysis of two-dimensional piezoelectric semiconductor. Engineering Analysis with Boundary Elements 83:87–95

Zhang TY, Gao CF (2004) Fracture behaviors of piezoelectric materials. Theoretical and Applied Fracture Mechanics 41(1):339–379

Zhao MH, Pan YB, Fan CY, Xu GT (2016) Extended displacement discontinuity method for analysis of cracks in 2d piezoelectric semiconductors. International Journal of Solids and Structures 94-95:50–59

Zhao MH, Yang CH, Fan CY, Xu GT (2018) Extended displacement discontinuity method for analysis of penny-shaped cracks in three-dimensional thermal piezoelectric semiconductors. European Journal of Mechanics - A/Solids 70:23–36

Zhou YY, Chen WQ, Lu CF (2010) Semi-analytical solution for orthotropic piezoelectric laminates in cylindrical bending with interfacial imperfections. Composite Structures 92(4):1009–1018

Chapter 17
On the Quasi-Static Approximation to the Initial Traction Boundary Problem of Linear Elastodynamics

Robin J. Knops

Abstract Conditions are investigated that are sufficient for the quasi-static approximation to be valid in the initial traction boundary value problem of linear elastodynamics on a bounded three-dimensional spatial region. The approach consists of two main steps both of which involve trace inequalities to derive explicit estimates. The first establishes continuous dependence of the solution upon the inertia. The second treats continuous dependence of the inertia upon surface traction and body-force. Circumstances when the approximation is not valid are briefly discussed.

Keywords: Quasi-static approximation · Linear elasticity

17.1 Introduction

The main objective of this note is to establish conditions on selected data that justify the quasi-static approximation for classical linear elastodynamics. The quasi-static approximation to any dynamical system assumes that the inertia, or accleration, is negligible compared to the displacement, displacement gradient (strains) and velocity. Yet precise conditions under which inertia can be neglected are seldom stated in the literature. Conditions that produce asymptotic vanishing of inertia are impractical. Of interest are conditions that ensure the inertia is suitably small for all time or after a short interval once motion has commenced. Dissipative systems are of obvious significance and include theories of linear thermoelastodynamics (Boley and Weiner, 1960; Carlson, 1972) and linear viscous elasticity (Eringen, 1967). The formation of shock waves in the solution to the nonlinear wave equation or in that of elastodynamics are further examples of dissipative behaviour for which the inertia

Robin J. Knops
The Maxwell Institute of Mathematical Sciences and School of Mathematical and Computing Sciences, Heriot-Watt University, Edinburgh EH14 4AS, Scotland, UK
e-mail: r.j.knops@hw.ac.uk

B. E. Abali et al. (eds.), *New Achievements in Continuum Mechanics and Thermodynamics*, Advanced Structured Materials 108,
https://doi.org/10.1007/978-3-030-13307-8_17

can be expected to become negligible in finite or infinite time. Vanishing of inertia in systems without intrinsic dissipation must rely upon suitable initial conditions, source terms and boundary data.

Initial boundary value problems for the linear wave equation in two and three spatial dimensions (see, for example, Egorov and Shubin, 1998) demonstrate that rapid oscillatory initial conditions violate the quasi-static approximation. Similar examples in linear elastodynamics include those discussed in Shield and Green (1963); Shield (1965) and emphasise the relation with Liapunov stability. For the initial boundary value problem subject to arbitrary initial data the solution to the nonlinear wave equation can be unstable, as illustrated among others by Glassey (1973); Knops et al (1974)) and Levine (1974a,b) and quoted references. Counterexamples are provided also by metastable systems and in particular the Schaffer-Shnirelmann phenomenon (Schaffer, 1993; Shnirelmann, 1997) in which a fluid, after a period during which it is stationary, exhibits motion without input of energy. More generally, the quasi-static approximation may be valid for some geometries but not others. Furthermore, weak solutions may behave differently to strong or classical solutions. These and other examples reinforce the need for a precise statement of conditions under which the quasi-static approximation is valid.

The literature on this question is not vast and is mainly confined to discussion of particular problems. Notable contributions include those by Day (1981, 1982, 1985) mainly for the one dimensional initial boundary value problem of linear homogeneous isotropic uncoupled thermoelastodynamics subject to square integrability conditions. Elements of his method occur in the present development. Other techniques include scaling arguments that examine the relative magnitude of the nondimensionalised inertia, displacement, and velocity, as illustrated in the familiar derivation of Stokes flow in fluid mechanics (see for example Childress, 2009). Singular perturbations applied to first order are used in linear three dimensional isotropic thermoelastodynamics by Esham and Weinacht (1994, 1999) to obtain pointwise estimates that establish the uniform exponential decay of the inertia. Dafermos (1968, 1976) considers linear anisotropic thermoelasticity in three dimensions and apart from discussing existence and uniqueness, proves that the temperature gradient and specific entropy decay asymptotically to zero, while the displacement tends to zero except for certain regions and boundary conditions where convergence is to an undamped oscillation. Additional detailed results are due to Lebeau and Zuazua (1999). A unified account may be found in the book by Jiang and Racke (2000). Shearing motions in certain viscous materials are explored by Saccomandi and co-authors (Gilchrist et al, 2013; Pucci and Saccomandi, 2010, 2011; Saccomandi and Vergori, 2016). It is shown that a boundary layer exists in which the inertia does not initially decay although it does eventually tend to zero.

The present study adopts a different method to those just described. The approach is motivated by proofs of continuous data dependence in the theory of partial differential equations. Pointwise behaviour is not considered. Instead, integral measures are introduced to establish various mean square estimates required in the validation of the quasi-static approximation. The argument has two components both of which employ trace and other standard inequalities. The first supposes that a solution exists to the quasi-static approxmimation, and examines conditions sufficient for the difference between the exact and quasi-static approximate solution to depend continuously in mean square measure upon the inertia. Initial and traction boundary data are assumed to be the same for both the exact and approximate systems, although it is comparatively straightforward to include initial conditions that differ. The second component argument identifies conditions sufficient for continuous dependence of the inertia upon surface traction and body-force together with their time derivatives up to fourth order. The combination of both components determines conditions under which the inertia becomes vanishingly small, and just as importantly, shows how the error due to the quasi-static approximation vanishes with the magnitude of the inertia.

Notation and the trace inequalities are introduced in Sect. 17.2, while the exact initial traction boundary value problem of linear elastodynamics is stated in Subsect. 17.3.1. Subsection 17.3.2 derives a conservation law for the exact system and discusses why the resulting inequality, although of intrinsic interest, is unsatisfactory for the present study. The quasi-static approximation is formulated in Subsect. 17.4.1 and a corresponding conservation law derived in Subsect. 17.4.2. The first component argument for validation of the quasi-static approximation is considered in Sect. 17.5 where continuous dependence of the displacement upon the inertia is established. Section 17.6, which completes the analysis, deals with the second component of the argument and concerns the proof that the inertia depends continuously upon the surface traction, body-force and their time derivatives to fourth order. The treatment, similar to a corresponding development by Knops and Quintanilla (2018), appeals to the quasi-static approximate solution together with another conservation law to construct an effective bound which, under suitably prescribed surface traction and body-force, implies that the inertia is small for all time. Consequently, conditions are obtained that justify the quasi-static approximation. Sect. 17.7 contains brief concluding remarks.

A classical solution is assumed to exist on the finite time interval $[0, T)$, although it is often convenient to suppose that T is unbounded. The conventions are adopted of summation over repeated suffixes and of the comma notation to indicate partial differentiation. Suffixes range over 1, 2, 3 except for the indices t and η which are reserved for the time variable. Scalar, vector, and tensor quantities are not typographically distinguished.

17.2 Preliminaries

17.2.1 Notation

We consider a bounded region $\Omega \subset \mathbb{R}^n$, where $n = 2, 3$, with Lipschitz boundary $\partial\Omega$ on which the unit outward normal is the vector field N. A spatial point in either Ω or $\partial\Omega$ is specified by the vector position x, while the time variable is given by

$t \geq 0$. A common Cartesian orthogonal set of coordinates is used throughout unless otherwise stated

The region Ω is occupied by an anisotropic nonhomogeneous compressible linear elastic material whose elastic modulus tensor has components c_{ijkl} that satisfy the symmetries

$$c_{ijkl} = c_{klij} = c_{jikl}, \tag{17.1}$$

and is positive-definite in the sense that there exists a positive constant c_0 such that

$$c_0 \psi_{ij} \psi_{ij} \leq c_{ijkl} \psi_{ij} \psi_{kl}, \qquad \forall\, \psi_{ij} = \psi_{ji}, \tag{17.2}$$

for all symmetric second order tensors ψ. Relaxation of assumption (17.2) is required when similar problems to those discussed here are considered in the theory of small elastic deformaations superposed upon large elastic deformations; see for example, Green et al (1952).

17.2.2 Inequalities

The standard Schwarz and Young's (arithmetic–geometric mean) inequalities are repeatedly used in what follows. Also employed are trace and Rellich-Korn inequalities proved by Bramble and Payne (1962) but stated here for reference.

Theorem 17.1 (Bramble and Payne (1962)). *Let the coordinate origin be in Ω and let $B(0, a) \subset \Omega$ be the ball of radius a and centre at the origin. Let u be a continuously differentiable vector field defined on $\overline{\Omega}$ where an overbar denotes closure of the region. Then subject to the normalisation*

$$\oint_{\partial B(0,a)} u_i \, dS = \oint_{\partial B(0,a)} (x_i u_j - x_j u_i) \, dS = 0, \qquad i, j = 1, 2, 3, \tag{17.3}$$

where dS denotes the element of surface integration, it follows from (Bramble and Payne, 1962, Sect. 2) that

$$\oint_{\partial\Omega} u_i u_i \, dS \leq C_1 \int_\Omega c_{ijkl} e_{ij} e_{kl} \, dx = C_1 \int_\Omega c_{ijkl} u_{i,j} u_{k,l} \, dx \tag{17.4}$$

and

$$\int_\Omega u_i u_i \, dx \leq C_2 \int_\Omega c_{ijkl} e_{ij} e_{kl} \, dx = C_2 \int_\Omega c_{ijkl} u_{i,j} u_{k,l} \, dx, \tag{17.5}$$

where

$$e_{ij} = \frac{1}{2} \left(u_{i,j} + u_{j,i} \right), \tag{17.6}$$

and C_1, C_2 are computable positive constants.

The proof of both inequalities depends upon Rellich identities and additional vector functions selected to satisfy certain bounds. Several such functions are constructed by Bramble and Payne (1962).

17.3 Exact Initial Boundary Value Problem

17.3.1 Statement of Problem

The linear elastic body is in motion subject to prescribed time dependent body- force $f(x,t)$ per unit mass, time dependent surface traction $T(x,t)$, $x \in \partial\Omega$, and Cauchy initial data. Let $w(x,t)$, $(x,t) \in \Omega \times [0,T)$ be the corresponding displacement vector field. The initial boundary value problem governing the motion is then given by the system

$$\begin{aligned} (c_{ijkl}(x)w_{k,l}(x,t))_{,j} + \rho f_i(x,t) &= \rho \ddot{w}_i(x,t), && (x,t) \in \Omega \times [0,T), \\ (c_{ijkl}(x)w_{k,l}(x,t))\,N_j &= T_i(x,t), && (x,t) \in \partial\Omega \times [0,T), \\ w_i(x,0) &= w_i^{(0)}(x), && x \in \Omega \\ \dot{w}_i(x,0) &= w_i^{(1)}(x), && x \in \Omega, \end{aligned} \tag{17.7}$$

where a superposed dot indicates differentiation with respect to time, and the mass density $\rho(x)$ is assumed to satisfy the bounds

$$0 < \underline{\rho} \leq \rho(x) \leq \bar{\rho}, \qquad x \in \Omega, \tag{17.8}$$

for specified positive constants $\underline{\rho}$, $\bar{\rho}$.

It is supposed that the traction and initial data are compatible at $t = 0$ so that

$$\begin{aligned} c_{ijkl}(x)w_{k,l}^{(0)}(x)N_j &= T_i(x,0), && x \in \Omega, \\ c_{ijkl}(x)w_{k,l}^{(1)}(x)N_j &= \dot{T}_i(x,0), && x \in \Omega. \end{aligned} \tag{17.9}$$

Remark 17.1 (Reduction to initial homogeneous traction boundary value problem). The initial traction boundary value problem (17.7) can be reduced to an initial homogeneous traction boundary value problem using the following standard procedure. Let $w_i^* \in C_2(\Omega \times [0,T))$ be any twice continuously differentiable functions that satisfy

$$(c_{ijkl}w_{k,l}^*)N_j = T_i, \qquad (x,t) \in \partial\Omega \times [0,T).$$

Set

$$w_i^{**}(x,t) = w_i(x,t) - w_i^*(x,t), \qquad (x,t) \in \bar{\Omega} \times [0,T],$$

to obtain the system

$$\left(c_{ijkl}w_{k,l}^{**}\right)_{,j} + \rho f_i^* = \rho\ddot{w}_i^{**}, \qquad (x,t) \in \Omega \times [0,T),$$
$$c_{ijkl}w_{k,l}^{**}N_j = 0, \qquad (x,t) \in \partial\Omega \times [0,T),$$

$$w_i^{**}(x,0) = w_i^{(0)}(x) - w_i^*(x,0), \quad \dot{w}_i^{**}(x,0) = w_i^{(1)}(x) - \dot{w}_i^*(x,0), \quad x \in \Omega,$$

where

$$\rho f_i^*(x,t) = \rho f_i(x,t) + \left(c_{ijkl}w_{k,l}^*(x,t)\right)_{,j} - \rho\ddot{w}_i^*(x,t), \qquad (x,t) \in \Omega \times [0,T).$$

A non-zero surface traction, however, is retained in order to emphasise the explicit effect of surface traction upon the inertia and more generally upon the quasi-static approximation.

17.3.2 Conservation Law for Exact Problem

We derive a conservation law that establishes a weak form of continuous dependence of the inertia upon the initial and boundary data and body-force in appropriate function spaces.

Multiplication of (17.7) by the velocitiy $\dot{w}_i$ and integration by parts gives

$$\begin{aligned}\int_{\Omega(t)} \left(c_{ijkl}w_{i,j}\dot{w}_{k,l} + \rho\dot{w}_i\ddot{w}_i\right) dx &= \oint_{\partial\Omega(t)} T_i\dot{w}_i\, dS + \int_{\Omega(t)} \rho f_i\dot{w}_i\, dx \\ &= \frac{d}{dt}\oint_{\partial\Omega(t)} T_i w_i\, dS - \oint_{\partial\Omega(t)} \dot{T}_i w_i\, dS \\ &\quad + \int_{\Omega(t)} \rho f_i\dot{w}_i\, dx,\end{aligned} \tag{17.10}$$

where the subscript $\Omega(t)$ indicates that terms in the corresponding integrand are evaluated at time t.

Let $E(t)$ be defined by $(17.12)_2$ below. Integration of $(17.10)_1$ with respect to time gives the well known conservation law

$$E(t) = E(0) + \int_0^t \oint_{\partial\Omega(\eta)} T_i w_{i,\eta}\, dSd\eta + \int_0^t \int_{\Omega(\eta)} \rho f_i w_{i,\eta}\, dxd\eta. \tag{17.11}$$

Before discussing implications of the second identity $(17.10)_2$, we introduce further notation and derive an embedding inequality.

We set

$$K(t)=\frac{1}{2}\int_{\Omega(t)}\rho\dot{w}_i\dot{w}_i\,dx,\ V(t)=\frac{1}{2}\int_{\Omega(t)}c_{ijkl}w_{i,j}w_{k,l}\,dx,\ E(t)=V(t)+K(t),$$
$$A=E(0)-\oint_{\partial\Omega(0)}T_iw_i^{(0)}\,dS,\quad H(t)=\int_{\Omega(t)}\rho f_if_i\,dx,\quad I(t)=\int_0^t H(\eta)\,d\eta,$$
$$J(t)=\oint_{\partial\Omega(t)}T_iT_i\,dS,\qquad L(t)=\int_0^t\oint_{\partial\Omega(\eta)}T_{i,\eta}T_{i,\eta}\,dSd\eta. \tag{17.12}$$

The initial surface traction $T_i(x,0)$ appearing in $(17.12)_3$ is given in terms of $w_i^{(0)}(x)$ by the compatibility relation (17.9).

A Sobolev embedding theorem next provides an upper bound for the quantity $J(t)$ in terms of $L(t)$. The estimate is derived directly by defining

$$\widehat{T}_i(x,t)=T_i(x,t)-T_i(x,0) \tag{17.13}$$

and noting that

$$\widehat{T}_i(x,0)=0,$$
$$\frac{d}{dt}\widehat{T}_i(x,t)=\frac{d}{dt}T_i(x,t). \tag{17.14}$$

In consequence, we have

$$J(t)=\oint_{\partial\Omega(t)}T_iT_i\,dS\leq 2\left[\oint_{\partial\Omega(t)}\widehat{T}_i\widehat{T}_i\,dS+\oint_{\partial\Omega(0)}T_iT_i\,dS\right], \tag{17.15}$$

and

$$\oint_{\partial\Omega(t)}\widehat{T}_i\widehat{T}_i\,dS=2\int_0^t\oint_{\partial\Omega(\eta)}\widehat{T}_i\widehat{T}_{i,\eta}\,dSd\eta$$
$$\leq 2\left(\int_0^t\oint_{\partial\Omega(\eta)}\widehat{T}_i\widehat{T}_i\,dSd\eta\int_0^t\oint_{\partial\Omega(\eta)}T_{i,\eta}T_{i,\eta}\,dSd\eta\right)^{1/2}, \tag{17.16}$$

which on integration gives

$$\left(\int_0^t\oint_{\partial\Omega(\eta)}\widehat{T}_i\widehat{T}_i\,dSd\eta\right)^{1/2}\leq\int_0^t\left(\int_0^\eta\oint_{\partial\Omega(\tau)}T_{i,\tau}T_{i,\tau}\,dSd\tau\right)^{1/2}d\eta$$
$$\leq\left[\int_0^t h(\eta)\,d\eta\int_0^t h^{-1}(\eta)\left(\int_0^\eta\oint_{\partial\Omega(\tau)}T_{i,\tau}T_{i,\tau}\,dSd\tau\right)d\eta\right]^{1/2}, \tag{17.17}$$

where $h(t)$ is an arbitrary integrable function together with its inverse on $[0,T)$. On substitution of $(17.17)_2$ in (17.16) we obtain

$$\oint_{\partial\Omega(t)}\widehat{T}_i\widehat{T}_i\,dS\leq 2\left[L(t)\int_0^t h(\eta)\,d\eta\int_0^t h^{-1}(\eta)L(\eta)d\eta\right]^{1/2}, \tag{17.18}$$

and the notation $(17.12)_5$ is used. Insertion of (17.18) into (17.15) gives the required Sobolev embedding inequality.

Now consider $(17.10)_2$. Integration with respect to time followed by application of Schwarz's inequality and inequality (17.4) subject to (17.3), yields

$$\begin{aligned} E(t) &= A + \oint_{\partial\Omega(t)} T_i w_i \, dS - \int_0^t \oint_{\partial\Omega(\eta)} T_{i,\eta} w_i \, dS d\eta + \int_0^t \int_{\Omega(\eta)} \rho f_i w_{i,\eta} \, dx d\eta \\ &\leq A + \left(J(t) \oint_{\partial\Omega(t)} w_i w_i \, dS \right)^{1/2} + \left(L(t) \int_0^t \oint_{\partial\Omega(\eta)} w_i w_i \, dx d\eta \right)^{1/2} \\ &+ \left(I(t) \int_0^t \int_{\Omega(\eta)} \rho w_{i,\eta} w_{i,\eta} \, dx d\eta \right)^{1/2} \\ &\leq A + (2C_1 J(t) V(t))^{1/2} + \left(2C_1 L(t) \int_0^t V(\eta) \, d\eta \right)^{1/2} \\ &+ \left(2I(t) \int_0^t K(\eta) \, d\eta \right)^{1/2} \\ &\leq A + \gamma_1^{-1} C_1 J(t) + \gamma_1 V(t) + \left(2C_1 L(t) \int_0^t V(\eta) \, d\eta \right)^{1/2} \\ &+ \left(2I(t) \int_0^t K(\eta) \, d\eta \right)^{1/2}, \end{aligned} \tag{17.19}$$

where γ_1 is a positive constant to be chosen and Young's inequality is applied to obtain the last line.

Select γ_1 to satisfy

$$(1 - \gamma_1) \equiv \gamma_2^{-1} < 1, \tag{17.20}$$

and set

$$B(t) = \gamma_1^{-1} C_1 J(t) + C_1 L(t) + I(t) \geq 0, \qquad G(t) = \int_0^t E(\eta) \, d\eta. \tag{17.21}$$

so that after further application of Young's inequality, we obtain from (17.19) the differential inequality

$$\dot{G}(t) \leq \gamma_2 \left(A + B(t) + G(t) \right). \tag{17.22}$$

which upon integration, or by Gronwell's inequality, gives the bounds

$$\begin{aligned} G(t) &\leq \gamma_2 \int_0^t \exp\{\gamma_2 (t - \eta)\} \left(A + B(\eta) \right) d\eta \\ E(t) &\leq \gamma_2 \left(A + B(t) \right) + \gamma_2^2 \int_0^t \exp\{\gamma_2 (t - \eta)\} \left(A + B(\eta) \right) d\eta. \end{aligned} \tag{17.23}$$

The estimates (17.23) are effective only when the surface traction $T_i(x,t)$ and body-force $f_i(x,t)$ are such that the right sides are bounded for all time. Sufficient conditions are that $A \leq 0$ and not only that the surface traction, its time derivative, and body-force are square integrable on their respective domains of definition but as $t \to \infty$, there also holds

$$\begin{aligned} B(t) &= O(\exp-(nt)), \qquad 2n > \gamma_2, \\ \int_0^t B(\eta)\, d\eta &= O(\exp-(nt)), \qquad 2n > \gamma_2. \end{aligned} \tag{17.24}$$

The right sides become asymptotically unbounded when initial conditions imply $A > 0$.

It follows from (17.2),(17.5), and $(17.12)_{1,2}$ that

$$\int_{\Omega(t)} w_i w_i\, dx \leq 2C_2 V(t) \leq 2C_2 E(t), \tag{17.25}$$

which in conjunction with estimates (17.23) indicates how the displacement and velocity depend upon the surface traction and body-force. It is of interest also to employ these estimates to derive similar information regarding dependence on the inertia. For this purpose, the displacement is assumed to be three times continuously differentiable with respect to space and time. The previous argument can then be applied to the system (17.7) after appropriate differentiation with respect to time. The process yields, for example, a bound for

$$E_1(t) = \frac{1}{2}\int_{\Omega(t)} \left(\rho \ddot{w}_i \ddot{w}_i + c_{ijkl}\dot{w}_{i,j}\dot{w}_{k,l}\right) dx,$$

and consequently the inertia on appeal to (17.5). However, as just remarked, these bounds are of limited practical interest. In later sections, improved estimates are obtained which require the relevant functions to be only square integrable and not necessarily to satisfy asymptotic behaviour corresponding to (17.24).

17.4 Quasi-static Approximation

17.4.1 Formulation

The quasi-static approximation to the exact problem specified by (17.7) assumes that the inertia decays faster than either the strain or velocity and consequently may be omitted. The displacement vector field $v(x,t)$, however, remains dependent upon the time variable now regarded as a parameter. Accordingly, the quasi-static approximaation corresponding to (17.7) is given by the system

$$\begin{aligned} \left(c_{ijkl}v_{k,l}\right)_{,j} + \rho f_i &= 0, \qquad x \in \Omega, \\ c_{ijkl}v_{k,l}N_j &= T_i(x,t), \quad (x,t) \in \partial\Omega \times [0,T), \\ v_i(x.0) &= w_i^{(0)}(x), \quad \dot{v}_i(x,0) = w_i^{(1)}(x), \quad x \in \Omega., \end{aligned} \tag{17.26}$$

where the body-force, surface traction, and initial conditions of the exact problem are retained for convenience. Extension of the present arguments to the general quasi-static initial boundary problem is comparatively easy.

17.4.2 Conservation Laws

Certain properties are established including a conservation law for the quasi-static approximation system as just specified.

Multiplication of (17.26) by the field v_i normalised by (17.3), followed by an integration by parts, appeal to Schwarz's inequality, and inequalities (17.4) and (17.5) successively yields

$$\begin{aligned}\int_\Omega c_{ijkl}v_{i,j}v_{k,l}\,dx &= \oint_{\partial\Omega} T_i v_i\,dS + \int_\Omega \rho f_i v_i\,dx\\ &\le \left(\oint_{\partial\Omega} T_iT_i\,dS \oint_{\partial\Omega} v_iv_i\,dS\right)^{1/2}\\ &+ \left(\int_\Omega \rho f_i f_i\,dx \int_\Omega \rho v_i v_i\,dx\right)^{1/2}\\ &\le \left[\left(C_1\oint_{\partial\Omega} T_iT_i\,dS\right)^{1/2} + \left(\bar{\rho}C_2\int_\Omega \rho f_i f_i\,dx\right)^{1/2}\right]^2.\end{aligned} \tag{17.27}$$

The bound (17.5) then leads to

$$\begin{aligned}\int_{\Omega(t)} v_iv_i\,dx \le C_2\Bigg[&\left(C_1\oint_{\partial\Omega(t)} T_iT_i\,dS\right)^{1/2}\\ &+\left(\bar{\rho}C_2\int_{\Omega(t)} \rho f_i f_i\,dx\right)^{1/2}\Bigg]^2.\end{aligned} \tag{17.28}$$

We remark that initial data are absent from both estimates (17.27) and (17.28). After a time integration of (17.27), the same calculation gives

$$\begin{aligned}\int_0^t\int_{\Omega(\eta)} v_iv_i\,dxd\eta \le C_2\Bigg[&\left(C_1\int_0^t\oint_{\partial\Omega(\eta)} T_iT_i\,dS\eta\right)^{1/2}\\ &+\left(\bar{\rho}C_2\int_0^t\int_{\Omega(\eta)} \rho f_i f_i\,dxd\eta\right)^{1/2}\Bigg]^2.\end{aligned} \tag{17.29}$$

which is required in Sect. 17.6.

17.5 Continuous Dependence on Inertia

This Section examines the accuracy of the quasi-static approximation for the problem under consideration by demonstrating that subject to the normalisation (17.3), the difference displacement depends continuously upon the inertia in mean-square norm. For this purpose, set

$$u_i(x,t) = w_i(x,t) - v_i(x,t), \qquad (x,t) \in \overline{\Omega} \times [0,T), \tag{17.30}$$

and by subtraction of (17.26) respectively from (17.7) obtain the system

$$\begin{aligned} (c_{ijkl}u_{k.l})_{,j} &= \rho\ddot{w}_i, \qquad (x,t) \in \Omega \times [0,T),\\ c_{ijkl}u_{k,l}N_j &= 0, \qquad (x,t) \in \partial\Omega \times [0,T),\\ u_i(x,0) &= \dot{u}_i(x,0) = 0, \qquad x \in \Omega. \end{aligned} \tag{17.31}$$

Multiplication of $(17.31)_1$ by u_i, a spatial integration by parts, followed by Schwarz's inequality and inequality (17.5) gives

$$\begin{aligned} \int_{\Omega(t)} c_{ijkl}u_{i,j}u_{k,l}\,dx &= -\int_{\Omega(t)} \rho\ddot{w}_i u_i\,dx\\ &\leq \bar{\rho}C_2 \int_{\Omega(t)} \rho\ddot{w}_i\ddot{w}_i\,dx. \end{aligned} \tag{17.32}$$

A second application of (17.5) then leads to the desired continuous dependence estimate:

$$\int_{\Omega(t)} u_i u_i\,dx \leq \bar{\rho}C_2^2 \int_{\Omega(t)} \rho\ddot{w}_i\ddot{w}_i\,dx. \tag{17.33}$$

Remark 17.2. Inequalities (17.32) and (17.33) are explicitly independent of data including initial data.

It is therefore demonstrated that in the $L_2(\Omega)$ norm the difference displacement u depends continuously upon the inertia $\ddot{w}$. In order, however, to confirm the validity of the quasi-static approximation, conditions must be determined under which the inertia becomes, or is, small.

In the next section, we discuss such conditions and examine continuous dependence of the inertia upon the body-force and surface traction in appropriate function spaces.

17.6 Continuous Dependence of Inertia on Prescribed Surface Tractions and Body-force

The influence of data upon inertia can be determined entirely within the context of the exact problem. Indeed, for sufficiently smooth solutions the result is easily derived from inequalities (17.23) provided the expressions $(17.12)_{4,5}$ for $H(t)$, $J(t)$, $L(t)$ are replaced by

$$\begin{aligned} H_1(t) &= \int_{\Omega(t)} \rho f_{i,\eta\eta} f_{i,\eta\eta}\, dx, \\ J_1(t) &= \oint_{\partial\Omega(t)} T_{i,\eta\eta} T_{i,\eta\eta}\, dx, \\ L_1(t) &= \int_0^t \oint_{\partial\Omega(\eta)} T_{i,\eta\eta\eta} T_{i,\eta\eta\eta}\, dS d\eta. \end{aligned} \tag{17.34}$$

Nevertherless, introduction of the quasi-static approximate displacement v enables improved bounds to be obtained. An analogous derivation is developed by Knops and Quintanilla (2018).

On using the previous notation, we have

$$\ddot{w}_i = \ddot{u}_i + \ddot{v}_i$$

which leads to the inequality

$$\int_{\Omega(t)} \rho \ddot{w}_i \ddot{w}_i\, dx \leq 2 \left(\int_{\Omega(t)} \rho \ddot{u}_i \ddot{u}_i\, dx + \int_{\Omega(t)} \rho \ddot{v}_i \ddot{v}_i\, dx. \right). \tag{17.35}$$

Upper bounds are now sought for each integral on the right of (17.35) in terms of the time dependent surface traction and body-force. For this purpose, we note that under sufficient differentiability, a bound for the second integral is easily obtained from the inequality corresponding to (17.28). A similar calculation yields

$$\begin{aligned} \int_{\Omega(t)} \rho \ddot{v}_i \ddot{v}_i\, dx &\leq \bar{\rho} \int_{\Omega(t)} \ddot{v}_i \ddot{v}_i\, dx \\ &\leq \bar{\rho} C_2 \left[(C_1 J_1(t))^{1/2} + (\bar{\rho} C_2 H_1(t))^{1/2} \right]^2. \end{aligned} \tag{17.36}$$

Treatment of the first integral on the right of (17.35) consists in deriving a preliminary bound for the integral

$$K_1(t) = \frac{1}{2} \int_{\Omega(t)} \rho \dot{u}_i \dot{u}_i\, dx. \tag{17.37}$$

The desired estimate is then obtained on replacing first time derivatives of u and other terms in the bound for $K_1(t)$ by their second time derivatives.

The system (17.31) can be rewritten as

$$\begin{aligned} (c_{ijkl} u_{k,l})_{,j} &= \rho \ddot{u}_i + \rho \ddot{v}_i, \qquad (x,t) \in \Omega \times [0,T), \\ c_{ijkl} u_{k,l} N_j &= 0, \qquad (x,t) \in \partial\Omega \times [0,T), \\ u_i(x,0) &= \dot{u}_i(x,0) = 0, \qquad x \in \Omega. \end{aligned} \tag{17.38}$$

Note that $(17.38)_1$ is assumed to hold at $t = 0$.

Let

$$
\begin{aligned}
V_1(x) &= \frac{1}{2}\int_{\Omega(t)} c_{ijkl}u_{i,j}u_{k,l}\,dx \geq 0,\\
E_1(t) &= K_1(t) + V_1(t),
\end{aligned}
\tag{17.39}
$$

so that $E_1(0) = 0$ by $(17.38)_3$. Multiplication of $(17.38)_1$ by $\dot{u}$ and integration both by parts and with respect to time gives

$$
\begin{aligned}
E_1(t) &= -\int_0^t\int_{\Omega(\eta)} \rho v_{i,\eta\eta}u_{i,\eta}\,dxd\eta\\
&\leq \left[2\int_0^t\int_{\Omega(\eta)} \rho v_{i,\eta\eta}v_{i,\eta\eta}\,dxd\eta \int_0^t K_1(\eta)\,d\eta\right]^{1/2},
\end{aligned}
\tag{17.40}
$$

where Schwarz's inequality is used. But by (17.39) and (17.2) we have

$$
K_1(t) \leq E_1(t),
$$

and insertion into (17.40) succeeded by a time integration gives

$$
\int_0^t K_1(\eta)\,d\eta \leq \frac{1}{2}\left[\int_0^t ||v(\eta)||\,d\eta\right]^2,
\tag{17.41}
$$

where

$$
||v(t)||^2 = \int_0^t\int_{\Omega(\eta)} \rho v_{i,\eta\eta}v_{i,\eta\eta}\,dxd\eta.
\tag{17.42}
$$

The norm (17.42) is bounded by means of inequality (17.29) suitably adapted to second time derivatives. We have

$$
\begin{aligned}
||v(t)||^2 &\leq \bar{\rho}\int_0^t\int_{\Omega(\eta)} v_{i,\eta\eta}v_{i,\eta\eta}\,dxd\eta\\
&\leq \bar{\rho}C_2\left[(C_1J_2(t))^{1/2} + (\bar{\rho}C_2H_2(t))^{1/2}\right]^2.
\end{aligned}
\tag{17.43}
$$

where

$$
\begin{aligned}
J_2(t) &= \int_0^t J_1(\eta)\,d\eta,\\
H_2(t) &= \int_0^t H_1(\eta)\,d\eta.
\end{aligned}
\tag{17.44}
$$

In consequence, by virtue of (17.40) and (17.41) the required preliminary estimate for $K_1(t)$ is obtained from that for $E_1(t)$ and is given by

$$
\begin{aligned}
K_1(t) \leq E_1(t) &\leq ||v(t)| \int_9^t ||v(\eta)||\, d\eta \\
&\leq (\bar{\rho} C_2) \left[(C_1 J_2(t))^{1/2} + (\bar{\rho} C_2 H_2(t))^{1/2} \right] \times \\
&\times \left[\int_0^t (C_1 J_2(\eta))^{1/2}\, d\eta + \int_0^t (\bar{\rho} C_2 H_2(\eta))^{1/2}\, d\eta \right].
\end{aligned} \tag{17.45}
$$

The estimate $(17.45)_1$ is meaningful only when the acceleration $\ddot{v}_i$ belonging to the quasi-static approximation is mean-square integrable over space-time, and moreover the mean-square norm $||v(t)||$ is integrable over time. These assumptions are satisfied on imposing appropriate restrictions on the norms $J_2(t)$ and $H_2(t)$ and therefore on the surface traction and body-force.

It is now possible to bound the mean-square integral of $\ddot{u}_i$ appearing in (17.35) subject to sufficient smoothness and conditions for higher time derivatives of u to vanish at $t = 0$. Thus, in an obvious notation, we assume

$$
\begin{aligned}
f_{i,tt}(x,0) = f_{i,ttt}(x,0) = 0. \qquad & x \in \Omega, \\
T_{i,tt}(x,0) = T_{i,ttt}(x,0) = 0, \qquad & x \in \partial\Omega.
\end{aligned} \tag{17.46}
$$

Moreover, assume that the differentiated forms of the quasi-static approximation equation (17.26) hold at $t = 0$. Then $v_{i,tt}(x,0)$ and $v_{i,ttt}(x,0)$ satisfy homogeneous boundary value problems and from the Kirchhoff uniqueness theorem of linear elastostatics we conclude that

$$
v_{i,tt}(x,0) = v_{i,ttt}(x,0) = 0, \qquad x \in \bar{\Omega}. \tag{17.47}
$$

Furthermore, on differentiation of $(17.31)_1$ and use of $(17.31)_3$, we have

$$
w_{i,tt}(x,0) = w_{i,ttt}(x,0) = 0, \qquad x \in \Omega, \tag{17.48}
$$

and therefore it follows that

$$
u_{i,tt}(x,0) = u_{i,ttt}(x,0) = 0, \qquad x \in \Omega. \tag{17.49}
$$

Differentiation of the system (17.26) gives

$$
\begin{aligned}
\left(c_{ijkl}\ddot{u}_{k,l}\right)_{,j} = \rho u_{i,tttt} + \rho v_{i,tttt}, \qquad & (x,t) \in \Omega \times [0,T), \\
c_{ijkl}\ddot{u}_{k,l}N_j = 0, \qquad (x,t) \in \partial\Omega \times [0,T), &
\end{aligned} \tag{17.50}
$$

to which are adjoined the initial conditions (17.49). This system is analogous to (17.38) and consequently similar operations may be applied.

We have from (17.5) that

$$
\int_{\Omega(t)} \ddot{u}_i \ddot{u}_i \, dx \leq 2C_2 V_2(t)
$$

where

$$V_2(t) = \frac{1}{2}\int_{\Omega(t)} c_{ijkl}\ddot{u}_{i,j}\ddot{u}_{k,l}\,dx. \tag{17.51}$$

Let us also put

$$K_2(t) = \int_{\Omega(t)} \rho\ddot{u}_i\ddot{u}_i\,dx, \qquad E_2(t) = K_2(t) + V_2(t), \qquad t \in [0,T). \tag{17.52}$$

The argument leading to $(17.45)_{1,2}$ suitably modified and applied to the system (17.49) and (17.50) yields

$$\begin{aligned}\int_{\Omega(t)} \rho\ddot{u}_i\ddot{u}_i &\le 2\bar{\rho}C_2V_2(t) \le 2\bar{\rho}C_2E_2(t)\\ &\le 2\bar{\rho}C_2\|v_{,tt}(t)\|\int_0^t \|v_{,\eta\eta}(\eta)\|\,d\eta.\end{aligned} \tag{17.53}$$

But corresponding to (17.29) we have

$$\|v_{,tt}(t)\| \le (\bar{\rho}C_2)^{1/2}\left[(C_1J_3(t))^{1/2} + (\bar{\rho}C_2H_3(t))^{1/2}\right], \tag{17.54}$$

where

$$\begin{aligned}J_3(t) &= \int_0^t \oint_{\partial\Omega(\eta)} T_{i,\eta\eta\eta}T_{i,\eta\eta\eta}\,dSd\eta,\\ H_3(t) &= \int_0^t \int_{\Omega(\eta)} \rho f_{i,\eta\eta\eta}f_{i,\eta\eta\eta}\,dxd\eta.\end{aligned} \tag{17.55}$$

Consequently, we have constructed a mean-square bound for the difference acceleration $\ddot{u}_i$ in terms of the surface traction and body-force similar to the estimate $(17.45)_2$, provided the time derivatives occurring in (17.55) are restricted to ensure that J_3 and H_3 exist and are integrable for all time.

Finally, the required continuous data dependence of the inertia is proved on insertion of (17.53) and (17.36) into (17.35) and use of (17.54). We obtain

$$\begin{aligned}\int_{\Omega(t)} \rho\ddot{w}_i\ddot{w}_i\,dx &\le 2\left[2\bar{\rho}C_2\|v_{,tt}\|\int_0^t \|v_{,\eta\eta}\|\,d\eta + \int_{\Omega(t)} \rho\ddot{v}_i\ddot{v}_i\,dx\right]\\ &\le (2\bar{\rho}C_2)^2\left\{\left[(C_1J_3(t))^{1/2} + (\bar{\rho}C_2H_3(t))^{1/2}\right]\right.\\ &\quad\times\left.\left[\int_0^t (C_1J_3(\eta))^{1/2} + \int_0^t (\bar{\rho}C_2H_3(\eta))^{1/2}\,d\eta\right]\right\}\\ &\quad + (2\bar{\rho}C_2)\left[(C_1J_1(t))^{1/2} + (\bar{\rho}C_2H_1(t))^{1/2}\right]^2\end{aligned} \tag{17.56}$$

Simplified continuous data estimates, which, however, to be meaningful require prescription of certain asymptotic behaviour, may be derived from (17.56). Consider the second term on the right of $(17.53)_2$. Schwarz's inequality gives

$$\int_0^t ||v_{,\eta\eta}(\eta)||\,d\eta \le \left[t\int_0^t ||v_{,\eta\eta}(\eta)||^2\,d\eta\right]^{1/2}$$
$$= \left[t\int_0^t \int_{\Omega(\eta)} (t-\eta)\rho v_{i,\eta\eta}v_{i,\eta\eta}\,dxd\eta\right]^{1/2}$$
$$\le t||v_{,tt}(t)||.$$

Substitution in (17.53) gives

$$\begin{aligned}\int_{\Omega(t)} \rho\ddot{u}_i\ddot{u}_i\,dx &\le 2\bar{\rho}C_2E_2(t)\\ &\le 2t\bar{\rho}C_2||v_{,tt}(t)||^2\\ &\le t\bar{\rho}C_2\left[(C_1J_3(t))^{1/2}+(\bar{\rho}C_2H_3(t))^{1/2}\right]^2\\ &\le 2t\bar{\rho}C_2\left[C_1J_3(t)+\bar{\rho}C_2H_3(t)\right], \qquad t\ge 0.\end{aligned} \tag{17.57}$$

The continuous data dependence estimate (17.56) becomes

$$\begin{aligned}\int_{\Omega(t)} \rho\ddot{w}_i\ddot{w}_i\,dx &\le 2\left[2t\bar{\rho}C_2||v_{,tt}||^2+\int_{\Omega(t)}\rho\ddot{v}_i\ddot{v}_i\,dx\right]\\ &\le 2\bar{\rho}C_2\left[2t\bar{\rho}C_2\left((C_1J_3(t))^{1/2}+(\bar{\rho}H_3(t))^{1/2}\right)^2\right.\\ &\left.+\left((C_1J_1(t))^{1/2}+(\bar{\rho}C_2H_1(t))^{1/2}\right)^2\right]\\ &\le 4\bar{\rho}C_2\left[2t\bar{\rho}C_2\left(C_1J_3(t)+\bar{\rho}C_2H_3(t)\right)+\left(C_1J_1(t)+\bar{\rho}C_2H_1(t)\right)\right].\end{aligned} \tag{17.58}$$

To be of practical interest, the data terms in (17.58) must be such that J_1, J_3, H_1, and H_3 are integrable and moreover as $t\to\infty$ satisfy the asymptotic behaviour specified by

$$\begin{aligned} J_1(t) &= O(1), & H_1(t) &= O(1),\\ J_3(t) &= O(t^{-(1+\epsilon)}), & H_3(t) &= O(t^{-(1+\epsilon)}),\end{aligned} \tag{17.59}$$

for $\epsilon > 0$

We have constructed a bound for the inertia in terms of the surface traction and body-force provided these functions are of sufficient smoothness and consistent with (17.59). Initial data do not explicitly appear. These results improve upon the bounds corresponding to (17.23) since they require only that the relevant time derivatives of the traction and body-force are square integrable over space time.

These remarks may be illustrated by supposing that

$$T_i(x,t) = X_i(x)Z(t), \qquad (x,t)\in\partial\Omega\times[0,T), \tag{17.60}$$

so that

$$\int_0^t \int_{\partial\Omega(\eta)} T_{i,\eta\eta} T_{i,\eta\eta} \, dS d\eta = \int_{\partial\Omega} X_i(x) X_i(x) \, dx \int_0^t Z_{,\eta\eta}(\eta) Z_{\eta\eta}(\eta) \, d\eta.$$

Let the second derivative of Z be square integrable and suppose that

$$\int_{\partial\Omega} X_i(x) X_i(x) \, dS \leq \epsilon_1 \tag{17.61}$$

for small ϵ_1. Similar treatment of other terms in the estimates then leads to the conclusion that the inertia is small for all time.

17.7 Concluding Remarks

The discussion presented in the preceding Sections is comparatively straightforward and is intended to clarify conditions under which the quasi-static approximation is valid both for the particular initial traction boundary value problem under consideration and more generally. In this respect, it is important to identify data that ensure the inertia in suitable measure is sufficiently small, if not for all time, then at least after an interval sufficiently short to be of practical use. The second crucial element in the argument establishes continuous dependence of the difference solution on the inertia in an appropriate function space. The calculations are based on standard inequalities, but as previously stated other methods of validating the quasi-static approximation, especially in linear thermoelastodynamics, employ scaling arguments and singular perturbation techniques. See, for esample, Esham and Weinacht (1994, 1999); pointwise behaviour is also investigated by Lebeau and Zuazua (1999).

The quasi-static estimates require various orders of the time derivative of the surface traction and body-force to be bounded but not these quantities themselves. They are, however, subject to restrictions imposed by the embedding inequality.

Explicit examples of surface tractions that lead to meaningful bounds are not presented here, partly because emphasis is directed to producing conditions that justify the quasi-static approximation. Nevertheless, surface tractions for which the present bounds are not effective may still be compatible with the quasi-static approximation.

It is obvious that there is scope for further exploration of the topic for both special problems and fundamental principles.

Acknowledgements The author acknowledges valuable discussions with Professor C M Dafermos at an early stage of the study reported here.

References

Boley BA, Weiner JH (1960) Theory of Thermal Stresses. John Wiley & Sons, New York London Sydney

Bramble JH, Payne LE (1962) Some inequalities for vector functions with applications in elasticity. Arch Rational Mech Anal 11:16–26
Carlson DE (1972) Linear Thermoelasticity. In: Flügge S (ed) Handbuch der Physik, vol VIa/2, Springer-Verlag Berlin Heidelberg New York, pp 297–346
Childress S (2009) An Introduction to Theoretical Fluid Mechanics. In: Courant Lecture Notes in Mathematics, American Mathematical Society, vol 19
Dafermos CM (1968) On the existence and the asymptotic stability of solutions to the equations of linear thermoelasticity. Arch Rational Mech Anal 29:241–271
Dafermos CM (1976) Contraction semigroups and trend to equilibrium in continuum mechanics. In: Lecture Notes in Mathematics, vol 503, Springer-Verlag Berlin Heidelberg, pp 295–306
Day WA (1981) Justification of the uncoupled and quasi-static approximation in the problem of dynamic thermoelasticity. Arch Rational Mech Anal 77:387–396
Day WA (1982) Further justification of the uncoupled and quasi-approximations in thermoelasticity. Arch Rational Mech Anal 79:85–95
Day WA (1985) Heat Conduction within Linear Thermoelasticity, Springer Tracts in Natural Philosophy, vol 30. Springer-Verlag Berlin
Egorov Y, Shubin MA (1998) Foundations of the Classical Theory of Partial Differential Equations. Springer Berlin Heidelberg New York
Eringen AC (1967) Mechanics of Continua. John Wiley & Co. New York London Sydney
Esham BF, Weinacht RJ (1994) Singular perturbations and the coupled/quasi-static approximation in linear thermoelasticity. SIAM J Math Anal 25(6):1521–1536
Esham BF, Weinacht RJ (1999) Limitation of the coupled/quasi-static approximation in multi-dimensional linear thermoelasticity. Applicable Analysis 73(1-2):72–87
Gilchrist MD, Rashid B, Murphy JG, Saccomandi G (2013) Quasi-static deformations of biological soft tissue. Math Mech Solids 18:622–633
Glassey R (1973) Blow-up theories for nonlinear wave equations. Math Zeit 132:183–203
Green AE, Rivlin RS, Shield RT (1952) General theory of small elastic deformations superposed upon large elastic deformations. Proc Roy Soc A211:128–154
Jiang S, Racke R (2000) Evolution Equations in Thermoelasticity, Monographs and Surveys in Pure and Applied Mathematics, vol 112. Chapman & Hall/CRC Boca Raton
Knops RJ, Quintanilla R (2018) Quasi-static approximations in linearised thermoelastodynamics. J Thermal Stresses (to appear)
Knops RJ, Levine HA, Payne LE (1974) Nonexistence, instability and growth theorems for solutions to an abstact nonlinear equation with applications to elastodynamics. Arch Rational Mech Anal 35:52–72
Lebeau G, Zuazua E (1999) Decay rates for the three-dimensional linear system of thermoelasticity. Arch Rational Mech Anal 148:179–231
Levine HA (1974a) Instability and nonexistence of global solutions to nonlinear wave equations of the form $\mathrm{P}u_{tt} = -au + \mathcal{F}(\mathrm{u})$. Trans Amer Math Soc 194:1–21
Levine HA (1974b) Some additional remarks on the nonexistence of global solutions to nonlinear wave equations. SIAM J Appl Math 5:138–146
Pucci E, Saccomandi G (2010) On a special class of nonlinear viscoelastic solids. Math Mech Solids 15:803–811
Pucci E, Saccomandi G (2011) On the nonlinear theory of viscoelasticity of differential type. Math Mech Solids 17:624–630
Saccomandi G, Vergori L (2016) Large time approximation for shearing motions. SIAM J Appl Math 78:1964–1983
Schaffer V (1993) An inviscid flow with compact support in space-time. J Geom Anal 3(4):343–401
Shield RT (1965) On the stability of linear continuous systems. Zeit Angew Math Phys 16:649–686
Shield RT, Green AE (1963) On certain methods in the stability theory of continuous systems. Arch Rational Mech Anal 12:354–560
Shnirelmann A (1997) On the uniqueness of weak solutions of the Euler equation. Comm Pure and Appl Math 50(12):1261–1286

Chapter 18
Delamination Buckling in Composite Plates: an Analytical Approach to Predict Delamination Growth

Anton Köllner, Fabian Forsbach & Christina Völlmecke

Abstract An analytical modelling approach is presented which is capable of determining the post-buckling responses as well as the onset of delamination growth of multi-layered composite plates with an embedded circular delamination. In order to overcome current drawbacks of analytical models regarding embedded delaminations, the model employs a problem description in cylindrical coordinates and a novel geometric representation of delamination growth in conjunction with a RAYLEIGH-RITZ formulation and the so-called crack-tip element analysis. The modelling approach is applied to study the compressive response of composite plates with thin-film delaminations loaded under radial compressive strain. Post-buckling responses and the onset of delamination growth are determined for several layups. The results are in very good agreement with finite element simulations while requiring low computational cost.

Keywords: Delamination buckling · Energy release rate · Composites · Plates · Delamination

18.1 Introduction

Delamination buckling is a well-known failure mode in layered slender structures which has attracted a lot of interest since the pioneering work of KACHANOV (Kachanov, 1976) and CHAI et al. (Chai and Babcock, 1985; Chai et al, 1981). Owing to its relevance, particularly for the aircraft industry (Baker and Murray, 2016; Butler et al, 2012), the problem of delaminated composite structures loaded un-

Anton Köllner · Fabian Forsbach · Christina Völlmecke
Technische Universität Berlin, Institute of Mechanics, Stability and Failure of Functionally Optimized Structures Group, Einsteinufer 5, 10587 Berlin, Germany,
e-mail: anton.koellner@tu-berlin.de, fabianfo@mailbox.tu-berlin.de,
christina.voellmecke@tu-berlin.de

B. E. Abali et al. (eds.), *New Achievements in Continuum Mechanics and Thermodynamics*, Advanced Structured Materials 108,
https://doi.org/10.1007/978-3-030-13307-8_18

der in-plane compression represents an area of ongoing research (Chen et al, 2018; Köllner and Völlmecke, 2018; Ouyang et al, 2018). Significant progress has recently been made regarding analytical modelling approaches (Köllner et al, 2018; Köllner and Völlmecke, 2017a,b, 2018) providing insight into the interaction of stability and material failure by determining the post-buckling behaviour during delamination growth and investigating the effect of damage types, dimensions and locations. The effect of delamination location (Ipek et al, 2018; Nilsson et al, 2001), layups (i.e. anisotropy of the sublaminates Butler et al, 2012), local-global buckling (Rhead et al, 2017) and stiffeners (Ouyang et al, 2018) has also been investigated in experimental studies. On the other hand, current numerical studies (Abir et al, 2017; Sun and Hallet, 2018; Tan et al, 2016) mainly investigate the compressive response of certain configurations of damaged composite panels, where damage originated from out-of-plane impact scenarios.

Regarding the evaluation of the compressive strength of delaminated composite panels, the accurate prediction of the onset of delamination growth is important. Analytical models considering embedded delaminations are hitherto not capable of determining the energy release rate along the boundary, which is required to determine the onset of delamination growth precisely. Therefore, the current work aims at improving the capabilities of analytical modelling approaches further by resolving one of the major drawbacks regarding the application of analytical descriptions to embedded delaminations: the prediction of delamination growth by an increase in the initial radius (circular delaminations) (Bottega and Maewal, 1983) or in the major and minor axis (elliptical delaminations), which is commonly referred to as *global* approach.

However, except for certain configurations of the initial delamination (cf. Köllner and Völlmecke, 2018), the global description does not allow for an accurate prediction of the onset of delamination growth (applied load, displacement field, shape of growth) or requires simple model reductions such that only the load causing growth can be approximated (Butler et al, 2012). The current modelling approach considers delamination growth along the boundary of the delamination, thus delamination growth is not associated with a complete disbond of the boundary, which is referred to as *local* approach. This is enabled by using cylindrical coordinates (r, φ, z) as well as a geometric representation of the newly generated delamination area. Despite the resulting dependence of the stiffness tensor on the angle φ, the total potential energy, the equilibrium equations and the energy release rate can be determined analytically yielding an efficient engineering tool to adequately predict post-buckling responses and the onset of delamination growth in multi-layered composite panels with embedded delaminations.

18.2 Model Description

The geometric model and the geometric representation of delamination growth are shown in Fig. 18.1. The circular plate has a radius R^* and a thickness t. The depth of

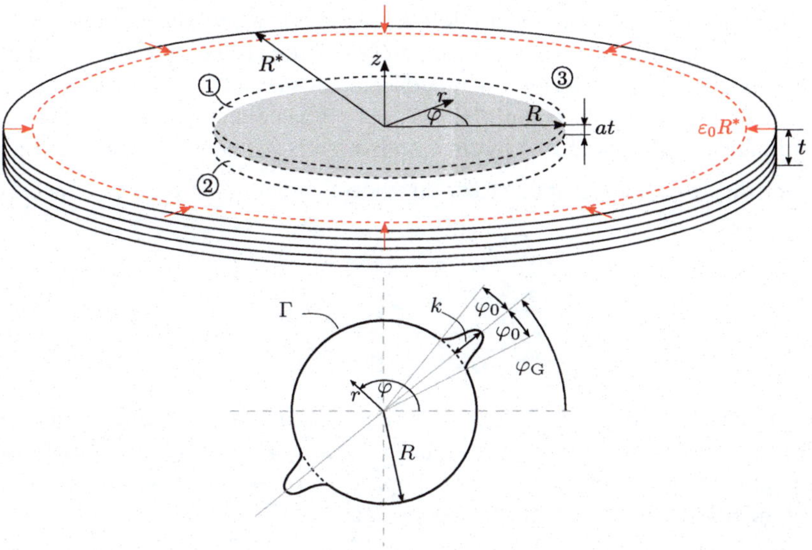

Fig. 18.1 Geometric model of the delaminated plate (top), visualization of the geometric representation of local growth (bottom).

the delamination is defined by the parameter a. Since, in the current work, thin-film delaminations are studied, delaminations complying with $a < 0.1$ are considered. The initial radius of the circular delamination is denoted by R. The plate is subjected to an compressive in-plane radial strain ε_0.

The plate is subdivided into three parts. Parts ① and ② describe the upper and lower sublaminate respectively; part ③ represents the intact region of the plate. It is further assumed that $R^* \gg R$, such that, owing to the thin-film assumption, only the upper delaminated region ① undergoes buckling (out-of-plane deflections).

Delamination growth is modelled with the aid of a trigonometric function added to the given initial radius R in the region where delamination growth is present. Therefore, three parameters φ_G, k and φ_0 are introduced representing the direction of delamination growth, the amplitude of the newly generated delamination and the span of growth respectively. Thus, the boundary of the delamination Γ can be defined as

$$\Gamma = \begin{cases} R + k\cos^2\left(\frac{\varphi - \varphi_G}{\varphi_0}\frac{\pi}{2}\right) & \text{for } \varphi_G - \varphi_0 \leq \varphi \leq \varphi_G + \varphi_0 \\ R, & \text{elsewhere} \end{cases}, \tag{18.1}$$

where, owing to the symmetry of the problem, half of the plate can be considered, i.e. $0 \leq \varphi \leq \pi$. The trigonometric description used is in good agreement with experimental observations made regarding embedded circular delaminations (cf. Nilsson et al (2001)).

The Classical Laminate Theory (Reddy, 2004) is employed since out-of-plane shear effects are deemed small for the laminates considered (the thin-film assumption). The post-buckling behaviour is modelled with the aid of a RAYLEIGH–RITZ formulation where the displacement field is approximated using a set of generalized coordinates q_i. However, as aforementioned, parts ② and ③ experience no out-of-plane displacement, such that their displacement field $\boldsymbol{u}^{(i)}$ can be defined as

$$\begin{aligned} u^{(i)}(r,\varphi) &= \varepsilon_0 r, \\ v^{(i)}(r,\varphi) &= 0, \\ w^{(i)}(r,\varphi) &= 0, \end{aligned} \tag{18.2}$$

where u, v and w are the radial, circumferential and out-of-plane displacements respectively, ε_0 is the loading parameter and $i = 2, 3$. The displacement field of the upper sublaminate is approximated by employing a series of axisymmetric and non-axisymmetric continuous shape functions, thus

$$\begin{aligned} u^{(1)}(r,\varphi) &= \varepsilon_0 r + \sum_{m=1}^{M^u}\sum_{n=0}^{N^u} \sin\left(m\pi\frac{r}{\Gamma}\right)\left(a^u_{mn}\sin(2n\varphi) + b^u_{mn}\cos(2n\varphi)\right), \\ v^{(1)}(r,\varphi) &= \sum_{m=1}^{M^v}\sum_{n=1}^{N^v} \sin(m\pi\frac{r}{\Gamma})\left(a^v_{mn}\sin(2n\varphi) + b^v_{mn}\cos(2n\varphi)\right), \\ w^{(1)}(r,\varphi) &= \sum_{m=1}^{M^w_1} c^w_m\left(\cos\left(m\pi\frac{r}{\Gamma}\right) + (-1)^{m+1}\right) \\ &+ \sum_{m=1}^{M^w_2}\sum_{n=1}^{N^w}\sum_{o=1}^{O^w} \sin\left(m\pi\frac{r}{\Gamma}\right)\sin\left(n\pi\frac{r}{\Gamma}\right)\left(a^w_{mno}\sin(2o\varphi)\right. \\ &+ \left. b^w_{mno}\cos(2o\varphi)\right), \end{aligned} \tag{18.3}$$

where a^u_{mn}, b^u_{mn}, a^v_{mn}, b^v_{mn}, c^w_m, a^w_{mno} and b^w_{mno} are sets of generalized coordinates which will subsequently be summarized in the set q_i. Eqs. (18.2) and (18.3) comply with the geometric boundary conditions:

$$\begin{aligned} u^{(i)}(r=\Gamma,\varphi) &= u^{(1)}(r=\Gamma,\varphi) &&= \varepsilon_0\Gamma, \\ v^{(i)}(r=\Gamma,\varphi) &= v^{(1)}(r=\Gamma,\varphi) &&= 0, \\ w^{(i)}(r=\Gamma,\varphi) &= w^{(1)}(r=\Gamma,\varphi) &&= 0, \\ \nabla_j w^{(i)}(r=\Gamma,\varphi) &= \nabla_j w^{(1)}(r=\Gamma,\varphi) &&= 0, \end{aligned} \tag{18.4}$$

with $\nabla_j = \{\frac{\partial}{\partial r}, \frac{1}{r}\frac{\partial}{\partial\varphi}\}$ and $i = 2, 3$.

The amount of generalized coordinates required to adequately model the post-buckling responses varies strongly with the layup of the upper sublaminate as well as the delamination depth. Therefore, with the aid of a parametric study, 84 generalized coordinates corresponding to $M^u = 8$, $N^u = 3$, $M^v = N^v = 3$, $M^w_1 = 4$, $M^w_2 = N^w = 2$, $O^w = 1$ (cf. Eq. (18.3)) have been determined to provide satisfactory results, where certain configurations such as unidirectional layups and deeper delaminations (within the thin-film range) may only require 10 generalized coordinates.

18.3 Energy Formalism

18.3.1 Total potential energy principle

Owing to the description of the given problem in cylindrical coordinates, the well-known in-plane (A_{IJ}), coupling (B_{IJ}) and bending (D_{IJ}) stiffness matrices comprised within the CLASSICAL LAMINATE THEORY ($\{I, J\} = \{1, 2, 6\}$) are rewritten employing the coordinate transformation illustrated in Fig. 18.2 (from the local fibre coordinate system ($\boldsymbol{e_1}, \boldsymbol{e_2}, \boldsymbol{e_3}$) to the cylindrical coordinate system ($\boldsymbol{e_r}, \boldsymbol{e_\varphi}, \boldsymbol{e_z}$)).

With the assumption of plane stress, the reduced transformed stiffness matrix $[\bar{Q}]$ can be expressed in terms of the reduced stiffness matrix $[Q]$ of the respective unidirectional layers (assumed to be transversally isotropic) of the laminate, the fibre orientation angle θ and the angle φ, thus

$$[\bar{Q}](\varphi) = [K][Q][K]^{-\mathrm{T}}, \text{ with} \tag{18.5}$$

$$[K] = \begin{bmatrix} \cos^2\omega & \sin^2\omega & 2\sin\omega\cos\omega \\ \sin^2\omega & \cos^2\omega & -2\sin\omega\cos\omega \\ -\sin\omega\cos\omega & \sin\omega\cos\omega & \cos^2\omega - \sin^2\omega \end{bmatrix}, \tag{18.6}$$

$$[Q] = \begin{bmatrix} Q_{11} & Q_{12} & 0 \\ Q_{12} & Q_{22} & 0 \\ 0 & 0 & Q_{66} \end{bmatrix}, \text{ and} \tag{18.7}$$

$$\omega = \varphi - \theta. \tag{18.8}$$

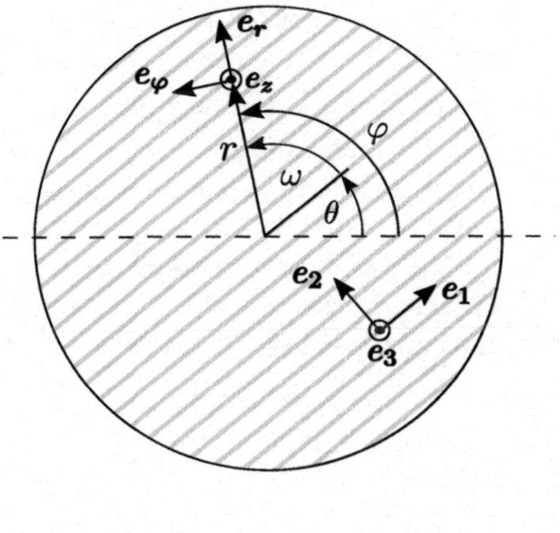

Fig. 18.2 Coordinate transformation from the local fibre coordinate system ($\boldsymbol{e_1}, \boldsymbol{e_2}, \boldsymbol{e_3}$) to the cylindrical coordinate system ($\boldsymbol{e_r}, \boldsymbol{e_\varphi}, \boldsymbol{e_z}$).

In order to model the post-buckling response, non-linear strains associated with the out-of-plane displacement (i.e. VON KÁRMÁN strains, see Reddy, 2004) are considered in the modelling approach, thus

$$\begin{pmatrix} \varepsilon_{rr} \\ \varepsilon_{\varphi\varphi} \\ 2\varepsilon_{r\varphi} \end{pmatrix} = \begin{pmatrix} \varepsilon_1 \\ \varepsilon_2 \\ \varepsilon_6 \end{pmatrix} = \{\varepsilon^0\} + z\{\kappa\}$$
$$= \begin{pmatrix} \dfrac{\partial u}{\partial r} + \dfrac{1}{2}\left(\dfrac{\partial w}{\partial r}\right)^2 \\ \dfrac{1}{r}\dfrac{\partial v}{\partial \varphi} + \dfrac{u}{r} + \dfrac{1}{2}\left(\dfrac{1}{r}\dfrac{\partial w}{\partial \varphi}\right)^2 \\ \dfrac{1}{r}\dfrac{\partial u}{\partial \varphi} + \dfrac{\partial v}{\partial r} - \dfrac{v}{r} + \dfrac{1}{r}\dfrac{\partial w}{\partial \varphi}\dfrac{\partial w}{\partial r} \end{pmatrix} + \begin{pmatrix} -\dfrac{\partial^2 w}{\partial r^2} \\ -\dfrac{1}{r^2}\dfrac{\partial^2 w}{\partial \varphi^2} - \dfrac{1}{r}\dfrac{\partial w}{\partial r} \\ -\dfrac{2}{r}\dfrac{\partial^2 w}{\partial r \partial \varphi} + \dfrac{2}{r^2}\dfrac{\partial w}{\partial \varphi} \end{pmatrix}, \tag{18.9}$$

where $\{\varepsilon^0\}$ and $\{\kappa\}$ are the membrane strains and the curvatures, respectively.

The strain energy W_s is determined by integrating the strain energy density,

$$w_\mathrm{s} = \frac{1}{2}\bar{Q}_{IJ}\varepsilon_I\varepsilon_J, \tag{18.10}$$

over the volume, yielding

$$W_\mathrm{s} = \frac{1}{2}\int_\varphi\int_r \left(\varepsilon_I^0 A_{IJ}\varepsilon_J^0 + 2\varepsilon_I^0 B_{IJ}\kappa_J + \kappa_I D_{IJ}\kappa_J\right) r\mathrm{d}r\,\mathrm{d}\varphi, \tag{18.11}$$

where the displacement field defined in Eqs. (18.2) and (18.3) as well as Eq. (18.9) are employed, with $W_\mathrm{s} = W_\mathrm{s}^{①} + W_\mathrm{s}^{②} + W_\mathrm{s}^{③}$. It should be noted that, in Eq. (18.11), the in-plane (A_{IJ}), coupling (B_{IJ}) and bending stiffness (D_{IJ}) matrices depend on the angle φ. Owing to the displacement controlled problem description, the strain energy is the governing functional. Thus, the post-buckling response can be determined by the well-known variational principle

$$\delta W_\mathrm{s}(q_i) = \frac{\partial W_\mathrm{s}}{\partial q_i}\delta q_i = 0 \quad \text{yielding} \quad \frac{\partial W}{\partial q_i} = 0, \tag{18.12}$$

where the set of non-linear algebraic equations is solved using the NEWTON–RAPHSON method. Owing to the presence of the delamination, an initial imperfection in the form of a small out-of-plane displacement of the upper sublaminate (amplitude of $t/1000$) is commonly assumed modelling delamination buckling (Sheinman et al, 1998). The energy contributions associated with the imperfection are deducted from Eq. (18.11) (cf. Köllner, 2017). The strain energy (Eq. (18.11)) as well as the equilibrium equations (18.12) are determined analytically.

18.3.2 Energy Release Rate

With the aid of the equilibrium solution $q_i(\varepsilon_0)$ obtained from Eq. (18.12), the energy release rate G for delamination growth can be calculated as (cf. Fig. 18.1):

$$G = -\frac{\partial W_\mathrm{s}}{\partial A_\mathrm{del}} \quad \text{with} \quad A_\mathrm{del} = 2 \int\limits_{\varphi_\mathrm{G}-\varphi_0}^{\varphi_\mathrm{G}-\varphi_0+\pi} \int\limits_0^{\Gamma} r\mathrm{d}r\,\mathrm{d}\varphi = R\pi + \frac{1}{4}k\varphi_0\,(8R+3k)\,. \tag{18.13}$$

Eq. (18.13) can be rewritten, since the onset of delamination growth is determined by a change of the amplitude k of the newly generated delamination area for a certain span φ_0, thus

$$G(\varphi_\mathrm{G},\varphi_0) = -\frac{1}{\frac{\partial A_\mathrm{del}}{\partial k}} \frac{\partial W_\mathrm{s}}{\partial k}\bigg|_{k=0}\,. \tag{18.14}$$

Equation (18.14) has to be evaluated for all possible φ_0 (span of delamination growth), i.e. $0 \leq \varphi_0 \leq \pi/2$. Maximizing Eq. (18.14) with respect to φ_0 yields the energy release rate along the boundary of the delamination (φ_G):

$$G(\varphi_\mathrm{G}) = \max_{\varphi_0 \in (0,\frac{\pi}{2})} \left(-\frac{1}{\frac{\partial A_\mathrm{del}}{\partial k}} \frac{\partial W_\mathrm{s}}{\partial k}\bigg|_{k=0} \right). \tag{18.15}$$

Even though the calculation of the energy release rate along the boundary of embedded delaminations constitutes a significant advancement in analytical modelling approaches, it should be noted that Eq. (18.15) provides the total amount of the energy release rate. Particularly for embedded delaminations, delamination growth is governed by mode mixture, which is not considered in Eq. (18.15). Therefore, mode mixture is determined by evaluating the force and moment resultants along the boundary of the delamination in conjunction with employing the crack-tip element analysis as described in Schapery and Davidson (1990). Such a crack-tip element, adjusted for the given problem of thin-film delaminations, is illustrated in Fig. 18.3.

In Fig. 18.3, a one-dimensional representation of the crack-tip element is shown. The thin-film assumption is enforced by the supports added to the bottom of the plate. Following Davidson et al (2000, 1995), the lengths d and e as well as the width of the element (annulus) are small enough such that geometric nonlinearities are negligible as well as force and moment resultants remain uniform within the element. As done in Davidson et al (1995), the force and moment resultants $n_{\varphi\varphi}$ and $m_{\varphi\varphi}$ are omitted assuming that they do not affect the crack-tip element (state of plane strain in the width direction of the element). Moreover, it has been shown (e.g. Nilsson et al, 2001) that Mode III remains negligible for the delaminations studied, thus the shear components $n_{r\varphi}$ and $m_{r\varphi}$ are subsequently also omitted.

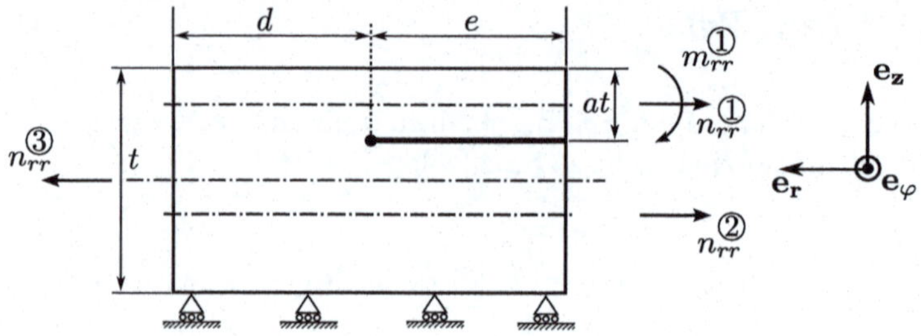

Fig. 18.3 Crack-tip element for the circular composite plate with a thin-film delamination.

With the aid of a free body diagram of the upper sublaminate illustrated in Fig. 18.4, the crack-tip force n_c and moment m_c can be determined:

$$n_\mathrm{c} = -n_{rr}^{①} + \tilde{n}_{rr}^{③}, \tag{18.16}$$

$$m_\mathrm{c} = -m_{rr}^{①} + n_\mathrm{c}\left(\frac{at}{2}\right), \tag{18.17}$$

with $\tilde{n}_{rr}^{③} = \varepsilon_0 \left(A_{11}^{①} + A_{12}^{①} \right)$.

The energy release rate G employing the crack-tip forces and resultants as well as the concept of virtual crack closure (Krueger, 2004) can be calculated as

$$G = \frac{1}{2d}\left(n_\mathrm{c}\Delta u + m_\mathrm{c}\Delta\beta\right), \tag{18.18}$$

where $\Delta u = u^{①} - u^{②}$ and $\Delta\beta = \beta^{①}$ are the differences in the displacement of the crack surfaces in the radial direction and in the rotation around the φ-axis respectively, i.e.

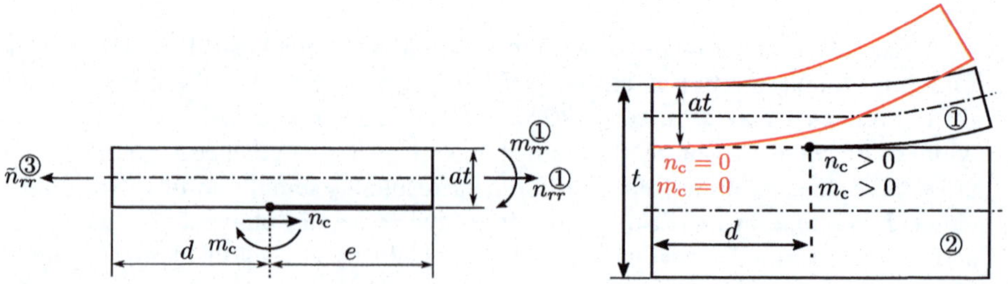

Fig. 18.4 Free-body diagram of the upper sublaminate.

$$
\begin{aligned}
u^{①} &= d\left(\varepsilon_{rr}^{0} - \kappa_{rr}\tfrac{at}{2}\right)^{①}, \\
u^{②} &= d\,\varepsilon_{rr}^{0②}, \\
\beta^{①} &= d\,\kappa_{rr}^{①}.
\end{aligned}
\tag{18.19}
$$

The parameters Δu and $\Delta\beta$ can also be expressed by the inverted constitutive relation using the crack tip force $n_{\rm c}$ and moment $m_{\rm c}$ in conjunction with the compliances (Schapery and Davidson, 1990),

$$
\begin{bmatrix} a_{11} & b_{11} \\ b_{11} & d_{11} \end{bmatrix}^{(i)} = \left(\begin{bmatrix} A_{11} & B_{11} \\ B_{11} & D_{11} \end{bmatrix}^{(i)} \right)^{-1}, \tag{18.20}
$$

yielding

$$
\begin{aligned}
\Delta u/d &= \left(a_{11}^{①} + a_{11}^{②} - b_{11}^{①} at + d_{11}^{①}\left(\tfrac{at}{2}\right)^{2}\right) n_{\rm c} + \left(b_{11}^{①} - d_{11}^{①}\tfrac{at}{2}\right) m_{\rm c} \\
&= c_{11}\, n_{\rm c} + c_{12}\, m_{\rm c}, \\
\Delta\beta/d &= \left(b_{11}^{①} - d_{11}^{①}\tfrac{at}{2}\right) n_{\rm c} + d_{11}^{①}\, m_{\rm c} \\
&= c_{12}\, n_{\rm c} + c_{22}\, m_{\rm c}\,.
\end{aligned}
\tag{18.21}
$$

With the parameters c_{11}, c_{12} and c_{22} given by Eq. (18.21), the mode mixture between mode I and mode II can be calculated by determining the phase angle $\Psi = \tan^{-1}\sqrt{G_{\rm II}/G_{\rm I}}$, i.e.

$$
\Psi = \tan^{-1}\left(\frac{\sqrt{c_{11}} n_{\rm c} \cos(\Omega) + \sqrt{c_{22}} m_{\rm c} \sin(\Omega + \Gamma)}{-\sqrt{c_{11}} n_{\rm c} \sin(\Omega) + \sqrt{c_{22}} m_{\rm c} \cos(\Omega + \Gamma)} \right), \tag{18.22}
$$

as given in Davidson et al (2000), where $\Gamma = \sin^{-1}(c_{12}(c_{11}c_{22})^{-1/2})$ and Ω is the mode-mix parameter. Note that employing the CLASSICAL LAMINATE THEORY, the parameter Ω cannot be determined analytically for thin-film delaminated multi-layered plates; experimental (Davidson et al, 2000) or numerical studies (Schapery and Davidson, 1990) are required. In the current work, Ω is determined with the aid of a finite element simulation (cf. Table 18.1) and remains constant for thin-film delaminations (cf. Davidson et al, 2000).

In order to determine the critical energy release rate $G_{\rm c}$, Eq. (18.22) is used in a crack growth criterion provided by Hutchinson and Suo (1992), i.e.

$$
G_{\rm c} = G_{\rm I}\left(1 + \tan^{2}\left((1-\lambda)\,\Psi\right)\right), \tag{18.23}
$$

with

$$
\lambda = 1 - \frac{2}{\pi}\tan^{-1}\left(\frac{G_{\rm c}^{\rm II} - G_{\rm c}^{\rm I1}}{G_{\rm c}^{\rm I}}\right),
$$

where G_c^I and G_c^{II} are the critical energy release rates for mode I and II respectively.

18.4 Results

The capabilities of the modelling approach are presented in two ways. First, the post-buckling behaviour of a unidirectional laminate with varying delamination depth is studied (Fig. 18.5). Second, the effect of the layup (i.e. angle orientation) on the behaviour of the energy release rate and thus the onset of delamination growth is analysed (Fig. 18.6). A multi-layered composite plate made of 40 CFRP plies is investigated. The material parameters and the dimensions of the plate are provided in Table 18.1. The results obtained are compared with FE simulations performed in Abaqus using SR4 elements.

Table 18.1 Dimensions and material parameters of the circular plate.

E_{11}	137.90 GPa	G_c^I	0.19 N/mm
E_{22}	8.98 GPa	G_c^{II}	0.63 N/mm
G_{12}	7.20 GPa	R	5 mm
ν_{12}	0.3	t	3.556 mm
$^a\ \nu_{23}$	0.5	t_{ply}	0.0889 mm
$^a\ R^*$	50 mm	Ω	58°

^{a} parameters used for FEM only.

In Fig. 18.5, the post-buckling response of a circular plate with a unidirectional ($[0^\circ_{40}]$) layup and a circular delamination ($R = 5\,\text{mm}$) for three different delamination depths ($a = \{1/40, 2/40, 3/40\}$) is shown. The post-buckling behaviour is analysed in terms of applied compressive strain against midpoint deflection (top in Fig. 18.5) and compressive force acting on the upper sublaminate against the end-shortening of the plate (bottom in Fig.18.5). Normalization is performed against the buckling load and strain of a respective intact plate with the radius R^*. The midpoint deflection is normalized against the total thickness of the plate t. As expected, Fig. 18.5 shows that with larger delamination depths (a) the buckling load increases and the midpoint deflection during the post-buckling response decreases. This behaviour is verified by the FEM showing very good agreement with the analytical modelling approach. The onset of delamination growth is visualized in Fig. 18.5 by diamond symbols; filled diamonds for the current model and non-filled diamonds the for the FEM. Analysing the force against end-shortening behaviour, it can be seen that delamination growth occurs earlier during the post-buckling response with increasing delamination depth. The prediction of the onset of delamination growth is also in very good agreement with the FEM.

The reason for the accurate prediction of delamination growth in Fig. 18.5 is presented in Fig. 18.6 illustrating the out-of-plane displacements of the delaminated region (top row in Fig. 18.6), the behaviour of the energy release rate along the

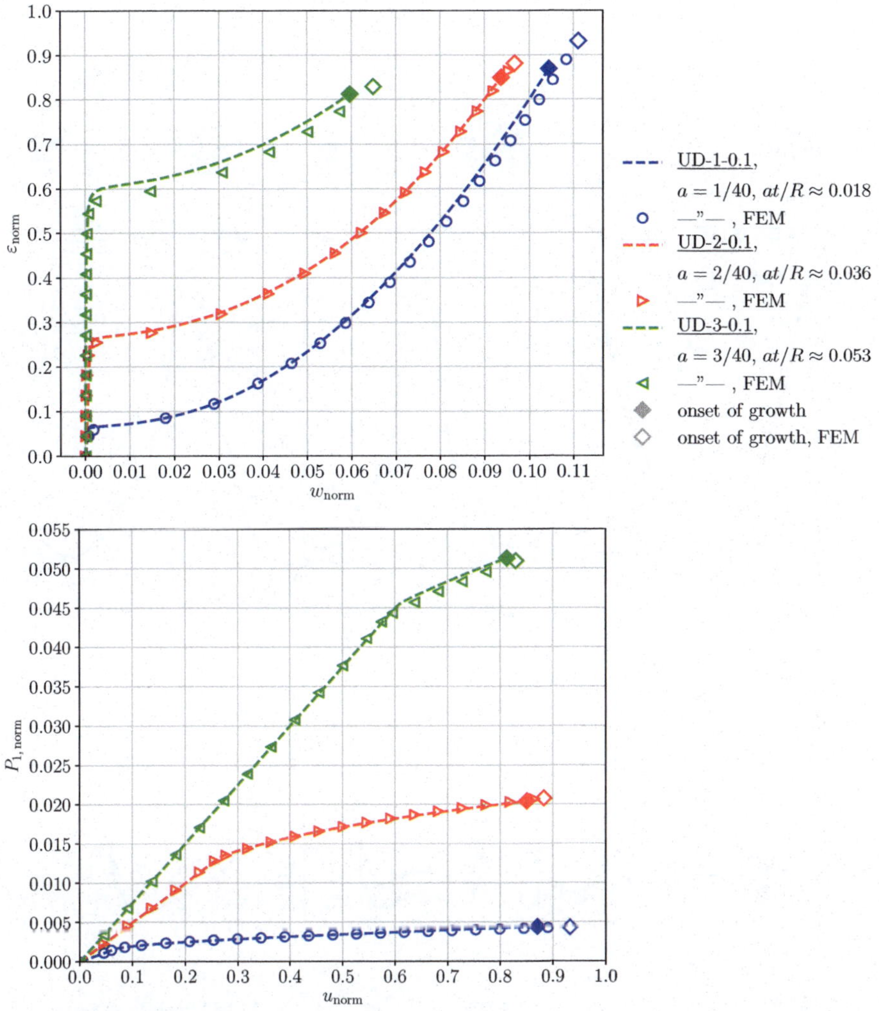

Fig. 18.5 Top: normalized applied compressive strain $\varepsilon_{\mathrm{norm}}$ vs. normalized midpoint deflection w_{norm}; bottom: normalized compressive force P_{norm} vs. normalized end-shortening u_{norm}.

boundary of the delamination (middle row in Fig. 18.6) and the span of delamination growth as well as the phase angle along the boundary (bottom row in Fig. 18.6). A delamination depth of $a = 3/40$ is chosen and three different layups of the upper sublaminate are analysed: **a)** unidirectional $[0^\circ_3]$, **b)** cross-ply $[90^\circ/0^\circ/90^\circ]$ and **c)** an arbitrary angle layup $[45^\circ/0^\circ/45^\circ]$.

Owing to the local geometric representation presented in Sect. 18.2, the behaviour of the energy release rate along the boundary can be analysed (cf. second row of Fig. 18.6). The energy release rate is normalized against the respective criti-

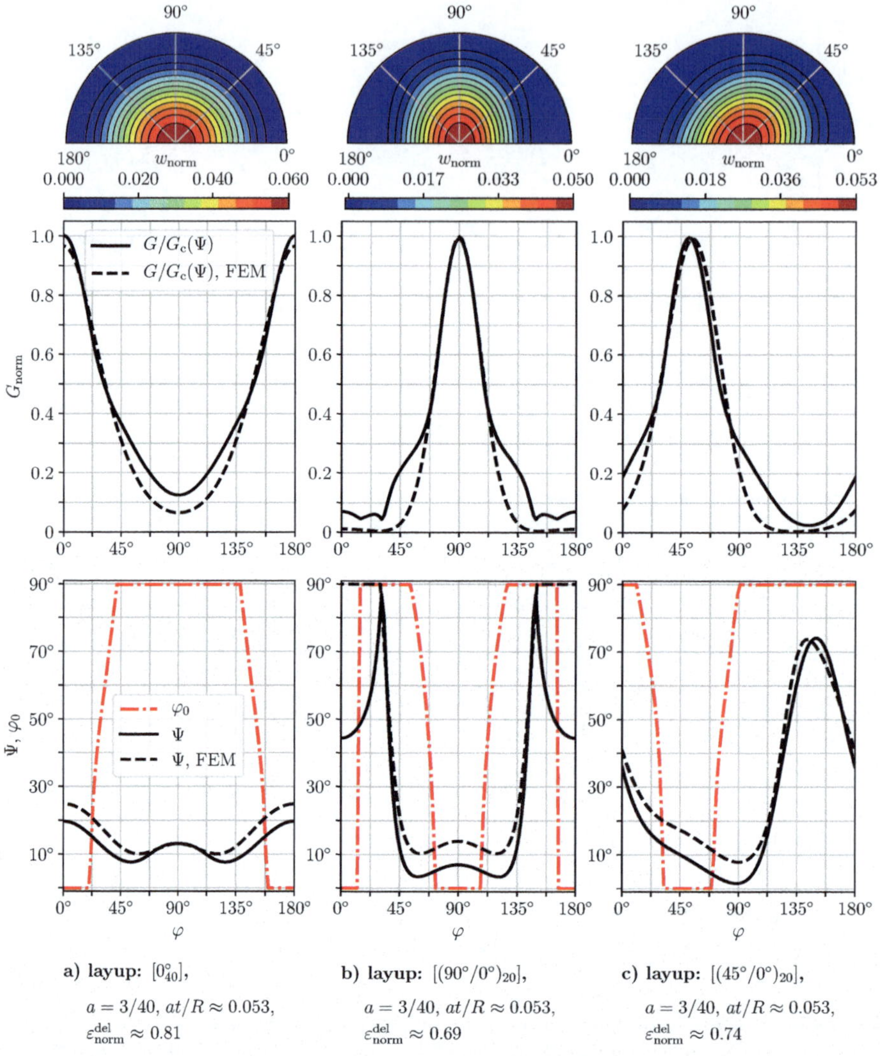

Fig. 18.6 Normalized out-of-plane displacements of the delaminated region w_{norm} (top), normalized energy release rate G_{norm} along the boundary (middle), span of delamination growth φ_0 and phase angle Ψ along the boundary (bottom); at the onset of growth, $\varepsilon^{\mathrm{del}}_{\mathrm{norm}}$; layups studied: **a)** $[0^\circ_3]$, **b)** $[90^\circ/0^\circ/90^\circ]$, **c)** $[45^\circ/0^\circ/45^\circ]$.

cal energy release rate that depends on the phase angle illustrated in the bottom row of Fig. 18.6. Thus, where $G_{\mathrm{norm}} = 1$ is reached along the boundary, delamination growth occurs. As can be seen in Fig. 18.6, the direction of growth is strongly dependent on the layup of the laminate, where growth for the $[0^\circ_3]$ layup is initiated in the 0° direction, for [90°/0°/90°] in the 90° direction and for [45°/0°/45°] growth is shifted to approximately 51°.

The normalized applied strain causing delamination growth ($\varepsilon_{\text{norm}}^{\text{del}}$) is provided at the bottom of Fig. 18.6. The unidirectional layup (a) requires the highest applied strain to cause delamination growth. This is related with the phase angle Ψ at the location of the boundary experiencing growth. In growth direction, the unidirectional layup shows the highest value of the phase angle (20°), whereas the layups [90°/0°/90°] and [45°/0°/45°] indicate angles of approximately 7° and 9°, respectively. Larger phase angles, representing larger mode II contributions, increase the critical energy release rate and therefore larger levels of load input are required to reach the respective critical value. The FEM shows qualitatively the same behaviour with small quantitative deviations in the phase angle.

In the bottom row of Fig. 18.6, besides the phase angle, the span of initial delamination growth φ_0 determined by maximizing Eq. (18.14) is plotted along the boundary. As expected, for all cases investigated the initial span of delamination growth tends to zero indicating a localized delamination growth pattern that corresponds well with the FEM where initial delamination growth is given by disbonding of a single node.

18.5 Conclusions

An analytical modelling approach for predicting post-buckling responses and the onset of delamination growth of multi-layered composite plates with a circular delamination has been presented. For the first time, local delamination growth has adequately been modelled by means of a (semi-)analytical approach. This has been enabled by a geometric representation of the newly generated delamination area and a problem description using cylindrical coordinates. Studies employing (semi-)analytical models have hitherto considered delamination growth in a global manner, i.e. growing major and/or minor axis of a circular(elliptical) delamination, which either only applies to certain configurations or yields significant overestimations of the applied load required to cause delamination growth. Thus, with the aid of the modelling approach presented in this work, the capabilities of (semi-)analytical approaches towards a structural stability analysis of delaminated composite structures have been improved significantly.

Despite using cylindrical coordinates as well as the geometric representation of the boundary of the delamination, the total potential energy, the equilibrium equations and the energy release rate have been determined analytically. Post-buckling responses have been determined by only solving once a set of non-linear algebraic equations. As a consequence, efficient parametric studies are enabled which has been demonstrated, in the current work, by studying the effect of varying delamination depths (cf. Fig. 18.5). The adequate prediction of the onset of delamination growth has been enabled by the analysis of the energy release rate along the entire boundary of the delamination, which hitherto could not be done by semi-analytical modelling approaches. Mode mixture has been considered by employing the crack-tip element analysis, in which the mode mix parameter Ω has been determined with the aid of a finite element simulation. Since Ω mainly depends on the geometry (cf. Davidson et al, 2000), the parameter remains constant for all cases investigated, i.e. for thin-film delaminations, which has been validated by experimental studies in Davidson et al (2000) investigating beam-like structures.

In summary, with the modelling approach developed, a major drawback in semi-analytical models for delamination buckling of embedded delaminations has been overcome, viz. delamination growth can be modelled along the entire boundary. Thus, post-buckling responses of delaminated composite plates considering delamination growth can be determined adequately.

References

Abir MR, Tay TE, Ridga M, Lee HP (2017) Modelling damage growth in composites subjected to impact and compression after impact. Compos Struct 168:13–25

Baker AA, Murray LS (2016) Composite Materials for Aircraft Structures. AIAA Education Series

Bottega WJ, Maewal A (1983) Post-buckled propagation model for compressive fatigue of impact damaged laminates. J Appl Mech 50(1):184–189

Butler R, Rhead AT, Liu W, Kontis N (2012) Compressive strength of delaminated aerospace composites. Phil Trans R Soc Lond A 370:1759–1779

Chai H, Babcock CD (1985) Two-dimensional modelling of compressive failure in delaminated laminates. J Compos Mater 19:67–98

Chai H, Babcock CD, Knauss WG (1981) One dimensional modelling of failure in laminated plates by delamination buckling. Int J Solids Struct 17(11):1069–1083

Chen X, Wu Z, Nie G, Weaver P (2018) Buckling analysis of variable angle tow composite plates with a through-the-width or an embedded rectangular delamination. Int J Solids Struct 138:166–180

Davidson BD, Hurang H, Schapery RA (1995) An analytical crack-tip element for layered elastic structures. J Appl Mech-T ASME 62:294–305

Davidson BD, Gharibian SJ, Yu L (2000) Evaluation of energy release rate-based approaches for predicting delamination growth in laminated composites. Int J Fract 105(4):343–365

Hutchinson JW, Suo Z (1992) Mixed mode cracking in layered materials. Adv Appl Mech 29:63–191

Ipek G, Arman Y, Celik A (2018) The effect of delamination size and location to buckling behavior of composite materials. Composites Part B 155:69–76

Kachanov LM (1976) Separation failure of composite materials. Poly Mech 12(5):812–815

Köllner A (2017) An analytical framework for the structural stability analysis of damageable structures and its application to delaminated composites. Dissertation, Technische Universität Berlin

Köllner A, Völlmecke C (2017a) An analytical framework to extend the general structural stability analysis by considering certain inelastic effects—theory and application to delaminated composites. Compos Struct 170:261–270

Köllner A, Völlmecke C (2017b) Buckling and postbuckling behaviour of delaminated composite struts. Int J Comput Meth Eng Sci Mech 18(1):25–33

Köllner A, Völlmecke C (2018) Post-buckling behaviour and delamination growth characteristics of delaminated composite plates. Compos Struct 203:777–788

Köllner A, Kashtalyan M, Guz I, Völlmecke C (2018) Modelling cracked cross-ply laminates with delamination buckling. Key Eng Mater 774:60–65

Krueger R (2004) Virtual crack closure technique: History, approach, and applications. Appl Mech Rev 57(2):109–143

Nilsson KF, Asp LE, Alpman JE, Nystedt L (2001) Delamination buckling and growth for delaminations at different depths in a slender composite panel. Int J Solids Struct 38:3039–3071

Ouyang T, Sun W, Guan Z, Tan R, Li Z (2018) Experimental study on delamination growth of stiffened composite panels in compression after impact. Compos Struct 206:791–800

Reddy JN (2004) Mechanics of laminated composite plates and shells: theory and analysis. CRC press

Rhead AT, Butler R, Hunt GW (2017) Compressive strength of composite laminates with delamination-induced interaction of panel and sublaminate buckling modes. Compos Struct 171:326–334

Schapery RA, Davidson BD (1990) Prediction of energy release rate for mixed-mode delamination using classical plate theory. Appl Mech Rev 43(5):281–287

Sheinman I, Kardomateas GA, Pelegri AA (1998) Delamination growth during pre- and post-buckling phases of delaminated composite laminates. Int J Solids Struct 35(1-2):19–31

Sun XC, Hallet SR (2018) Failure mechanisms and damage evolution of laminated composites under compression after impact (cai): experimental and numerical study. Composites Part A 104:41–59

Tan W, Falzon BG, Chiu LNS, Price M (2016) Predicting low velocity impact damage and compression-after-impact(cai) behaviour of composite laminates. Composites Part A 71:212–226

Chapter 19
Dynamical Vector Fields on Pantographic Sheet: Experimental Observations

Marco Laudato, Fabio Di Cosmo, Rafał Drobnicki & Peter Göransson

Abstract In this work, we will present and discuss some experimental observations of the dynamical displacement vector field on a pantographic sheet. We will sketch the experimental setup and we will qualitatively describe the observed behavior for a set of relevant frequencies.

Keywords: Pantographic Material · Metamaterial · Digital Imaging Correlation · Dynamics

Marco Laudato
Dipartimento di Ingegneria e Scienze dell'Informazione e Matematica, Università degli Studi dell'Aquila. Via Vetoio, Coppito, 67100 L'Aquila, Italy.
e-mail: marco.laudato@graduate.univaq.it

Fabio Di Cosmo
Dipartimento di Ingegneria Civile, Edile, Ambientale e Architettura Università degli Studi dell'Aquila, L'Aquila, Italy & International Research Center M&MoCS, Università degli Studi dell'Aquila. Via Giovanni Gronchi 18 - Zona industriale di Pile, 67100 L'Aquila, Italy,
e-mail: fabiodicosmo@gmail.com

Rafał Drobnicki
Department of Machinery Design and Biomedical Engineering, Warsaw University of Technology, Warsaw, Poland & Dipartimento di Ingegneria e Scienze dell'Informazione e Matematica, Università degli Studi dell'Aquila, Via Vetoio, Coppito, 67100 L'Aquila, Italy,
e-mail: draf1992@gmail.com

Peter Göransson
KTH Royal Institute of Technology, Marcus Wallenberg Laboratory for Sound and Vibration Research, SE-100 44, Stockholm, Sweden,
e-mail: pege@kth.se

B. E. Abali et al. (eds.), *New Achievements in Continuum Mechanics and Thermodynamics*, Advanced Structured Materials 108,
https://doi.org/10.1007/978-3-030-13307-8_19

19.1 Introduction

Mechanical metamaterials are usually defined as mechanical systems that exhibit an unconventional deformation behavior (Del Vescovo and Giorgio, 2014; Barchiesi et al, 2018c; dell'Isola et al, 2016a). In particular, for these systems, such exotic behavior is usually the result of a complex interaction between the elements of their microstructure (dell'Isola et al, 2016d; Laudato and Di Cosmo, 2018; Golaszewski et al, 2018). From the mathematical point of view, a metamaterial is described in terms of two (or more) length scales of description measuring the ratio of macroscopic length to microscopic lengths. The behavior observed at the larger scale is the result of the collective behavior of the smaller one. In a more formal way, the macroscopic (continuum) model of a mechanical metamaterial is obtained by means of homogenization of the (usually discrete) microscopic system for, both, statistical distributed microstructures (De Masi and Olla, 2015; De Masi et al, 2015; Carinci et al, 2014b,a; Chatzigeorgiou et al, 2014; Saeb et al, 2016; Caprino et al, 1993; Esposito and Pulvirenti, 2004; Pulvirenti, 1996) and ordered microstructures (Pideri and Seppecher, 1997; Cecchi and Rizzi, 2001; Javili et al, 2013a,b, 2012).

An interesting example of mechanical metamaterial is the so-called pantographic material (Alibert et al, 2003; dell'Isola et al, 2016c; Barchiesi and Placidi, 2017; Alibert and Della Corte, 2015; Eremeyev et al, 2017; Boutin et al, 2017). This system is characterized by a microstructure made of arrays of (usually straight) fibers. These arrays lie on different planes which are connected by a set of pivots in the points where the fibers intersect. In this paper, however, we will focus on a particular kind of pantographic material: the so-called pantographic sheet. Its microstructure is a rectangular lattice of two arrays of fibers oriented, with respect to the longest side, with an angle of $\pi/4$ (see Fig. 19.1). Moreover, the cylindrical pivots connecting the two arrays are perfect i.e., in principle, frictionless.

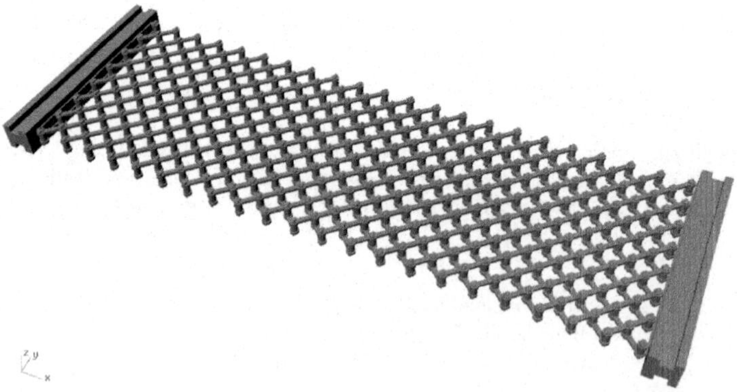

Fig. 19.1 A SolidWork CAD model of pantographic sheet.

Pantographic materials have been, and they still are, the object of investigation for their interesting features (for a general review, see dell'Isola et al (2018)) that can be, in a simplistic way, summarized in two main aspects. The first is related to the mathematical modeling of these structures. Indeed, to properly model the behavior of this kind of materials, one has to postulate an analytical form of the deformation energy which depends on the second gradient of the displacement field (Abali et al, 2017; Ganzosch et al, 2018; Yang et al, 2018; Rinaldi and Placidi, 2013; Cuomo et al, 2017; Andreaus et al, 2016; Giorgio, 2016; Placidi et al, 2015; Ferretti et al, 2014; Seppecher, 2002; Eremeyev and Altenbach, 2014; Auffray et al, 2015; Madeo et al, 2012; Battista et al, 2016; Placidi, 2014; Yang and Misra, 2012). In a wider perspective, pantographic materials can be classified as *generalized continua* (Trinh et al, 2012; dell'Isola et al, 2015b; Balobanov and Niiranen, 2018; Carcaterra et al, 2015; Misra and Poorsolhjouy, 2015; Yang et al, 2011), i.e. mechanical systems that cannot be fully described by the usual approach in which deformation measures depend only on the first gradient of the displacement field. This is a whole research field which is traced back to the Italian mathematician Gabrio Piola (dell'Isola et al, 2014a; Rahali et al, 2015; dell'Isola et al, 2014b, 2019), and represents a formidable challenge for both mathematicians and engineers. The second interesting aspect is related to possible technological applications. Indeed, although pantographic systems are usually made of linear elastic materials with elongation at break of 20% of relative elongation (in the case of polyamide, for instance), the particular arrangement of their microstructure is able to widen the relative elongation for the first rupture up to 30% and they are able to store an amount of energy until the final rupture which is of the same order of magnitude of the energy required for the first one (Barchiesi et al, 2018a; Placidi et al, 2018a,b; Turco et al, 2016a, 2017a; Misra et al, 2018; Placidi and Barchiesi, 2018; Turco et al, 2016d; dell'Isola et al, 2016b; Andreaus et al, 2018; Ganzosch et al, 2016; Scerrato et al, 2016b,a; Giorgio et al, 2016; Barchiesi et al, 2018b; Turco et al, 2016c).

In general, mechanical metamaterials exhibit "exotic" behaviors when observed in their dynamical regime (Deymier, 2013)]. Classical examples of such behaviors are the so-called cloak-effects, as well as the presence of a band-gap in the dispersion relations, which can be potentially exploited to control and guide waves propagation in such materials. This effects are typical of microstructured continua since they can be usually traced back to resonance phenomena in their microstructure (di Cosmo et al, 2018; Madeo et al, 2013; Berezovski et al, 2015). Pantographic materials are not an exception, and several numerical studies on their dynamical behavior have been performed in the last years (Giorgio et al, 2017b; dell'Isola et al, 2015a; Placidi et al, 2014; Madeo et al, 2014; Giorgio et al, 2017a). However, in their microstructure, resonance phenomena are not expected. The "exotic" behavior, in this case, will be due to the *complexity* of the microstructure which results, from the viewpoint of the model, in the occurrence of the second gradient of the displacement field. This is the main difference which makes the dynamical behavior of these materials worthy of interest.

To the best of the authors' knowledge, experimental observations of the dynamic behavior of pantographic materials, aimed at confirming the results of these huge numerical and mathematical efforts, have not been published in the open literature. In this paper, we will present a new set of observations relative to the experiment described in Laudato et al (2018). In particular, after having sketched the description of the experimental setup in Sec. 19.2, we will qualitatively describe the dynamical displacement vector field of the pantographic sheet in Sec. 19.3. Finally, in Sec. 19.4 we will discuss some future perspectives.

19.2 Setup of the experiment

In order to observe the displacement field of the pantographic sample in time, an experiment has been set up using an electrodynamic shaker to put the sample in motion and a pair of high-speed cameras to record the moving sample. The specimens under analysis are made of polyamide and have been 3D printed at Warsaw University of Technology. The measurements have been performed at Marcus Wallenberg Laboratory for Sound and Vibration at KTH Royal Institute of Technology.

In this experiment, one of the two short ends of the sample is clamped to the ground (assumed infinitely stiff) while its other end is solidly connected to the shaker by means of a clamping device manufactured specifically for this purpose. The shaker is suspended in soft rubber springs, and mounted collinear to the sample in its longitudinal direction.

The driving excitation signal is sinusoidal and the shaker moves along sample's axis. To avoid the occurrence of buckling instabilities, the sample is mounted in a pre-stretched configuration. With respect to this configuration, the sample oscillates by imposing a time-varying displacement at one of its ends, while its other end is fixed.

The displacement of the sample is recorded for different excitation frequencies (spanned both in ascending and descending fashion), different amplitudes, and two sample orientations. A delay is imposed between the start of the mechanical excitation and the triggering of the cameras to record only the steady-state condition, skipping the hysteresis behavior. Observing the behavior of the sample at different excitation amplitudes allows determining whether the sample behaves as a linear elastic system, and spanning the excitation frequencies in different order is done in order to examine any viscous behavior. Testing the sample in two different orientations yield the same results, due to the symmetries of the samples tested.

The actual determination of the displacement field is by means of three dimensional (3D) digital image correlation (DIC) (Hild and Roux, 2006). The nature of the method requires the sample to have a random speckle pattern on it, such that the algorithm may uniquely identify and track the different subsets of such dense random pattern to ultimately resolve the displacement field. In light of this, matte black and white aerosol paint has been deposited on one of the sample faces, effectively realizing a random pattern usable by the DIC algorithm. The quality of the analysis depends strongly on lighting conditions: while the matte paint helps in avoiding reflections (which may affect only one of the two camera images, effectively making it harder for the DIC algorithm to match features), effectively having stereo images means that the dynamic range of the intensity values is reduced by a normalization process necessary to compare the two video streams. For such purpose, four halogen spotlights are used to illuminate the sample from for distinct directions – top, bottom, left and right. This also minimizes the shades that struts closer to the foreground cast on neighbouring struts, and allows for the use of short exposure times that practically eliminate motion blur (see Fig. 19.2).

Fig. 19.2 3D reconstruction of the camera setup operated by the DIC software, overlaid to a picture of the real setup.

19.3 Qualitative analysis of the results

In this section we present and discuss qualitatively the measurements' results. In particular, we describe the dynamical displacement vector fields of a pantographic sheet under uniaxial sinusoidal oscillation of its top end. We focus on the configurations relative to maximum displacement at 4 relevant frequencies: 30 Hz, 100 Hz, 140 Hz, and 200 Hz. The amplitudes of the imposed displacement relative to this set of frequencies range from 7.23×10^{-2} mm for 200 Hz to 2.76 mm for 30 Hz. The behavior of the specimen appears to be linear elastic for all the investigated frequencies since all the points of the specimen that we have analyzed oscillate at the same frequency of the shaker with a constant phase delay in time. However, a careful analysis of the observed elastic behavior, in terms of phase delay and transfer functions, will be the object of future investigations.

The resulting displacement vector fields are displayed in Figs. 19.3-19.6. The arrows represent the in-plane displacement whereas their color represents the amplitude of the out-of-plane displacement, both computed with respect to the prestretched configuration. For each vector field, the length of the arrows is normalized with respect to the maximum observed length. Analogously, for the out-of-plane displacements the values relative to the color palette are normalized with respect to the maximum observed amplitude in this direction. Since we are interested in a qualitative description of the topology of such vector fields, however, we have not added any reference length on the figures in order to avoid any misleading unnecessary detail.

From this figure it is possible to remark two interesting features, relative to in-plane and out-of-plane displacements, respectively:

1. The in-plane displacement vector field is curved. The position and the number of curvature centers appear to be dependent on the frequency. In particular, at higher frequencies (see for instance Fig. 19.6, right panel) we can spot three centers of curvature and the curvature itself appears to be proportional to the frequency. Moreover, for higher frequencies, it is possible to observe the presence of vortexes (see Fig. 19.5, right panel).
2. The specimen tends to rotate around its vertical middle axis giving rise to out-of-plane displacements. The amplitude of this perturbation depends inversely on the frequency. Indeed, for lower frequencies (see Fig. 19.3), the peak to peak amplitude of out-of-plane oscillations is around 2 mm (to be compared with the specimen's thickness which is 6 mm), whereas for higher frequencies they can be considered negligible since the corresponding peak to peak amplitude is $\approx 60\mu m$.

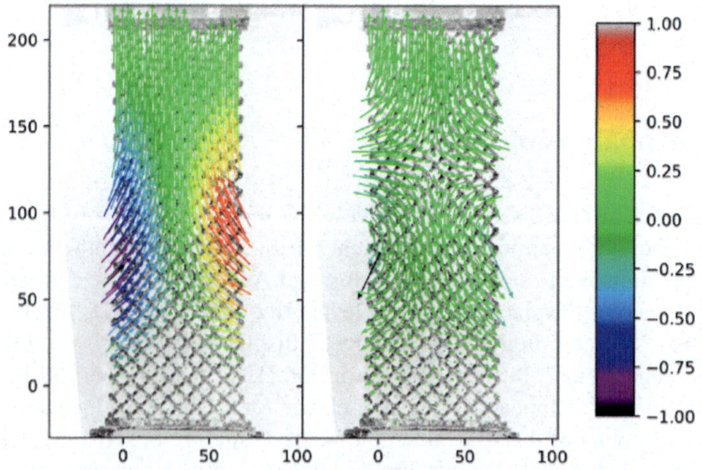

Fig. 19.3 Displacement vector field of a pantographic sheet at maximum and minimum peak amplitudes of the periodic oscillation, at 30 Hz, respectively. The arrows represent in-plane displacement while the color represents the amplitude of the out-of-plane oscillation. The scales are expressed in millimeter.

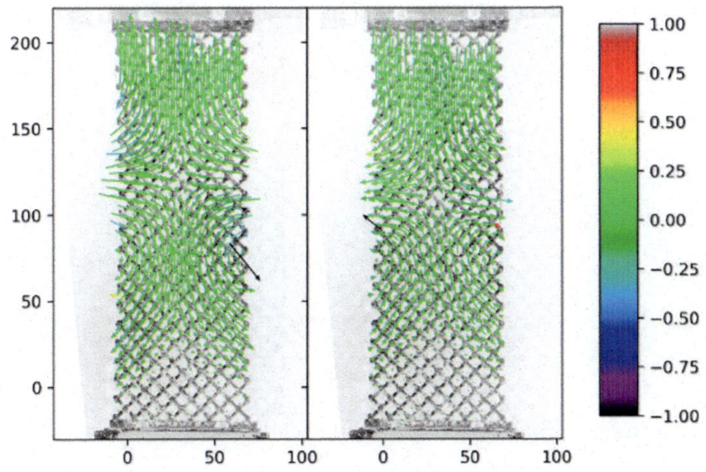

Fig. 19.4 Displacement vector field of the same pantographic sheet, at 100 Hz.

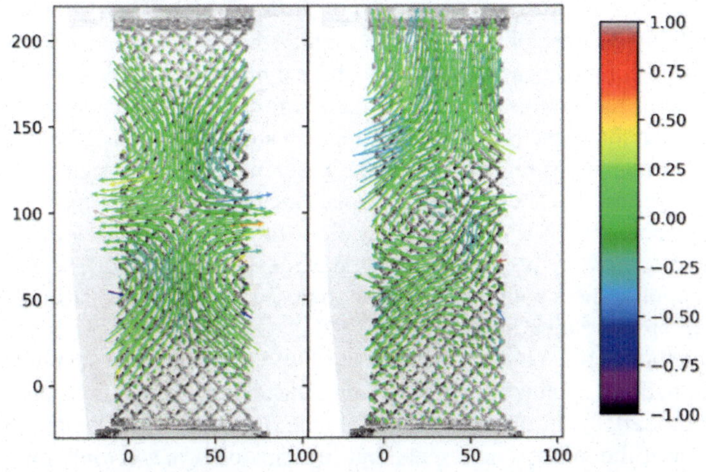

Fig. 19.5 Displacement vector field of the same pantographic sheet, at 140 Hz.

19.4 Conclusions and perspectives

In this paper, experimental observations of the dynamical displacement vector field of a pantographic sheet under uniaxial periodic oscillation is presented. The measurement procedure and the following data analysis have been described in details. The results have been represented in terms of vector fields relative to maximum displacement and they have been qualitatively described.

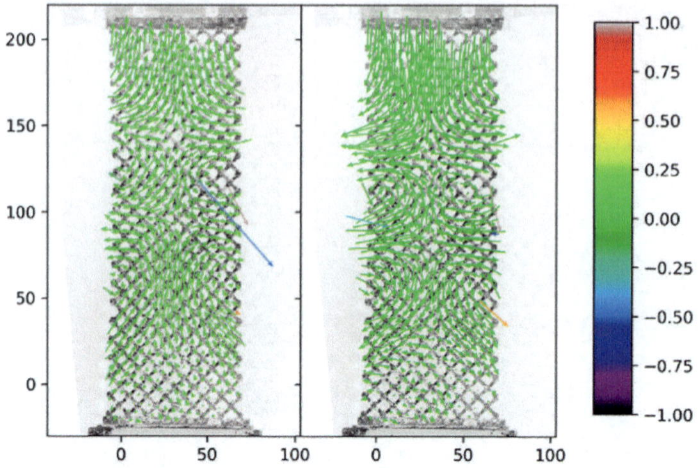

Fig. 19.6 Displacement vector field of the same pantographic sheet, at 200 Hz.

Despite the qualitative nature of the presented results, they can be considered as the starting point for a complete characterization of the dynamical behavior of pantographic material. Indeed, and this will be the aim of a forthcoming paper, the first step will be the comparison of the results from numerical simulations of a 2D model with the presented results, in the higher frequencies case. Indeed, in this regime, we have observed that the out-of-plane amplitude displacement is small with respect to the thickness of the pantographic sheet which may allow for a description in terms of a two-dimensional model. Moreover, the study of phase delay and amplitude of each point of the pantographic sheet with respect to the imposed boundary displacement will be the topic of a forthcoming investigation.

Another possible interesting path is the comparison of the presented results with the numerical simulations obtained by Hencky-type discrete model (Turco et al, 2016b, 2017b; Turco, 2018). In this approach, the material is described neither at the microscopic nor at the macroscopic scale, but in terms of a *mesoscopic* finite-dimensional discrete system which is able to capture the main features of the microstructure despite its discrete nature. The main advantage of such approach is that the computational costs are highly reduced (Turco et al, 2018). Of course, although the discrete model is directly inspired by the microstructure, it represents an arbitrary choice and the presented measurements can be a possible significant benchmark for the validity of such assumption.

Finally, from the experimental point of view, the next step will be the investigation of higher amplitudes, in which non-linear effects due to the complexity of the microstructure (modeled in terms of the second gradient) will become more relevant.

Acknowledgements The authors want to thanks Luca Manzari for the precious help in the production of the figures.

References

Abali BE, Müller WH, dell'Isola F (2017) Theory and computation of higher gradient elasticity theories based on action principles. Archive of Applied Mechanics 87(9):1495–1510

Alibert JJ, Della Corte A (2015) Second-gradient continua as homogenized limit of pantographic microstructured plates: a rigorous proof. Zeitschrift für angewandte Mathematik und Physik

Alibert JJ, Seppecher P, dell'Isola F (2003) Truss modular beams with deformation energy depending on higher displacement gradients. Mathematics and Mechanics of Solids 8(1):51–73

Andreaus U, dell'Isola F, Giorgio I, Placidi L, Lekszycki T, Rizzi N (2016) Numerical simulations of classical problems in two-dimensional (non) linear second gradient elasticity. International Journal of Engineering Science 108:34–50

Andreaus U, Spagnuolo M, Lekszycki T, Eugster SR (2018) A ritz approach for the static analysis of planar pantographic structures modeled with nonlinear euler–bernoulli beams. Continuum Mechanics and Thermodynamics pp 1–21

Auffray N, dell'Isola F, Eremeyev V, Madeo A, Rosi G (2015) Analytical continuum mechanics à la Hamilton–Piola least action principle for second gradient continua and capillary fluids. Mathematics and Mechanics of Solids 20(4):375–417

Balobanov V, Niiranen J (2018) Locking-free variational formulations and isogeometric analysis for the timoshenko beam models of strain gradient and classical elasticity. Computer Methods in Applied Mechanics and Engineering 339:137–159

Barchiesi E, Placidi L (2017) A review on models for the 3d statics and 2d dynamics of pantographic fabrics. In: Wave Dynamics and Composite Mechanics for Microstructured Materials and Metamaterials, Springer, pp 239–258

Barchiesi E, dell'Isola F, Laudato M, Placidi L, Seppecher P (2018a) A 1d continuum model for beams with pantographic microstructure: Asymptotic micro-macro identification and numerical results. In: Advances in Mechanics of Microstructured Media and Structures, Springer, pp 43–74

Barchiesi E, Ganzosch G, Liebold C, Placidi L, Grygoruk R, Müller WH (2018b) Out-of-plane buckling of pantographic fabrics in displacement-controlled shear tests: experimental results and model validation. Continuum Mechanics and Thermodynamics pp 1–13

Barchiesi E, Spagnuolo M, Placidi L (2018c) Mechanical metamaterials: a state of the art. Mathematics and Mechanics of Solids p 1081286517735695

Battista A, Rosa L, dell'Erba R, Greco L (2016) Numerical investigation of a particle system compared with first and second gradient continua: Deformation and fracture phenomena. Mathematics and Mechanics of Solids p 1081286516657889

Berezovski A, Giorgio I, Della Corte A (2015) Interfaces in micromorphic materials: Wave transmission and reflection with numerical simulations. Mathematics and Mechanics of Solids p 1081286515572244

Boutin C, Giorgio I, Placidi L, et al (2017) Linear pantographic sheets: Asymptotic micro-macro models identification. Mathematics and Mechanics of Complex Systems 5(2):127–162

Caprino S, Esposito R, Marra R, Pulvirenti M (1993) Hydrodynamic limits of the vlasov equation. Communications in partial differential equations 18(5):805–820

Carcaterra A, dell'Isola F, Esposito R, Pulvirenti M (2015) Macroscopic description of microscopically strongly inhomogenous systems: A mathematical basis for the synthesis of higher gradients metamaterials. Archive for Rational Mechanics and Analysis pp 1–24

Carinci G, De Masi A, Giardinà C, Presutti E (2014a) Hydrodynamic limit in a particle system with topological interactions. Arabian Journal of Mathematics 3(4):381–417

Carinci G, De Masi A, Giardinà C, Presutti E (2014b) Super-hydrodynamic limit in interacting particle systems. Journal of Statistical Physics 155(5):867–887
Cecchi A, Rizzi N (2001) Heterogeneous elastic solids: a mixed homogenization-rigidification technique. International journal of solids and structures 38(1):29–36
Chatzigeorgiou G, Javili A, Steinmann P (2014) Unified magnetomechanical homogenization framework with application to magnetorheological elastomers. Mathematics and Mechanics of Solids 19(2):193–211
di Cosmo F, Laudato M, Spagnuolo M (2018) Acoustic metamaterials based on local resonances: Homogenization, optimization and applications. In: Generalized Models and Non-classical Approaches in Complex Materials 1, Springer, pp 247–274
Cuomo M, dell'Isola F, Greco L, Rizzi N (2017) First versus second gradient energies for planar sheets with two families of inextensible fibres: investigation on deformation boundary layers, discontinuities and geometrical instabilities. Composites Part B: Engineering 115:423–448
De Masi A, Olla S (2015) Quasi-static hydrodynamic limits. Journal of Statistical Physics 161(5):1037–1058
De Masi A, Galves A, Löcherbach E, Presutti E (2015) Hydrodynamic limit for interacting neurons. Journal of Statistical Physics 158(4):866–902
Del Vescovo D, Giorgio I (2014) Dynamic problems for metamaterials: review of existing models and ideas for further research. International Journal of Engineering Science 80:153–172
dell'Isola F, Andreaus U, Placidi L (2014a) At the origins and in the vanguard of peridynamics, non-local and higher-gradient continuum mechanics: An underestimated and still topical contribution of Gabrio Piola. Mathematics and Mechanics of Solids p 1081286513509811
dell'Isola F, Maier G, Perego U, Andreaus U, Esposito R, Forest S (2014b) The complete works of Gabrio Piola: volume I. Cham, Switzerland: Springer
dell'Isola F, Giorgio I, Andreaus U (2015a) Elastic pantographic 2d lattices: a numerical analysis on static response and wave propagation. Proceedings of the Estonian Academy of Sciences 64(3):219–225
dell'Isola F, Seppecher P, Della Corte A (2015b) The postulations á la D'alembert and á la Cauchy for higher gradient continuum theories are equivalent: a review of existing results. Proc R Soc A 471(2183):20150,415
dell'Isola F, Bucci S, Battista A (2016a) Against the fragmentation of knowledge: The power of multidisciplinary research for the design of metamaterials. In: Advanced Methods of Continuum Mechanics for Materials and Structures, Springer, pp 523–545
dell'Isola F, Cuomo M, Greco L, Della Corte A (2016b) Bias extension test for pantographic sheets: numerical simulations based on second gradient shear energies. Journal of Engineering Mathematics pp 1–31
dell'Isola F, Giorgio I, Pawlikowski M, Rizzi N (2016c) Large deformations of planar extensible beams and pantographic lattices: Heuristic homogenization, experimental and numerical examples of equilibrium. Proceedings of the Royal Society A: Mathematical, Physical and Engineering Sciences 472(2185)
dell'Isola F, Steigmann D, Della Corte A (2016d) Synthesis of fibrous complex structures: Designing microstructure to deliver targeted macroscale response. Applied Mechanics Reviews 67(6):21–pages
dell'Isola F, Seppecher P, Alibert JJ, et al (2018) Pantographic metamaterials: an example of mathematically driven design and of its technological challenges. Continuum Mechanics and Thermodynamics pp 1–34
dell'Isola F, Maier G, Perego U, Andreaus U, Esposito R, Forest S (2019) The complete works of gabrio piola: volume ii. Cham, Switzerland: Springer
Deymier P (2013) Acoustic Metamaterials and Phononic Crystals. Springer Series in Solid-State Sciences, Springer Berlin Heidelberg
Eremeyev VA, Altenbach H (2014) Equilibrium of a second-gradient fluid and an elastic solid with surface stresses. Meccanica 49(11):2635–2643
Eremeyev VA, dell'Isola F, Boutin C, Steigmann D (2017) Linear pantographic sheets: existence and uniqueness of weak solutions. Journal of Elasticity pp 1–22

Esposito R, Pulvirenti M (2004) From particles to fluids. Handbook of mathematical fluid dynamics 3:1–82

Ferretti M, Madeo A, dell'Isola F, Boisse P (2014) Modeling the onset of shear boundary layers in fibrous composite reinforcements by second-gradient theory. Zeitschrift für angewandte Mathematik und Physik 65(3):587–612

Ganzosch G, dell'Isola F, Turco e, Lekszycki T, Müller W (2016) Shearing tests applied to pantographic structures. Acta Polytechnica CTU Proceedings 7:1–6

Ganzosch G, Hoschke K, Lekszycki T, Giorgio I, Turco E, Müller W (2018) 3d-measurements of 3d-deformations of pantographic structures. Technische Mechanick 38(3):233–245

Giorgio I (2016) Numerical identification procedure between a micro-Cauchy model and a macro-second gradient model for planar pantographic structures. Zeitschrift fur Angewandte Mathematik und Physik 67(4)

Giorgio I, Della Corte A, dell'Isola F, Steigmann DJ (2016) Buckling modes in pantographic lattices. Comptes rendus Mecanique 344(7):487–501

Giorgio I, Della Corte A, dell'Isola F (2017a) Dynamics of 1d nonlinear pantographic continua. Nonlinear Dynamics 88(1):21–31

Giorgio I, Rizzi N, Turco E (2017b) Continuum modelling of pantographic sheets for out-of-plane bifurcation and vibrational analysis. Proceedings of the Royal Society of London Series A 473:20170,636

Golaszewski M, Grygoruk R, Giorgio I, Laudato M, Cosmo FD (2018) Metamaterials with relative displacements in their microstructure: technological challenges in 3d printing, experiments and numerical predictions. Continuum Mechanics and Thermodynamics DOI 10.1007/s00161-018-0692-0

Hild F, Roux S (2006) Digital image correlation: from displacement measurement to identification of elastic properties–a review. Strain 42(2):69–80

Javili A, McBride A, Steinmann P (2012) Thermomechanics of solids with lower-dimensional energetics: On the importance of surface, interface, and curve structures at the nanoscale. A unifying review. Applied Mechanics Reviews 65:010,802–1–31

Javili A, Chatzigeorgiou G, Steinmann P (2013a) Computational homogenization in magneto-mechanics. International Journal of Solids and Structures 50(25):4197–4216

Javili A, McBride A, Mergheim J, Steinmann P, Schmidt U (2013b) Micro-to-macro transitions for continua with surface structure at the microscale. International Journal of Solids and Structures 50(16):2561–2572

Laudato M, Di Cosmo F (2018) Euromech 579 arpino 3–8 april 2017: Generalized and microstructured continua: new ideas in modeling and/or applications to structures with (nearly)inextensible fibers—a review of presentations and discussions. Continuum Mechanics and Thermodynamics

Laudato M, Manzari L, Barchiesi E, Di Cosmo F, Göransson P (2018) First experimental observation of the dynamical behavior of a pantographic metamaterial. Mechanics Research Communications 94:125–127

Madeo A, George D, Lekszycki T, Nierenberger M, Rémond Y (2012) A second gradient continuum model accounting for some effects of micro-structure on reconstructed bone remodelling. Comptes Rendus Mécanique 340(8):575–589

Madeo A, Neff P, Ghiba ID, Placidi L, Rosi G (2013) Wave propagation in relaxed micromorphic continua: modeling metamaterials with frequency band-gaps. Continuum Mechanics and Thermodynamics pp 1–20

Madeo A, Della Corte A, Greco L, Neff P (2014) Wave propagation in pantographic 2D lattices with internal discontinuities. Accepted for publication in Proceedings of the Estonian Academy of Sciences

Misra A, Poorsolhjouy P (2015) Identification of higher-order elastic constants for grain assemblies based upon granular micromechanics. Math Mech Complex Syst 3(3):285–308

Misra A, Lekszycki T, Giorgio I, Ganzosch G, Müller WH, dell'Isola F (2018) Pantographic metamaterials show atypical poynting effect reversal. Mechanics Research Communications 89:6–10

Pideri C, Seppecher P (1997) A second gradient material resulting from the homogenization of an heterogeneous linear elastic medium. Continuum Mechanics and Thermodynamics 9(5):241–257

Placidi L (2014) A variational approach for a nonlinear one-dimensional damage-elasto-plastic second-gradient continuum model. Continuum Mechanics and Thermodynamics

Placidi L, Barchiesi E (2018) Energy approach to brittle fracture in strain-gradient modelling. Proc R Soc A 474(2210):20170,878

Placidi L, Rosi G, Giorgio I, Madeo A (2014) Reflection and transmission of plane waves at surfaces carrying material properties and embedded in second-gradient materials. Mathematics and Mechanics of Solids 19(5):555–578

Placidi L, Andreaus U, Della Corte A, Lekszycki T (2015) Gedanken experiments for the determination of two-dimensional linear second gradient elasticity coefficients. Zeitschrift für angewandte Mathematik und Physik 66(6):3699–3725

Placidi L, Barchiesi E, Misra A (2018a) A strain gradient variational approach to damage: a comparison with damage gradient models and numerical results. Mathematics and Mechanics of Complex Systems 6(2):77–100

Placidi L, Misra A, Barchiesi E (2018b) Simulation results for damage with evolving microstructure and growing strain gradient moduli. Continuum Mechanics and Thermodynamics pp 1–21

Pulvirenti M (1996) Kinetic limits for stochastic particle systems. Lecture Notes in Mathematics-Springer-Verlag

Rahali Y, Giorgio I, Ganghoffer J, dell'Isola F (2015) Homogenization à la Piola produces second gradient continuum models for linear pantographic lattices. International Journal of Engineering Science 97:148–172

Rinaldi A, Placidi L (2013) A microscale second gradient approximation of the damage parameter of quasi-brittle heterogeneous lattices. ZAMM - Zeitschrift für Angewandte Mathematik und Mechanik/Journal of Applied Mathematics and Mechanics

Saeb S, Steinmann P, Javili A (2016) Aspects of computational homogenization at finite deformations: A unifying review from reuss' to voigt's bound. Applied Mechanics Reviews 68(5):050,801

Scerrato D, Giorgio I, Rizzi N (2016a) Three-dimensional instabilities of pantographic sheets with parabolic lattices: numerical investigations. Zeitschrift fur Angewandte Mathematik und Physik 67(3)

Scerrato D, Zhurba Eremeeva I, Lekszycki T, Rizzi N (2016b) On the effect of shear stiffness on the plane deformation of linear second gradient pantographic sheets. ZAMM-Journal of Applied Mathematics and Mechanics/Zeitschrift für Angewandte Mathematik und Mechanik

Seppecher P (2002) Second-gradient theory: application to cahn-hilliard fluids. In: Continuum Thermomechanics, Springer, pp 379–388

Trinh DK, Janicke R, Auffray N, Diebels S, Forest S (2012) Evaluation of generalized continuum substitution models for heterogeneous materials. International Journal for Multiscale Computational Engineering 10(6)

Turco E (2018) Discrete is it enough? the revival of Piola–Hencky keynotes to analyze three-dimensional elastica. Continuum Mechanics and Thermodynamics pp 1–19

Turco E, Barcz K, Rizzi N (2016a) Non-standard coupled extensional and bending bias tests for planar pantographic lattices. Part II: comparison with experimental evidence. Zeitschrift fur Angewandte Mathematik und Physik 67(5)

Turco E, dell'Isola F, Cazzani A, Rizzi N (2016b) Hencky-type discrete model for pantographic structures: numerical comparison with second gradient continuum models. Zeitschrift fur Angewandte Mathematik und Physik 67(4):1–28

Turco E, dell'Isola F, Rizzi N, Grygoruk R, Müller W, Liebold C (2016c) Fiber rupture in sheared planar pantographic sheets: Numerical and experimental evidence. Mechanics Research Communications 76:86–90

Turco E, Golaszewski M, Cazzani A, Rizzi NL (2016d) Large deformations induced in planar pantographic sheets by loads applied on fibers: experimental validation of a discrete lagrangian model. Mechanics Research Communications 76:51–56

Turco E, Golaszewski M, Giorgio I, D'Annibale F (2017a) Pantographic lattices with non-orthogonal fibres: experiments and their numerical simulations. Composites Part B: Engineering 118:1–14

Turco E, Golaszewski M, Giorgio I, Placidi L (2017b) Can a hencky-type model predict the mechanical behaviour of pantographic lattices? In: Mathematical Modelling in Solid Mechanics, Springer, pp 285–311

Turco E, Misra A, Pawlikowski M, dell'Isola F, Hild F (2018) Enhanced Piola–Hencky discrete models for pantographic sheets with pivots without deformation energy: numerics and experiments. International Journal of Solids and Structures

Yang H, Ganzosch G, Giorgio I, Abali BE (2018) Material characterization and computations of a polymeric metamaterial with a pantographic substructure. Zeitschrift für angewandte Mathematik und Physik 69(4):105

Yang Y, Misra A (2012) Micromechanics based second gradient continuum theory for shear band modeling in cohesive granular materials following damage elasticity. International Journal of Solids and Structures 49(18):2500–2514

Yang Y, Ching W, Misra A (2011) Higher-order continuum theory applied to fracture simulation of nanoscale intergranular glassy film. Journal of Nanomechanics and Micromechanics 1(2):60–71

Chapter 20
Numerical Solution of the Tri-harmonic KIRCHHOFF Plate Equation Resulting from a Strain Gradient Theory

Christian Liebold & Belal M. Dawwas

Abstract A second gradient continuum theory is formulated based on second gradients of displacements. For a reduction of additional material parameters, the modified strain gradient model is used and a partial differential equation of rank six is developed using the KIRCHHOFF plate assumptions. The solutions of the governing tri-harmonic plate bending equation incoorperate size-effects. Balance equations are presented and higher-order stress-strain relations are derived. In order to account for second gradients of displacements, which manifest themselves in the higher-order terms of a strain energy density, a C^1–continuous displacement field is preferable. So-called HERMITE finite element formulations allow for merging gradients between elements and are used to achieve global C^1–continuity of the solution. Element stiffness matrices as well as the global stiffness matrix are developed for a lexicographical order of nodes and for equidistantly distributed elements. The convergence, the C^1–continuity, and the size effect are demonstrated.

Keywords: Second gradient elasticity · Size-effect · Hermite finite elements · Continuum mechanics · Computational mechanics · Tri-harmonic equation

20.1 Introduction

Materials with intrinsic micro or nano-structure may show size-dependent material behavior, which is reflected, *e. g.*, in a stiffer elastic response to external forces, when the size of the material body is reduced. A quantitative understanding of a size effect is of great importance when modeling Micro- and Nano-Electro-Mechanical Systems (MEMS/NEMS). Driven by the miniaturization as an improvement of the

---

Christian Liebold & Belal M. Dawwas
Technische Universität Berlin, Chair of Continuum Mechanics and Constitutive Theory, Einsteinufer 5, 10587 Berlin, Germany,
e-mail: christian.liebold@mail.de, belal.m.dawwas@campus.tu-berlin.de

B. E. Abali et al. (eds.), *New Achievements in Continuum Mechanics and Thermodynamics*, Advanced Structured Materials 108,
https://doi.org/10.1007/978-3-030-13307-8_20

performance of MEMS, the requirement of reliability in simulation techniques increases. Experimental validation for size effects is given in, *e. g.*, Cuenot et al (2004); Lam et al (2003); Li et al (2010); McFarland and Colton (2005). Materials which are modeled this way are referred to as "non-simple materials of the gradient type". This is accurate, for example, for polymers at a small scale, or even fibre-reinforced materials (Giorgio et al, 2018). In Sect. 20.2 the present work deals with the KIRCHHOFF plate assumptions as well as the *Modified Strain Gradient theory* (MSG) developed by, *e. g.*, Mindlin and Tiersten (1962); Toupin (1962), since conventional continuum theories based on the CAUCHY continuum are not able to predict size effects. As a result, a tri-harmonic partial differential equation for plates is derived. Their solution for a boundary value problem of a rectangular plate under a uniform load is numerically investigated in Sect. 20.3. The application of conventional Finite Element (FE) strategies may lead to inaccurate results, if finite element formulations are used, which only fulfill global C^0–continuity. The scope of this work is, to develop a FE formulation based on HERMITE polynomials in order to account for C^1–continuity of the solution for the tri-harmonic plate equation.

20.2 The Tri-harmonic Plate Equation

20.2.1 Modified Strain Gradient Theory

The present work is based on one of the three reduced forms of the strain gradient energy density for small deformations, as postulated by Mindlin and Tiersten (1962). Because of the later on modification of this theory by an introduction of a rotational degree of freedom, the resulting theory is addressed as *modified strain gradient theory* here, different to the common name *modified (indeterminate) couple stress theory* frequently to be found in the litarature (Eremeyev and dell'Isola, 2018). The fact, that the rotational degree of freedom in the kinematical description of the continuum is replaced by a second gradient of displacement (valid for solids under small translational and rotational deformation only), supports the naming here. In what follows, the EINSTEIN summation convention is used on repeated indices. Spatial partial derivatives in the Cartesian coordinate system are denoted by comma-separated indices. MINDLIN's second form of a linear isotropic strain energy density originally reads:

$$
\begin{aligned}
u^{\mathrm{SG}} &= \alpha_1 \varepsilon_{ij}\varepsilon_{ij} + \alpha_2 \varepsilon_{kk}\varepsilon_{mm} \\
&\quad + \beta_1 \eta_{ijk}\eta_{ijk} + \beta_2 \eta_{iik}\eta_{jjk} + \beta_3 \eta_{iik}\eta_{kjj} + \beta_4 \eta_{ijj}\eta_{ikk} + \beta_5 \eta_{ijk}\eta_{kji},
\end{aligned} \tag{20.1}
$$

where α_1 and α_2 denote the conventional elastic constants for isotropic materials, $\beta_{1,\dots,5}$ are the additional material constants accompanied with the five irreducible parts of the strain gradient tensor $\eta_{ijk} = \varepsilon_{kj,i}$ (Lazar, 2016). The formulation of the *modified* strain gradient energy density is derived from MINDLIN's second form by using the macroscopic vector of rotation $\varphi_i = \frac{1}{2}\epsilon_{ijk}u_{k,j}$, applicable to solids under small deformations. ϵ_{ijk} denotes the LEVI-CIVITA symbol. Taking into account the balance of spin (Abali et al, 2015, 2017), the irreducible parts of η_{ijk} reduce to three. Based on Fleck and Hutchinson (1997), the independent expressions of η_{ijk} are introduced and the second order displacement gradient is decomposed into a symmetric part η^{S}_{ijk} and a remaining part η^{R}_{ijk}. The remaining part is not necessarily anti-symmetric, like the work of Fleck and Hutchinson (1997) supposed, *c.f.* Eq. (20.2). Figure 20.1 shows the scheme of decomposition, where:

$$\begin{aligned} \eta_{ijk} &= \eta^{\mathrm{S}}_{ijk} + \eta^{\mathrm{R}}_{ijk} , \\ \eta^{\mathrm{S}}_{ijk} &= \frac{1}{3}(u_{k,ij} + u_{i,jk} + u_{j,ki}) , \\ \eta^{\mathrm{R}}_{ijk} &= \frac{2}{3}(\epsilon_{ikl}\bar{\eta}_{lj} + \epsilon_{jkl}\bar{\eta}_{li}) + \epsilon_{kjl}\bar{\eta}_{li} . \end{aligned} \tag{20.2}$$

$\bar{\eta}_{ij} = \varphi_{j,i}$ is the gradient of rotation, which is decomposed into its symmetric and anti-symmetric part, χ^{S}_{ij} and χ^{A}_{ij}, respectively:

$$\chi^{\mathrm{A}}_{ij} = \frac{1}{2}\left(\varphi_{i,j} - \varphi_{j,i}\right), \quad \chi^{\mathrm{S}}_{ij} = \frac{1}{2}\left(\varphi_{i,j} + \varphi_{j,i}\right). \tag{20.3}$$

The tensor η^{S}_{ijk} is further decomposed into its spherical and deviatoric part, $\eta^{(0)}_{ijk}$ and $\eta^{(1)}_{ijk}$, *c.f.*, Fig. 20.1 . The quantity $\eta^{(0)}_{ijk}$ is related to χ^{A}_{ij} and the dilatation gradient $\varepsilon_{mm,i}$ in the following manner:

$$\begin{aligned} \eta^{(0)}_{ijk} &= \frac{1}{5}\left(\delta_{ij}\eta^{\mathrm{S}}_{mmk} + \delta_{jk}\eta^{\mathrm{S}}_{mmi} + \delta_{ki}\eta^{\mathrm{S}}_{mmj}\right) , \\ \eta^{\mathrm{S}}_{mmi} &= \varepsilon_{mm,i} + \frac{2}{3}\epsilon_{iln}\chi^{\mathrm{A}}_{ln} , \\ \eta^{(1)}_{ijk} &= \eta^{\mathrm{S}}_{ijk} - \eta^{(0)}_{ijk} . \end{aligned} \tag{20.4}$$

χ^{A}_{ij} is a power conjugated measure for an antisymmetric couple stress tensor. For static problems it can be assumed, that the couple stress tensor μ_{ij} will be symmetric only. Because of that, χ^{A}_{ij} does not influence the strain energy, as it is motivated in Liebold and Müller (2013); Yang et al (2002) and further examined in Münch et al (2015).

Based on Fleck and Hutchinson (1997); Liebold and Müller (2017), the formulation of the modified strain gradient energy density reads:

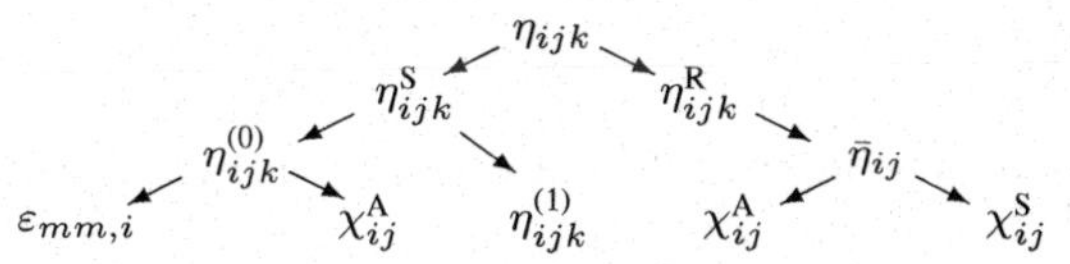

Fig. 20.1 Scheme of decomposition

$$u^{\text{MSG}} = G\varepsilon_{ij}\varepsilon_{ij} + \frac{\lambda}{2}\varepsilon_{kk}\varepsilon_{ii} + G\ell_0^2\varepsilon_{mm,i}\varepsilon_{kk,i} + G\ell_1^2\eta_{ijk}^{(1)}\eta_{ijk}^{(1)} + G\ell_2^2\chi_{ij}^{\text{S}}\chi_{ij}^{\text{S}} \,. \tag{20.5}$$

G and λ are LAMÉ's constants, whereas ℓ_0, ℓ_1 and ℓ_2 denote the three additional material length scale parameters given in the dimension of a length. The multiplication of the higher-order terms by G is arbitrary. The classical strain-energy Eugster and C. (2017) is extended here. Without further reasoning ℓ_0, ℓ_1 and ℓ_2 are set to be equal to ℓ:

$$\boxed{u^{\text{MSG}} = G\varepsilon_{ij}\varepsilon_{ij} + \frac{\lambda}{2}\varepsilon_{kk}\varepsilon_{ii} + G\ell^2\left(\varepsilon_{mm,i}\varepsilon_{kk,i} + \eta_{ijk}^{(1)}\eta_{ijk}^{(1)} + \chi_{ij}^{\text{S}}\chi_{ij}^{\text{S}}\right)} \,. \tag{20.6}$$

The strain and the higher-order strain tensors are:

$$\begin{aligned} \varepsilon_{ij} &= \frac{1}{2}\left(u_{i,j} + u_{j,i}\right) \,, \\ \chi_{ij}^{\text{S}} &= \frac{1}{4}\left(\epsilon_{ilk}u_{k,lj} + \epsilon_{jlk}u_{k,li}\right) \,, \\ \eta_{ijk}^{(1)} &= \frac{1}{3}(u_{k,ij}+u_{i,jk}+u_{j,ki}) - \frac{1}{15}\left[\delta_{ij}(u_{k,mm}+2u_{m,mk})\right. \\ &\quad \left. +\delta_{jk}(u_{i,mm}+2u_{m,mi}) + \delta_{ki}(u_{j,mm}+2u_{m,mj})\right] \,, \end{aligned} \tag{20.7}$$

given in terms of the displacement field $u_i(\mathbf{x})$.

20.2.2 KIRCHHOFF Plate assumptions

The present work investigates the following restrictions to the displacement field $u_i(\mathbf{x})$ in order to derive the Partial Differential Equation (PDE) of the system and their weak form by the help of variational calculus. The so-called KIRCHHOFF-LOVE model of plates is a two-dimensional mathematical model for thin plates subjected to forces and moments, *c.f.* Fig. 20.2. The assumptions are, that:

- the smallest diameter d is much larger then the thickness t,
- the mid-surface plane is the only deformation plane,
- forces are prependicular to the deformation plane,
- straight lines normal to the mid-surface remain straight and normal after deformation,
- the thickness of the plate does not change during the deformation
- the deformations are small.

Fig. 20.2 Outline of a Kirchhoff-Love plate

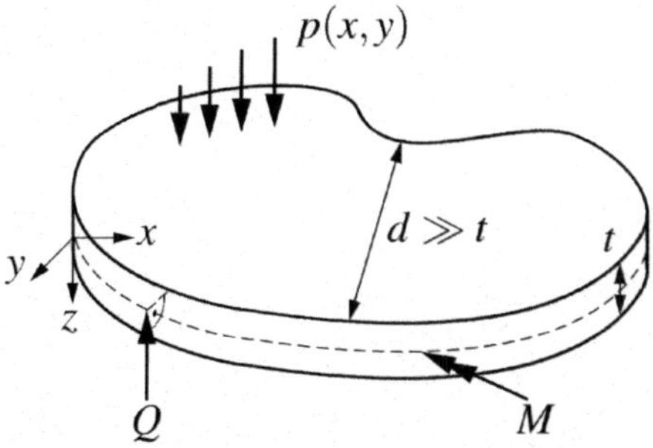

The displacement field of a Kirchhoff-Love plate reads:

$$u_x = -z\frac{\partial w(x,y)}{\partial x}\,, \quad u_y = -z\frac{\partial w(x,y)}{\partial y}\,, \quad u_z = w(x,y) \tag{20.8}$$

where $w(x,y)$ is the bending plane, $p(x,y)$ the load distribution, Q the boundary force and M the boundary moment.

20.2.3 Variation of the Modified Strain Energy of a Kirchhoff Plate

The strain energy density of the problem is derived by evaluating the prescribed displacement field Eq. (20.8) in combination with u^{MSG}, Eq. (20.6). Partial derivatives are denoted by subscripts in the following manner:

$$\frac{\partial(\cdot)}{\partial x} = (\cdot)_x\,, \quad \frac{\partial(\cdot)}{\partial y} = (\cdot)_y\,, \quad \frac{\partial^2(\cdot)}{\partial x \partial y} = (\,)_{xy}\,. \tag{20.9}$$

The first term in Eq. (20.6) becomes

$$\varepsilon_{ij}\varepsilon_{ij} = (zw_{xx})^2 + 2\,(zw_{xy})^2 + (zw_{yy})^2\,, \tag{20.10}$$

the second one

$$\varepsilon_{kk}\varepsilon_{ii} = (zw_{xx})^2 + 2z^2 w_{xx}w_{yy} + (zw_{yy})^2\,, \tag{20.11}$$

the third one

$$\begin{aligned}\varepsilon_{mm,i}\varepsilon_{kk,i} = {} & (zw_{xxx})^2 + (zw_{yyy})^2 + (zw_{yyx})^2 + (zw_{xxy})^2 + \\ & + 2\,(zw_{xxx})\,(zw_{yyx}) + 2\,(zw_{xxy})\,(zw_{yyy}) + 2w_{xx}w_{yy} + \\ & + w_{xx}^2 + w_{yy}^2\,,\end{aligned} \tag{20.12}$$

the fourth one

$$\eta^{(1)}_{ijk}\eta^{(1)}_{ijk} = \frac{2}{5}\left(zw_{xxx}\right)^2 + \frac{2}{5}\left(zw_{yyy}\right)^2 + \frac{12}{5}\left(zw_{yyx}\right)^2 + \frac{12}{5}\left(zw_{xxy}\right)^2 - \frac{6}{5}zw_{xxx}\,zw_{yyx} - \frac{6}{5}zw_{yyy}\,zw_{xxy} + \frac{4}{15}w_{xx}^2 + \frac{4}{15}w_{yy}^2 - \frac{2}{15}w_{xx}w_{yy} + \frac{2}{3}w_{yy}^2\,, \tag{20.13}$$

and the last one

$$\chi^{\mathrm{S}}_{ij}\chi^{\mathrm{S}}_{ij} = \frac{1}{2}\left(w_{xx} + w_{yy}\right)^2\,. \tag{20.14}$$

In summary, the modified strain energy density is:

$$\begin{aligned} u^{\mathrm{MSG}} &= \frac{7}{5}G\ell^2 z^2\left[w_{xxx}^2 + w_{yyy}^2\right] + \frac{17}{5}G\ell^2 z^2\left[w_{yyx}^2 + w_{xxy}^2\right] \\ &+ \frac{4}{5}G\ell^2 z^2\left[w_{xxx}\,w_{yyx} + w_{xxy}\,w_{yyy}\right] \\ &+ \left(Gz^2 + \frac{\lambda}{2}z^2 + \frac{53}{30}G\ell^2\right)\left[w_{xx}^2 + w_{yy}^2\right] \\ &+ \left(z^2\lambda + \frac{13}{5}G\ell^2\right) w_{xx}\,w_{yy} + \left(2Gz^2 + \frac{2}{3}G\ell^2\right) w_{yx}^2\,. \end{aligned} \tag{20.15}$$

For a proof of concept, this work restricts to a square plate of the length L and of thickness t. Then, the variation of the strain energy is derived as follows:

$$\begin{aligned} \delta W^{\mathrm{MSG}} &= \int\limits_V \delta u^{\mathrm{MSG}}\,\mathrm{d}V \\ &= \int\limits_{-\frac{t}{2}}^{+\frac{t}{2}}\int\limits_0^L\int\limits_0^L \Bigg(\frac{14}{5}G\ell^2 z^2\left[w_{xxx}\delta w_{xxx} + w_{yyy}\delta w_{yyy}\right] \\ &+ \frac{34}{5}G\ell^2 z^2\left[w_{yyx}\delta w_{yyx} + w_{xxy}\delta w_{xxy}\right] \\ &+ \frac{4}{5}G\ell^2 z^2\left[\delta w_{xxx}\,w_{yyx} + w_{xxx}\,\delta w_{yyx} + \delta w_{xxy}\,w_{yyy} + w_{xxy}\,\delta w_{yyy}\right] \\ &+ \left(2Gz^2 + \lambda z^2 + \frac{53}{15}G\ell^2\right)\left[w_{xx}\delta w_{xx} + w_{yy}\delta w_{yy}\right] \\ &+ \left(z^2\lambda + \frac{13}{5}G\ell^2\right)\left[\delta w_{xx}\,w_{yy} + w_{xx}\,\delta w_{yy}\right] \\ &+ \left(4Gz^2 + \frac{4}{3}G\ell^2\right) w_{yx}\delta w_{yx}\Bigg)\,\mathrm{d}x\mathrm{d}y\mathrm{d}z\,. \end{aligned} \tag{20.16}$$

Employing the rules of variational calculus and multiple application of two-dimensional partial integration, δW^{MSG} becomes:

$$
\begin{aligned}
\delta W^{\mathrm{MSG}} &= -\frac{t^3}{12}\int_0^L\int_0^L \delta w\Bigg[\frac{14}{5}G\ell^2\left(w_{xxxxxx}+w_{yyyyyy}\right) \\
&+\frac{42}{5}G\ell^2\left(w_{xxxxyy}+w_{yyyyxx}\right) \\
&-\left(2G+\lambda+\frac{212}{5}G\frac{\ell^2}{t^2}\right)\left(w_{xxxx}+w_{yyyy}+2w_{xxyy}\right)\Bigg]\,\mathrm{d}x\mathrm{d}y \\
&+\frac{Gt^3}{12}\int_0^L \Lambda_1(x,y)\Big|_{y=0}^{y=L}\,\mathrm{d}x+\frac{Gt^3}{12}\int_0^L \Lambda_2(x,y)\Big|_{x=0}^{x=L}\,\mathrm{d}y\,,
\end{aligned}
\tag{20.17}
$$

with Λ_1 and Λ_2 being the boundary relations:

$$
\begin{aligned}
\Lambda_1(x,y) &= \left(\frac{7}{5}\ell^2 w_{yyy}+\frac{2}{5}\ell^2 w_{xxy}\right)\delta w_{yy}+\left(\frac{17}{5}\ell^2 w_{xxy}+\frac{2}{5}\ell^2 w_{yyy}\right)\delta w_{xx} \\
&+\Bigg[\left(\frac{\lambda}{G}+\frac{78}{5}\frac{\ell^2}{t^2}\right)w_{xx}+\left(1+\frac{\lambda}{2G}+\frac{106}{5}\frac{\ell^2}{t^2}\right)w_{yy}-\frac{7}{5}\ell^2 w_{yyyy} \\
&-\frac{17}{5}\ell^2 w_{yyxx}-\frac{2}{5}\ell^2\left(w_{xxxx}+w_{xxyy}\right)\Bigg]\delta w_y+\left(1+4\frac{\ell^2}{t^2}\right)w_{xy}\,\delta w_x \\
&+\Bigg[\frac{7}{5}\ell^2 w_{yyyyy}+\frac{17}{5}\ell^2 w_{yyyxx}+\frac{2}{5}\ell^2\left(w_{xxxxy}+w_{xxyyy}\right) \\
&-\left(\frac{\lambda}{G}+\frac{78}{5}\frac{\ell^2}{t^2}\right)w_{xxy} \\
&-\left(1+\frac{\lambda}{2G}+\frac{106}{5}\frac{\ell^2}{t^2}\right)w_{yyy}-\left(1+4\frac{\ell^2}{t^2}\right)w_{yxx}\Bigg]\delta w\,,
\end{aligned}
\tag{20.18}
$$

and

$$
\begin{aligned}
\Lambda_2(x,y) &= \left(\frac{7}{5}\ell^2 w_{xxx}+\frac{2}{5}\ell^2 w_{yyx}\right)\delta w_{xx}+\left(\frac{17}{5}\ell^2 w_{yyx}+\frac{2}{5}\ell^2 w_{xxx}\right)\delta w_{yy} \\
&+\Bigg[\left(\frac{\lambda}{G}+\frac{78}{5}\frac{\ell^2}{t^2}\right)w_{yy}+\left(1+\frac{\lambda}{2G}+\frac{106}{5}\frac{\ell^2}{t^2}\right)w_{xx}-\frac{7}{5}\ell^2 w_{xxxx} \\
&-\frac{17}{5}\ell^2 w_{yyxx}-\frac{2}{5}\ell^2\left(w_{xxxx}+w_{xxyy}\right)\Bigg]\delta w_y+\left(1+4\frac{\ell^2}{t^2}\right)w_{xy}\,\delta w_x \\
&+\Bigg[\frac{7}{5}\ell^2 w_{xxxxx}+\frac{17}{5}\ell^2 w_{xxxyy}+\frac{2}{5}\ell^2\left(w_{yyyyx}+w_{yyxxx}\right) \\
&-\left(\frac{\lambda}{G}+\frac{78}{5}\frac{\ell^2}{t^2}\right)w_{yyx}-\left(1+\frac{\lambda}{2G}+\frac{106}{5}\frac{\ell^2}{t^2}\right)w_{xxx} \\
&-\left(1+4\frac{\ell^2}{t^2}\right)w_{xyy}\Bigg]\delta w\,.
\end{aligned}
\tag{20.19}
$$

In fact, the formulations Λ_1 and Λ_2 at the boundaries would allow to identify boundary-forces and -moments, when considering the work done by the external loads. In the present work, the external loads are limited to the application of a pressure distribution $p(x, y)$.

20.2.4 The Governing Tri-harmonic Plate Equation

To derive the respective partial differential equation for static problems, the difference of the virtual strain energy and the virtual work done by the external loads δA, assumed as:

$$\begin{aligned}\delta A = &\int_0^L \int_0^L p(x,y)\delta w \,\mathrm{d}x\,\mathrm{d}y \\ &+ \int_0^L \Big[Q\delta w - M^y \delta w_x - M^x \delta w_y + M^{Hy}\delta w_{xx} + M^{Hx}\delta w_{yy}\Big]_{x=0}^{x=L} \mathrm{d}x \\ &+ \int_0^L \Big[Q\delta w - M^y \delta w_x - M^x \delta w_y + M^{Hy}\delta w_{xx} + M^{Hx}\delta w_{yy}\Big]_{x=0}^{x=L} \mathrm{d}y \,,\end{aligned} \tag{20.20}$$

has to be minimized, where M^x and M^y are classical moments, which affect the deflection angle (first derivative) of w at the boundaries. M^{Hx} and M^{Hy} are higher order moments, affecting the curvature (second derivative) in the respective direction at the boundaries. An energy minimization yields in:

$$\delta W^{\mathrm{MSG}} - \delta A = \delta \Pi \,, \qquad \delta \Pi \to 0 \qquad \Rightarrow \qquad \delta W^{\mathrm{MSG}} = \delta A \,, \tag{20.21}$$

and by comparison of Eq. (20.17) and (20.20) the *tri-harmonic plate equation* arises:

$$\boxed{D\Delta\Delta w(x,y) - H\Delta\Delta\Delta w(x,y) = p(x,y)} \,, \tag{20.22}$$

using the Laplacian $\Delta(\cdot) = (\cdot)_{xx} + (\cdot)_{yy}$. Similarities can be drawn to the result of Kotchergenko (2015). In the present work, the plate stiffnesses are:

$$D = \frac{t^3}{12}\left(2G + \lambda + \frac{212}{5}G\frac{\ell^2}{t^2}\right) \,, \qquad H = \frac{7}{60}Gt^3\ell^2 \,. \tag{20.23}$$

In the limit case to the conventional continuum theory, if $\ell = 0$, Eq. (20.22) will turn into the classical KIRCHHOFF plate equation.

20.2.5 A Navier-Solution with Fourier-Series

Combining the continuous ansatz for *simply supported rectangular plates* after Navier (Becker and Gross, 2002):

$$w_{\text{ana}}^{\text{SG}}(x,y) = \sum_{m=1}^{\infty}\sum_{n=1}^{\infty} w_{mn} \sin\left(\frac{m\pi x}{L}\right) \sin\left(\frac{n\pi y}{L}\right) \tag{20.24}$$

and rewriting the load-function as follows:

$$p(x,y) = \sum_{m=1}^{\infty}\sum_{n=1}^{\infty} p_{mn} \sin\left(\frac{m\pi x}{L}\right) \sin\left(\frac{n\pi y}{L}\right) , \tag{20.25}$$

$$p_{mn} = \frac{4}{L^2}\int_0^L\int_0^L p(x,y) \sin\left(\frac{m\pi x}{L}\right) \sin\left(\frac{n\pi y}{L}\right) \mathrm{d}x\mathrm{d}y , \tag{20.26}$$

a solution for the equation (20.22) is given in form of a series in Eq. (20.24), having the coefficients:

$$\begin{aligned} w_{mn} = p_{mn}\Bigg\{ D\left[\left(\frac{m\pi}{L}\right)^4 + \left(\frac{n\pi}{L}\right)^4 + 2\left(\frac{n\pi}{L}\right)^2\left(\frac{m\pi}{L}\right)^2\right] \\ + H\left[\left(\frac{m\pi}{L}\right)^6 + \left(\frac{n\pi}{L}\right)^6 + 3\left(\frac{m\pi}{L}\right)^4\left(\frac{n\pi}{L}\right)^2 + 3\left(\frac{m\pi}{L}\right)^2\left(\frac{n\pi}{L}\right)^4\right]\Bigg\}^{-1} . \end{aligned} \tag{20.27}$$

For a constant distribution of loads $p(x,y)$, a sufficient convergence of this series is achieved by taking about 50 terms into account for m and n.

20.3 A C^1–continuous Finite Element Approach

20.3.1 The Weak Form of the PDE

The weak form of the tri-harmonic plate equation, Eq. (20.22), is in a sense already given with the variation of the strain gradient energy density in Eq. (20.16), only by replacing the variational terms by the independent test-functions $v(x,y)$:

$$
\begin{aligned}
&\overbrace{\int_V p(x,y)\, v(x,y)\mathrm{d}V}^{f(v)} \\
&\underbrace{\begin{aligned}
= \int_V & \frac{14}{5} G\ell^2 z^2 \left[w_{xxx}v_{xxx} + w_{yyy}v_{yyy}\right] + \frac{34}{5} G\ell^2 z^2 \left[w_{yyx}v_{yyx} + w_{xxy}v_{xxy}\right] \\
&+ \frac{4}{5} G\ell^2 z^2 \left[v_{xxx}\, w_{yyx} + w_{xxx}\, v_{yyx} + v_{xxy}\, w_{yyy} + w_{xxy}\, v_{yyy}\right] \\
&+ \left(2Gz^2 + \lambda z^2 + \frac{53}{15} G\ell^2\right) \left[w_{xx}v_{xx} + w_{yy}v_{yy}\right] \\
&+ \left(z^2\lambda + \frac{13}{5} G\ell^2\right) \left[v_{xx}\, w_{yy} + w_{xx}\, v_{yy}\right] + \left(4Gz^2 + \frac{4}{3} G\ell^2\right) w_{yx}v_{yx}\mathrm{d}V
\end{aligned}}_{a(v,w)}
\end{aligned}
\tag{20.28}
$$

The left-hand side is the so-called bi-linear form $a(v,w)$ of the PDE. The right-hand side $f(v)$ is formed, using the variation of the work done by external loads (Eq. 20.20) with the same replacement. It can be seen, that the weak form is only a third-order partial differential equation in both, the test- and trial-functions, $v(x,y)$ and $w(x,y)$.

20.3.2 Two Dimensional HERMITE *Finite Element Formulation*

The requirements for a straight-forward finite element solution for the present weak form are: (i), that the basis functions for the GALERKIN discretization are at least three times differentiable and (ii), that the global behavior of the solution fulfills C^1–continuity. The latter is needed due to the fact, that the strain energy used in Eq. (20.6) evaluates second derivatives of displacements and in the end of the test- and trial-functions, too. To guarantee, that the second derivatives in each element will be well connected to the global behavior, the first derivatives need to be continuous inbetween the neighboring elements. Both requirements are fulfilled using so-called HERMITE finite element formulations, which consist of the HERMITE polynomials, as plotted in Fig. (20.3):

$$
\begin{aligned}
H_1(\zeta) &= 2\zeta^3 - 3\zeta^2 + 1\,, \qquad & H_2(\zeta) &= \zeta^3 - 2\zeta^2 + \zeta\,, \\
H_3(\zeta) &= -2\zeta^3 + 3\zeta^2\,, \qquad & H_4(\zeta) &= \zeta^3 - \zeta^2,
\end{aligned}
\tag{20.29}
$$

which are linearly superposed and multiplicatively connected to form either 1D or 2D test- and trail-functions v^{e} and w^{e} per element, e,

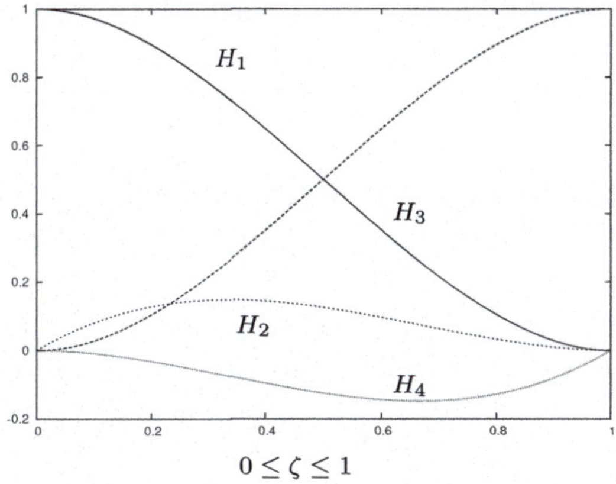

Fig. 20.3 Plot of the Hermite polynomials

$$v^{\mathrm{e}}(\zeta,\xi) = \sum_{\alpha=1}^{4}\sum_{\beta=1}^{4} H_\alpha(\zeta)H_\beta(\xi) \,, \quad w^{\mathrm{e}}(\zeta,\xi) = \sum_{\delta=1}^{4}\sum_{\gamma=1}^{4} c^{\mathrm{e}}_{\delta\gamma} H_\delta(\zeta)H_\gamma(\xi) \,, \tag{20.30}$$

where $c^{\mathrm{e}}_{\delta\gamma}$ denote the coefficients (the unknowns) to be calculated to form the solution. Such element formulations are called Bogner–Fox–Schmit elements Bogner et al (1965). Two exemplary combinations $\Phi_1 = H_1 H_1$ and $\Phi_2 = H_1 H_2$ are given in Fig. (20.4). Φ_1 will directly influence the value of deflection at the node position (0,0), whereas Φ_2 will influence the first derivative in the ζ-direction. Equation (20.30) is used in the following condensed form:

$$v^{\mathrm{e}}(\zeta,\xi) = \sum_{i=1}^{16} \Phi_i(\zeta,\xi) \,, \quad w^{\mathrm{e}}(\zeta,\xi) = \sum_{i=1}^{16} c^{\mathrm{e}}_i \Phi_i(\zeta,\xi) \,, \tag{20.31}$$

where the assignment of the combinations for $\Phi_i \to H_\alpha H_\beta$ is: $(i \to \alpha\beta)$ $1 \to 11$, $2 \to 12$, $3 \to 13$, $4 \to 14$, $5 \to 21$, $6 \to 22$, $7 \to 23$, $8 \to 24$, $9 \to 31$, $10 \to 32$, $11 \to 33$, $12 \to 34$, $13 \to 41$, $14 \to 42$, $15 \to 43$ and $16 \to 44$. c^{e}_i represent the 16 coefficients per element, of which four of them directly represent the value of deflections at the four nodes, eight of them represent the first derivatives

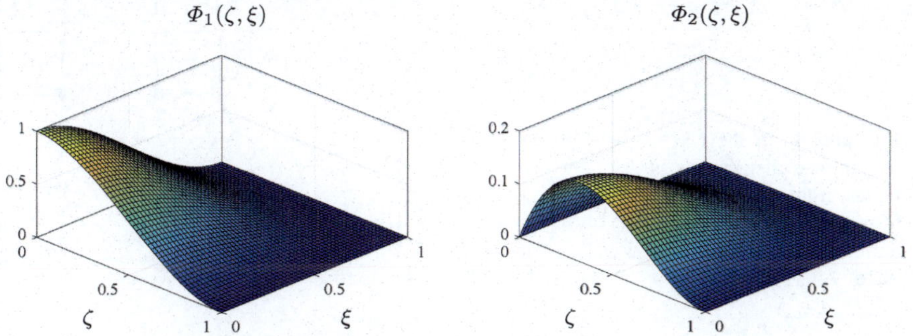

Fig. 20.4 Exemplary two-dimensional Hermite polynomials acting on the node (0,0)

in each direction at the nodes and four of them carry the information of the so-called internal- (or bubble-) modes.

20.3.3 The Element and Global Stiffness Matrix and Realization of the Boundary Condition

The element stiffness matrix is established by inserting the test- and trial-functions into the integral weak form. Doing so, Eq. (20.28) is rewritten:

$$\mathbf{K}^{\mathrm{e}}\mathbf{c}^{\mathrm{e}} = \mathbf{f}^{\mathrm{e}}\,, \qquad \mathbf{c}^{\mathrm{e}} = [c_1^{\mathrm{e}},\, c_2^{\mathrm{e}},\, \ldots, c_{16}^{\mathrm{e}}\,]^{\mathrm{T}}\,, \tag{20.32}$$

such that the left-hand side is represented by $\mathbf{K}^{\mathrm{e}}\mathbf{c}^{\mathrm{e}}$, with $\mathbf{K}^{\mathrm{e}}$ being the element stiffness matrix and $\mathbf{c}^{\mathrm{e}}$ the vector of coefficients, and the right-hand side is expressed by thev vector $\mathbf{f}^{\mathrm{e}}$. Using the element stiffness components:

$$K^{\mathrm{e}}_{i|j} = a^{\mathrm{e}}(\Phi_i, \Phi_j)\,, \tag{20.33}$$

in which a^{e} denotes the bilinearform in the integration domain of a single element, the weak form Eq. (20.28) per element can be expressed as:

$$\begin{bmatrix} K^{\mathrm{e}}_{1|1} & K^{\mathrm{e}}_{1|2} & \cdots & K^{\mathrm{e}}_{1|16} \\ K^{\mathrm{e}}_{2|1} & K^{\mathrm{e}}_{2|2} & \cdots & K^{\mathrm{e}}_{2|16} \\ \vdots & \vdots & \ddots & \vdots \\ K^{\mathrm{e}}_{16|1} & K^{\mathrm{e}}_{16|2} & \cdots & K^{\mathrm{e}}_{16|16} \end{bmatrix} \begin{bmatrix} c^{\mathrm{e}}_1 \\ c^{\mathrm{e}}_2 \\ \vdots \\ c^{\mathrm{e}}_{16} \end{bmatrix} = \begin{bmatrix} \int_{V^{\mathrm{e}}} p(\zeta,\xi)\,\Phi_1\,\mathrm{d}V^{\mathrm{e}} \\ \int_{V^{\mathrm{e}}} p(\zeta,\xi)\,\Phi_2\,\mathrm{d}V^{\mathrm{e}} \\ \vdots \\ \int_{V^{\mathrm{e}}} p(\zeta,\xi)\,\Phi_{16}\,\mathrm{d}V^{\mathrm{e}} \end{bmatrix}. \tag{20.34}$$

In a next step the global stiffness matrix is developed. To achieve a sparse band matrix, which will have advantages in dissolving large systems, the lexicographical distribution of nodes is used here, *c.f.* Fig. (20.5). Here, the nodes at the physical boundary of the plate are highlighted to be remembered for the assignement of Dirichlet boundary conditions, whereas node number "5" is used to demonstrate,

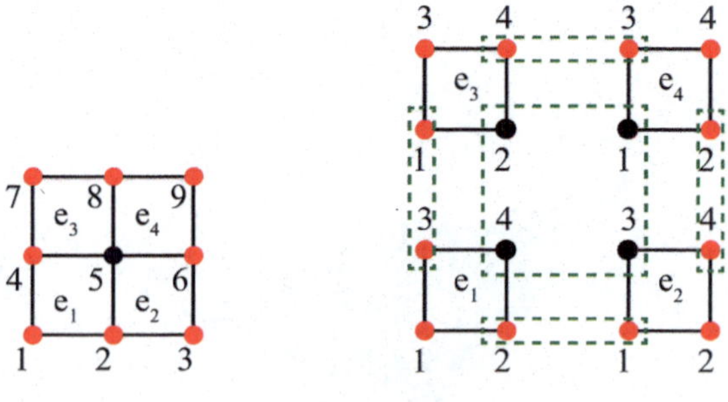

Fig. 20.5 Left: the lexicographical distribution of nodes; right: the numbering of nodes per element e_N for a minimal discretization of a plate

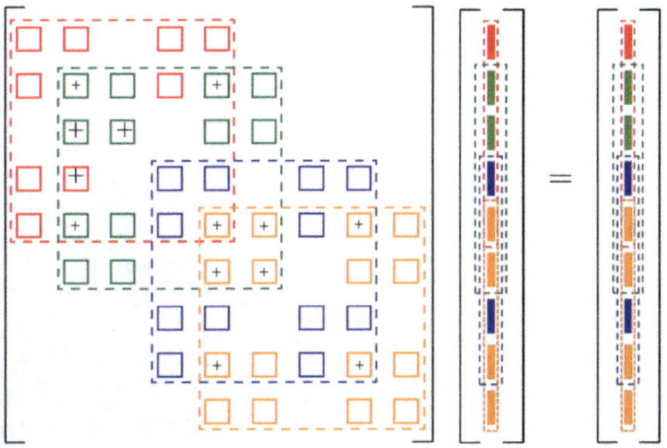

Fig. 20.6 Scheme of the composition of the global matrix equation

that its value of deflection as well as the derivatives belong to all four neighboring elements at the same time. This is achieved by assembling the global stiffness matrix by overlapping the element matrices in this way, such that the respective values of deflection as well as the derivatives are multiplied with the same coefficients c, respectively. Therefor, the element's matrices are split into 4×4 submatrices, in which the components will represent the deflection, the derivatives and the internal mode for a single node of an element. Figure (20.6) demontrates the scheme of composition of the global system of equations. This scheme represents the global linear algebraic equation:

$$\mathbf{K}^{\mathrm{G}}\mathbf{c}^{\mathrm{G}} = \mathbf{f}^{\mathrm{G}} \,, \tag{20.35}$$

where the dashed lines in Fig. (20.6) declare an assignment to single elements and squares to the 4×4 submatrices. The Dirichlet boundary condition for a simply supported plate is realized by a direct manipulation of the coefficients of the respective boundary nodes. For the reason that the deflection at the boundary nodes are set equal to zero, the respective columns are removed in the global stiffness matrix. The reduced system of equations, denoted by the superscript S, is solved with the *backslash* operator of `Matlab` (MathWorks, Inc.):

$$\mathbf{c}^{\mathrm{S}} = \mathbf{K}^{\mathrm{S}} \backslash \mathbf{f}^{\mathrm{S}} \,, \tag{20.36}$$

using a banded solver. The procedure described here is numerically rather fast and in general applicable to different situations or geometries.

20.4 Results

For the numerical simulations and tests for convergence and size effect, a squared micro-plate of the length L and thickness t is used. Figure (20.7) represents the deflection of the plate, which is simply supported at all edges and loaded by the constant distribution of force $p(x, y)$. Table 20.1 gives the material and geometry

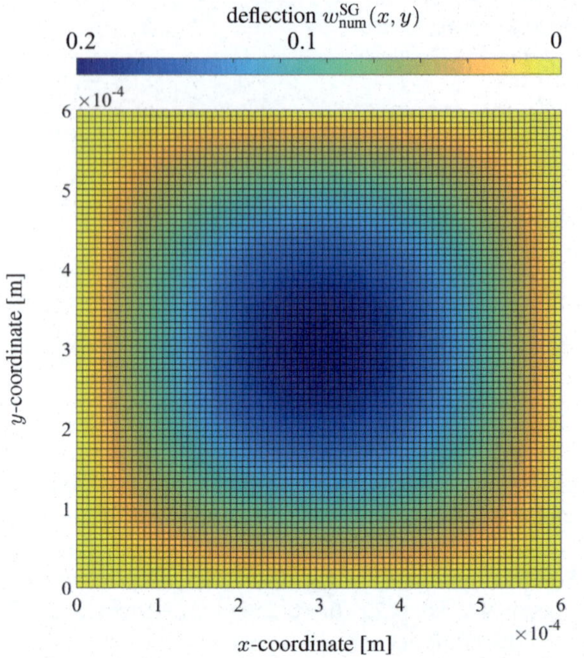

Fig. 20.7 Deflection of the midplane of a square plate, $t = 30\,\mu\text{m}$, $L = 20t$, $\ell = 10\,\mu\text{m}$, NumOfEl=6400

parameters, which are quite realistic when modeling a small polymer plate (Chong, 2002; Kong et al, 2009; Lam et al, 2003; Nikolov et al, 2007). $w_{\text{num}}^{\text{SG}}(x, y)$ denotes the numerical result of the deflection of the plate at the coordinate x and y w.r.t. the Second Gradient (SG) continuum approach, whereas $w_{\text{ana}}^{\text{SG}}$ denotes the "‘analytical result"’ from the truncated FOURIER-series, *c.f.* Eq. (20.24). $w^{\text{class}}(x, y)$ will represent the solution of plate-bending for the classical CAUCHY-continuum theory. Figure (20.8) shows the behavior of the numerical solution of the midpoint of the plate while refining the mesh of elements.

20.4.1 Concerning the Convergence

To demonstrate the convergence behavior of the present numerical approach, solutions with different sizes of equidistantly distrubuted elements were conducted. The global number of degrees of freedoms (DOF) is used to compare different solutions.

Table 20.1 Thickness t, lengths L, elastic modulus E, material length scale parameter ℓ and the distributed load $p(x, y)$ used for the plate simulations

t	L	E	ℓ	$p(x, y)$
$30\,\mu\text{m}$	$20\,t$	$3.8\,\text{GPa}$	$10\,\mu\text{m}$	10 MPa

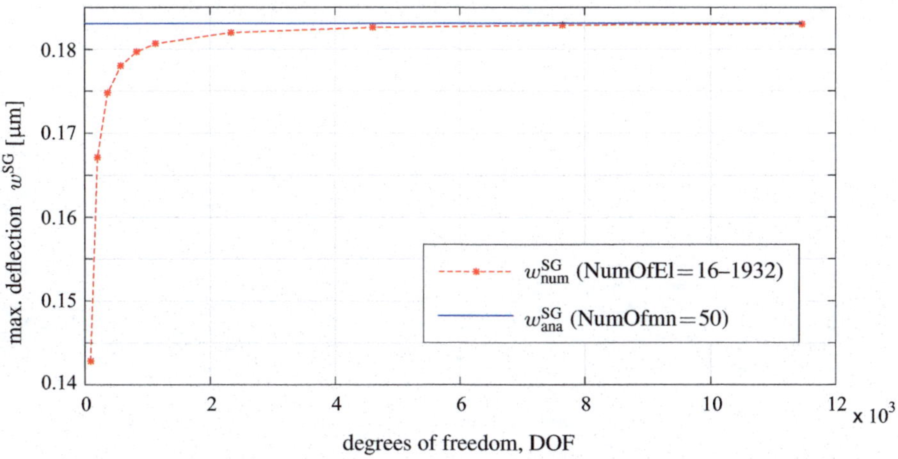

Fig. 20.8 Convergence of the deflections of the center point compared to the analytical value for $t = 30\,\mu\text{m}$

For the beginning, Fig. (20.8) gives the absolute deflection of the center point $(\frac{L}{2}, \frac{L}{2})$ of the plate for different mesh sizes, in comparison to the constant analytical value.

It has to be remarked, that with the same set of parameters used in Fig. (20.8), the classical value (where ℓ is equal to zero) is more than twice as large as the second gradient (SG) solution. In a next investigation, in Fig. (20.9), the error in percent between the numerical and the Fourier solution is charted for different degrees of freedom. The error between the numerical and the analytical values for the deflection of the center point is calculated by:

$$\text{error} = \left| \frac{w^{SG}_{\text{num}}}{w^{SG}_{\text{ana}}} - 1 \right| \times 100\,. \tag{20.37}$$

In the logarithmic plot of the error we clearly observe a quite constant rate of convergence. From a certain number of degrees of freedom on, the convergence rate seems to increase. In the authors opinion, this fact is due to the truncation of the Fourier solution at m=50, which is taken as the reference value. It can be assumed, that the numerical solution passes the truncated Fourier solution at a further point of DOFs. Beside this assumption, it has to be taken into account, that the computational errors, which go along with these very large numbers of calculation steps, will add up and will shift the solution for very large DOFs. In summary, however, from a numerical point of view, the approach shown is extremely robust and well-built, reaching an error of $0.1\,\%$ quite easily.

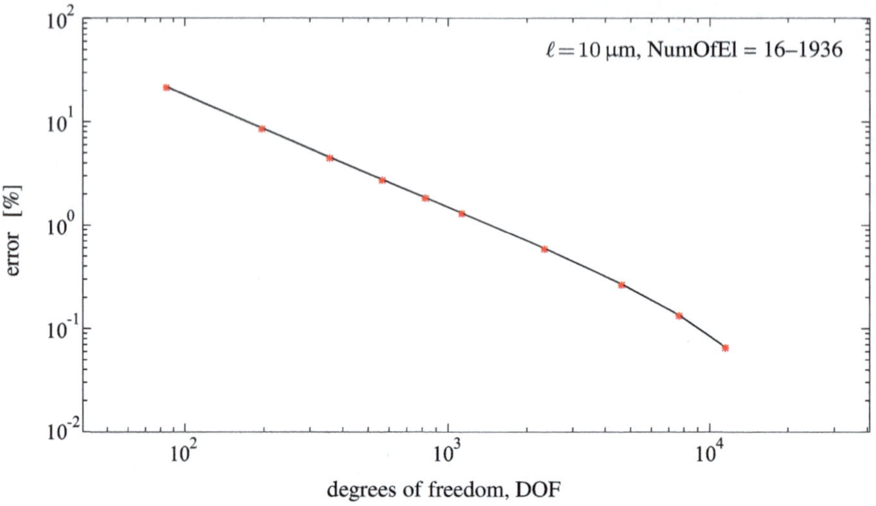

Fig. 20.9 Convergence of the error between numerical and analytical deflections of the midpoint for $t = 30$ µm in a logarithmic plot

20.4.2 Results for the Size Effect

For the analysis of the size dependent behavior of the solutions, the ratio of both – the numerical as well as the FOURIER solution – to the result of the classical continuum theory (without any length scale influence) is calculated for different sizes of the plate, *c.f.* Fig. (20.10). The ratio of the length to the thickness is constant. The numerical values in Fig. (20.10) are calculated for a more or less coarse mesh using 16 elements, which includes a constant error of about 21 %.

20.4.3 Analysis of the C^1–continuity

To demonstrate the intended C^1–continuity of the present solution, a cascade of results along a center line $(x, \frac{L}{2})$ is established: (i) for the z-deflections, see Fig. (20.11), (ii) the first derivatives in the x-direction, see Fig. (20.12) an (iii) the second derivatives in the x-direction, see Fig. (20.13).

The plots in Figs. (20.11–20.13) are based on the post-prozessed data for the element's solutions. In each interval, the slope of the weighted HERMITE element function, Eq. $(20.30)_2$, is plotted independently. The element's solutions for the 0th and 1st derivatives are continuously connected between the elements. The first derivative at a node of one element equals the first derivative of the neighbouring element, and so on. A different picture is drawn, when looking at the second derivatives. Fig. (20.13) demonstrates by the jumps between the element solutions, that the sec-

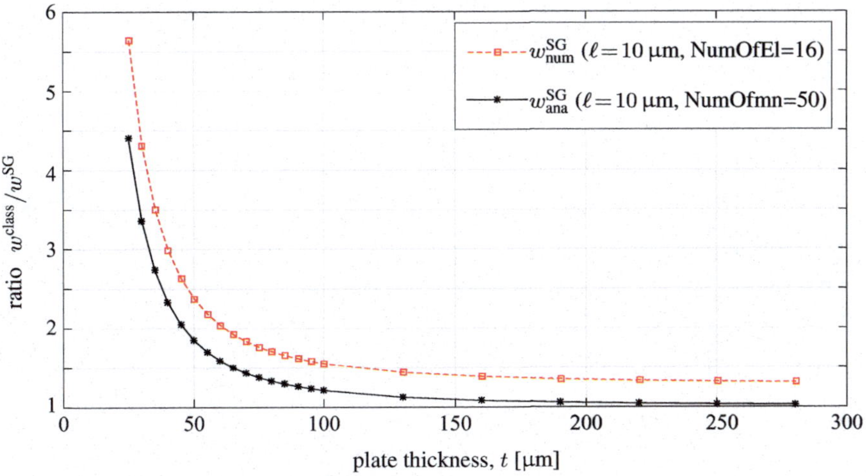

Fig. 20.10 Size effect for a square plate, $L = 20t$; numerical vs. analytical results of the deflection of the midpoint $w^{\text{SG}}(\frac{L}{2}, \frac{L}{2})$ in relation to the classical result w^{class}

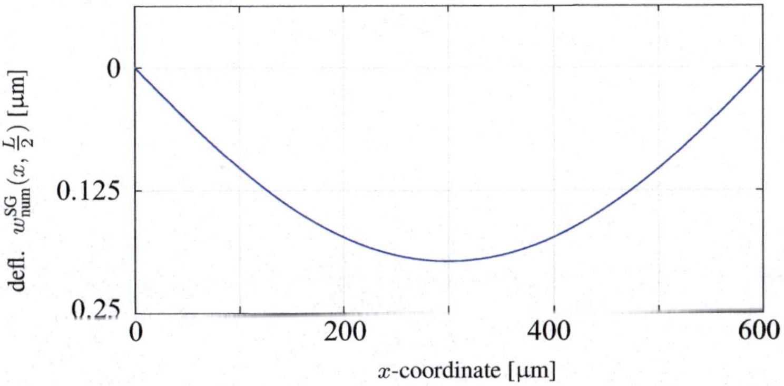

Fig. 20.11 Deflection of the center line of a square plate, t=30 µm, L=20t, ℓ=10 µm, NumOfEl = 80×80

ond derivative at a node of one element does not equal the second derivative of the neighbouring element. This behavior of the numerical solution suggests, that the intended C^1–continuity is fulfilled.

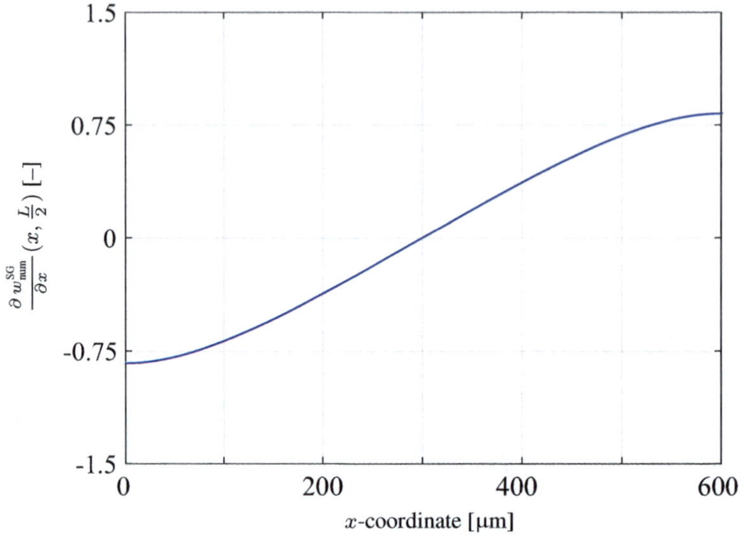

Fig. 20.12 First derivative of the deflection of the center line, $t = 30$ μm, $L = 20t$, $\ell = 10$ μm, NumOfEl = 80×80

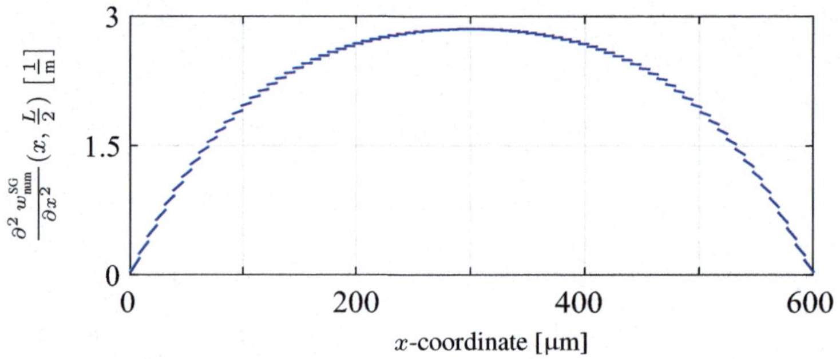

Fig. 20.13 Second derivative of the deflection of the center line, $t = 30$ μm, $L = 20t$, $\ell = 10$ μm, NumOfEl = 80×80

20.5 Conclusions

A modified second gradient continuum theory of elasticity was elaborated. The restriction on the displacement field of a KIRCHHOFF-LOVE plate was carried out in order to derive the corresponding partial differential equation (the *tri-harmonic equation*) and its weak form, respectively. In order to keep the first derivative of the solution continuous, the problem was discretized using HERMITE polynomials, of which the so-called BOGNER–FOX–SCHMIT elements consist of. So far, the present resluts are restricted to equidistantly distributed quad element meshes. It is further accompanied by a large number of element coefficients in comparison to a conventional FEM. The elaborated FE approximations show a size effect, as expected from the higher-order theory, as well as convergence in terms of increasing degrees of freedoms in the mesh discretization. This will allow to simulate the elastostatic problem of KIRCHHOFF-LOVE plates in arbitrary geometries for micromechanical applications, when considering a higher-order material behavior.

References

Abali BE, Müller WH, Eremeyev VA (2015) Strain gradient elasticity with geometric nonlinearities and its computational evaluation. Mechanics of Advanced Materials and Modern Processes 1(4):2–11

Abali BE, Müller WH, dell'Isola F (2017) Theory and computation of higher gradient elasticity theories based on action principles. Archive of Applied Mechanics 87(9):1495–1510, DOI 10.1007/s00419-017-1266-5

Becker W, Gross D (2002) Mechanik elastischer Körper und Strukturen. Springer-Verlag, Berlin Heidelberg

Bogner FK, Fox RL, A SL (1965) The generation of interelement compatible stiffness and mass matrices by the use of interpolation formulas. In: Proceedings of the Conference on Matrix Methods in Structural Mechanics, Wright-Patterson Air Force Base, Ohio, pp 397–444

Chong CM (2002) Experimental investigation and modeling of size effect in elasticity. PhD thesis, Hong Kong University of Science and Technology

Cuenot S, Fretigny C, Demoustier-Champagne S, Nysten B (2004) Surface tension effect on the mechanical properties of nanomaterials measured by atomic force microscopy. Physical Review B 69:01–05

Eremeyev VA, dell'Isola F (2018) A note on reduced strain gradient elasticity. Generalized Models and Non-classical Approaches in Complex Materials 1:301–310

Eugster SR, C G (2017) On the notion of stress in classical continuum mechanics. Mathematics and Mechanics of Complex Materials 5

Fleck NA, Hutchinson JW (1997) Strain gradient plasticity. In: W HJ, Y WT (eds) Advances in Applied Mechanics, Academic Press, New York, pp 295–361

Giorgio I, dell'Isola F, Steigmann DJ (2018) Axisymmetric deformations of a 2nd grade elastic cylinder. Mechanics Research Communications 94:45–48

Kong S, Zhou S, Nie Z, Wang K (2009) Static and dynamic analysis of micro beams based on strain gradient elasticity theory. International Journal of Engineering Science 47:487–498

Kotchergenko ID (2015) The tri-harmonic plate bending equation. WIT Transactions on Modelling and Simulation 59:367–378

Lam DCC, Yang F, Chong CM, Wang J, Tong P (2003) Experiments and theory in strain gradient elasticity. J Mech Phys Sol 51(8):1477–1508

Lazar M (2016) Irreducible decomposition of strain gradient tensor in isotropic strain gradient elasticity. ZAMM - Journal of Applied Mathematics and Mechanics / Zeitschrift für Angewandte Mathematik und Mechanik 96(11):1291–1305

Li XF, Wang BL, Lee KY (2010) Size effect in the mechanical response of nanobeams. J of Adv Research in Mech Eng 1(1):4–16

Liebold C, Müller WH (2013) Measuring material coefficients of higher gradient elasticity by using AFM techniques and Raman-spectroscopy. In: Altenbach H, Forest S, Krivtsov A (eds) Generalized continua as models for materials, Advanced Structured Materials, 22, pp 255–271

Liebold C, Müller WH (2017) 1D Hermite elements for C1–continuous solutions in second gradient elasticity. Acta Polytechnica CTU Proceedings 7:33–37

McFarland AW, Colton JS (2005) Role of material microstructure in plate stiffness with relevance to microcantilever sensors. Journal of Micromechanics and Microengineering 15(5):1060–1067
Mindlin RD, Tiersten HF (1962) Effects of couple-stresses in linear elasticity. ARMA 11:415–448
Münch I, Neff P, Madeo A, Ghiba ID (2015) The modified indeterminate couple stress model: Why yang et al.'s arguments motivating a symmetric couple stress tensor contain a gap and why the couple stress tensor may be chosen symmetric nevertheless. preprint, arXiv, 1512.02053
Nikolov S, Han CS, Raabe D (2007) On the origin of size effects in small-strain elasticity of solid polymers. International Journal of Solids and Structures 44:1582–1592
Toupin RA (1962) Elastic materials with couple-stresses. ARMA 11:385–414
Yang F, Chong CM, Lam DCC, Tong P (2002) Couple stress based strain gradient theory for elasticity. International Journal of Solids and Structures 39(10):2731–2743

Chapter 21
Implications of the Lagrange Identity in Thermoelasticity of Dipolar Bodies

Marin Marin, Andreas Öchsner & Sorin Vlase

Abstract This paper is concerned with the mixed initial-boundary value problem in the context of the theory of thermoelasticity of dipolar bodies. We prove a uniqueness theorem and some continuous dependence theorems without recourse to any energy conservation law, or to any boundedness assumptions on the thermoelastic coefficients. This was possible due to the use of Lagrange's identity. Because of the flexibility of this identity, we also avoid the use of positive definiteness assumptions on the thermoelastic coefficients.

Keywords: Dipolar bodies · Lagrange identity · Uniqueness · Continuous dependence results

21.1 Introduction

One of the first initiators of generalized theories of thermoelasticity was Eringen. For instance, the studies of Eringen (1972, 1990, 1999) were dedicated to the theory of microstretch elastic bodies, this theory being a generalization of the micropolar

Marin Marin
Department of Mathematics and Computer Science, Transilvania University of Brasov, 500093 Brasov, Romania,
e-mail: m.marin@unitbv.ro

Andreas Öchsner
Faculty of Mechanical Engineering, Esslingen University of Applied Sciences, 73728 Esslingen, Germany,
e-mail: andreas.oechsner@gmail.com

Sorin Vlase
Department of Mechanical Engineering, Transilvania University of Brasov, 500093 Brasov, Romania,
e-mail: svlase@unitbv.ro

B. E. Abali et al. (eds.), *New Achievements in Continuum Mechanics and Thermodynamics*, Advanced Structured Materials 108,
https://doi.org/10.1007/978-3-030-13307-8_21

theory and a special case of the micromorphic theory. We must outline that a body is a microstretch continuum if the directors are constrained to have only breathing-type microdeformations. Also, specific for this theory is that each material point is endowed with three deformable directors. The material points of a microstretch solid can stretch and contract independently of their translations and rotations. The purpose of all non-classic theories is to eliminate discrepancies between classical elasticity and experiments, since the classical elasticity failed to predict acceptable results when the effects of material microstructure were known to contribute significantly to the body's overall deformations. This occurs, for example, in the case of granular bodies with large molecules (e.g. polymers), graphite or human bones.

Lately, the number of studies dedicated to bodies with microstructure has increased steadily. Different problems of these environments have been addressed in Abbas (2014b,a); Othman (2003); Sharma and Othman (2007); Marin (1997, 1998); Marin and Öchsner (2017).

The theory of dipolar structure occupies a privileged place between the theories that are dedicated to the microstructure. In this theory any material point has the degrees of freedom of a solid. To see the importance of the dipolar structure of materials, it is enough to analyze the importance given to this structure by some of the outstanding researchers. For example, the studies of Mindlin (1964) and Green and Rivlin (1964) are very significant from this point of view. It is important to note that these researchers have approached multipolar structures in many of their studies, and the dipolar structure appears as a particular case. For instance, Gurtin published a few articles in which he systematized the multipolar structures. The work of Fried and Gurtin (2007) is significant in this respect, where Fried and Gurtin formulated integral statements of the force balance and energy balance for an interface between a body and its environment. It is known that this theory proposes twelve degrees of freedom for a material point: nine micro-deformations and three (classical) translations. Another restriction imposed by this structure is that all material points of the body have only homogeneous deformations. Taking into account the fact that there are many published articles dedicated to the dipolar structure, it can be inferred that the theory of this type of structure is quite adequate for many applications of continuum mechanics. These cases are becoming increasingly important in the design and manufacture of modern day advanced materials, as small-scale effects become paramount in the prediction of the overall mechanical behavior of these materials. Other intended applications of this theory are composite materials reinforced with chopped fibers and various porous materials (Altenbach and Öchsner, 2010). In the following we will use Lagrange's known identity that will facilitate both the obtaining of uniqueness results and those of the continuous dependence on solutions. It is worth noting that previous papers on uniqueness and continuous dependence in elasticity or the thermoelasticity have been based almost exclusivelly on the assumptions that the elasticity tensor or thermoelastic coefficients are positive definite (see for instance, Wilkes, 1980). Also, in other papers, the authors recourse to an energy conservation law, in order to derive the uniqueness or continuous dependence of solutions. For instance, an uniqueness result was indicated in Green and Laws (1972) by supplementing the restrictions arising from thermodynamics with certain definiteness assumptions. In a more general context, some results of existence and uniqueness can be found in the book Marin and Öchsner (2018).

We will examine by a new approach the mixed initial-boundary value problem in the context of thermoelasticity of microstretch solids. We establish the uniqueness and continuous dependence of solutions with respect to body forces, dipolar body force and heat supply. We also deduce the continuous dependence of solutions of our problem with respect to initial data and, at last, with respect to thermoelastic coefficients. The results are obtained for bounded regions of the Euclidian three-dimensional space. We point out, again, that the results are obtained without recourse to an energy conservation law or to any boundedness assumptions on the thermoelastic coefficients. Also, we avoid the use of definiteness assumptions on the thermoelastic coefficients. It can be concluded that our results on continuous dependence of solutions are natural generalizations of the results of Wilkes from classical linear thermoelasticity, see Wilkes (1980) and Marin (2010).

21.2 Basic Equations

In what follows, a bounded region B of the three-dimensional Euclidian space R^3 is occupied by a microstretch elastic body, referred to the reference configuration and a fixed system of rectangular Cartesian axes. Let $\bar{B}$ denote the closure of B and call ∂B the boundary of the domain B. The border ∂B is a piecewise smooth surface and we designated by n_i the components of the outward unit normal to the surface ∂B. Letters in boldface stand for vector fields. We use the notation v_i to designate the components of the vector $\mathbf{v}$ in the underlying rectangular Cartesian coordinates frame. Superposed dots stand for the material time derivative. We shall employ the usual summation and differentiation conventions: the subscripts are understood to range over integers $(1, 2, 3)$. Summation over repeated subscripts is implied and subscripts preceded by a comma denote the partial differentiation with respect to the corresponding Cartesian coordinate.

The spatial argument and the time argument of a function will be omitted when there is no likelihood of confusion. We refer the motion of the body to a fixed system of rectangular Cartesian axes $Ox_i,\ i = 1, 2, 3$.

The behavior of this kind of body is characterized by means of the variables

$$u_i(x,t),\ \varphi_{ij}(x,t),\ \theta(x,t),\ (x,t) \in B \times [0,\infty), \tag{21.1}$$

where u_i are the components of the displacement vector, φ_{ij} are the components of the dipolar displacement tensor and θ is the difference from the reference temperature, where we choose the absolute temperature T_0 as the reference temperature.

We will use three strain tensors ε_{ij}, γ_{ij} and χ_{ijk} which are defined by means of the geometric equations:

$$\varepsilon_{ij} = \frac{1}{2}\left(u_{i,j} + u_{j,i}\right),\ \gamma_{ij} = u_{j,i} - \varphi_{ij},\ \chi_{ijk} = \varphi_{jk,i}. \tag{21.2}$$

The equations of motion from thermoelasticity of dipolar bodies are (see Marin and Öchsner, 2017):

$$
\begin{aligned}
&\left(\tau_{ij}+\sigma_{ij}\right)_{,j}+\varrho f_i=\varrho \ddot{u}_i,\\
&\mu_{ijk,i}+\sigma_{jk}+\varrho g_{jk}=I_{kr}\ddot{\varphi}_{jr}.
\end{aligned} \tag{21.3}
$$

The equation of energy is given by (see Eringen, 1999):

$$
\varrho T_0\dot{\eta}=q_{i,i}+\varrho r. \tag{21.4}
$$

Being in the context of a linear theory, this requires a quadratic form for the Helmholtz's free energy per mass with regards to its independent constitutive variables. We denote by Ψ the Helmholtz's free energy per mass and according to the principle of conservation of energy we can expand in series the Helmholtz's free energy per mass about the reference configuration so that it can be written in the following form

$$
\begin{aligned}
\Psi=&\frac{1}{2}C_{ijmn}\varepsilon_{ij}\varepsilon_{mn}+G_{ijmn}\varepsilon_{ij}\gamma_{mn}+F_{ijmnr}\varepsilon_{ij}\chi_{mnr}\\
&+\frac{1}{2}B_{ijmn}\gamma_{ij}\gamma_{mn}+D_{ijmnr}\gamma_{ij}\chi_{mnr}\\
&+\frac{1}{2}A_{ijkmnr}\chi_{ijk}\chi_{mnr}-a_{ij}\varepsilon_{ij}\theta-b_{ij}\gamma_{ij}\theta-c_{ijk}\chi_{ijk}\theta-\frac{1}{2}c\theta^2.
\end{aligned} \tag{21.5}
$$

As a consequence, this form of Ψ is used in the inequality of entropy in order to obtain the above equations of motion. Also, in the same manner we deduce the constitutive equations that give the expressions for the stress measures in terms of the strain measures. The components of the stress tensor are denoted by τ_{ij}, σ_{ij} and μ_{ijk}. There is a correlation between the strain tensors and stress tensors. From (21.5) we obtain

$$
\begin{aligned}
\tau_{ij}&=\frac{\partial\Psi}{\partial\varepsilon_{ij}}=C_{ijmn}\varepsilon_{mn}+G_{mnij}\gamma_{mn}+F_{mnrij}\chi_{mnr}-a_{ij}\theta,\\
\sigma_{ij}&=\frac{\partial\Psi}{\partial\gamma_{ij}}=G_{ijmn}\varepsilon_{mn}+B_{ijmn}\gamma_{mn}+D_{ijmnr}\chi_{mnr}-b_{ij}\theta,\\
\mu_{ijk}&=\frac{\partial\Psi}{\partial\chi_{ijk}}=F_{ijkmn}\varepsilon_{mn}+D_{mnijk}\gamma_{mn}+A_{ijkmnr}\chi_{mnr}-c_{ijk}\theta,\\
\eta&=-\frac{\partial\Psi}{\partial\theta}=a_{ij}\varepsilon_{ij}+b_{ij}\gamma_{ij}+c_{ijk}\chi_{ijk}+c\theta,
\end{aligned} \tag{21.6}
$$

where η is the entropy per unit mass.

The following symmetry relations are assumed to hold true in the domain Ω

$$
C_{ijmn}=C_{mnij},\ B_{ijmn}=B_{mnij},\ A_{ijkmnr}=A_{mnrijk},\ a_{ij}=a_{ji}. \tag{21.7}
$$

In the above equations we have used the following notations:

- f_i the components of the body force;
- g_{ij} the components of the dipolar body force;
- ϱ is the reference constant mass density;
- $I_{ij} = I_{ji}$ are the coefficients of microinertia;
- r is the heat supply per unit mass;
- q_i are the components of the heat flux vector.

For an anisotropic and homogeneous dipolar thermoelastic material, the components of the heat flux q_i can be expressed in the form

$$q_i = k_{ij}\theta_{,j}, \tag{21.8}$$

where k_{ij} is the thermal conductivity tensor. One can assume that a positive constant λ_0 exists such that

$$I_{ij}\xi_i\xi_j \geq \lambda_0\xi_i\xi_i, \ \forall \xi_i.$$

Also, the Second Law of Thermodynamics implies that:

$$k_{ij}\xi_i\xi_j \geq 0, \ \forall \xi_i.$$

We denote by t_i the components of surface traction, m_{jk} the components of the surface couple, p the microsurface traction and q the heat flux. These quantities are defined by

$$t_i = (\tau_{ji} + \sigma_{ji})\, n_j, \ m_{jk} = \mu_{ijk}n_i, \ q = q_i n_i,$$

at regular points of the surface ∂B.
Here, n_i are the components of the outward unit normal of the surface ∂B.
We now complete the initial boundary value problem by considering the following initial conditions:

$$\begin{aligned} u_i(x,0) &= u_{0i}(x), \ \dot{u}_i(x,0) = u_{1i}(x), \\ \varphi_{ij}(x,0) &= \varphi_{0ij}(x), \ \dot{\varphi}_{ij}(x,0) = \varphi_{1ij}(x), \ x \in \bar{B} \\ \theta(x,0) &= \theta_0(x), \end{aligned} \tag{21.9}$$

and the following given boundary conditions

$$\begin{aligned} &u_i = \bar{u}_i \text{ on } \partial B_1 \times [0,t_0), && t_i = \bar{t}_i \text{ on } \partial B_1^c \times [0,t_0), \\ &\varphi_{ij} = \bar{\varphi}_{ij} \text{ on } \partial B_2 \times [0,t_0), && m_{jk} = \bar{m}_{jk} \text{ on } \partial B_2^c \times [0,t_0), \\ &\theta = \bar{\theta} \text{ on } \partial B_3 \times [0,t_0), && q = \bar{q} \text{ on } \partial B_3^c \times [0,t_0), \end{aligned} \tag{21.10}$$

where t_0 is some instant that may be infinite. Also, the surfaces ∂B_1, ∂B_2 and ∂B_3 with respective complements ∂B_1^c, ∂B_2^c and ∂B_c^c are subsets of the surface ∂B such that

$$\begin{aligned} \partial B_1 \cap \partial B_1^c = \partial B_2 \cap \partial B_2^c = \partial B_3 \cap \partial B_3^c = \emptyset, \\ \partial B_1 \cup \partial B_1^c = \partial B_2 \cup \partial B_2^c = \partial B_3 \cup \partial B_3^c = \partial B. \end{aligned}$$

The quantities $u_{0i},\ u_{1i},\ \varphi_{0i},\ \varphi_{1i},\ \omega_0,\ \omega_1,\ \theta_0,\ \bar{u}_i,\ \bar{t}_i,\ \bar{\varphi}_i,\ \bar{m}_i,\ \bar{\omega}, \bar{p}, \bar{\theta}$ and $\bar{q}$ from (21.9) and (21.10) are prescribed and regular functions in their domains.
We will assume the following regularity assumptions:

- all constitutive coefficients are continuously differentiable functions on $\bar{B}$;
- $\varrho,\ I_{jk}$ are continuous functions on $\bar{B}$;
- $f_i,\ g_{jk}$ and r are continuous on $\bar{B} \times [0, t_1)$;
- $u_{0i},\ u_{1i},\ \varphi_{0ij},\ \varphi_{1ij}$ and θ_0 are continuous functions on $\bar{B}$;
- the functions $\bar{u}_i,\ \bar{\varphi}_{ij}$ and $\bar{\theta}$ are continuous on $\partial B_1 \times [0, t_1)$, $\partial B_2 \times [0, t_1)$ and $\partial B_3 \times [0, t_1)$, respectively;
- the functions $\bar{t}_i,\ \bar{m}_{jk}$ and $\bar{q}$ are piecewise regular on $\partial B_1^c \times [0, t_1)$, $\partial B_2^c \times [0, t_1)$ and $\partial B_3^c \times [0, t_1)$, respectively and continuous in time.

Taking into account the constitutive equations (21.6) and the geometric equations (21.2), the system of equations (21.3) and (21.4) becomes

$$\begin{aligned}
\varrho \ddot{u}_i &= \left[\left(C_{ijmn} + G_{ijmn} \right) u_{n,m} + \left(G_{mnij} + B_{ijmn} \right) \left(u_{n,m} - \varphi_{mn} \right) \right. \\
&\quad \left. + \left(F_{mnrij} + D_{ijmnr} \right) \varphi_{nr,m} - \left(a_{ij} + b_{ij} \right) \theta \right]_{,j} + \varrho f_i, \\
I_{kr} \ddot{\varphi}_{jr} &= \left[F_{ijkmn} u_{n,m} + D_{mnijk} \left(u_{n,m} - \varphi_{mn} \right) + A_{ijkmnr} \varphi_{nr,m} - c_{ijk} \theta \right]_{,i} \\
&\quad + G_{jkmn} u_{m,n} + B_{jkmn} (u_{n,m} - \varphi_{mn}) + D_{jkmnr} \varphi_{nr,m} - b_{jk} \theta + \varrho g_{jk}, \\
\varrho r + k_{ij} \theta_{,ji} &= T_0 \left[c \dot{\theta} + a_{ij} \dot{u}_{i,j} + b_{ij} \left(\dot{u}_{j,i} - \dot{\varphi}_{ij} \right) + c_{ijk} \dot{\varphi}_{ij,k} \right].
\end{aligned} \tag{21.11}$$

An ordered array $(u_i,\ \varphi_{ij},\ \theta)$ is a solution of the mixed initial boundary value problem of the theory of thermoelasticity of dipolar bodies in the cylinder $\Omega_0 = B \times [0, t_0)$ if it satisfies the system of equations (21.11) for all $(x, t) \in \Omega_0$, the boundary conditions (21.10) and the initial conditions (21.9).

21.3 Main Result

Consider the functions $\alpha(t, x)$ and $\beta(t, x)$ being twice continuously differentiable with respect to the time variable t. Clearly, we have the following identity

$$\frac{d}{dt} \left(\alpha \dot{\beta} - \dot{\alpha} \beta \right) = \dot{\alpha} \dot{\beta} + \alpha \ddot{\beta} - \ddot{\alpha} \beta - \dot{\alpha} \dot{\beta} = \alpha \ddot{\beta} - \ddot{\alpha} \beta.$$

Because there is no likelihood of confusion, the spatial argument and the time argument of the functions $\alpha(t, x)$ and $\beta(t, x)$ are omitted, for the sake of simplicity.
Instead of the functions $\alpha(t, x)$ and $\beta(t, x)$ we will use, in the above identity, the functions $u_i(x, t)$ and $v_i(x, t)$, which are, also, assumed to be twice continuously differentiable with respect to the time variable. The identity thus obtained is integrated over the domain B so that we are led to the following well known Lagrange's identity:

$$\int_B \varrho(x)[u_i(x,t)\dot{v}_i(x,t) - \dot{u}_i(x,t)v_i(x,t)\mathrm{d}V \tag{21.12}$$
$$= \int_0^t \int_B \varrho(x)\left[u_i(x,s)\ddot{v}_i(x,s) - \ddot{u}_i(x,s)v_i(x,s)\right]\mathrm{d}V\mathrm{d}s$$
$$+ \int_B \varrho(x)\left[u_i(x,0)\dot{v}_i(x,0) - \dot{u}_i(x,0)v_i(x,0)\right]\mathrm{d}V.$$

Let us consider two solutions $\left(u_i^{(\alpha)},\ \varphi_{ij}^{(\alpha)},\ \theta^{(\alpha)}\right)$, $(\alpha = 1,\ 2)$ of the mixed initial boundary value problem defined by the equations (11) and the conditions (21.9) and (21.10) which correspond to the same boundary data and same initial data, but to different body forces, dipolar couples and heat supplies,

$$\left(f_i^{(\alpha)},\ g_{jk}^{(\alpha)},\ r^{(\alpha)}\right),\ (\alpha = 1,\ 2),$$

respectively.

For the difference of the above two solutions, we will use the following notations:

$$v_i = u_i^{(2)} - u_i^{(1)},\ \psi_{ij} = \varphi_{ij}^{(2)} - \varphi_{ij}^{(1)},\ \chi = \theta^{(2)} - \theta^{(1)}. \tag{21.13}$$

We are now in a position to prove our first result. This will be the basis for all subsequent results.

Theorem 21.1. *For the differences (v_i, ψ_{ij}, χ) of two solutions of our mixed initial boundary value problem the Lagrange identity becomes:*

$$2\int_B [\varrho v_i(t)\dot{v}_i(t) + I_{jk}\psi_{jr}(t)\dot{\psi}_{kr}(t)]\mathrm{d}V$$
$$+ \int_B \frac{1}{T_0}k_{ij}\left(\int_0^t \chi_{,i}(\xi)d\xi\right)\left(\int_0^t \chi_{,j}(\xi)d\xi\right)\mathrm{d}V$$
$$= \int_0^t \mathrm{d}s\int_B \varrho\left[v_i(2t-s)\mathcal{F}_i(s) - v_i(s)\mathcal{F}_i(2t-s)\right]\mathrm{d}V$$
$$+ \int_0^t \mathrm{d}s\int_B \varrho\left[\psi_{jk}(2t-s)\mathcal{G}_{jk}(s) - \psi_{jk}(s)\mathcal{G}_{jk}(2t-s)\right]\mathrm{d}V \tag{21.14}$$
$$+ \int_0^t \mathrm{d}s\int_B \frac{1}{T_0}\left[\chi(s)\int_0^{2t-s}\mathcal{P}(\xi)d\xi - \chi(2t-s)\int_0^s \mathcal{P}(\xi)d\xi\right]\mathrm{d}V,$$

where we have used the notations

$$\mathcal{F}_i = f_i^{(2)} - f_i^{(1)},\ \mathcal{G}_{jk} = g_{jk}^{(2)} - g_{jk}^{(1)},\ \mathcal{P} = r^{(2)} - r^{(1)}. \tag{21.15}$$

Proof. Of course, the mixed problem defined by (21.11), (21.9) and (21.10) is linear such that we can deduce that the differences $(v_i,\ \psi_{ij},\ \chi)$ from (21.13) represent the solution of a mixed initial boundary value problem analogous to (21.11), (21.9) and (21.10) in which the loads are $(\mathcal{F}_i,\ \mathcal{G}_{jk},\ \mathcal{P})$ from (21.15), but the corresponding initial conditions and the corresponding boundary conditions become homogeneous.

If we use the substitution

$$u_i(x,s)\to v_i(x,s),\ v_i(x,s)\to v_i(x,2t-s),\ s\in[0,2t],\ t\in[0,\frac{t_1}{2}),$$

after some straightforward calculus, the identity (21.12) becomes

$$2\int_B \varrho v_i(t)v_i(t)\mathrm{d}V=\int_0^t \mathrm{d}s\int_B \varrho\left[v_i(2t-s)\ddot{v}_i(s)-\ddot{v}_i(2t-s)v_i(s)\right]\mathrm{d}V. \quad (21.16)$$

In deducting equality (21.16) we took into account that the initial and boundary data are null.
The inertial terms on the right-hand side of the relation (21.16) can be eliminated by means of the equations of motion for the differences $(v_i,\ \psi_{ij},\ \chi)$.
First, in view of the equation $(21.3)_1$, we have

$$\begin{aligned}&\varrho\left[v_i(2t-s)\ddot{v}_i(s)-\ddot{v}_i(2t-s)v_i(s)\right]=\\&\quad=\left[v_i(2t-s)\left(\tau_{ji}+\sigma_{ji}\right)(s)-v_i(s)\left(\tau_{ji}+\sigma_{ji}\right)(2t-s)\right]_{,j}+\\&\qquad+\left[\left(\tau_{ji}+\sigma_{ji}\right)(2t-s)v_{j,i}(s)-\left(\tau_{ji}+\sigma_{ji}\right)(s)v_{j,i}(2t-s)\right]+\\&\qquad\quad+\varrho\left[\mathcal{F}_i(s)v_i(2t-s)-\mathcal{F}_i(2t-s)v_i(s)\right].\end{aligned}$$

If we use the geometric equations (21.4), the above equation becomes

$$\begin{aligned}\varrho[v_i(2t-s)\ddot{v}_i(s)\ &-\ddot{v}_i(2t-s)v_i(s)]\\&=\left[v_i(2t-s)\left(\tau_{ji}+\sigma_{ji}\right)(s)-v_i(s)\left(\tau_{ji}+\sigma_{ji}\right)(2t-s)\right]_{,j}\\&+\varrho\left[\mathcal{F}_i(s)v_i(2t-s)-\mathcal{F}_i(2t-s)v_i(s)\right]\\&=\left(C_{ijrs}+G_{ijrs}\right)\left[\varepsilon_{rs}(2t-s)\varepsilon_{ij}(s)-\varepsilon_{rs}(s)\varepsilon_{ij}(2t-s)\right]\\&+\left(G_{ijrs}+B_{ijrs}\right)\left[\gamma_{rs}(2t-s)\varepsilon_{ij}(s)-\gamma_{rs}(s)\varepsilon_{ij}(2t-s)\right]\\&+\left(F_{ijkrs}+D_{ijkrs}\right)\left[\gamma_{rs}(2t-s)\chi_{ijk}(s)-\gamma_{rs}(s)\chi_{ij}(2t-s)\right]\\&+\left(a_{ij}+b_{ij}\right)\left[\chi(2t-s)\varepsilon_{ij}(s)\right.\\&\left.-\chi(s)\varepsilon_{ij}(2t-s)\right].\end{aligned} \quad (21.17)$$

Based on the symmetry conditions of the coefficients I_{ij}, it is easy to see that:

$$I_{jk}\frac{d}{\mathrm{d}t}\left[\omega_{jr}(t)\dot{\psi}_{kr}(t)-\dot{\omega}_{jr}(t)\psi_{kr}(t)\right]=I_{jk}\left[\omega_{jr}(t)\ddot{\psi}_{kr}(t)-\ddot{\omega}_{jr}(t)\psi_{kr}(t)\right].$$

Now, we integrate this equality on cylinder $B\times[0,t]$ and, recalling that the differences satisfy null initial data, we are led to the following identity:

$$\begin{aligned}&\int_B I_{jk}\left[\omega_{jr}(t)\dot{\psi}_{kr}(t)-\dot{\omega}_{jr}(t)\psi_{kr}(t)\right]\mathrm{d}V=\\&\quad=\int_0^t\left\{\int_B I_{jk}\left[\omega_{jr}(s)\ddot{\psi}_{kr}(s)-\ddot{\omega}_{jr}(s)\psi_{kr}(s)\right]\mathrm{d}V\right\}\mathrm{d}s.\end{aligned}$$

By using the substitution:

$$\omega_{ij}(s) \to \psi_{ij}(2t-s), \; s \in [0, 2t], \; t \in [0, \frac{t_1}{2}),$$

previous identity becomes:

$$2\int_B I_{kj}\psi_{jr}(t)\dot{\psi}_{kr}(t)\mathrm{d}V = \int_0^t \left\{ \int_B I_{jk} \left[\psi_{jr}(2t-s)\ddot{\psi}_{kr}(s) - \ddot{\psi}_{jr}(2t-s)\psi_{kr}(s) \right] \mathrm{d}V \right\} \mathrm{d}s. \tag{21.18}$$

The inertial terms on the right-hand side of the relation (21.18) can be eliminated by means of the equations of motion for the differences $(v_i, \; \psi_{ij}, \; \chi)$.
So, in view of the equation $(21.3)_2$, we have

$$\begin{aligned}
I_{jk} & \left[\psi_{jr}(2t-s)\ddot{\psi}_{kr}(s) - \ddot{\psi}_{jr}(2t-s)\psi_{kr}(s) \right] \\
&= \psi_{jr}(2t-s)[\mu_{jri,i}(s) + \varrho\mathcal{G}_{jr}(s) + \sigma_{jr}(s)] \\
&- \psi_{jr}(s)[\mu_{jri,i}(2t-s) + \varrho\mathcal{G}_{jr}(2t-s) + \sigma_{jr}(2t-s)] \\
&= [\psi_{jk}(2t-s)\mu_{jki}(s) - \psi_{jk}(s)m_{jki}(2t-s)]_{,i} \\
&- \psi_{jk,i}(2t-s)\mu_{jki}(s) + \psi_{jk,i}(s)\mu_{jki}(2t-s) \\
&+ \psi_{jk}(2t-s)\sigma_{jk}(s) - \psi_{jk}(s)\sigma_{jk}(2t-s) \\
&+ \varrho\left[\psi_{jk}(2t-s)\mathcal{G}_{jk}(s) - \psi_{jk}(s)\mathcal{G}_{jk}(2t-s)\right].
\end{aligned}$$

With the help of the geometric equations (21.2), the above identity receives the form:

$$\begin{aligned}
I_{jk} & \left[\psi_{jr}(2t-s)\ddot{\psi}_{kr}(s) - \ddot{\psi}_{jr}(2t-s)\psi_{kr}(s) \right] \\
&= \psi_{jr}(2t-s)[\mu_{jri,i}(s) + \varrho\mathcal{G}_{jr}(s) + \sigma_{jr}(s)] \\
&- \psi_{jr}(s)[\mu_{jri,i}(2t-s) + \varrho\mathcal{G}_{jr}(2t-s) + \sigma_{jr}(2t-s)] \\
&= [\psi_{jk}(2t-s)\mu_{jki}(s) - \psi_{jk}(s)\mu_{jki}(2t-s)]_{,i} \\
&- \chi_{ijk}(2t-s)\mu_{jki}(s) + \chi_{ijk}(s)\mu_{jki}(2t-s) \\
&+ \psi_{jk}(2t-s)\sigma_{jk}(s) - \psi_{jk}(s)\sigma_{jk}(2t-s) \\
&+ \varrho\left[\psi_{jk}(2t-s)\mathcal{G}_{jk}(s) - \psi_{jk}(s)\mathcal{G}_{jk}(2t-s)\right].
\end{aligned}$$

If we take into account the constitutive equations (21.6), from the above identity we are led to:

$$
\begin{aligned}
&I_{jk}\left[\psi_{jr}(2t-s)\ddot{\psi}_{kr}(s)-\ddot{\psi}_{jr}(2t-s)\psi_{kr}(s)\right] \\
&\quad=\left[\psi_{jk}(2t-s)\mu_{jki}(s)-\psi_{jk}(s)\mu_{jki}(2t-s)\right]_{,i} \\
&\quad+\psi_{jk}(2t-s)\sigma_{jk}(s)-\psi_{jk}(s)\sigma_{jk}(2t-s) \\
&\quad+\varrho\left[\psi_{jk}(2t-s)\mathcal{G}_{jk}(s)-\psi_{jk}(s)\mathcal{G}_{jk}(2t-s)\right] \\
&\quad=F_{ijkrs}\left[\varepsilon_{rs}(2t-s)\chi_{ijk}(s)-\varepsilon_{rs}(s)\chi_{ijk}(2t-s)\right] \\
&\quad+D_{ijkrs}\left[\gamma_{rs}(2t-s)\chi_{ijk}(s)-\gamma_{rs}(s)\chi_{ijk}(2t-s)\right] \\
&\quad+b_{ij}\left[\chi(2t-s)\gamma_{ji}(s)-\chi(s)\gamma_{ji}(2t-s)\right].
\end{aligned}
\tag{21.19}
$$

Now we integrate the equalities (21.17) and (21.19) over B and add the resulting relations, term by term. After that we use the divergence theorem and take into account that the differences $(v_i,\ \psi_{ij},\ \chi)$ satisfy the null initial data and null boundary data. In this way, we obtain the identity

$$
\begin{aligned}
&\int_B\left[\varrho v_i\dot{v}_i+I_{jk}\psi_{jr}\dot{\psi}_{kr}\right]\mathrm{d}V \\
&\quad=\int_0^t\left\{\int_B\varrho\left[\mathcal{F}_i(s)v_i(2t-s)+\mathcal{G}_{jk}(s)\psi_{jk}(2t-s)\right.\right. \\
&\quad\left.\left.-\ \mathcal{F}_i(2t-s)v_i(s)-\mathcal{G}_{jk}(2t-s)\psi_{jk}(s)\right]\mathrm{d}V\right\}\mathrm{d}s \\
&\quad=\int_0^t\left\{\int_B a_{ij}\left[\varepsilon_{ij}(2t-s)\chi(s)-\varepsilon_{ij}(s)\chi(2t-s)\right]\right\}\mathrm{d}s\mathrm{d}V \\
&\quad+\int_0^t\left\{\int_B b_{ij}\left[\gamma_{ij}(2t-s)\chi(s)-\gamma_{ij}(s)\chi(2t-s)\right]\mathrm{d}V\right\}\mathrm{d}s \\
&\quad+\int_0^t\left\{\int_B c_{ijk}\left[\chi_{ijk}(2t-s)\chi(s)-\chi_{ijk}(s)\chi(2t-s)\right]\mathrm{d}V\right\}\mathrm{d}s \\
&\quad+\int_0^t\left\{\int_B c\chi(2t-s)\chi(s)\mathrm{d}V\right\}\mathrm{d}s.
\end{aligned}
\tag{21.20}
$$

Considering the constitutive equation $(21.6)_4$, it is easy to deduce the equality

$$
\begin{aligned}
&a_{ij}\left[\varepsilon_{ij}(2t-s)\chi(s)-\varepsilon_{ij}(s)\chi(2t-s)\right] \\
&+b_{ij}\left[\gamma_{ij}(2t-s)\chi(s)-\gamma_{ij}(s)\chi(2t-s)\right] \\
&+c_{ijk}\left[\chi_{ijk}(2t-s)\chi(s)-\chi_{ijk}(s)\chi(2t-s)\right] \\
&+c\chi(2t-s)\chi(s)=\eta(2t-s)\chi(s)-\eta(s)\chi(2t-s).
\end{aligned}
\tag{21.21}
$$

For the term from the right-hand side of the equality (21.21) we can get the following evaluation:

$$
\begin{aligned}
\eta(2t-s)\chi(s) &- \eta(s)\chi(2t-s) \\
&= \frac{1}{T_0}k_{ij}\left[\chi(s)\int_0^{2t-s}\chi_{,j}(\tau)\mathrm{d}\tau - \chi(2t-s)\int_0^{s}\chi_{,j}(\tau)\mathrm{d}\tau\right]_{,i} \\
&+ \frac{1}{T_0}k_{ij}\left[\chi_{,i}(2t-s)\int_0^{s}\chi_{,j}(\tau)\mathrm{d}\tau - \chi_{,i}(s)\int_0^{2t-s}\chi_{,j}(\tau)\mathrm{d}\tau\right] \quad (21.22) \\
&+ \frac{\varrho}{T_0}\left[\chi(s)\int_0^{2t-s}\mathcal{P}(\tau)\mathrm{d}\tau - \chi(2t-s)\int_0^{s}\mathcal{P}(\tau)\mathrm{d}\tau\right]
\end{aligned}
$$

Taking into account that k_{ij} is a symmetric tensor, we can write

$$
\begin{aligned}
&\int_0^t \mathrm{d}s\int_B \frac{1}{T_0}k_{ij}\frac{d}{\mathrm{d}s}\left[\left(\int_0^s \chi_{,i}(\tau)\mathrm{d}\tau\right)\left(\int_0^{2t-s}\chi_{,j}(\tau)\mathrm{d}\tau\right)\right]\mathrm{d}V \\
&= \int_B \frac{1}{T_0}k_{ij}\int_0^t \frac{d}{ds}\left[\left(\int_0^s \chi_{,i}(\tau)\mathrm{d}\tau\right)\left(\int_0^{2t-s}\chi_{,j}(\tau)\mathrm{d}\tau\right)\right]\mathrm{d}s\,\mathrm{d}V \quad (21.23) \\
&= \int_B \frac{1}{T_0}k_{ij}\left(\int_0^t \chi_{,i}(\tau)\mathrm{d}\tau\right)\left(\int_0^t \chi_{,j}(\tau)\mathrm{d}\tau\right)\mathrm{d}V
\end{aligned}
$$

If we integrate by parts, we get the next equality

$$
\begin{aligned}
&\int_0^t \mathrm{d}s\int_B \frac{1}{T_0}k_{ij}\left[\chi_{,i}(s)\int_0^{2t-s}\chi_{,j}(\tau)\mathrm{d}\tau - \chi_{,j}(2t-s)\int_0^{s}\chi_{,i}(\tau)\mathrm{d}\tau\right]\mathrm{d}V \\
&= \int_0^t \mathrm{d}s\int_B \frac{1}{T_0}k_{ij}\frac{d}{ds}\left[\left(\int_0^s \chi_{,i}(\tau)\mathrm{d}\tau\right)\left(\int_0^{2t-s}\chi_{,j}(\tau)\mathrm{d}\tau\right)\right]\mathrm{d}V \quad (21.24)
\end{aligned}
$$

Considering the identities (21.22) and (21.23) we are led to the following equality:

$$
\begin{aligned}
&\int_0^t \mathrm{d}s\int_B \frac{1}{T_0}k_{ij}\left[\chi_{,i}(2t-s)\int_0^{s}\chi_{,j}(\tau)\mathrm{d}\tau - \chi_{,i}(2t-s)\int_0^{2t-s}\chi_{,j}(\tau)\mathrm{d}\tau\right]\mathrm{d}V \\
&= -\int_B \frac{1}{T_0}k_{ij}\left[\left(\int_0^s \chi_{,i}(\tau)\mathrm{d}\tau\right)\left(\int_0^{2t-s}\chi_{,j}(\tau)\mathrm{d}\tau\right)\right]\mathrm{d}V. \quad (21.25)
\end{aligned}
$$

Finally, by taking into account the identities (21.16), (21.18), (21.20) and (21.25), applying the divergence theorem and using the null initial data and null boundary data, we obtain the identity (21.14) and the proof of Theorem 21.1 is concluded. ■

The identity (21.14) constitutes the basis on which we shall prove the uniqueness and the continuous dependence results. We proceed first to obtain the uniqueness of the solution of the mixed initial boundary value problem defined by the system of equations (21.11), the initial conditions (21.9) and the boundary conditions (21.10).

Theorem 21.2. *Let us suppose that the symmetry relations (21.7) are satisfied and the conductivity tensor k_{ij} is positive definite, that is, there exists a positive constant k_0 such that*

$$k_{ij}x_i x_j \geq k_0 x_i x_i, \ \forall x_i.$$

If ∂B_3 is not empty or $c(x) \neq 0$ on B, then the mixed initial boundary value problem in thermoelastodynamics of dipolar bodies admits only one solution.

Proof. Suppose by absurd that our mixed problem has two solutions

$$\left(u_i^{(\alpha)},\ \varphi_{ij}^{(\alpha)},\ \theta^{(\alpha)}\right), \quad \alpha = 1, 2.$$

Also, we assume that these solutions correspond to the same boundary data, the same initial data, the same body force, the same dipolar body force and the same heat supply.

Consider the difference of these two solutions:

$$v_i = u_i^{(2)} - u_i^{(1)},\ \psi_{ij} = \varphi_{ij}^{(2)} - \varphi_{ij}^{(1)},\ \chi = \theta_i^{(2)} - \theta_i^{(1)}. \tag{21.26}$$

Then we must prove that

$$v_i(x,t) = 0,\ \psi_{ij}(x,t) = 0,\ \chi(x,t) = 0,\ \forall (x,t) \in B \times [0, t_1). \tag{21.27}$$

Because of the linearity of our problem we deduce that the differences $(v_i,\ \psi_{ij},\ \chi)$ from (21.27) are also a solution of the problem, but in a particular case, namely, with null body force, null dipolar body force and null heat supply. In this case the identity (21.14) receives the following simple form:

$$2\int_B \left[\varrho v_i(t)\dot{v}_i(t) + I_{jk}\psi_{jr}(t)\dot{\psi}_{kr}(t)\right] \mathrm{d}V +$$
$$+\int_B \frac{1}{T_0} k_{ij} \left(\int_0^t \chi_{,i}(\tau)\mathrm{d}\tau\right)\left(\int_0^t \chi_{,j}(\tau)\mathrm{d}\tau\right) \mathrm{d}V = 0.$$

For a fixed $s \in [0, t_1/2)$, we integrate this equality on the interval $[0, s]$ so that we obtain

$$\int_B \varrho v_i(s) v_i(s) \mathrm{d}V + \int_B I_{ij}\psi_i(s)\psi_i(s)\mathrm{d}V + \int_B J\delta^2 \mathrm{d}V +$$
$$+\int_0^s \int_B \frac{1}{T_0} k_{ij} \left(\int_0^\xi \chi_{,i}(\tau)\mathrm{d}\tau\right)\left(\int_0^\xi \chi_{,j}(\tau)\mathrm{d}\tau\right) \mathrm{d}V\, d\xi = 0.$$

Based on the properties of ϱ, I_{ij} and k_{ij}, from the above identity we are led to

$$v_i(x,t) = 0,\ \psi_{ij}(x,t) = 0,\ \chi_{,i}(x,t) = 0,\ \forall (x,t) \in B \times [0, t_1/2). \tag{21.28}$$

In the case that ∂B_3 is not empty, taking into account the null boundary conditions (21.7), from (21.28) we deduce that (21.27) holds. In the case that $c(x) \neq 0$, we write the equation of energy for the differences and obtain $\dot{\chi} = 0$. Taking into account the null initial conditions, that is, χ vanishes initially, we obtain again that (21.26) holds true. If we suppose that t_1 is infinite, then the proof of Theorem 21.2 ends.

If t_1 is finite, then we can do the substitution

$$v_i(x,t_1/2) \to 0,\ \dot{v}_i(x,t_1/2) \to 0,\ \psi_{ij}(x,t_1/2) \to 0,\ \dot{\psi}_{ij}(x,t_1/2) \to 0,$$
$$\chi(x,t_1/2) \to 0,$$

and we can repeat the above procedure with new null initial conditions on the interval $[t_1/2, t_1/2 + t_1/4]$. In this way, the conclusion (21.27) is extended on $B \times [0, 3t_1/4)$, and so on. As such, the conclusion (21.27) is obtained on $B \times [0, t_1)$ and the proof of Theorem 21.2 is concluded. ■

Now, we state the second main result, namely a continuous dependence theorem with regard to the body forces, dipolar body forces and heat supplies for the solution of the mixed initial boundary value problem defined by the system of equations (21.11), the initial conditions (21.9) and the boundary conditions (21.10).

In this regard, let us consider $\left(u_i^{(\alpha)},\ \varphi_{ij}^{(\alpha)},\ \theta^{(\alpha)}\right)$, $(\alpha = 1, 2)$, two solutions of our mixed problem which correspond to the same initial data and boundary data, but to different body force, dipolar body force and to different heat supply, $\left(\mathcal{F}_i^{(\alpha)},\ \mathcal{G}_i^{(\alpha)},\ \mathcal{L}^{(\alpha)},\ \mathcal{P}^{(\alpha)}\right)$, $(\alpha = 1, 2)$, where $\mathcal{F}_i = f_i^2 - f_i^1$, $\mathcal{G}_i = g_i^2 - g_i^1$, $\mathcal{L} = l^2 - l^1$, $\mathcal{P} = r^2 - r^1$. Also, we need to suppose that there exists $t_* \in (0, t_1)$ such that

$$\int_0^{t_*}\int_B \varrho\mathcal{F}_i(t)\mathcal{F}_i(t)\mathrm{d}V\mathrm{d}t \le M_1^2,\ \int_0^{t_*}\int_B \varrho\mathcal{G}_i(t)\mathcal{G}_i(t)\mathrm{d}V\mathrm{d}t \le M_2^2,$$
$$\int_0^{t_*}\int_B \frac{\varrho}{T_0}\left(\int_0^t \mathcal{P}^2(\tau)\mathrm{d}\tau\right)^2\mathrm{d}V\mathrm{d}t \le M_3^2,\ \int_0^{t_*}\int_B \varrho u_i(t)u_i(t)\mathrm{d}V\mathrm{d}t \le K^2, \tag{21.29}$$
$$\int_0^{t_*}\int_B I_{ij}\varphi_i(t)\varphi_j(t)\mathrm{d}V\mathrm{d}t \le M^2,\ \int_0^{t_*}\int_B \frac{\varrho}{T_0}\chi^2(t)\mathrm{d}V\mathrm{d}t \le Q^2.$$

The result is obtained on the compact subintervals of the interval $[0, t_1)$.

Theorem 21.3. *Suppose that the assumptions from Theorem 21.2 are satisfied and the conditions (21.29) take places. Then there exists $s \in [0, t_1/2)$ so that we have the following estimate:*

$$\begin{aligned}
&\int_B \left[\varrho u_i(s)u_i(s) + I_{jk}\varphi_{jr}(s)\varphi_{kr}(s)\right]\mathrm{d}V \\
&+ \int_0^s\int_B \frac{1}{T_0}k_{ij}\left(\int_0^t \chi_{,i}(\tau)\mathrm{d}\tau\right)\left(\int_0^t \chi_{,j}(\tau)\mathrm{d}\tau\right)\mathrm{d}V\mathrm{d}t \\
&\le t_*K\left[\int_0^{t_*}\int_B \varrho\mathcal{F}_i(t)\mathcal{F}_i(t)\mathrm{d}V\mathrm{d}t\right]^{1/2} \\
&+ t_*M\left[\int_0^{t_*}\int_B \varrho\mathcal{G}_{ij}(t)\mathcal{G}_{ij}(t)\mathrm{d}V\mathrm{d}t\right]^{1/2} \\
&+ t_*Q\left[\int_0^{t_*}\int_B \frac{\varrho}{T_0}\left(\int_0^t \mathcal{P}(\tau)\mathrm{d}\tau\right)^2\mathrm{d}V\mathrm{d}t\right]^{1/2}.
\end{aligned} \tag{21.30}$$

Proof. Our approach will be based on the identity (21.14). For each of the three integrals from the right-hand side of this identity we will use the Schwarz's inequality. For example, for the first integral we have

$$\begin{aligned}
&\int_0^t \int_B \varrho u_i(2t-s)\mathcal{F}_i(s)\mathrm{d}V\mathrm{d}s \\
&\leq \left[\int_0^t \int_B \varrho \mathcal{F}_i(s)\mathcal{F}_i(s)\mathrm{d}V\mathrm{d}s\right]^{1/2} \left[\int_0^{t_*} \int_B \varrho u_i(2t-s)u_i(2t-s)\mathrm{d}V\mathrm{d}s\right]^{1/2} \\
&= \left[\int_0^t \int_B \varrho \mathcal{F}_i(s)\mathcal{F}_i(s)\mathrm{d}V\mathrm{d}s\right]^{1/2} \left[\int_t^{2t} \int_B \varrho u_i(s)u_i(s)\mathrm{d}V\mathrm{d}s\right]^{1/2} \qquad (21.31) \\
&\leq K\left[\int_0^{t_*} \int_B \varrho \mathcal{F}_i(s)\mathcal{F}_i(s)\mathrm{d}V\mathrm{d}s\right]^{1/2}
\end{aligned}$$

To obtain the last inequality, we used the substitution $2t-s \to s$.
Analogous inequalities are obtained for the other two integrals from the right-hand side of this identity (21.14). If we integrate the inequality (21.31) and the other two resulting inequalities over $[0,s]$, $s \in [0, t_*/2]$, we are led to the inequality (21.30), which ends the proof of Theorem 21.3. ■

In the following theorem, we will prove also a continuous dependence result, but with regards to the initial data. The demonstration will be facilitated by the use of the estimate (21.30). For this purpose, let us consider

$$\left(u_i^1,\, \varphi_i^1,\, \theta^1\right), \qquad \left(u_i^1+v_i,\, \varphi_i^1+\psi_i,\, \theta^1+\chi\right), \qquad (21.32)$$

i.e. two solutions of the mixed initial boundary value problem, which correspond to two different initial data,

$$\begin{aligned}
&\left(u_{0i}^1,\, u_{1i}^1,\, \varphi_{0ij}^1,\, \varphi_{1ij}^1,\, \theta^1\right), \\
&\left(u_{0i}^1+a_i^0,\, u_{1i}^1+a_i^1,\, \varphi_{0ij}^1+b_{ij}^0,\, \varphi_{1ij}^1+b_{ij}^1,\, \theta^1+c^0\right), \qquad (21.33)
\end{aligned}$$

but to the same body force, dipolar body force and heat supply and to the same boundary data. The above perturbations $\left(a_i^0,\, a_i^1,\, b_{ij}^0,\, b_{ij}^1,\, c^0\right)$ must be bounded, that is, there exist the constants M_4, M_5 and M_6 such that

$$\begin{aligned}
&\int_B \varrho\left(a_i^0 a_i^0 + a_i^1 a_i^1\right)\mathrm{d}V \leq M_4^2, \\
&\int_B \varrho\left(b_{ij}^0 b_{ij}^0 + b_{ij}^1 b_{ij}^1\right)\mathrm{d}V \leq M_5^2, \\
&\int_B \frac{T_0}{\varrho}\eta_0^2\mathrm{d}V \leq M_6^2,
\end{aligned}$$

where η_0 has the following meaning:

$$\eta_0(x) = a_{ij}(x)a^0_{j,i}(x) + b_{ij}(x)\left[a^0_{j,i}(x) - b^0_{ij}(x)\right]$$
$$+c_{ijk}(x)b^0_{ij,k}(x) + \frac{c}{T_0}c^0(x).$$

With the help of the differences between the two solutions $(v_i,\ \psi_{ij},\ \chi)$ we define the functions $V_i(x,t)$, $\Psi_{ij}(x,t)$ and $\Theta(x,t)$ by

$$\begin{aligned} V_i(x,t) &= \int_0^t \int_0^s v_i(x,\tau)\mathrm{d}\tau\mathrm{d}s, \\ \Psi_{ij}(x,t) &= \int_0^t \int_0^s \psi_{ij}(x,\tau)\mathrm{d}\tau\mathrm{d}s, \\ \Theta(x,t) &= \int_0^t \int_0^s \chi(x,\tau)\mathrm{d}\tau\mathrm{d}s. \end{aligned} \tag{21.34}$$

Theorem 21.4. *If the functions $(U_i, \Phi_i, \Psi_i, \Theta)$ satisfy the conditions (21.29), then we have the next estimate:*

$$\begin{aligned} &\int_b \left[\varrho V_i(t)V_i(t) + I_{jk}\Psi_{jr}(t)\Psi_{kr}(t)\right]\mathrm{d}V \\ &+\int_0^t \int_B \frac{1}{T_0}k_{ij}\left(\int_0^s \Theta_{,i}(\tau)\mathrm{d}\tau\right)\left(\int_0^s \Theta_{,j}(\tau)\mathrm{d}\tau\right)\mathrm{d}V\mathrm{d}s \\ &\leq t_* K\left[(t_*+t_*/2)\int_B \varrho a^0_i a^0_i \mathrm{d}V + (t_*^2/2+t_*^3/3)\int_B \varrho a^1_i a^1_i \mathrm{d}V\right]^{1/2} \\ &+t_* M\left[(t_*+t_*/2)\int_B I_{jk}b^0_{jr}b^0_{kr}\mathrm{d}V + (t_*^2/2+t_*^3/3)\int_B I_{jk}b^1_{jr}b^1_{kr}\mathrm{d}V\right]^{1/2} \\ &+t_*^{7/2}Q\frac{1}{\sqrt{20}}\left(\int_B \frac{T_0}{\varrho}\eta_0^2\mathrm{d}V\right)^{1/2}. \end{aligned} \tag{21.35}$$

Proof. Clearly, from (21.30), by integrating by parts, it is easy to deduce

$$\begin{aligned} V_i(x,t) &= \int_0^t (t-s)v_i(x,s)\mathrm{d}s, \\ \Psi_{ij}(x,t) &= \int_0^t (t-s)\psi_{ij}(x,s)\mathrm{d}s, \\ \Theta(x,t) &= \int_0^t (t-s)\chi(x,s)\mathrm{d}s. \end{aligned}$$

We can observe that the difference functions $(v_i,\ \psi_i,\ \chi)$ satisfy the equations of motion (21.3) and the equation of energy (21.4), corresponding to null loads

$$f_i = g_{ij} = r = 0.$$

Also, it is easy to deduce that the difference functions satisfy the following initial conditions:

$$v_i(x,0) = a_i^0(x),\ \dot{v}_i(x,0) = a_i^1(x),\ \psi_{ij}(x,0) = b_{ij}^0(x),$$
$$\dot{\psi}_{ij}(x,0) = b_{ij}^1(x),\ \chi(x,0) = d^0(x),\ \forall x \in B.$$

As a consequence, let us observe that the functions (V_i, Ψ_{ij}, Θ) from (21.34) satisfy the equations of motion (21.3), corresponding the following loads:

$$f_i(x,t) = a_i^0(x) + ta_i^1(x), \quad g_{ij}(x,t) = b_{ij}^0(x) + tb_{ij}^1(x),$$

and the equation of energy (21.4), corresponding the following heat flux:

$$r(x,t) = T_0 t\Big[a_{ij}(x)a_{j,i}^0(x) + b_{ij}(x)\left(a_{j,i}^0(x)\right.$$
$$\left. - b_{ij}^0(x)\right) + c_{ijk}(x)b_{ij,k}^0(x) + \frac{c}{T_0}c^0(x)\Big].$$

We take into account the above specifications in the inequality (21.30), so that we are led to the estimate (21.35), so that the proof Theorem 21.4 is concluded. ■

Our last result is also a result of the continuous dependence of the solution to the problem (21.11), (21.9) and (21.10) with respect to the thermoelastic coefficients. The result is also based on the estimate (21.30) from Theorem 21.3.

For this purpose, as above, we will consider two solutions of the form (21.32), which correspond to the same body force, dipolar body force and heat supply, to the same boundary data and to the same initial data, but to different thermoelastic coefficients:

$$\left(C_{ijmn}^{(1)}, G_{ijmn}^{(1)}, F_{ijmnr}^{(1)}, B_{ijmn}^{(1)}, D_{ijmnr}^{(1)}, A_{ijkmnr}^{(1)}, a_{ij}^{(1)}, b_{ij}^{(1)}, c_{ijk}^{(1)}, k_{ij}^{(1)}, c^{(1)}\right)$$
$$\left(C_{ijmn}^{(1)} + \mathcal{C}_{ijmn}, G_{ijmn}^{(1)} + \mathcal{G}_{ijrs}, F_{ijmnr}^{(1)} + \mathcal{F}_{ijmnr},\right.$$
$$B_{ijmn}^{(1)} + \mathcal{B}_{ijrs}, D_{ijmnr}^{(1)} + \mathcal{D}_{ijmnr}, A_{ijkmnr}^{(1)} + \mathcal{A}_{ijkmnr},$$
$$\left. a_{ij}^{(1)} + \alpha_{ij}, b_{ij}^{(1)} + \beta_{ij}, c_{ijk}^{(1)} + \delta_{ijk}, k_{ij}^{(1)} + \kappa_{ij}, c^{(1)} + \zeta\right).$$

Assume that the perturbations $(v_i,\ \psi_{ij},\ \chi)$ satisfy the restrictions (21.29).

Theorem 21.5. *Any solution $(u_i,\ \varphi_{ij},\ \theta)$ of the initial boundary value problem defined by (21.11), (21.9) and (21.10) that satisfies the condition*

$$\int_0^{t_*}\int_B \left(u_{i,j}u_{i,j} + u_{i,jk}u_{i,jk} + \dot{u}_{i,j}\dot{u}_{i,j} + \varphi_{ij,k}\varphi_{ij,k} + \varphi_{ij,km}\varphi_{ij,km}\right.$$
$$\left. + \dot{\varphi}_{ij,k}\dot{\varphi}_{ij,k} + \theta_{,j}\theta_{,j} + \theta_{,jk}\theta_{,jk} + \dot{\theta}^2\right) \mathrm{d}V\mathrm{d}s \le M_7^2,$$

depends continuously on the thermoelastic coefficients with regards to the measure

$$\int_B \left[\varrho v_i(t) v_i(t) + I_{jk}\psi_{jr}(t)\psi_{kr}(t)\right] \mathrm{d}V +$$
$$+ \int_0^t \int_B \frac{1}{T_0} k_{ij} \left(\int_0^s \chi_{,i}(\tau)\mathrm{d}\tau \right) \left(\int_0^s \chi_{,j}(\tau)\mathrm{d}\tau \right) \mathrm{d}V\mathrm{d}s,$$

on the interval $[0, t_*/2]$.

Proof. A straightforward calculation proves that the perturbations $(v_i,\ \psi_{ij},\ \chi)$ of two solutions verify the equations of motion and the equation of energy with the following body force, dipolar body force and heat supply:

$$f_i = \Big[\left(\mathcal{C}_{ijmn} + \mathcal{C}_{ijmn}\right) \varepsilon^{(2)}_{mn} + \left(\mathcal{G}_{ijmn} + \mathcal{B}_{ijmn}\right) \gamma^{(2)}_{mn} +$$
$$+ \left(\mathcal{F}_{ijmnr} + \mathcal{D}_{ijmnr}\right) \chi^{(2)}_{mnr} - \left(\alpha_{ij} + \beta_{ij}\right) \theta^{(2)} \Big]_{,j},$$
$$g_{jk} = \Big[\mathcal{F}_{ijkmn}\varepsilon^{(2)}_{mn} + \mathcal{D}_{mnijk}\gamma^{(2)}_{mn} + \mathcal{A}_{ijkmnr}\chi^{(2)}_{mnr} - \delta_{ijk}\theta^{(2)} \Big]_{,i}$$
$$+ \mathcal{G}_{jkmn}\varepsilon^{(2)}_{mn} + \mathcal{B}_{jkmn}\gamma_{mn} + \mathcal{D}_{jkmnr}\chi^{(2)}_{mnr} - \beta_{ij}\theta^{(2)}$$
$$\frac{r}{T_0} = \alpha_{ij}\varepsilon^{(2)}_{ij} + \beta_{ij}\gamma^{(2)}_{ij} + \delta_{ijk}\chi^{(2)}_{ijk} + \frac{\zeta}{T_0}\theta^{(2)} - \frac{1}{T_0}\left(\kappa_{ij}\theta^{(2)}_{,j}\right)_{,i}.$$

In this way, we have transformed the problem into an analogous one to that from Theorem 21.4. Therefore, we can use the estimates (21.35) and (21.30) to obtain the continuous dependence result with regards to body force, dipolar body force and heat supply. ■

21.4 Conclusion

As can be seen from our above-mentioned approaches, it was possible to demonstrate both the theorem of uniqueness of the solution and three results of the continuous dependence of the solution in low hypotheses. Due to the use of Lagrange's identity, it was possible to obtain these results without recourse to any conservation law or to any boundedness assumptions on the thermoelastic coefficients. In other studies devoted to mixed initial boundary value problems, similar to that consisting of (21.11), (21.9) and (21.10), the theorems regarding the uniqueness of the solution or the continuous dependence of the solutions are proven in more restrictive conditions. As an example, Iesan and Quintanilla (2005) have obtained the existence of the solution by assuming that the Helmholtz's free energy per mass is a positive definite quadratic form. It can be concluded that our results on the continuous dependence of solutions are natural generalizations of the results of Wilkes (1980) and Marin (2010).

References

Abbas IA (2014a) Eigenvalue approach for an unbounded medium with a spherical cavity based upon two-temperature generalized thermoelastic theory. Journal of Mechanical Science and Technology 28(10):4193 – 4198

Abbas IA (2014b) A gn model based upon two-temperature generalized thermoelastic theory in an unbounded medium with a spherical cavity. Applied Mathematics and Computation 245:108 – 115

Altenbach H, Öchsner A (eds) (2010) Cellular and porous materials in structures and processes, CISM Courses and Lectures, vol 521. Springer, Wien

Eringen AC (1972) Theory of micromorphic materials with memory. International Journal of Engineering Science 10(7):623 – 641

Eringen AC (1990) Theory of thermo-microstretch elastic solids. International Journal of Engineering Science 28(12):1291 – 1301

Eringen AC (1999) Microcontinuum Field Theories, vol I. Foundations and Solids. Springer, New York

Fried E, Gurtin ME (2007) Thermomechanics of the interface between a body and its environment. Continuum Mechanics and Thermodynamics 19(5):253–271

Green AE, Laws N (1972) On the entropy production inequality. Archive for Rational Mechanics and Analysis 45:47–53

Green AE, Rivlin RS (1964) Multipolar continuum mechanics. Archive for Rational Mechanics and Analysis 17(2):113–147

Iesan D, Quintanilla R (2005) Thermal stresses in microstretch bodies. International Journal of Engineering Science 43(11):885–907

Marin M (1997) Cesaro means in thermoelasticity of dipolar bodies. Acta Mechanica 122(1-4):155–168

Marin M (1998) Contributions on uniqueness in thermoelastodynamics of bodies with voids. Revista Ciencias Matematicas (Havana) 16(2):101–109

Marin M (2010) Lagrange identity method for microstretch thermoelastic materials. Journal of Mathematical Analysis and Applications 363(1):275 – 286

Marin M, Öchsner A (2017) The effect of a dipolar structure on the holder stability in green-naghdi thermoelasticity. Continuum Mechanics and Thermodynamics 29(6):1365–1374

Marin M, Öchsner A (2018) Complements of Higher Mathematics. Springer, Cham

Mindlin RD (1964) Micro-structure in linear elasticity. Archive for Rational Mechanics and Analysis 16:51–78

Othman MIA (2003) State-space approach to generalized thermoelasticity plane waves with two relaxation times under dependence of the modulus of elasticity on reference temperature. Canadian Journal of Physics 81(12):1403–1418

Sharma JN, Othman MIA (2007) Effect of rotation on generalized thermo-viscoelastic rayleighâ€“lamb waves. International Journal of Solids and Structures 44(13):4243 – 4255

Wilkes N (1980) Continuous dependence and instability in linear thermoelasticity. SIAM Journal on Mathematical Analysis 11(2):292–299

Chapter 22
Theory and Computation of Nonlinear Damage Accumulation for Lifetime Prediction

Anton Matzenmiller & Ulrich Kroll

Abstract Nonlinear damage accumulation is modelled for the lifetime prediction in order to capture the loading sequence effect, which is the influence of the chronological order of the loading values on the lifetime. The prediction results from the solution of the damage evolution equation, which is defined according to the theory of continuum damage mechanics and applied together with a cohesive zone model for structural adhesive joints. The damage model consists of a creep and fatigue damage part, both taking into account the influence of the mean stress and the load multiaxiality on the predicted time to rupture. The analytical investigation of the model shows the meaning of the model parameters and propose their identification by means of tests with static and constant amplitude loading. In order to capture the loading sequence effect by nonlinear damage accumulation, the fatigue damage part is enhanced with a factor, which influences the predicted lifetime due to variable amplitude loading in the case of pure fatigue damage, while the prediction for constant amplitude loading is unaffected. The influences of the enhancement on the predicted lifetime and the damage evolution are discussed. The comparison of lifetimes with numerical predictions proves the validity of the proposed approach.

Keywords: Damage mechanics · Lifetime prediction · Adhesive joints

22.1 Introduction

Components in engineering applications suffer sustained mechanical service loading. In the particular case, where two constant values exist for each the local maxima and minima, which e. g. is the case for the harmonic, sawtooth wave or triangle

Anton Matzenmiller · Ulrich Kroll
Institute of Mechanics, Department of Mechanical Engineering, University of Kassel, Mönchebergstraße 7, 34125 Kassel, Germany,
e-mail: amat@uni-kassel.de, ulrich.kroll@uni-kassel.de

B. E. Abali et al. (eds.), *New Achievements in Continuum Mechanics and Thermodynamics*, Advanced Structured Materials 108,
https://doi.org/10.1007/978-3-030-13307-8_22

wave function, service loading is called constant amplitude (CA) loading, otherwise variable amplitude (VA) loading. Fatigue damage is contributed due to each loading cycle, which may be defined by the load reversals of two neighbouring local maxima or minima of the service loading and the corresponding values in between. In high cycle fatigue (HCF), fatigue damage is accumulated over a large number of loading cycles – usually more than several tens of thousands up to millions – and causes the gradual degradation of the integrity (load bearing capacity) at a material point. After a certain time, the material lifetime, the complete integrity of the material point is lost and a local crack initiates. The loading continues and the crack grows, while other cracks may initiate and grow at different material points. As a consequence of this process of degradation, after some time, the structural lifetime, there remains no load bearing capacity of the structure, which leads to fatal failure of the component. Therefore, the operation time of the component must not exceed its lifetime, which has to be ensured by lifetime prediction.

The lifetime prediction for components and structures is generally performed by use of technical codes and guidelines, e. g. Normenausschuss Bauwesen (NABau) im DIN (2010, 2011); Rennert et al (2012) for steel and aluminium components. These guidelines propose empirical methods, which are highly adapted to particular applications and make use of a variety of simplifying assumptions for the process of fatigue. Most of the assumptions apply superposition, which leads to linearity. Therefore, nonlinear phenomena are not captured, although they may have a great influence on the material and the structural lifetime. One of these nonlinear phenomena is the loading sequence effect, which is the influence of the chronological order of the loading values on the lifetime. The sequence effect is modelled with so called nonlinear damage accumulation, for which the damage increments cannot be easily superimposed, which is usually the case in common procedures for lifetime prediction. Nevertheless, several lifetime prediction methods introduce influencing factors for the consideration of nonlinear phenomena for the particular component, material and application, e. g. influencing factors for temperature, surface condition, loading sequence, mean stress, frequency, multiaxiality etc. Such methods cannot be easily transferred to different materials and structures and are not able to be applied in general. As a consequence, different lifetime prediction methods have been proposed for various materials, components and applications. Furthermore, because of this lack of generality, there exist no commonly accepted lifetime prediction methods for a number of joining techniques such as adhesive bonding.

Continuum damage mechanics (CDM) strives to overcome this shortcoming. In contrast to conventional methods, the lifetime prediction for the structure is an outcome of the lifetime prediction for each material point and the consideration of the whole loading process. The representation of all material phenomena results from the definition of the constitutive equations for the stress and the internal variables in the framework of continuum mechanics. Thereby, one internal variable represents material damage, which evolution equation is adapted to the description of the fatigue process in order to predict the material lifetime.

The lifetime prediction with CDM started with the approach in Kachanov (1958) for the prediction of the creep rupture time of brittle materials in the uniaxial case by means of the definition of the so called continuity, which stands for structural integrity. In Lemaitre and Chaboche (1975), first, damage is defined as the contrary variable to the continuity. Second, the approach in Kachanov (1958) is extended for nonlinear creep damage accumulation and, third, a creep-fatigue damage model is proposed, for which the formalism of the creep damage approach in Kachanov (1958) is transferred to fatigue damage. The creep-fatigue damage model in Lemaitre and Chaboche (1975) is a differential equation in terms of differential damage, time and loading cycles. The model extension for multiaxial loading and plastic damage is proposed in Lemaitre (1979), where the inclusion of the damage theory into thermodynamics of irreversible processes is also addressed. The predicted creep-fatigue damage interaction of the proposed theory is presented in Lemaitre and Plumtree (1979); Cailletaud and Levaillant (1984) and Cailletaud et al (1984), where two-level loadings with pure creep and pure fatigue levels are considered. In Chaboche (1981), the theory of CDM and its application for lifetime prediction are reviewed, accompanied by further studies of creep-fatigue damage interaction and one of the first approaches for the application of CDM for anisotropic damage, which has been also initially investigated in Murakami and Ohno (1981). In Chaboche (1978); Lemaitre (1984), an enhancement of the model in Lemaitre and Chaboche (1975) for the consideration of nonproportional loading is presented. The developed methods and approaches of CDM for lifetime prediction are reviewed again in Lemaitre (1984); Chaboche (1987); Krajcinovic and Lemaitre (1987); Chaboche (1988a,b). In Chaboche and Lesne (1988), the main features of the approach in Lemaitre and Chaboche (1975) are reviewed and discussed. The lifetime prediction with CDM has been further developed in Paas et al (1993) and applied in Lemaitre and Doghri (1994). All the results mentioned before are part of the monographs Lemaitre and Chaboche (1990); Lemaitre (1996); Lemaitre and Desmorat (2005). In Lemaitre and Desmorat (2005), the extension of the proposed approaches with criterions suitable for fatigue damage evolution due to nonproportional loading is mentioned. A short review of the theory and a discussion of improvements for future investigations are presented in Chaboche (2003). In Chaboche (2011), several models for nonlinear damage accumulation are discussed, including the model in Lemaitre and Chaboche (1975) and Chaboche and Lesne (1988).

In CDM, the solution of the damage differential equation results in the predicted number of cycles until rupture for a given stress level. This result is called stress-number (S-N) model, which e. g. takes the form of the Basquin equation (Basquin, 1910) and represents the influence of the amplitude of mechanical CA loading on the lifetime. Other influences of CA loading are related to mean stress, frequency, difference of tension and compression as well as multiaxiality. As in the case of the influence of the amplitude, all these influences are considered by the solution of the damage differential equation. Even the influence of nonproportional (out of phase) loading can be taken into account, which may result from different phases of the stress components. For VA loading, it is well known that also the load-level sequence has an effect on the lifetime (Lemaitre and Chaboche, 1990). The modelling of this effect by nonlinear damage accumulation is of great importance for the warranty of fatigue durability due to lifetime prediction, since linear damage accumulation generally leads to a significant overestimation of the lifetime for complex VA loading cases, based on load spectra and standardised loadings from measurements during service (Chaboche, 2011, p. 50). This observation is also made for adhesives, see Erpolat et al (2004). Hence, the loading sequence effect must be considered in order to prevent the overestimation of the lifetime of adhesively bonded components and possible disastrous consequences.

In this contribution, a creep-fatigue damage model is presented and its consideration of the loading sequence effect by nonlinear damage accumulation is explained in detail. Although the model at hand is applied through a cohesive zone model for the lifetime prediction of adhesively bonded joints, the general characteristics for the consideration of the loading sequence effect can be directly transferred and applied for the lifetime prediction of various materials.

22.2 Modelling of Damage Growth

The theory of CDM is transferred to the cohesive zone model, relating the separation $\boldsymbol{\Delta}$ via the constitutive equation to traction $\mathbf{t} = \left[t_{\mathrm{t}}\ t_{\mathrm{b}}\ t_{\mathrm{n}}\right]^{\mathsf{T}}$, which consists of the tangential t_{t}, binormal t_{b} and normal stress component t_{n} and is calculated according to effective stress concept (Rabotnov, 1963, 1969) for the multiaxial case (Murakami and Ohno, 1981; Lemaitre and Chaboche, 1990):

$$\mathbf{t} = (1 - D)\tilde{\mathbf{t}}\ ,\ D \in [0, 1]\ . \tag{22.1}$$

The effective traction $\tilde{\mathbf{t}} = \tilde{\mathbf{t}}(\boldsymbol{\Delta})$ represents the material behaviour without consideration of damage D. Every damage free model can be used for the effective traction $\tilde{\mathbf{t}}$, e. g. constitutive equations for a (visco-)elastic-(visco-)plastic cohesive model. The damage free state is characterised by $D = 0$. Mechanical loading causes initiation and growth of voids and, thus, increase of damage, which results in $D > 0$. If mechanical loading is further applied, then this process continues until local rupture at $D = 1$. For this damage evolution, a differential equation must be defined, which has to be suitable for the particular case of application. The additive split of the damage increment $\mathrm{d}D$ is proposed in Lemaitre and Chaboche (1975) for lifetime prediction: damage consists of the creep $\mathrm{d}D_{\mathrm{c}}$ and fatigue damage part $\mathrm{d}D_{\mathrm{f}}$, both caused by sustained loading:

$$\mathrm{d}D = \mathrm{d}D_{\mathrm{c}} + \mathrm{d}D_{\mathrm{f}}\ . \tag{22.2}$$

Based on this idea, Eq. (22.2) is reformulated in Matzenmiller and Kurnatowski (2012); Kroll and Matzenmiller (2017) in order to define a differential equation in time according to the general framework of continuum mechanics with internal variables (Truesdell and Toupin, 1960; Coleman and Gurtin, 1967):

$$\dot{D} = \dot{D}_{\mathrm{c}} + \dot{D}_{\mathrm{f}}\ . \tag{22.3}$$

For the lifetime prediction in the case of long-term sustained static and cyclic loading, the creep $\dot{D}_c$ and fatigue damage evolution $\dot{D}_f$ must be specified. In view of the additive split of creep and fatigue damage in Eq. (22.3), creep damage should primarily evolve due to creep loading while fatigue damage should mainly evolve due to loading cycles.

22.2.1 Creep Damage Evolution

For the creep damage evolution $\dot{D}_c$ in Eq. (22.3), the following model is proposed in Matzenmiller and Kurnatowski (2012) and based on the uniaxial version in Kachanov (1958):

$$\dot{D}_c = \frac{1}{c_0}\left(\frac{\langle \sigma_{eqc} - \sigma_{dc} \rangle}{\sigma_{ref}(1-D)}\right)^n \quad , \quad c_0 = 1\,\mathrm{s}\,. \tag{22.4}$$

Creep damage evolves due to the following equivalent stress, which depends on the tractions in Eq. (22.1) and reads

$$\sigma_{eqc} = \sqrt{\langle b_{1c} t_n^2 + b_{2c} t_n + t_t^2 + t_b^2 \rangle}\,, \tag{22.5}$$

where positive tractions $t_i > 0$, $i = \mathrm{t}, \mathrm{b}, \mathrm{n}$ are assumed for simplicity. Hence, alternating stress and pressure are not considered in this contribution but are addressed in Kroll and Matzenmiller (2017); Kroll (2018). The parameters b_{1c} and b_{2c} in Eq. (22.5) take into account the multiaxiality of the loading (Kroll and Matzenmiller, 2017; Kroll, 2018). The constant c_0 in Eq. (22.4) is introduced for consistent units. The MACAULAY operator $\langle x \rangle = (x + |x|)/2$ in Eq. (22.4) results in no creep damage evolution, if the loading is below the creep limit σ_{dc}. The meaning of the two remaining creep damage model parameters n and σ_{ref} is demonstrated by means of creep loading, for which the tractions in Eq. (22.1) are constant: $t_i = \mathrm{const.}$, $i = \mathrm{t}, \mathrm{b}, \mathrm{n}$. If creep loading results in pure creep damage in Eq. (22.3), i. e. $\dot{D}_f = 0$, $D = D_c$, then Eq. (22.4) yields

$$\dot{D}_c = \frac{1}{c_0}\left(\frac{\langle \sigma_{eqc} - \sigma_{dc} \rangle}{\sigma_{ref}(1-D_c)}\right)^n \,. \tag{22.6}$$

If separation and integration are applicable, then the characteristics of the damage differential equation for lifetime prediction arise from three solutions, which are obtained by use of different limits.

For the first solution of Eq. (22.6), separation and integration of the damage equation (22.6) from zero damage $D = 0$ at time $t = 0$ until total failure $D = 1$ at rupture time t_R for creep loading $\sigma_{eqc} = \mathrm{const.} > \sigma_{dc}$ results in the following expression, cf. Kachanov (1958):

$$\int_0^1 (1 - D_\mathrm{c})^n \mathrm{d}D_\mathrm{c} = \int_0^{t_\mathrm{R}} \frac{1}{c_0} \left(\frac{\sigma_\mathrm{eqc} - \sigma_\mathrm{dc}}{\sigma_\mathrm{ref}} \right)^n \mathrm{d}t \Rightarrow t_\mathrm{R} = \frac{c_0}{n+1} \left(\frac{\sigma_\mathrm{ref}}{\sigma_\mathrm{eqc} - \sigma_\mathrm{dc}} \right)^n . \tag{22.7}$$

Application of the logarithm and rearrangement of terms yields the double logarithmic straight line

$$\ln\left(\sigma_\mathrm{eqc} - \sigma_\mathrm{dc}\right) = -\frac{1}{n} \ln \frac{t_\mathrm{R}}{c_0} + \ln \frac{\sigma_\mathrm{ref}}{(n+1)^{\frac{1}{n}}} . \tag{22.8}$$

The meaning of the creep damage parameters n and σ_ref becomes apparent in Eq. (22.8), which is illustrated in Fig. 22.1(a): Parameter n determines the slope of the double logarithmic straight line and σ_ref stands for the ordinate value for fixed n.

The second solution of Eq. (22.6) is performed for a virgin material with $D = 0$ at $t = 0$ until damage D at time t for creep loading $\sigma_\mathrm{eqc} = \text{const.}$, $\sigma_\mathrm{eqc} > \sigma_\mathrm{dc}$, see Lemaitre and Chaboche (1975):

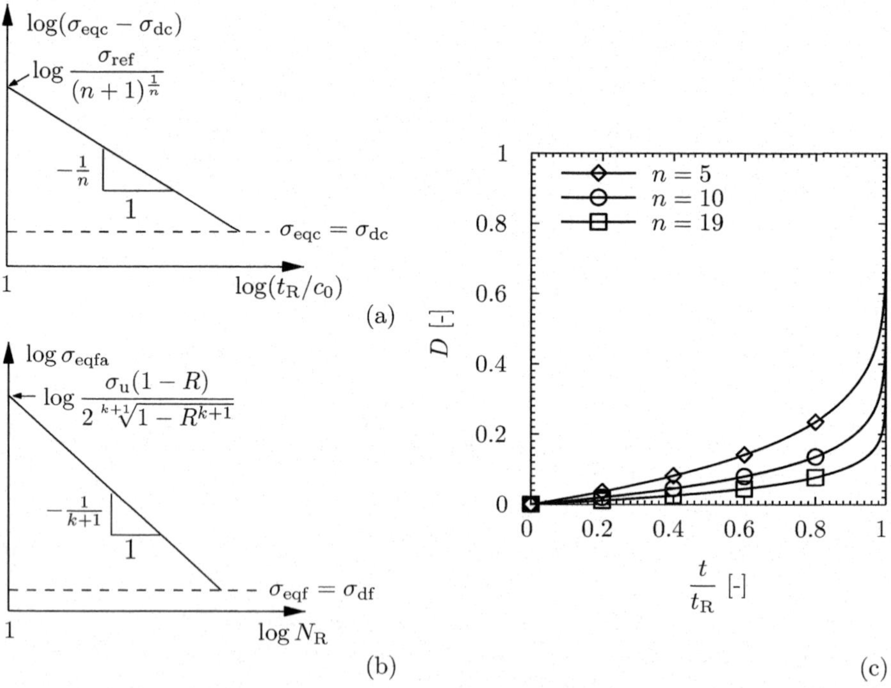

Fig. 22.1 Influence of parameters: (a) influence of creep damage parameters on predicted time to rupture; (b) influence of fatigue damage parameters on predicted number of cycles to rupture; (c) influence of parameter n on damage evolution over normalised time according to Eq. (22.9)

$$\int_0^D (1-D_\mathrm{c})^n \,\mathrm{d}D_\mathrm{c} = \int_0^t \frac{1}{c_0}\left(\frac{\sigma_\mathrm{eqc}-\sigma_\mathrm{dc}}{\sigma_\mathrm{ref}}\right)^n \mathrm{d}\tilde{t} \Rightarrow D = 1-\left(1-\frac{t}{t_\mathrm{R}}\right)^{\frac{1}{n+1}} . \tag{22.9}$$

Equation (22.7) has been used for the substitution of rupture time t_R in Eq. (22.9), which is illustrated in Fig. 22.1(c): The parameter n influences the evolution of damage D over normalised time t/t_R between the fixed start and end point at $D = 0$ and $D = 1$. In contrast, the parameters σ_dc and σ_ref do not influence the curve at all.

The third solution of Eq. (22.6) is obtained by separation and integration from damage D_{i-1} at time t_{i-1} to new damage state D_i due to loading $\sigma_{\mathrm{eqc},i} = \mathrm{const.}$, acting over time Δt_i (Lemaitre and Plumtree, 1979):

$$\int_{D_{i-1}}^{D_i} (1-D_\mathrm{c})^n \,\mathrm{d}D_\mathrm{c} = \int_{t_{i-1}}^{t_{i-1}+\Delta t_i} \frac{1}{c_0}\left(\frac{\sigma_{\mathrm{eqc},i}-\sigma_\mathrm{dc}}{\sigma_\mathrm{ref}}\right)^n \mathrm{d}t , \tag{22.10}$$

$$D_i = 1-\left((1-D_{i-1})^{n+1} - \frac{\Delta t_i}{t_{\mathrm{R}i}}\right)^{\frac{1}{n+1}} . \tag{22.11}$$

In Eq. (22.11), $t_{\mathrm{R}i}$ denotes the time to rupture, if creep loading with $\sigma_{\mathrm{eqc},i}$ is applied from the undamaged state until total failure, i. e. Eq. (22.7) with $t_{\mathrm{R}i}$ instead of t_R and $\sigma_{\mathrm{eqc},i}$ instead of σ_eqc. Eq. (22.11) represents the actual damage value D_i after the so called load level or load block $i = 1, ..., K$ of a K-level or K-block creep loading sequence in form of a recurrence relation.

22.2.2 *Fatigue Damage Evolution*

The following model for fatigue damage evolution $\dot{D}_\mathrm{f}$ is based on the approach in terms of loading cycles in Lemaitre (1979) and proposed in Matzenmiller and Kurnatowski (2012) for the approach in Eq. (22.3):

$$\dot{D}_\mathrm{f} = \left(\frac{\langle\sigma_\mathrm{eqf}-\sigma_\mathrm{df}\rangle}{(\sigma_\mathrm{u}-\sigma_\mathrm{df})(1-D)}\right)^k \frac{\langle\dot{\sigma}_\mathrm{eqf}\rangle}{\sigma_\mathrm{u}-\sigma_\mathrm{df}} . \tag{22.12}$$

The equivalent stress depends on the tractions in Eq. (22.1) and is defined as

$$\sigma_\mathrm{eqf} = \sqrt{\langle b_{1\mathrm{f}} t_\mathrm{n}^2 + b_{2\mathrm{f}} t_\mathrm{n} + t_\mathrm{t}^2 + t_\mathrm{b}^2\rangle} , \tag{22.13}$$

whereby positive tractions $t_\mathrm{i} > 0$, $i = \mathrm{t, b, n}$ are considered for simplicity. As mentioned before in Sect. 22.2.1, alternating stress and pressure are not considered here but are addressed in Kroll and Matzenmiller (2017); Kroll (2018). Note that $\dot{D}_\mathrm{f} = 0$, if $\dot{\sigma}_\mathrm{eqf} \le 0$, which includes creep loading $t_\mathrm{i} = \mathrm{const.}$, $i = \mathrm{t, b, n}$, yielding pure creep damage evolution, which has been assumed for Eq. (22.6). The parameters $b_{1\mathrm{f}}$ and $b_{2\mathrm{f}}$ in Eq. (22.13) take the multiaxiality of the loading into consideration (Kroll and Matzenmiller, 2017; Kroll, 2018). Obviously, the parameter σ_{df} in Eq. (22.12) represents the fatigue limit. The meaning of the two remaining fatigue damage parameters k and σ_{u} becomes apparent, if pure fatigue damage is assumed, i. e. $\dot{D}_{\mathrm{c}} = 0$, $D = D_{\mathrm{f}}$. Then, Eqs. (22.3) and (22.12) are represented by

$$(1 - D_{\mathrm{f}})^k \mathrm{d}D_{\mathrm{f}} = \frac{1}{(\sigma_{\mathrm{u}} - \sigma_{\mathrm{df}})^{k+1}} \langle \sigma_{\mathrm{eqf}} - \sigma_{\mathrm{df}} \rangle^k \langle \mathrm{d}\sigma_{\mathrm{eqf}} \rangle \,, \tag{22.14}$$

where separation of variables has been applied. As in the previous Sect. 22.2.1, the characteristics of the damage equation (22.14) result from three solutions in form of integrations with different limits.

The first solution results from the integration of damage free material with $D = 0$ until rupture at $D = 1$ due to the periodic loading $\sigma_{\mathrm{eqf}}(t) = \sigma_{\mathrm{eqf}}(t + T)$ with smallest period T as well as local and global minimum $\sigma_{\mathrm{eqfmin}} = \min \sigma_{\mathrm{eqf}}$ and maximum $\sigma_{\mathrm{eqfmax}} = \max \sigma_{\mathrm{eqf}}$. An example of such a loading is the harmonic function

$$\sigma_{\mathrm{eqf}} = \sigma_{\mathrm{eqfm}} + \sigma_{\mathrm{eqfa}} \sin(2\pi f t) \tag{22.15}$$

with mean stress σ_{eqfm}, stress amplitude σ_{eqfa} and frequency $f = 1/T$. Consequently, Eq. (22.14) can be integrated over period T, which corresponds to the integration over a stress cycle, consisting of the stress values within the periodic time. The stress cycle results in a damage increment ΔD, thus, Eq. (22.14) becomes

$$\int\limits_{D}^{D+\Delta D} (1 - D_{\mathrm{f}})^k \mathrm{d}D_{\mathrm{f}} = \frac{1}{(\sigma_{\mathrm{u}} - \sigma_{\mathrm{df}})^{k+1}} \oint\limits_{\sigma_{\mathrm{eqf}}} \langle \sigma_{\mathrm{eqf}} - \sigma_{\mathrm{df}} \rangle^k \langle \mathrm{d}\sigma_{\mathrm{eqf}} \rangle \,. \tag{22.16}$$

The Macaulay operator in Eq. (22.16) results in $\langle \mathrm{d}\sigma_{\mathrm{eqf}} \rangle = 0$, if σ_{eqf} decreases, which is the case for the integration from σ_{eqfmax} to σ_{eqfmin}. Additionally, the simplifying assumption $\sigma_{\mathrm{eqf}} > \sigma_{\mathrm{df}}$ is applied in the following. Furthermore, the number of periodic load repetitions N is introduced, which is called cycle number and described as dimensionless time (Paas et al, 1993). Hence, Eq. (22.16) becomes

$$\int\limits_{D}^{D+\Delta D} (1 - D_{\mathrm{f}})^k \mathrm{d}D_{\mathrm{f}} = \frac{1}{(\sigma_{\mathrm{u}} - \sigma_{\mathrm{df}})^{k+1}} \int\limits_{N}^{N+1} \int\limits_{\sigma_{\mathrm{eqfmin}}}^{\sigma_{\mathrm{eqfmax}}} (\sigma_{\mathrm{eqf}} - \sigma_{\mathrm{df}})^k \, \mathrm{d}\sigma_{\mathrm{eqf}} \mathrm{d}\tilde{N} \,. \tag{22.17}$$

For simplicity and without loss of generality, it is assumed that rupture occurs immediately after a certain cycle. Then, Eq. (22.17) results in the following expression for the number of cycles to rupture N_{R}:

$$\int_0^1 (1-D_\mathrm{f})^k \mathrm{d}D_\mathrm{f} = \frac{1}{(\sigma_\mathrm{u}-\sigma_\mathrm{df})^{k+1}} \int_0^{N_\mathrm{R}} \int_{\sigma_\mathrm{eqfmin}}^{\sigma_\mathrm{eqfmax}} (\sigma_\mathrm{eqf}-\sigma_\mathrm{df})^k \,\mathrm{d}\sigma_\mathrm{eqf}\mathrm{d}\tilde{N}, \quad (22.18)$$

$$N_\mathrm{R} = \frac{(\sigma_\mathrm{u}-\sigma_\mathrm{df})^{k+1}}{(\sigma_\mathrm{eqfmax}-\sigma_\mathrm{df})^{k+1} - (\sigma_\mathrm{eqfmin}-\sigma_\mathrm{df})^{k+1}}\,. \quad (22.19)$$

In the case of cyclic loading given by Eq. (22.15), the load ratio $R = \sigma_\mathrm{eqfmin}/\sigma_\mathrm{eqfmax}$ may be introduced. If $\sigma_\mathrm{df} = 0$, Eq. (22.19) is equivalent to the most common S-N model known as the Basquin equation (Basquin, 1910):

$$\ln\sigma_\mathrm{eqfa} = -\frac{1}{k+1}\ln N_\mathrm{R} + \ln\frac{\sigma_\mathrm{u}(1-R)}{2\sqrt[k+1]{1-R^{k+1}}}\,. \quad (22.20)$$

The influence of mean stress on the rupture time is considered by the creep damage part in Eq. (22.4) and also by the fatigue damage part in Eq. (22.12), since the stress amplitude for a given number of cycles to rupture depends on the load ratio R in Eq. (22.20) for pure fatigue damage. The following solution for damage over normalised loading cycles is almost similar to Eq. (22.9): The integration from the damage free state $D = 0$ at cycle $\tilde{N} = 0$ until damage D at cycle $\tilde{N} = N$ for constant amplitude loading in Eq. (22.15) with $\sigma_\mathrm{eqf} > \sigma_\mathrm{df}$ results in

$$\int_0^D (1-D_\mathrm{f})^k \mathrm{d}D_\mathrm{f} = \int_0^N \int_{\sigma_\mathrm{eqfmin}}^{\sigma_\mathrm{eqfmax}} \frac{(\sigma_\mathrm{eqf}-\sigma_\mathrm{df})^k}{(\sigma_\mathrm{u}-\sigma_\mathrm{df})^{k+1}} \mathrm{d}\sigma_\mathrm{eqf}\mathrm{d}\tilde{N} \Rightarrow D = 1-\left(1-\frac{N}{N_\mathrm{R}}\right)^{\frac{1}{k+1}}. \quad (22.21)$$

Equation (22.19) has been used for the substitution of the number of loading cycles until rupture N_R. Damage in Eq. (22.21) is almost similar to the expression in Eq. (22.9). Hence, parameter k has the same influence as parameter n, illustrated in Fig. 22.1(c).

The third solution is obtained by integration from damage D_{i-1} at cycle N_{i-1} to new damage state D_i due to constant amplitude loading with minimum $\sigma_{\mathrm{eqfmin},i}$ and maximum $\sigma_{\mathrm{eqfmax},i}$ over ΔN_i cycles (Lemaitre and Plumtree, 1979):

$$\int_{D_{i-1}}^{D_i} (1-D_\mathrm{f})^k \mathrm{d}D_\mathrm{f} = \frac{1}{(\sigma_\mathrm{u}-\sigma_\mathrm{df})^{k+1}} \int_{N_{i-1}}^{N_{i-1}+\Delta N_i} \int_{\sigma_{\mathrm{eqfmin},i}}^{\sigma_{\mathrm{eqfmax},i}} (\sigma_\mathrm{eqf}-\sigma_\mathrm{df})^k \,\mathrm{d}\sigma_\mathrm{eqf}\mathrm{d}\tilde{N}, \quad (22.22)$$

$$D_i = 1-\left((1-D_{i-1})^{k+1} - \frac{\Delta N_i}{N_{\mathrm{R}i}}\right)^{\frac{1}{k+1}}. \quad (22.23)$$

The number of cycles to rupture $N_{\mathrm{R}i}$ in Eq. (22.23) denotes the lifetime, if constant amplitude loading with minimum $\sigma_{\mathrm{eqfmin},i}$ and maximum $\sigma_{\mathrm{eqfmax},i}$ is applied from the damage free state until rupture, i. e. Eq. (22.19) with $N_{\mathrm{R}i}$ instead of N_R, $\sigma_{\mathrm{eqfmin},i}$ instead of σ_eqfmin and $\sigma_{\mathrm{eqfmax},i}$ instead of σ_eqfmax. Similar to Eq. (22.11), which results from pure creep damage, Eq. (22.23) is also a recurrence relation for the calculation of pure fatigue damage for K-level constant amplitude loading.

22.3 Damage Accumulation

If a certain load level of a K-level loading is applied, then the time spent on that load level results in a corresponding damage increment. Damage accumulation is the model property, which defines, how a certain damage increment is contributed to the actual amount of damage. If the damage accumulation contains the sum of the damage increments, then the damage accumulation is called linear, otherwise nonlinear. The property of linearity refers to the commutativity of the sum, which is equal to the superposition of the damage increments regardless of their chronological appearance. Thus, a model, which exhibits linear damage accumulation, does not account for the loading sequence effect, while a model with nonlinear damage accumulation does. This is illustrated by means of the two-level creep loading until rupture in Fig. 22.2.

The loadings σ_1 and σ_2 in Fig. 22.2(a) lead to the rupture times t_{R1} and t_{R2}. In the top illustration of Fig. 22.2(b), σ_1 is applied from $t = 0$ until $t = \Delta t_1 = 0.5 t_{R1}$, followed by σ_2 from $t = \Delta t_1$ until rupture, which is observed after $\Delta t_2 = 0.5 t_{R2}$, so $t_{R12} = \Delta t_1 + \Delta t_2$. This is a High-Low (HL) creep loading sequence, since $\sigma_1 > \sigma_2$. In the case of linear damage accumulation, the same rupture time is observed, if the chronological order of the loadings is interchanged, see the Low-High (LH) creep loading scenario in the bottom illustration of Fig. 22.2(b). The loading sequence effect is not captured, since the HL and LH loading result in the same rupture time: $t_{R12} = t_{R21}$. This fact is represented by the ROBINSON rule for linear creep damage accumulation (Robinson, 1938) with $K = 2$ in the case of Fig. 22.2(b):

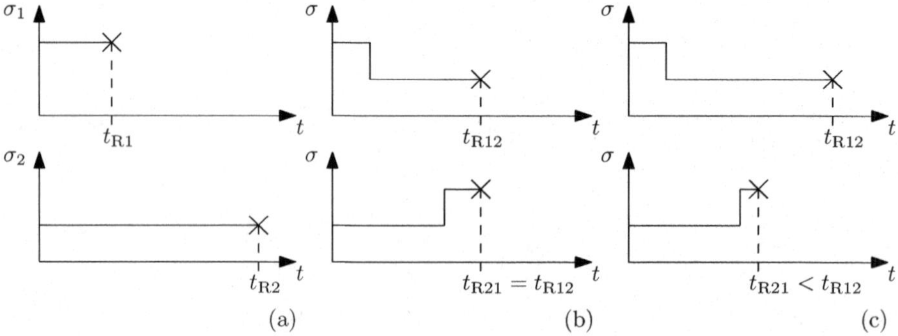

Fig. 22.2 Illustration of damage accumulation by means of two level loading until rupture ($\times$): (a) considered loadings σ_1 and σ_2 leading to rupture times t_{R1} and t_{R2}; (b) example for linear damage accumulation; (c) example for nonlinear damage accumulation

$$\sum_{i=1}^{K} \frac{\Delta t_i}{t_{\mathrm{R}i}} = 1\,. \tag{22.24}$$

The time spent on load level i is denoted with Δt_i, while $t_{\mathrm{R}i}$ is the rupture time in the case, where the loading of load level i is applied as one-level loading from the undamaged material at $t = 0$ until rupture. Fig. 22.2(c) shows nonlinear damage accumulation, where the rupture times for both scenarios are not equal. A loading sequence effect is observed and Eq. (22.24) is not fulfilled anymore: The sum of creep life ratios is not always equal to one, see e.g. Pavlou (2001), where LH creep loading test data are shown, which correspond to the bottom illustration in Fig. 22.2(c).

The theory of creep damage accumulation is directly transferred to fatigue loading. Linear fatigue damage accumulation is represented by the PALMGREN–MINER rule (Palmgren, 1924; Miner, 1945), where ΔN_i denotes the number of cycles spent on load level i and $N_{\mathrm{R}i}$ represents the number of cycles to rupture in the case, where the loading of load level i is applied as one-level loading from $t = 0$ until rupture:

$$\sum_{i=1}^{K} \frac{\Delta N_i}{N_{\mathrm{R}i}} = 1\,. \tag{22.25}$$

As the ROBINSON rule in Eq. (22.24), the PALMGREN–MINER rule in Eq. (22.25) also does not take the loading sequence effect into consideration.

22.3.1 Creep and Fatigue Damage Accumulation

In the following, the accumulation of damage is presented due to the evolutions in the cases of pure creep and fatigue given in Eqs. (22.6) and (22.14). Damage due to creep loading over time Δt_1 is according to Eq. (22.9) and Eq. (22.11) with $i = 1$ and $D_0 = 0$

$$D_1 = 1 - \left(1 - \frac{\Delta t_1}{t_{\mathrm{R}1}}\right)^{\frac{1}{n+1}}\,. \tag{22.26}$$

Equation (22.11) with $i = 2$ is applied for the case, where a second creep loading follows and acts over time Δt_2:

$$D_2 = 1 - \left((1 - D_1)^{n+1} - \frac{\Delta t_2}{t_{\mathrm{R}2}}\right)^{\frac{1}{n+1}}\,. \tag{22.27}$$

Inserting of Eq. (22.26) into (22.27) results in

$$D_2 = 1 - \left(1 - \sum_{i=1}^{2} \frac{\Delta t_i}{t_{\mathrm{R}i}}\right)^{\frac{1}{n+1}}\,. \tag{22.28}$$

This process can be continued, which results in the amount of pure creep damage D_j after load level j:

$$D_j = 1 - \left(1 - \sum_{i=1}^{j} \frac{\Delta t_i}{t_{\mathrm{R}i}}\right)^{\frac{1}{n+1}} . \tag{22.29}$$

Occurrence of rupture after K load levels results the ROBINSON rule in Eq. (22.24):

$$D_K = 1 = 1 - \left(1 - \sum_{i=1}^{K} \frac{\Delta t_i}{t_{\mathrm{R}i}}\right)^{\frac{1}{n+1}} \Leftrightarrow \sum_{i=1}^{K} \frac{\Delta t_i}{t_{\mathrm{R}i}} = 1 . \tag{22.30}$$

Thus, the damage evolution equation (22.6) represents linear damage accumulation and cannot take the loading sequence effect into consideration.

The same result is obtained for the case of pure fatigue damage by consideration of Eq. (22.14). Damage after load level j is

$$D_j = 1 - \left(1 - \sum_{i=1}^{j} \frac{\Delta N_i}{N_{\mathrm{R}i}}\right)^{\frac{1}{k+1}} , \tag{22.31}$$

where Eqs. (22.21) and (22.23) have been taken into account and a similar procedure to Eqs. (22.26) to (22.29) is applied. Equation (22.31) has the same form as Eq. (22.29). Consequently, if rupture occurs after the application of K load levels, then

$$D_K = 1 = 1 - \left(1 - \sum_{i=1}^{K} \frac{\Delta N_i}{N_{\mathrm{R}i}}\right)^{\frac{1}{k+1}} \Leftrightarrow \sum_{i=1}^{K} \frac{\Delta N_i}{N_{\mathrm{R}i}} = 1 , \tag{22.32}$$

which is the PALMGREN–MINER rule, given by Eq. (22.25).

According to the results in Eqs. (22.30) and (22.32), the damage evolution equations for pure creep and pure fatigue damage exhibit linear damage accumulation. This is a result of the separability of the differential equations: All damage differential equations, which are separable, lead to linear damage accumulation (Ostergren and Krempl, 1979; Todinov, 2001). A model with nonlinear damage evolution as in Eqs. (22.9) or (22.21) does not automatically represent nonlinear damage accumulation (Chaboche and Lesne, 1988; Lemaitre and Chaboche, 1990).

22.3.2 *Modelling of Nonlinear Damage Accumulation*

Since separability of the damage equation leads to linear damage accumulation, a model must not be separable in order to represent nonlinear damage accumulation and to account for the loading sequence effect. In the previous section, only pure creep and pure fatigue damage evolution given by Eqs. (22.6) and (22.14) have been considered. But, the differential equation (22.3) with the approaches in Eqs. (22.4) and (22.12) is not separable for $n \neq k$ and, in this case, represents nonlinear damage accumulation through nonlinear damage interaction.

However, in practice, two circumstances lead to linear or almost linear damage accumulation of the model given by Eqs. (22.3), (22.4) and (22.12). First, the result of the parameter identification and numerical optimisation for the prediction in the case of creep and CA loading may be $n = k$, see Kroll and Matzenmiller (2017) or Cavdar et al (2018), which leads to separability and linear damage accumulation. Second, even if the identification and optimisation for other test data results in $n \neq k$, the nonlinearity of damage interaction and of the resulting damage accumulation may only weakly pronounced, see Kroll and Matzenmiller (2016). This is explained by negligible creep damage: The magnitude of the factor in (22.12) with the time derivative of the equivalent fatigue stress is of second order for usual HCF loading, e. g. Eq. (22.15) with $\sigma_{\mathrm{eqfm}} = \sigma_{\mathrm{eqfa}} = f = 10$ and pure shear $t_{\mathrm{n}} = 0$, $\sigma_{\mathrm{eqc}} = \sigma_{\mathrm{eqf}}$. In addition, if the terms with the exponent n and k in Eqs. (22.4) and (22.12) have the same orders of magnitude as a result of the identification, see Kroll and Matzenmiller (2015, 2016, 2017) and Kroll (2018), then creep damage evolution appears to be negligible compared to fatigue damage evolution: $\dot{D}_{\mathrm{c}} \approx 0$. But $\dot{D}_{\mathrm{c}} = 0$ is the condition to match Eqs. (22.3), (22.4), (22.12) with Eq. (22.14), which is separable and represents linear damage accumulation according to Eq. (22.32). Thus, if creep damage is negligible compared to fatigue damage for usual HCF loading, then the model Eqs. (22.3), (22.4) and (22.12) represent linear damage accumulation.

As a consequence, it appears reasonable to set the focus on the modelling of nonlinear fatigue damage accumulation, which is related to the fatigue damage evolution in Eq. (22.12). The following approach is proposed in Kroll and Matzenmiller (2017):

$$\dot{D}_{\mathrm{f}} = \frac{\left(1-(1-D)^{k+1}\right)^{\alpha}}{1-\alpha}\left(\frac{\langle\sigma_{\mathrm{eqf}}-\sigma_{\mathrm{df}}\rangle}{(\sigma_{\mathrm{u}}-\sigma_{\mathrm{df}})(1-D)}\right)^{k}\frac{\langle\dot{\sigma}_{\mathrm{eqf}}\rangle}{\sigma_{\mathrm{u}}-\sigma_{\mathrm{df}}}\,. \tag{22.33}$$

By contrast with Eq. (22.12), the fatigue damage evolution in Eq. (22.33) has an additional factor, which contains the variable α for nonlinear fatigue damage accumulation. In the following, α is assumed to be a function, which is constant for any integration over a stress cycle, resulting in a damage increment according to Eqs. (22.16), (22.17):

$$\alpha = \alpha(\sigma_{\mathrm{eqfmin}}, \sigma_{\mathrm{eqfmax}})\,. \tag{22.34}$$

The following approach for α in Kroll and Matzenmiller (2017) is based on the proposals in Chaboche and Lesne (1988) and Do et al (2015):

$$\alpha = \alpha_{\mathrm{p2}}\left(\left\langle 1-\alpha_{\mathrm{p1}}\left\langle\frac{\sigma_{\mathrm{eqfa}}-\sigma_{\mathrm{df}}}{\tau_{\mathrm{u}}-\sigma_{\mathrm{eqfmax}}}\right\rangle-\alpha_{\mathrm{p3}}\right\rangle+\alpha_{\mathrm{p3}}\right)\,. \tag{22.35}$$

In Eq. (22.35), α_{p1}, α_{p2} and α_{p3} are parameters. The first parameter α_{p1} controls the damage interaction (Chaboche and Lesne, 1988), which is not addressed in this contribution. The switching variable α_{p2} is used in order to set $\alpha_{\mathrm{p2}} = 0 \to \alpha = 0$, which results in linear fatigue damage accumulation, see Subsect. 22.3.3. The third parameter α_{p3} represents a lower boundary for α for the stabilisation of the numerical treatment (Kroll and Matzenmiller, 2017).

As in Subsect. 22.2.2, three integrations of the damage equation (22.33) will be presented for pure fatigue damage $D = D_{\mathrm{f}}$. Thus, Eq. (22.33) becomes

$$\mathrm{d}D_{\mathrm{f}} = \frac{\left(1-(1-D_{\mathrm{f}})^{k+1}\right)^{\alpha}}{1-\alpha}\left(\frac{\langle\sigma_{\mathrm{eqf}}-\sigma_{\mathrm{df}}\rangle}{(\sigma_{\mathrm{u}}-\sigma_{\mathrm{df}})(1-D_{\mathrm{f}})}\right)^{k}\frac{\langle\mathrm{d}\sigma_{\mathrm{eqf}}\rangle}{\sigma_{\mathrm{u}}-\sigma_{\mathrm{df}}}\,. \qquad (22.36)$$

The first solution is obtained by separation and integration until rupture according to Eqs. (22.15) to (22.18) for CA fatigue loading, for which α is constant due to Eq. (22.34):

$$\int\limits_0^1 \frac{(1-\alpha)(1-D_{\mathrm{f}})^k}{\left(1-(1-D_{\mathrm{f}})^{k+1}\right)^{\alpha}}\mathrm{d}D_{\mathrm{f}} = \int\limits_0^{N_{\mathrm{R}}}\int\limits_{\sigma_{\mathrm{eqfmin}}}^{\sigma_{\mathrm{eqfmax}}}\left(\frac{\sigma_{\mathrm{eqf}}-\sigma_{\mathrm{df}}}{\sigma_{\mathrm{u}}-\sigma_{\mathrm{df}}}\right)^{k}\frac{\mathrm{d}\sigma_{\mathrm{eqf}}}{\sigma_{\mathrm{u}}-\sigma_{\mathrm{df}}}\mathrm{d}N, \qquad (22.37)$$

$$\frac{1}{k+1}\int\limits_0^1 \frac{1-\alpha}{\tilde{D}^{\alpha}}\mathrm{d}\tilde{D} = \int\limits_0^{N_{\mathrm{R}}}\int\limits_{\sigma_{\mathrm{eqfmin}}}^{\sigma_{\mathrm{eqfmax}}}\left(\frac{\sigma_{\mathrm{eqf}}-\sigma_{\mathrm{df}}}{\sigma_{\mathrm{u}}-\sigma_{\mathrm{df}}}\right)^{k}\frac{\mathrm{d}\sigma_{\mathrm{eqf}}}{\sigma_{\mathrm{u}}-\sigma_{\mathrm{df}}}\mathrm{d}N, \qquad (22.38)$$

$$\Rightarrow N_{\mathrm{R}} = \frac{(\sigma_{\mathrm{u}}-\sigma_{\mathrm{df}})^{k+1}}{(\sigma_{\mathrm{eqfmax}}-\sigma_{\mathrm{df}})^{k+1}-(\sigma_{\mathrm{eqfmin}}-\sigma_{\mathrm{df}})^{k+1}}\,, \qquad (22.39)$$

where the following substitution has been used:

$$\tilde{D} = 1-(1-D_{\mathrm{f}})^{k+1}\quad,\quad \mathrm{d}\tilde{D} = (k+1)(1-D_{\mathrm{f}})^{k}\mathrm{d}D_{\mathrm{f}}\,. \qquad (22.40)$$

Since the numbers of cycles to rupture in Eqs. (22.19) and (22.39) are equal, the term with α has no influence on the lifetime for CA loading.

According to the limits for integration in Eq. (22.21), the integration of Eq. (22.36) from the virgin state until a certain amount of damage D for CA loading together with the substitution in Eq. (22.40) and constant α due to Eq. (22.34) yields

$$\int\limits_0^D \frac{(1-\alpha)(1-D_{\mathrm{f}})^k}{\left(1-(1-D_{\mathrm{f}})^{k+1}\right)^{\alpha}}\mathrm{d}D_{\mathrm{f}} = \int\limits_0^{N}\int\limits_{\sigma_{\mathrm{eqfmin}}}^{\sigma_{\mathrm{eqfmax}}}\left(\frac{\langle\sigma_{\mathrm{eqf}}-\sigma_{\mathrm{df}}\rangle}{\sigma_{\mathrm{u}}-\sigma_{\mathrm{df}}}\right)^{k}\frac{\mathrm{d}\sigma_{\mathrm{eqf}}}{\sigma_{\mathrm{u}}-\sigma_{\mathrm{df}}}\mathrm{d}\tilde{N}, \qquad (22.41)$$

$$D = 1-\left(1-\left(\frac{N}{N_{\mathrm{R}}}\right)^{\frac{1}{1-\alpha}}\right)^{\frac{1}{k+1}}\,. \qquad (22.42)$$

As can be seen from the comparison of Eq. (22.42) with Eq. (22.21), in addition to the parameter k, the variable α has also an influence on the course of damage over the cycle ratio.

The last integration is performed for a damage increment according to the limits in Eq. (22.22) and by use of the substitution in Eq. (22.40), which results in the following expression for actual damage D_i, where $\alpha_i = \alpha_i(\sigma_{\mathrm{eqfmin},i}, \sigma_{\mathrm{eqfmax},i})$ is constant for every load level i, cf. Eq. (22.34):

$$\frac{1}{k+1}\int_{\tilde{D}_{i-1}}^{\tilde{D}_i} \frac{1-\alpha_i}{\tilde{D}^{\alpha_i}}\mathrm{d}\tilde{D} = \int_{N_{i-1}}^{N_{i-1}+\Delta N_i}\int_{\sigma_{\mathrm{eqfmin},i}}^{\sigma_{\mathrm{eqfmax},i}} \left(\frac{\sigma_{\mathrm{eqf}}-\sigma_{\mathrm{df}}}{\sigma_{\mathrm{u}}-\sigma_{\mathrm{df}}}\right)^k \frac{\mathrm{d}\sigma_{\mathrm{eqf}}}{\sigma_{\mathrm{u}}-\sigma_{\mathrm{df}}}\mathrm{d}N\,, \tag{22.43}$$

$$D_i = 1 - \left(1 - \left[\left(1-[1-D_{i-1}]^{k+1}\right)^{1-\alpha_i} + \frac{\Delta N_i}{N_{\mathrm{R}i}}\right]^{\frac{1}{1-\alpha_i}}\right)^{\frac{1}{k+1}}. \tag{22.44}$$

The variable α_i in Eq. (22.44) has an additional influence on the actual amount of damage compared to Eq. (22.23), as observed by comparing of Eqs. (22.42) and (22.21).

22.3.3 Discussion of Modelling Approach

In the following, the damage accumulation behaviour of Eq. (22.36) will be analysed as already performed for Eq. (22.14) in Subsect. 22.3.1. According to Eqs. (22.44) and (22.42), damage due to CA fatigue loading over ΔN_1 loading cycles from the undamaged state to damage D_i with $i = 1$ and $D_0 = 0$ is

$$D_1 = 1 - \left(1 - \left(\frac{\Delta N_1}{N_{\mathrm{R1}}}\right)^{\frac{1}{1-\alpha_1}}\right)^{\frac{1}{k+1}}. \tag{22.45}$$

If further loading with a second load level is applied, according to Eq. (22.44), damage after this second load level then is

$$D_2 = 1 - \left(1 - \left[\left(1-[1-D_1]^{k+1}\right)^{1-\alpha_2} + \frac{\Delta N_2}{N_{\mathrm{R2}}}\right]^{\frac{1}{1-\alpha_2}}\right)^{\frac{1}{k+1}}. \tag{22.46}$$

Insertion of Eq. (22.45) into Eq. (22.46) results in

$$D_2 = 1 - \left(1 - \left[\left(\frac{\Delta N_1}{N_{\mathrm{R1}}}\right)^{\frac{1-\alpha_2}{1-\alpha_1}} + \frac{\Delta N_2}{N_{\mathrm{R2}}}\right]^{\frac{1}{1-\alpha_2}}\right)^{\frac{1}{k+1}}. \tag{22.47}$$

If the second load level is applied until rupture, then $D_2 = 1$ in Eq. (22.47), becoming

$$\left(\frac{\Delta N_1}{N_{\text{R}1}}\right)^{\frac{1-\alpha_2}{1-\alpha_1}} + \frac{\Delta N_2}{N_{\text{R}2}} = 1. \tag{22.48}$$

If the second and third level do not result in rupture, then insertion of Eq. (22.47) into Eq. (22.44) for $i = 3$ yields the following amount of damage:

$$D_3 = 1 - \left(1 - \left[\left[\left(\frac{\Delta N_1}{N_{\text{R}1}}\right)^{\frac{1-\alpha_2}{1-\alpha_1}} + \frac{\Delta N_2}{N_{\text{R}2}}\right]^{\frac{1-\alpha_3}{1-\alpha_2}} + \frac{\Delta N_3}{N_{\text{R}3}}\right]^{\frac{1}{1-\alpha_3}}\right)^{\frac{1}{k+1}}. \tag{22.49}$$

If rupture occurs after load level $K = 3$, then $D_3 = 1$ and Eq. (22.49) becomes

$$\left[\left(\frac{\Delta N_1}{N_{\text{R}1}}\right)^{\frac{1-\alpha_2}{1-\alpha_1}} + \frac{\Delta N_2}{N_{\text{R}2}}\right]^{\frac{1-\alpha_3}{1-\alpha_2}} + \frac{\Delta N_3}{N_{\text{R}3}} = 1. \tag{22.50}$$

The preceding steps can be performed for a block loading with arbitrary number of load levels. Damage after load level j is

$$D_j = 1 - \left(1 - \left[\left[\dots\left[\left[\left(\frac{\Delta N_1}{N_{\text{R}1}}\right)^{\frac{1-\alpha_2}{1-\alpha_1}} + \frac{\Delta N_2}{N_{\text{R}2}}\right]^{\frac{1-\alpha_3}{1-\alpha_2}} + \frac{\Delta N_3}{N_{\text{R}3}}\right]^{\frac{1-\alpha_4}{1-\alpha_3}} \dots \right.\right.\right.$$
$$\left.\left.\left. + \frac{\Delta N_{j-1}}{N_{\text{R}j-1}}\right]^{\frac{1-\alpha_j}{1-\alpha_{j-1}}} + \frac{\Delta N_j}{N_{\text{R}j}}\right]^{\frac{1}{1-\alpha_j}}\right)^{\frac{1}{k+1}}. \tag{22.51}$$

If rupture occurs after load level K, then $D_j = D_K = 1$ in Eq. (22.51), which becomes

$$\left[\dots\left[\left[\left(\frac{\Delta N_1}{N_{\text{R}1}}\right)^{\frac{1-\alpha_2}{1-\alpha_1}} + \frac{\Delta N_2}{N_{\text{R}2}}\right]^{\frac{1-\alpha_3}{1-\alpha_2}} + \frac{\Delta N_3}{N_{\text{R}3}}\right]^{\frac{1-\alpha_4}{1-\alpha_3}} \dots \right.$$
$$\left. + \frac{\Delta N_{K-1}}{N_{\text{R}K-1}}\right]^{\frac{1-\alpha_K}{1-\alpha_{K-1}}} + \frac{\Delta N_K}{N_{\text{R}K}} = 1. \tag{22.52}$$

The difference of Eq. (22.52) compared to the PALMGREN–MINER rule in Eq. (22.25) is, that the cycle ratios are not commutatively superimposed, which is nonlinearity of damage accumulation in the sense of noncommutativity of the chronological order of the load levels.

This property becomes apparent, if the PALMGREN–MINER rule in Eq. (22.25) is interpreted as a chain formed by the cycle ratios as summands, then the chain links can be arbitrarily interchanged without any change of the result. In the simplest case of two load levels until rupture, this means commutativity of $\Delta N_1/N_{\text{R}1} + \Delta N_2/N_{\text{R}2} = \Delta N_2/N_{\text{R}2} + \Delta N_1/N_{\text{R}1} = 1$. In the case of Eq. (22.52), the chain links cannot be interchanged without any change of the result. Consider for example the particular case of $K = 2$ load levels in Eq. (22.48), which is equivalent to

$$\frac{\Delta N_1}{N_{\mathrm{R1}}} = \left(1 - \frac{\Delta N_2}{N_{\mathrm{R2}}}\right)^{\frac{1-\alpha_1}{1-\alpha_2}}, \tag{22.53}$$

where the loading $\sigma_{\mathrm{eqf},1}$ with $\sigma_{\mathrm{eqfmin},1}$ and $\sigma_{\mathrm{eqfmax},1}$ of level one is applied first, followed by the loading $\sigma_{\mathrm{eqf},2}$ with $\sigma_{\mathrm{eqfmin},2}$ and $\sigma_{\mathrm{eqfmax},2}$ of level two until rupture. If the chronological order of the load levels is interchanged, then the result is different from Eq. (22.53):

$$\left(\frac{\Delta N_2}{N_{\mathrm{R2}}}\right)^{\frac{1-\alpha_1}{1-\alpha_2}} + \frac{\Delta N_1}{N_{\mathrm{R1}}} = 1 \Leftrightarrow \frac{\Delta N_1}{N_{\mathrm{R1}}} = 1 - \left(\frac{\Delta N_2}{N_{\mathrm{R2}}}\right)^{\frac{1-\alpha_1}{1-\alpha_2}}. \tag{22.54}$$

The same observation holds for the general case with K-level loading in Eq. (22.52), which, therefore, considers the loading sequence effect. Linear damage accumulation results from the case $\alpha_i = \alpha_j$ for all i, j or particularly $\alpha = 0$, then Eq. (22.52) becomes the PALMGREN–MINER rule in Eq. (22.25).

Although nonlinear damage accumulation can be further investigated by means of the general case in Eq. (22.52), the particular case $K = 2$ is much simpler, represented by Eqs. (22.53) and (22.54). If $\alpha_{\mathrm{p2}} = 1$ and

$$1 - \alpha_{\mathrm{p1}} \left\langle (\sigma_{\mathrm{eqfa}} - \sigma_{\mathrm{df}})/(\tau_{\mathrm{u}} - \sigma_{\mathrm{eqfmax}}) \right\rangle > \alpha_{\mathrm{p3}},$$

then the insertion of the approach for α given by Eq. (22.35) into Eqs. (22.53) and (22.54) results in the following expressions, which are independent of the parameter α_{p1} due to the division in the exponent:

$$\frac{\Delta N_1}{N_{\mathrm{R1}}} = \left(1 - \frac{\Delta N_2}{N_{\mathrm{R2}}}\right)^{\frac{\left\langle (\sigma_{\mathrm{eqfa},1} - \sigma_{\mathrm{df}})/(\tau_{\mathrm{u}} - \sigma_{\mathrm{eqfmax},1}) \right\rangle}{\left\langle (\sigma_{\mathrm{eqfa},2} - \sigma_{\mathrm{df}})/(\tau_{\mathrm{u}} - \sigma_{\mathrm{eqfmax},2}) \right\rangle}}, \tag{22.55}$$

$$\frac{\Delta N_1}{N_{\mathrm{R1}}} = 1 - \left(\frac{\Delta N_2}{N_{\mathrm{R2}}}\right)^{\frac{\left\langle (\sigma_{\mathrm{eqfa},1} - \sigma_{\mathrm{df}})/(\tau_{\mathrm{u}} - \sigma_{\mathrm{eqfmax},1}) \right\rangle}{\left\langle (\sigma_{\mathrm{eqfa},2} - \sigma_{\mathrm{df}})/(\tau_{\mathrm{u}} - \sigma_{\mathrm{eqfmax},2}) \right\rangle}}. \tag{22.56}$$

The same observation is made for the damage accumulation by Eq. (22.52). Hence, in the case of pure fatigue, the parameter α_{p1} in Eq. (22.35) does not influence the damage accumulation behaviour, but only influences the course of damage over the cycle ratio for one-level loading until rupture, which becomes apparent by insertion of Eq. (22.35) into Eq. (22.42).

22.4 Parameter Identification

Creep loading is the special case of CA loading, where the global extrema coincide. Thus, the parameter identification starts with the direct determination of the creep damage parameters σ_{dc}, σ_{ref} and n by means of creep tests until rupture as illustrated in Fig. 22.1(a), where pure shear is considered: $t_{\mathrm{n}} = t_{\mathrm{b}} = 0$, $\sigma_{\mathrm{eqc}} = \sigma_{\mathrm{eqf}} = t_{\mathrm{t}}$. Afterwards, the fatigue damage parameters σ_{df}, σ_{u} and k are identified directly by means of tests with CA pure shear fatigue loading as shown in Fig. 22.1(b). Since the illustration in Fig. 22.1(b) is only true for negligible creep damage, the fatigue damage parameters σ_{u} and k need to be numerically optimised. However, as mentioned in Subsect. 22.3.2, creep damage is expected to be negligible for usual HCF loading and the values of the identified parameters are expected to be marginally changed by the optimisation. After the determination of the damage model parameters by means of creep and CA fatigue tests with pure shear, the shear-tension interaction parameters $b_{1\mathrm{c}}$, $b_{2\mathrm{c}}$, $b_{1\mathrm{f}}$ and $b_{2\mathrm{f}}$ are directly identified and numerically optimised by means of multiaxial creep and CA fatigue tests. The validity of the parameter identification has been shown in Kroll and Matzenmiller (2015, 2016, 2017) and Kroll (2018).

As pointed out in Subsect. 22.3.2, the modelling approach for nonlinear fatigue damage accumulation has no influence on the lifetime prediction for CA loading in the case of pure fatigue, see Eq. (22.39). Thus, material dependent parameters in the approach for the function $\alpha = \alpha(\sigma_{\mathrm{eqfmin}}, \sigma_{\mathrm{eqfmax}})$ in Eq. (22.34) may be determined by means of one of the following two sets of test data: The first set contains the results of two-level loading tests, for which the parameters in the function α have to be determined in order to fit Eqs. (22.53) and (22.54) with the test data. Thereby, the parameters to be identified must have an influence on the rupture time, contrary to parameter α_1 in the approach for α in Eq. (22.35), see Eqs. (22.55) and (22.56). An alternative way for the identification procedure is the evaluation of the damage values over the cycle ratio as the second set of test data. In this case, the parameters in α have to be identified in order to fit Eq. (22.42) with the data points. Unfortunately, damage is an internal variable and cannot be measured directly, but only indirectly by means of several methods, see Lemaitre and Dufailly (1987). The main difficulties are the reasonable choice and the reliable detection of the quantity, which is supposed to represent damage best for the particular case of application.

For the previous explained direct identification of the parameters in α, creep damage must be negligible compared to fatigue damage, which is the necessary condition for the application of the Eqs. (22.36) to (22.54). If this is not the case, then the fatigue damage parameters σ_{u}, σ_{df} and k as well as the parameters in the function α have to be numerically and simultaneously optimised in a last step by means of S-N curves and rupture times from two-level loadings or indirect damage measurements.

The tests for the parameter identification must provide an almost homogeneous state of stress in the bonding layer. Therefore, the adhesive layer thickness must be very thin in order to describe its constitutive behaviour by a cohesive zone model, see e.g. Su et al (2004). This applies to structural adhesives, which usually have a bonding layer thickness between 0.1 and 1 mm. Second, the geometries of the adherends as well as the load application in the test setup must be appropriate such that peeling and inhomogeneous shear loading is minimised. Examples are the specimens with single lap and butt joints and the corresponding test setups in Schlimmer et al (2012); Schneider et al (2012) and Cavdar and Meschut (2017) providing almost homogeneous shear-, tension- and combined shear and tension in the thin structural adhesive bonding layer.

22.5 Application to Lifetime Prediction for Adhesive Joints

In the following, the application in Kroll and Matzenmiller (2017) and Kroll (2018) of the proposed damage model is presented for the lifetime prediction of butt-bonded thin steel tubes under two-level torsional loading with force control until rupture. The bonding layer consists of the thermosetting, one-component, ductile modified epoxy structural adhesive BETAMATE™1496V and has a thickness of 0.3 mm only. Therefore, it is modelled as a cohesive zone, which suffers pure shear stress t_t due to the torsional loading of the specimen. The preparation of the adherends, the bonding procedure and the test setup are detailly described in Cavdar and Meschut (2017).

The test results in Fig. 22.3 show nonlinear damage accumulation due to LH loading, while the data points for HL loading almost coincide with the dashed line, representing linear damage accumulation. The location of some data points outside the unit square may be explained by the scatter of the data, because this phenomenon exists for both the LH and HL sequence. Another explanation refers to an effect of the firstly applied load level, which in some situations, mostly for LH loading, increases the fatigue life for the subsequently applied load level. For an amount of loading cycles below $\Delta N_2 = 0.5 N_{R2}$, this effect may even increase the original CA fatigue life.

Since the model equations in this contribution do not account for scatter of data and the previously mentioned effect of lifetime increase, only the unit square is il-

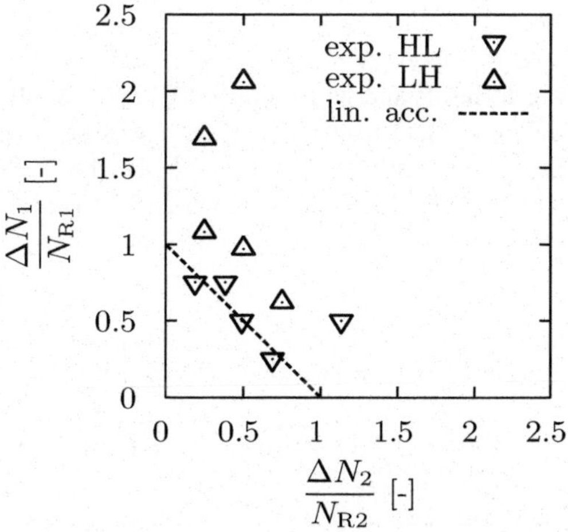

Fig. 22.3 Experimental test data (Cavdar and Meschut, 2017) due to High-Low (HL) and Low-High (LH) two-level shear loading with $f = 10$ Hz; number of cycles to rupture for high loading (in MPa) $t_{t1} = 16.97 + 13.88 \sin(2\pi f t)$ is $N_{R1} = 5663$ and for low loading (in MPa) $t_{t2} = 13.2 + 10.8 \sin(2\pi f t)$ is $N_{R2} = 713246$

lustrated in Fig. 22.4, where lifetime predictions are compared with the test results. The predictions denoted with ODE1, ODE2 and ODE3 result from the implicit nu-

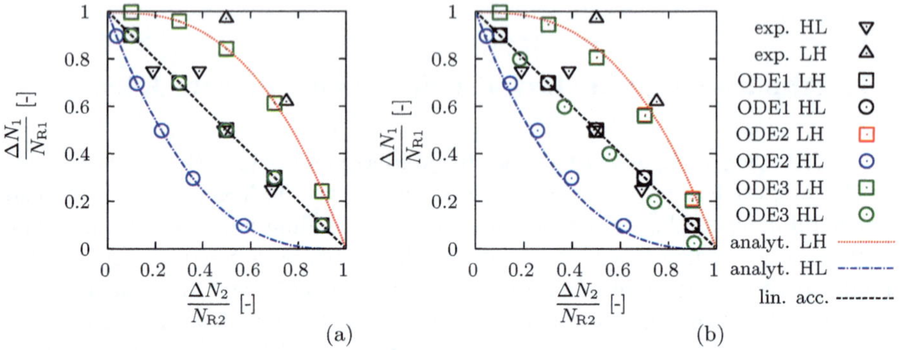

Fig. 22.4 Experimental test data and prediction for two-level loading in Fig. 22.3 for (a) pure fatigue ($\dot{D}_c = 0$) and (b) creep-fatigue; ODE1: Eqs. (22.3), (22.4), (22.5), (22.12), (22.13); ODE2: Eqs. (22.3), (22.4), (22.5), (22.13) (22.33), (22.35); ODE3: Eqs. (22.3), (22.4), (22.5), (22.13), (22.35), (22.57), (22.58); analyt. HL: Eq. (22.55); analyt. LH: Eq. (22.56)

merical solution of the corresponding damage evolution equation, for which the two step backward differentiation formula together with the second order finite difference approximation are applied, see Kroll and Matzenmiller (2015). Besides the presented model equations in the previous sections, the following expressions are additionally considered for ODE3:

$$\dot{D}_{\text{f}} = \frac{\left(1-(1-D)^{k+1}\right)^{\alpha_{\min}}}{1-\alpha_{\min}} \left(\frac{\langle\sigma_{\text{eqf}}-\sigma_{\text{df}}\rangle}{(\sigma_{\text{u}}-\sigma_{\text{df}})(1-D)}\right)^{k} \frac{\langle\dot{\sigma}_{\text{eqf}}\rangle}{\sigma_{\text{u}}-\sigma_{\text{df}}}, \tag{22.57}$$

$$\alpha_{\min}(t) = \min_{0\leq\tau\leq t} \alpha(\tau) \; . \tag{22.58}$$

The damage model parameters in Tabelle 22.1 are identified and optimised according to Kroll and Matzenmiller (2017); Kroll (2018) as explained in Sect. 22.4 by means of data from tests of the steel tube specimen under torsional creep and CA fatigue loading, see Cavdar and Meschut (2017). The parameters in the equivalent stresses given by Eqs. (22.5) and (22.13) are irrelevant in the following due to pure shear loading. The factor for nonlinear damage accumulation is active for all numer-

Table 22.1 Identified damage model parameters in Kroll and Matzenmiller (2017); Kroll (2018) by means of test data in Cavdar and Meschut (2017)

σ_{dc} [MPa]	σ_{ref} [MPa]	n [-]	σ_{df} [MPa]	σ_{u} [MPa]	k [-]	α_{p1} [-]	α_{p2} [-]	α_{p3} [-]	τ_{u} [MPa]
0	51	19	0	49	19	1	1	-10	39

ical computations: $\alpha_{\mathrm{p2}} = 1$. Since no data for the course of damage over cycles or time are considered, the identification of parameter α_{p1} is obsolete, thus: $\alpha_{\mathrm{p1}} = 1$. The parameter $\alpha_{\mathrm{p3}} = -10$ in Eq. (22.35) is applied for numerical stability, see Kroll and Matzenmiller (2017) and Kroll (2018) for details. The ultimate shear strength $\tau_{\mathrm{u}} = 39$ MPa is identified by means of torsion tests of the steel tube specimen under quasistatic loading Cavdar and Meschut (2017).

The amplitude and maximum of σ_{eqf} in Eq. (22.35) are computed from stress values in the last half cycle, which consists of the most recently passed local minimum and maximum stress as well as the values in between. A detailed description and an alternative formulation are given in Kroll and Matzenmiller (2017); Kroll (2018).

Due to the consideration of pure fatigue damage in Fig. 22.4(a), the numerical predictions for ODE1, ODE2 and ODE3 match with the analytical expressions in the previous sections. Since ODE1 becomes a separable differential equation for pure fatigue damage, it represents linear damage accumulation and the predictions match with the dashed black line, represented by Eq. (22.32) for $K = 2$. The predictions with ODE2 match with Eqs. (22.55) and (22.56), which verifies the numerical solution. The predictions with ODE2 are in better agreement with the test data compared to ODE1 for the LH sequence. But they are worse for the HL sequence. As ODE3 matches with the dotted red line and with the black dashed line, it provides the best results for the representation of the damage accumulation of the structural adhesive at hand. Thereby, Eq. (22.58) ensures α to be constant for the HL scenario in order to match the prediction with the dashed black line for linear accumulation, since $\alpha_1 = -1.252$ and $\alpha_2 = 0.169$, cf. Fig. 22.3.

In Fig. 22.4(b), ODE1 represents nonlinear damage accumulation again because of $n = k$, which is a result of the identification for creep and CA fatigue loading, see Table 22.1. The predictions with ODE2 differ from the curves represented by Eqs. (22.55) and (22.56), which shows the influence of creep damage. The predictions with ODE3 are still in best agreement with the test data.

22.6 Conclusion

In this contribution, nonlinear damage accumulation is modelled by a damage differential equation with a creep and fatigue part for the consideration of the loading sequence effect. Three integrations of the differential equation are presented for one-level loading, where different integration limits are applied in order to get three results: the time to rupture and the course of damage over time, both for the initially undamaged state, as well as the damage increment due to a load level. These three results are used to demonstrate the damage accumulation behaviour for pure creep and fatigue damage. They represent the Robinson and Palmgren–Miner rules of linear damage accumulation, if the underlying damage differential equation is separable and confirm that nonlinear damage evolution not automatically implies nonlinear damage accumulation.

The magnitude of creep damage is low compared to fatigue damage for usual HCF loading, which justifies the consideration of pure fatigue damage for the modelling of nonlinear damage accumulation. An adaption is proposed of the fatigue damage part by a factor, which contains the approach for variable α and influences the course of damage over time and the lifetime prediction due to VA loading, but not the prediction due to CA loading. The definition of the variable α ensures separability of the damage equation for each load level in the case of pure fatigue damage, but inseparability in general. Thus, nonlinear damage accumulation can be modelled independently from the S-N approach for CA loading. This is shown by the three integrations of the damage differential equation with different limits, providing a procedure for the identification of the parameters in the function for α.

Two particular approaches for α are validated by means of the comparison of numerical lifetime predictions with test data for an adhesive layer subjected to shear loading. Since the predictions are in good agreement with the test data, it is concluded that the loading sequence effect is well captured for pure shear loading.

Nonlinear creep damage accumulation is not considered, but it can be modelled similarly to the fatigue damage accumulation (Kroll, 2018). Although the influences of mean stress, pressure, multiaxiality and nonproportionality on the lifetime for CA loading and the corresponding considerations by the damage model are addressed in Kroll and Matzenmiller (2015, 2017) and Kroll (2018), they have to be experimentally investigated for VA loading of structural adhesives in order to further validate the proposed approach for the variable α according to the presented theory. Therefore, multiaxial tests with two-level loading and different mean stresses as well as indirect damage measurement for one-level loading have to be performed.

References

Basquin OH (1910) The exponential law of endurance tests. ASTM 10:625–630

Cailletaud G, Levaillant C (1984) Creep-fatigue life prediction: what about initiation? Nucl Eng Des 84:279–292

Cailletaud G, Nouailhas D, Grattier J, Levaillant C, Mottot M, Tortel J, Escavarage C, Héliot J, Kang S (1984) A review of creep-fatigue life prediction methods: identification and extrapolation to long term and low strain cyclic loading. Nucl Eng Des 83:267–278

Cavdar S, Meschut G (2017) Analyse der Schwingfestigkeit geklebter Stahlverbindungen unter mehrkanaliger Belastung. Tech. rep., Forschungsvereinigung Stahlanwendung e.V., number = FOSTA-Report P1028, chap 11-12, Düsseldorf

Cavdar S, Kroll U, Meschut G, Matzenmiller A (2018) Experimental characterization and numerical lifetime prediction of adhesively bonded joints under multiaxial fatigue loading. In: Proc Adh Soc 2018

Chaboche JL (1978) Description thermodynamique et phénoménologique de la viscoplasticité cyclique avec endommagement. Dissertation, Université Pierre et Marie Curie, Paris

Chaboche JL (1981) Continuous damage mechanics — A tool to describe phenomena before crack initiation. Nucl Eng Des 64(2):233–247

Chaboche JL (1987) Continuum damage mechanics: present state and future trends. Nucl Eng Des 105:19–33

Chaboche JL (1988a) Continuum damage mechanics: Part I - general concepts. J Appl Mech 55:59–64

Chaboche JL (1988b) Continuum damage mechanics: Part II - damage growth, crack initiation, and crack growth. J Appl Mech 55:65–72

Chaboche JL (2003) Damage Mechanics. In: Karihaloo B, Knauss WG (eds) Fundamental Theories and Mechanisms of Failure, Elsevier, Comprehensive Structural Integrity, vol 2, pp 213–284

Chaboche JL (2011) Cumulative Damage. In: Bathias C, Pineau A (eds) Fatigue of Materials and Structures, John Wiley & Sons

Chaboche JL, Lesne PM (1988) A non-linear continous fatigue damage model. Fatigue Fract Eng Mater Struct 11(1):1–17

Coleman BD, Gurtin ME (1967) Thermodynamics with internal state variables. J Chem Phys 47(2):597–613

Do VNV, Lee CH, Chang KH (2015) High cycle fatigue analysis in presence of residual stresses by using a continuum damage mechanics model. Int J Fatigue 70:51–62

Erpolat S, Ashcroft IA, Crocombe AD, Abdel-Wahab MM (2004) A study of adhesively bonded joints subjected to constant and variable amplitude fatigue. Int J Fatigue 26(11):1189–1196

Kachanov LM (1958) On rupture time under conditions of creep. Izv Akad Nauk Sssr Otdelenie, Otd Tech 8(2631):26–31, in Russian. English translation (1999) Rupture time under creep conditions. Int J Fract 97:11–18.

Krajcinovic D, Lemaitre J (1987) Continuum Damage Mechanics: Theory and Applications. Springer

Kroll U (2018) Modellierung der Schädigungsentwicklung und Lebensdauerprognose für Stahlklebverbindungen unter hochzyklischer Ermüdungsbelastung. Dissertation, Fachbereich Maschinenbau, Universität Kassel

Kroll U, Matzenmiller A (2015) Parameter identification of a damage model for the lifetime prediction of adhesively bonded joints. In: Saanouni K (ed) Damage Mechanics: Theory, Computation and Practice, Trans Tech Publications, Appl Mech Mater, vol 784, pp 300–307

Kroll U, Matzenmiller A (2016) On nonlinear damage accumulation and creep-fatigue interaction of a damage model for the lifetime prediction of adhesively bonded joints. PAMM 16(1):151–152

Kroll U, Matzenmiller A (2017) Analyse der Schwingfestigkeit geklebter Stahlverbindungen unter mehrkanaliger Belastung. Tech. Rep. FOSTA-Report P1028, chap 13-15, Forschungsvereinigung Stahlanwendung e.V., Düsseldorf

Lemaitre J (1979) Damage Modeling for Prediction of Plastic or Creep Fatigue Failure in Structures. In: Proc IASMiRT 1979 (Trans. 5th Intl. Conf. Struct. Mech. React. Tech.)

Lemaitre J (1984) How to use damage mechanics. Nucl Eng Des 80:233–245

Lemaitre J (1996) A Course on Damage Mechanics. Springer

Lemaitre J, Chaboche JL (1975) A Non-Linear Model of Creep-Fatigue Damage Cumulation and Interaction. In: Hult J (ed) Mechanics of Visco-Elastic Media and Bodies, Springer, pp 291–301

Lemaitre J, Chaboche JL (1990) Mechanics of Solid Materials. Cambridge University Press

Lemaitre J, Desmorat R (2005) Engineering Damage Mechanics. Springer

Lemaitre J, Doghri I (1994) Damage 90: a post processor for crack initiation. Comput Methods Appl Mech Eng 115(3–4):197–232

Lemaitre J, Dufailly J (1987) Damage measurements. Eng Fract Mech 28(5–6):643–661

Lemaitre J, Plumtree A (1979) Application of damage concepts to predict creep-fatigue failures. J Eng Mater Technol 101(3):284–292

Matzenmiller A, Kurnatowski B (2012) Schwingfestigkeitsauslegung von geklebten Stahlbauteilen des Fahrzeugbaus unter Belastung mit variablen Amplituden. Tech. Rep. FOSTA-Report P796, pp. 51–101, 133–138, 210–215, Forschungsvereinigung Stahlanwendung e.V., Verlag und Vertriebsgesellschaft mbH, Düsseldorf

Miner MA (1945) Cumulative damage in fatigue. J Appl Mech 67:A159–A164

Murakami S, Ohno N (1981) A Continuum Theory of Creep and Creep Damage. In: Ponter ARS, Hayhurst DR (eds) Creep in Structures, Springer, pp 422–444

Normenausschuss Bauwesen (NABau) im DIN (2010) Eurocode 3: Bemessung und Konstruktion von Stahlbauten - Teil 1-8: Bemessung von Anschlüssen. DIN Deutsches Institut für Normung e. V.

Normenausschuss Bauwesen (NABau) im DIN (2011) Eurocode 9: Bemessung und Konstruktion von Aluminiumtragwerken - Teil 1-3: Ermüdungsbeanspruchte Tragwerke. DIN Deutsches Institut für Normung e. V.

Ostergren WJ, Krempl E (1979) A uniaxial damage accumulation law for time-varying loading including creep-fatigue interaction. J Press Vessel Technol 101:118–124

Paas M, Schreurs P, Brekelmans W (1993) A continuum approach to brittle and fatigue damage: theory and numerical procedures. Int J Solids Struct 30(4):579–599

Palmgren A (1924) Die Lebensdauer von Kugellagern. Zeitschrift des Vereines Deutscher Ingenieure (VDI Zeitschrift) 68(14):339–341

Pavlou D (2001) Creep life prediction under stepwise constant uniaxial stress and temperature conditions. Eng Struct 23(6):656–662

Rabotnov YN (1963) On the Equation of State of Creep. Proc IMechE 1963 (Joint Intl Conf Creep) 178(68):2–117–2–122

Rabotnov YN (1969) Creep rupture. In: Hetenyi M, Vincenti M (eds) Applied Mechanics – Proceedings of the XII International Congress on Applied Mechanics, Springer, pp 342–349

Rennert R, Kullig E, Vormwald M, Esderts A, Siegele D (2012) Rechnerischer Festigkeitsnachweis für Maschinenbauteile aus Stahl, Eisenguss und Aluminiumwerkstoffen: FKM-Richtlinie, 6th edn. Forschungskuratorium Maschinenbau e. V. (FKM), VDMA Verlag

Robinson E (1938) Effect of temperature variation on the creep strength of steels. Trans ASME 160:253–259

Schlimmer M, Hahn O, Hennemann OD, Mihm KM, Jendrny J, Teutenberg D, Brede M, Nagel C (2012) Methodenentwicklung zur Berechnung und Auslegung geklebter Stahlbauteile im Fahrzeugbau bei schwingender Beanspruchung. Tech. Rep. FOSTA-Report P653, Forschungsvereinigung Stahlanwendung e.V., Düsseldorf

Schneider B, Kehlenbeck H, Nagel C (2012) Schwingfestigkeitsauslegung von geklebten Stahlbauteilen des Fahrzeugbaus unter Belastung mit variablen Amplituden. Tech. Rep. FOSTA-Report P796, chap 2, Forschungsvereinigung Stahlanwendung e.V., Düsseldorf

Su C, Wei YJ, Anand L (2004) An elastic–plastic interface constitutive model: application to adhesive joints. Int J Plast 20(12):2063–2081

Todinov M (2001) Necessary and sufficient condition for additivity in the sense of the Palmgren–Miner rule. Comput Mater Sci 21(1):101–110

Truesdell C, Toupin RA (1960) The Classical Field Theories. In: Flügge S (ed) Principles of Classical Mechanics and Field Theory, Handbuch der Physik, vol 3/I, Springer

Chapter 23
A Non-equilibrium Approach Concerning Thermostatics of Schottky Systems

Wolfgang Muschik

Abstract Non-equilibrium processes in Schottky systems generate by projection and relaxation reversible accompanying processes which serve as thermostatic approximations of different accuracy. The compatibility of the accompanying processes with the non-equilibrium ones according to the embedding theorem is shortly discussed.

Keywords: Schottky systems · Contact quantities · Non-equilibrium entropy · Thermostatic approach · Embedding theorem

23.1 Introduction

Once in 1982 after having written his Diplomarbeit (Müller, 1982), we were interested in the following problem (Müller and Muschik, 1983; Muschik and Müller, 1983): In contrast to Rational Thermodynamics—which starts with “bodies”—we considered an “open” multicomponent discrete (Schottky) system for deriving the usual balance equations of mass, momentum, energy and entropy. Of course, evident was that these balance equations are not influenced by closing or opening of the considered discrete system, but the purpose of our procedure was to define quantities, such as the stress tensor, as measurable contact quantities, thus avoiding the more formal definitions of the mixture theory (Müller, 1973).

Also evident was at that time, that discrete systems play a prominent role for the description of technical devices because of their often complicated geometry which prevents the application of a field formulation on account of unknown boundary conditions of the fields. Thus, a global description as discrete systems in contact with

Wolfgang Muschik
Institut für Theoretische Physik, Technische Universität Berlin, Hardenbergstr. 36, D-10623 Berlin, Germany
e-mail: muschik@physik.tu-berlin.de

B. E. Abali et al. (eds.), *New Achievements in Continuum Mechanics and Thermodynamics*, Advanced Structured Materials 108,
https://doi.org/10.1007/978-3-030-13307-8_23

their environments is adequate. Favourable to this description is that thermodynamics of discrete systems is well developed (Muschik, 1993) and thermodynamical concepts—such as exchanges, partitions, state spaces and processes—can be easily formulated. An other reason is that these technical non-equilibrium devices are often successfully treated by use of thermostatic procedures. The question arises: what is the link between non-equilibrium systems and their thermostatic descriptions?

The paper is organized as follows: First of all, some facts of open non-equilibrium Schottky systems are repeated, such as state spaces, processes, First and Second Law, contact and exchange quantities. Subsequently, the thermostatic approximation and its consequences are considered.

23.2 Schottky Systems

A system[1] $\mathcal{G}$, described as undecomposed and homogeneous, which is separated by a partition $\partial\mathcal{G}$ from its environment[2] $\mathcal{G}^\square$ is called a *Schottky system* (Schottky, 1929), if the interaction between $\mathcal{G}$ and $\mathcal{G}^\square$ through $\partial\mathcal{G}$ can be described by

$$\text{heat exchange } \overset{\bullet}{Q}, \quad \text{power exchange } \overset{\bullet}{W}, \quad \text{and material exchange } \overset{\bullet}{\boldsymbol{n}}{}^{e}. \tag{23.1}$$

The power exchange is related to the work variables $\boldsymbol{a}$ of the system

$$\overset{\bullet}{W} = \boldsymbol{A} \cdot \overset{\bullet}{\boldsymbol{a}} \,. \tag{23.2}$$

Here, $\boldsymbol{A}$ are the generalized forces which are as well known as the work variables. Kinetic and potential energy are constant and therefore out of scope. $\overset{\bullet}{Q}$ is measurable by calorimetry and the time rate of the mole numbers due to material exchange $\overset{\bullet}{\boldsymbol{n}}{}^{e}$ by weigh.

23.2.1 State Spaces and Processes

A large state space[3] $\mathcal{Z}$ (Muschik, 1990a) is decomposed into its equilibrium subspace containing the equilibrium variables $\boldsymbol{\Omega}$ and into the non-equilibrium part containing the variables $\boldsymbol{\Xi}$ which are not included in the equilibrium subspace

$$\boldsymbol{Z} = (\boldsymbol{\Omega}, \boldsymbol{\Xi}) \in \mathcal{Z}. \tag{23.3}$$

---

[1] $\mathcal{G} \subset \mathbb{R}^3$ may be a control volume or a "body" without material transport through its surface.

[2] The environment is a heat reservoir, that means, its temperature is not influenced by the exchanges with the system.

[3] Two kinds of state spaces should be distinguished: small and large ones. Small state spaces require constitutive equations, which contain histories of the state space variables, whereas large state spaces avoid such histories.

The *states of equilibrium* $\boldsymbol{\Omega}$ are defined by time independent states of an isolated Schottky system and are determined by the *Zeroth Law*: The state space of a thermal homogeneous Schottky system in equilibrium is spanned by the work variables, the mole numbers and the internal energy

$$\boldsymbol{\Omega} = (\boldsymbol{a}, \boldsymbol{n}, U) \quad \longrightarrow \quad \boldsymbol{Z} = (\boldsymbol{a}, \boldsymbol{n}, U, \boldsymbol{\Xi}). \tag{23.4}$$

A projection $\mathcal{P}$ is introduced which projects the non-equilibrium state $\boldsymbol{Z}$ onto the equilibrium subspace (Muschik, 1993)

$$\mathcal{P}\boldsymbol{Z} = \mathcal{P}\Big(\boldsymbol{a}, \boldsymbol{n}, U, \boldsymbol{\Xi}\Big) = \boldsymbol{\Omega} = \Big(\boldsymbol{a}, \boldsymbol{n}, U\Big) =: \boldsymbol{Z}^* \in \mathcal{P}\mathcal{Z} \tag{23.5}$$

whose projected states are marked by *. If the considered system is in equilibrium, the non-equilibrium variables become dependent on the equilibrium ones

$$\boldsymbol{Z}^{eq} := \Big(\mathcal{P}\boldsymbol{Z}, \boldsymbol{\Xi}(\mathcal{P}\boldsymbol{Z})\Big) =: \mathcal{P}^{eq}\boldsymbol{Z} \in \mathcal{P}^{eq}\mathcal{Z}, \qquad \mathcal{P}^{eq}\boldsymbol{Z} \neq \mathcal{P}\boldsymbol{Z}. \tag{23.6}$$

The projection $\mathcal{P}$ cuts the non-equilibrium variables, whereas $\mathcal{P}^{eq}$ takes their values at equilibrium into account which are not zero in general.

A process, defined as a trajectory on the non-equilibrium state space,

$$\boldsymbol{Z}(t) = \Big(\boldsymbol{a}, \boldsymbol{n}, U, \boldsymbol{\Xi}\Big)(t), \quad t = \text{time} \tag{23.7}$$

generates by projection a trajectory on the equilibrium subspace

$$\mathcal{P}\boldsymbol{Z}(t) = \Big(\boldsymbol{a}, \boldsymbol{n}, U\Big)(t) = \boldsymbol{Z}^*(t) \tag{23.8}$$

which is different from the "equilibrium process"

$$\boldsymbol{Z}^{eq}(t) = \Big(\mathcal{P}\boldsymbol{Z}, \boldsymbol{\Xi}(\mathcal{P}\boldsymbol{Z})\Big)(t) = \Big(\boldsymbol{a}, \boldsymbol{n}, U, \boldsymbol{\Xi}(\boldsymbol{a}, \boldsymbol{n}, U)\Big)(t) \tag{23.9}$$

which is generated by isolation (the equilibrium variables are constant during the irreversible relaxation to equilibrium) of the Schottky system at time t and relaxation into the corresponding equilibrium state keeping the "time" as path parameter of $\boldsymbol{Z}^{eq}(t)$.

Both trajectories—$\mathcal{P}\boldsymbol{Z}(t)$ and $\mathcal{P}^{eq}\boldsymbol{Z}(t)$—are induced by the original irreversible process $\boldsymbol{Z}$. Consequently, they depend on the same equilibrium variables $\Big(\boldsymbol{a}, \boldsymbol{n}, U\Big)$ as $\boldsymbol{Z}$, but in a different way as a comparison of (23.8) with (23.9) depicts. Trajectories on the equilibrium subspace are denoted as *reversible processes*, a bit strange denotation because no "process" with progress in time takes place on the equilibrium subspace. The "time" in (23.8) and (23.9) is generated by projection and relaxation, representing the path parameter along the reversible processes. Both trajectories are also denoted as *accompanying processes* of $\boldsymbol{Z}$(t) (Keller, 1971). Although not existing in nature, reversible processes serve as mathematical closing of the set of "real" (irreversible) processes which are defined as trajectories on the non-equilibrium state space.

23.2.2 The First Law

Up to now, the internal energy was introduced in $(23.4)_1$ as one variable of the equilibrium subspace of a thermally homogeneous[4] Schottky system. The connection between the time rate of the internal energy of the system and the exchange quantities through $\partial\mathcal{G}$ is established by the *First Law*

$$\overset{\bullet}{U} = \overset{\bullet}{Q} + \boldsymbol{h}\cdot \overset{\bullet}{\boldsymbol{n}}{}^{e} + \overset{\bullet}{W} \tag{23.10}$$

which states that the internal energy U of the system should be conserved in isolated Schottky systems.[5] The second term of the RHS of (23.10) originates from the fact that the heat exchange has to be redefined for open systems ($\overset{\bullet}{\boldsymbol{n}}{}^{e} \neq \boldsymbol{0}$) (W. Muschik and S. Gümbel, 1969). Here, $\boldsymbol{h}$ are the molar enthalpies of the chemical components in $\mathcal{G}$. The modified heat exchange which is combined with the material exchange appearing in the First Law (23.10) was used by Haase (1969).

23.2.3 Entropy Rate and Second Law

Considering a discrete system $\mathcal{G}$, a quantity $\boldsymbol{J}$ of $\mathcal{G}$ is called *balanceable*, if its time rate can be decomposed into an exchange $\boldsymbol{\Psi}$ with the environment and a production $\boldsymbol{R}$ in the interior of the system

$$\overset{\bullet}{\boldsymbol{J}} = \boldsymbol{\Psi} + \boldsymbol{R}, \qquad \boldsymbol{\Psi} = \boldsymbol{\Phi} + \varphi\, \overset{\bullet}{\boldsymbol{n}}{}^{e}. \tag{23.11}$$

The exchange is composed of its conductive part $\boldsymbol{\Phi}$ and its convective part $\varphi\, \overset{\bullet}{\boldsymbol{n}}{}^{e}$.

Doubtless, a non-equilibrium entropy of a Schottky system is a balanceable quantity[6]. Presupposing that the power exchange does not contribute to the entropy rate, if the heat exchange is taken into consideration, we obtain according to (23.1) a balance equation

$$\overset{\bullet}{S}(\boldsymbol{Z}) = \frac{1}{\Theta}\overset{\bullet}{Q} + \boldsymbol{s}\cdot \overset{\bullet}{\boldsymbol{n}}{}^{e} + \Sigma. \tag{23.12}$$

Here, $\overset{\bullet}{Q}$ and $\overset{\bullet}{\boldsymbol{n}}{}^{e}$ are the exchange quantities, $\boldsymbol{s}$ the molar entropy, whereas Σ is the entropy production which is not negative according to the *Second Law*

---

[4] Thermally homogeneous means: there are no adiabatic partitions in the interior of the system.

[5] Irreversible processes in isolated systems do not change their internal energy.

[6] about the existence of a "non-equilibrium" entropy see Muschik (2018)

$$\Sigma \geq 0 \longrightarrow \overset{\bullet}{S} \geq \frac{1}{\Theta}\overset{\bullet}{Q} + \boldsymbol{s}\cdot\overset{\bullet}{\boldsymbol{n}}{}^{e} \longrightarrow 0 \geq \oint \left[\frac{1}{\Theta}\overset{\bullet}{Q} + \boldsymbol{s}\cdot\overset{\bullet}{\boldsymbol{n}}{}^{e}\right] dt. \tag{23.13}$$

If the considered Schottky system is adiabatically unique[7] (Muschik, 2009), the entropy rate $\overset{\bullet}{S}$ is the time derivative of a state space function entropy $S(\boldsymbol{Z})$ according to (23.4).

Presupposing that all chemical components in $\mathcal{G}$ have the same temperature Θ, the non-equilibrium molar entropies $\boldsymbol{s}$ in (23.12) are (Kestin, 1979)

$$\boldsymbol{s} = \frac{1}{\Theta}\Big(\boldsymbol{h} - \boldsymbol{\mu}\Big). \tag{23.14}$$

Consequently taking (23.10) and (23.14) into account, the entropy rate (23.12) becomes

$$\overset{\bullet}{S} = \frac{1}{\Theta}\Big(\overset{\bullet}{U} - \boldsymbol{A}\cdot\overset{\bullet}{\boldsymbol{a}} - \boldsymbol{\mu}\cdot\overset{\bullet}{\boldsymbol{n}}{}^{e}\Big) + \Sigma. \tag{23.15}$$

Because the external mole number rates $\overset{\bullet}{\boldsymbol{n}}{}^{e}$ are no state variables[8], but the mole numbers themselves are included in the equilibrium subspace $(23.4)_1$ according to the Zeroth Law, the missing term for generating the mole numbers in (23.15) is hidden in the entropy production

$$\Sigma = -\frac{1}{\Theta}\boldsymbol{\mu}\cdot\overset{\bullet}{\boldsymbol{n}}{}^{i} + \Sigma^0, \tag{23.16}$$

and (23.15) results in

$$\overset{\bullet}{S}(\boldsymbol{Z}) = \frac{1}{\Theta}\Big(\overset{\bullet}{U} - \boldsymbol{A}\cdot\overset{\bullet}{\boldsymbol{a}} - \boldsymbol{\mu}\cdot\overset{\bullet}{\boldsymbol{n}}\Big) + \Sigma^0, \tag{23.17}$$

if the internal mole number changes due to chemical reactions

$$\overset{\bullet}{\boldsymbol{n}}{}^{i} = \overset{\bullet}{\boldsymbol{n}} - \overset{\bullet}{\boldsymbol{n}}{}^{e} \tag{23.18}$$

are taken into account. Because the bracket in (23.17) contains only rates of equilibrium variables, those of the non-equilibrium state variables appear in the entropy production Σ^0 which is established below.

---

[7] Adiabatic uniqueness means that the final equilibrium state of an irreversible relaxation of an isolated system is unique, a statement which is not satisfied in Stochastical Thermodynamics.

[8] because they depend on the environment

23.3 Contact Quantities

23.3.1 *Defining Inequalities*

The non-equilibrium quantities Θ and $\boldsymbol{s}$ in (23.12) are contact quantities (Muschik, 2009; Muschik and Brunk, 1975; Muschik, 1977; Muschik and Brunk, 1977) which are determined by *defining inequalities*[9] (Muschik, 2009, 2018)

$$\left(\frac{1}{\Theta}-\frac{1}{T^{\square}}\right)\overset{\bullet}{Q}\overset{*}{\geq} 0,\quad \left(\frac{\boldsymbol{h}}{\Theta}-\frac{\boldsymbol{h}^{\square}}{T^{\square}}\right)\cdot\overset{\bullet}{\boldsymbol{n}}{}^{e}\overset{*}{\geq} 0,\quad \left(\frac{\boldsymbol{\mu}^{\square}}{T^{\square}}-\frac{\boldsymbol{\mu}}{\Theta}\right)\cdot\overset{\bullet}{\boldsymbol{n}}{}^{e}\overset{*}{\geq} 0. \quad (23.19)$$

In more detail: the non-equilibrium Schottky systen $\mathcal{G}$ is in contact with a thermally homogeneous equilibrium environment $\mathcal{G}^{\square}$ of the thermostatic temperature $T^{\square}$, the molar equilibrium enthalpy $\boldsymbol{h}^{\square}$ and the equilibrium chemical potential $\boldsymbol{\mu}^{\square}$. All the chemical components in $\mathcal{G}^{\square}$ have the same temperature $T^{\square}$.

We now consider special equilibrium environments which cause that certain non-equilibrium contact rates vanish

$$\mathcal{G}^{\square}_{\odot}:\quad \overset{\bullet}{Q}=0,\longrightarrow\ \Theta\ =\ T^{\square}_{\odot}, \quad (23.20)$$

$$\mathcal{G}^{\square}_{nj}:\quad \overset{\bullet}{n}_j=0,\longrightarrow\ h_j\ =\ \frac{T^{\square}_{\odot}}{T^{\square}}h^{\square}_j,\quad \mu_j\ =\ \frac{T^{\square}_{\odot}}{T^{\square}}\mu^{\square}_j. \quad (23.21)$$

For deriving $(23.20)_2$ and $(23.21)_{2,3}$ from (23.19), the following proposition (Muschik, 1984) is used:

$$f(\boldsymbol{X})\cdot\boldsymbol{X}\ \geq\ \boldsymbol{0}\ (\text{for all } \boldsymbol{X}\wedge f \text{ continuous at } \boldsymbol{X}=\boldsymbol{0})\ \Longrightarrow\ f(\boldsymbol{0})=\boldsymbol{0}. \quad (23.22)$$

Consequently, the non-equilibrium contact quantities *contact temperature* Θ, *molar non-equilibrium enthalpies* $\boldsymbol{h}$ and *non-equilibrium chemical potentials* $\boldsymbol{\mu}$ are defined by equilibrium quantities of special contacting equilibrium environments according to (23.20) and (23.21). The contact temperature is a non-equilibrium analogue to the thermostatic equilibrium temperature in the following sense: The system's contact temperature is that thermostatic temperature of the system's equilibrium environment for which the net heat exchange between the system and this environment through an inert partition vanishes by change of sign according to (23.21). In equilibrium, the contact temperature changes into the thermostatic temperature of the system which is equal to that of the environment.

[9] * means a setting

23.3.2 Internal Energy and Contact Temperature

As easily to demonstrate, contact temperature Θ and internal energy U are independent of each other in non-equilibrium. For this purpose, a rigid partition $\partial\mathcal{G}$ ($\overset{\bullet}{\boldsymbol{a}}\equiv \mathbf{0}$) between the Schottky system $\mathcal{G}$ and its equilibrium environment $\mathcal{G}^\square$ is chosen which is impervious to matter ($\overset{\bullet}{\boldsymbol{n}}{}^e \equiv 0$) and a time-dependent environment temperature $T^\square(t)$ which is always set equal to the value of the momentary contact temperature $\Theta(t)$ of $\mathcal{G}$, resulting according to $(23.19)_1$ and (23.10) in

$$T^\square(t) \stackrel{*}{=} \Theta(t) \quad\longrightarrow \overset{\bullet}{Q} = 0 \longrightarrow \overset{\bullet}{U} = 0. \tag{23.23}$$

Because Θ is time-dependent and U is constant, totally different from thermostatics, both quantities are independent of each other.

Because the contact temperature is independent of the internal energy, it represents an additional variable which is included in Σ^0. The choice of further non-equilibrium variables depends on the system in consideration. Here, *internal variables* $\boldsymbol{\xi}$ are chosen because they allow a great flexibility of describing non-equilibria (Muschik, 1990b; Maugin and Muschik, 1994). Consequently, the created non-equilibrium state space and the entropy production caused by the contact temperature and the internal variables are

$$\boldsymbol{Z} \;=\; (\boldsymbol{a},\boldsymbol{n},U,\Theta,\boldsymbol{\xi}), \qquad \Sigma^0 = \alpha\,\overset{\bullet}{\Theta} + \boldsymbol{\beta}\cdot\overset{\bullet}{\boldsymbol{\xi}} \geq 0. \tag{23.24}$$

According to (23.9), the contact temperature is in equilibrium a function of the internal energy, the work variables and the mole numbers. That is the "thermal equation of state."

23.4 Thermostatic Approach for Schottky Systems

The entropy "rate" of the accompanying process $\mathcal{P}\boldsymbol{Z}$ of $\boldsymbol{Z}$ is according to (23.5) and (23.17)

$$\overset{\bullet}{S}{}^*(\mathcal{P}\boldsymbol{Z}) \;=\; \frac{1}{T^*}\Big(\overset{\bullet}{U} - \boldsymbol{A}^*\cdot\overset{\bullet}{\boldsymbol{a}} - \boldsymbol{\mu}^*\cdot\overset{\bullet}{\boldsymbol{n}}\Big), \qquad \boldsymbol{A}^* := \boldsymbol{A}(\mathcal{P}\boldsymbol{Z}),\ \boldsymbol{\mu}^* := \boldsymbol{\mu}(\mathcal{P}\boldsymbol{Z}), \tag{23.25}$$

Evident is, that the equilibrium subspace quantities T^*, $\boldsymbol{A}^*$ and $\boldsymbol{\mu}^*$ are different from the non-equilibrium ones Θ, $\boldsymbol{A}$ and $\boldsymbol{\mu}$ in (23.17). The entropy rate belonging to the accompanying process (23.9) is according to (23.24)

$$\overset{\bullet}{S}(\mathcal{P}^{eq}\boldsymbol{Z}) = \overset{\bullet}{S}(\mathcal{P}\boldsymbol{Z},\Xi(\mathcal{P}\boldsymbol{Z})) = \frac{1}{T^{eq}}\Big(\overset{\bullet}{U} - \boldsymbol{A}^{eq}\cdot\overset{\bullet}{\boldsymbol{a}} - \boldsymbol{\mu}^{eq}\cdot\overset{\bullet}{\boldsymbol{n}}\Big)$$
$$=: \overset{\bullet}{S}{}^{eq}(\mathcal{P}\boldsymbol{Z}), \tag{23.26}$$
$$\overset{\bullet}{S}(\mathcal{P}^{eq}\boldsymbol{Z}) = \Big(\frac{\partial S(\boldsymbol{Z}^{eq})}{\partial U} + \frac{\partial S(\boldsymbol{Z}^{eq})}{\partial \Xi}\cdot\frac{\partial \Xi}{\partial U}\Big)\overset{\bullet}{U} + \tag{23.27}$$
$$+\Big(\frac{\partial S(\boldsymbol{Z}^{eq})}{\partial \boldsymbol{a}} + \frac{\partial S(\boldsymbol{Z}^{eq})}{\partial \Xi}\cdot\frac{\partial \Xi}{\partial \boldsymbol{a}}\Big)\cdot\overset{\bullet}{\boldsymbol{a}} +$$
$$+\Big(\frac{\partial S(\boldsymbol{Z}^{eq})}{\partial \boldsymbol{n}} + \frac{\partial S(\boldsymbol{Z}^{eq})}{\partial \Xi}\cdot\frac{\partial \Xi}{\partial \boldsymbol{n}}\Big)\cdot\overset{\bullet}{\boldsymbol{n}}, \tag{23.28}$$

and the connection to the original irreversible process $\mathcal{T}$ is given analogous to $(23.25)_2$ by

$$\boldsymbol{A}^{eq} := \boldsymbol{A}(\mathcal{P}^{eq}\boldsymbol{Z}),\ \boldsymbol{\mu}^{eq} := \boldsymbol{\mu}(\mathcal{P}^{eq}\boldsymbol{Z}). \tag{23.29}$$

A comparison of (23.25) with (23.26) to (23.29) results in the following facts: The irreversible process $\mathcal{T}$ generates an accompanying process on the equilibrium subspace

$$\mathcal{T}:\ \boldsymbol{Z}(t) \longrightarrow \mathcal{R}:\ \mathcal{P}\boldsymbol{Z}(t) = (\boldsymbol{a},\boldsymbol{n},U)(t). \tag{23.30}$$

This accompanying process is charged with different equilibrium entropies

$$S^*(\mathcal{P}\boldsymbol{Z}) \neq S^{eq}(\mathcal{P}\boldsymbol{Z}) \tag{23.31}$$

dependent on the specially chosen thermostatic approximation: S^* not considering any irreversibility and S^{eq} taking it partly into account by the equilibrium values of the non-equilibrium variables. Interesting is that different equilibrium entropies can be defined on the common equilibrium subspace depending on the chosen thermostatic approximation of the original irreversible process.

23.5 The Embedding Theorem

Because equilibrium and non-equilibrium entropies have to be compatible with each other, we have to demand the *embedding theorem*: the non-equilibrium entropy rate integrated along an irreversible process $\mathcal{T}$ starting and ending in equilibrium states—A^* and B^* or A^{eq} and B^{eq}—has the same value as the equilibrium entropy difference between these two equilibrium states calculated along the corresponding accompanying process $\mathcal{R}$

$$\mathcal{T}^*\int_{A*}^{B*}\overset{\bullet}{S}(\boldsymbol{Z})dt = \mathcal{R}\int_{A*}^{B*}\overset{\bullet}{S}{}^*(\mathcal{P}\boldsymbol{Z})dt = S^*(B^*) - S^*(A^*), \tag{23.32}$$
$$\mathcal{T}^{eq}\int_{Aeq}^{Beq}\overset{\bullet}{S}(\boldsymbol{Z})dt = \mathcal{R}\int_{Aeq}^{Beq}\overset{\bullet}{S}{}^{eq}(\mathcal{P}\boldsymbol{Z})dt = S^{eq}(B^{eq}) - S^{eq}(A^{eq}) \tag{23.33}$$

This yields according to $(23.13)_{2,3}$

$$\oint \overset{\bullet}{S}(\boldsymbol{Z})dt = 0, \qquad \oint \overset{\bullet}{S}{}^{*}(\mathcal{P}\boldsymbol{Z})dt \qquad \oint \overset{\bullet}{S}{}^{eq}(\mathcal{P}\boldsymbol{Z})dt = 0. \tag{23.34}$$

According to (23.17) and $(23.25)_1$, the embedding theorems (23.32) and (23.33) results in

$$\begin{aligned}
\mathcal{T}^{*}/\mathcal{R}\int_{A*}^{B*}\Big[\overset{\bullet}{S}(\boldsymbol{Z}) - \overset{\bullet}{S}{}^{*}(\mathcal{P}\boldsymbol{Z})\Big]dt &= 0 = \\
&= \mathcal{T}^{*}/\mathcal{R}\int_{A*}^{B*}\Big[\Big(\frac{1}{\Theta}-\frac{1}{T^{*}}\Big)\overset{\bullet}{U} - \Big(\frac{\boldsymbol{A}}{\Theta}-\frac{\boldsymbol{A}^{*}}{T^{*}}\Big)\cdot\overset{\bullet}{\boldsymbol{a}} - \\
&\quad -\Big(\frac{\boldsymbol{\mu}}{\Theta}-\frac{\boldsymbol{\mu}^{*}}{T^{*}}\Big)\cdot\overset{\bullet}{\boldsymbol{n}} + \Sigma^{0}\Big]dt,
\end{aligned} \tag{23.35}$$

$$\begin{aligned}
\mathcal{T}^{eq}/\mathcal{R}\int_{Aeq}^{Beq}\Big[\overset{\bullet}{S}(\boldsymbol{Z}) - \overset{\bullet}{S}(\mathcal{P}^{eq}\boldsymbol{Z})\Big]dt &= 0 = \\
&= \mathcal{T}^{eq}/\mathcal{R}\int_{Aeq}^{Beq}\Big[\Big(\frac{1}{\Theta}-\frac{1}{T^{eq}}\Big)\overset{\bullet}{U} - \Big(\frac{\boldsymbol{A}}{\Theta}-\frac{\boldsymbol{A}^{eq}}{T^{eq}}\Big)\cdot\overset{\bullet}{\boldsymbol{a}} - \\
&\quad -\Big(\frac{\boldsymbol{\mu}}{\Theta}-\frac{\boldsymbol{\mu}^{eq}}{T^{eq}}\Big)\cdot\overset{\bullet}{\boldsymbol{n}} + \Sigma^{0}\Big]dt.
\end{aligned} \tag{23.36}$$

Every equilibrium entropy has to be defined uniquely, that means process independently on the equilibrium subspace. If the considered system is adiabatically unique (Muschik, 2009), to each equilibrium entropy non-equilibrium entropies can be uniquely defined by use of the entropy production. By construction, these non-equilibrium entropies satisfy the embedding theorem which represents a constraint for the contact quantities. Here, the procedure is the other way round: the non-equilibrium process induces the thermostatic approximation by accompanying reversible processes.

23.6 Summary

A non-equilibrium discrete Schottky system is considered which is approximately described as undecomposed and homogeneous and which is contacted with an equilibrium environment by an undecomposed partition permeable to heat, power and material. A non-equilibrium state space of the Schottky system is introduced being composed of the equilibrium subspace and of non-equilibrium state variables, the contact temperature and a set of internal variables. Every non-equilibrium process generates by projection or by relaxation onto the equilibrium subspace a reversible accompanying process on which equilibrium entropies are defined as approximations of different accuracy of the original irreversible process. The compatibility of these equilibrium entropies with the non-equilibrium entropy of the original irreversible process has to be tested by using the embedding theorem.

References

Haase R (1969) Thermodynamics of Irreversible Processes. § 1.7, Addison-Wesley, Reading Ma.

Keller JU (1971) Ein Beitrag zur Thermodynamik fluider Systeme. I. Physica 53(4):602 – 620

Kestin J (1979) A Course in Thermodynamics, vol I, sect.13.6. Hemisphere Pub. Corp., Washington/London

Maugin GA, Muschik W (1994) Thermodynamics with internal variables. Part I. General Concepts. J Non-Equilib Thermodyn 19(3):217 – 249

Müller I (1973) Thermodynamik –Grundlagen der Materialtheorie–. Bertelsmann Universitätsverlag, Düsseldorf 1973

Müller WH (1982) Mischungstheorien der Vieltemperatursysteme: Darstellung und Vergleich. Diplomarbeit, Inst. Theor. Physik, TU-Berlin

Müller WH, Muschik W (1983) Bilanzgleichungen offener mehrkomponentiger Systeme I. Massen- und Impulsbilanzen. J Non-Equilib Thermodyn 8:22–46

Muschik W (1977) Empirical foundation and axiomatic treatment of non-equilibrium temperature. Archive for Rational Mechanics and Analysis 66(4):379–401

Muschik W (1984) Fundamental remarks on evaluating dissipation inequalities. In: Casas-Vazquez J, Jou D, Lebon G (eds) Recent Developments in Nonequilibrium Thermodynamics, Springer, Berlin, Lecture Notes in Physics, vol 199, pp 388–397

Muschik W (1990a) Aspects of Non-Equilibrium Thermodynamics. sect. 1.2, World Scientific, Singapore

Muschik W (1990b) Internal variables in non-equilibrium thermodynamics. J Non-Equilib Thermodyn 15(2):127 – 137

Muschik W (1993) Fundamentals of non-equilibrium thermodynamics. In: Muschik W (ed) Non-Equilibrium Thermodynamics with Application to Solids, Springer, Wien, CISM Courses and Lectures, vol 336, pp 1–63

Muschik W (2009) Contact quantities and non-equilibriun entropy of discrete systems. J Non-Equilib Thermodyn 34(1):75 – 92

Muschik W (2018) Second Law and Non-Equilibrium Entropy of Schottky Systems –Doubts and Verification–. Entropie 20(10), article No. 740

Muschik W, Brunk G (1975) Temperatur und Irreversibilität in der Rationalen Mechanik. ZAMM 55(4):T102–T105

Muschik W, Brunk G (1977) A concept of non-equilibrium temperature. International Journal of Engineering Science 15(6):377 – 389

Muschik W, Müller WH (1983) Bilanzgleichungen offener mehrkomponentiger Systeme II. Energie- und Entropiebilanz. J Non-Equilib Thermodyn 8:47–66

Schottky W (1929) Thermodynamik. Erster Teil, § 1, Springer, Berlin

W Muschik and S Gümbel (1969) Does Clausius' inequality analogue exists for open discrete systems? J Non-Equilib Thermodyn 24(1):97 – 106

Chapter 24
On the Temperature Gradient in the Standard Troposphere

Ingo Müller & Wolf Weiss

Abstract The temperature gradient in the upper troposphere is $-0.65\frac{\mathrm{K}}{100\mathrm{m}}$. We suggest that the thermal diffusion factor of air in the gravitational field determines this gradient.

Keywords: Troposphere · Thermal diffusion · Temperature gradient

24.1 Introduction

The troposphere up to a height of ca. 10 km consists of two layers: The lower layer, up to a maximal height of ca. 2 km in our latitudes and on sunny summer days, is strongly agitated by convective thermal up- and down-draughts. The temperature gradient in that range is $-1\frac{\mathrm{K}}{100\mathrm{m}}$ provided that the air is dry which we shall assume here. That value is well-understood and exhaustively explained in the meteorological literature: It is due to the adiabatic expansion of the rising air "bubbles".

Above that range the normal condition of the troposphere is a state of rest, and the temperature gradient has a constant value of $-0.65\frac{\mathrm{K}}{100\mathrm{m}}$ as far up as the troposphere reaches[1]. To be sure, there may be violent motion in all directions – horizontal and vertical – even in the upper troposphere. It occurs in thunderstorms and hurricanes. But those are local and transient phenomena and they do not concern us, because the vast majority of the tropospheric air above a height of ca. 2 km is stationary most of the time; it is essentially at rest, and has the small temperature gradient $-0.65\frac{\mathrm{K}}{100\mathrm{m}}$. We ask why there is a temperature gradient in that range at all, and why the gradient is independent of the height?

Ingo Müller · Wolf Weiss
Technische Universität Berlin, Germany
e-mail: ingo.mueller@alumni.tu-berlin.de, wolf.weiss@alumni.tu-berlin.de

[1] This gradient is part of the *standard atmosphere* (Wikipedia, 2019) abstracted by meteorologists from their observations.

B. E. Abali et al. (eds.), *New Achievements in Continuum Mechanics and Thermodynamics*, Advanced Structured Materials 108,
https://doi.org/10.1007/978-3-030-13307-8_24

Another somewhat surprising observation concerns the composition of air: 21% oxygen and 79% of the lighter nitrogen[2]. One might have expected that those ratios vary with the height – more of the heavy oxygen near the bottom (say). This is not the case, however. Indeed, samples gathered by airplanes, rockets, and balloons from all heights of the troposphere have the same composition, see Chapman and Cowling (1970, p. 257), Roedel and Wagner (2011, p. 71), and Ishidoya et al (2012).

We combine the governing thermodynamic equations for a mixture of ideal gases with the observed conditions in the upper troposphere and come to the conclusion that the reason for the temperature gradient is related to thermal diffusion. This conclusion may have been reached by meteorologists before; if so, we are unaware where and when this has happened. The literature known to us provides information about the observed conditions, but it does not offer an explanation, e.g. see Manabe and Strickler (1964, p. 363).

24.2 Equations of Balance and Constitutive Relations

For a mixture of ν non reacting constituents $\alpha = 1, 2, ..., \nu$ the equations of balance of masses and momenta of the constituents and of energy of the mixture as a whole read

$$\begin{aligned} \frac{\partial \varrho^\alpha}{\partial t} + \frac{\partial \varrho^\alpha v_i^\alpha}{\partial x_i} &= 0 \\ \frac{\partial \varrho^\alpha v_j^\alpha}{\partial t} + \frac{\partial (\varrho^\alpha v_j^\alpha v_i^\alpha - t_{ij}^\alpha)}{\partial x_i} &= m_j^\alpha + \varrho^\alpha f_j \\ \varrho \frac{\partial \varepsilon}{\partial t} + \varrho v_i \frac{\partial \varepsilon}{\partial x_i} + \frac{\partial q_i}{\partial x_i} &= t_{ij} \frac{\partial v_i}{\partial x_j} \end{aligned} \tag{24.1}$$

The customary notation is employed here: ϱ^α, v_i^α, t_{ij}^α, m_i^α are the partial mass densities, velocities, stresses and the inter-constituent forces respectively of the constituents.

$$\sum_{\alpha=1}^{\nu} m_i^\alpha = 0$$

holds, because the momentum of the mixture as a whole is conserved. f_i is the specific force of gravitation so that

$$f_i = (0, 0, -g) = (0, 0, -9.81 \frac{\mathrm{m}}{\mathrm{s}^2})$$

holds, if gravity acts opposite to the x_3-axis. ϱ, v_i, t_{ij}, ε, and q_i in $(24.1)_3$ are defined by

[2] We ignore the presence of argon, because it is small, merely 1%.

$$\varrho = \sum_{\alpha=1}^{\nu} \varrho^\alpha, \; v_i = \sum_{\alpha=1}^{\nu} \frac{\varrho^\alpha}{\varrho} v_i^\alpha, \; t_{ij} = \sum_{\alpha=1}^{\nu} \left(t_{ij}^\alpha - \varrho^\alpha u_i^\alpha u_j^\alpha\right),$$
$$\varepsilon = \sum_{\alpha=1}^{\nu} \frac{\varrho^\alpha}{\varrho} \left(\varepsilon^\alpha + \frac{u_i^\alpha u_i^\alpha}{2}\right), \; q_i = \sum_{\alpha=1}^{\nu} \left(q_i^\alpha + \varrho^\alpha \left(\varepsilon^\alpha + \frac{u_j^\alpha u_j^\alpha}{2}\right) u_i^\alpha - t_{ij}^\alpha u_j^\alpha\right) \tag{24.2}$$

where ε^α and q_i^α are the specific internal energies of the constituents and their heat fluxes.

$$u_i^\alpha = v_i^\alpha - v_i$$

is called the diffusion velocity; it is also – sometimes – useful to introduce the diffusion fluxes

$$J_i^\alpha = \varrho^\alpha u_i^\alpha,$$

whose sum over α obviously vanishes. If derivation of the equations (24.1) and (24.2) is needed, it may be obtained from Müller (1985, p. 172 ff.).

In a binary mixture the fields to be determined are $\varrho^1(x_k, t)$, $\varrho^2(x_k, t)$, $v_i^1(x_k, t)$, $v_i^2(x_k, t)$, and $T(x_k, t)$ or, more conveniently

$$\varrho(x_k, t), \; v_j(x_k, t), \; T(x_k, t), \; c \equiv \varrho^1/\varrho \text{ and } J_i(x_k, t) = \varrho^1 u_i^1. \tag{24.3}$$

c is called the concentration of constituent 1. For that choice of fields the appropriate equations follow from (24.2). They have the forms

$$\begin{aligned}
\frac{\partial \varrho}{\partial t} + \frac{\partial \varrho v_i}{\partial x_i} &= 0, \\
\varrho \frac{\partial v_j}{\partial t} + \varrho v_i \frac{\partial v_j}{\partial x_i} - \frac{\partial t_{ij}}{\partial x_i} &= \varrho f_j, \\
\varrho \frac{\partial \varepsilon}{\partial t} + \varrho v_i \frac{\partial \varepsilon}{\partial x_i} + \frac{\partial q_i}{\partial x_i} &= t_{ij} \frac{\partial v_i}{\partial x_j}, \\
\varrho \frac{\partial c}{\partial t} + \varrho v_i \frac{\partial c}{\partial x_i} + \frac{\partial J_i}{\partial x_i} &= 0, \\
\frac{\partial J_i}{\partial t} + \frac{\partial}{\partial x_j}\left(J_i v_j + \frac{1}{\varrho c} J_i J_j + c t_{ij} - t_{ij}^1\right) &= -J_j \frac{\partial v_i}{\partial x_j} + t_{ij} \frac{\partial c}{\partial x_j} + m_i^1.
\end{aligned} \tag{24.4}$$

This latter form is most appropriate for our purposes. Although it is formally more complicated than (24.1), – particularly $(24.4)_5$ –, the two sets are completely equivalent for binary mixtures. And obviously neither set is closed. In order to close (24.4) we need constitutive equations which relate the constitutive quantities

$$t_{ij}, \; t_{ij}^1, \; m_i^1, \; \varepsilon, \text{ and } q_i \tag{24.5}$$

to the fields (24.3) in a manner dependent on the nature of the constituents. For instance, in a typical case, the constitutive quantities (24.5) at point x_i and time t may depend on the values of the fields (24.3) at that point and time, and on the gradients and rates of change of the fields there.

The evaluation of the constitutive functions is the subject of thermodynamics of mixtures. It relies heavily on the entropy principle. We do not go into this here. Suffice it to say that the interaction force m_i^1 and the flux of internal energy q_i are assumed to be given – in terms of the relative velocity of the constituents and of the temperature gradient – by linear constitutive relations of the forms

$$\begin{aligned} m_i^1 &= m_V\left(v_i^1 - v_i^2\right) + m_T\frac{\partial T}{\partial x_i}, \\ q_i &= q_V\left(v_i^1 - v_i^2\right) + q_T\frac{\partial T}{\partial x_i}. \end{aligned} \tag{24.6}$$

One might think that the gradients of ϱ^α – or the gradients of ϱ and c – should also be represented on the right hand sides of (24.6) along with $v_i^1 - v_i^2$ and $\partial T/\partial x_i$. However, such terms are forbidden by the entropy principle, at least for mixtures of ideal gases, see again Müller (1985).

In the early days of theories of mixtures of gases the equations of balance (24.4) and the constitutive relations (24.6) were derived from the kinetic theory of gases. The latter ones came out as first approximations in a complicated iterative scheme – the Chapman-Enskog method, see (Chapman and Cowling, 1970, p. 140). While the present macroscopic theory is conceptually simpler than the kinetic theory, we cannot ignore the earlier work, if only because it has set the standards in the field and has fixed the nomenclature. Thus our coefficients m_V and m_T are related to the diffusion coefficient D and to the thermal diffusion factor α – both defined in the kinetic theory – by

$$m_V = -kT\frac{n^1 n^2}{n}\frac{1}{D} \quad \text{and} \quad m_T = k\frac{n^1 n^2}{n}\alpha. \tag{24.7}$$

$k = 1.38 \cdot 10^{-23}\mathrm{J/K}$ is the Boltzmann constant, and n^1, n^2, and $n = n^1 + n^2$ are number densities of molecules. In the kinetic theory it is customary and appropriate to choose n^α as variables, or n and $X = n^1/n$, rather than ϱ^α, or ϱ and c. X is called the mol fraction of constituent α. D and α have been measured and tables of their values are available, see again Chapman and Cowling (1970). Actually D and α can also be *calculated* in the kinetic theory from an assumed model of molecular interaction; the agreement between measured and calculated values is fairly good, if not perfect.

Now, however, the foregoing arguments about m_i^1 and q_i refer to an isotropic fluid and it may well be argued that air in the atmosphere is not isotropic, because there is a preferred direction, the vertical direction, defined by gravitation. We may ask the question: Does a horizontal temperature gradient create the same horizontal interaction force m_i^1 as a vertical gradient of the same magnitude does for a vertical interaction force? We allow for a negative answer to this question by splitting the equations for m_i^1 and q_i into two parts, one part for the horizontal components 1, 2 and the other one for the vertical component. Thus (24.6) is replaced by

$$
\begin{aligned}
m_i^1 &= -\frac{kT}{D^{\perp}}\frac{n_1 n_2}{n}\left(v_i^1 - v_i^2\right) + k\frac{n_1 n_2}{n}\alpha^{\perp}\frac{\partial T}{\partial x_i}, \\
q_i^1 &= q_V^{\perp}\left(v_i^1 - v_i^2\right) + q_T^{\perp}\frac{\partial T}{\partial x_i} \quad (i=1,2) \\
m_3^1 &= -\frac{kT}{D^{\parallel}}\frac{n_1 n_2}{n}\left(v_i^1 - v_i^2\right) + k\frac{n_1 n_2}{n}\alpha^{\parallel}\frac{\partial T}{\partial x_3}, \\
q_3^1 &= q_V^{\parallel}\left(v_i^1 - v_i^2\right) + q_T^{\parallel}\frac{\partial T}{\partial x_3}.
\end{aligned}
\tag{24.8}
$$

The coefficients $D^{\perp}$, $\alpha^{\perp}$, $q_V^{\perp}$, $q_T^{\perp}$ in the horizontal components may differ from $D^{\parallel}$, $\alpha^{\parallel}$, $q_V^{\parallel}$, $q_T^{\parallel}$ in the vertical component. And we call $\alpha^{\perp}$ and $\alpha^{\parallel}$ the thermal diffusion factors.

24.3 Application to the Troposphere

As discussed in the introduction we shall consider the upper part of the troposphere to be essentially stationary with both constituents – O_2 and N_2 – at rest, in stark contrast to the name which literally means *sphere of change* [3]. For stationary processes and with $v_i = 0$ and $J_i = 0$ the equations $(24.4)_{1,4}$ are identically satisfied and the equations $(24.4)_{2,3,5}$ reduce to

$$
\begin{aligned}
-\frac{\partial t_{ij}}{\partial x_i} &= \varrho f_j \\
\frac{\partial\left(c t_{ij} - t_{ij}^1\right)}{\partial x_j} &= t_{ij}\frac{\partial c}{\partial x_j} + m_i^1, \\
\frac{\partial q_i}{\partial x_i} &= 0.
\end{aligned}
\tag{24.9}
$$

We insert $t_{ij} = -nkT\delta_{ij}$ and $t_{ij}^1 = -n^1 kT\delta_{ij}$ – appropriate to the ideal gas character of air – as well as (24.8), or what is left of (24.8) in the case of rest. Thus we obtain[4]

$$
\begin{aligned}
\frac{\partial nT}{\partial x_i} &= 0 \quad (i = 1, 2), \\
\frac{\partial nT}{\partial x_3} &= -n\frac{\left(\mu^2 + \left(\mu^1 - \mu^2\right) X\right) g}{k}
\end{aligned}
\tag{24.10}
$$

[3] Weather phenomena occur in the troposphere. Hence the name, notwithstanding the observation that the bulk of the upper troposphere – *above the clouds* – is at rest most of the time.

[4] From here on ϱ and c are replaced by n and X, because that makes the equations a little simpler. The reason why thermodynamicists prefer ϱ over n is that ϱv_i is the density of a conserved quantity, the momentum, which $n v_i$ is not, and that the gravitational force density is proportional to ϱ not n. On the other hand, physicists prefer n over ϱ, because n can be counted – in principle – while ϱ has to be weighed. When all is said and done, however, the choice comes down to a matter of taste.

$$\alpha^{\perp}\frac{\partial \ln T}{\partial x_i}=\frac{\partial}{\partial x_i}\left(\ln\frac{X}{1-X}\right)\quad (i=1,2)\,,$$
$$\alpha^{\|}\frac{\partial \ln T}{\partial x_3}=\frac{\partial}{\partial x_3}\left(\ln\frac{X}{1-X}\right)-\frac{\mu^2-\mu^1}{kT}g \tag{24.11}$$

$$\sum_{i=1}^{2}\frac{\partial}{\partial x_i}\left(q_T^{\perp}\frac{\partial T}{\partial x_i}\right)+\frac{\partial}{\partial x_i}\left(q_T^{\|}\frac{\partial T}{\partial x_3}\right)=0, \tag{24.12}$$

where μ^{α} are the molecular masses of the constituents. It is thus clear that it was prudent to distinguish between $\alpha^{\perp}$ and $\alpha^{\|}$. Indeed, if the air were isotropic in the presence of gravitation – so that $\alpha^{\perp}$ were equal to $\alpha^{\|}$ – the two equations (24.11) could not both be satisfied.

The equations (24.10) through (24.12) still permit many solutions for $n(x_i)$, $X(x_i)$, and $T(x_i)$, even though the air is at rest and without diffusion. Let us consider two of those solutions:

If T is constant, (24.12) is identically satisfied and we conclude from $(24.10)_1$ and $(24.11)_1$ that n and X can only depend on x_3. And $(24.11)_2$ implies that

$$X(x_3)=\frac{X(0)\exp\left(\frac{\mu^2-\mu^1}{kT}gx_3\right)}{1-X(0)\left(1-\exp\left(\frac{\mu^2-\mu^1}{kT}gx_3\right)\right)} \tag{24.13}$$

holds, whence $n(x_3)$ may easily be calculated by integration of $(24.11)_2$. It follows that the temperature in the troposphere cannot be constant, because it has been observed – as reported in the introduction – that the composition of the air does not change with height.

Now then, let us investigate a case which *is* consistent with those observations in the upper troposphere: If T depends only on x_3, we conclude from (24.12) that its gradient must be constant and observations have shown that the constant gradient has the value $-0.65\frac{\text{K}}{100\text{m}}$. Also no height-dependence of X has been observed. Therefore $(24.11)_2$ reduces to – with $\mu^1=32\mu^H$, $\mu^2=28\mu^H$ for oxygen and nitrogen ($\mu^H=1.67\cdot 10^{-27}$kg) –

$$-\alpha^{\|}0.65\frac{\text{K}}{100\text{m}}=4\frac{g}{k/\mu^H},\quad \text{hence}\quad \alpha^{\|}=-0.73. \tag{24.14}$$

We conclude that the conditions in the standard troposphere satisfy the equations (24.4) and (24.8) provided that the vertical thermo-diffusion factor $\alpha^{\|}$ has the value -0.73. And we suggest that this value be confirmed in the laboratory. That should be easy for a person with the proper equipment: All it takes is to set up an upward temperature gradient in a box, – so as to avoid convection –, wait for stationarity, and measure the eventual mol fractions on top and at the bottom.

Incidentally, the value $\alpha^{\|} = -0.73$ is about 40 times bigger in magnitude than the measured value $\alpha^{\perp} = 0.018$, see Chapman and Cowling (1970, p. 278), – apart from having a different sign[5].

24.4 Discussion

The foregoing considerations do not explain, of course, how the conditions of the standard troposphere come about. They merely explain how the fields $n(x_i)$, $X(x_i)$, and $T(x_i)$ are related in a stationary state of rest. And, if $X(x_i)$ is constant, – as observed –, they permit the calculation of the temperature field as a consequence of thermal diffusion.

Thermal diffusion is, perhaps, the most esoteric one of the transport processes in gases. Thus for instance it has no clear-cut simple explanation in terms of molecular motion as diffusion itself has, or thermal conduction, or viscous friction. Chapman and Cowling (1961) in the 2nd edition of their book deplore the lack of such an "elementary" interpretation.[6]

Still, thermal diffusion is on everybody's mind in the context of the separation columns for the enrichment of isotopes capable of nuclear fission. This may well be the most common application of thermal diffusion as a contributing effect in a complicated process of convection, diffusion and thermal conduction. An early paper (Furry et al, 1939) by Furry, Jones, and Onsager deals with the problem. This work is not relevant to our investigation, however, because the temperature gradient is set up horizontally, so that our equation $(24.11)_2$ is absent; also there is convective motion.

More recently, there is a sizable community interested in thermal diffusion as the driving force for the spreading of pollutants in air and the segregation of polymer molecules of different sizes in liquids. Thus there are regular international meetings on the subject, a fairly recent one in Bonn, Germany. The proceedings (Wiegand et al, 2008) are available with dozens of contributions on thermal diffusion in simple fluids, polymer melts, chemically reacting fluids, porous materials, thin channels, with and without a gravitational field.

Among those contributions, Galliero and Montel (2008) come closest to our investigation. It is concerned with the gravitational segregation in oil reservoirs and how this may be affected by thermal diffusion. In the article our equation (24.11) – with a single value for $\alpha^{\perp}$ and $\alpha^{\|}$, however – is integrated and the result is used for the calculation of the vertical gradient of the concentration. Such a gradient of concentration does not exist in the atmosphere, see above.

In another interesting contribution, Wiegand et al (2008) use molecular dynamics to describe atmospheric pollution, and they in fact calculate the air temperature in

---

[5] $\alpha^{\perp}$ is positive, if constituent 1 has a bigger molecular mass than constituent 2: $\mu^1 > \mu^2$.

[6] See Chapman and Cowling (1961, p. 143 and pp. 399-401). Ominously the authors have dropped these remarks in the 1970 edition of the book. By that time it may have become inappropriate, perhaps, to write in a book about what one does *not* know.

the troposphere. It comes out linear, but the gradient is too big. Moreover, they give a somewhat hasty interpretation of the temperature gradient.

Thus it appears, – unlikely as this may seem –, that nobody before us has identified thermal diffusion as the dominating phenomenon in the upper troposphere.

24.5 Remark on the Lack of Isotropy of Air in a Gravitational Field

Looking back on the arguments of Sect. 24.2 we recognize that it is the replacement of (24.6) by (24.8) that has made it possible to have a vertical temperature gradient in the troposphere without a gradient of composition, cf. (24.11). Therefore let us take a critical view at the equations (24.8). There are two peculiarities involved.

First of all, the equations (24.8) depend on the choice of coordinates: horizontal axes x_1, x_2 and vertical axis x_3. That choice is dictated by having the gravitational field f_j point opposite to the x_3-axis. If we wish to make the equations independent of the choice of coordinates, it suffices to replace the scalar coefficients m_V through q_T in (24.6) – generically we may call them c – by tensors of the form

$$c_{ij} = c^{\perp}\delta_{ij} + \left(c^{\parallel} - c^{\perp}\right)\frac{f_i f_j}{f^2}. \tag{24.15}$$

With those tensorial coefficients the interaction force m_i^1 reads

$$\begin{aligned} m_i^1 = &\left(m_V^{\perp}\delta_{ij} + \left(m_V^{\parallel} - m_V^{\perp}\right)\frac{f_i f_j}{f^2}\right)\left(v_j^1 - v_j^2\right) \\ &+ \left(m_T^{\perp}\delta_{ij} + \left(m_T^{\parallel} - m_T^{\perp}\right)\frac{f_i f_j}{f^2}\right)\frac{\partial T}{\partial x_j} \end{aligned} \tag{24.16}$$

which clearly reduces to the two equations (24.8) for m_i^1, if we choose

$$f_i = (0, 0, -g).$$

The second peculiarity is more subtle and it concerns a basic tenet of continuum thermodynamics. Indeed, continuum mechanics and thermodynamics stipulate that the governing equations should be strictly divided into balance laws and constitutive equations: While the former ones are universal and may contain inertial and gravitational forces, the latter ones should not be influenced by such forces; they should merely represent the properties of the materials under consideration, here air. This distinction found its expression in the *principle of material frame indifference*. Constitutive equations like (24.16) are clearly incompatible with that axiom.

It is true that material frame indifference is a somewhat precarious proposition according to molecular theories. Thus the kinetic theory of gases contradicts the principle: According to the kinetic theory the heat flux is no longer opposite to the temperature gradient. Rather it has a component perpendicular to grad$T$ due to the Coriolis force on the molecular motion, see Müller (1972). However, quantitatively this is a small effect, and it is therefore mostly ignored.

In our case it is possible to save material frame indifference by replacing f_i in (24.16) by

$$\frac{1}{\mid \text{grad}p \mid} \frac{\partial p}{\partial x_i}$$

so that the equation reads

$$\begin{aligned} m_i^1 = & \left(m_V^{\perp} \delta_{ij} + \left(m_V^{\parallel} - m_V^{\perp} \right) \frac{\frac{\partial p}{\partial x_i} \frac{\partial p}{\partial x_j}}{\mid \text{grad}p \mid^2} \right) \left(v_i^1 - v_j^2 \right) \\ & + \left(m_T^{\perp} \delta_{ij} + \left(m_T^{\parallel} - m_T^{\perp} \right) \frac{\frac{\partial p}{\partial x_i} \frac{\partial p}{\partial x_j}}{\mid \text{grad}p \mid^2} \right) \frac{\partial T}{\partial x_j}. \end{aligned} \tag{24.17}$$

Since the pressure p, as part of the stress t_{ij}, is a constitutive quantity, there is no contradiction with material frame indifference. And, if stationarity prevails, and v_i and J_i both vanish, the pressure gradient points opposite to f_i, and (24.17) reduces to (24.16). This scheme, however, seems a somewhat artificial recipe for the rescue of material frame indifference. It may be better to rethink the principle fundamentally in its application to bodies in a gravitational field, or inertial force fields.

References

Chapman S, Cowling TG (1961) The Mathematical Theory of Non-uniform Gases, 2nd edn. Cambridge University Press, Cambridge UK

Chapman S, Cowling TG (1970) The Mathematical Theory of Non-uniform Gases, 3rd edn. Cambridge University Press, Cambridge UK

Furry WH, Jones RC, Onsager L (1939) On the theory of isotope separation by thermal diffusion. Physical Review 55(11):1083

Galliero G, Montel F (2008) Non-isothermal gravitational segregation by molecular dynamics simulations. Thermal Equilibrium Proceedings of the 8th International Meeting on Thermodiffusion 3

Ishidoya S, Aoki S, Goto D, Nakazawa T, Taguchi S, Patra PK (2012) Time and space variations of the O2/N2 ratio in the troposphere over Japan and estimation of the global CO2 budget for the period 2000–2010. Tellus B: Chemical and Physical Meteorology 64(1):18,964

Manabe S, Strickler RF (1964) Thermal equilibrium of the atmosphere with a convective adjustment. Journal of the Atmospheric Sciences 21(4):361–385

Müller I (1972) On the frame dependence of stress and heat flux. Archive for Rational Mechanics and Analysis 45(4):241–250

Müller I (1985) Thermodynamics. Pitman Publishing Program, London

Roedel W, Wagner T (2011) Physik unserer Umwelt: Die Atmosphäre. Springer, Heidelberg

Wiegand S, Köhler W, Dhont JKGe (2008) Thermal Equilibrium. Proceedings of the 8th International Meeting on Thermodiffusion. Schriften des Forschungszentrums Jülich Vol 3
Wikipedia (2019) International standard atmosphere

Chapter 25
A Brief History of Mechanical Stress and the Method of Experimental Micromechanics with the Raman Microprobe

Giuseppe Pezzotti

Abstract The Raman microprobe is a unique tool for experimental stress analyses at the microscopic level, capable to overcome the drawbacks of other probes including low spatial resolution and lack of tensorial stress deconvolution. The presence of elastic stress in the lattice of crystalline materials results in shifts in frequency of Raman bands, which in turn obey individual dependencies, related to the specific molecular vibrations that they represent. More importantly, the observed frequency shifts depend on the reciprocal orientation between different stress components and the vibrating lattice. We present here some basic algorithms for stress assessments with the Raman micro-probe and validate them with some examples of practical engineering applications. We also introduce some details for three-dimensional measurement procedures using a confocal microprobe through a characterization of the probe itself. The high spatial resolution, the complete contactless nature, and the possibility to deconvolute the six independent components of the stress tensor from the Raman spectrum are key features, which place the Raman method at the frontiers of modern micromechanics.

Keywords: Experimental mechanics · Raman microprobe · Stress analysis · Micromechanics

---

Giuseppe Pezzotti
Ceramic Physics Laboratory, Kyoto Institute of Technology, Sakyo-ku, Matsugasaki, Kyoto 606-8585, Japan, Department of Orthopedic Surgery, Tokyo Medical University, 6-7-1 Nishi-Shinjuku, Shinjuku-ku, 160-0023 Tokyo, Japan, The Center for Advanced Medical Engineering and Informatics, Osaka University, 2-2 Yamadaoka, Suita, Osaka 565-0854, Japan & Department of Immunology, Graduate School of Medical Science, Kyoto Prefectural University of Medicine, Kamigyo-ku, 465 Kajii-cho, Kyoto 602-8566, Japan,
e-mail: pezzotti@kit.ac.jp

B. E. Abali et al. (eds.), *New Achievements in Continuum Mechanics and Thermodynamics*, Advanced Structured Materials 108,
https://doi.org/10.1007/978-3-030-13307-8_25

25.1 Introduction

25.1.1 A bit of History on the Concepts of Stress and Strain

When Hooke conceived his principle of elasticity in 1658, he certainly created a new concept but conspicuously failed in finding simple words to explain it. Despite the mathematical link between stress and strain was recorded in history as the Hooke's law, the cause-effect mechanism between tension and deformation was left obscure and unexplained in the Hooke's description. In the 17th century, the word "stress" did not even exist to promptly convey the concept that we all nowadays understand and use in many diverse circumstances. The situation was probably worsened by the fact that Hooke, in a lack of definitions and suitable math, created an *anagram* to describe the new concept: *CEIIINOSSSTTUU* (deciphered in Latin as *"ut tension sic vis"*, namely, "like the distension so the force"): conceptual confusion followed for a couple of centuries. About 150 years later, a definition of modulus of elasticity was reported by Thomas Young to the British Admiralty as *"a fundamental concept in engineering which would enable preventing ships from sinking and buildings and bridges from falling down"*. However, his definition did not help much in shedding conceptual clarity onto the fundamental concepts of stress and strain: *"The modulus of elasticity of any substance is a column of the same substance, capable of producing a pressure on its base which is to the weight causing a certain degree of compression as the length of the substance is to the diminution of its length."* (Young, 1807) The reply of the British Admiralty, not surprisingly, was: *"Though science is much respected by their Lordships and your paper is much esteemed, it is too learned in short it is not understood."* (Müller, 2014) Nearly two centuries after the Hooke's intuition, Cauchy succeeded in rigorously defining the concepts of stress and strain as second-rank tensors. Building upon this progress, in 1826, Navier elegantly located their correct proportionality relationship (yet in usage nowadays). Since then, because of the matching contributions of four brilliant minds, we have indeed learnt how to *"prevent ships from sinking and bridging from falling down"*. However, a number of circumstances in modern mechanics at the microscopic scale yet challenge such historical findings. The concepts of stress and strain, posited by Cauchy and Navier about two centuries ago, withstand the test of time, but need more specific definitions and new approaches to be predictively applied.

25.1.2 Residual Stress Ghosts from the Industrial World

In the experimental practice, mechanical stresses are often highly graded in space and might appear as ghostly entities because their magnitude depends on the scale at which they are measured: the larger the scale of measurements, the smaller the measured stress magnitude. An internal state of stress can develop within a solid as a consequence of an externally applied force (referred to as "applied stresses"). However, another important category of stresses exists in a solid even in absence of any externally applied force: the so-called "residual stresses" (Hosford, 2010). Devices experience residual stresses since their manufacturing cycle, and the magnitude of such stresses could be as high as the GPa level. Residual stresses become involved with any manufactured device any time different materials or phases with different physical properties (e.g., elastic modulus and/or thermal expansion coefficient) are "forced" to join, link or bond to each other, or even when a monolithic material is exposed to local gradients of temperature or repeated loading cycles. No matter how precise and sophisticated a manufacturing process could be conceived, there will always be the possibility that residual stresses silently remain hidden within the as-manufactured device. The residual stress field shall eventually overlap externally applied stresses and stresses related to functional fatigue during service, significantly degrading the reliability of the device and reducing its lifetime. So, re-examining *ad litteram* the Hooke's definition *"ut tension sic vis"* in the case of residual stress brings us no visible *tension* and no apparent *vis*, just a ghostly entity... although the device might yet fail because of it. Residual stresses are usually highly graded at the microscopic scale, for example, across grain boundaries, interfaces, or nearby micro-cracks, edges, holes, or notches. Accordingly, attempts to probe them at the tens of micron might result in apparently stress-free outputs. Invisible to the naked eye and hidden to dimensional inspections, microscopically graded stresses might thus elude X-ray or neutron scattering probes because those probes average over too large volumes. Popular examples of modern devices prone to residual stresses are: multilayered capacitors (ceramic and metal parts bonded together), and femoral heads in artificial joints (bearing dome and internal taper). The uncontrolled presence of microscopic residual stresses has long represented a curse for device manufacturers.

25.1.3 The Need for a Microscopic and Contactless Probe for Stress

Technologists have long relied on finite element modelling (FEM) for predicting internal stress and counteracting structural failures in devices (Rohlmann et al, 1980). Despite the usefulness of this method, limitations due to its purely computational nature might arise, which might include an insufficient understanding of the physical phenomena involved with the computations and/or an arbitrary/inappropriate choice of boundary conditions. For this reason, precise and flexible experimental stress probes have long been searched for. Leaving aside center-hole drilling, which has widely been used but requires material removal being, non-destructive stress probes are based on different physical principles: diffractive methods, such as X-ray and neutron diffraction directly measure the lattice displacements due to strain, while the ultrasonic method relies on variations in velocity of ultrasonic waves upon stress magnitude. Note that each of the above-mentioned methods presents advantages and disadvantages (not discussed here; for a review, see Withers and Bhadeshia

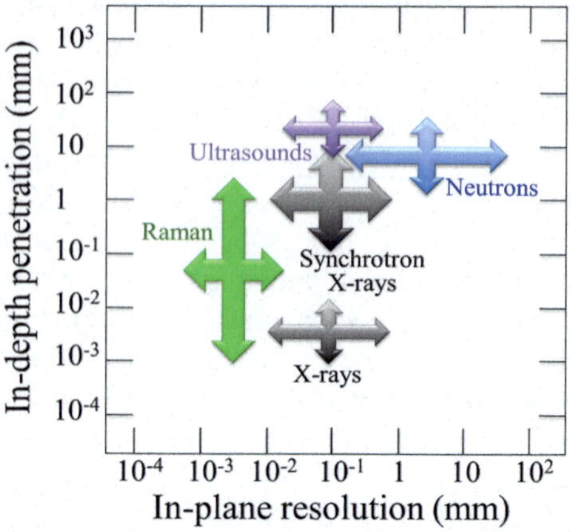

Fig. 25.1 Comparison of in-plane and in-depth spatial resolution among various stress measurement methods

(2001)). However, all the above methods have a common shortcoming in their spatial resolution, which can hardly be pushed toward the single micrometer. Figure 25.1 shows an indicative plot, which summarizes the in-plane and in-depth spatial resolution of different stress probes. From the plot, it is clear that none of the above-mentioned probes matches the needs of modern miniaturized devices.

The Raman probe exploits molecular vibrations and its spatial resolution is dictated by the wavelength of the laser used to excite the Raman scattered light. With using conventional lasers in the visible, a spatial resolution in the order of the single micrometer can be easily achieved. The Raman probe is Gaussian in nature and its response function can be easily calibrated according to defocusing procedures on each given material (Atkinson and Jain, 1999). This enables to enhance spatial resolution beyond the diffraction limit. The Raman frequency is quite sensitive to stress and the availability of several bands in the Raman spectrum allows deconvoluting the local stress tensor with a single measurement. Information of crystallographic orientation is also contained in the collected spectrum, provided that preliminary calibrations are made on the investigated material (Pezzotti, 2013, 2017; Pezzotti and Zhu, 2015). In the following sections, we will first show the basics of spatially and tensor-deconvoluted stress assessments by Raman spectroscopy, and then describe two applications of this technique to modern devices.

25.2 Working Algorithms for Stress Assessments by Raman Spectroscopy

25.2.1 *Step 1: Extracting Crystallographic Information with Polarized Probes*

The inherent properties of a phonon with respect to Raman scattering are described by a second-rank Raman tensor, $\mathfrak{R}$, namely a quantity that is represented by a 3x3 matrix of Raman tensor elements (RTE) (Loudon, 1964). The tensor, $\mathfrak{R}$, is related to an expansion of the polarizability tensor in powers of the scalar amplitude of the vibrational displacement of the lattice (to the first order). Therefore, the polarizability tensor and the Raman tensors are simply related through a direct proportionality relationship. Upon exploiting symmetry properties of the phonons, one can deduce the number of independent components of the Raman tensor, which proves always smaller than the total number of nine elements of a second-rank tensor. In the case of the A_1 phonon in a tetragonal crystal of $BaTiO_3$, for example, the number of independent components is only two, given the symmetry in the crystal basal plane, and reduces to one for the B_1 phonon. A more complex and (less symmetric) structure, such as a monoclinic lattice, possesses three and two independent Raman tensor elements for the A_g and B_g phonons, respectively. The Raman tensor formalism has a direct physical meaning. If neither the energy of the incoming nor that of the scattered light is in resonance with an electronic interband excitation of the solid, the Raman tensor represents the partial derivative of the dielectric tensor of the material with respect to the phonon normal coordinate of the vibration for a given frequency of the incoming laser (Cardona and Giintherodt, 1982). This tensorial formalism is based on the simple concept that different phonon branches in a crystal will correspond to different symmetries of vibration and will, thus, be conditional on irreducible representations of the space group of the studied crystal lattice. The selection rules for Raman-active phonons can be determined by standard methods of group theory, while the intensities of Raman-scattered radiations can be rationalized according to the directions of polarized incoming/scattered monochromatic light with respect to the principal axes of the investigated crystal (Loudon, 1964). In other words, the experimental collection of polarized Raman spectra from known crystallographic planes might serve as a means for providing physical insight into the actual symmetry properties of phonon branches or, vice versa, the knowledge of the selection rules for the investigated crystal can be applied to quantitative assessments of unknown crystallographic directions. Polarized Raman experiments can be defined according to the so-called Porto formalism in Porto and Krishnan (1967), namely by means of two distinct systems of Cartesian axes (associated to the incoming and scattered radiation), and described by a total of four rotational indexes. The formalism, expressed as $i(kl)j$, means that the incident light is propagated along the i direction with its electric vector in the k direction, while the Raman scattered light is collected from the j direction with the analyzer so placed that it passes light with the electric vector aligned along the l direction. The symbols outside and inside the brackets refer to directions of light propagation and electric vector, respectively. The relative intensities of a given Raman mode can be described, as follows:

$$I \propto \left|\mathrm{e}_i \mathfrak{R} \mathrm{e}_s\right|^2 \tag{25.1}$$

where I is the scattered Raman intensity, $\mathfrak{R} \equiv \mathfrak{R}_{\mathrm{y}}$ is the second-rank Raman tensor; and e_{i} and e_{s} are the unit polarization vectors of the electric field for incident and scattered light, respectively. In the present experiments, the polarization of the incident light was fixed (i.e., parallel to the y-axis), while both parallel and cross polarization were applied to the Raman scattered light in a backscattered geometry. Such configurations correspond to $z(yy)\bar{z}$ and $z(xy)\bar{z}$, respectively (cf. Fig. 25.2 for our choice of Cartesian axes and Euler angles for the vibrational modes generated in different crystal structures). The configurations expressed as $z(yy)\bar{z}$ and $z(xy)\bar{z}$ will be henceforth simply denoted as "parallel" and "cross" polarization configuration, respectively. The unit polarization vectors can then be explicitly expressed in Cartesian coordinates:

$$\mathrm{e}_{i\ xyz} = \begin{pmatrix} 0 & 1 & 0 \end{pmatrix}, \qquad \mathrm{e}^{\parallel}_{i\ xyz} = \begin{pmatrix} 0 \\ 1 \\ 0 \end{pmatrix}, \qquad \mathrm{e}^{\perp}_{i\ xyz} = \begin{pmatrix} 1 \\ 0 \\ 0 \end{pmatrix} \tag{25.2}$$

where the superscripts, $\parallel$ and $\perp$, refer to parallel and cross configurations, respectively. Taking now into account the matrix of Euler angles, Φ_{xyz} and its inverse $\tilde{\Phi}_{\mathrm{xyz}}$, which enable one transforming the Cartesian system of coordinates associated to the crystallographic frame into that of the laboratory frame, the back-scattered intensity

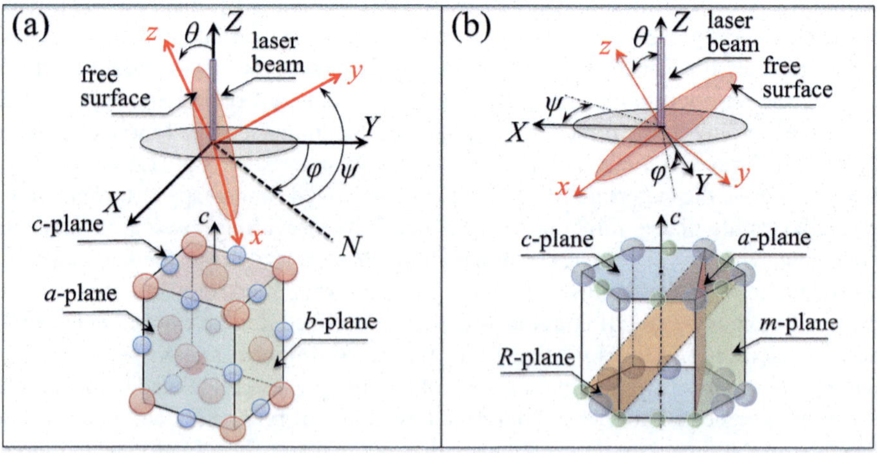

Fig. 25.2 Definition of Cartesian axes and Euler angles for (a) tetragonal and (b) trigonal (corundum) structures

of any Raman band can be expressed as:

$$\mathfrak{R}_{xyz} = \Phi_{xyz}\mathfrak{R}\tilde{\Phi}_{xyz} \tag{25.3}$$

where the transformation matrix of Euler angles and its inverse are given as

$$\Phi_{xyz} = \begin{pmatrix} \cos\theta\cos\varphi\cos\psi - \sin\varphi\sin\psi & \cos\theta\sin\varphi\cos\psi + \cos\varphi\sin\psi & -\sin\theta\cos\psi \\ -\sin\varphi\cos\psi - \cos\theta\cos\varphi\sin\psi & \cos\varphi\sin\psi - \cos\theta\sin\varphi\sin\psi & \sin\theta\sin\psi \\ \sin\theta\cos\varphi & \sin\theta\sin\varphi & \cos\theta \end{pmatrix} \tag{25.4}$$

$$\tilde{\Phi}_{xyz} = \begin{pmatrix} \cos\theta\cos\varphi\cos\psi - \sin\varphi\sin\psi & -\sin\varphi\cos\psi - \cos\theta\cos\varphi\sin\psi & \sin\theta\cos\varphi \\ \cos\theta\sin\varphi\cos\psi + \cos\varphi\sin\psi & \cos\varphi\cos\psi - \cos\theta\sin\varphi\sin\psi & \sin\theta\sin\varphi \\ -\sin\theta\cos\psi & \sin\theta\sin\psi & \cos\theta \end{pmatrix} \tag{25.5}$$

with $0 \leq \theta \leq \pi$, $0 \leq \varphi \leq 2\pi$, and $0 \leq \psi \leq 2\pi$. Euler angles (θ, φ, ψ), Cartesian coordinate crystal system, $X_{\text{cry}}Y_{\text{cry}}Z_{\text{cry}}$, and laboratory system coordinates, $x_{\text{lab}}y_{\text{lab}}z_{\text{lab}}$, as defined for the tetragonal and trigonal crystallographic structures, are shown in Fig. 25.2. Substituting for Eqs. (25.3)-(25.5) into Eq. (25.1), and introducing the Raman tensor, $\mathfrak{R}$, pertaining to each of the investigated crystal structures, it is possible to obtain a series of independent periodic equations in three Euler angles, which represent an expansion of the so-called Raman selection rules of crystals and fully describe the angular dependences of the intensity of the Raman modes for any given crystal structure. Accordingly, an explicit morphology for such "extended" Raman selection rules can be worked out and experimentally validated for any available crystal. As an example, we consider here the tetragonal structure of barium titanate ($BaTiO_3$), which belongs to the C_{4v} (in Schönflies notations) or $4mm$ (in Hermann–Mauguin notations) point group. The irreducible representation of the optical modes, according to group theory, is expressed as (Loudon, 1964; DiDomenico Jr et al, 1968):

$$\Gamma = 3T_{1u} + T_{2u} \tag{25.6}$$

Each of the T_{1u} modes in the irreducible representation splits into $A_1 + E$ modes. These modes can be classified as longitudinal optic (LO) or transverse optic (TO), according to the propagation direction of the scattered light with respect to incoming polarization. This physical circumstance explains why we can observe four modes: A_1(TO), A_1(LO), E(TO), and E(LO) from each T_{1u} mode. T_{2u} is usually referred to as the "silent mode," although it represents a mixed $B_1 + E$ mode. For this reason, it is no longer be silent. However, the spectral separation between its components is small, which impedes their separate measurement (El Marssi et al, 2003). The second-rank tensors, pertaining to $BaTiO_3$ crystal, has been calculated by Loudon (1964) and, for the active Raman phonon modes A_1, B_1, and E can be expressed, as follows:

$$\mathrm{R}_{A_1}=\begin{bmatrix}a\,0\,0\\0\,a\,0\\0\,0\,b\end{bmatrix};\ \mathrm{R}_{B_1}=\begin{bmatrix}c&0&0\\0&-c&0\\0&0&0\end{bmatrix};\ \mathrm{R}_{E(X)}=\begin{bmatrix}0\,0\,e\\0\,0\,0\\e\,0\,0\end{bmatrix};\ \mathrm{R}_{E(Y)}=\begin{bmatrix}0\,0\,0\\0\,0\,e\\0\,e\,0\end{bmatrix} \tag{25.7}$$

where the constants a, b, c, and e represent the four independent Raman tensor elements that describe the Raman response of the tetragonal structure. General angular dependencies on Euler angles of the intensities of different Raman modes in a tetragonal structure can be explicitly given according to Eqs. (25.1), (25.5), and (25.7), as follows:

$$\begin{aligned}I^{||}_{A_1}&=\Psi^{||}\left|b\sin^2\theta\sin^2\psi+a(-\cos\psi\sin\varphi-\cos\theta\cos\varphi\sin\psi)^2\right.\\&\left.+a(\cos\varphi\cos\psi-\cos\theta\sin\varphi\sin\psi)^2\right|^2+\mathrm{H}^{||}\end{aligned} \tag{25.8}$$

$$\begin{aligned}I^{||}_{B_1}&=\Psi^{||}\left|c(\cos\psi\sin\varphi+\cos\theta\cos\varphi\sin\psi)^2\right.\\&\left.-c(\cos\varphi\cos\psi-\cos\theta\sin\varphi\sin\psi)^2\right|^2+\mathrm{H}^{||}\end{aligned} \tag{25.9}$$

$$I^{||}_{E(X)}=\Psi^{||}|2e\sin\theta\sin\psi(\cos\psi\sin\varphi+\cos\theta\cos\varphi\sin\psi)|^2+\mathrm{H}^{||} \tag{25.10}$$

$$I^{||}_{E(Y)}=\Psi^{||}|2e\sin\theta\sin\psi(\cos\varphi\cos\psi-\cos\theta\sin\varphi\sin\psi)|^2+\mathrm{H}^{||} \tag{25.11}$$

$$I^{\perp}_{A_1}=\Psi^{\perp}\left|(a-b)\cos\psi\sin^2\theta\sin\psi\right|^2+\mathrm{H}^{\perp} \tag{25.12}$$

$$I^{\perp}_{B_1}=\Psi^{\perp}\left|\frac{c}{2}(2\cos\theta\cos2\psi\sin2\varphi+(1+\cos^2\theta)\cos2\varphi\sin2\psi)\right|^2+\mathrm{H}^{\perp} \tag{25.13}$$

$$I^{\perp}_{E(X)}=\Psi^{\perp}|e\sin\theta(\cos2\psi\sin\varphi+\cos\theta\cos\varphi\sin2\psi)|^2+\mathrm{H}^{\perp} \tag{25.14}$$

$$I^{\perp}_{E(Y)}=\Psi^{\perp}|e\sin\theta(\cos\varphi\sin2\psi-\cos\theta\sin\varphi\sin2\psi)|^2+\mathrm{H}^{\perp} \tag{25.15}$$

where the parameters $\Psi^{||}$, $\Psi^{\perp}$, $H^{||}$, $H^{\perp}$ are instrumental constants that depend on the optical circuit in the Raman setup and can be preliminary calibrated. They might be different for different Raman bands. Note also that the above constants have a physical nature, but also contain values characteristic of the measurement device (i.e., in particular of the optical circuit adopted). Therefore, the numerical values of the measured constants might differ when measured with different instruments (and/or under even slightly different conditions of the same optical circuit) and need to be calibrated case by case. The Raman intensity of the E mode consists of two distinct components, $E(X)$ and $E(Y)$ (cf. Eqs. (14.14), (25.11) and Eqs. (25.14), (25.15)). The related bands almost fully overlap and only a cumulative Raman emission can be experimentally measured. Such a cumulative emission can be described according to the following equation:

$$I_E=\mathrm{x}I_{E(X)}+(1-\mathrm{x})I_{E(Y)} \tag{25.16}$$

which is valid for both parallel and cross configurations. The weight parameter, x, can be assumed equal to 0.5, if band symmetry is observed, which is often the case.

The Raman spectrum of the tetragonal structure of $BaTiO_3$ presents two bands at around 520 (A_1(TO)) and 720 cm^{-1} (A_1(LO)), which represent the transverse optic mode of the Ti-O stretching vibration in TiO_6 octahedra and its longitudinal optic pair, respectively. An additional band located at around 280 cm^{-1}, also belongs to the A_1(TO) mode, but arises from O-Ti-O bending vibrations. A sharp band located at around 310 cm^{-1} arises from a mix mode B_1 + E(TO) and originates from Ti-O_3 torsional vibrations. Equations (14.12)-(25.16) can thus be applied to individual planes of the tetragonal structure for selected Euler angles to locate four independent working equations needed to experimentally measure the Raman tensor elements in Eq. (14.11). One of the possible choices of working equations by fixing the out-of-plane Euler angles and rotating the Raman probe within the selected crystallographic plane is, as follows:

For the a-plane: $\theta = \frac{\pi}{2}$; $\varphi = 0$

$$RI^{\parallel}_{270/305} = \frac{I^{\parallel}_{A_1}}{I^{\parallel}_{B_1+E}} = \frac{(a\cos^2\psi + b\sin^2\psi)^2}{\cos^2\psi(c^2 + 2e^2\sin^2\psi)} \tag{25.17}$$

$$RI^{\perp}_{270/305} = \frac{I^{\perp}_{A_1}}{I^{\perp}_{B_1+E}} = \frac{2(a-b)^2\sin 2\psi}{c^2 + 2e^2 - (c^2 - 2e^2)\cos 4\psi} \tag{25.18}$$

For the c-plane: $\theta = 0$; $\varphi = 0$

$$RI^{\parallel}_{720/305} = \frac{I^{\parallel}_{A_1+E}}{I^{\parallel}_{B_1+E}} = \frac{a^2}{\left|c(\sin^2\psi - \cos^2\psi)\right|^2} \tag{25.19}$$

$$RI^{\perp}_{720/305} = \frac{I^{\perp}_{A_1+E}}{I^{\perp}_{B_1+E}} = \frac{1}{\left|2c\cos\psi\sin\psi\right|^2} \tag{25.20}$$

where RI represents relative intensity ratios of the bands located by the subscripts. Figure 25.3(a) shows parallel- and cross-polarized Raman spectra collected on the a- and c-plane of $BaTiO_3$ single-crystal together with a deconvolutive fitting of the spectra according to the above-indicated vibrational modes (our choice of Cartesian axes and Euler angles for the tetragonal structure are shown in Fig. 25.2(a)). From the observed periodicity trends obtained upon in-plane rotation (shown in Fig. 25.3(b)), it can be seen that the bands from the A mode are greatly enhanced when the c-axis of the tetragonal structure is perpendicular to the laser polarization vector. Conversely, they undergo an abrupt intensity decrease when the c-axis is parallel to the laser polarization vector. The experimentally determined angular dependence of the intensities of various Raman bands of tetragonal $BaTiO_3$ can be fit to linear combinations of Eqs. (25.17)-(25.20) in order to retrieve the four elements of the Raman tensor of $BaTiO_3$, as follows: a=0.318±0.041, b=-0.871±0.062; c=0.290±0.009; and, e=0.817±0.072. With the precise knowledge of the Raman tensor elements, one can now use Eqs. (14.12)-(25.16) to locate the set of three unknown Euler angles at any location of a $BaTiO_3$ sample. Similar treatments starting from the irreducible representation of the optical modes, according

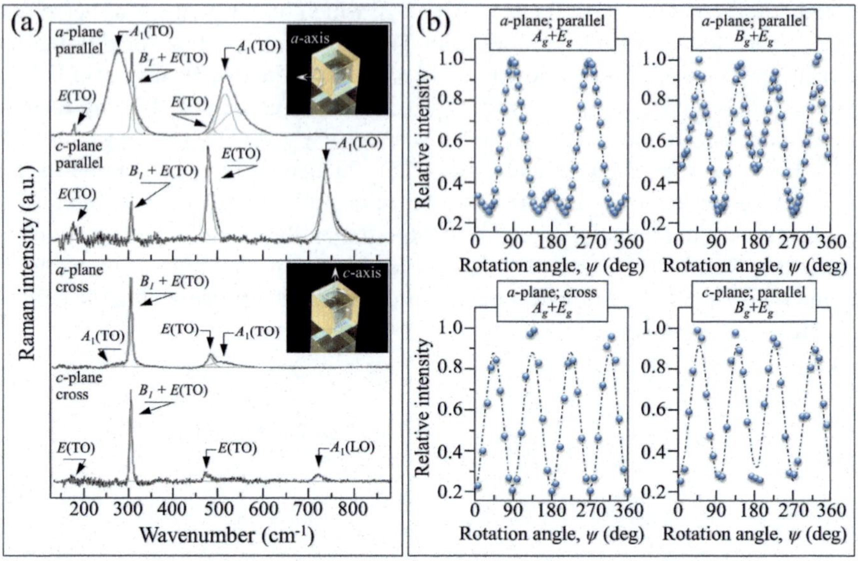

Fig. 25.3 (a) Polarized Raman spectra on different crystallographic planes of tetragonal $BaTiO_3$; and, (b) in-plane angular dependencies of their Raman intensities on selected crystallographic planes (see labels in inset; broken lines are least-square fitting curves to the experimental data dots)

to group theory, could be worked out for different crystallographic structures (Pezzotti, 2013, 2017; Pezzotti and Zhu, 2015). Of interest in this paper are the RTE of the trigonal (corundum) structure of Al_2O_3. The trigonal structure possesses 4 independent RTE, which were precisely determined, as follows (Pezzotti and Zhu, 2015): a=-0.8±0.001, b=0.2±0.001; c=0.015±0.03; and, d=-0.49±0.011.

25.2.2 Deconvoluting the Stress Tensor Elements

The so-called "secular equation" links the strain tensor to the shift in Raman frequency of the vibrational modes available to a given crystal structure. The secular equation can be directly derived from the dynamic equation (in its linear approximation for small deformations) by introducing the variance with strain due to anharmonic effects in the lattice (Ganesan et al, 1970). The phonon deformation potentials (PDP) are constants, which describe the stiffness of crystal lattice vibrations. They represent a set of proportionality parameters to individual components of the strain tensor in the secular equation. Altered values of phonon frequencies under a given strain tensor can be retrieved by diagonalization of the secular equation, whose solution leads to a set of eigenvalues, which relate the Raman frequency to the strain tensor components. The PDP constants thus represent a link between the strain tensor components and the changes induced on phonon frequency, the same way the elastic compliance tensor components link the strain components to stresses (i.e., through the generalized Hooke's law). When a general stress field, σ_{kl}, (with $k, l = x, y, z$) is applied to a single-crystal along arbitrary crystallographic axes, the configuration of the stress tensor expressed in any orthogonal reference frame (i.e., the laboratory reference frame) can be transformed into an equivalent tensor in the crystal reference frame, σ_{ij}, (with $i, j = 1, 2, 3$) by using tensorial rules of coordinate transformation, as follows:

$$\begin{bmatrix} \sigma_{11} & \sigma_{12} & \sigma_{13} \\ \sigma_{21} & \sigma_{22} & \sigma_{23} \\ \sigma_{31} & \sigma_{32} & \sigma_{33} \end{bmatrix} = \begin{bmatrix} \Phi_{11} & \Phi_{12} & \Phi_{13} \\ \Phi_{21} & \Phi_{22} & \Phi_{23} \\ \Phi_{31} & \Phi_{32} & \Phi_{33} \end{bmatrix} \begin{bmatrix} \sigma_{xx} & \sigma_{xy} & \sigma_{xz} \\ \sigma_{yx} & \sigma_{yy} & \sigma_{yz} \\ \sigma_{zx} & \sigma_{zy} & \sigma_{zz} \end{bmatrix} \begin{bmatrix} \tilde{\Phi}_{11} & \tilde{\Phi}_{12} & \tilde{\Phi}_{13} \\ \tilde{\Phi}_{21} & \tilde{\Phi}_{22} & \tilde{\Phi}_{23} \\ \tilde{\Phi}_{31} & \tilde{\Phi}_{32} & \tilde{\Phi}_{33} \end{bmatrix} \quad (25.21)$$

where the transformation matrix, Φ_{ij}, and its inverse, $\tilde{\Phi}_{ij}$, in the crystal reference frame are given by Eqs. (14.9) and (14.10) in terms of three Euler angles. Under a given elastic stress field, the Raman frequencies of each individual vibrational mode in the crystal will be altered due to the anharmonic nature of atomic interactions. It follows that, for any j-th vibrational mode associated with a frequency, ω_j, an effective force constant, K_j, can be defined, which is the second derivative of the crystal potential energy with respect to mode normal coordinates. The change in force constants under stress/strain can be then represented with a matrix, $[\Delta K] = \Delta K_{ij} = K_{ij} - K_{ij}^{(0)}$, whose size is in turn determined by the degeneracy characteristics of the Raman mode. The eigenvalues, λ, of the $[\Delta K]$ matrix are related to the alteration, $\Delta\omega$, of a Raman frequency, ω, under stress with respect to the unstressed frequency, ω_0, as follows:

$$\lambda - \omega^2 - \omega_0^2 \approx 2\omega_0(\omega - \omega_0) = 2\omega_0\Delta\omega \quad (25.22)$$

and can be obtained by solving the crystal-structure dependent equation:

$$\det\{[\Delta K] - \lambda[I]\} = 0 \quad (25.23)$$

where $[I]$ is a unit diagonal matrix. The symmetry of the crystal under stress is determined by the symmetry elements common to both the unstrained crystal and the strain ellipsoid, namely the geometric representation of the three-dimensional state of strain that develops during a homogeneous deformation. Each vibrational mode is then associated to a stress-free frequency, ω_0, and to an effective force constant, K_{ij} (i.e., the second derivative of the crystal potential energy with respect to normal mode coordinates). The PDP constants, K_{ijkl}, are the components of a fourth-rank tensor, connecting the strain tensor to the changes of the elements of the matrix of phonon force constants, ΔK_{ij}, as follows:

$$\Delta K_{ij} = K_{ij} - K_{ij}^{(0)} = K_{ijkl}\varepsilon_{kl} \quad (25.24)$$

The dynamical equation that relates each non-degenerate optical phonon of the crystal structure to terms linear in strain has the following form:

$$K_{ij} = K_{ij}^{(0)} + \sum_{k,l=1}^{3} \varepsilon_{kl} K_{ij,kl}^{(\varepsilon)} = \omega_0^2 \delta_{ij} + \sum_{k,l=1}^{3} \varepsilon_{kl} \frac{\partial K_{ij}}{\partial \varepsilon_{kl}} \tag{25.25}$$

where δ_{ij} is the Kronecker delta in its usual meaning and the strain tensor, ϵ_{ij}, along the principal crystallographic axes of a general crystal is related to the stress tensor, σ_{kl}, through the usual equation:

$$\varepsilon_{ij} = S_{ijkl} \sigma_{kl} \tag{25.26}$$

where S_{ijkl} is the matrix of elastic compliance constants. A quantity referred to as the perturbing potential, V, which is the summation of terms linear in strain (directly related to the basis functions of the crystal symmetry), can be also defined as:

$$V = \sum_{i,j=1}^{3} V_{ij} \varepsilon_{ij} \tag{25.27}$$

where the constants V_{ij}, are the PDP constants of the crystal, namely the elements of a (symmetric) second-rank tensor. A comparison between Eqs. (25.25) and (25.27) clarifies the relationship between effective force constants, V_{ij}, and the perturbing potential, V. The general procedure in Raman assessments of stress tensors, thus, will consist in setting first the total change in energy of the vibrational state of the crystal according to Eq. (25.27) and to the basis functions characteristic of the studied crystal structure (i.e., known from group theory) (DiDomenico Jr et al, 1968). Then, the changes in phonon force constants, ΔK_{ij}, can be made explicit and the contributions of different vibrational modes separated by means of Eq. (25.25). Finally, rearranging from Eqs. (25.22) and (25.23), working equations for the spectral shifts of each vibrational mode can be obtained as a function of the PDP constants and of strain (or stress through Eq. (25.26)) tensor components. The obtained equations fully describe the shift of individual Raman bands in response to a given tensorial stress/strain state. The tensorial relationship between strain and stress, applied along the principal crystallographic axes for the trigonal structure of sapphire, can explicitly be expressed, as follows:

$$\begin{pmatrix} \varepsilon_{11} \\ \varepsilon_{22} \\ \varepsilon_{33} \\ 2\varepsilon_{23} \\ 2\varepsilon_{13} \\ 2\varepsilon_{12} \end{pmatrix} = \begin{bmatrix} S_{11} & S_{12} & S_{13} & S_{14} & 0 & 0 \\ S_{12} & S_{11} & S_{13} & -S_{14} & 0 & 0 \\ S_{13} & S_{13} & S_{33} & 0 & 0 & 0 \\ S_{14} & -S_{14} & 0 & S_{44} & 0 & 0 \\ 0 & 0 & 0 & 0 & S_{44} & 2S_{14} \\ 0 & 0 & 0 & 0 & 2S_{14} & S_{66} \end{bmatrix} \begin{pmatrix} \sigma_{11} \\ \sigma_{22} \\ \sigma_{33} \\ \sigma_{23} \\ \sigma_{13} \\ \sigma_{12} \end{pmatrix} \tag{25.28}$$

where S_{ij} are the elastic compliance constants of the sapphire crystal, with values reported as: $S_{11} = 2.3 \times 10^{-12}\text{Pa}^{-1}$, $S_{12} = -0.7 \times 10^{-12}\text{Pa}^{-1}$, $S_{13} = -0.4 \times 10^{-12}\text{Pa}^{-1}$, $S_{33} = 2.2 \times 10^{-12}\text{Pa}^{-1}$, $S_{44} = 6.8 \times 10^{-12}\text{Pa}^{-1}$, and $S_{14} = 0.5 \times 10^{-12}\text{Pa}^{-1}$ (Goto et al, 1989). The perturbing potential, V, for the trigonal (corundum) structure can be deduced from the irreducible representation of the sapphire crystal, according to group theory: $\Gamma = 2A_{1g} + 2A_{1u} + 3A_{2g} + 3A_{2u} + 5E_g + 5E_u$. The components of the tensor operator, V_{ij}, of the trigonal structure can be grouped as follows:

$$V = \sum_{i,j} V_{ij}\varepsilon_{ij} = \frac{1}{2}(V_{11} + V_{22})(\varepsilon_{11} + \varepsilon_{22}) + V_{33}\varepsilon_{33} + \frac{1}{2}[(V_{11} - V_{22})(\varepsilon_{11} - \varepsilon_{22}) + V_{12}\varepsilon_{12}] + 2V_{13}\varepsilon_{13} + 2V_{23}\varepsilon_{23} \tag{25.29}$$

In Eq. (25.29), $(V_{11} + V_{22})$ and V_{33} belong to A_{1g}, $(V_{11} - V_{22})$ and V_{13} belong to $E_g(X)$, while V_{12} and V_{23} belong to $E_g(Y)$, according to the basis functions appropriate for the representations of the A_{1g} and E_g modes of the D_{3d} point group. Under stress, the energy of a vibrational state belonging to A_{1g} will shift by an amount:

$$\frac{1}{2}\langle A_{1g}|V_{11} + V_{22}|A_{1g}\rangle(\varepsilon_{11} + \varepsilon_{22}) + \langle A_{1g}|V_{33}|A_{1g}\rangle\varepsilon_{33} \tag{25.30}$$

with the remaining terms vanishing according to the orthogonality theorem. Consequently, for the non-degenerated A_{1g} mode, Eq. (25.23) can be rewritten as:

$$\left|K_1^A\varepsilon_{11} + K_2^A\varepsilon_{22} + K_3^A\varepsilon_{33} - \lambda\right| = 0 \tag{25.31}$$

According to Eqs. (25.22) and (25.30), the spectral shift of the A_{1g} mode is, therefore:

$$\Delta\omega_A = \omega_A - \omega_{0,A} = \frac{1}{2\omega_{0,A}}(K_1^A\varepsilon_{11} + K_2^A\varepsilon_{22} + K_3^A\varepsilon_{33}) \tag{25.32}$$

where $K_1^A = K_2^A = \frac{1}{2}\langle A_{1g} \mid V_{11} + V_{22} \mid A_{1g}\rangle$ and $K_3^A = \langle A_{1g} \mid V_{33} \mid A_{1g}\rangle$ are the PDP constants. Equations (25.28) and (25.32) can then be re-arranged to give:

$$\begin{aligned}\Delta\omega_A = &\frac{1}{2\omega_{0,A}}\big[(K_1^A S_{11} + K_2^A S_{12} + K_3^A S_{13})\sigma_{11} \\ &+ (K_2^A S_{11} + K_1^A S_{12} + K_3^A S_{13})\sigma_{22} \\ &+ (K_1^A S_{13} + K_2^A S_{13} + K_3^A S_{33})\sigma_{33}\big]\end{aligned} \tag{25.33}$$

According to Eqs. (25.24), (25.25), (25.21), and (25.33), the spectral shift of the Raman mode in response to stress in the laboratory reference frame can explicitly be expressed as:

$$
\begin{aligned}
\Delta\omega_A = \frac{1}{2\omega_{0,A}}\Big\{&\big[(K_1^A S_{11} + K_2^A S_{12} + K_3^A S_{13})(\cos\theta\cos\varphi\cos\psi - \sin\psi\sin\varphi)^2 \\
&+ (K_2^A S_{11} + K_1^A S_{12} + K_3^A S_{13})(\cos\psi\sin\varphi + \cos\theta\cos\varphi\sin\psi)^2 \\
&+ (K_1^A S_{13} + K_2^A S_{13} + K_3^A S_{33})\sin^2\theta\cos^2\varphi\big]\sigma_{xx} \\
&+ \big[(K_1^A S_{11} + K_2^A S_{12} + K_3^A S_{13})(-\sin 2\varphi\sin^2\psi + \cos^2\theta\sin 2\varphi\cos^2\psi \\
&+ \cos\theta\cos 2\varphi\sin 2\psi) + (K_2^A S_{11} + K_1^A S_{12} + K_3^A S_{13})(\cos^2\theta\sin 2\varphi\sin^2\psi \\
&- \sin 2\varphi\cos^2\psi - \cos\theta\cos 2\varphi\sin 2\psi) \\
&+ (K_1^A S_{13} + K_2^A S_{13} + K_3^A S_{13})\cos^2\theta\sin 2\varphi\big]\sigma_{xy} \\
&+ \big[(K_1^A S_{11} + K_2^A S_{12} + K_3^A S_{13})(\sin\varphi\sin\theta\sin 2\psi - \cos\varphi\sin 2\theta\cos^2\psi) \\
&+ (K_2^A S_{11} + K_1^A S_{12} + K_3^A S_{13})(-\sin\varphi\sin\theta\sin 2\psi - \cos\varphi\sin 2\theta\sin^2\psi) \\
&+ (K_1^A S_{13} + K_2^A S_{13} + K_3^A S_{33})(\cos\varphi\sin 2\theta\big]\sigma_{xz} \\
&+ \big[(K_1^A S_{11} + K_2^A S_{12} + K_3^A S_{13})(\cos\theta\sin\varphi\cos\psi + \sin\psi\cos\varphi)^2 \\
&+ (K_2^A S_{11} + K_1^A S_{12} + K_3^A S_{13})(-\cos\theta\cos\varphi\sin\psi + \cos\psi\sin\varphi)^2 \\
&+ (K_1^A S_{13} + K_2^A S_{13} + K_3^A S_{33})\sin^2\theta\sin^2\varphi\big]\sigma_{yy} \\
&+ \big[(K_1^A S_{11} + K_2^A S_{12} + K_3^A S_{13})(-\cos\varphi\sin\theta\sin 2\psi - \sin\varphi\sin 2\theta\cos^2\psi) \\
&+ (K_2^A S_{11} + K_1^A S_{12} + K_3^A S_{13})(-\cos\varphi\sin\theta\sin 2\psi - \sin\varphi\sin 2\theta\sin^2\psi) \\
&+ (K_1^A S_{13} + K_2^A S_{13} + K_3^A S_{33})\sin\varphi\sin 2\theta\big]\sigma_{yz} \\
&+ [(K_1^A S_{11} + K_2^A S_{12} + K_3^A S_{13})\sin^2\theta\cos^2\psi \\
&+ (K_2^A S_{11} + K_1^A S_{12} + K_3^A S_{13})\sin^2\theta\sin^2\psi \\
&+ (K_1^A S_{13} + K_2^A S_{13} + K_3^A S_{33})]\sigma_{zz}\Big\}
\end{aligned}
\tag{25.34}
$$

In order to express in close form the elements of the matrix V_{ij} for the doubly degenerate states X and Y of the E_g mode, a decomposition of the products of the basis functions of the class D_{3d} should preliminary be performed. Accordingly, the secular equation for the E_g mode can be expressed in terms of PDP constants, K_1^E, K_2^E, K_3^E and K_4^E, as follows:

$$
\begin{vmatrix}
K_1^E(\varepsilon_{11}+\varepsilon_{22}) + K_2^E\varepsilon_{33} + K_3^E(\varepsilon_{11}-\varepsilon_{22}) + K_4^E\varepsilon_{13} - \lambda & K_3^E\varepsilon_{12} + K_4^E\varepsilon_{23} \\
K_3^E\varepsilon_{12} + K_4^E\varepsilon_{23} & K_1^E(\varepsilon_{11}+\varepsilon_{22}) + K_2^E\varepsilon_{33} - K_3^E(\varepsilon_{11}-\varepsilon_{22}) - K_4^E\varepsilon_{13} - \lambda
\end{vmatrix} = 0
\tag{25.35}
$$

where

$$K_1^E = \frac{1}{2}\langle\psi(A_{1g})|V_{11} + V_{22}\rangle \tag{25.36}$$

$$K_2^E = \langle\psi(A_{1g})|V_{33}\rangle \tag{25.37}$$

$$K_3^E = \frac{1}{2}\langle\psi_X(E_g)|V_{11} - V_{22}\rangle = \frac{1}{2}\langle\psi_Y(E_g)|V_{12}\rangle \tag{25.38}$$

$$K_4^E = 2\langle\psi_X(E_g)|V_{13}\rangle = 2\langle\psi_Y(E_g)|V_{23}\rangle \tag{25.39}$$

According to Eqs. (25.22) and (25.35), the spectral shift can be expressed as:

$$\Delta\omega_{E(X,Y)} = \omega_{E(X,Y)} - \omega_{0,E(X,Y)} = \frac{1}{2\omega_{0,E(X,Y)}}\left[K_1^E(\epsilon_{11}+\epsilon_{22}) + K_2^E\epsilon_{33} \pm \sqrt{[K_3^E(\epsilon_{11}-\epsilon_{22}) + K_4^E\epsilon_{13}]^2 + (K_3^E\epsilon_{12}+K_4^E\epsilon_{23})^2}\right] \tag{25.40}$$

By considering the relationship between strain and stress in Eq. (25.28), Eq. (25.40) can be rewritten, as follows:

$$\Delta\omega_{E(X,Y)} = \frac{1}{2\omega_{0,E}}\Bigg\{\left[K_1^E(S_{11}+S_{12}) + K_2^E S_{13}\right](\sigma_{11}+\sigma_{22}) + (2K_1^E S_{13} + K_2^E S_{33})\sigma_{33} \pm \Bigg\{\left[K_3^E(S_{11}-S_{12})(\sigma_{11}-\sigma_{22}) + K_4^E(S_{44}\sigma_{13}+2S_{14}\sigma_{12})\right]^2 + \left\{2K_3^E\left[S_{14}\sigma_{13}+(S_{11}-S_{12})\sigma_{12}\right] + K_4^E\left[S_{14}(\sigma_{11}-\sigma_{22}) + S_{44}\sigma_{23}\right]\right\}^2\Bigg\}^{\frac{1}{2}}\Bigg\} \tag{25.41}$$

The overall peak shift of the combined signal can be set as the average of individual peak-shifts weighted by their relative intensities:

$$\Delta\bar{\omega} = \frac{\sum_{i=1}^{2}(I_i\Delta\omega_i)}{\sum_{i=1}^{2} I_i} \tag{25.42}$$

For the cases of cross and parallel polarization, Eq. (25.42) can then be re-written as:

$$\Delta\bar{\omega}_E^{\perp\parallel} = \frac{\Delta\omega_X I_{E_g(X)}^{\perp\parallel} + \Delta\omega_Y I_{E_g(Y)}^{\perp\parallel}}{I_{E_g(X)}^{\perp\parallel} + I_{E_g(Y)}^{\perp\parallel}} \tag{25.43}$$

which can in turn be expanded into:

$$\Delta\bar{\omega}_E^{\perp\parallel} = \frac{1}{2\omega_{0,E}}\Bigg\{\left[K_1^E(S_{11}+S_{12}) + K_2^E S_{13}\right](\sigma_{11}+\sigma_{22}) + (2K_1^E S_{13} + K_2^E S_{33})\sigma_{33} + \frac{I_{E_g(X)}^{\perp\parallel} - I_{E_g(Y)}^{\perp\parallel}}{2\omega_{0,E}(I_{E_g(X)}^{\perp\parallel} - I_{E_g(Y)}^{\perp\parallel})}\Bigg\{\left[K_3^E(S_{11}-S_{12})(\sigma_{11}-\sigma_{22}) + K_4^E(S_{44}\sigma_{13}+2S_{14}\sigma_{12})\right]^2 + \left\{2K_3^E\left[S_{14}\sigma_{13}+(S_{11}-S_{12})\sigma_{12}\right] + K_4^E[S_{14}(\sigma_{11}-\sigma_{22}) + S_{44}\sigma_{23}]\right\}^2\Bigg\}^{\frac{1}{2}}\Bigg\} \tag{25.44}$$

or, in the laboratory reference frame:

$$
\begin{aligned}
\Delta\bar{\omega}_E^{\perp\parallel} = {} & \frac{1}{2\omega_{0,E}}\Bigg\{\Big[K_1^E(S_{11}+S_{12})+K_2^E S_{13}\Big]\Big[\sigma_{xx}(\cos^2\theta\cos^2\varphi+sin^2\varphi)+\sigma_{xy}(-sin2\varphi+cos^2\theta\sin 2\varphi)\\
&+\sigma_{xz}(-cos\varphi\sin 2\theta)+\sigma_{yy}(cos^2\theta+1)(sin^2\varphi\cos^2\psi+cos^2\varphi\sin^2\psi)+\sigma_{yz}(-\sin\varphi\sin 2\theta)+\sigma_{zz}sin^2\theta\Big]\\
&+(2K_1^E S_{13}+K_2^E S_{33})\Big[\sigma_{xx}\sin^2\theta\cos^2\varphi+\sigma_{xy}\cos^2\theta\sin 2\varphi+\sigma_{xz}\cos\varphi\sin 2\theta+\sigma_{yy}\sin^2\theta\sin^2\varphi\\
&+\sigma_{yz}\sin\varphi\sin 2\theta+\sigma_{zz}\cos^2\theta\Big]\\
&+\frac{I_{E_g(X)}^{\parallel\perp}-I_{E_g(Y)}^{\parallel\perp}}{I_{E_g(X)}^{\parallel\perp}+I_{E_g(Y)}^{\parallel\perp}}\times\Bigg\{\{K_3^E(S_{11}-S_{12})\{\sigma_{xx}(\cos^2\theta\cos^2\varphi\cos 2\psi-\cos 2\psi\sin^2\varphi-\sin 2\varphi\cos\theta\sin 2\psi)\\
&+\sigma_{xy}(\sin 2\varphi\cos 2\psi(1+\cos^2\theta)+2\cos\theta\cos 2\varphi\sin 2\psi)\\
&+\sigma_{xz}(2sin\varphi\sin\theta\sin 2\psi-\cos\varphi\sin 2\theta\cos 2\psi)\\
&+\sigma_{yy}\Big[(\cos^2\theta-1)(\sin^2\varphi\cos^2\psi+\sin^2\psi\cos^2\varphi)+\cos\theta\sin 2\psi\sin 2\varphi\Big]\\
&+\sigma_{yz}(-2cos\varphi\sin\theta\sin 2\psi-\sin\varphi\sin 2\theta\cos 2\psi)+\sigma_{zz}\sin^2\theta\cos 2\psi\\
&+K_4^E\{S_{44}\Big\{\sigma_{xx}\Big[\cos\theta\sin\theta\cos^2\varphi\cos\psi-\sin\theta\cos\varphi\sin\varphi\sin\psi\Big]\\
&+\sigma_{xy}\Big[\cos\theta\sin\theta\sin 2\varphi\cos\psi+\sin\theta\cos 2\varphi\sin\psi\Big]+\sigma_{xz}(\cos 2\theta\cos\varphi\cos\psi-\cos\theta\sin\varphi\sin\psi)\\
&+\sigma_{yy}\Big[\cos\theta\sin\theta\sin^2\varphi\cos\psi+\sin\theta\sin\varphi\cos\varphi\sin\psi\Big]+\sigma_{yz}(\cos 2\theta\sin\varphi\cos\psi+\cos\theta\cos\varphi\sin\psi)\\
&-\sigma_{zz}\cos\theta\sin\theta\cos\psi\Big\}+2S_{14}\Big\{\sigma_{xx}\Big[(\sin^2\varphi-\cos^2\theta\cos^2\varphi)\sin\psi\cos\psi-\cos\theta\sin\varphi\cos\varphi\cos 2\psi\Big]\\
&+\sigma_{xy}[\cos\theta\cos 2\varphi\cos 2\psi-\sin 2\varphi\sin\psi\cos\psi(1+\cos^2\theta)]\\
&+\sigma_{xz}(\cos\varphi\sin 2\theta\sin\psi\cos\psi+\sin\varphi\sin\theta\cos 2\psi)+\\
&+\sigma_{yy}\Big[\cos\theta\sin\varphi\cos\varphi\cos 2\psi+\sin\psi\cos\psi(\cos^2\varphi-\cos^2\theta\sin^2\varphi)\Big]\\
&+\sigma_{yz}(\sin\varphi\sin 2\theta\sin\psi\cos\psi-\cos\varphi\sin\theta\cos 2\psi)\\
&-\sigma_{zz}\sin^2\theta\sin\psi\cos\psi\Big\}\Big\}^2+\Big\{2K_3^E\Big[S_{14}\Big\{\sigma_{xx}\Big[\cos\theta\sin\theta\cos^2\varphi\cos\psi-\sin\theta\cos\varphi\sin\varphi\sin\psi\Big]\\
&+\sigma_{xy}\Big[cos\theta\sin\theta\sin 2\varphi\cos\psi+\sin\theta\cos 2\varphi\sin\psi\Big]+\sigma_{xz}(\cos 2\theta\cos\varphi\cos\psi-\cos\theta\sin\varphi\sin\psi)\\
&+\sigma_{yy}\Big[\cos\theta\sin\theta\sin^2\varphi\cos\psi+\sin\theta\sin\varphi\cos\varphi\sin\psi\Big]+\sigma_{yz}(\cos 2\theta\sin\varphi\cos\psi+\cos\theta\cos\varphi\sin\psi)\\
&-\sigma_{zz}\cos\theta\sin\theta\cos\psi\Big\}\\
&(S_{11}-S_{12})\Big\{\sigma_{xx}\Big[(\sin^2\varphi-\cos^2\theta\cos^2\varphi)\sin\psi\cos\psi-\cos\theta\sin\varphi\cos\varphi\cos 2\psi\Big]\\
&+\sigma_{xy}\Big[\cos\theta\cos 2\varphi\cos 2\psi-\sin 2\varphi\sin\psi\cos\psi(1+\cos^2\theta)\Big]\\
&+\sigma_{xz}(\cos\varphi\sin 2\theta\sin\psi\cos\psi+\sin\varphi\sin\theta\cos 2\psi)\\
&+\sigma_{yy}\Big[\cos\theta\sin\varphi\cos\varphi\cos 2\psi+\sin\psi\cos\psi(\cos^2\varphi-\cos^2\theta\sin^2\varphi)\Big]\\
&+\sigma_{yz}(\sin\varphi\sin 2\theta\sin\psi\cos\psi-\cos\varphi\sin\theta\cos 2\psi)-\sigma_{zz}\sin^2\theta\sin\psi\cos\psi\Big\}\Big]\\
&+K_4^E\Big\{S_{14}\Big\{\sigma_{xx}(\cos^2\theta\cos^2\varphi\cos 2\psi-\cos 2\psi\sin^2\varphi-\sin 2\varphi\cos\theta\sin 2\psi)\\
&+\sigma_{xy}(\sin 2\varphi\cos 2\psi(1+\cos^2\theta)+2\cos\theta\cos 2\varphi\sin 2\psi)+\sigma_{xz}(2\sin\varphi\sin\theta\sin 2\psi-cos\varphi\sin 2\theta\cos 2\psi)\\
&+\sigma_{yy}\Big[(\cos^2\theta-1)(\sin^2\varphi\cos^2\psi+\sin^2\psi\cos^2\varphi)+\cos\theta\sin 2\psi\sin 2\varphi\Big]\\
&+\sigma_{yz}(-2\cos\varphi\sin\theta\sin 2\psi-\sin\varphi\sin 2\theta\cos 2\psi)+\sigma_{zz}\sin^2\theta\cos 2\psi\Big\}\\
&+S_{44}\Big\{\sigma_{xx}\Big[-\cos\theta\sin\theta\cos^2\varphi\sin\psi-\sin\theta\cos\varphi\sin\varphi\cos\psi\Big]+\sigma_{xy}\Big[-cos\theta\sin\theta\sin 2\varphi\sin\psi+sin\theta\cos 2\varphi\cos\psi\Big]\\
&+\sigma_{xz}(-\cos 2\theta\cos\varphi\sin\psi-\cos\theta\sin\varphi\cos\psi)+\sigma_{yy}\Big[-\cos\theta\sin\theta\sin^2\varphi\sin\psi+\sin\theta\sin\varphi\cos\varphi\cos\psi\Big]\\
&+\sigma_{yz}(-cos2\theta\sin\varphi\sin\psi+\cos\theta\cos\varphi\cos\psi)+\sigma_{zz}\cos\theta\sin\theta\sin\psi\Big\}\Big\}\Big\}^2\Big\}^{\frac{1}{2}}\Big\}\Big\}
\end{aligned}
\tag{25.45}
$$

As described in in the previous section, local information of crystal orientation in a sample with the corundum structure (i.e., the local values of the Euler angles, $(\theta\varphi\psi)$ can be retrieved, based on the knowledge of the RTE constants, from the angular dependencies of experimentally obtained polarized Raman intensities. Provided that also the PDP constants of sapphire become known, a tensorial algorithm based on Eqs. (25.34) and (25.45) applied to different bands could be built up, with the related Euler angles determined from best fitting the measured intensity variations as a function of in-plane rotation angle. A simple approach to PDP calibrations consists in generating a known uniaxial stress field along a known direction of the investigated crystal, for example, in a four-point flexure jig. The jig, equipped with a load-cell, is then placed under the Raman microprobe, and Raman line scans can be performed along its thickness to reveal spectral shifts along the bar thickness. Figure 25.4(a) shows a schematic draft of the bending calibration setup with the selected Euler angles and laboratory Cartesian system. The PDP calibration method using the four-point flexure configuration involves both compressive and tensile uniaxial stresses, whose maxima, $\pm\sigma_{max}$, are of the same magnitude. The uniaxial (elastic) stress field varies linearly along the bar thickness, as a function of the abscissa, y, as follows:

$$\sigma_{\mathrm{xx}}(y) = \frac{2y}{h}\sigma_{max} = \frac{2y}{h} \times \frac{3P(L-I)}{2wh^2} \tag{25.46}$$

where P is the (known) applied load, w and h are the width and the thickness of the bending bar, respectively, and L and l are the large and the small span of the bending jig, respectively (cf. Fig. 25.4(a)). Figure 25.5(a) shows a series of polarized Raman spectra taken on different crystallographic planes. Upon applying a given bending load, P, a series of spectra can be collected at locations along the specimen

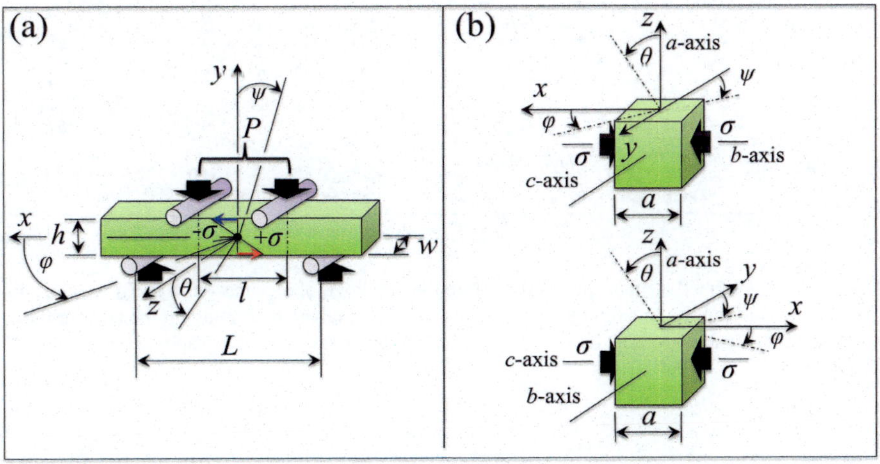

Fig. 25.4 (a) Bending and (b) compression calibration setups and the related definition of Cartesian coordinates and Euler angles

thickness, as a function of the abscissa, y. Since the uniaxial stress, σ_{xx} , is linked to the abscissa, y, through Eq. (25.46), the experimentally collected dependence, $\Delta\omega(y)$, can then be converted into a dependence, $\Delta\omega(\pm\sigma_{xx})$. For this simple case of uniaxial stress, Eqs. (25.34) and (25.45), for A_{1g} and E_g modes, respectively, can be greatly simplified to become:

$$\Delta\omega_A^{\|\perp} = \Delta\omega_A = \frac{1}{2\omega_{0,A}}\Big\{\big[(K_1^A S_{11} + K_2^A S_{12} + K_3^A S_{13})(\cos\theta\cos\varphi\cos\psi - \sin\psi\sin\varphi)^2 + (K_2^A S_{11} K_{11}^A S_{12} + K_3^A S_{13})(\cos\psi\sin\varphi + \cos\theta\cos\varphi\sin\psi)^2 + (K_1^A S_{13} K_2^A S_{13} + K_3^A S_{33})\sin^2\theta\cos^2\varphi\big]\sigma_{\mathrm{xx}}\Big\} \tag{25.47}$$

and

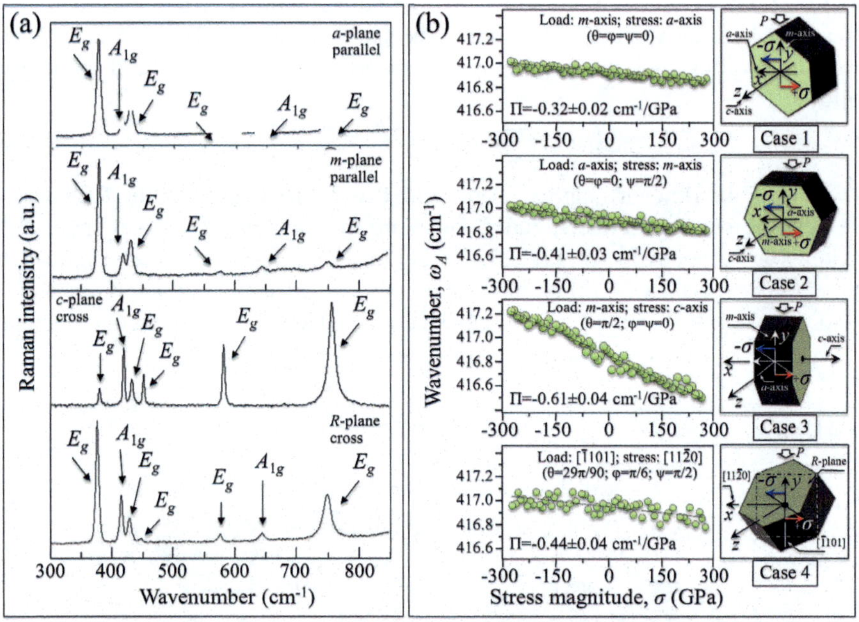

Fig. 25.5 (a) Polarized Raman spectra on different crystallographic planes of trigonal Al_2O_3; and, (b) bending calibration plots (and their best linear fits) as obtained by impinging the laser probe upon different crystallographic faces of a sapphire single-crystal (crystallographic planes as defined in Fig. 25.2(b))

$$
\begin{aligned}
\Delta\bar{\omega}_E^{\|\perp} = \frac{\sigma_{xx}}{2\omega_{0,E}} & \Bigg\{ \Bigg[K_1^E(S_{11}+S_{12}) + K_2^E S_{13} \Bigg] (\cos^2\theta\cos^2\varphi + \sin^2\varphi) + (2K_1^E S_{13} + K_2^E S_{33})\sin^2\theta\cos^2\varphi \\
& + \frac{I_{E_{g(X)}}^{\|\perp} - I_{E_{g(Y)}}^{\|\perp}}{I_{E_{g(X)}}^{\|\perp} + I_{E_{g(Y)}}^{\|\perp}} \times \Bigg\{ \Bigg\{ K_3^E(S_{11}-S_{12})(\cos^2\theta\cos^2\varphi\cos 2\psi - \cos 2\psi\sin^2\varphi - \sin 2\varphi\cos\theta\sin 2\psi) \\
& + K_4^E \Bigg\{ S_{44}(\cos\theta\sin\theta\cos^2\varphi\cos\psi - \sin\theta\cos\varphi\sin\varphi\sin\psi) \\
& + 2S_{14} \Bigg[(\sin^2\varphi - \cos^2\theta\cos^2\varphi)\sin\psi\cos\psi - \cos\theta\sin\varphi\cos\varphi\cos 2\psi \Bigg] \Bigg\} \Bigg\}^2 \\
& + \Bigg\{ 2K_3^E \Bigg\{ S_{14}(\cos\theta\sin\theta\cos^2\varphi\cos\psi - \sin\theta\cos\varphi\sin\varphi\sin\psi) \\
& + (S_{11}-S_{12}) \Bigg[(\sin^2\varphi - \cos^2\theta\cos^2\varphi)\sin\psi\cos\psi - \cos\theta\sin\varphi\cos\varphi\cos 2\psi \Bigg] \Bigg\} \\
& + K_4^E \Bigg[S_{14}(\cos^2\theta\cos^2\varphi\cos 2\psi - \cos 2\psi\sin^2\varphi - \sin 2\varphi\cos\theta\sin 2\psi) \\
& + S_{44}(-\cos\theta\sin\theta\cos^2\varphi\sin\psi - \sin\theta\cos\varphi\sin\varphi\cos\psi) \Bigg] \Bigg\}^2 \Bigg\}^{\frac{1}{2}} \Bigg\}
\end{aligned}
\tag{25.48}
$$

Under the assumption of a linearly elastic deformation of the bending bar, the plot of spectral shifts, $\Delta\omega(\sigma_{xx})$, in the interval, $-\sigma_{max} \leq \sigma_{xx} \leq \sigma_{max}$, is linear. The slope of such a plot, Π , depends on the vibrational mode analyzed and is related to a specific combination of PDP constants, as governed by Eqs. (25.47) and (25.48). It should be noted that, as far as the corundum structure is concerned, the full set of PDP constants includes 7 independent values, namely 3 and 4 constants for the A_{1g} and the E_g mode, respectively. In other words, one needs to carry out at least four uniaxial bending experiments along different crystallographic directions. Four different bending calibration tests with a uniaxial stress field were made in 4 different configurations, as follows: (Case 1) loading direction, m-axis; stress direction, a-axis ($\theta = \varphi = \psi = 0$); (Case 2) loading direction, a-axis; stress direction, m-axis ($\theta = \varphi = 0; \psi = \frac{\pi}{2}$); (Case 3) loading direction, m-axis; stress direction, c-axis ($\theta = \frac{\pi}{2}; \varphi = \psi = 0$); and, (Case 4) loading along direction, $[\bar{1}101]$; stress direction, $[11\bar{2}0]$ ($\theta = \frac{29\pi}{90}; \varphi = \frac{\pi}{6}; \psi = \frac{\pi}{2}$). The four cases mentioned above are schematically depicted in Fig. 25.5(b) with reference to the orientation of the corundum structure depicted in Fig. 25.2(b). The equations derived from Eqs. (25.47) and (25.48), which are pertinent to the above 4 cases and give the predicted slopes of the plots $\Delta\omega(\sigma_{xx})$, can be listed, as follows:

Case 1:

$$
\Delta\omega_A^{\|\perp} = \frac{K_1^A S_{11} + K_2^A S_{12} + K_3^A S_{13}}{2\omega_{0,A}} \sigma_{xx} \tag{25.49}
$$

$$
\Delta\bar{\omega}_E^{\|\perp} = \frac{\sigma_{xx}}{2\omega_{0,E}} \Bigg\{ K_1^E(S_{11}+S_{12}) + K_2^E S_{13} + \frac{I_{E_{g(X)}}^{\|\perp} - I_{E_{g(Y)}}^{\|\perp}}{I_{E_{g(X)}}^{\|\perp} + I_{E_{g(Y)}}^{\|\perp}} \times \sqrt{\left(K_3^E(S_{11}-S_{12})^2 + (K_4^E S_{14} \right)^2} \Bigg\} \tag{25.50}
$$

Case 2:

$$
\Delta\omega_A^{\|\perp} = \frac{K_2^A S_{11} + K_1^A S_{12} + K_3^A S_{13}}{2\omega_{0,A}} \sigma_{xx} \tag{25.51}
$$

$$\Delta\bar{\omega}_{E}^{\parallel\perp}=\frac{\sigma_{xx}}{2\omega_{0,E}}\left\{K_1^E(S_{11}+S_{12})+K_2^E S_{13}+\frac{I_{E_g(X)}^{\parallel\perp}-I_{E_g(Y)}^{\parallel\perp}}{I_{E_g(X)}^{\parallel\perp}+I_{E_g(Y)}^{\parallel\perp}}\times\sqrt{\left(K_3^E(S_{11}-S_{12}\right)^2+(K_4^E S_{14})^2}\right\} \tag{25.52}$$

Case 3:

$$\Delta\omega_{A}^{\parallel\perp}=\frac{K_1^A S_{13}+K_2^A S_{13}+K_3^A S_{33}}{2\omega_{0,A}}\sigma_{xx} \tag{25.53}$$

$$\Delta\bar{\omega}_{E}^{\parallel\perp}=\frac{\sigma_{xx}}{2\omega_{0,E}}(2K_1^E S_{13}+K_2^E S_{33}) \tag{25.54}$$

Case 4:

$$\Delta\omega_{A}^{\parallel\perp}=\frac{\sigma_{xx}}{8\omega_{0,A}}\left[(K_1^A S_{11}+K_2^A S_{12}+K_3^A S_{13})+3(K_1^A S_{12}+K_2^A S_{11}+K_3^A S_{13})\cos^2\frac{29\pi}{90}\right.$$
$$\left.+3(K_1^A S_{13}+K_2^A S_{13}+K_3^A S_{33})\sin^2\frac{29\pi}{90}\right] \tag{25.55}$$

$$\Delta\bar{\omega}_{E}^{\parallel\perp}=\frac{\sigma_{xx}}{8\omega_{0,E}}\left\{\left[K_1^E(S_{11}+S_{12})+K_2^E S_{13}\right]\left(3\cos^2\frac{29\pi}{90}+1\right)+3(2K_1^E S_{13}+K_2^E S_{33})\sin^2\frac{29\pi}{90}\right.$$
$$+\frac{I_{E_g(X)}^{\parallel\perp}-I_{E_g(Y)}^{\parallel\perp}}{I_{E_g(X)}^{\parallel\perp}+I_{E_g(Y)}^{\parallel\perp}}\times\left\{\left[K_3^E(S_{11}-S_{12})\left(1-3\cos^2\frac{29\pi}{90}\right)+\sqrt{3}K_4^E\left(2S_{14}\cos\frac{29\pi}{90}-S_{44}\sin\frac{29\pi}{90}\right)\right]^2\right.$$
$$+\left\{2\sqrt{3}K_3^E\left[S_{14}\sin\frac{29\pi}{90}+(S_{11}-S_{12})\cos\frac{29\pi}{90}\right]\right.$$
$$\left.\left.\left.+K_4^E\left[S_{14}\left(1-3\cos^2\frac{29\pi}{90}\right)-3S_{44}\cos\frac{29\pi}{90}\sin\frac{29\pi}{90}\right]\right\}^2\right\}^{\frac{1}{2}}\right\} \tag{25.56}$$

According to the equations given above, the PDP constants of the detected vibrational bands were extracted from four independent sets of calibration data in bending configuration on single-crystalline sapphire samples uniaxially loaded along different crystallographic directions. Figures 25.5(b) show bending calibration plots of spectral shifts for the 417 cm^{-1} A_{1g} vibrational mode as a function of uniaxial stress magnitude for the five different samples (Cases 1-4), in parallel polarization configuration. Linear fitting according to the least square method enabled us to retrieve the values of the slopes, Π , which are explicitly given in inset to Fig. 25.5(b). Accordingly, a system of working equations could be obtained from Eqs. (25.49), (25.51), (25.53), and (25.55), as follows:

$$\begin{cases} K_1^A S_{11}+K_2^A S_{12}+K_3^A S_{13}=-267 \\ K_2^A S_{11}+K_1^A S_{12}+K_3^A S_{13}=-345 \\ K_1^A S_{13}+K_2^A S_{13}+K_3^A S_{33}=-1339 \\ \left(S_{11}+3S_{12}\cos^2\frac{29\pi}{90}+3S_{13}\sin^2\frac{29\pi}{90}\right)K_1^A+\left(S_{12}+3S_{11}\cos^2\frac{29\pi}{90}+3S_{13}\sin^2\frac{29\pi}{90}\right)K_2^A \\ \qquad +\left(S_{13}+S_{13}\cos^2\frac{29\pi}{90}+3S_{33}\sin^2\frac{29\pi}{90}\right)K_3^A=-1467 \end{cases} \tag{25.57}$$

All the obtained experimental plots were linear in the stress interval investigated, but exhibited different slopes. Similarly, for the 378 cm^{-1} E_g mode, calibration plots obtained on the sapphire single-crystals for Cases 1-4 (not shown). A system of independent equations could then be obtained, according to Eqs. (25.50), (25.52), (25.54) and (25.56), as follows:

$$\begin{cases} K_1^E(S_{11}+S_{12})+K_2^E S_{13}+\sqrt{\left[K_3^E(S_{11}-S_{12})\right]^2+(K_4^E S_{14})^2}=-367 \\ 2K_1^E S_{13}+K_2^E S_{33}=-266 \\ \left[K_1^E(S_{11}+S_{12})+K_2^E S_{13}\right](3\cos^2\frac{29\pi}{90}+1)+3(2K_1^E S_{13}+K_2^E S_{33})\sin^2\frac{29\pi}{90} \\ \quad -\frac{1}{2}\times\left\{\left[K_3^E(S_{11}-S_{12})\left(1-3\cos^2\frac{29\pi}{90}\right)+\sqrt{3}K_4^E\left(2S_{14}\cos\frac{29\pi}{90}-S_{44}\sin\frac{29\pi}{90}\right)\right]^2\right. \\ \quad +\left\{2\sqrt{3}K_3^E\left[S_{14}\sin\frac{29\pi}{90}+(S_{11}-S_{12})\cos\frac{29\pi}{90}\right]\right. \\ \quad \left.\left.+K_4^E\left[S_{14}\left(1-3\cos^2\frac{29\pi}{90}\right)-3S_{44}\cos\frac{29\pi}{90}\sin\frac{29\pi}{90}\right]\right\}^2\right\}^{\frac{1}{2}}=-1316 \\ 2\left[K_1{}^E(S_{11}+S_{12})+K_2^E S_{13}\right]+2K_1^E S_{13}+K_2^E S_{33})=-1021 \end{cases} \tag{25.58}$$

The last one in Eq. (25.58) was obtained from the bending on a polycrystalline alumina. Upon solving the two independent systems of Eqs. (25.57) and (25.58), the two sets of PDP values, $K_i^A(i = 1,2,3)$ and $K_j^A(j = 1,2,3,4)$ could be determined, which corresponds to the 417 cm^{-1} band of the A_{1g} mode and to the 378 cm^{-1} band of the E_g mode, respectively. Similar procedures were also performed on the shifts recorded for the band located at around 645 cm^{-1} (belonging to the A_{1g} vibrational mode), and for the three bands detected at around 430, 578, and 750 cm^{-1} (all belonging to the E_g vibrational mode). The values of PDP constants obtained for all the above-mentioned Raman bands are listed in Table 25.1.

Table 25.1 PDP constants for the trigonal structure of Al_2O_3

Band	Mode	K_1 ($\times 10^3$cm^{-2})	K_2 ($\times 10^3$cm^{-2})	K_3 ($\times 10^3$cm^{-2})	K_4 ($\times 10^3$cm^{-2})
378	E_g	-293±22	-227±24	-3.2±0.5	-8.5±1
417	A_{1g}	-396±19	-406±6	-777±18	-
430	E_g	-664±67	-613±74	-112±15	-30±5
576	E_g	-976±89	-721±76	-259±38	-16±3
645	A_{1g}	-610±69	-380±50	-764±73	-
750	E_g	-2117±319	-1193±166	-165±28	-35±4

A similar procedure can be applied to a tetragonal crystal of $BaTiO_3$ with using the appropriate secular equation and elastic constants in order to obtain the PDP constants (Pezzotti, 2013). Given the small dimensions of the available $BaTiO_3$ crystal, we used calibrations in compression (Fig. 25.4(b)) or exploited the equibiaxial stress field developed at a crack tip (Pezzotti, 2013). The PDP values retrieved for the tetragonal $BaTiO_3$ crystal are listed in Table 25.2.

Table 25.2 PDP constants for the tetragonal structure of $BaTiO_3$

Band	Mode	a (Uniax. compress.)	a (Biax. tens.)	b (Uniax. compress.)	b (Biax. tens.)	c (Biax. tens.)
270	$A_1(TO_2)$	838±39	987±101	458±38	606±60	-
520	$A_1(TO_2)$	1016±195	1224±250	424±93	623±221	-
490	$E(TO_4)$	-	2306±426	-	-1622±214	250±50

25.2.3 Deconvoluting the Raman Probe in Space

Raman spectra collected at given geometrical locations (x_0, y_0, z_0) contain spectral contributions originating from portions of the crystal belonging to the finite probe volume. Accordingly, a probe deconvolution procedure is needed, which is based on the introduction of a probe response function (PRF). The PRF describes the intensity of light scattered from a given point (x, y, z) when the incident laser beam is focused at the point (x_0, y_0, z_0). A suitable mathematical form of PRF in three dimensions can be given as follows (Atkinson and Jain, 1999):

$$G(x,y,z,x_0,y_0,z_0) \propto exp\left[-2\frac{(x-x_0)^2+(y-y_0)^2}{R^2}\right] \times \left[\frac{p^2}{(z-z_0)^2+p^2}e^{-2\alpha_{\text{eff}}z}\right] \tag{25.59}$$

where R is the waist diameter of the laser probe in its focal plane, p is the in-depth probe response parameter, which for an unfocused beam tends to infinity, and α_{eff} is the effective absorption coefficient of the material at the incident wavelength. The values of these parameters for the investigated crystals were determined in previous studies in Pezzotti (2013); Pezzotti and Zhu (2015) as: $BaTiO_3$: p=15.3 and 13.0 µm (for $a-$ and $c-$plane, respectively), R=3.0 and 5.0 µm (for a- and c-plane, respectively), and $\alpha_{\text{eff}} = 3.4 \times 10^{-3}$ and 2.0×10^{-3} µm^{-1} (for $a-$ and $c-$plane, respectively); and, Al_2O_3: p=3.6 µm, R=2.0 µm, and α_{eff}=0.030 µm^{-1}. In order to obtain a complete description of the observed Raman spectrum, $I_{\text{obs}}(\omega)$, a convolution of infinitesimal spectral contributions originating from each point of the probe volume in the sample must be considered, as described by the following equation:

$$I_{\text{obs}}(\omega) \propto \int_{-\infty}^{+\infty}\int_{-\infty}^{+\infty}\int_{0}^{+\infty} I(\omega)e^{\left[-2\frac{(x-x_0)^2+(y-y_0)^2}{R^2}\right]}\frac{p^2}{p^2+(z-z_0)^2}e^{-2\alpha_{\text{eff}}z}dxdydz \tag{25.60}$$

where $I(\omega)$ is the local Raman line shape, which is in turn a function of the local stress field in the case of an inhomogeneous stress distribution inside the probe volume. If the spectral shift of a selected Raman band in presence of stress, $\Delta\omega = \omega - \omega_0$, is negligible with respect to the band width, (i.e., $(\omega-\omega_0)^2 << N^2$), the observed band shift can be expressed to a degree of precision by an average shift weighted by the effective scattered intensity at the irradiated point. Accordingly, the observed band shift, $\Delta\omega_{\text{obs}}(x_0, y_0, z_0)$, can be calculated from the PRF as a weighted average of infinitesimal (local) band shifts, $\Delta\omega(x, y, z)$, emitted from different parts of the probe, as follows:

$$\Delta\omega_{\text{obs}}(x_0, y_0, z_0) = \frac{\int_{-\infty}^{+\infty}\int_{-\infty}^{+\infty}\int_{0}^{+\infty} G(x, y, z, x_0, y_0, z_0)\Delta\omega(x, y, z)dxdydz}{\int_{-\infty}^{+\infty}\int_{-\infty}^{+\infty}\int_{0}^{+\infty} G(x, y, z, x_0, y_0, z_0)dxdydz} \tag{25.61}$$

Note that a deconvolution procedure using the above equation involves the solution of an integral equation whose unknown function, $\Delta\omega(x, y, z)$, lies inside a triple integral. The solution for such equation is not unique unless the "character" of the unknown function can be guessed *a priori* or the morphology of the (experimentally retrieved) convoluted function can be assumed to yet retain the "character" of the native function and be used as a trial function. One usually accepts that the character of the deconvolute function should coincide with that of the experimentally retrieved function before deconvolution.

25.3 Applications of Raman Stress Analysis to Modern Devices

25.3.1 Miniaturized Multilayered Ceramic Condensers

$BaTiO_3$ ceramic with its high dielectric constant is the most common dielectric constituent of commercial multilayer ceramic capacitors (MLCC). However, this material is highly anisotropic, and also prone to crystallographic instability (domain switching) and phase transitions during service. The most recent technological developments to-wards miniaturization of MLCC dielectric devices have foreseen a significant decrease in layer thickness (down to few micrometers) for both ceramic dielectric layers and internal metallic Ni electrodes Kishi et al (2003). The purpose of doing so is to increase the MLCC capacitance by increasing the number of the dielectric layers. In such a miniaturized context and given the high anisotropy in dielectric constant of the $BaTiO_3$ crystallites, a control of the texture of crystallographic domains becomes crucial for the performance of the device and decisive for its reliability. Significant variations in electric and mechanical properties of ferroelectric materials could arise from the formation of residual stress-driven domain textures. The formation of such textures is indeed a consequence of the mechanical residual stresses developed in the dielectric MLCC layers upon manufacturing, which are partly released by the domain formation. Residual stresses are mainly due to the high thermal expansion mismatch between Ni metal and $BaTiO_3$ ceramic (den Toonder et al, 2003). Common reasons of failure stem from the presence of internal defects introduced during manufacturing (Nakano et al, 2003), residual stress fields experienced during firing or at various stages of assembly (Saito and Chazono, 2003; Uchino et al, 1989), and mechanical, thermal and/or electric fatigue during operation (Nomura et al, 1995). In all the above cases (and their combinations), internal cracks generate across the dielectric layers, which represent short/leakage current paths.

Given the importance of domain structures and residual stress fields in MLCC, several methods have been proposed for their detection and visualization. They include optical (polarized) microscopy, scanning force microscopy, and neutron diffraction (for a review see Potnis et al (2011)). However, Raman spectroscopy has some advantages on those methods: it enables to reach single-micronscale spatial resolution in the $BaTiO_3$ dielectric layers and to concurrently measure the tensor-resolved stress state. We employed the polarized Raman algorithms given in the previous sections for a quantitative characterization of local domain textures and residual stresses in the $BaTiO_3$ dielectric layers of a miniaturized MLCC device. Figure 25.6(a) shows a schematic draft of a MLCC device less than one millimeter in total thickness, and an optical micrograph of the $BaTiO_3$ interlayer 5 μm in thickness, which was probed with a Raman probe size of a single micron and a step of 0.1 μm in an area $5 \times 5\ \mu m^2$ (2500 measurement points). In Fig. 25.6(b), maps of domain texture in the investigated area are given in terms of three Euler angles in space as defined in Fig. 25.6(a). The local orientation of the domain c-axis was obtained by locally rotating in-plane (angle ψ) the polarized Raman probe at each measurement point, and then computing the values of three Euler angles by best fitting the experimentally obtained curves according to Eqs. (14.12)-(25.16) (based on the knowledge of the RTE for the $BaTiO_3$ crystal as determined in Subsect.

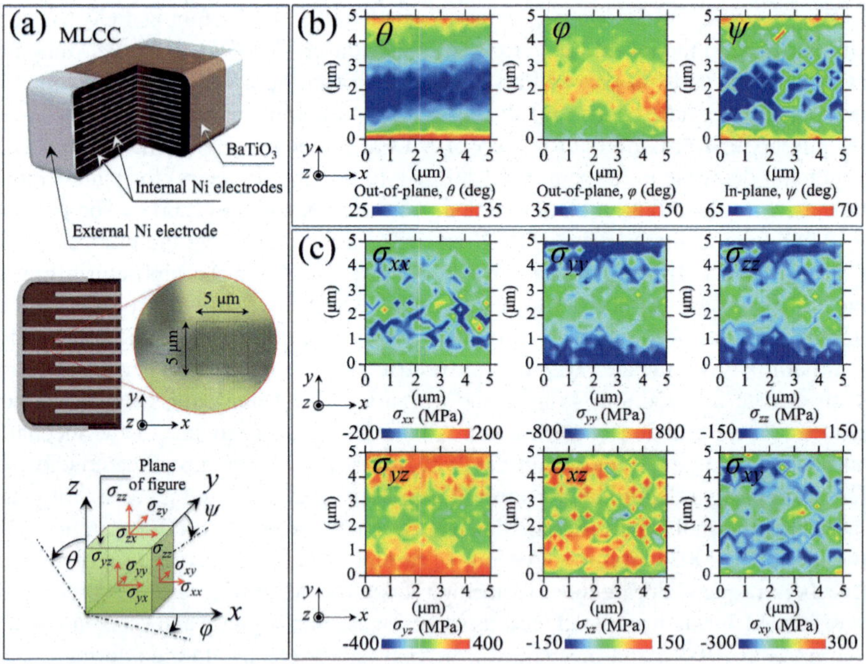

Fig. 25.6 (a) Schematic draft of MLCC miniaturized device, its internal electrode structure (with micrograph of the investigated interlayer area), and the related definition of Cartesian coordinates and Euler angles; (b) maps of three Euler angles in space in the interlayer area shown in the micrograph in (a); and, (c) maps of six independent stress components in space in the interlayer area shown in the micrograph in (a)

25.2.1). Figure 25.6(c) shows maps of the six independent stress tensor components in the same zone of the dielectric layer, according to the Cartesian axes given in Fig. 25.6(a)). The tensor-deconvoluted stress values were obtained according to the PDP constants in Table 25.2 and a treatment of the secular equation for the tetragonal structure (Pezzotti, 2013) similar to that described in Sect. 25.2.2 for the trigonal structure. Figures 25.6(b) and (c) contain fundamental information that clarifies the formation of crystallographic domains and their interactions with the residual stress field. The orientation of the domain structures experienced a clear gradient with increasing distance from the Ni-electrode area, the closer the distance from the electrode, the more homogeneous and marked the orientation pattern. Far from being randomly distributed, both in-plane and out-of-plane orientation angles appeared to obey a symmetric configuration with respect to the thickness direction (y-axis). The crystallographic structure of the dielectric interlayer was affected by steeper out-of-plane angular gradients as compared to the in-plane angle, although such fluctuations occurred within relatively narrow angular intervals of 10deg-15deg. Note that there was a clear correspondence between the geometry of the domains and the distributions of residual stresses; both hydrostatic and deviatoric components being mainly relaxed at the central parts were the "switched" domain structure appeared the most homogeneous. In other words, the present experiments demonstrated that the formation of domain textures occurs in order to relax manufacturing residual stresses due to thermal expansion mismatch between dielectrics and electrodes. The evolution of switching in the $BaTiO_3$ inter-layers under an electric field applied to the internal electrodes can be rationalized and computed by appropriate modeling in Choudhury et al (2007), but the domain conditions become difficult to predict when affected by thermal fatigue because of the complex nature of the generated residual stress fields. The impact of domain textures on the dielectric performance has been discussed in detail in a previous publication by Okai et al (2011). It was found that a non-random angular distribution of the domain c-axis, e.g., a strongly aligned one along a given axis of the MLCC device, significantly alters the effective value of dielectric constant as compared to a random angular population. In particular, the dielectric response of the MLCC could actually be lower than that expected for the same design with a random structure (i.e., in which all the directions in space exist with the same probability and thus the dielectric constant is simply the average of those along the a- and c-axis). The polarized Raman probe thus becomes a precious and unique instrument in microelectronics for understanding the evolution of domain orientation during service, thus leading to improved reliability and elongated lifetimes in MLCC devices.

25.3.2 Ceramic Femoral Heads in Artificial Hip Joint

Another field in which polarized Raman spectroscopy could suitably be applied in micromechanical assessments is joint orthopedics. A stunning finding is that a residual stress state remains stored onto the surface of ceramic femoral head retrievals, which thus retain "memory" of the sliding conditions *in vivo*, even if the sliding counterpart is a softer material (i.e., polyethylene). This circumstance is important in forensics, because it gives us a chance to interpret eventual complications and malfunctioning that might have occurred *in vivo* after the component has been extracted from the patient's body. Moreover, abnormal kinematics (coupled with material degradation) is hardly reproducible by *in vitro* simulations or obviously explainable through computational approaches. Raman spectroscopic algorithms contribute to link the joint performance with the micromechanical features that occur in real *in vivo* situations. Contact stresses play a fundamental role in the long term clinical performance of the artificial hip joint prostheses (Brand et al, 2001), because surface wear and the generation of polyethylene wear particles is recognized as one of the major factors contributing to loosening and failure of the prosthesis in the longer term. High wear rate and subsequently the partial or total penetration of the ceramic femoral head into the soft polyethylene acetabular liner can also cause abnormal biomechanics, such as joint migration, impingement, and fixation loosening. Accordingly, the establishment of a spectroscopic link between the residual stress states stored onto the ceramic head surface and the loading conditions *in vivo* enables acquiring precious biomechanical information, which contributes to improve the design of long-lasting artificial hip joints.

In this section, the principal stress magnitude was computed (after experimentally measuring the local magnitudes of the full set of stress components) and mapped on short (2 y and 6 mo), medium (8 y and 2 mo), and long-term (19 y) Al_2O_3 femoral head retrievals, which served in the human body coupled with polyethylene liners. Figure 25.7(a) shows a schematic draft of a femoral head sliding against an acetabular liner under body weight. Swing angles are shown in the cases of normal and abnormal gait, the latter case being a consequence of the previously mentioned phenomenon of "head migration." Head migration might occur during service due to an excess of wear and deformation of the polyethylene acetabular liner (Fig. 25.7(b)). Under such circumstances, the head occupies a restricted zone of the liner and slides with a limited angular swing (cf. Figs. 25.7(a) and (b)). Such an altered kinematics produces increased contact stresses, especially when the ceramic head directly impinges against the metallic cup that supports the polyethylene liner (Khanna and Beaulé, 2014). As a consequence of such an abnormal situation, metallic debris can be found on the ceramic surface, as shown in the scanning electron micrograph and in the *in toto* fractional map of metal stain in Figs. 25.7(a) and (b), respectively. In the bottom part of Fig. 25.7(a), the gait trajectory is shown together with the main path of sliding, and the paths of maximum shear and maximum pressure. The specific zone where all those paths intercept is subjected to formidable stress fields during gait. Under normal conditions, the residual stress field tends to concentrate at two specific zones, which correspond to the apex of the gait trajectory (cf. bottom part of Fig. 25.7(a)). During the joint lifetime, the stresses are initially of a tensile nature in these zones but, in the long range, they turn into compressive ones because the effect of repeated pressure prevails on the initial surface microdamage (compare maps of the trace of principal residual stresses for Case 1 and Case 3 in Fig. 25.7(c)). Case 2 (also in Fig. 25.7(c)) is a typical case of abnormal kinematics for the hip joint, with a number of distinct impingement zones (corresponding to a number of different spots of concentrated residual stress) in addition to the two physiological ones at the apex of the gait trajectory. Such a detrimental situation usually arises form recurrent hip sub-luxation/dislocations and/or from malpositioning during surgery. An in-depth tensor-resolved stress analysis is shown in Fig. 25.7(c) with a plot as a function of the in-depth abscissa, z, of the six independent stress components at the conjunction of maximum pressure and shear trajectories. The location of the Raman measurement is shown in the micrograph of Fig. 25.7(a) together with the Cartesian axes and the selected polarization direction. After determining the crystallographic orientation of the Al_2O_3 grain according to the extended selection rules and the knowledge of the RTE (see Sect. 25.2.1), both hydrostatic and deviatoric stress components were computed from experimental Raman band shifts, according to the algorithm explained in Sect. 25.2.2 (Eqs. (25.47)-(25.58)). The confocal probe was shifted along the in-depth direction, *z*, and the obtained trend then deconvoluted in space using the Raman probe response function discussed in Sect.

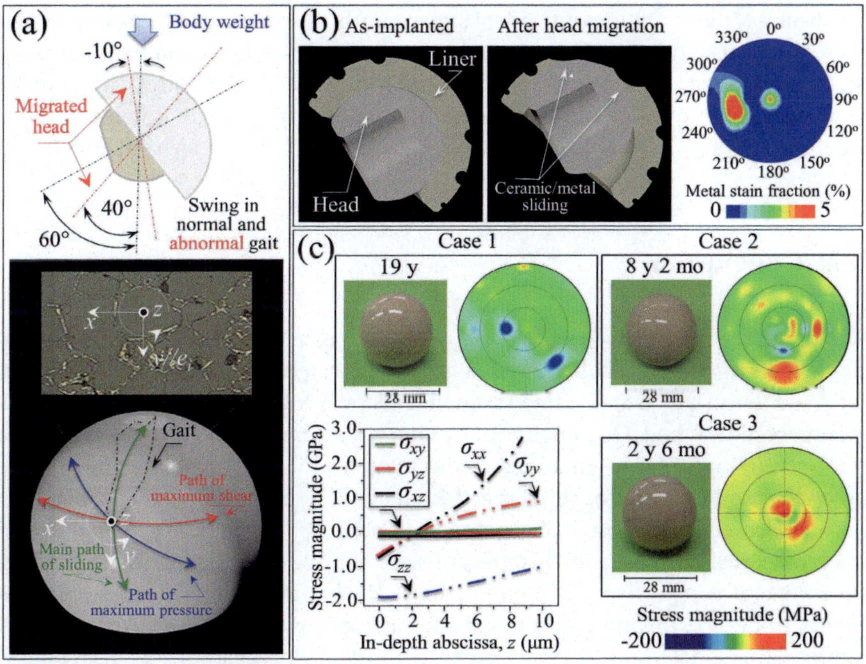

Fig. 25.7 (a) Schematic draft of artificial hip joint, swing angles under normal and abnormal gait, and the paths of maximum pressure and shear stress (with micrograph of the investigated microstructural area with the related Cartesian coordinates); (b) schematic draft of femoral head migration and *in toto* map of metal contamination; and, (c) *in toto* maps of the trace of principal stress tensor for three retrievals exposed *in vivo* for different periods of time (see labels), and dependence on the in-depth z-axis for the six independent stress components in space at the location of maximum pressure and shear stress paths shown in the micrograph in (a)

25.2.3 (Eq. (25.61)). From the measured in-depth trends, we could locate the following main items, based on the hypothesis that the residual stresses actually reflect the stress state developed during service on the surface of the ceramic bearing:

(i) The hydrostatic stress tensor component, σ_{zz} , was always compressive in nature because caused by body weight. The magnitude of this residual stress component was quite high, but it matched predictions according to Hertzian contact against the metallic support of the acetabular liner (Zhu et al, 2017);
(ii) The in-plane hydrostatic components, σ_{xx} and σ_{yy} are believed to mainly arise from partial lateral rotations (usually referred to as micro-displacement) of the femoral head during gait. The former experienced a pronounced in-depth variation and changed from compressive into tensile along the z-axis, while the latter followed a similar trend, but experienced a lower tensile magnitude along the subsurface.
(iii) The magnitudes of all the contact shear stress components were negligibly small. Shear stresses might have been relaxed by grain pulling out (as also visible in the micrograph of Fig. 25.7(a));

In substance, the evolution of residual stress motifs in space could be examined with micrometric resolution by Raman spectroscopy. Complications and malfunctioning became visualized *post mortem* through analyzing the residual stress fields stored on the surface of retrieved femoral heads. Raman spectroscopy thus helps unveiling peculiar micromechanical features as they occur during real *in vivo* situations.

25.4 Conclusion

This paper briefly reviewed the history of stress, posited the importance of expanding the concepts to residual stresses, and provided algorithms to measure their spatially and tensorially deconvoluted components by a completely non-contact Raman spectroscopic method. Quantitative algorithms to assess domain textures and single-grain crystallographic orientations at the microscopic level in three Euler angles were also reported. Confocal/polarized Raman analyses were found capable to advance understanding of basic phenomena behind the performance of both electronic and biomedical devices, and examples of such analyses were explicitly shown. In a more general perspective, the computational Raman algorithms could inspire new criteria in design, quality control, and lifetime of new devices, while concurrently incorporating a deterministic monitoring of residual stress parameters, a so far conspicuously neglected task that Raman spectroscopy could effectively accomplish.

Acknowledgements The content of Sect. 25.1 was inspired by the book written by Prof. W. H. Müller (Müller, 2014) and by his lecture given at the Kyoto Institute of Technology during the academic year 2016-2017. Through the development of tensor-resolved stress analyses by Raman micro-probe spectroscopy, the author was greatly inspired by lectures given by Prof. W. H. Müller at the Berlin Technical University, which he had a chance to assist during an invited professorship stay in Berlin during the year 2007. The author also gratefully thanks Prof. W. Zhu and Prof. E.

Marin at the Kyoto Institute of Technology for their support in this research, for revising the first draft of the manuscript, and for helping with formatting the manuscript.

References

Atkinson A, Jain S (1999) Spatially resolved stress analysis using raman spectroscopy. Journal of Raman Spectroscopy 30(10):885–891

Brand RA, Iglič A, Kralj-Iglič V (2001) Contact stresses in the human hip: implications for disease and treatment. Hip International 11(3):117–126

Cardona M, Giintherodt G (1982) Light Scntteing in Solids I-VIII. Springer-Verlag, Berlin

Choudhury S, Li Y, Krill Iii C, Chen L (2007) Effect of grain orientation and grain size on ferroelectric domain switching and evolution: Phase field simulations. Acta Materialia 55(4):1415–1426

DiDomenico Jr M, Wemple S, Porto S, Bauman R (1968) Raman spectrum of single-domain BaTiO3. Physical Review 174(2):522

El Marssi M, Le Marrec F, Lukyanchuk I, Karkut M (2003) Ferroelectric transition in an epitaxial barium titanate thin film: Raman spectroscopy and X-ray diffraction study. Journal of Applied Physics 94(5):3307–3312

Ganesan S, Maradudin A, Oitmaa J (1970) A lattice theory of morphic effects in crystals of the diamond structure. Annals of Physics 56(2):556–594

Goto T, Anderson OL, Ohno I, Yamamoto S (1989) Elastic constants of corundum up to 1825 K. Journal of Geophysical Research: Solid Earth 94(B6):7588–7602

Hosford WF (2010) Mechanical behavior of materials. Cambridge University Press

Khanna V, Beaulé PE (2014) Defining structural abnormalities of the hip joint at risk of degeneration. Journal of Hip Preservation Surgery 1(1):12–20

Kishi H, Mizuno Y, Chazono H (2003) Base-metal electrode-multilayer ceramic capacitors: past, present and future perspectives. Japanese Journal of Applied Physics 42(1R):1

Loudon R (1964) The Raman effect in crystals. Advances in Physics 13(52):423–482

Müller WH (2014) An expedition to continuum theory. Springer

Nakano Y, Nomura T, Takenaka T (2003) Residual stress of multilayer ceramic capacitors with Ni-electrodes (Ni-MLCCs). Japanese Journal of Applied Physics 42(9S):6041

Nomura T, Kawano N, Yamamatsu J, Arashi T, Nakano Y, Sato A (1995) Aging behavior of ni-electrode multilayer ceramic capacitors with X7R characteristics. Japanese Journal of Applied Physics 34(9S):5389

Okai K, Zhu W, Pezzotti G (2011) Domain structures in multilayer ceramic capacitors studied by polarized raman spectroscopy. Physica status solidi (a) 208(5):1132–1140

Pezzotti G (2013) Raman spectroscopy of piezoelectrics. Journal of Applied Physics 113(21):8_1

Pezzotti G (2017) Raman spectroscopy of biomedical polyethylenes. Acta Biomaterialia 55:28–99

Pezzotti G, Zhu W (2015) Resolving stress tensor components in space from polarized Raman spectra: polycrystalline alumina. Physical Chemistry Chemical Physics 17(4):2608–2627

Porto S, Krishnan R (1967) Raman effect of corundum. The Journal of Chemical Physics 47(3):1009–1012

Potnis PR, Tsou NT, Huber JE (2011) A review of domain modelling and domain imaging techniques in ferroelectric crystals. Materials 4(2):417–447

Rohlmann A, Bergmann G, Kölbel R (1980) Finite element analysis, its limitations and the relevance of its results to orthopaedic surgery as shown by simultaneous experimental measurements (author's transl). Zeitschrift fur Orthopadie und ihre Grenzgebiete 118(1):122–131

Saito K, Chazono H (2003) Stress and electrical field responses of X5R type multilayer ceramic capacitor with Ni internal electrode. Japanese Journal of Applied Physics 42(9S):6045

den Toonder JM, Rademaker CW, Hu CL (2003) Residual stresses in multilayer ceramic capacitors: measurement and computation. Journal of Electronic Packaging 125(4):506–511

Uchino K, Sadanaga E, Hirose T (1989) Dependence of the crystal structure on particle size in barium titanate. Journal of the American Ceramic Society 72(8):1555–1558

Withers PJ, Bhadeshia H (2001) Residual stress. part 1–measurement techniques. Materials Science and Technology 17(4):355–365

Young T (1807) A course of lectures on natural philosophy and the mechanical arts: in two volumes, vol 2. Johnson

Zhu W, Marin E, Sugano N, Pezzotti G (2017) Tensor-resolved raman spectroscopic analysis of wear-induced residual stress fields in long-term alumina hip-joint retrievals. Journal of the Mechanical Behavior of Biomedical Materials 66:201–210

Chapter 26
Analytical Solutions of 2-dimensional Second Gradient Linear Elasticity for Continua with Cubic-D_4 Microstructure

Luca Placidi, Giuseppe Rosi & Emilio Barchiesi

Abstract We consider in this paper analytical solutions for some remarkable cases and for a linear anisotropic D4 second gradient elastic model. The purpose is that of constitutive parameter identification. In general, analytical solutions are considered less important than in the past due to fast numerical tools and to the fact that they are generally very difficult to achieve. However, they still play very important roles such as for numerical comparison and for dealing with pathological mechanical systems.

Keywords: Analytical solution · Metamaterial · Inverse analysis · Parameter determination

26.1 Introduction

Metamaterials (see the reviews Del Vescovo and Giorgio (2014); Barchiesi et al (2018)) represent a very important challenge for engineering design and modeling. Indeed, from the point of view of numerical simulations, metamaterials are intrinsically complex to model and require high computational power Bilotta and Turco (2009); Cazzani and Ruge (2012); Garusi et al (2004); Hendy and Turco (2008).

Luca Placidi
International Telematic University Uninettuno, Faculty of Engineering, Corso Vittorio Emanuele II, 39, Rome, Italy,
e-mail: luca.placidi@uninettunounirsity.net

Giuseppe Rosi
Université Paris-Est, Laboratoire Modélisation et Simulation Multi Echelle, MSME UMR 8208 CNRS, 61 av du Général de Gaulle, 94010 Créteil Cedex, France,
e-mail: giuseppe.rosi@u-pec.fr

Emilio Barchiesi
Università degli Studi di Roma "La Sapienza" Dipartimento di Ingegneria Strutturale e Geotecnica, Via Eudossiana 18, 00184 Rome, Italy,
e-mail: BarchiesiEmilio@gmail.com

B. E. Abali et al. (eds.), *New Achievements in Continuum Mechanics and Thermodynamics*, Advanced Structured Materials 108,
https://doi.org/10.1007/978-3-030-13307-8_26

Discrete models, because of the increasing performances of modern computers, are becoming a suitable choice for capturing all the important features of complex multi-scale or microstructured mechanical systems (see Baraldi et al (2013); Rahali et al (2015); Placidi et al (2017a); Goda et al (2012) for numerical and theoretical results). Still, the development of continuum models is fundamental because of the possibility of having both analytical, in some simplified cases, and numerical solutions to be used in optimization processes. As it is well known, Classic Cauchy continuum theory is not rich enough to be used in the framework of metamaterial modelling if the geometry of the microstructure is not explicitly described (like in, e.g., Franciosi et al (2018)). Indeed, the size of the microstructural features (see, e.g., AminPour and Rizzi (2016); Aminpour and Rizzi (2015, 2016); Aminpour et al (2014) for nano-sized objects) is so small that the increased degrees of freedom results in a high computational cost and the effective geometry of the resulting body is complex (see e.g pantographic structures dell'Isola et al (2018); Scerrato et al (2016b) or truss structures Alibert et al (2003); Seppecher et al (2011)). Thus, numerical simulations, within the standard 3D Cauchy elasticity model, of the exact geometry of a 3D body goes generally beyond the present hardware capabilities. Starting with works of Piola dell'Isola et al (2015, 2016, 2019), when numerical simulations could not clearly be performed with the aid of computers, the necessity to find new, generalized, models able to deal with complex microstructures and from a continuum point of view is strongly felt by the scientific community. Higher order gradient continua Auffray et al (2015a) fulfill the above mentioned characteristics, via e.g. an homogenization criterion Cecchi and Rizzi (2001); Dos Reis and Ganghoffer (2012); Goda et al (2014); Rahali et al (2015), and second gradient dell'Isola et al (2009, 2008); Mindlin (1964); Pideri and Seppecher (1997); Sansour and Skatulla (2009); Terravecchia et al (2014); Selvadurai (1973) 2D elastic materials, in general anisotropic Auffray et al (2015b); dell'Isola and Steigmann (2015); Indelicato and Albano (2009); Steigmann (2009); Steigmann and dell'Isola (2015); Walpole (1984) ones, are the subject of this contribution. It is a matter of fact that the higher order continua can be seen as a specialization of micromorphic/microstructured continua Misra and Poorsolhjouy (2015); Misra and Singh (2015); Misra and Huang (2012); Misra et al (2015); Misra and Singh (2014), where the kinematical descriptors added to the Cauchy model are independent of the classic ones, by means of certain assumptions. Even if generalized continuum models have proven to be more efficient, as we have discussed above, one of the main issue concerning such models is the lack of procedures for estimating their constitutive coefficients. In this paper, analytical solutions will be derived for the purpose of constitutive parameter identification.

From a numerical point of view, difficulties mainly consist in conceiving new numerical schemes Atluri and Cazzani (1995); Greco and Cuomo (2015); Solaria et al (1997); Andreaus et al (2018) for finite element simulations suitable for higher order continua. In particular higher degree of continuity Bilotta et al (2010) of the basis functions must be guaranteed Cuomo et al (2014); Greco and Cuomo (2014); Hughes et al (2014). Among others, the fundamental reasons justifying these difficulties are related, e.g., to the inclusion of impact behaviors Andreaus et al (2013);

Bersani et al (2013); Kezmane et al (2017), of instabilities Luongo and D'annibale (2012); Piccardo et al (2015); Rizzi and Varano (2011a,b); Ruta et al (2008); Scerrato et al (2016a); Spagnuolo and Andreaus (2018) and/or of surface effects Altenbach et al (2010); dell'Isola and Rotoli (1995); Rosi et al (2017) and of damage or plastic behaviour D'Annibale and Luongo (2013); Goda et al (2012, 2014); Placidi et al (2018).

In general, analytical solutions are considered less important than in the past due to the new numerical tools dell'Isola and Seppecher (1995). Besides, they are generally very difficult to achieve. However, they still play the very important role for numerical comparison and for dealing with pathological mechanical systems Bersani et al (2016). In other words, in these cases, they are useful for the verification of those new numerical solution procedures, for which the use of the analytical solutions as a benchmark is needed.

In this paper, analytical solutions will be defined for the purposes of convenient constitutive parameter identification Placidi et al (2017a); Turco (2013) and will be exploited for that objective in a further contribution.

26.2 Outline of the Model

We briefly recall, in this Section, the main facts about the linear two-dimensional elastic second gradient D4 cubic continuum model employed in this paper. The model has been developed in more detail in Placidi et al (2017a). The reference configuration $\mathcal{B}$ is a 2-dimensional body, where X denotes the coordinates of its material points. The internal energy density functional $U(\mathbf{G}, \nabla\mathbf{G})$ is a function of the deformation matrix $\mathbf{G} = \left(\mathbf{F}^T\mathbf{F} - I\right)/2$ and of its gradient $\nabla\mathbf{G}$. $\mathbf{F} = \nabla\chi$ is the deformation gradient, where χ is the placement function, $\mathbf{F}^T$ is the transpose of $\mathbf{F}$, and ∇ is the gradient operator. The energy functional $\mathcal{E}(\mathbf{u}(\cdot))$ depends on the displacement $\mathbf{u} = \chi - \mathbf{X}$ and includes two contributions, that are the internal and the external energies,

$$\mathcal{E}(\mathbf{u}(\mathbf{X})) = \int_{\mathcal{B}} \left[U(\mathbf{G}, \nabla\mathbf{G}) - \mathbf{b}^{ext} \cdot \mathbf{u} - \mathbf{m}^{ext} \cdot \nabla\mathbf{u}\right] dA \tag{26.1}$$
$$- \int_{\partial\mathcal{B}} \left[\mathbf{t}^{ext} \cdot \mathbf{u} + \tau^{ext} \cdot \left[(\nabla\mathbf{u})\,\mathbf{n}\right]\right] ds - \int_{[\partial\partial\mathcal{B}]} \mathbf{f}^{ext} \cdot \mathbf{u}$$

where the dot is the scalar product between vectors or tensors (in the case $\mathbf{m}^{ext} \cdot \nabla\mathbf{u}$) and n is the unit external normal. $\mathbf{b}^{ext}$ and $\mathbf{m}^{ext}$ are the external body force and double force, respectively, per unit area. $\mathbf{t}^{ext}$ and τ^{ext} are the external force and double force, respectively and per unit length. $\mathbf{f}^{ext}$ is the external concentrated force, that is applied on the set of vertices $[\partial\partial\mathcal{B}]$. The boundary $\partial\mathcal{B}$ is assumed to be the union of m regular parts Σ_c (with $c = 1, \ldots, m$) and the so-called boundary of the boundary $[\partial\partial\mathcal{B}]$ is assumed to be the union of the corresponding m vertex-points $\mathcal{V}_c$ (with $c = 1, \ldots, m$) with coordinates X^c.

The most general second gradient quadratic function representing the internal energy density is

$$U(\mathbf{G}, \nabla\mathbf{G}) = \hat{U}(\epsilon, \eta) = \frac{1}{2} C_{IJ} \epsilon_I \epsilon_J + \frac{1}{2} A_{\alpha\beta} \eta_\alpha \eta_\beta \tag{26.2}$$

where, the indexes I and J vary from 1 to 3, the indexes α and β vary from 1 to 6, ϵ_I is the I-th component of the following column-vector ϵ

$$\epsilon = \begin{pmatrix} G_{11} \\ G_{22} \\ \sqrt{2} G_{12} \end{pmatrix}, \tag{26.3}$$

η_α is the α-th component of the following column-vector η

$$\eta = \begin{pmatrix} G_{11,1} \\ G_{22,1} \\ \sqrt{2} G_{12,2} \\ G_{22,2} \\ G_{11,2} \\ \sqrt{2} G_{12,1} \end{pmatrix}, \tag{26.4}$$

C_{IJ} is the IJ-th component of the 3×3 matrix $\mathbf{C}$

$$\mathbf{C} = \begin{pmatrix} c_{11} & c_{12} & 0 \\ c_{12} & c_{11} & 0 \\ 0 & 0 & c_{33} \end{pmatrix}, \tag{26.5}$$

for a D4 material, see also Auffray et al (2015b). $A_{\alpha\beta}$ is the $\alpha\beta$-th component of the matrix $\mathbf{A}$

$$\mathbf{A} = \begin{pmatrix} a_{11} & a_{12} & a_{13} & 0 & 0 & 0 \\ a_{12} & a_{22} & a_{23} & 0 & 0 & 0 \\ a_{13} & a_{23} & a_{33} & 0 & 0 & 0 \\ 0 & 0 & 0 & a_{11} & a_{12} & a_{13} \\ 0 & 0 & 0 & a_{12} & a_{22} & a_{23} \\ 0 & 0 & 0 & a_{13} & a_{23} & a_{33} \end{pmatrix}, \tag{26.6}$$

Stress and hyper stress tensors are,

$$S_{ij} = \frac{\partial U}{\partial G_{ij}}, \qquad T_{ijh} = \frac{\partial U}{\partial G_{ij,h}}, \tag{26.7}$$

26.3 Some Explicit Computations for the Identification Procedure

26.3.1 Stress and Hyperstress in Terms of the Displacement Field

Some explicit computations, which are instrumental for the identification procedure, will be shown in this Section. Thus, the stress components read from $(26.7)_1$, (26.2), (26.3) and (26.5):

$$S_{11} = c_{11}G_{11} + c_{12}G_{22}, \tag{26.8}$$

$$S_{12} = S_{21} = c_{33}G_{12}, \tag{26.9}$$

$$S_{22} = c_{12}G_{11} + c_{11}G_{22}. \tag{26.10}$$

The hyperstress components read from $(26.7)_2$, (26.2), (26.4) and (26.6):

$$T_{111} = a_{11}G_{11,1} + a_{12}G_{22,1} + \sqrt{2}a_{13}G_{12,2}, \tag{26.11}$$

$$T_{112} = a_{12}G_{22,2} + a_{22}G_{11,2} + \sqrt{2}a_{23}G_{12,1}, \tag{26.12}$$

$$T_{121} = T_{211} = \frac{\sqrt{2}}{2}a_{13}G_{22,2} + \frac{\sqrt{2}}{2}a_{23}G_{11,2} + a_{33}G_{12,1}, \tag{26.13}$$

$$T_{122} = T_{212} = \frac{\sqrt{2}}{2}a_{13}G_{11,1} + \frac{\sqrt{2}}{2}a_{23}G_{22,1} + a_{33}G_{12,2}, \tag{26.14}$$

$$T_{221} = a_{12}G_{11,1} + a_{22}G_{22,1} + \sqrt{2}a_{23}G_{12,2}, \tag{26.15}$$

$$T_{222} = a_{11}G_{22,2} + a_{12}G_{11,2} + \sqrt{2}a_{13}G_{12,1}. \tag{26.16}$$

By replacing the components of the strain and of the strain gradient tensors in terms of the displacement field yields,

$$G_{11} = u_{1,1}, G_{12} = G_{21} = \tfrac{1}{2}\left(u_{1,2} + u_{2,1}\right), G_{22} = u_{2,2}, \tag{26.17}$$

$$G_{11,1} = u_{1,11}, G_{11,2} = u_{1,12}, \quad G_{22,1} = u_{2,12}, G_{22,2} = u_{2,22}, \tag{26.18}$$

$$G_{12,1} = G_{21,1} = \tfrac{1}{2}\left(u_{1,12} + u_{2,11}\right), G_{12,2} = G_{21,2} = \tfrac{1}{2}\left(u_{1,22} + u_{2,12}\right). \tag{26.19}$$

Insertion of (26.17-26.19) into (26.8-26.16), gives the stress and the hyperstress fields in terms of the displacement fields, i.e.,

$$S_{11} = c_{11}u_{1,1} + c_{12}u_{2,2}, \tag{26.20}$$

$$S_{12} = S_{21} = \frac{1}{2}c_{33}\left(u_{1,2} + u_{2,1}\right), \tag{26.21}$$

$$S_{22} = c_{11}u_{2,2} + c_{12}u_{1,1}, \tag{26.22}$$

and

$$T_{111} = a_{11}u_{1,11} + a_{12}u_{2,12} + \frac{a_{13}}{\sqrt{2}}\left(u_{1,22} + u_{2,12}\right), \quad (26.23)$$

$$T_{112} = a_{12}u_{2,22} + a_{22}u_{1,12} + \frac{a_{23}}{\sqrt{2}}\left(u_{1,12} + u_{2,11}\right), \quad (26.24)$$

$$T_{121} = T_{211} = \frac{\sqrt{2}}{2}a_{13}u_{2,22} + \frac{\sqrt{2}}{2}a_{23}u_{1,12} + \frac{1}{2}a_{33}\left(u_{1,12} + u_{2,11}\right), \quad (26.25)$$

$$T_{122} = T_{212} = \frac{\sqrt{2}}{2}a_{13}u_{1,11} + \frac{\sqrt{2}}{2}a_{23}u_{2,12} + \frac{1}{2}a_{33}\left(u_{1,22} + u_{2,12}\right), \quad (26.26)$$

$$T_{221} = a_{12}u_{1,11} + a_{22}u_{2,12} + \frac{a_{23}}{\sqrt{2}}\left(u_{1,22} + u_{2,12}\right), \quad (26.27)$$

$$T_{222} = a_{11}u_{2,22} + a_{12}u_{1,12} + \frac{a_{13}}{\sqrt{2}}\left(u_{1,12} + u_{2,11}\right). \quad (26.28)$$

26.3.2 Partial Differential Equations and Boundary Conditions

The system of PDEs for anisotropic D_4 elastic second gradient materials has been deduced by the stationary principle in Placidi et al (2017b,c,a). Here, it is made explicit the first partial differential equation,

$$c_{11}u_{1,11} + \frac{1}{2}c_{33}\left(u_{1,22} + u_{2,12}\right) + c_{12}u_{2,12} = \quad (26.29)$$
$$= a_{11}u_{1,1111} + \sqrt{2}\left(a_{13} + a_{23}\right)\left(\frac{1}{2}u_{2,1222} + \frac{1}{2}u_{2,1112} + u_{1,1122}\right) + a_{22}u_{1,1122}$$
$$+a_{12}\left(u_{2,1222} + u_{2,1112}\right) + a_{33}\left(u_{1,2222} + u_{1,1122} + u_{2,1222} + u_{2,1112}\right)$$
$$-b_1^{ext} + m_{11,1}^{ext} + m_{12,2}^{ext}$$

The other partial differential equation, because of the D_4-symmetry, is simply given by an interchange of the indices 1 and 2 in (i) the displacement field $\mathbf{u}$, in (ii) its derivatives, in (iii) the external force per unit area $\mathbf{b}^{ext}$, in (iv) the external double force per unit area $\mathbf{m}^{ext}$ and in (v) its derivatives (not in the indices of the constitutive coefficients of eqns. (26.5) and (26.6)!).

The rectangular geometry of the homogeneous 2-dimensional second gradient elastic body we will here investigate has been pictured in Fig. 26.1. The external force and double force field per unit length have been already characterized in Placidi et al (2017b,c,a) for each sides of the rectangle of Fig. 26.1. In the rest of this subsection we will make these forces and double forces in terms stress and hyperstress tensors.

The characterization of side S is done as follows

$$t_1 = t_1^S = S_{11} - T_{112,2} - T_{111,1} - T_{121,2} \tag{26.30}$$

$$t_2 = t_2^S = S_{21} - T_{211,1} - T_{212,2} - T_{221,2}, \tag{26.31}$$

$$\tau_1 = \tau_1^S = T_{111}, \tag{26.32}$$

$$\tau_2 = \tau_2^S = T_{211}. \tag{26.33}$$

The characterization of side Q is, onthe other side, as foollows

$$t_1 = t_1^Q = -S_{11} + T_{112,2} + T_{111,1} + T_{121,2} \tag{26.34}$$

$$t_2 = t_2^Q = -S_{21} + T_{211,1} + T_{212,2} + T_{221,2}, \tag{26.35}$$

$$\tau_1 = \tau_1^Q = T_{111} \tag{26.36}$$

$$\tau_2 = \tau_2^Q = T_{211}. \tag{26.37}$$

We remark that, because of symmetry reasons, t_1^Q in (26.34) and t_2^Q in (26.35) are the opposite of t_1^S in (26.30) and of t_2^S in (26.31), respectively, and that τ_1^Q in (26.36) and τ_2^Q in (26.37) are the same of τ_1^S in (26.32) and of τ_2^S in (26.33), respectively.

The characterization of side R is done as follows

$$t_1 = t_1^R = S_{12} - T_{121,1} - T_{112,1} - T_{122,2} \tag{26.38}$$

$$t_2 = t_2^R = S_{22} - T_{221,1} - T_{212,1} - T_{222,2}, \tag{26.39}$$

$$\tau_1 = \tau_1^R = T_{122}, \tag{26.40}$$

$$\tau_2 = \tau_2^R = T_{222}. \tag{26.41}$$

We remark that, again for the same reasons, t_1^R in (26.38) and t_2^R in (26.39) are the same of t_2^S in (26.31) and of t_1^S in (26.30), respectively, by changing the indices 1 and 2. Besides, because of symmetry reasons, τ_1^R in (26.40) and τ_2^R in (26.41) are the same of τ_2^S in (26.32) and of τ_1^S in (26.33), respectively, by changing the indices 1 and 2.

Finally, the characterization of side T is, on the other side, done as follows

$$t_1 = t_1^T = -S_{12} + T_{121,1} + T_{112,1} + T_{122,2}, \tag{26.42}$$

$$t_2 = t_2^T = -S_{22} + T_{221,1} + T_{212,1} + T_{222,2}, \tag{26.43}$$

$$\tau_1 = \tau_1^T = T_{122}, \tag{26.44}$$

$$\tau_2 = \tau_2^T = T_{222}. \tag{26.45}$$

We again remark that t_1^T in (26.42) and t_2^T in (26.43) are the opposite of t_1^R in (26.38) and of t_2^R in (26.39), respectively, and that τ_1^T in (26.44) and τ_2^T in (26.45) are the same of τ_1^R in (26.40) and of τ_2^R in (26.41), respectively.

The external force on vertices are characterized in Placidi et al (2017b,c,a) for each vertices. For vertices $\mathcal{V}_1$ and $\mathcal{V}_3$ we have

$$f_\alpha = -T_{\alpha 21} - T_{\alpha 12}, \qquad \alpha = 1, 2 \tag{26.46}$$

For vertices $\mathcal{V}_2$ and $\mathcal{V}_4$, on the other side, we have

$$f_\alpha = T_{\alpha 21} + T_{\alpha 12}, \qquad \alpha = 1, 2 \tag{26.47}$$

26.4 Analytical Solutions of Homogeneous Second Gradient Model

In this section we explore, in each subsection, analytical solutions of this homogeneous 2-dimensional second gradient elastic model. In particular, force and double force for each sides and per unit length, have been calculated as well as the vertex-concentrated forces.

26.4.1 Simple Tension Test

The following displacement field is considered,

$$u_1 = 0, \qquad u_2 = \frac{\delta}{20}\left(-1 + \frac{X_2}{l}\right). \tag{26.48}$$

This is achieved, from (26.29), by neglecting distributed forces and double forces ($\mathbf{b}^{ext} = \mathbf{m}^{ext} = \mathbf{0}$) and by imposing the following set of kinematical boundary conditions for the macroscopic second gradient model (Fig. 26.1). At Q, that is at $X_1 = 0$ and $\forall X_2 \in [-l, l]$

$$u_1(0, X_2) = u_{1,1}(0, X_2) = 0, u_2(0, X_2) = \frac{\delta}{20}\left(-1 + \frac{X_2}{l}\right), u_{2,1}(0, X_2) = 0.$$

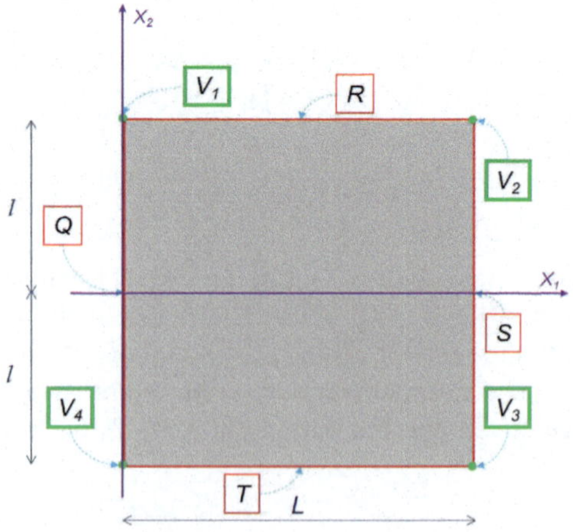

Fig. 26.1 Nomenclature and geometry of the homogeneous 2-dimensional second gradient elastic body.

At R, that is $\forall X_1 \in [0, L]$ and at $X_2 = l$,

$$u_1(X_1, l) = u_{1,2}(X_1, l) = u_2(X_1, l) = 0, \quad u_{2,2}(X_1, l) = \frac{\delta}{20l}.$$

At S, that is at $X_1 = L$ and $\forall X_2 \in [-l, l]$

$$u_1(L, X_2) = u_{1,1}(L, X_2) = 0, u_2(L, X_2) = \frac{\delta}{20}\left(-1 + \frac{X_2}{l}\right), u_{2,1}(L, X_2) = 0.$$

At T, that is $\forall X_1 \in [0, L]$ and at $X_2 = -l$,

$$u_1(X_1, -l) = u_{1,2}(X_1, -l) = 0, \quad u_2(X_1, -l) = -\frac{\delta}{10}, \quad u_{2,2}(X_1, -l) = \frac{\delta}{20l}.$$

From Placidi et al (2017b,c, 2015, 2017a) force and double force at each side have been calculated

$$t_1^{S,ext} = c_{12}\frac{\delta}{20l}, \quad t_2^{S,ext} = 0, \quad \tau_1^{S,ext} = 0, \quad \tau_2^{S,ext} = 0, \tag{26.49}$$

$$t_1^{Q,ext} = -c_{12}\frac{\delta}{20l}, \quad t_2^{Q,ext} = 0, \quad \tau_1^{Q,ext} = 0, \quad \tau_2^{Q,ext} = 0, \tag{26.50}$$

$$t_1^{R,ext} = 0, \quad t_2^{R,ext} = c_{11}\frac{\delta}{20l}, \quad \tau_1^{R,ext} = 0, \quad \tau_2^{R,ext} = 0, \tag{26.51}$$

$$t_1^{T,ext} = 0, \quad t_2^{T,ext} = -c_{11}\frac{\delta}{20l}, \quad \tau_1^{T,ext} = 0,, \quad \tau_2^{T,ext} = 0, \tag{26.52}$$

as well as vertex forces,

$$\left(f_j^{ext}\right)_{\mathcal{V}_i} = 0, \quad i = 1, 2, 3, 4, \quad j = 1, 2$$

26.4.2 Simple Shear Test

The following displacement field is considered,

$$u_1 = \frac{\delta}{20l}(X_2 + l), \qquad u_2 = 0. \tag{26.53}$$

This is achieved by considering no distributed forces and double forces

$$\mathbf{b}^{ext} = \mathbf{m}^{ext} = \mathbf{0}$$

and by imposing the following set of kinematical boundary conditions for the macroscopic second gradient model of Fig. 26.1.

At Q, that is at $X_1 = 0$ and $\forall X_2 \in [-l, l]$

$$u_1(0, X_2) = \frac{\delta}{20l}(X_2 + l), u_{1,1}(0, X_2) = 0, u_2(0, X_2) = 0, u_{2,1}(0, X_2) = 0.$$

At R, that is $\forall X_1 \in [0, L]$ and at $X_2 = l$,

$$u_1(X_1, l) = \frac{\delta}{10}, \quad u_{1,2}(X_1, l) = u_2(X_1, l) = 0, \quad u_{2,2}(X_1, l) = 0.$$

At S, that is at $X_1 = L$ and $\forall X_2 \in [-l, l]$

$$u_1(L, X_2) = \frac{\delta}{20l}(X_2 + l), u_{1,1}(L, X_2) = 0, u_2(L, X_2) = 0, u_{2,1}(L, X_2) = 0.$$

At T, that is $\forall X_1 \in [0, L]$ and at $X_2 = -l$,

$$u_1(X_1, -l) = u_{1,2}(X_1, -l) = 0, \quad u_2(X_1, -l) = 0, \quad u_{2,2}(X_1, -l) = 0.$$

From Placidi et al (2017b,c, 2015, 2017a) force and double force at each side have been calculated

$$t_1^{S,ext} = 0, \quad t_2^{S,ext} = c_{33}\frac{\delta}{40l}, \quad \tau_1^{S,ext} = 0, \quad \tau_2^{S,ext} = 0, \tag{26.54}$$

$$t_1^{Q,ext} = 0, \quad t_2^{Q,ext} = -c_{33}\frac{\delta}{40l}, \quad \tau_1^{Q,ext} = 0, \quad \tau_2^{Q,ext} = 0, \tag{26.55}$$

$$t_1^{R,ext} = c_{33}\frac{\delta}{40l}, \quad t_2^{R,ext} = 0, \quad \tau_1^{R,ext} = 0, \quad \tau_2^{R,ext} = 0, \tag{26.56}$$

$$t_1^{T,ext} = -c_{33}\frac{\delta}{40l}, \quad t_2^{T,ext} = 0, \quad \tau_1^{T,ext} = 0, \quad \tau_2^{T,ext} = 0, \tag{26.57}$$

as well as vertex forces,

$$\left(f_j^{ext}\right)_{V_i} = 0, \quad i = 1, 2, 3, 4, \quad j = 1, 2$$

26.4.3 Heavy Sheet

Let ρ be the mass per unit area and A the total area of the second gradient homogeneous elastic body and ρ_m be the mass per unit area and A_m the total area of the non-homogeneous first gradient elastic body. Thus, the correct identification between the two mass densities is as follows

$$\rho_m = \rho \frac{A}{A_m}$$

The following displacement field is considered,

$$u_1 = 0, \qquad u_2 = \frac{\rho g \left(X_2 - l\right)\left(3l + X_2\right)}{2c_{11}}. \tag{26.58}$$

This is achieved by the following distributed forces and double forces

$$b_1^{ext} = m_{ij}^{ext} = 0, \quad \forall i, j, \quad b_2^{ext} = -g\rho$$

the following set of kinematical boundary conditions for the macroscopic second gradient model of Fig. 26.1.

At Q, that is at $X_1 = 0$ and $\forall X_2 \in [-l, l]$

$$\begin{aligned} u_1\left(0, X_2\right) &= u_{1,1}\left(0, X_2\right) = 0, \\ u_2\left(0, X_2\right) &= \frac{\rho g \left(X_2 - l\right)\left(3l + X_2\right)}{2c_{11}}, \\ u_{2,1}\left(0, X_2\right) &= 0. \end{aligned}$$

At R, that is $\forall X_1 \in [0, L]$ and at $X_2 = l$,

$$u_1\left(X_1, l\right) = u_{1,2}\left(X_1, l\right) = u_2\left(X_1, l\right) = 0, \quad u_{2,2}\left(X_1, l\right) = \frac{2\rho g l}{c_{11}}.$$

At S, that is at $X_1 = L$ and $\forall X_2 \in [-l, l]$

$$\begin{aligned} u_1\left(L, X_2\right) &= u_{1,1}\left(L, X_2\right) = 0, \\ u_2\left(L, X_2\right) &= \frac{\rho g \left(X_2 - l\right)\left(3l + X_2\right)}{2c_{11}}, \\ u_{2,1}\left(L, X_2\right) &= 0. \end{aligned}$$

At T, that is $\forall X_1 \in [0, L]$ and at $X_2 = -l$,

$$u_1\left(X_1, -l\right) = u_{1,2}\left(X_1, -l\right) = 0, \quad u_2\left(X_1, -l\right) = -\frac{2\rho g l^2}{c_{11}}, \quad u_{2,2}\left(X_1, -l\right) = 0.$$

From Placidi et al (2017b,c, 2015, 2017a) force and double force at each side have been calculated

$$\begin{aligned} t_1^{S,ext} &= c_{12} \frac{\rho g \left(X_2 + l\right)}{c_{11}}, \quad t_2^{S,ext} = 0, \\ \tau_1^{S,ext} &= 0, \quad \tau_2^{S,ext} = \frac{\sqrt{2}}{2} a_{13} \frac{\rho g}{c_{11}}. \end{aligned} \tag{26.59}$$

$$t_1^{Q,ext} = -c_{12}\frac{\rho g\,(X_2+l)}{c_{11}}, \quad t_2^{Q,ext}=0,$$
$$\tau_1^{Q,ext}=0, \quad \tau_2^{Q,ext}=\frac{\sqrt{2}}{2}a_{13}\frac{\rho g}{c_{11}}. \tag{26.60}$$

$$t_1^{R,ext}=0, \quad t_2^{R,ext}=2\rho g\,l, \quad \tau_1^{R,ext}=0, \quad \tau_2^{R,ext}=a_{11}\frac{\rho g}{c_{11}}. \tag{26.61}$$

$$t_1^{T,ext}=0, \quad t_2^{T,ext}=0, \quad \tau_1^{T,ext}=0, \quad \tau_2^{T,ext}=a_{11}\frac{\rho g}{c_{11}}. \tag{26.62}$$

as well as vertex forces,

$$\left(f_1^{ext}\right)_{\mathcal{V}_i} = -\left(a_{12}+\frac{\sqrt{2}}{2}a_{13}\right)\frac{\rho g}{c_{11}}, \quad i=1,3 \tag{26.63}$$
$$\left(f_1^{ext}\right)_{\mathcal{V}_i} = \left(a_{12}+\frac{\sqrt{2}}{2}a_{13}\right)\frac{\rho g}{c_{11}}, \quad i=2,4 \tag{26.64}$$
$$\left(f_2^{ext}\right)_{\mathcal{V}_i}=0, \quad i=1,2,3,4. \tag{26.65}$$

26.4.4 Non-conventional Bending

The following displacement field is considered,

$$u_1=0, \qquad u_2=-B\frac{X_1^2}{2}, \tag{26.66}$$

where B is a small constant with the physical dimension of the inverse of a length. This is achieved by the following distributed forces and double forces

$$b_1^{ext}=m_{ij}^{ext}=0, \quad \forall i,j, \quad b_2^{ext}=-\frac{B}{2}c_{33}$$

and set of kinematical boundary conditions for the macroscopic second gradient model of Fig. 26.1.
At Q, that is at $X_1=0$ and $\forall X_2\in[-l,l]$

$$u_1\left(0,X_2\right)=u_{1,1}\left(0,X_2\right)=0, \quad u_2\left(0,X_2\right)=0, \quad u_{2,1}\left(0,X_2\right)=0.$$

At R, that is $\forall X_1\in[0,L]$ and at $X_2=l$,

$$u_1\left(X_1,l\right)=u_{1,2}\left(X_1,l\right)=0, \quad u_2\left(X_1,l\right)=-\frac{B\,X_1^2}{2}, \quad u_{2,2}\left(X_1,l\right)=0.$$

At S, that is at $X_1 = L$ and $\forall X_2 \in [-l, l]$

$$u_1(L, X_2) = u_{1,1}(L, X_2) = 0, \quad u_2(L, X_2) = -\frac{B\,L^2}{2}, \quad u_{2,1}(L, X_2) = -B\,L.$$

At T, that is $\forall X_1 \in [0, L]$ and at $X_2 = -l$,

$$u_1(X_1, -l) = u_{1,2}(X_1, -l) = 0, \quad u_2(X_1, -l) = -\frac{B\,X_1^2}{2}, \quad u_{2,2}(X_1, -l) = 0.$$

From Placidi et al (2017b,c, 2015, 2017a) force and double force at each side have been calculated

$$t_1^{S,ext} = 0, t_2^{S,ext} = -\frac{B}{2} X_1 c_{33}, \tau_1^{S,ext} = 0, \tau_2^{S,ext} = -\frac{B}{2} a_{33}, \tag{26.67}$$

$$t_1^{ext,Q} = 0, t_2^{Q,ext} = \frac{B}{2} X_1 c_{33}, \tau_1^{Q,ext} = 0, \tau_2^{Q,ext} = -\frac{B}{2} a_{33}, \tag{26.68}$$

$$t_1^{ext,R} = -\frac{B}{2} X_1 c_{33}, t_2^{ext,R} = 0, \tau_1^{R,ext} = 0, \tau_2^{R,ext} = -\frac{\sqrt{2}}{2} a_{13}\,B, \tag{26.69}$$

$$t_1^{ext,T} = \frac{B}{2} X_1 c_{33}, t_2^{ext,T} = 0, \tau_1^{T,ext} = 0, \tau_2^{T,ext} = -\frac{\sqrt{2}}{2} a_{13}\,B, \tag{26.70}$$

as well as vertex forces,

$$\left(f_1^{ext}\right)_{V_i} = B\left(\frac{\sqrt{2}}{2} a_{23} + \frac{1}{2} a_{33}\right), \quad i = 1, 3 \tag{26.71}$$

$$\left(f_1^{ext}\right)_{V_i} = -B\left(\frac{\sqrt{2}}{2} a_{23} + \frac{1}{2} a_{33}\right), \quad i = 2, 4 \tag{26.72}$$

$$\left(f_2^{ext}\right)_{V_i} = 0, \quad i = 1, 2, 3, 4. \tag{26.73}$$

26.4.5 Trapezoidal Case

Let us take into account the following displacement field

$$u_1 = 0, \qquad u_2 = T\,X_1 X_2, \tag{26.74}$$

where T is a small constant with the physical dimension of the inverse of a length. This is achieved by the following distributed forces and double forces

$$b_2^{ext} = m_{ij}^{ext} = 0, \quad \forall i, j, \quad b_1^{ext} = T\left(c_{12} + \frac{1}{2} c_{33}\right)$$

and set of kinematical boundary conditions for the macroscopic second gradient model of Fig. 26.1.

At Q, that is at $X_1 = 0$ and $\forall X_2 \in [-l, l]$

$$u_1(0, X_2) = u_{1,1}(0, X_2) = 0, \quad u_2(0, X_2) = 0, \quad u_{2,1}(0, X_2) = T X_2.$$

At R, that is $\forall X_1 \in [0, L]$ and at $X_2 = l$,

$$u_1(X_1, l) = u_{1,2}(X_1, l) = 0, \quad u_2(X_1, l) = T X_1 X_2, \quad u_{2,2}(X_1, l) = T X_1.$$

At S, that is at $X_1 = L$ and $\forall X_2 \in [-l, l]$

$$u_1(L, X_2) = u_{1,1}(L, X_2) = 0, \quad u_2(L, X_2) = T L X_2, \quad u_{2,1}(L, X_2) = T X_2.$$

At T, that is $\forall X_1 \in [0, L]$ and at $X_2 = -l$,

$$u_1(X_1, -l) = u_{1,2}(X_1, -l) = 0, \; u_2(X_1, -l) = -T\, l X_1, \; u_{2,2}(X_1, -l) = T X_1.$$

From Placidi et al (2017b,c, 2015, 2017a) force and double force at each side have been calculated,

$$\begin{aligned}
t_1^{ext,S} &= c_{12} T X_1,\; t_2^{ext,S} = \frac{1}{2} T\, c_{33} X_2, \\
\tau_1^{S,ext} &= T \left(a_{12} + \frac{\sqrt{2}}{2} a_{13} \right),\; \tau_2^{S,ext} = 0, \\
t_1^{ext,Q} &= -T\, c_{12} X_1,\; t_2^{ext,Q} = -\frac{1}{2} T\, c_{33} X_2 \\
\tau_1^{Q,ext} &= T \left(a_{12} + \frac{\sqrt{2}}{2} a_{13} \right),\; \tau_2^{Q,ext} = 0, \\
t_1^{ext,R} &= \frac{1}{2} T\, c_{33} X_2,\; t_2^{ext,R} = T\, c_{11} X_1, \\
\tau_1^{R,ext} &= \frac{T}{2} \left(\sqrt{2} a_{23} + a_{33} \right),\; \tau_2^{R,ext} = 0, \\
t_1^{ext,T} &= -\frac{1}{2} T\, c_{33} X_2,\; t_2^{ext,T} = -T\, c_{11} X_1, \\
\tau_1^{T,ext} &= \frac{T}{2} \left(\sqrt{2} a_{23} + a_{33} \right),\; \tau_2^{T,ext} = 0,
\end{aligned}$$

as well as vertex forces,

$$\left(f_1^{ext} \right)_{\mathcal{V}_i} = 0, \quad i = 1, 2, 3, 4 \tag{26.75}$$

$$\left(f_2^{ext} \right)_{\mathcal{V}_i} = -T \left(a_{22} + \sqrt{2} a_{23} + \frac{1}{2} a_{33} \right), \quad i = 1, 3 \tag{26.76}$$

$$\left(f_2^{ext} \right)_{\mathcal{V}_i} = T \left(a_{22} + \sqrt{2} a_{23} + \frac{1}{2} a_{33} \right), \quad i = 2, 4. \tag{26.77}$$

26.5 Conclusion

In this paper we have derived analytical solutions for a second gradient elastic model with a D_4 cubic symmetry in the static case. For this model, in 2D, nine constitutive coefficients are needed, that is six more than the standard first gradient case, where only three constitutive coefficients are needed. Thus, identification analysis is not an easy task and analytical solutions can concur in achieving the searched parameter identification. Such analytical solutions are computed for simple mechanical tests, namely: simple tension test, simple shear test, heavy sheet, non-conventional bending and the trapezoidal case.

These solutions will be used in a further contribution for the convenient identification of the nine coefficients (Abali et al, 2015). In fact, it is a matter of facts that if we can measure, for the four cited experiments, the force $\mathbf{t}$ and the double force τ per unit length and the force $\mathbf{f}$ at the edges, then the identification of the material coefficients could be done as shown in Table 26.1.

Table 26.1 Material coefficients

Experiment	Identified constants	Equations for the identification
Simple tension	c_{11} and c_{12}	(26.51) and (26.49)
Simple shear	c_{33}	(26.54)
Heavy sheet	a_{11}, a_{12}, and a_{13}	(26.61), (26.63), and (26.59)
Non-conventional bending	a_{23}, and a_{33}	(26.71) and (26.68)
Trapezoidal case	a_{22}	(26.76)

Acknowledgements Authors thank the *Laboratoire International Associé Coss&Vita* for the support *via* "Fédération Francilienne de Mécanique, CNRS FR2609". Giuseppe Rosi has been supported by the CNRS through the PEPS program "HipImpact".

References

Abali BE, Müller WH, Eremeyev VA (2015) Strain gradient elasticity with geometric nonlinearities and its computational evaluation. Mechanics of Advanced Materials and Modern Processes 1(1):1–11

Alibert JJ, Seppecher P, dell'Isola F (2003) Truss modular beams with deformation energy depending on higher displacement gradients. Mathematics and Mechanics of Solids 8(1):51–73

Altenbach H, Eremeev V, Morozov N (2010) On equations of the linear theory of shells with surface stresses taken into account. Mechanics of Solids 45(3):331–342

Aminpour H, Rizzi N (2015) On the continuum modelling of carbon nano tubes. Civil-Comp Proceedings 108

Aminpour H, Rizzi N (2016) On the modelling of carbon nano tubes as generalized continua. In: Generalized Continua as Models for Classical and Advanced Materials, Springer, pp 15–35

AminPour H, Rizzi N (2016) A one-dimensional continuum with microstructure for single-wall carbon nanotubes bifurcation analysis. Mathematics and Mechanics of Solids 21(2):168–181

Aminpour H, Rizzi N, Salerno G (2014) A one-dimensional beam model for single-wall carbon nano tube column buckling. Civil-comp proceedings 106

Andreaus U, Chiaia B, Placidi L (2013) Soft-impact dynamics of deformable bodies. Continuum Mechanics and Thermodynamics 25(2-4):375–398

Andreaus U, Spagnuolo M, Lekszycki T, Eugster SR (2018) A ritz approach for the static analysis of planar pantographic structures modeled with nonlinear euler–bernoulli beams. Continuum Mechanics and Thermodynamics pp 1–21

Atluri S, Cazzani A (1995) Rotations in computational solid mechanics. Archives of Computational Methods in Engineering 2(1):49–138

Auffray N, dell'Isola F, Eremeyev VA, Madeo A, Rosi G (2015a) Analytical continuum mechanics à la Hamilton–Piola least action principle for second gradient continua and capillary fluids. Mathematics and Mechanics of Solids 20(4):375–417

Auffray N, Dirrenberger J, Rosi G (2015b) A complete description of bi-dimensional anisotropic strain-gradient elasticity. International Journal of Solids and Structures 69:195–206

Baraldi D, Reccia E, Cazzani A, Cecchi A (2013) Comparative analysis of numerical discrete and finite element models: the case of in-plane loaded periodic brickwork. Composites: Mechanics, Computations, Applications: An International Journal 4(4)

Barchiesi E, Spagnuolo M, Placidi L (2018) Mechanical metamaterials: a state of the art. Mathematics and Mechanics of Solids p 1081286517735695

Bersani AM, Giorgio I, Tomassetti G (2013) Buckling of an elastic hemispherical shell with an obstacle. Continuum Mechanics and Thermodynamics 25(2-4):443–467

Bersani AM, Della Corte A, Piccardo G, Rizzi NL (2016) An explicit solution for the dynamics of a taut string of finite length carrying a traveling mass: the subsonic case. Zeitschrift für angewandte Mathematik und Physik 67(4):108

Bilotta A, Turco E (2009) A numerical study on the solution of the cauchy problem in elasticity. International Journal of Solids and Structures 46(25-26):4451–4477

Bilotta A, Formica G, Turco E (2010) Performance of a high-continuity finite element in three-dimensional elasticity. International Journal for Numerical Methods in Biomedical Engineering 26(9):1155–1175

Cazzani A, Ruge P (2012) Numerical aspects of coupling strongly frequency-dependent soil–foundation models with structural finite elements in the time-domain. Soil Dynamics and Earthquake Engineering 37:56–72

Cecchi A, Rizzi NL (2001) Heterogeneous elastic solids: a mixed homogenization-rigidification technique. International journal of solids and structures 38(1):29–36

Cuomo M, Contrafatto L, Greco L (2014) A variational model based on isogeometric interpolation for the analysis of cracked bodies. International Journal of Engineering Science 80:173–188

D'Annibale F, Luongo A (2013) A damage constitutive model for sliding friction coupled to wear. Continuum Mechanics and Thermodynamics 25(2-4):503–522

Del Vescovo D, Giorgio I (2014) Dynamic problems for metamaterials: review of existing models and ideas for further research. International Journal of Engineering Science 80:153–172

dell'Isola F, Rotoli G (1995) Validity of laplace formula and dependence of surface tension on curvature in second gradient fluids. Mechanics research communications p 6

dell'Isola F, Seppecher P (1995) The relationship between edge contact forces, double forces and interstitial working allowed by the principle of virtual power. Comptes rendus de l'Académie des sciences Série IIb, Mécanique, physique, astronomie p 7

dell'Isola F, Steigmann D (2015) A two-dimensional gradient-elasticity theory for woven fabrics. Journal of Elasticity 118(1):113–125

dell'Isola F, Gouin H, Seppecher P (2008) Radius and surface tension of microscopic bubbles by second gradient theory. arXiv preprint arXiv:08080312

dell'Isola F, Gouin H, Rotoli G (2009) Nucleation of spherical shell-like interfaces by second gradient theory: numerical simulations. arXiv preprint arXiv:09061897

dell'Isola F, Andreaus U, Placidi L (2015) At the origins and in the vanguard of peridynamics, non-local and higher-gradient continuum mechanics: An underestimated and still topical contribution of Gabrio Piola. Mathematics and Mechanics of Solids 20(8):887–928 dell'Isola F, Della Corte A, Esposito R, Russo L (2016) Some cases of unrecognized transmission of scientific knowledge: from antiquity to Gabrio Piola's peridynamics and generalized continuum theories. In: Generalized continua as models for classical and advanced materials, Springer, pp 77–128

dell'Isola F, Seppecher P, Alibert JJ, Lekszycki T, Grygoruk R, Pawlikowski M, Steigmann D, Giorgio I, Andreaus U, Turco E, et al (2018) Pantographic metamaterials: an example of mathematically driven design and of its technological challenges. Continuum Mechanics and Thermodynamics pp 1–34

dell'Isola F, Andreaus U, Cazzani A, Barchiesi E (2019) Introductory remarks about the volume ii of the complete works of Gabrio Piola. In: The Complete Works of Gabrio Piola: Volume II, Springer, pp 1–22

Dos Reis F, Ganghoffer J (2012) Construction of micropolar continua from the asymptotic homogenization of beam lattices. Computers & Structures 112:354–363

Franciosi P, Spagnuolo M, Salman OU (2018) Mean green operators of deformable fiber networks embedded in a compliant matrix and property estimates. Continuum Mechanics and Thermodynamics pp 1–32

Garusi E, Tralli A, Cazzani A (2004) An unsymmetric stress formulation for reissner-mindlin plates: A simple and locking-free rectangular element. International Journal of Computational Engineering Science 5(03):589–618

Goda I, Assidi M, Belouettar S, Ganghoffer J (2012) A micropolar anisotropic constitutive model of cancellous bone from discrete homogenization. Journal of the mechanical behavior of biomedical materials 16:87–108

Goda I, Assidi M, Ganghoffer JF (2014) A 3d elastic micropolar model of vertebral trabecular bone from lattice homogenization of the bone microstructure. Biomechanics and modeling in mechanobiology 13(1):53–83

Greco L, Cuomo M (2014) An implicit g1 multi patch b-spline interpolation for kirchhoff–love space rod. Computer Methods in Applied Mechanics and Engineering 269:173–197

Greco L, Cuomo M (2015) Consistent tangent operator for an exact kirchhoff rod model. Continuum Mechanics and Thermodynamics 27(4-5):861–877

Hendy CR, Turco E (2008) Numerical validation of simplified theories for design rules of transversely stiffened plate girders. Struct Eng pp 37–42

Hughes TJ, Evans JA, Reali A (2014) Finite element and nurbs approximations of eigenvalue, boundary-value, and initial-value problems. Computer Methods in Applied Mechanics and Engineering 272:290–320

Indelicato G, Albano A (2009) Symmetry properties of the elastic energy of a woven fabric with bending and twisting resistance. Journal of Elasticity 94(1):33

Kezmane A, Chiaia B, Kumpyak O, Maksimov V, Placidi L (2017) 3d modelling of reinforced concrete slab with yielding supports subject to impact load. European Journal of Environmental and Civil Engineering 21(7-8):988–1025

Luongo A, D'annibale F (2012) Bifurcation analysis of damped visco-elastic planar beams under simultaneous gravitational and follower forces. International Journal of Modern Physics B 26(25):1246,015

Mindlin RD (1964) Micro-structure in linear elasticity. Archive for Rational Mechanics and Analysis 16(1):51–78

Misra A, Huang S (2012) Micromechanical stress–displacement model for rough interfaces: Effect of asperity contact orientation on closure and shear behavior. International Journal of Solids and Structures 49(1):111–120

Misra A, Poorsolhjouy P (2015) Micro-macro scale instability in 2d regular granular assemblies. Continuum Mechanics and Thermodynamics 27(1-2):63–82

Misra A, Singh V (2014) Nonlinear granular micromechanics model for multi-axial rate-dependent behavior. International Journal of Solids and Structures 51(13):2272–2282

Misra A, Singh V (2015) Thermomechanics-based nonlinear rate-dependent coupled damage-plasticity granular micromechanics model. Continuum Mechanics and Thermodynamics 27(4-5):787–817

Misra A, Parthasarathy R, Singh V, Spencer P (2015) Micro-poromechanics model of fluid-saturated chemically active fibrous media. ZAMM-Journal of Applied Mathematics and Mechanics/Zeitschrift für Angewandte Mathematik und Mechanik 95(2):215–234

Piccardo G, Pagnini LC, Tubino F (2015) Some research perspectives in galloping phenomena: critical conditions and post-critical behavior. Continuum Mechanics and Thermodynamics 27(1-2):261–285

Pideri C, Seppecher P (1997) A second gradient material resulting from the homogenization of an heterogeneous linear elastic medium. Continuum Mechanics and Thermodynamics 9(5):241–257

Placidi L, Andreaus U, Della Corte A, Lekszycki T (2015) Gedanken experiments for the determination of two-dimensional linear second gradient elasticity coefficients. Zeitschrift für angewandte Mathematik und Physik 66(6):3699–3725

Placidi L, Andreaus U, Giorgio I (2017a) Identification of two-dimensional pantographic structure via a linear d4 orthotropic second gradient elastic model. Journal of Engineering Mathematics 103(1):1–21

Placidi L, Barchiesi E, Battista A (2017b) An inverse method to get further analytical solutions for a class of metamaterials aimed to validate numerical integrations. In: Mathematical Modelling in Solid Mechanics, Springer, pp 193–210

Placidi L, Barchiesi E, Della Corte A (2017c) Identification of two-dimensional pantographic structures with a linear d4 orthotropic second gradient elastic model accounting for external bulk double forces. In: Mathematical Modelling in Solid Mechanics, Springer, pp 211–232

Placidi L, Barchiesi E, Misra A (2018) A strain gradient variational approach to damage: a comparison with damage gradient models and numerical results. Mathematics and Mechanics of Complex Systems 6(2):77–100

Rahali Y, Giorgio I, Ganghoffer J, dell'Isola F (2015) Homogenization à la Piola produces second gradient continuum models for linear pantographic lattices. International Journal of Engineering Science 97:148–172

Rizzi N, Varano V (2011a) On the postbuckling analysis of thin-walled frames. In: Thirteenth international conference on civil, structural and environmental engineering computing. Civil-Comp Press

Rizzi NL, Varano V (2011b) The effects of warping on the postbuckling behaviour of thin-walled structures. Thin-Walled Structures 49(9):1091–1097

Rosi G, Placidi L, Nguyen VH, Naili S (2017) Wave propagation across a finite heterogeneous interphase modeled as an interface with material properties. Mechanics Research Communications 84:43–48

Ruta G, Varano V, Pignataro M, Rizzi N (2008) A beam model for the flexural–torsional buckling of thin-walled members with some applications. Thin-Walled Structures 46(7-9):816–822

Sansour C, Skatulla S (2009) A strain gradient generalized continuum approach for modelling elastic scale effects. Computer Methods in Applied Mechanics and Engineering 198(15-16):1401–1412

Scerrato D, Giorgio I, Rizzi NL (2016a) Three-dimensional instabilities of pantographic sheets with parabolic lattices: numerical investigations. Zeitschrift für angewandte Mathematik und Physik 67(3):53

Scerrato D, Zhurba Eremeeva IA, Lekszycki T, Rizzi NL (2016b) On the effect of shear stiffness on the plane deformation of linear second gradient pantographic sheets. ZAMM-Journal of Applied Mathematics and Mechanics/Zeitschrift für Angewandte Mathematik und Mechanik 96(11):1268–1279

Selvadurai A (1973) Plane strain problems in second-order elasticity theory. International Journal of Non-Linear Mechanics 8(6):551–563

Seppecher P, Alibert JJ, dell'Isola F (2011) Linear elastic trusses leading to continua with exotic mechanical interactions. In: Journal of Physics: Conference Series, IOP Publishing, vol 319, p 012018

Solaria G, Pagnini L, Piccardo G (1997) A numerical algorithm for the aerodynamic identification of structures. Journal of wind engineering and industrial aerodynamics 69:719–730

Spagnuolo M, Andreaus U (2018) A targeted review on large deformations of planar elastic beams: extensibility, distributed loads, buckling and post-buckling. Mathematics and Mechanics of Solids p 1081286517737000

Steigmann DJ (2009) Linear theory for the bending and extension of a thin, residually stressed, fiber-reinforced lamina. International Journal of Engineering Science 47(11-12):1367–1378

Steigmann DJ, dell'Isola F (2015) Mechanical response of fabric sheets to three-dimensional bending, twisting, and stretching. Acta Mechanica Sinica 31(3):373–382

Terravecchia S, Panzeca T, Polizzotto C (2014) Strain gradient elasticity within the symmetric BEM formulation. Frattura ed Integrità Strutturale 8(29):61–73

Turco E (2013) Identification of axial forces on statically indeterminate pin-jointed trusses by a nondestructive mechanical test. The Open Civil Engineering Journal 7:50–57

Walpole L (1984) Fourth-rank tensors of the thirty-two crystal classes: multiplication tables. Proc R Soc Lond A 391(1800):149–179

Chapter 27
Gradient Theory of Adhesion and Tabor Parameter

Valentin L. Popov

Abstract There are two basic approaches to mathematical description of media having microstructure, or more general, one or more intrinsic characteristic lengths. The first approach is to consider the underlying structure explicitly. The opposite possibility is to try to simulate the medium still as a homogeneous one but having either additional degrees of freedom or being characterized by some intrinsic characteristic length. This second way is the way used in a wide spectrum of micropolar and gradient approaches.
In the present paper, the philosophy of the gradient approaches is applied to the problem of adhesion. Adhesive forces have some range of action which naturally introduces a length parameter into the system. Relative role which plays this characteristic length in adhesive behavior is governed by a parameter introduced 1977 by Tabor. However, the interaction range is only one possible characteristic length in adhesion. Other lengths may be connected with spatial heterogeneity of the surfaces at the micro scale. We discuss the possibility of describing adhesion with a finite characteristic length in the framework of a continuum theory introducing proper gradient expansion. The result is both encouraging and disappointing: The gradient theory provides a description of adhesion with finite length which is much simpler compared with known explicit approaches. However, it remains phenomenological, so that proper physical interpretation of the characteristic length is not straightforward.

Keywords: Adhesive forces · Characteristic length · Gradient theory

---

Valentin L. Popov
Technische Universität Berlin
Department of System Dynamics and Friction Physics, 10623 Berlin, Germany,
e-mail: v.popov@tu-berlin.de

B. E. Abali et al. (eds.), *New Achievements in Continuum Mechanics and Thermodynamics*, Advanced Structured Materials 108,
https://doi.org/10.1007/978-3-030-13307-8_27

27.1 Introduction

1971 Johnson, Kendall and Roberts (JKR) published their classical paper on the theory of adhesion of elastic bodies having shape of paraboloids of revolution in Johnson et al (1971). Four years later, an alternative theory was published by Derjaguin et al (1975) followed soon by a paper of Tabor (1977) who demonstrated that both theories are limiting cases of very short range (JKR) and long range (DMT) adhesive forces. He introduced a parameter, known as "Tabor parameter" which basically compares the characteristic action range of adhesive forces with the length of the adhesive "neck."

The JKR theory is applicable well for macroscopic bodies. It is based on a simple physical concept of energy balance as first formulated by Griffith (1921) in his theory of brittle cracks. The simplicity of the theoretical ideas behind and the possibility of using the energy-based methods for analyzing adhesive contacts made JKR theory very popular. However, for micromechanical devices a finite range of adhesive forces may play an essential role. That is why the DMT theory and its generalizations, especially to contacts of rough surfaces, continue attracting vivid interest of researchers as in Ciavarella (2017). An important contribution for understanding the transition between both limiting cases of JKR and DMT provided the theory of Maugis (1992) who considered adhesive contact problem of parabolic bodies with very simple interaction potential (with adhesive pressure remaining constant up to some critical distance and vanishing abruptly at larger distances.) This simple interaction made it possible to provide an almost completely analytical solution.

The JKR solution for adhesive contact of axially symmetric bodies can alternatively be derived in the framework of the formalism of the Method of Dimensionality Reduction (MDR) described in Popov and Heß (2015). In this method, the three-dimensional contact is first mapped onto a contact problem of a modified plane profile with a linear elastic foundation (Winkler foundation, a series of independent springs). After solution of the contact problem for this simple system, the exact three-dimensional solution can be easily restored using the rules of the Method of Dimensionality Reduction. In the present paper we will proceed from the formulation of the JKR theory via MDR and suggest a generalization to adhesion exhibiting a characteristic length scale.

The paper is organized as follows: In Sect. 27.2 we shortly recapitulate the main steps of the method of dimensionality reduction. In Sect. 27.3 we suggest a generalization of the simple adhesive MDR to adhesion with finite length scale. Sect. 27.4 concludes the paper.

27.2 Method of Dimensionality Reduction Formulation of Johnson-Kendall-Roberts Theory

We consider the frictionless normal contact between two elastic bodies with the elasticity moduli E_1 and E_2, and Poisson's ratios ν_1 and ν_1. The axisymmetric difference between the profiles will be written as $\tilde{z} = f(r)$, where r is the polar radius in the contact plane. As is well known since Hertz, this contact problem is equivalent to the contact of a rigid indenter with the profile $\tilde{z} = f(r)$ and an elastic half-space with the effective elasticity modulus E^* as in Hertz (1882):

$$\frac{1}{E^*} = \frac{1-\nu_1^2}{E_1} + \frac{1-\nu_2^2}{E_2} . \tag{27.1}$$

As already mentioned in the Introduction, the contact of any given axially symmetric bodies can be solved very easily with the so-called "Method of Dimensionality Reduction." Solving the contact problem in the framework of the MDR requires two preparatory steps:

I. Firstly, the three-dimensional elastic bodies are replaced by a Winkler foundation: a linear arrangement of elements with independent degrees of freedom with arbitrary but sufficiently small distance Δx between the elements, as shown in Fig. 27.1. In the case of elastic bodies, the foundation consists of linear elastic springs with a normal stiffness

$$\Delta k_z = E^* \Delta x , \tag{27.2}$$

whereby E^* is given by Eq. (27.1).

II. In the second step, the three-dimensional profile $\tilde{z} = f(r)$ (Fig. 27.2, left) is transformed to a plane profile $g(x)$ (Fig. 27.2, right) according to "MDR-Transform"

$$g(x) = |x| \int_0^{|x|} \frac{f'(r)}{\sqrt{x^2 - r^2}} \, \mathrm{d}r . \tag{27.3}$$

The further calculation procedure consists of the following steps:

III. The profile $g(x)$ is now pushed into the one-dimensional elastic foundation defined according to (27.2) until a contact radius a is reached. Until this point, the adhesion will not be considered. This process is depicted in Fig. 27.3 on the left.

IV. After that the indenter is lifted up. It is assumed that all springs involved in the contact adhere to the indenter—the contact radius thus remains constant. In this

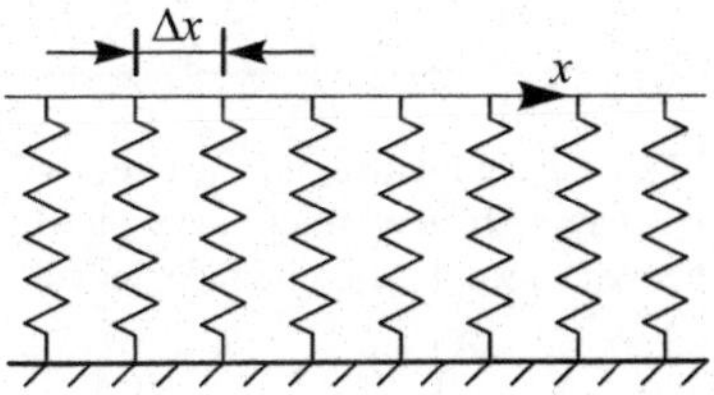

Fig. 27.1 One-dimensional elastic foundation (Winkler foundation).

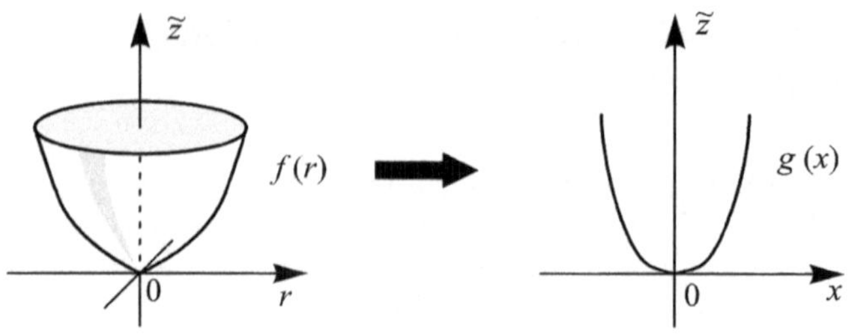

Fig. 27.2 With in the MDR, the three-dimensional profile $f(r)$ is transformed to a plane profile $g(x)$ according to the transformation rule (27.3).

process, the springs at the edge experience the maximum increase in tension. Upon reaching the the critical value of elongation of the outer springs

$$\Delta l(a) = \sqrt{\frac{2\pi a \Delta\gamma}{E^*}} \tag{27.4}$$

they detach (Fig. 27.3 right). $\Delta\gamma$ is here the specific work of separation per unit area. This criterion was discovered by Heß (2011) and is known as *rule of Hess*. A simple derivation of this criterion can be found in Popov et al (2018). The displacement of the outer springs is in the critical state negative, with the absolute value equal to the critical value (27.4):

$$w(a) = d - g(a) = -\Delta l(a) \tag{27.5}$$

where d is the indentation depth. It follows that

$$d = g(a) - \Delta l(a)\,. \tag{27.6}$$

This equation determines the relation between the contact radius and the indentation depth in an adhesive contact.

The normal force F_N is given by the sum of all springs in contact:

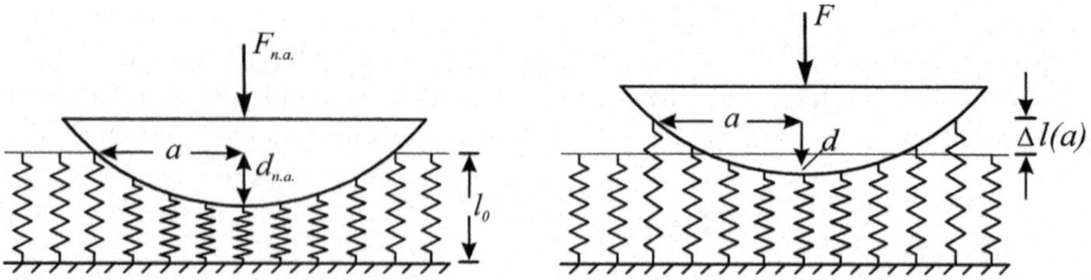

Fig. 27.3 Qualitative representation of the indentation and lifting process of a parabolic planar indenter with an elastic foundation which exactly models the properties of the adhesive contact between a rigid parabolic indenter and an elastic half-space.

$$F_N = 2E^* \int_0^a \left[d - g(x)\right] \mathrm{d}x \, . \tag{27.7}$$

Note that solution of the described replacement model coincides exactly with the solution of the corresponding three-dimensional problem (proofs can be found in Popov and Heß (2015) and Popov et al (2018).)

The criterion (27.4) can be also formulated as a critical value of the stress intensity factor. Indeed, by defining the linear force density

$$q_z(x) = \frac{\Delta F_N(x)}{\Delta x} = E^* w(x) = E^* \begin{cases} d - g(x)\,, & |x| < a \\ 0\,, & |x| > a \end{cases}, \tag{27.8}$$

with $w(x)$ being the vertical displacement of springs in the MDR model, one can restore the stress distribution in the original three-dimensional problem by using Equation (27.7)

$$\sigma_{zz}(r) = -r(r) = \frac{1}{\pi} \int_r^\infty \frac{q_z'(x)}{\sqrt{x^2 - r^2}} \mathrm{d}x \, . \tag{27.9}$$

If the force density has a jump, the corresponding stress distribution will have a typical for cracks square-root singularity.

The absolute stability of the contact is determined by the relations (27.10)

$$\left.\frac{\mathrm{d}g(a)}{\mathrm{d}a}\right|_{a=a_c} = \xi \sqrt{\frac{\pi \Delta\gamma}{2E^* a_c}}\,, \xi = \begin{cases} 3\,, & \text{force-control trials,} \\ 1\,, & \text{displacement-control trials.} \end{cases} \tag{27.10}$$

where the body shape enters only over the derivative of the effective MDR-transformed profile $g(x)$ at the edge of the contact.

It is important to stress that the whole essential information about the contact properties of any profile is completely contained in the effective profile $g(x)$ which directly determines both the current contact radius while its derivative determining the condition for the absolute stability.

27.3 Generalization of Method of Dimensionality Reduction for Adhesion with Finite Length Scale

If the adhesion problem contains a finite length, then the formulation of the adhesion criterion should be changed in a way allowing for definition of a characteristic length. At this point we introduce a corresponding formulation *formally*, without discussion of physical mechanisms behind. A detailed analysis of the physical meaning of the characteristic length which will be introduced below will be subject of later research.

Our suggestion is to replace the equilibrium condition (27.5) by a generalized condition

$$\left[w^2 - L^2\left(\frac{\mathrm{d}w}{\mathrm{d}x}\right)^2\right]_{x=a} = \Delta l(a)^2 \;, \tag{27.11}$$

where L is some characteristic length associated with adhesive interaction. In the following we analyze the properties of adhesive contact with the above generalized law of adhesion. Criterion (27.11) can be written explicitly as

$$\left(d - g(a)\right)^2 - L^2\left(\frac{\mathrm{d}g(a)}{\mathrm{d}a}\right)^2 = \frac{2\pi a \Delta\gamma}{E^*} \;. \tag{27.12}$$

Consider as example a parabolically shaped body with the profile $f(r) = r^2/(2R)$ where R is the radius of curvature. The corresponding MDR-modified profile is, according to (27.3), $g(x) = x^2/R$ and criterion (27.12) takes the form

$$\left(d - \frac{a^2}{R}\right)^2 - 4\frac{L^2}{R^2}a^2 = \frac{2\pi a \Delta\gamma}{E^*} \;. \tag{27.13}$$

The relation of the indentation depth and contact radius is thus given by

$$d = \frac{a^2}{R} - \sqrt{\frac{2\pi a \Delta\gamma}{E^*} + 4\frac{L^2}{R^2}a^2} \tag{27.14}$$

while the dependency of the normal force on the contact radius being determined by the equation

$$\begin{aligned} F_N &= 2E^* \int_0^a \left[d - g(x)\right] \mathrm{d}x = 2E^*\left(ad - \frac{a^3}{3R}\right) \\ &= 2E^*\left(\frac{2}{3}\frac{a^3}{R} - a\sqrt{\frac{2\pi a \Delta\gamma}{E^*} + 4\frac{L^2}{R^2}a^2}\right) \;. \end{aligned} \tag{27.15}$$

After normalizing the quantities to the critical values of the force-controlled trial of the JKR approximation (critical values of the JKR theory)

$$\hat{a} := \frac{a}{a_c} \;, \; \hat{d} := \frac{d}{|d_c|} \;, \; \hat{F} := \frac{F_N}{|F_c|} \tag{27.16}$$

with

$$a_c = \left(\frac{9\pi R^2 \Delta\gamma}{8E^*}\right)^{1/3}, \quad d_c = -\frac{1}{4}\left(\frac{3\pi^2(\Delta\gamma)^2 R}{(E^*)^2}\right)^{1/3}, \quad F_c = -\frac{3}{2}\pi\Delta\gamma R \tag{27.17}$$

we can rewrite (27.14) and (27.15) in the form

$$\tilde{d} = 3\tilde{a}^2 - 4\sqrt{\tilde{a} + \Lambda^{-2}\tilde{a}^2} \;, \qquad \tilde{F}_N = \tilde{a}^3 - 2\tilde{a}\sqrt{\tilde{a} + \Lambda^{-2}\tilde{a}^2} \;, \tag{27.18}$$

where Λ is a dimensionless parameter defined by the following expression:

$$\Lambda = \left(\frac{\pi\Delta\gamma R^2}{192L^3E^*}\right)^{1/3}. \tag{27.19}$$

As L is a phenomenological parameter which till now does not have a clear physical meaning, it is logically to consider Λ just as a new fitting parameter. Note that $\Lambda^{3/2} \propto R$, and thus thus can be considered as directly related to the Tabor parameter (which is also proportional to the radius of curvature.)

We would like to stress here, that the adhesion law used in the present paper does not coincide with that of Maugis's model and corresponding relations should not be compared with those of the Maugis's theory. For very large values of Λ we recapitulate the JKR theory. For intermediate values of Λ the adhesion curve deforms similarly to prediction of Maugis's theory as in Maugis (1992). However, while Maugis's theory is in a sense "exact" (at least for the kind of interaction potential used by him), the present theory can be considered only for values of $\Lambda > 2$.

From a "philosophical point of view" restriction of the field of applicability of gradient theories by system sizes larger than (or comparable to) the characteristic length is a basic, intrinsic restriction. Indeed, the characteristic length intends to describe a "microstructure" of the system; this is obviously only possible as far as we are at the scale larger than the characteristic length of microstructure. From mathematical point of view, gradient theories are based on expansions of relevant quantities (stresses of energies) with respect to gradients and have sense only as long as the characteristic length of the system remains smaller than the characteristic size of the system.

Figure 27.4 presents dependencies of normal force on the indentation depth obtained by parametric plot of Eqs. (27.18) and (27.19) in the interval from $\Lambda = \infty$ (JKR limit) to a critical value of $\Lambda \approx 2$ roughly corresponding to the limiting case of DMT. Smaller values of Λ correspond to no physical reality.

27.4 Conclusion

The charm of gradient theories consists in their simplicity and the drawback in their phenomenological character and limited range of applicability. Basing on the gradient expansions, this class of theories is applicable only as long as the characteristic length due to gradient expansion is much smaller than the relevant system lengths. In the case of adhesion, this means that the gradient theory can picture the limiting case of JKR and the case of moderate Tabor parameters but not the limiting case of extremely small Tabor parameters. For the case of moderate Tabor parameters, the gradient theory provides a very convenient, easy-to-derive and easy-to-use approximation.

To the best knowledge of the author, this is the first attempt of considering scale effects in adhesion using gradient approaches. Further developments surely may lead to more universal approaches with proper physical interpretation of the underlying length scales.

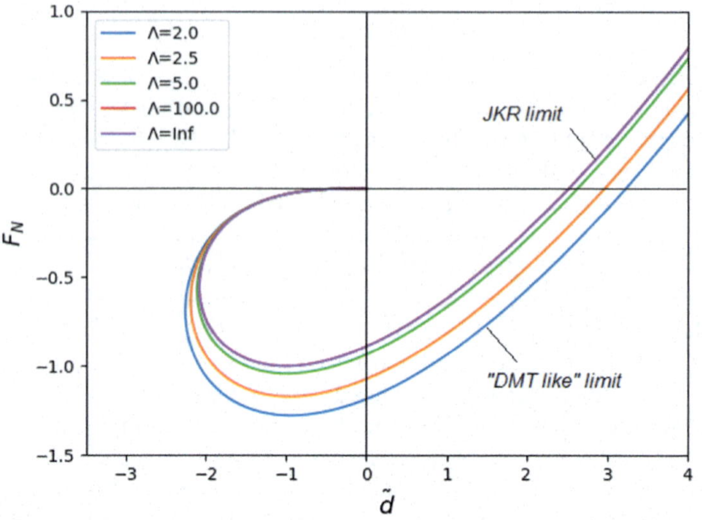

Fig. 27.4 Dependencies of the normal force on the indentation depth according to Eqs. (27.18) and (27.19) for a series of values of parameter Λ.

Acknowledgements The author acknowledges financial support of the Deutsche Forschungsgemeinschaft (DFG PO 810-55-1) and thanks M. Popov for valuable discussions.

References

Ciavarella M (2017) On the use of dmt approximations in adhesive contacts, with remarks on random rough contacts. Tribology International 114:445–449

Derjaguin B, Muller V, Toporov YP (1975) Effect of contact deformations on the adhesion of particles. Journal of Colloid and Interface Science 53(2):314–326

Griffith AA (1921) The phenomena of rupture and flow in solids. Phil Trans R Soc Lond A 221(582-593):163–198

Hertz H (1882) Über die Berührung fester elastischer Körper. Journal für die reine und angewandte Mathematik 92:156–171

Heß M (2011) Über die exakte Abbildung ausgewählter dreidimensionaler Kontakte auf Systeme mit niedrigerer räumlicher Dimension. Cuvillier Verlag Göttingen

Johnson KL, Kendall K, Roberts A (1971) Surface energy and the contact of elastic solids. Proc R Soc Lond A 324(1558):301–313

Maugis D (1992) Adhesion of spheres: the JKR-DMT transition using a Dugdale model. Journal of Colloid and Interface Science 150(1):243–269

Popov VL, Heß M (2015) Method of dimensionality reduction in contact mechanics and friction. Springer

Popov VL, Heß M, Willert E (2018) Handbuch der Kontaktmechanik: Exakte Lösungen axialsymmetrischer Kontaktprobleme. Springer-Verlag

Tabor D (1977) Surface forces and surface interactions. Journal of Colloid and Interface Science 58(1):2–13

Chapter 28
Cavity Flow of a Micropolar Fluid - a Parameter Study

Wilhelm Rickert & Sebastian Glane

Abstract This paper presents a parameter study of the flow of a micropolar fluid. The underlying equations and the choice of boundary conditions are discussed. Two flow situations are considered: Couette flow as a reference problem and the lid-driven cavity problem. The governing equations are specialized for the case of two-dimensional flow and discussed in dimensionless form. Several dimensionless parameters common in the theory of micropolar fluids are identified and their impact on the solutions is analyzed using the finite element method.

Keywords: Micropolar fluid theory · Microstructured material · Lid-driven cavity problem · Forced convection

28.1 Introduction

Generalized continuum theories have gained high interest in continuum mechanics and material modeling. Among Cosserat elasticity and gradient theories for solids and the Ericksen-Leslie theory for liquid crystals, Eringen's micropolar theory is one of the representatives of this field, see Ariman et al (1973) for an extensive review and Maugin (2011) for a historical discussion. Since its introduction by Eringen (1964), the theory of micropolar fluids has been widely applied, for example in the modeling of blood, Ariman et al (1973), particle suspensions, liquid crystals, lubrication, Prakash and Sinha (1975), and colloidal suspensions, Eringen (1991). In a series of papers, Müller and Vilchevskaya extended micropolar theory and investigated the production of microinertia, see Müller and Vilchevskaya (2017); Müller et al (2017); Müller and Vilchevskaya (2018); Vilchevskaya and Müller (2018). This extension may be applied to problems with microstructural changes.

Wilhelm Rickert · Sebastian Glane
Technische Universität Berlin, Institute of Mechanics, Einsteinufer 5, 10587 Berlin, Germany,
e-mail: rickert@tu-berlin.de, glane@tu-berlin.de

B. E. Abali et al. (eds.), *New Achievements in Continuum Mechanics and Thermodynamics*, Advanced Structured Materials 108,
https://doi.org/10.1007/978-3-030-13307-8_28

In this study the effect of micropolar material parameters is investigated through parameter variations using a numerical model. The flow problem considered in the parameter study is the lid-driven cavity problem. This is a (benchmark) problem in fluid mechanics. Many authors report numerical solutions for the flow of a Navier–Stokes fluid in a two or three dimensional cavity using different numerical techniques, see Botella and Peyret (1998); Bruneau and Saad (2006); Cortes and Miller (1994); Erturk and Gökccöl (2008); Freitas et al (1985); Nallasamy and Prasad (1977). Furthermore, this problem was studied for a thermally-driven flow of a buoyant fluid using the Boussinesq approximation, Iwatsu and Hyun (1995); Kareem et al (2016); Le Quere et al (1981), for nanofluids, Tiwari and Das (2007), and also for multiphase flow, Anders and Weinberg (2011); Chakravarthy and Ottino (1996). In context of micropolar fluids, thermally-driven convection was studied for cavities with different geometrical shapes, Aydin and Pop (2007); Bourantas and Loukopoulos (2014); Gibanov et al (2016a,b); Hsu and Chen (1996); Jena and Bhattacharyya (1986); Sheremet et al (2017), and including electromagnetic fields, Türk and Tezer-Sezgin (2017). Most of the papers on the micropolar cavity problem are concerned with the influence of micropolar parameters on the onset of convection (critical Rayleigh number), on the heat transfer and effects of the "vortex viscosity" on the heat generation. However, a parameter study neglecting thermal effects and only focusing on the mechanical behavior has not yet been conducted to the best of the authors' knowledge.

Following a presentation of the theory of micropolar fluids in Sect. 28.2, two flow situations are considered: Couette flow as a reference problem and the lid-driven cavity flow. The governing equations are specialized for these two-dimensional problems in Sect. 28.3 and subsequently solved numerically using the finite element method. The numerical procedure is described in Sect. 28.4 and a convergence analysis is performed based on the analytical solution for the Couette flow. The parameter study is presented and discussed in Sect. 28.5 before a conclusion is given in Sect. 28.6.

28.2 Theory of Micropolar Fluids

In context of a generalized continuum theory, the balance equations of mass, momentum and energy are supplemented by additional balances for the fields of angular velocity and moment of inertia. This introduces additional flux terms as well as production terms in the balance equations and the constitutive equations are modified in order to account for effects associated to the additional fields. Moreover, additional constitutive equations for the coupled stress and the production terms are required. In this paper, the theory of micropolar fluids is employed, in which an additional independent rotational degree of freedom, namely the angular velocity field[1], $\boldsymbol{\omega}(\boldsymbol{x},t)$, is introduced. In this framework, the microinertia tensor, $\boldsymbol{J}(\boldsymbol{x},t)$, is

[1] The angular velocity field is sometimes also referred to as the microgyration vector, *cf.* Eringen (2001).

an additional field accounting for the inertia of the material against microrotation. Below, we present the equations governing micropolar fluids and neglect microinertia effects later on.

28.2.1 Governing Local Balance Equations

In spatial description, the governing local equations are the balances of mass, linear momentum, moment of inertia and spin (Müller and Vilchevskaya, 2018)

$$\frac{\mathrm{d}\rho}{\mathrm{d}t} = -\rho(\nabla \cdot \boldsymbol{v}) \,, \quad \rho\frac{\mathrm{d}\boldsymbol{v}}{\mathrm{d}t} = \nabla \cdot \boldsymbol{\sigma} + \rho \boldsymbol{f} \,, \quad \frac{\mathrm{d}\boldsymbol{J}}{\mathrm{d}t} = \boldsymbol{\omega} \times \boldsymbol{J} - \boldsymbol{J} \times \boldsymbol{\omega} + \boldsymbol{\chi} \,,$$
$$\rho\frac{\mathrm{d}}{\mathrm{d}t}(\boldsymbol{J} \cdot \boldsymbol{\omega}) = \nabla \cdot \boldsymbol{\mu} + \boldsymbol{\sigma} \cdot\cdot \overset{(3)}{\boldsymbol{\epsilon}} + \rho \boldsymbol{m} + \rho \boldsymbol{\chi} \cdot \boldsymbol{\omega} \,, \tag{28.1}$$

where ρ is the density, $\boldsymbol{v}$ the velocity, $\boldsymbol{\sigma}$ the Cauchy stress tensor, $\boldsymbol{f}$ the specific body force, $\boldsymbol{J}$ the inertia tensor, $\boldsymbol{\mu}$ the couple stress tensor, $\boldsymbol{m}$ the specific volume couple, $\boldsymbol{\chi}$ the symmetric production of moment of inertia, and $\overset{(3)}{\boldsymbol{\epsilon}}$ the complete anti-symmetric tensor (density) of rank three. The standard scalar product of two tensors of second rank can be evaluated via $\boldsymbol{A} \cdot\cdot \boldsymbol{B} = A_{ij}B_{ij}$, where Einstein's summation convention applies and an orthonormal coordinate system is used. Furthermore, cross products of a second-rank tensor $\boldsymbol{A}$ and a vector $\boldsymbol{b}$ are given by $\boldsymbol{A} \times \boldsymbol{b} = (\boldsymbol{A} \otimes \boldsymbol{b}) \cdot\cdot \overset{(3)}{\boldsymbol{\epsilon}}$ and $\boldsymbol{b} \times \boldsymbol{A} = \overset{(3)}{\boldsymbol{\epsilon}} \cdot\cdot (\boldsymbol{b} \otimes \boldsymbol{A})$.[2]

This set of equations accounts for microstructural changes, because a production term, $\boldsymbol{\chi}$, is present in balance of the moment of inertia tensor. While such a production is not present in early works on micropolar fluids, see Eringen (1964, 1966), it was later introduced by Eringen (1985) to model a sticking of fluid to suspended rigid particles, see Eringen (1991, 1985); Zhilin (2006). Other examples for microstructural changes such as the crushing of particles, Glane et al (2017); Vilchevskaya and Müller (2018), the expansion of pressurized spherical particles, Müller and Vilchevskaya (2018), or the orientation and elongation of charged particles in an electric field, Müller and Vilchevskaya (2018), were proposed in a series of papers. Therein, different types of constitutive equations for the production term were studied. According to Zhilin (2006), the production term can be neglected, if the particles are considered rigid on the microscale. In summary, the question of whether or not the production term shall be included is a question of modeling of a specific fluid.

[2] In an orthonormal coordinate system, the components of this product may also be expressed as:

$$(\boldsymbol{A} \times \boldsymbol{b})_{ij} = A_{ik} b_l \varepsilon_{klj} \,.$$

28.2.2 Constitutive Laws and Field Equations

The local balances stated in Eq. (28.1) need to be supplemented by suitable constitutive equations. In this study, the quantities $\boldsymbol{\chi}$, $\boldsymbol{g}$, and $\boldsymbol{m}$ are neglected later and linear constitutive relations for the flux terms $\boldsymbol{\sigma}$ and $\boldsymbol{\mu}$ are employed. For isotropic micropolar fluids the following constitutive relations were derived in, *e.g.*, Cowin (1974); Eringen (2001); Zhilin (2006) and are applied subsequently:

- The Cauchy stress tensor is given by

$$\boldsymbol{\sigma} = (-p + \lambda \nabla \cdot \boldsymbol{v})\mathbf{1} + 2\mu \boldsymbol{D} - 2\tau(\boldsymbol{W} + \overset{(3)}{\boldsymbol{\epsilon}} \cdot \boldsymbol{\omega}) \,, \tag{28.2}$$

  where $\boldsymbol{D}$ is the symmetric and $\boldsymbol{W}$ is the skew-symmetric part of the velocity gradient $\boldsymbol{v} \otimes \nabla$. In Eq. (28.2), p denotes the thermodynamic pressure, λ and μ the volume and shear viscosity, respectively, and τ is an additional viscosity associated to rotational shear, which also has the unit $\mathrm{kg\,m^{-1}\,s^{-1}}$. The vorticity, $\boldsymbol{w} = \nabla \times \boldsymbol{v}/2$, may be referred to as regional angular velocity, Cowin (1974), and is the axial vector of the skew-symmetric part of the velocity gradient with $\boldsymbol{W} = -\overset{(3)}{\boldsymbol{\epsilon}} \cdot \boldsymbol{w}$. Thus, the last term in Eq. (28.2) is a difference between the local angular velocity and the vorticity. Therefore, we may refer to τ as rotational shear viscosity. Note that, in the limiting case $\tau \to 0$, the constitutive relation of an ordinary Navier–Stokes fluid is recovered.
- The couple stress tensor can be written as:

$$\boldsymbol{\mu} = \alpha(\nabla \cdot \boldsymbol{\omega})\mathbf{1} + 2\beta \boldsymbol{Q} - 2\gamma \boldsymbol{R} \,, \tag{28.3}$$

  where $\boldsymbol{Q}$ is the symmetric and $\boldsymbol{R}$ the skew-symmetric part of the angular velocity gradient $\boldsymbol{\omega} \otimes \nabla$. Here, α, β and γ are generalized viscosities that have the unit $\mathrm{kg\,m\,s^{-1}}$. The constitutive relations for the Cauchy stress and couple stress tensor is similar. However, it should be noted that the couple stress tensor, $\boldsymbol{\mu}$, does not have a direct functional dependency on the velocity field, which is not the case for the Cauchy stress tensor, $\boldsymbol{\sigma}$.

In the following, we restrict the analysis to constant material coefficients. Insertion of the constitutive equations specified above gives rise to the following field equations:

$$\begin{aligned}
0 &= \frac{\mathrm{d}\rho}{\mathrm{d}t} + \rho(\nabla \cdot \boldsymbol{v}) \,, \quad \frac{\mathrm{d}\boldsymbol{J}}{\mathrm{d}t} = \boldsymbol{\omega} \times \boldsymbol{J} - \boldsymbol{J} \times \boldsymbol{\omega} + \boldsymbol{\chi} \,, \\
\rho \frac{\mathrm{d}\boldsymbol{v}}{\mathrm{d}t} &= -\nabla p + (\lambda + \mu - \tau)\nabla[\nabla \cdot \boldsymbol{v}] + (\mu + \tau)\Delta \boldsymbol{v} + 2\tau \nabla \times \boldsymbol{\omega} + \rho \boldsymbol{f} \,, \\
\rho \frac{\mathrm{d}}{\mathrm{d}t}(\boldsymbol{J} \cdot \boldsymbol{\omega}) &= (\alpha + \beta - \gamma)\nabla[\nabla \cdot \boldsymbol{\omega}] + (\beta + \gamma)\Delta \boldsymbol{\omega} - 4\tau \boldsymbol{\omega} + \\
&\quad + 2\tau \nabla \times \boldsymbol{v} + \rho \boldsymbol{m} + \rho \boldsymbol{\chi} \cdot \boldsymbol{\omega} \,,
\end{aligned} \tag{28.4}$$

where Δ is the Laplace operator.

According to Müller and Vilchevskaya (2018), the production of moment of inertia, χ, must be interpreted as an additional constitutive quantity. However, we restrict our investigations to rigid spherical particles (on the mesoscale). As a consequence, the moment of inertia tensor is a constant spherical tensor, *i.e.*, $\boldsymbol{J} = J_0\mathbf{1}$, the production of moment of inertia vanishes and Eq. $(28.4)_2$ is fulfilled trivially because $\boldsymbol{\omega} \times \boldsymbol{J} - \boldsymbol{J} \times \boldsymbol{\omega}$ vanishes. Additionally we assume that the micropolar fluid is incompressible, *i.e.*, the mass density is a constant, ρ_0. Moreover, we assume that the specific body force, $\boldsymbol{f}$, and the specific volume couple, $\boldsymbol{m}$, are negligible. Under these assumptions, Eqs. (28.4) reduce to:

$$\begin{aligned} \nabla \cdot \boldsymbol{v} = 0\,, \quad \rho_0 \frac{\mathrm{d}\boldsymbol{v}}{\mathrm{d}t} &= -\nabla p + (\mu + \tau)\Delta \boldsymbol{v} + 2\tau \nabla \times \boldsymbol{\omega}\,, \\ \rho_0 J_0 \frac{\mathrm{d}\boldsymbol{\omega}}{\mathrm{d}t} &= (\alpha + \beta - \gamma)\nabla[\nabla \cdot \boldsymbol{\omega}] + (\beta + \gamma)\Delta \boldsymbol{\omega} + 2\tau \nabla \times \boldsymbol{v} - 4\tau \boldsymbol{\omega}\,. \end{aligned} \tag{28.5}$$

28.3 Problem Statement

In the following, two different problems of a stationary and two-dimensional flow of a micropolar fluid are considered. Figure 28.1 shows the two different problems. First, the Couette flow is considered as a reference problem, see Fig. 28.1a. For this problem, an analytical closed-form solution can be obtained for micropolar fluids, *cf.* Cowin (1974), which is subsequently used for a verification of the finite element method.

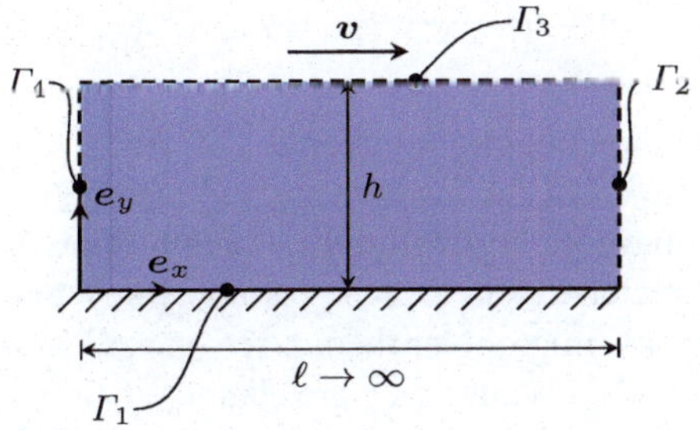

(a) Couette flow problem.

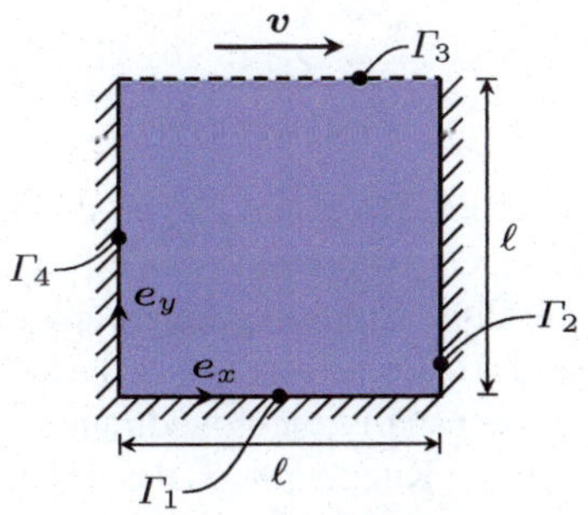

(b) Lid-driven cavity problem.

Fig. 28.1 Example problems.

The second problem is the lid-driven cavity problem, see Fig. 28.1b. At the top of the cavity, a "conveyor" modeled through purely tangential velocity forces a fluid motion in the underlying domain.

28.3.1 Dimensionless Equations

The parameters present in the equations for micropolar fluids are the generalized viscosities α, μ, τ, α, β, γ and the generalized inertias ρ_0, J_0. In order to determine the governing combinations of these material parameters, the problem is written in dimensionless form. The normalized quantities are introduced as:

$$\boldsymbol{x} = \ell_{\text{ref}}\tilde{\boldsymbol{x}} \,, \quad \boldsymbol{v} = v_{\text{ref}}\tilde{\boldsymbol{v}} \,, \quad t = t_{\text{ref}}\tilde{t} \,, \quad p = p_{\text{ref}}\tilde{p} \,, \quad \boldsymbol{\omega} = \omega_{\text{ref}}\tilde{\boldsymbol{\omega}} \,, \tag{28.6}$$

where symbols with a tilde are dimensionless. As usual in fluid mechanics, the time scale is chosen as the time scale of convective transport, *i.e.*, $t_{\text{ref}} = \ell_{\text{ref}}/v_{\text{ref}}$, and the pressure is normalized by $p_{\text{ref}} = \rho_0 v_{\text{ref}}^2$. Because of the fact that we will impose zero spin boundary conditions at walls, the angular velocity field, $\boldsymbol{\omega}$, is solely triggered by the velocity. It seems therefore reasonable to choose the reference angular velocity accordingly. Here, $\omega_{\text{ref}} = v_{\text{ref}}/\ell_{\text{ref}}$ is applied to emphasize the influence of the velocity. As a consequence, the following dimensionless system of equations arises:

$$\begin{aligned} &\tilde{\nabla} \cdot \tilde{\boldsymbol{v}} = 0 \,, \quad \frac{\mathrm{d}\tilde{\boldsymbol{v}}}{\mathrm{d}\tilde{t}} = -\tilde{\nabla}\tilde{p} + \frac{1}{(1-N^2)Re}\left(\tilde{\Delta}\tilde{\boldsymbol{v}} + 2N^2\tilde{\nabla} \times \tilde{\boldsymbol{\omega}}\right) \,, \\ &\frac{\mathrm{d}\tilde{\boldsymbol{\omega}}}{\mathrm{d}\tilde{t}} = \frac{1}{\Theta Re}\left(\frac{1}{M^2}\tilde{\nabla}[\tilde{\nabla} \cdot \tilde{\boldsymbol{\omega}}] + \frac{1}{L^2}\tilde{\Delta}\tilde{\boldsymbol{\omega}} + \frac{2N^2}{(1-N^2)}\left[\tilde{\nabla} \times \tilde{\boldsymbol{v}} - 2\tilde{\boldsymbol{\omega}}\right]\right) \,, \end{aligned} \tag{28.7}$$

with the characteristic numbers

$$\begin{aligned} Re &= \frac{\rho_0 v_{\text{ref}} \ell_{\text{ref}}}{\mu} \,, \quad L = \frac{\ell_{\text{ref}}}{l} \,, \quad M = \frac{\ell_{\text{ref}}}{m} \,, \quad N = \sqrt{\frac{\tau}{\mu + \tau}} \,, \\ \Theta &= \frac{J_0}{\ell_{\text{ref}}^2} \,, \quad l = \sqrt{\frac{\beta + \gamma}{\mu}} \,, \quad m = \sqrt{\frac{\alpha + \beta - \gamma}{\mu}} \,. \end{aligned} \tag{28.8}$$

Therein, Re is the Reynolds number, N is the coupling number, L, M and Θ are characteristic length scale parameters. While all of the aforementioned parameters are dimensionless, the parameters l and m have the dimension of a length. Thus L and M are referred to as length scale parameters. Note that the parameters N and L are usually employed in context of Eringen's micropolar theory, see, *e.g.*, Cowin (1974); Rueger and Lakes (2016); Singh (1982), because they are characteristic for solutions of common flow problem such as Hagen-Poiseuille flow, *cf.* Cowin (1974), or Couette flow, see Eq. (28.14).

The coupling parameter N may vary between zero and one, where the former corresponds to a vanishing influence of the mesoscopic scale and the latter represents a negligible macroscopic scale. Obviously, the limit value $N = 0$ can be achieved by considering an ordinary fluid. The case of $N = 1$, however, is a rather theoretical limit and only values smaller than one should be considered. The other length scale parameters such as L and M are positive numbers that may become infinite in the limit of a vanishing influence of the mesoscopic scale. In the other limit of a strong influence they tend to zero but are always greater than zero, because the shear viscosity, μ, does not vanish.

We proceed by employing the following two dimensional *ansatz* for the velocity and angular velocity to both problems:

$$\tilde{\boldsymbol{v}} = \tilde{v}_x(\tilde{x},\tilde{y})\boldsymbol{e}_x + \tilde{v}_y(\tilde{x},\tilde{y})\boldsymbol{e}_y \,, \quad \tilde{\boldsymbol{\omega}} = \tilde{\omega}(\tilde{x},\tilde{y})\boldsymbol{e}_z \,. \tag{28.9}$$

Note that the angular velocity field is solenoidal in the two-dimensional case and therefore the parameter M is not present in the equations specialized for two-dimensional flows. The simplified dimensionless system reads:

$$\tilde{\nabla} \cdot \tilde{\boldsymbol{v}} = 0 \,, \quad \tilde{\boldsymbol{v}} \cdot (\tilde{\nabla} \otimes \tilde{\boldsymbol{v}}) = -\tilde{\nabla}\tilde{p} + \frac{1}{(1-N^2)Re}\left(\tilde{\Delta}\tilde{\boldsymbol{v}} + 2N^2\tilde{\nabla} \times \tilde{\boldsymbol{\omega}}\right) ,$$

$$\tilde{\boldsymbol{v}} \cdot (\tilde{\nabla}\tilde{\omega}) = \frac{1}{\Theta Re}\left(\frac{1}{L^2}\tilde{\Delta}\tilde{\omega} + \frac{2N^2}{(1-N^2)}\left[(\tilde{\nabla} \times \tilde{\boldsymbol{v}}) \cdot \boldsymbol{e}_z - 2\tilde{\omega}\right]\right) . \tag{28.10}$$

In contrast to a Navier–Stokes fluid with the Reynolds number as a single characteristic parameter, there are four characteristic parameters for a micropolar fluid. For a discussion of the rough orders of magnitude of the dimensionless parameters in context with blood flow the reader is referred to Sect. 28.5.

28.3.2 Boundary Conditions and Boundary Value Problems

For the statement of a complete problem, the equations presented above need to be supplemented by boundary conditions. For ordinary fluids physically correct boundary conditions seem to be "intuitively" clear, so that it is customary to impose (say) no-slip boundary conditions for the velocity at solid walls, because fluids usually stick to walls. It turns out, that this is neither intuitively clear nor always true. Although the historical review in Day (1990) strongly suggests, that the no-slip boundary condition is applicable in many flow situations, there are papers, *e.g.*, Brenner (2011); Lauga et al (2007), stating, that there are several mechanisms, which can lead to slip effects. Lauga et al (2007) state, that although there are different mechanisms for slip at fluid-solid interfaces, a distinction is of no practical importance. Furthermore, the so-called apparent slip length ranges over several orders of magnitude up to hundreds of nanometers. However, because usual (macroscopic) experiments are of much larger dimensions, the effect of slip is negligible such that no-slip boundary conditions at solid walls are reasonably applicable in many flow situations - at least as a first approximation.

Analogously the question of correct boundary conditions for the angular velocity in the theory of micropolar fluids is still under debate, see Alizadeh et al (2011); Hogan and Henriksen (1989); Kirwan (1986); Kolpashchikov et al (1983); Łukaszewicz (1999); Silber et al (2007). There are several suggestions of boundary conditions for the angular velocity at solid walls, but there is no convincing argument for one of them to be always preferable. Most of the suggested conditions can be categorized in three groups and may be referred to as follows:

- stick (or slip) controlled, where the angular velocity is prescribed due to the state of the wall (and the fluid) but independently of the velocity,
- vorticity controlled, where the angular velocity is proportional to the vorticity of the fluid near the wall, or
- force controlled, where the term *force* also refers to generalized forces and the angular velocity is influenced through forces and moments acting on the surface of the fluid.

While the first condition is analogous to the (say) no-slip condition of the velocity, the same comments as above apply for the angular velocity. Łukaszewicz (1999) refers to this boundary condition as "physically clear (the viscous fluid sticks to the solid boundary)." However, another study considers slip conditions and confirms them by molecular dynamic simulations, Chakraborty and Chakraborty (2008).

Another boundary condition commonly used is to assume that the angular velocity is proportional to the vorticity, Kirwan (1986); Kolpashchikov et al (1983); Hogan and Henriksen (1989). The reasoning behind this is, that the microstructure near solid walls needs to become irrelevant such that the angular velocity is solely given through the vorticity.

A third type of boundary conditions considers the coupling of angular velocity, vorticity and possible moment tractions at interfaces, Aero et al (1965); Łukaszewicz (1999). However, the correct choice of boundary conditions involving tractions is still under debate even for ordinary fluids, Sani and Gresho (1994). The same applies for micropolar fluids. Insights of how to impose angular velocity boundary conditions experimentally is given in none of the references mentioned above, see also Rickert et al (2018) for a discussion.

In this paper, only geometric boundary conditions directly applied on the velocity or angular velocity field are considered. Furthermore, we will impose only no-slip boundary conditions, because they are experimentally confirmed, Day (1990), at least for ordinary fluids.

In the Couette flow problem, two parallel plates of infinite extent are moving relatively to each other. The upper plate moves at a prescribed velocity $\boldsymbol{v} = v_0 \boldsymbol{e}_x$ while the bottom plate is at rest. The angular velocity is assumed to vanish at both plates, because they are manufactured in such a way not to induce any microrotation. In order to mimic Couette flow, periodic boundary conditions are employed. Hence, the boundary value problem is to solve Eqs. (28.10) subjected to the following boundary conditions:

$$\begin{cases} \tilde{\omega}(\tilde{\boldsymbol{x}}) = 0\,, \quad \tilde{\boldsymbol{v}}(\tilde{\boldsymbol{x}}) = \boldsymbol{0}\,, & \tilde{\boldsymbol{x}} \in \Gamma_1\,, \\ \tilde{\omega}(\tilde{\boldsymbol{x}}) = 0\,, \quad \tilde{\boldsymbol{v}}(\tilde{\boldsymbol{x}}) = \boldsymbol{e}_x\,, & \tilde{\boldsymbol{x}} \in \Gamma_3\,, \\ \tilde{\boldsymbol{v}}(\tilde{x} = 0, \tilde{y}) = \tilde{\boldsymbol{v}}(\tilde{x} = \ell/h, \tilde{y})\,, & \\ \tilde{\omega}(\tilde{x} = 0, \tilde{y}) = \tilde{\omega}(\tilde{x} = \ell/h, \tilde{y})\,, & 0 < \tilde{y} < 1\,. \end{cases} \tag{28.11}$$

For the lid-driven cavity problem, it is also assumed that the angular velocity vanishes at the top. Furthermore, at all other boundaries the velocity and the angular velocity vanish. The complete boundary value problem consists of Eqs. (28.10) subjected to the following boundary conditions:

$$\begin{cases} \tilde{\omega}(\tilde{\boldsymbol{x}}) = 0 \,, & \tilde{\boldsymbol{v}}(\tilde{\boldsymbol{x}}) = \boldsymbol{e}_x \,, \quad \tilde{\boldsymbol{x}} \in \Gamma_3 \,, \\ \tilde{\omega}(\tilde{\boldsymbol{x}}) = 0 \,, & \tilde{\boldsymbol{v}}(\tilde{\boldsymbol{x}}) = \boldsymbol{0} \,, \quad \tilde{\boldsymbol{x}} \in (\Gamma_1 \cup \Gamma_2 \cup \Gamma_4) \,. \end{cases} \tag{28.12}$$

28.3.3 Reference Solution

For the Couette flow, an analytical solution is given in Cowin (1974) and derived in Rickert et al (2018). Cowin (1974) uses a semi-inverse *ansatz* for the stationary solution:

$$\boldsymbol{v} = v(y)\boldsymbol{e}_x \,, \qquad \boldsymbol{\omega} = \omega(y)\boldsymbol{e}_z \,, \qquad p = p(y) \,, \tag{28.13}$$

for which the convective terms drop out. Hence, the resulting problem is linear and the following solution can be derived:

$$\begin{aligned} \frac{v(\tilde{y})}{v_0} &= \frac{1}{2(1-P)}\left[2\tilde{y} - P\left(1 + \frac{\sinh(NL[2\tilde{y}-1])}{\sinh(NL)}\right)\right] \,, \\ \frac{\omega(\tilde{y})}{v_0/h} &= \frac{1}{2(1-P)}\left[\frac{\cosh(NL[2\tilde{y}-1])}{\cosh(NL)} - 1\right] \,, \quad P = \frac{N}{L}\tanh(NL) \,, \end{aligned} \tag{28.14}$$

where $\tilde{y} = y/h$ is the dimensionless vertical coordinate and for the dimensionless numbers $\ell_{\text{ref}} = h$ is applied. The solution does not depend on the parameter Θ, which describes the influence of the microinertia.

28.4 Numerical Treatment

We solve the resulting set of dimensionless partial differential equations (28.10) using the finite element method. A spatial discretization based on the stable P_2-P_1 Taylor-Hood element for velocity and pressure is employed, Taylor and Hood (1973). The scalar angular velocity is discretized using a P_1 element. We describe the derivation of the weak form for both of the proposed problems in the Appendix.

The FEniCS library, Alnæs et al (2015); Logg et al (2012), as applied for many problems in continuum mechanics (Abali, 2017), allows to solve the finite element problem once the weak form is implemented. The discrete weak form represents a set of nonlinear algebraic equations, which is commonly solved using Newton's method. However, for our stationary problem, there is no appropriate initial guess available. Due to this fact, the solution procedure is altered and based on a hybrid approach possessing a larger convergence radius (Elman et al, 2006). This hybrid approach first performs a Picard iteration and as a second step applies Newton's method. The main difference between the two steps lies in the linearization of the convective term, see Elman et al (2006).

The parameter study requires a series of subsequent simulations and while one parameter is varied, we use the previous solution as the initial guess. This reduces the number of iterations required. In this sense, our method can also be regarded as a continuation method. Nevertheless, we tested the hybrid approach without parameter continuation and the simulations converged over the entire range of tested parameters.

28.4.1 Convergence Analysis

In order to verify the implemented finite element code, we perform a convergence analysis based on the analytical solution of the Couette flow problem. The set of nominal values for the dimensionless parameters given in Sec. 28.5 is used. The error of the numerical solution obtained from the finite element program w.r.t. the analytical solution is computed using the following error measures:

$$e^{\text{abs}}(\psi) = \frac{\int_V \|\psi^{\text{ana}} - \psi^{\text{num}}\| \,\mathrm{d}V}{\int_V \mathrm{d}V}, \quad e^{\text{rel}}(\psi) = \frac{\int_V \|\psi^{\text{ana}} - \psi^{\text{num}}\| \,\mathrm{d}V}{\int_V \|\psi^{\text{ana}}\| \,\mathrm{d}V}, \tag{28.15}$$

where the $\|\cdot\|$ is the absolute value if ψ is a scalar and the Eulidean norm if ψ is a vector. Here, e^{abs} and e^{rel} refer to the absolute and relative error respectively. Furthermore, ψ^{ana} and ψ^{num} denote the analytical and the numerical solution, respectively. A series of simulations on globally refined meshes is performed and the fineness of the mesh is characterized by the number of nodes, n, in the vertical direction of the channel ($\boldsymbol{e}_y$ direction). Although the mesh is not successively refined, the aspect ratio of the elements remains constant in the refinement process, which ensures that the quality of the elements is not deteriorated.

Figure 28.2 shows the results of the convergence analysis. The errors for the velocity (Fig. 28.2a) as well as for the angular velocity (Fig. 28.2b) decrease monotonically for the relative and the absolute error as the number of nodes is increased. Both regression lines have a slope of -2 in the double logarithmic chart, which indicates that the method is of second order. In conclusion, our finite element implementation to simulate the stationary flow of a micropolar fluid is convergent and therefore it is considered as a reliable tool to assess the influence of the model parameters in more complex flow situations. The largest number of nodes in the vertical direction applied in the convergence analysis is 800, which corresponds to $645\,284$ degrees of freedom in total.

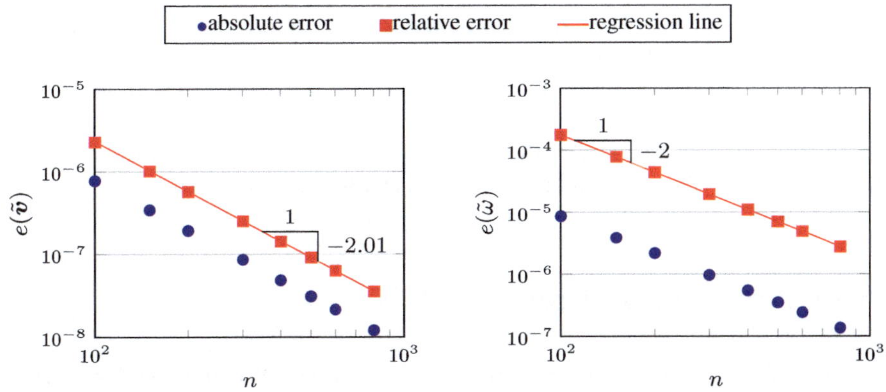

(a) Errors w.r.t. the velocity field.

(b) Errors w.r.t. the angular velocity field.

Fig. 28.2 Convergence results for the Couette flow problem. The legend above applies for both diagrams.

28.5 Results and Discussion

In this section, the results of the parameter study for the lid-driven cavity problem shown in Fig. 28.1b are presented. The computations are performed on an equidistant grid using 100 points in horizontal and vertical direction. For the parameter analysis a set of nominal values is selected and only one parameter at a time is varied in a given range. The following nominal values are chosen: $Re = 200$, $L = 0.4$, $N = 0.25$, and $\Theta = 10^{-9}$. This choice as well as the parameter ranges will be motivated by the following considerations.

In context of blood flow with a viscosity of $\mu = 3 \times 10^{-3}\,\mathrm{kg\,m^{-1}\,s^{-1}}$ to $4 \times 10^{-3}\,\mathrm{kg\,m^{-1}\,s^{-1}}$ (Popel and Pittman, 2000), the nominal Reynolds number corresponds to a mean velocity of $5.6 \times 10^{-2}\,\mathrm{m\,s^{-1}}$ to $7.6 \times 10^{-2}\,\mathrm{m\,s^{-1}}$ for a reference length of 1 cm. If the reference length is 1 mm, the mean velocity ranges from $5.6 \times 10^{-1}\,\mathrm{m\,s^{-1}}$ to $7.6 \times 10^{-1}\,\mathrm{m\,s^{-1}}$. Comparing these values with the ones given in Lieber (2000) suggests that the Reynolds number is slightly too high when considering arterioles. However, blood flow is not stationary and the Reynolds number, which depends on the type of the blood vessel and ranges from approx. 0.005 to 7000, exceeds six orders of magnitude (Caro et al, 2012). Therefore, the Reynolds number is varied over a large range in this study.

Based on the viscosities for human blood specified in Kang and Eringen (1976) and Papautsky et al (1999), $N = 0.25$ and $L = 0.4$ is obtained as a rough estimate for a reference length of 1 cm. The length scale parameter L decreases linearly, if the reference length decreases, but it strongly depends upon the viscosities and may therefore vary over a larger range. In order to obtain an estimate for Θ, erythrocytes are considered as a principle of component of blood. Applying $\ell_{\mathrm{ref}} = 0.1$ mm to 1 mm for the diameter of small arteries or terminal branches (Schneck, 2000), and $d \approx 7.5\,\mu\mathrm{m}$ for the diameter of erythrocytes, Schneck (2000), gives $J_0 \approx 10^{-12}\,\mathrm{m}^2$

and $\Theta = 10^{-6}$ to 10^{-5}. This number would of course increase to 10^{-2} for arterioles ($d < 100\,\mu\text{m}$) or even to 1 for capillaries ($d < 10\,\mu\text{m}$), see Schneck (2000).

In view of the facts outlined above, the Reynolds number is varied from the 0.01 to 5000. The coupling parameter N is varied from 0 to 0.95. The length scale L is varied from 10^{-3} to 10^{2}, where the latter represents a rather extreme case. Finally, the parameter Θ characterizing microinertia is varied from 10^{-9} to 1.

28.5.1 Vertical and Horizontal Profiles

The results are presented by using line plots of the velocity and angular velocity along the vertical and horizontal middle lines of the cavity, see dashed lines in Figs. 28.3. In Fig. 28.3a an example of the streamline of the flow of an ordinary Navier–Stokes fluid is given. There is a main eddy located close to the center and two so-called Moffatt eddies are present at the two bottom corners, Moffatt (1964). The eddy structure of this problem was explored in detail for Navier–Stokes fluids. It was shown that another eddy occurs close to the top-left corner and secondary corner eddies are present for higher Reynolds numbers (Shankar, 1993; Shen, 1991).

Figures 28.4 and 28.5 show the profiles for varying values of N. Considering the horizontal and vertical profiles of the angular velocity, it is evident that the larger the value of N the larger the amplitudes of the angular velocity. The angular velocity is mainly negative along the horizontal and vertical profile for small to moderate values of N. There are small domains at the boundaries, where the angular velocity is positive.[3] These domains grow for large values of N. This suggests that the parameter N gradually changes the spatial structure of the angular velocity field, whereas the qualitative behavior is not altered. The influence of the parameter N on the velocity profiles is not so strong when compared to the angular velocity. Slight changes are visible, which are more prominent in the vertical profile.

Figures 28.6 and 28.7 show the profiles for varying values of L. In the profiles of the angular velocity the formation of a very thin boundary layer is visible—the larger L the thiner the boundary layer. The boundary layer is most prominent at the top of the cavity, whereas there is no layer at the bottom. At the left and right walls, boundary layers are also present. Although the amplitudes differ only by an order of magnitude compared to those obtained for large values of N, the qualitative structure of the profiles for the angular velocity is very different when L is varied. Comparing the angular velocity profiles for varying L among each other shows, that apart from a scaling a marginal shift in the spatial structure is visible. The zero crossings move gradually, if L varies. For the influence of the parameter L on the velocity profiles, the same applies as for the parameter N—a significant change was not observed.

Figures 28.8 and 28.9 show the profiles for varying values of Re. Increasing the Reynolds number above approximately 100 results in changes of the velocity

---

[3] These domains are not really visible due to the scaling.

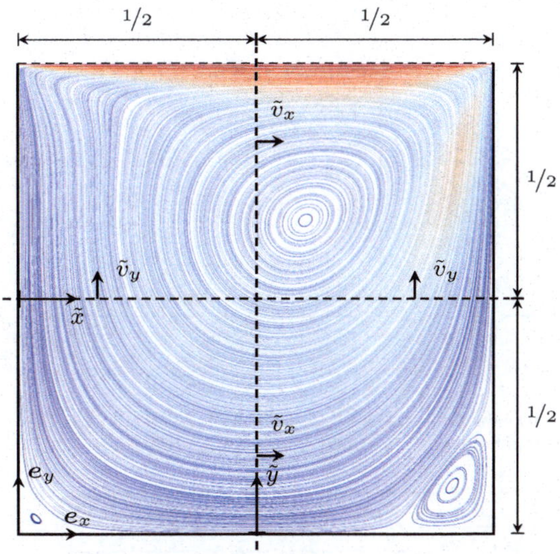

(a) Streamline plot with dashed lines for profiles.

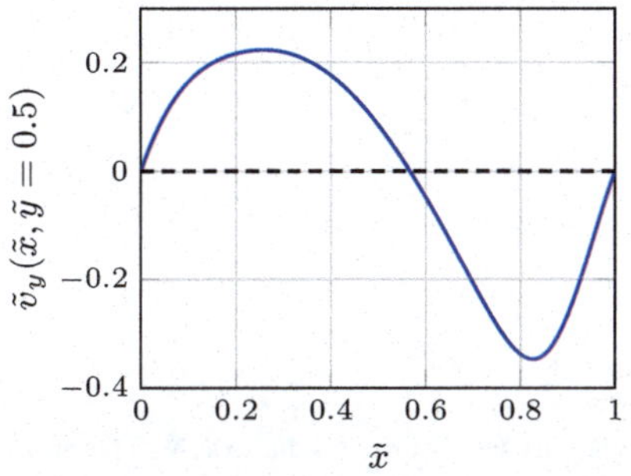

(b) Velocity profile along $\tilde{x}$-axis.

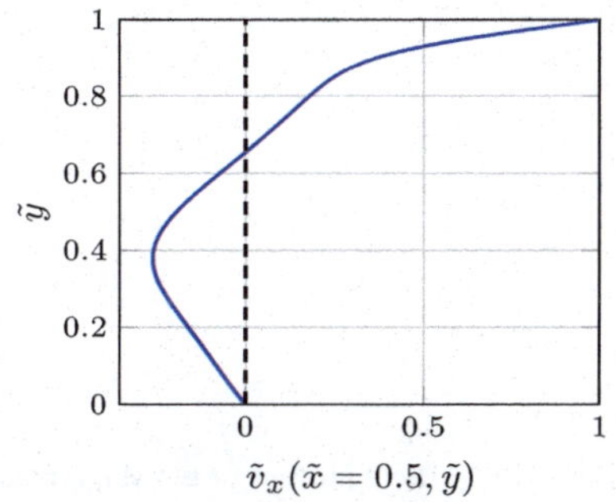

(c) Velocity profile along $\tilde{y}$-axis.

Fig. 28.3 Streamlines (a) and profiles of the velocity (b, c) for the lid-driven cavity problem of a Navier–Stokes fluid with $Re = 200$. Color indicates the magnitude of the velocity.

profiles. If the Reynolds number is below approximately 10, the profiles of the velocity are identical for the horizontal and vertical direction regardless of the value of Re. The same holds true for the angular velocity. At higher Reynolds numbers, a boundary layer develops and the center of the main vortex (zero crossing of the velocity in the vertical profile) moves downwards. Regarding the vertical profiles, there is a correlation of the angular velocity with the velocity—the minimum of the angular velocity is roughly located where the velocity is zero. This effect is also visible in the right inset of Fig. 28.8 in terms of a decrease of the angular velocity at $\tilde{x} \approx 0.5$ for $Re \geq 200$. We stress that this decrease is not due to a change of the underlying solution because the angular velocity is linked to the velocity field

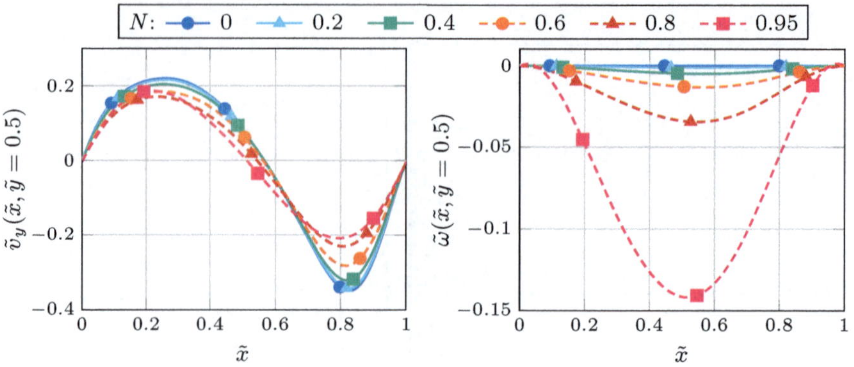

Fig. 28.4 Horizontal profiles of the velocity and angular velocity for different values of N. The other parameters Re, L, and Θ have nominal values.

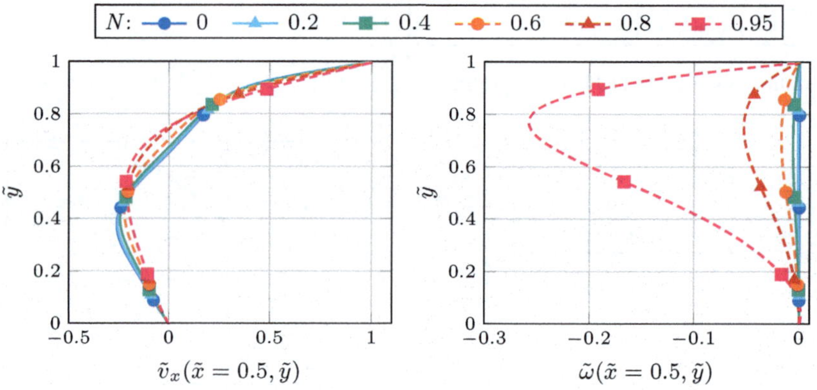

Fig. 28.5 Vertical profiles of the velocity and angular velocity for different values of N. The other parameters Re, L, and Θ have nominal values.

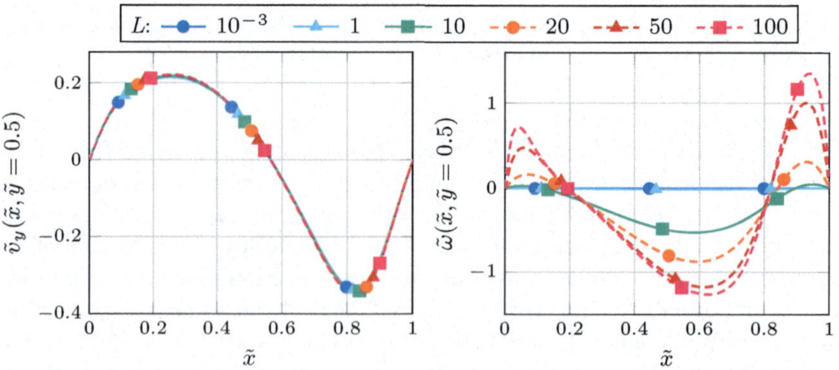

Fig. 28.6 Horizontal profiles of the velocity and angular velocity for different values of L. The other parameters Re, N, and Θ have nominal values.

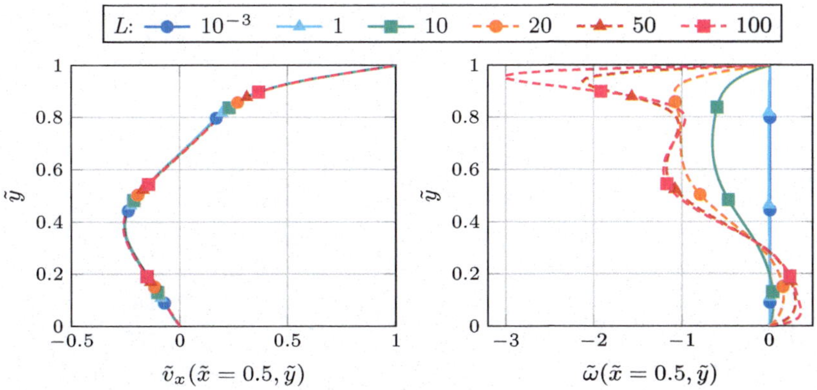

Fig. 28.7 Vertical profiles of the velocity and angular velocity for different values of L. The other parameters Re, N, and Θ have nominal values.

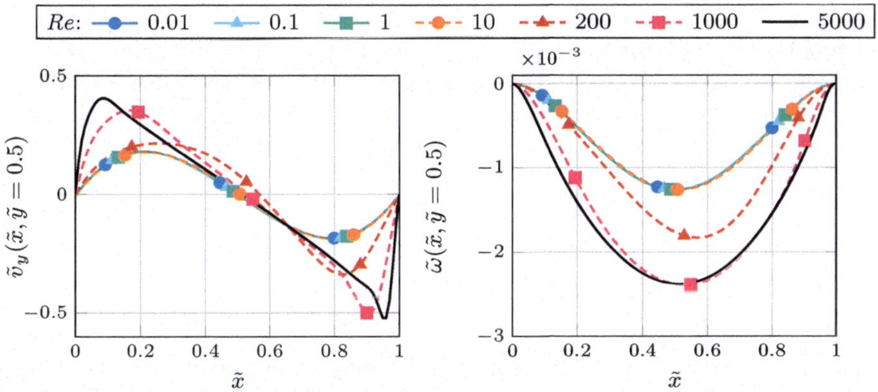

Fig. 28.8 Horizontal profiles of the velocity and angular velocity for different values of Re. The other parameters L, N, and Θ have nominal values.

for chosen set of parameters. Finally, note that the angular velocity in Figs. 28.8 and 28.9 is three orders of magnitudes smaller than for example in Figs. 28.6 and 28.7.

The last parameter, which was varied, is the microinertia parameter Θ. An influence of the microinertia parameter was not observed in the velocity field for the set of nominal values chosen in this study. This is expected because Θ is not present in the balance of linear momentum. The angular velocity field is also not significantly altered for small and moderate values of Θ.

In summary, the approach of varying one parameter while the others parameter keep their nominal values has shown that the microinertia has the smallest effect on the structure of the solution. However, this is only true for the considered case and parameter set. The velocity field changes the most, if the Reynolds number is increased. But smaller alterations are also observable in the velocity field, if the

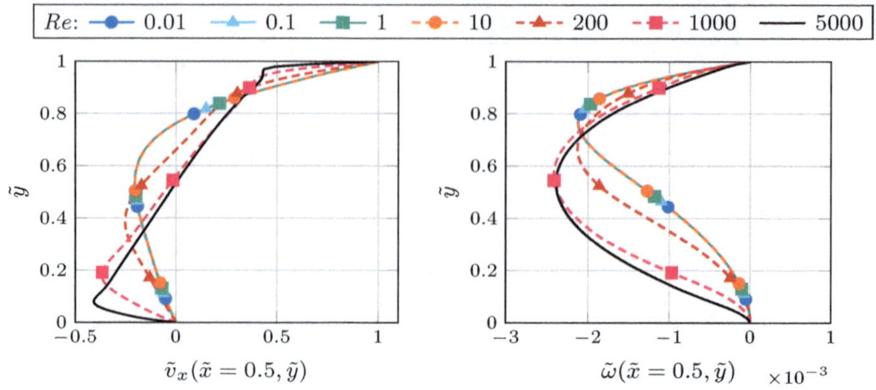

Fig. 28.9 Vertical profiles of the linear and angular velocity for different values of Re. The other parameters L, N, and Θ have nominal values.

micropolar viscosity parameters L and N are changed. The micropolar viscosity parameters L and N both yield to the changes in the angular velocity field. The largest amplitudes associated with these parameters are approximately 0.3 and 3 respectively. In a broad sense, the parameter N scales the solution, whereas increasing values of the parameter L result in a stronger boundary layer.

28.5.2 Analysis of the Angular Velocity Field

Figure 28.10 shows the angular velocity for a representative selection of parameters combinations. In Figs. 28.10a and 28.10b, the microrotation is confined to the top layer of the cavity and the spatial structure is strongly connected to the velocity field, see Fig. 28.3a. Comparing Figs. 28.10a to 28.10b suggests that the region at the top gets wider and thiner if N is increased. In Figs. 28.10c and 28.10d, boundary layers become apparent. They are strongest close to the top-right corner of the cavity and the angular velocity penetrates deeper regions of the cavity.

An explanation might be that for the considered parameter set the convective term in Eq. (28.10)$_3$ can be dropped because the product ΘRe is small. This gives the following spin balance:

$$\frac{1}{L^2}\tilde{\Delta}\tilde{\omega} + \frac{2N^2}{(1-N^2)}\left[(\tilde{\nabla} \times \tilde{\boldsymbol{v}}) \cdot \boldsymbol{e}_z - 2\tilde{\omega}\right] = 0\,, \tag{28.16}$$

which is interpreted as a Helmholtz equation for $\tilde{\omega}$ with the source term $\tilde{\nabla} \times \tilde{\boldsymbol{v}}$. If the parameter L is large, the diffusive term is negligible because the factor $N^2/(1-N^2)$ is of order 1 for $N < 0.9$. In the limit, we can make the approximation $\tilde{\omega} \approx (\tilde{\nabla} \times \tilde{\boldsymbol{v}})/2 \cdot \boldsymbol{e}_z$, which means that the angular velocity $\tilde{\omega}$ is directly coupled to the vorticity. This approximation was roughly confirmed for $L = 50$ by comparing the

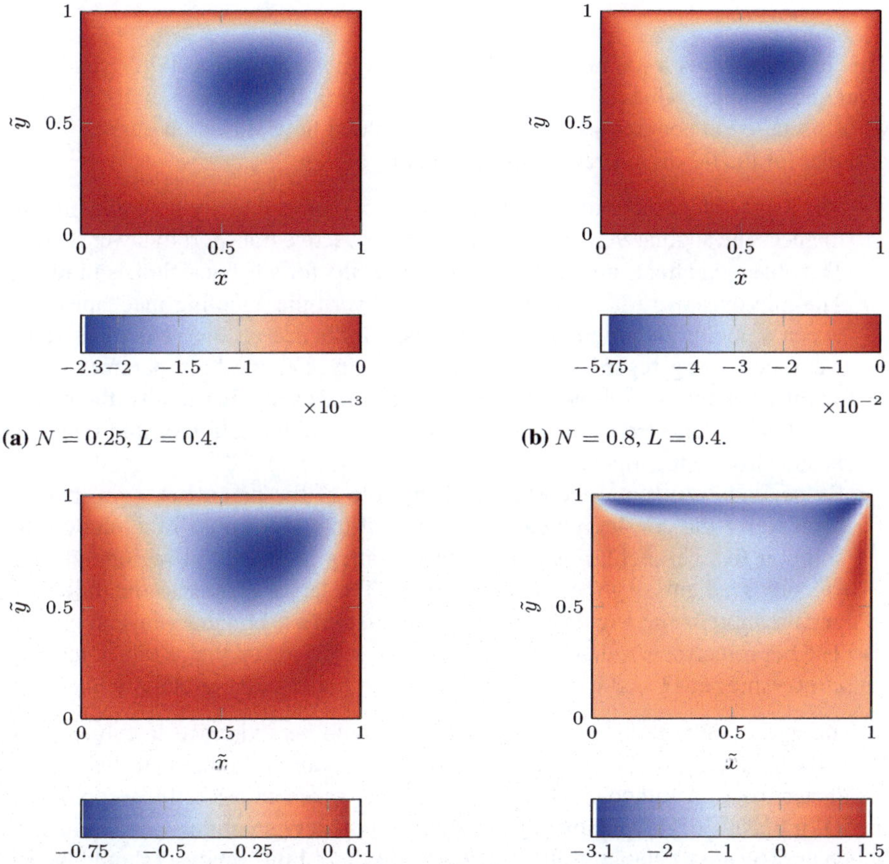

(a) $N = 0.25$, $L = 0.4$.

(b) $N = 0.8$, $L = 0.4$.

(c) $N = 0.25$, $L = 10$.

(d) $N = 0.25$, $L = 50$.

Fig. 28.10 Angular velocity field $\tilde{\omega}$ for different values of N and L. The other parameters have nominal values.

spatial structure of the angular velocity field with the one of the vorticity using our numerical results. However, for moderate values of L, the terms governed by the coupling parameter N are balanced with the diffusion term. This seems to weaken the coupling mechanism because strong gradients close to the walls and corners are smoothed by diffusion. In conclusion, the findings suggest the dominant parameter for the lid-driven cavity problem is the micropolar viscosity parameter L.

28.6 Conclusion

This paper presents a parameter study of a micropolar fluid in a lid-driven cavity using the nominal values: $Re = 200$, $L = 0.4$, $N = 0.25$, and $\Theta = 10^{-9}$, which were discussed in context with the example of blood flow. Based on our results, we find that for the lid-driven cavity flow problem:

- The Reynolds number has the strongest effect on the velocity field and only influences the angular velocity by enhanced convective transport to lower regions. The observed effects are analogous to the results for a Navier–Stokes fluid.
- The microviscosity parameters L and N govern the coupling mechanism between velocity and angular velocity. Their influence on the velocity is rather weak but strong regarding the angular velocity. For the both parameters, the angular velocity is confined in the top half of the cavity. Especially, the parameter L was identified as the one responsible for the formation of thin boundary layers close to the top-right corner.
- By scaling arguments, we explained the role of the parameter L in the spin balance for the considered case. We showed that low values of L smooth the solution for the angular velocity, whose source is located at regions of large velocity gradients, *e.g.*, at boundary layers. The larger L the less the diffusivity and the more is the angular velocity confined to these boundary layers.
- The microinertia parameter Θ has almost no influence on the results. Only very large values of Θ lead to visible alterations in the angular velocity field.

Our study points to several aspects, which could be addressed in future work. First, we have only varied one parameter at a time, which limits us to lines in the parameter space. A full coverage of the parameter space allows to fully explore the mutual influence of the parameters. Secondly, stationary solutions were computed, which may be not be stable at all. Because of this, a stability analysis should be performed by either integrating the equations in time using the precomputed stationary solution as an initial condition or by performing a linear perturbation analysis. Third, the example of buoyancy driven flow in the cavity should be studied to further explore the impact of the parameters identified in this study. This could also incorporate microstructural changes due to temperature or pressure fluctuations, which were proposed in Müller and Vilchevskaya (2017).

Appendix: Weak Forms

For the finite element formulation of Eqs. (28.10), weak forms of the differential equations need to be derived. Subsequently, the following short hand notation for inner products in the volume and on the boundary will be applied:

$$(\boldsymbol{A}, \boldsymbol{B})_{\Omega} = \int\limits_{\Omega} \boldsymbol{A} \star \boldsymbol{B} \, \mathrm{d}V \,, \quad \langle \boldsymbol{A}, \boldsymbol{B} \rangle_{\Gamma} = \int\limits_{\Gamma} \boldsymbol{A} \star \boldsymbol{B} \, \mathrm{d}A \,,$$

where $\boldsymbol{A} \star \boldsymbol{B}$ represents the contraction of two tensors $\boldsymbol{A}$ and $\boldsymbol{B}$ of arbitrary rank to a scalar. Here, Ω denotes the domain of integration with the boundary $\Gamma = \partial\Omega$. With the help of the operators

$$\begin{aligned}
\mathcal{A}(\boldsymbol{\phi},\boldsymbol{\psi}) &= (\tilde{\nabla}\otimes\boldsymbol{\phi},\tilde{\nabla}\otimes\boldsymbol{\psi})_{\Omega}, \quad & \mathcal{B}(\boldsymbol{\phi},\psi) &= (\tilde{\nabla}\cdot\boldsymbol{\phi},\psi)_{\Omega},\\
\mathcal{C}(\boldsymbol{\phi},\boldsymbol{\psi}) &= (\tilde{\boldsymbol{v}}\cdot(\tilde{\nabla}\otimes\boldsymbol{\phi}),\boldsymbol{\psi})_{\Omega}, \quad & \mathcal{D}_{\Gamma}(\boldsymbol{\phi},\boldsymbol{\psi}) &= \langle\boldsymbol{\phi},\boldsymbol{\psi}\times\boldsymbol{e}_z\rangle_{\Gamma},\\
\mathcal{D}(\phi,\boldsymbol{\psi}) &= (\phi,\tilde{\nabla}\cdot(\boldsymbol{\psi}\times\boldsymbol{e}_z))_{\Omega} = \left(\phi,\frac{\partial\psi_y}{\partial x}-\frac{\partial\psi_x}{\partial y}\right)_{\Omega}, & &
\end{aligned}$$

a multiplication of Eqs. (28.10) with test functions δp, $\delta\boldsymbol{v}$, $\delta\omega$, δJ and subsequent integration over the (dimensionless) domain gives rise to the following weak forms:

$$\begin{aligned}
&\mathcal{B}(\tilde{\boldsymbol{v}},\delta p) = 0\,,\\
&\mathcal{C}(\tilde{\boldsymbol{v}},\delta\boldsymbol{v}) = \mathcal{B}(\delta\boldsymbol{v},\tilde{p}) - \langle\tilde{p}\boldsymbol{n},\delta\boldsymbol{v}\rangle_{\Gamma} + \frac{2N^2}{(1-N^2)Re}\Big\{\mathcal{D}(\tilde{\omega},\delta\boldsymbol{v}) - \mathcal{D}_{\Gamma}(\tilde{\omega}\boldsymbol{n},\delta\boldsymbol{v})\Big\} +\\
&\qquad + \frac{1}{(1-N^2)Re}\Big\{\langle\boldsymbol{n}\cdot(\tilde{\nabla}\otimes\tilde{\boldsymbol{v}}),\delta\boldsymbol{v}\rangle_{\Gamma} - \mathcal{A}(\tilde{\boldsymbol{v}},\delta\boldsymbol{v})\Big\}\,,\\
&\mathcal{C}(\tilde{\omega},\delta\omega) = \frac{1}{L^2\Theta Re}\Big(\langle\boldsymbol{n}\cdot(\tilde{\nabla}\tilde{\omega}),\delta\omega\rangle_{\Gamma} - \mathcal{A}(\tilde{\omega},\delta\omega)\Big) +\\
&\qquad + \frac{2N^2}{(1-N^2)\Theta Re}\Big(\mathcal{D}(\delta\omega,\tilde{\boldsymbol{v}}) - 2(\tilde{\omega},\delta\omega)_{\Omega}\Big).
\end{aligned}$$

For the lid-driven cavity problem, pure Dirichlet boundary conditions are considered in form of no-slip conditions for both the velocity and angular velocity. In this case, the test functions δp, $\delta\boldsymbol{v}$ and $\delta\omega$ vanish at the boundary and therefore all boundary terms vanish. Hence, the simplified weak forms can be written as:

$$\begin{aligned}
&\mathcal{B}(\tilde{\boldsymbol{v}},\delta p) = 0\,,\\
&\mathcal{C}(\tilde{\boldsymbol{v}},\delta\boldsymbol{v}) = \mathcal{B}(\delta\boldsymbol{v},\tilde{p}) - \frac{1}{(1-N^2)Re}\Big(\mathcal{A}(\tilde{\boldsymbol{v}},\delta\boldsymbol{v}) - 2N^2\mathcal{D}(\tilde{\omega},\delta\boldsymbol{v})\Big)\,,\\
&\mathcal{C}(\tilde{\omega},\delta\omega) = \frac{1}{\Theta Re}\left[\frac{2N^2}{(1-N^2)}\Big(\mathcal{D}(\delta\omega,\tilde{\boldsymbol{v}}) - 2(\tilde{\omega},\delta\omega)_{\Omega}\Big) - \frac{1}{L^2}\mathcal{A}(\tilde{\omega},\delta\omega)\right].
\end{aligned}$$

The same holds for the Coutte flow except for the periodic boundaries Γ_2 and Γ_4. The surface integral related to the pressure, p, vanishes due to periodicity. The pressure gradient has the same values at Γ_2 and Γ_4, but the normal vectors have a different direction, *i.e.*, $\boldsymbol{n}_{\Gamma_2} = -\boldsymbol{n}_{\Gamma_4}$. Therefore, the simplified weak forms for the Couette flow can be expressed as:

$$
\begin{aligned}
\mathcal{B}(\tilde{\boldsymbol{v}}, \delta p) &= 0 \,, \\
\mathcal{C}(\tilde{\boldsymbol{v}}, \delta \boldsymbol{v}) &= \mathcal{B}(\delta \boldsymbol{v}, \tilde{p}) + \frac{2N^2}{(1-N^2)Re} \mathcal{D}(\tilde{\omega}, \delta \boldsymbol{v}) + \\
&\quad + \frac{1}{(1-N^2)Re} \Big\{ \langle \boldsymbol{n} \cdot (\tilde{\nabla} \otimes \tilde{\boldsymbol{v}}), \delta \boldsymbol{v} \rangle_{\Gamma} - \mathcal{A}(\tilde{\boldsymbol{v}}, \delta \boldsymbol{v}) \Big\} \,, \\
\mathcal{C}(\tilde{\omega}, \delta \omega) &= \frac{1}{L^2 \Theta Re} \Big(\langle \boldsymbol{n} \cdot (\tilde{\nabla} \tilde{\omega}), \delta \omega \rangle_{\Gamma} - \mathcal{A}(\tilde{\omega}, \delta \omega) \Big) + \\
&\quad + \frac{2N^2}{(1-N^2)\Theta Re} \Big(\mathcal{D}(\delta \omega, \tilde{\boldsymbol{v}}) - 2 \big(\tilde{\omega}, \delta \omega \big)_{\Omega} \Big).
\end{aligned}
$$

References

Abali BE (2017) Computational Reality, 1st edn. Advanced Structured Materials, Springer

Aero EL, Bulygin AN, Kuvshinskii EV (1965) Asymmetric hydromechanics. Journal of Applied Mathematics and Mechanics 29(2):333–346

Alizadeh M, Silber G, Nejad AG (2011) A continuum mechanical gradient theory with an application to fully developed turbulent flows. Journal of Dispersion Science and Technology 32(2):185–192

Alnæs MS, Blechta J, Hake J, Johansson A, Kehlet B, Logg A, Richardson C, Ring J, Rognes ME, Wells GN (2015) The fenics project version 1.5. Archive of Numerical Software 3(100)

Anders D, Weinberg K (2011) A variational approach to the decomposition of unstable viscous fluids and its consistent numerical approximation. ZAMM—Journal of Applied Mathematics and Mechanics / Zeitschrift für Angewandte Mathematik und Mechanik 91(8):609–629

Ariman T, Turk M, Sylvester N (1973) Microcontinuum fluid mechanic—A review. International Journal of Engineering Science 11(8):905–930

Aydin O, Pop I (2007) Natural convection in a differentially heated enclosure filled with a micropolar fluid. International Journal of Thermal Sciences 46(10):963–969

Botella O, Peyret R (1998) Benchmark spectral results on the lid-driven cavity flow. Computers & Fluids 27(4):421–433

Bourantas G, Loukopoulos V (2014) Modeling the natural convective flow of micropolar nanofluids. International Journal of Heat and Mass Transfer 68:35–41

Brenner H (2011) Beyond the no-slip boundary condition. Physical Review E 84(4):0463,091–8

Bruneau CH, Saad M (2006) The 2d lid-driven cavity problem revisited. Computers & Fluids 35(3):326–348

Caro CG, Pedley TJ, Schroter RC, Seed WA (2012) The mechanics of the circulation., 2nd edn. Cambridge University Press

Chakraborty D, Chakraborty S (2008) Towards a generalization of hydrodynamic boundary conditions for flows of fluids with homogeneous internally rotating structures. Physics Letters A 372(33):5462–5466

Chakravarthy V, Ottino J (1996) Mixing of two viscous fluids in a rectangular cavity. Chemical Engineering Science 51(14):3613–3622

Cortes AB, Miller JD (1994) Numerical experiments with the lid driven cavity flow problem. Computers & Fluids 23(8):1005–1027

Cowin SC (1974) The theory of polar fluids. In: Yih CS (ed) Advances in Applied Mechanics, vol. 14, Academic Press, Inc., New York, pp 279–347

Day MA (1990) The no-slip condition of fluid dynamics. Erkenntnis 33(3):285–296

Elman H, Silvester D, Wathen A (2006) Finite Elements and Fast Iterative Solvers: with Application in Incompressible Fluid Dynamics, Numerical Mathematics and Scientific Computation, vol 1. Oxford University Press

Eringen AC (1964) Simple microfluids. International Journal of Engineering Science 2(2):205–217

Eringen AC (1966) Theory of micropolar fluids. Journal of Mathematics and Mechanics 16(1):1–18

Eringen AC (1985) Rigid suspensions in viscous fluid. International Journal of Engineering Science 23(4):49–495

Eringen AC (1991) Continuum theory of dense rigid suspensions. Rheologica Acta 30(1):23–32

Eringen AC (2001) Microcontinuum Field Theories- II Fluent Media. Springer, Berlin, Heidelberg, New York

Erturk E, Gökccöl C (2008) Fourth-order compact formulation of navier-stokes equations and driven cavity flow at high reynolds numbers. International Journal for Numerical Methods in Fluids 50(4):421–436

Freitas CJJ, Street RL, Findikakis AN, Koseff JR (1985) Numerical simulation of three-dimensional flow in a cavity. International Journal for Numerical Methods in Fluids 5(6):561–575

Gibanov NS, Sheremet MA, Pop I (2016a) Free convection in a trapezoidal cavity filled with a micropolar fluid. International Journal of Heat and Mass Transfer 99:831–838

Gibanov NS, Sheremet MA, Pop I (2016b) Natural convection of micropolar fluid in a wavy differentially heated cavity. Journal of Molecular Liquids 221:518–525

Glane S, Rickert W, Müller WH, Vilchevskaya E (2017) Micropolar media with structural transformations: Numerical treatment of a particle crusher. In: Advanced problems in mechanics/ Proceedings of the XLV Summer school-conference, Russian Academy of Sciences, Institute for Problems in Mechanical Engineering, Peter the Great St.Petersburg Polytechnic University, pp 197–211

Hogan HA, Henriksen M (1989) An evaluation of a micropolar model for blood flow through an idealized stenosis. Journal of Biomechanics 22(3):211–218

Hsu TH, Chen CK (1996) Natural convection of micropolar fluids in a rectangular enclosure. International Journal of Engineering Science 34(4):407–415

Iwatsu R, Hyun JM (1995) Three-dimensional driven-cavity flows with a vertical temperature gradient. International Journal of Heat and Mass Transfer 38(18):3319–3328

Jena SK, Bhatacharyya SP (1986) The effect of microstructure on the thermal convection in a rectangular box of fluid heated from below. International Journal of Engineering Science 24(1):69–78

Kang CK, Eringen AC (1976) The effect of microstructure on the rheological properties of blood. Bulletin of Mathematical Biology 38(2):135–159

Kareem AK, Gao S, Ahmed AQ (2016) Unsteady simulations of mixed convection heat transfer in a 3d closed lid-driven cavity. International Journal of Heat and Mass Transfer 100:121–130

Kirwan AD (1986) Boundary conditions for micropolar fluids. International Journal of Engineering Science 24(7):1237–1242

Kolpashchikov VL, Migun NP, Prokhorenko PP (1983) Experimental determination of material micropolar fluid constants. International Journal of Engineering Science 21(4):405–411

Lauga E, Brenner M, Stone H (2007) Microfluidics: The no-slip boundary condition. In: Tropea C, Yarin AL, Foss JF (eds) Springer Handbook of Experimental Fluid Mechanics, Springer, Berlin, Heidelberg, New York, pp 1219–1240

Le Quere P, Humphrey JAC, Sherman FS (1981) Numerical calculation of thermally driven two-dimensional unsteady laminar flow in cavities of rectangular cross section. Numerical Heat Transfer 4(3):249–283

Lieber BB (2000) Arterial Macrocirculatory Hemodynamics. In: Bronzino JD (ed) The Biomedical Engineering Handbook, vol 1, 2nd edn, CRC Press, chap 30

Logg A, Wells GN, Hake J (2012) DOLFIN: a C++/Python Finite Element Library, Springer, chap 10, pp 173–225

Łukaszewicz G (1999) Micropolar fluids: Theory and applications. Springer Science & Business Media

Maugin GA (2011) A Historical Perspective of Generalized Continuum Mechanics. In: Altenbach H, Maugin GA, Erofeev V (eds) Mechanics of Generalized Continua, Springer, Berlin, Heidelberg, chap 1, pp 3–19

Moffatt HK (1964) Viscous and resistive eddies near a sharp corner. Journal of Fluid Mechanics 18(1):1–18

Müller WH, Vilchevskaya EN (2017) Micropolar theory from the viewpoint of mesoscopic and mixture theories. Physical Mesomechanics 20(3):263–279

Müller WH, Vilchevskaya EN (2018) Micropolar theory with production of rotational inertia: A rational mechanics approach. In: Altenbach H, Pouget J, Rousseau M, Collet B, Michelitsch T (eds) Generalized Models and Non-classical Approaches in Complex Materials 1, Advanced Structured Materials, Springer International Publishing, pp 581–606

Müller WH, Vilchevskaya EN, Weiss W (2017) Micropolar theory with production of rotational inertia: A farewell to material description. Physical Mesomechanics 20(3):250–262

Nallasamy M, Prasad KK (1977) On cavity flow at high reynolds numbers. Journal of Fluid Mechanics 79(2):391–414

Papautsky I, Brazzle J, Ameel T, Frazier AB (1999) Laminar fluid behavior in microchannels using micropolar fluid theory. Sensors and Actuators A: Physical 73(1-2):101–108

Popel AS, Pittman RN (2000) Mechanics and Transport in the Microcirculation. In: Bronzino JD (ed) The Biomedical Engineering Handbook, vol 1, 2nd edn, CRC Press, chap 31

Prakash J, Sinha P (1975) Lubrication theory for micropolar fluids and its application to a journal bearing. International Journal of Engineering Science 13(3):217–232

Rickert W, Müller WH, Vilchevskaya EN (2018) A note on couette flow of micropolar fluids according to eringen's theory. Mathematics and Mechanics of Complex Systems Submitted

Rueger Z, Lakes RS (2016) Experimental cosserat elasticity in open-cell polymer foam. Philosophical Magazine 96(2):93–111

Sani RL, Gresho PM (1994) Résumé and remarks on the open boundary condition minisymposium. International Journal for Numerical Methods in Fluids 18(10):983–1008

Schneck DJ (2000) An Outline of Cardiovascular Structure and Function. In: Bronzino JD (ed) The Biomedical Engineering Handbook, vol 1, 2nd edn, CRC Press, chap 1

Shankar PN (1993) The eddy structure in stokes flow in a cavity. Journal of Fluid Mechanics 250:371–383

Shen J (1991) Hopf bifurcation of the unsteady regularized driven cavity flow. Journal of Computational Physics 95(1):228–245

Sheremet MA, Pop I, Ishak A (2017) Time-dependent natural convection of micropolar fluid in a wavy triangular cavity. International Journal of Heat and Mass Transfer 105:610–622

Silber G, Janoske U, Alizadeh M, Benderoth G (2007) An application of a gradient theory with dissipative boundary conditions to fully developed turbulent flows. Journal of Fluids Engineering 129(5):643–651

Singh K (1982) Couette flow of microthermopolar fluids between two parallel plates. Acta Mechanica 43(1):1–13

Taylor C, Hood P (1973) A numerical solution of the navier-stokes equations using the finite element technique. Computers & Fluids 1(1):73–100

Tiwari RK, Das MK (2007) Heat transfer augmentation in a two-sided lid-driven differentially heated square cavity utilizing nanofluids. International Journal of Heat and Mass Transfer 50(9):2002–2018

Türk Ö, Tezer-Sezgin M (2017) FEM solution to natural convection flow of a micropolar nanofluid in the presence of a magnetic field. Meccanica 52(4):889–901

Vilchevskaya EN, Müller WH (2018) Some remarks on recent developments in micropolar continuum theory. Journal of Physics: Conference Series 991(1):012,079

Zhilin PA (2006) A micro-polar theory for binary media with application to flow of fiber suspensions. In: Indeitsev DA, Ivanova EA, Krivtsov A (eds) Advanced Problems in Mechanics, vol 2, Russian Academy of Sciences, Insitute for Problems in Mechanical Engineering

Chapter 29
Graded Insulation to Improve High Pressure Resistance in Deepwater Flowlines: a Closed Form Analytical Elastic Solution

Roberta Sburlati & Maria Kashtalyan

Abstract In this paper, an insulated pipe system comprising the inner pipe, insulation and outer jacket is investigated in the context of elasticity theory with the view to establish whether introducing stiffness gradient in the insulation would improve performance of the pipe under external pressure. Closed form analytical solutions are derived for stresses and displacements in the pipe system. Comparative analysis of pipes with homogeneous and graded insulation is performed and beneficial effect of graded insulation on stresses and displacements in the pipe is established.

Keywords: Elasticity theory · Syntactic foam · Internal pressure · External pressure

29.1 Introduction

Oil and gas production in deep and ultra deepwater faces many technological challenges (Bruschi et al, 2015). High temperatures of extracted fluid combined with low external temperature and high hydrostatic pressure of the deepwater environment place simultaneous requirements for thermal insulation and mechanical integrity on the production infrastructure. Failure of subsea infrastructure can lead to significant environmental pollution (Drumond et al, 2018).

Over the past two decades, for fields with flow assurance challenges pipe-in-pipe systems have been developed in which the annular space between the inner and outer

Roberta Sburlati
Department of Civil, Chemical and Environmental Engineering (DICCA), Polytechnic School, University of Genova, Via Montallegro 1, 16145 Genova, Italy,
e-mail: roberta.sburlati@unige.it

Maria Kashtalyan
Centre for Micro-and Nanomechanics (CEMINACS), School of Engineering, University of Aberdeen, Fraser Noble Building, Aberdeen AB24 3UE, Scotland UK,
e-mail: m.kashtalyan@abdn.ac.uk

B. E. Abali et al. (eds.), *New Achievements in Continuum Mechanics and Thermodynamics*, Advanced Structured Materials 108,
https://doi.org/10.1007/978-3-030-13307-8_29

pipes is filled with insulation material to meet specific thermal requirements while the outer pipe is designed to withstand high external pressure dictated by the water depth and installation method (Bai and Bai, 2014). The capability to maintain the required temperature of the fluid inside the inner pipe can be further increased by using electrically heated pipe-in-pipe systems which offer enhanced flow assurance (Denniel et al, 2011; Denniel, 2015).

Lightweight alternatives to pipe-in-pipe systems based on steel pipes surrounded by insulation materials without the need for the outer steel pipe were considered by Grosjean et al (2009); Choqueuse et al (2010) and Bouchonneau et al (2010). A finite element model was developed for predicting thermomechanical behaviour of complex multi-layered pipes subjected to ultra-deepwater conditions, which was validated using experimental data for large-scale industrial prototype. Analytical solutions for stresses and displacements in heated and pressurised multi-layered pipes were developed by Vedeld and Sollund (2014) and Sollund et al (2014). Solution of Vedeld and Sollund (2014), which assumed uniform temperature distribution within each layer of the pipe, was subsequently refined by Yeo et al (2017), who found that the refined solution was able to produce more accurate predictions than the original solution of Vedeld and Sollund (2014). Analysis of coupled thermo-mechanical phenomena in such systems can be accomplished only by computational means (Abali, 2017).

Currently, insulation materials used in thermally insulated multilayered systems used in subsea applications include polymers and syntactic foams, which are polymer-matrix composites filled with hollow glass particles (microspheres, microballoons). They combine thermal insulation function with low buoyancy and high compressive strength (Lachambre et al, 2013). Properties of syntactic foams can be further enhanced by incorporating additional reinforcing phases such as nanoclay and nano- and micro-fibres (Gupta et al, 2013). Another way of enhancing properties of syntactic foams involves introducing a stiffness gradient within the syntactic foam, either by varying volume fraction of microballoons or by varying microballoon wall thickness. Varying volume fraction leads to a gradient in the coefficient of thermal expansion and moisture absorption, which results in warping when the material is exposed to changing temperature or moisture conditions. These disadvantages are absent in graded syntactic foams with constant volume fraction and varying wall thickness which were first proposed by Gupta (2007).

Investigation of the dependence of the elastic properties of syntactic foams on volume fraction and wall thickness over the full range of wall thicknesses was performed by Porfiri and Gupta (2009). It was found that while the Young's modulus of syntactic foams with very thin microballoons is lower than that of the matrix, thick-walled microballons can stiffen the matrix material. This means that graded syntactic foams with gradient in microbaloon wall thickness have the potential of performing both insulating and load-bearing functions.

Beneficial effect of stiffness gradient on stress and displacement fields in multilayered systems has been already established for coating/substrate systems (Kashtalyan and Menshykova, 2009; Sburlati et al, 2013) and sandwich pipes Sburlati and Kashtalyan (2016). Using thin functionally graded layer was also shown to reduce stresses in hollow pressurized cylinders (Sburlati, 2012) and spherical vessels (Atashipour and Sburlati, 2016) as well as around open holes (Sburlati et al, 2014). A versatile computational approach to modelling materials with gradients in elastic properties has been proposed by Abali et al (2012) and Abali et al (2014).

In this paper, the effect of stiffness gradient within the insulation layer on performance of thick-walled pipe subjected to internal and external pressure is investigated in the context of elasticity theory. A closed-form analytical solution to the problem is derived for a three-layer system comprising the inner pipe, insulation layer and outer jacket. On its basis, a comparative study of stress and displacement fields in pipes with homogeneous and graded insulation is then performed.

29.2 Analytical Modelling

29.2.1 Problem Formulation

Figure 29.1 shows cross-section of a three-layer system comprising the inner pipe, insulation and outer jacket, although industrial prototypes may include additional layers such as primer and adhesive layers (Bouchonneau et al, 2010). The system is referred to the cylindrical co-ordinate system, with z-axis directed along the pipe axis. The inner pipe has inner radius a and outer radius b and is modelled as homogeneous isotropic material with Young's modulus E_a and Poisson's ν_a.

The insulation has inner radius b and outer radius c and is modelled as graded isotropic material with constant Poisson's ratio ν_c and Young's modulus that is a linear function of the radius in the form

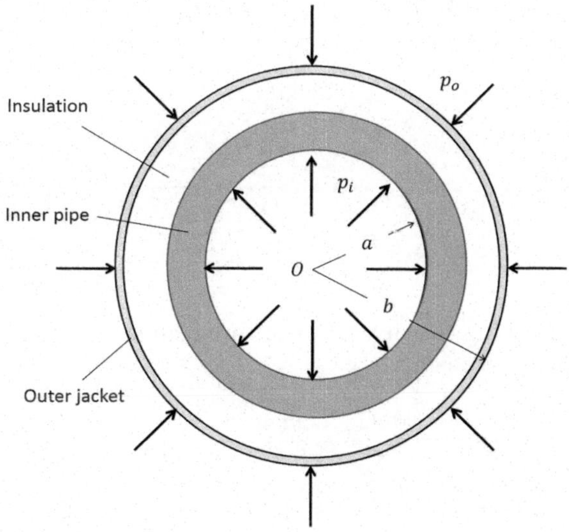

Fig. 29.1 Sketch of the pipe cross-section

$$E(r) = E_2 r + E_1 \tag{29.1}$$

If the Young's modulus values E_b at $r = b$ and E_c at $r = c$ are known then coefficients E_1 and E_2 are equal to

$$E_1 = \frac{E_c b - E_b c}{b - c}, \quad E_2 = -\frac{E_c - E_b}{b - c} \tag{29.2}$$

where

$$E(b) = E_2 b + E_1, \quad E(c) = E_2 c + E_1 \tag{29.3}$$

The Young's moduli E_b and E_c of the insulation material are generally not related to the Young's modulus E_a of the inner pipe, but are dictated by what it physically possible to achieve in a particular insulation material. A discussion of this is given in Sect. 29.3. The outer jacket has inner radius c and outer radius d and is modelled as homogeneous isotropic material with Young's modulus E_d and Poisson's ratio ν_d.

The system is subjected to inner pressure p_i and outer pressure p_o. Perfect bonding conditions are assumed between the pipe and insulation and the insulation and outer jacket, so that continuity conditions for stresses and displacements are fulfilled at the interfaces $r = b$ and $r = c$, respectively.

29.2.2 Method of Solution

Due to the axial symmetry of the system and applied loading, we can analyse it as plane strain elasticity problem. In this case, the equilibrium equations, strain-displacement equations and stress-strain relations have the form

$$\begin{aligned}
&\frac{d\sigma_r}{dr} + \frac{\sigma_r - \sigma_\theta}{r} = 0, \\
&\epsilon_r = \frac{du}{dr}, \quad \epsilon_\theta = \frac{u}{r}, \\
&\sigma_r = \frac{E(r)}{(1+\nu)(1-2\nu)} \left[(1-\nu)\,\epsilon_r + \nu\epsilon_\theta \right], \\
&\sigma_\theta = \frac{E(r)}{(1+\nu)(1-2\nu)} \left[(1-\nu)\,\epsilon_\theta + \nu\epsilon_r \right].
\end{aligned} \tag{29.4}$$

By using the displacement formulation, the above equations can be reduced to Navier equation

$$\frac{d^2u(r)}{dr^2} + \frac{1}{r}\frac{du(r)}{dr} - \frac{u(r)}{r^2} + \frac{1}{E(r)}\left(\frac{dE(r)}{dr}\right)\left(\frac{du(r)}{dr} + \frac{\nu}{(1-\nu)}\frac{u(r)}{r}\right) = 0. \tag{29.5}$$

The specific form of Navier equations depends on the function that describes variation of the Young's modulus with the radial co-ordinate within each component of the system. For the inner pipe and outer jacket, which are treated as homogeneous isotropic materials with constant Young's modulus, Navier equation is reduced to

$$\frac{d^2u(r)}{dr^2} + \frac{1}{r}\frac{du(r)}{dr} - \frac{u(r)}{r^2} = 0 \tag{29.6}$$

For the inner pipe, the solution of Navier equation can be written as

$$u(r) = A_1 r + \frac{A_2}{r} \tag{29.7}$$

Stresses in the inner pipe are then given as

$$\begin{aligned} \sigma_r(r) &= \frac{2E_a}{(1+\nu_a)(1-2\nu_a)}A_1 - \frac{E_a}{(1+\nu_a)r^2}A_2, \\ \sigma_\theta(r) &= \frac{2E_a}{(1+\nu_a)(1-2\nu_a)}A_1 + \frac{E_a}{(1+\nu_a)r^2}A_2. \end{aligned} \tag{29.8}$$

Similarly, for the outer jacket, the solutions of Navier equation can be written as

$$u(r) = C_1 r + \frac{C_2}{r} \tag{29.9}$$

Stresses in the outer jacket are then

$$\begin{aligned} \sigma_r(r) &= \frac{2E_a}{(1+\nu_a)(1-2\nu_a)}C_1 - \frac{E_a}{(1+\nu_a)r^2}C_2, \\ \sigma_\theta(r) &= \frac{2E_a}{(1+\nu_a)(1-2\nu_a)}C_1 + \frac{E_a}{(1+\nu_a)r^2}C_2. \end{aligned} \tag{29.10}$$

For the graded insulation, which is modelled as graded isotropic material with linear variation of the Young's modulus with the radial co-ordinate, Eq. (29.1), Navier equation becomes

$$\frac{d^2u(r)}{dr^2} + \frac{(\alpha-2r)}{(\alpha-r)r}\frac{du(r)}{dr} - \frac{4\alpha+(1-\beta^2)r}{4(\alpha-r)r^2}u(r) = 0 \tag{29.11}$$

where

$$\alpha = -\frac{E_1}{E_2}, \quad \beta = \frac{\sqrt{9\nu_c - 5}}{\sqrt{\nu_c - 1}}. \tag{29.12}$$

Solution to this equation can be found in terms of hypergeometric functions as

$$u(r) = B_1 r^{-(\beta+1)/2}\,\Theta_1(r) + B_2 r^{(\beta-1)/2}\,\Theta_2(r) \tag{29.13}$$

where

$$
\begin{aligned}
\Theta_1(r) &= {}_2\mathrm{F}_1\left(\left[\frac{\beta}{2}-\frac{1}{2},\frac{\beta}{2}+\frac{3}{2}\right],[\beta+1],\frac{\alpha}{r}\right),\\
\Theta_2(r) &= {}_2\mathrm{F}_1\left(\left[-\frac{\beta}{2}-\frac{1}{2},-\frac{\beta}{2}+\frac{3}{2}\right],[-\beta+1],\frac{\alpha}{r}\right).
\end{aligned}
\tag{29.14}
$$

Then stresses in the insulation can be expressed as

$$
\begin{aligned}
\sigma_r(r) &= w_{11}(r-\alpha)r^{-(3+\beta)/2}\,\Theta_1(r)\,B_1 + w_{12}(r-\alpha)r^{-(3-\beta)/2}\,\Theta_2(r)\,B_2\\
&\quad + w_{13}(r-\alpha)r^{-(5+\beta)/2}\,\Theta_3(r)\,B_1 + w_{14}(r-\alpha)r^{-(5-\beta)/2}\,\Theta_4(r)\,B_2,\\
\sigma_\theta(r) &= w_{21}(r-\alpha)r^{-(3+\beta)/2}\,\Theta_1(r)\,B_1 + w_{22}(r-\alpha)r^{-(3-\beta)/2}\,\Theta_2(r)\,B_2\\
&\quad + w_{23}(r-\alpha)r^{-(5+\beta)/2}\,\Theta_3(r)\,B_1 + w_{24}(r-\alpha)r^{-(5-\beta)/2}\,\Theta_4(r)\,B_2,
\end{aligned}
\tag{29.15}
$$

where

$$
\begin{aligned}
\Theta_3(r) &= {}_2\mathrm{F}_1\left(\left[\frac{\beta}{2}+\frac{5}{2},\frac{\beta}{2}+\frac{1}{2}\right],[\beta+2],\frac{\alpha}{r}\right),\\
\Theta_4(r) &= {}_2\mathrm{F}_1\left(\left[-\frac{\beta}{2}+\frac{1}{2},-\frac{\beta}{2}+\frac{5}{2}\right],[-\beta+2],\frac{\alpha}{r}\right).
\end{aligned}
\tag{29.16}
$$

and

$$
\begin{aligned}
w_{11} &= -\frac{(\beta-3)(\beta+3)^2\,E_2}{2(\beta+1)(\beta^2-7)}, \quad w_{12} = -\frac{(\beta-3)^2(\beta+3)\,E_2}{2(\beta-1)(\beta^2-7)},\\
w_{13} &= -\frac{(\beta-3)(\beta+3)^2\,E_2\,\alpha}{2(\beta+1)^2(\beta^2-7)}, \quad w_{14} = -\frac{(\beta-3)^2(\beta+3)\,E_2\,\alpha}{2(\beta-1)^2(\beta^2-7)},\\
w_{21} &= \frac{(\beta-3)(\beta+3)^2(\beta-1)\,E_2}{4(\beta+1)(\beta^2-7)}, \quad w_{22} = -\frac{(\beta-3)^2(\beta+3)(\beta+1)\,E_2}{4(\beta-1)(\beta^2-7)},\\
w_{23} &= \frac{(\beta-3)(\beta+3)^2(\beta^2-5)\,E_2\,\alpha}{8(\beta+1)^2(\beta^2-7)}, \quad w_{24} = -\frac{(\beta-3)^2(\beta^2-5)(\beta+3)\,E_2\,\alpha}{8(\beta-1)^2(\beta^2-7)}.
\end{aligned}
\tag{29.17}
$$

Arbitrary constants $A_1, A_2, B_1, B_2, C_1, C_2$ are determined by boundary conditions at the innermost and outermost radii of the insulated pipe system as well as continuity conditions at the interfaces.

They have been found as

$$
\begin{aligned}
A_1 &= \frac{(1+\nu_a)(2\nu_a-1)p_i}{E_a} - \frac{(2\nu_a-1)}{a^2}A_2,\\
C_1 &= \frac{(1+\nu_d)(2\nu_d-1)p_o}{E_d} - \frac{(2\nu_d-1)}{d^2}C_2,\\
B_1 &= \frac{b^{(\beta+3)/2}}{\Theta_1(b)}A_1 + \frac{b^{(\beta-1)/2}}{\Theta_1(b)}A_2 - \frac{b^{\beta}\,\Theta_2(b)}{\Theta_1(b)}B_2,\\
B_2 &= -\frac{c^{-\beta}\,\Theta_1(c)}{\Theta_2(c)}B_1 + \frac{c^{(3-\beta)/2}}{\Theta_2(c)}C_1 + \frac{c^{-(\beta+1)/2}}{\Theta_2(c)}C_2,
\end{aligned}
\tag{29.18}
$$

and

$$A_2 = -\frac{a_{10}a_{22} - a_{12}a_{20}}{a_{11}a_{22} - a_{12}a_{21}}, \qquad C_2 = \frac{a_{10}a_{21} - a_{11}a_{20}}{a_{11}a_{22} - a_{12}a_{21}}. \tag{29.19}$$

Quantities a_{ij} with $i = 0, 1, 2; j = 0, 1, 2$, involved in the above expressions, are listed in Appendix.

29.3 Results and Discussion

Similarly to Sburlati (2012), the correctness of the proposed analytical solution is verified by setting $E_a = E_b = E_c = E_d$ and establishing that it is possible to recover the results given by the well-known Lamé solution for a homogenous thick-walled pipe from the present solution. The developed solution is used to study the effect of insulation grading on the pipe performance under external and internal pressure.

The geometry of the insulated pipe system is set as follows: inner radius of the pipe $a = 0.1$ m outer radius of the pipe $b = 0.12$ m, inner radius of the jacket $c = 0.180$ m, outer radius of the jacket $d = 0.185$ m. The external pressure is taken as 30 MPa.

Material properties of the reference system with homogeneous insulation are adapted from Bouchonneau et al (2010) and are listed in Table 29.1.

It is evident from Table 29.1 that the Young's modulus of Polypropelene (PP) syntactic foam insulation is slightly lower than the modulus of the PP outer jacket (1.1 GPa versus 1.3 GPa). This indicates that wall thickness of microballoons in the insulation was very small. It is worth pointing out that when Porfiri and Gupta (2009) analysed the dependence of the elastic properties of syntactic foams on volume fraction and wall thickness over the full range of wall thicknesses, they found that the Young's modulus of syntactic foams with very thin microballoons is lower than that of the matrix, while thick microballons stiffen the matrix material.

The relationship between microballoon wall thickness and the Young's modulus of the resulting particle-reinforced composite, established by Porfiri and Gupta (2009), can be turned to advantage in graded syntactic foams. For instance, by keeping the volume fraction of microballoons constant (e.g. 60%) and varying wall thickness with the radius, one can produce a graded syntactic foam with the Young's modulus that varies from the modulus of the matrix E_m at the outer interface $r = c$ to 10 E_m at the inner interface $r = b$. The corresponding variation of the Poisson's

Table 29.1 System with homogeneous insulation

Component	Material	Young's modulus	Poisson's ratio
Inner pipe	Steel	218 GPa	0.33
Insulation	PP Syntactic Foam	1.1 GPa	0.32
Outer jacket	PP	1.3 GPa	0.40

ratio, according to Porfiri and Gupta (2009), would be from about $0.8\,\nu_m$ at $r = c$ to $0.7\,\nu_m$ at $r = b$; however, this variation is small and can be neglected to simplify the analysis.

Material properties of the system with graded insulation based on the above considerations have been proposed and are listed in Table 29.2.

The properties of the inner pipe and outer jacket are the same as in the reference system, whereas modulus of the insulation varies from 13 GPa at the interface with the inner pipe to 1.3 GPa at the interface with the outer jacket. Figure 29.2 shows variation of the normalised Young's modulus $\bar{E}(r) = E(r)/E(a)$ through the wall thickness of the pipe system.

Figures 29.3-29.5 show variation of respectively normalised radial displacement $\bar{u}(r) = u(r)/d$, normalised radial stress $\bar{\sigma}_r(r) = \sigma_r(r)/p_o$, and normalised hoop stress $\bar{\sigma}_\theta(r) = \sigma_\theta(r)/p_o$, through the wall thickness of the system, subjected to external pressure of 30 MPa. The results for the reference system with homogeneous insulation are shown in red, while the results for the system with graded insulation are shown in blue.

Under the external pressure, use of graded insulation leads to significant reduction of radial displacement in the insulation layer and the outer jacket (Fig. 29.3). The influence of graded insulation on the radial stress is comparatively small (Fig. 29.4). The hoop stress is reduced in pipe and jacket (Fig. 29.5); instead, the insulation redistribution of hoop stress is observed, with stress magnitude in the vicinity of the inner pipe increasing and in the vicinity of the outer jacket decreasing. In Table 29.3 we compare in details the values between graded and homogeneous insulation cases.

Table 29.2 System with graded insulation

Component	Material	Young's modulus	Poisson's ratio
Inner pipe	Steel	218 GPa	0.33
Insulation	PP Syntactic Foam ($r = b$)	13 GPa	0.32
	PP Syntactic Foam ($r = c$)	1.3 GPa	0.32
Outer jacket	PP	1.3 GPa	0.40

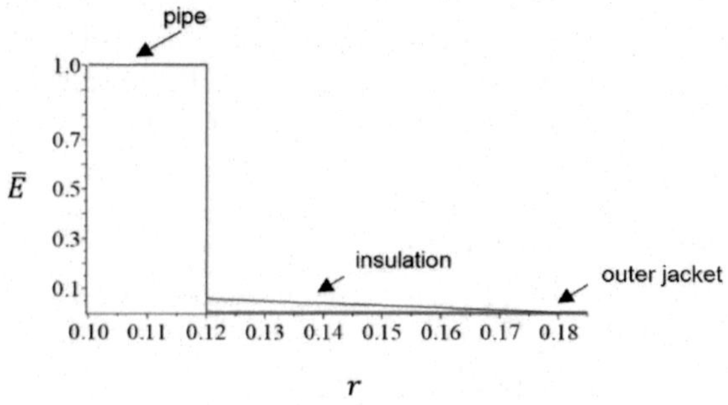

Fig. 29.2 Variation of the normalised Young's modulus through the thickness of the pipe system

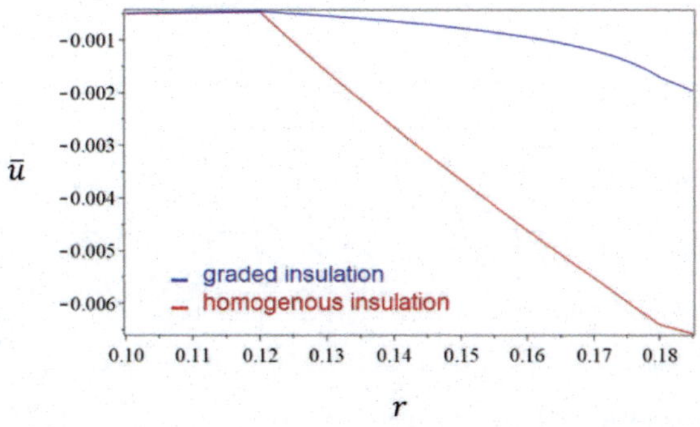

Fig. 29.3 Normalised radial displacement in the pipe system with and without graded insulation under external pressure

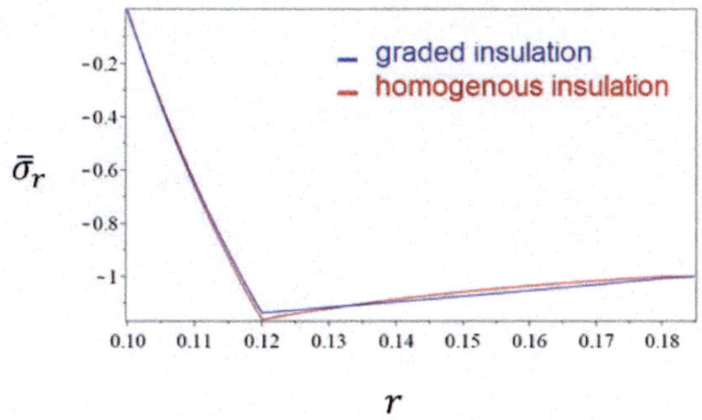

Fig. 29.4 Normalised radial stress in the pipe system with and without graded insulation under external pressure

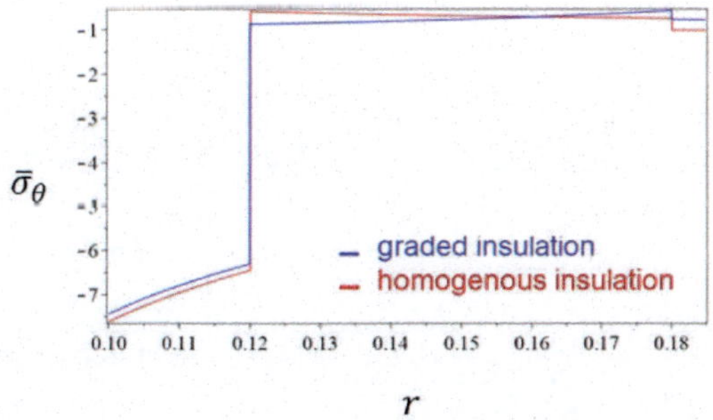

Fig. 29.5 Normalised hoop stress in the pipe system with and without graded insulation under external pressure

Figures 29.6-29.8 show variation of respectively radial displacement, radial stress $\bar{\sigma}_r(r) = \sigma_r(r)/p_i$, and hoop stress $\bar{\sigma}_\theta(r) = \sigma_\theta(r)/p_i$, through the wall thickness of the pipe, subjected to internal pressure of 10 MPa.

Under the internal pressure, use of graded insulation leads to significant reduction of radial displacement in both the insulation layer and the outer jacket (Fig. 29.6), and also to a slight increase of the radial stress (Fig. 29.7) in the pipe and the insulation layers in the vicinity of their interface. The hoop stress in the inner pipe is

Table 29.3 Comparison of hoop stress values in pipes with graded or homogeneous insulation under external pressure

Component	Location	Hoop stess (MPa)		Difference %
		Graded	Homogeneous	
Inner pipe	$r = a$	-223.18	-228.29	-2.24
	$r = b$	-189.08	-193.42	-2.24
Insulation	$r = b$	-26.25	-17.29	51.73
	$r = c$	-16.74	-22.18	-24.49
Outer jacket	$r = c$	-22.84	-30.18	-24.32
	$r = d$	-23.04	-30.17	-23.65

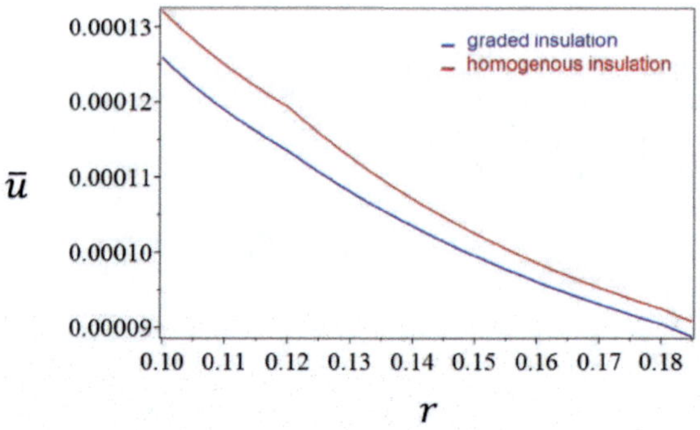

Fig. 29.6 Normalised radial displacement in the pipe system with and without graded insulation under internal pressure

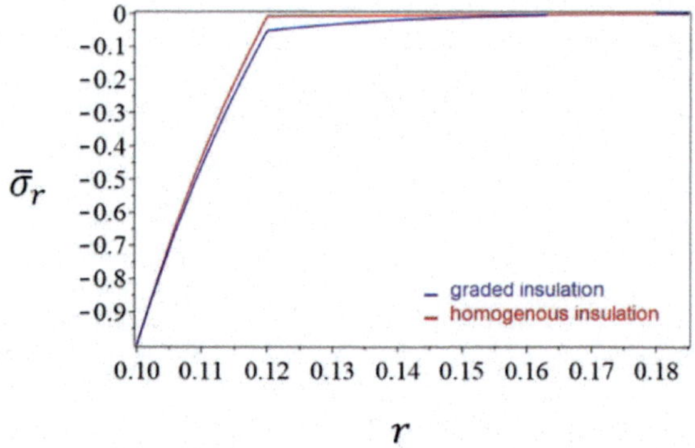

Fig. 29.7 Normalised radial stress in the pipe system with and without graded insulation under internal pressure

reduced, while an increase in the hoop stress is observed in the insulation layer, at the interface with the pipe (Fig. 29.8).

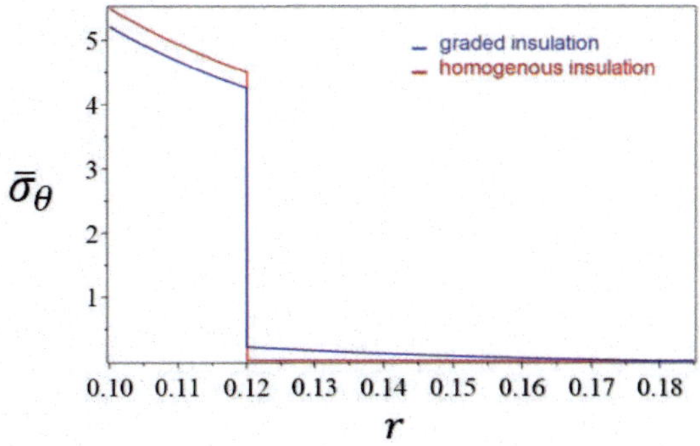

Fig. 29.8 Normalised hoop stress in the pipe system with and without graded insulation under internal pressure

29.4 Conclusions

In this paper, a closed form analytical solution for an insulated deepwater pipe system comprising an inner pipe, an insulation layer and an outer jacket is developed. Both the inner pipe and the outer jacket are modelled as homogeneous materials, while the insulation layer is assumed to contain a stiffness gradient of the type that can be realised in syntactic foams by varying wall thickness of glass microballoons while keeping their volume fraction constant. Based on the derived analytical solution, stress analysis of pipe systems with graded insulation subjected to external and internal pressure is performed and the beneficial effect of graded insulation on stresses and displacements in the system is established and quantified.

The developed closed-form analytical solution, while focusing on purely elastic behaviour and neglecting thermal effects, can be used as a valuable benchmark for validating models that attempt to take into account combined mechanical and thermal effects. Even in absence of stiffness gradients, such models often need to make simplifying assumptions to arrive at the results, as pointed out in our discussion of Vedeld and Sollund (2014) solution in Sect. 29.1. It is therefore helpful to have a benchmark elasticity solution that can be used to assess accuracy of approximate analytical and numerical models.

Acknowledgements Financial support of this research by the Royal Society of Edinburgh and the Italian Academy of Sciences through the Bilateral International Exchanges Programme is gratefully acknowledged.

Appendix

The coefficients of equations (29.19) are

$$a_{11} = \frac{(b-\alpha)\left((-2\,\nu_a+1)\,b^2+a^2\right)c^{\beta/2}\Theta_2\,(c)\,(w_{11}\Theta_1\,(b)\,b+w_{13}\Theta_3\,(b))}{a^2b^3\left(c^{-\beta/2}\Theta_2\,(b)\,b^{\beta}\Theta_1\,(c)-c^{\beta/2}\Theta_1\,(b)\,\Theta_2\,(c)\right)}$$
$$-\frac{(-b+\alpha)\left(b^2\,(2\,\nu_a-1)-a^2\right)b^{\beta}c^{-\beta/2}\Theta_1\,(c)\,(\Theta_2\,(b)\,w_{12}b+\Theta_4\,(b)\,w_{14})}{a^2b^3\left(c^{-\beta/2}\Theta_2\,(b)\,b^{\beta}\Theta_1\,(c)-c^{\beta/2}\Theta_1\,(b)\,\Theta_2\,(c)\right)}$$
$$-\frac{(a-b)\,(a+b)\,E_a}{(1+\nu_a)\,a^2b^2},$$

$$a_{12} = \frac{(b-\alpha)\left((2\,\nu_d-1)\,c^2-d^2\right)b^{\beta/2-3/2}\,(w_{11}-w_{12})\,\Theta_2\,(b)\,\Theta_1\,(b)}{\sqrt{c}d^2\left(c^{-\beta/2}\Theta_2\,(b)\,b^{\beta}\Theta_1\,(c)-c^{\beta/2}\Theta_1\,(b)\,\Theta_2\,(c)\right)}$$
$$-\frac{(b-\alpha)\left((2\,\nu_d-1)\,c^2-d^2\right)b^{\beta/2-5/2}\,(-\Theta_2\,(b)\,w_{13}\Theta_3\,(b)+\Theta_1\,(b)\,w_{14}\Theta_4\,(b))}{\sqrt{c}d^2\left(c^{-\beta/2}\Theta_2\,(b)\,b^{\beta}\Theta_1\,(c)-c^{\beta/2}\Theta_1\,(b)\,\Theta_2\,(c)\right)},$$

$$a_{21} = -\frac{(c-\alpha)\left(b^2\,(2\,\nu_a-1)-a^2\right)b^{\beta/2}\,(\Theta_4\,(c)\,\Theta_1\,(c)\,w_{14}-\Theta_3\,(c)\,\Theta_2\,(c)\,w_{13})}{\sqrt{b}c^{5/2}\left(c^{-\beta/2}\Theta_2\,(b)\,b^{\beta}\Theta_1\,(c)-c^{\beta/2}\Theta_1\,(b)\,\Theta_2\,(c)\right)a^2}$$
$$+\frac{(c-\alpha)\left(b^2\,(2\,\nu_a-1)-a^2\right)b^{\beta/2}\,(w_{11}-w_{12})\,\Theta_2\,(c)\,\Theta_1\,(c)}{\sqrt{b}c^{3/2}\left(c^{-\beta/2}\Theta_2\,(b)\,b^{\beta}\Theta_1\,(c)-c^{\beta/2}\Theta_1\,(b)\,\Theta_2\,(c)\right)a^2},$$

$$a_{22} = \frac{(\alpha-c)\left((2\,\nu_d-1)\,c^2-d^2\right)\left(c^{-\beta/2}\Theta_2\,(b)\,b^{\beta}\Theta_1\,(c)\,w_{11}-c^{\beta/2}\Theta_1\,(b)\,\Theta_2\,(c)\,w_{12}\right)}{\left(c^{-\beta/2}\Theta_2\,(b)\,b^{\beta}\Theta_1\,(c)-c^{\beta/2}\Theta_1\,(b)\,\Theta_2\,(c)\right)c^2d^2}$$
$$-\frac{(\alpha-c)\left((2\,\nu_d-1)\,c^2-d^2\right)\left(-\Theta_3\,(c)\,c^{-\beta/2}\Theta_2\,(b)\,b^{\beta}w_{13}+\Theta_4\,(c)\,c^{\beta/2}\Theta_1\,(b)\,w_{14}\right)}{c^3\left(c^{-\beta/2}\Theta_2\,(b)\,b^{\beta}\Theta_1\,(c)-c^{\beta/2}\Theta_1\,(b)\,\Theta_2\,(c)\right)d^2}$$
$$-\frac{E_d\,(c-d)\,(c+d)}{c^2d^2\,(1+\nu_d)}.$$

And, for inner pressure (i), the remaining coefficients assume the following form

$$a_{10}^{(i)} = \frac{(-b+\alpha)\,(1+\nu_a)\,(2\,\nu_a-1)\left(-b^{\beta}c^{-\beta/2}\Theta_1\,(c)\,\Theta_2\,(b)\,w_{12}+c^{\beta/2}\Theta_2\,(c)\,\Theta_1\,(b)\,w_{11}\right)p_i}{\left(-c^{-\beta/2}\Theta_2\,(b)\,b^{\beta}\Theta_1\,(c)+c^{\beta/2}\Theta_1\,(b)\,\Theta_2\,(c)\right)E_a}$$
$$+\frac{(-b+\alpha)\,(1+\nu_a)\,(2\,\nu_a-1)\left(-b^{\beta}c^{-\beta/2}\Theta_4\,(b)\,\Theta_1\,(c)\,w_{14}+c^{\beta/2}\Theta_3\,(b)\,\Theta_2\,(c)\,w_{13}\right)p_i}{b\left(-c^{-\beta/2}\Theta_2\,(b)\,b^{\beta}\Theta_1\,(c)+c^{\beta/2}\Theta_1\,(b)\,\Theta_2\,(c)\right)E_a}-p_i,$$

$$a_{20}^{(i)} = -\frac{(c-\alpha)\,(1+\nu_a)\,(2\,\nu_a-1)\,b^{3/2+\beta/2}\Theta_2\,(c)\,\Theta_1\,(c)\,(w_{11}-w_{12})\,p_i}{c^{3/2}\left(c^{-\beta/2}\Theta_2\,(b)\,b^{\beta}\Theta_1\,(c)-c^{\beta/2}\Theta_1\,(b)\,\Theta_2\,(c)\right)E_a}$$
$$-\frac{(\alpha-c)\,(1+\nu_a)\,(2\,\nu_a-1)\,b^{3/2+\beta/2}\,(\Theta_4\,(c)\,\Theta_1\,(c)\,w_{14}-\Theta_3\,(c)\,\Theta_2\,(c)\,w_{13})\,p_i}{c^{5/2}\left(c^{-\beta/2}\Theta_2\,(b)\,b^{\beta}\Theta_1\,(c)-c^{\beta/2}\Theta_1\,(b)\,\Theta_2\,(c)\right)E_a},$$

while, for outer pressure (o), we obtain

$$a_{10}^{(o)} = -\frac{(-b+\alpha)(1+\nu_d)(2\nu_d-1)\,b^{\beta/2}c^{3/2}\Theta_1(b)\,\Theta_2(b)\,(w_{11}-w_{12})\,p_o}{b^{3/2}\left(-c^{-\beta/2}\Theta_2(b)\,b^{\beta}\Theta_1(c)+c^{\beta/2}\Theta_1(b)\,\Theta_2(c)\right)E_d}$$
$$-\frac{(-b+\alpha)(1+\nu_d)(2\nu_d-1)\,b^{\beta/2-5/2}c^{3/2}\left(\Theta_2(b)\,w_{13}\Theta_3(b)-\Theta_1(b)\,w_{14}\Theta_4(b)\right)p_o}{\left(-c^{-\beta/2}\Theta_2(b)\,b^{\beta}\Theta_1(c)+c^{\beta/2}\Theta_1(b)\,\Theta_2(c)\right)E_d},$$

$$a_{20}^{(o)} = -\frac{(\alpha-c)(1+\nu_d)(2\nu_d-1)\left(-c^{-\beta/2}b^{\beta}\Theta_2(b)\,\Theta_1(c)\,w_{11}+c^{\beta/2}\Theta_2(c)\,\Theta_1(b)\,w_{12}\right)p_o}{\left(-c^{-\beta/2}\Theta_2(b)\,b^{\beta}\Theta_1(c)+c^{\beta/2}\Theta_1(b)\,\Theta_2(c)\right)E_d}$$
$$-\frac{(\alpha-c)(1+\nu_d)(2\nu_d-1)\left(-c^{-\beta/2}b^{\beta}\Theta_3(c)\,\Theta_2(b)\,w_{13}+c^{\beta/2}\Theta_4(c)\,\Theta_1(b)\,w_{14}\right)p_o}{c\left(-c^{-\beta/2}\Theta_2(b)\,b^{\beta}\Theta_1(c)+c^{\beta/2}\Theta_1(b)\,\Theta_2(c)\right)E_d}+p_o.$$

References

Abali BE (2017) Computational Reality, Solving Nonlinear and Coupled Problems in Continuum Mechanics. Advanced Structured Materials, Springer

Abali BE, Völlmecke C, Woodward B, Kashtalyan M, Guz I, Müller WH (2012) Numerical modeling of functionally graded materials using a variational formulation. Continuum Mechanics and Thermodynamics 24(4-6):377–390

Abali BE, Völlmecke C, Woodward B, Kashtalyan M, Guz I, Müller WH (2014) Three-dimensional elastic deformation of functionally graded isotropic plates under point loading. Composite Structures 118:367–376

Atashipour S, Sburlati R (2016) Electro-elastic analysis of a coated spherical piezoceramic sensor. Composite Structures 156:399–409

Bai Q, Bai Y (2014) Subsea pipeline design, analysis, and installation. Gulf Professional Publishing

Bouchonneau N, Sauvant-Moynot V, Choqueuse D, Grosjean F, Poncet E, Perreux D (2010) Experimental testing and modelling of an industrial insulated pipeline for deep sea application. Journal of Petroleum Science and Engineering 73(1-2):1–12

Bruschi R, Vitali L, Marchionni L, Parrella A, Mancini A (2015) Pipe technology and installation equipment for frontier deep water projects. Ocean Engineering 108:369–392

Choqueuse D, Davies P, Perreux D, Sohier L, Cognard JY (2010) Mechanical behavior of syntactic foams for deep sea thermally insulated pipeline. In: Applied Mechanics and Materials, Trans Tech Publ, vol 24, pp 97–102

Denniel S (2015) Pipe-in-pipe technology adapts to changing needs in deep and shallow water. Offshore 75(2):82–84

Denniel S, Bonneau P, Savy PA (2011) Electrically heated pipe-in-pipe- a qualified reelable technology to meet flow assurance challenges. Offshore 71(9):88–90

Drumond GP, Pasqualino IP, Pinheiro BC, Estefen SF (2018) Pipelines, risers and umbilicals failures: A literature review. Ocean Engineering 148:412–425

Grosjean F, Bouchonneau N, Choqueuse D, Sauvant-Moynot V (2009) Comprehensive analyses of syntactic foam behaviour in deepwater environment. Journal of Materials Science 44(6):1462–1468

Gupta N (2007) A functionally graded syntactic foam material for high energy absorption under compression. Materials Letters 61(4-5):979–982

Gupta N, Pinisetty D, Shunmugasamy VC (2013) Reinforced polymer matrix syntactic foams: effect of nano and micro-scale reinforcement. Springer Science & Business Media

Kashtalyan M, Menshykova M (2009) Effect of a functionally graded interlayer on three-dimensional elastic deformation of coated plates subjected to transverse loading. Composite Structures 89(2):167–176

Lachambre J, Maire E, Adrien J, Choqueuse D (2013) In situ observation of syntactic foams under hydrostatic pressure using x-ray tomography. Acta Materialia 61(11):4035–4043

Porfiri M, Gupta N (2009) Effect of volume fraction and wall thickness on the elastic properties of hollow particle filled composites. Composites Part B: Engineering 40(2):166–173

Sburlati R (2012) Analytical elastic solutions for pressurized hollow cylinders with internal functionally graded coatings. Composite Structures 94(12):3592–3600

Sburlati R, Kashtalyan M (2016) Elasticity analysis of sandwich pipes with functionally graded interlayers. European Journal of Mechanics-A/Solids 59:232–241

Sburlati R, Atashipour SR, Hosseini-Hashemi S (2013) Study on the effect of functionally graded coating layers on elastic deformation of thick circular plates: a closed-form elasticity solution. Composite Structures 99:131–140

Sburlati R, Atashipour SR, Atashipour SA (2014) Reduction of the stress concentration factor in a homogeneous panel with hole by using a functionally graded layer. Composites Part B: Engineering 61:99–109

Sollund HA, Vedeld K, Hellesland J (2014) Efficient analytical solutions for heated and pressurized multi-layer cylinders. Ocean Engineering 92:285–295

Vedeld K, Sollund HA (2014) Stresses in heated pressurized multi-layer cylinders in generalized plane strain conditions. International Journal of Pressure Vessels and Piping 120:27–35

Yeo W, Purbolaksono J, Aliabadi M, Ramesh S, Liew H (2017) Exact solution for stresses/displacements in a multilayered hollow cylinder under thermo-mechanical loading. International Journal of Pressure Vessels and Piping 151:45–53

Chapter 30
On Brake Pad Shim Characterization: a Homogenization Approach and Finite Element Analysis

Dominik Schmid, Nils Gräbner & Utz von Wagner

Abstract Brake squeal is a typical problem of "Noise, Vibration, Harshness" (NVH) phenomena in the automotive world leading to potential customer complaints. This high frequency noise in the audible frequency range of approximately 1 kHz to 15 kHz is induced by self excitation resulting from the frictional contact between brake pad and disk. A typical industrial countermeasure to address this problem is the mounting of thin composite structures consisting of elastomer and steel layers, so called shims, on the pad backplates. They are applied to increase the damping and to influence the vibration shapes.

The computational modeling of shims using Finite Elements is still a complex task and shows significant potential for improvement. To avoid problems resulting from element sizes of the partially very thin layers a classical homogenization theory from literature is considered. This homogenization approach maps shim properties in an improved manner which contributes to substantially smaller model sizes as well as less simulation effort and time. Therefore, analytical approaches for constrained layer damping structures are introduced and corresponding theoretical results are presented. To validate these theoretical results, experimental investigations are carried out on shims bonded to structures, especially steel plates and brake pads.

Keywords: Shims · Constrained layer damping · Homogenization · Finite Element simulation · NVH

Dominik Schmid · Nils Gräbner · Utz von Wagner
Chair of Mechatronics and Machine Dynamics, Technische Universität Berlin, 10587 Berlin, Germany,
e-mail: dominik.schmid@tu-berlin.de, nils.graebner@tu-berlin.de, utz.vonwagner@tu-berlin.de

B. E. Abali et al. (eds.), *New Achievements in Continuum Mechanics and Thermodynamics*, Advanced Structured Materials 108,
https://doi.org/10.1007/978-3-030-13307-8_30

30.1 Introduction

Numerous technical applications especially in lightweight structures nowadays require high stiffness characteristics with simultaneously high damping ratios in order to avoid undesired vibrations. A typical example for fulfilling such requirements is the application of composites. These structures meet the requirements by exhibiting good damping behavior without loosing stiffness (Marcelin et al, 1995). The application of such composites called shims on brake pad backplates is a typical countermeasure against undesired vibration phenomena like brake squeal. Brake squeal is a high frequency noise in the audible frequency range of approximately 1 kHz to 15 kHz based on self-excited vibrations caused by friction forces between pad and disk (Kinkaid et al, 2003; Cantoni et al, 2009). Shims are thin layer bonded structures consisting of viscoelastic elastomer layers and steel plates with high stiffness. Figure 30.1 shows the general set-up of a brake pad with shim consisting of the friction material being in contact with the disk during the braking process, the backplate and the shim, which is coupled to the backplate by an adhesive layer. Designing and selecting appropriate shims is still a major task to solve. There are plenty of experimental investigations required to find the right shim matching the individual noise problems of the respective brake. This includes the experimental investigation of the components as well as dynamometer tests of the entire brake and tests in the vehicle.

A standard industrial tool in the investigation of brake squeal is the so-called complex eigenvalue analysis (CEA). The CEA is based on large Finite Element models of the entire brake with disk, pads, caliper, carrier and the suspension. Equilibrium positions resulting from the applied brake torque are calculated by a static analysis and the equations of motion are linearized with respect to them (Gräbner et al, 2016). Gyroscopic terms and, due to the friction forces between disk and pads circulatoric terms, i.e. self excitation, is present. This may result in positive real parts of the eigenvalues, i.e. instability of the aforementioned equilibrium solution. Therefore the overall goal of CEA is to characterize the stability behavior of equilibrium solutions of brake systems as an indicator for possible onset of squeal. The accurate calculation of the eigenvalues anyway is a challenging problem (Gräbner et al, 2016) for that large gyroscopic-circulatory systems. The squealing itself is a limit cycle oscillation requiring to consider nonlinearities (Gräbner et al, 2014) and corresponding bifurcation behavior.

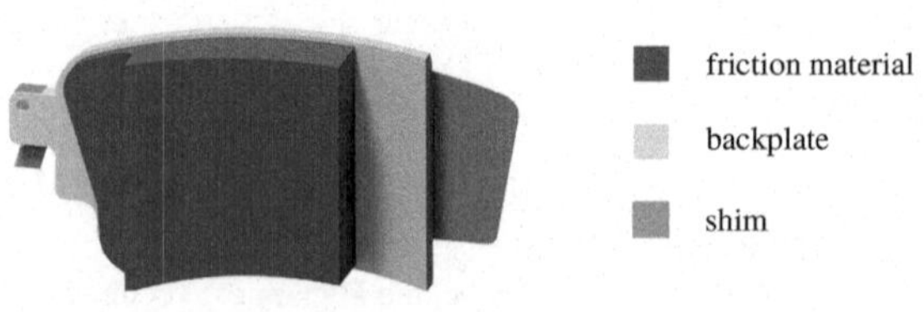

Fig. 30.1 Brake pad composition (Schmid, 2018a)

Damping is well known to influence and to be able to suppress self-excited vibrations, see e.g. (Gräbner et al, 2015) both in the linear as well as in the nonlinear case. On the other hand, damping is hard to identify and therefore in many applications only estimated or even neglected. Neglected or underestimated damping in general leads to overestimated eigenvalues with positive real parts and corresponding potential squeal frequencies. Therefore, modeling of damping, at least for components introduced by intention in order to damp vibrations like the shims, is a key issue for improved modeling of the system's behavior. Hereby, mapping the damping capacity of the shim's viscoelastic layer is an essential point. Compared to the overall high amount of literature on brake squeal including several review papers like the already mentioned in Kinkaid et al (2003) and Cantoni et al (2009) publications on modeling of shims are somewhat rare. Examples for FE investigations including shims are in Festjens et al (2012) and Kang (2012). Esgandari and Olatunbosun (2016) implemented Rayleigh damping in elastomer layers of shims, steel layers are still undamped. Rayleigh parameter are used based on previous investigations from Flint et al (2004). There are also some technical standards on shims like in SAE-J3001 (2011).

The problem in including the shims in FE-models are the thin layers in shims, which in general may have thicknesses in the range of 0.1 mm. Using element sizes in the same range with, in the FE sense "healthy", ratios of element dimensions would lead to element numbers which cannot be handled, if complete models of the entire brake are considered.

Therefore, the basic idea of the work described in this paper is, to homogenize the shim layered structure using classical theories and approaches. This is done in order to enable larger element sizes allowing for acceptable numbers of degrees of freedom, which can be handled in models of the entire brake. In this contribution shims with two thin elastomer layers enclosed by a metal core are examined in detail.

There is a large number of publications on the dynamics of composites combining metal and elastomer layers. One of the first theories has been published in the middle of the last century by Oberst describing unconstrained damping treatments (Oberst and Frankenfeld, 1952; Oberst et al, 1954). The purpose was to homogenize structures consisting of layers with different characteristics. As a result a single layer with equivalent mechanical properties describing stiffness and damping ratios is obtained. Characterizing system properties analytically Kerwin extended Oberst's theory considering an additional stiff top layer Kerwin Jr (1959) which in the following was called constrained layer. The energy dissipation for constrained layer compounds is mainly induced by shearing of the viscoelastic core material and exceeds the extensional damping of unconstrained damping treatments Ross et al (1959). The developed theory for free vibrations was introduced for fully coverage of the damping and constraining layer considering pinned-pinned boundary conditions. DiTaranto also addressed this problem and developed a sixth order equation for longitudinal displacement. He formulates the loss factor for coverage of the entire beam for any boundary condition (DiTaranto, 1965). The transverse displacement of beams with damping treatments has been published by Mead and Markuš for arbitrary boundary conditions and has been extended for forced vibrations (Mead and Markuš, 1969) and (Mead and Markuš, 1970). Using an energy approach an exact solution for the sixth order equation as well as numerical approaches were described by Rao (Rao, 1978).

The publications listed so far require full coverage of all layers. In general brake pads used in vehicles show only partial coverage of shims bonded to back plates. The more general set-up of only partial coverage has been investigated by Nokes describing damping of beams for any symmetrical boundary conditions requiring a centered constrained layer on the structure (Nokes and Nelson, 1968). Markuš (1974) dealed with the damping mechanism of beams and developed a theory for partially covered constrained layers predicting damping for any boundary conditions. Damping material calculation formulas of sandwich beams with partially covering damping layers have been presented by Sylwan achieving equal damping properties compared to full coverage (Sylwan, 1978). Moreover two approximate solutions and one exact method for the damping description of partially covered sandwich beams have been published by Lall et al (1988). Flint presents essential publications in his PhD thesis considering full coverage (Flint, 2002). An overview and classification of relevant surface damping treatments can be found in the books from Nashif et al (1985) and Sun and Lu (1995).

The aim of this contribution is to carry out Finite Element simulations of homogenized shims with a view to less experimental effort and better prediction quality. With this stiffness characteristics of shims and in particular loss factors of shims bonded to rectangular steel plates are determined analytically and applied to an Abaqus CAE model. The loss factors for full coverage are compared considering torsional and flexural mode shapes. The damping behavior of the viscoelastic core is implemented in a Finite Element model. As a further step towards the improved shim modeling, brake pads are examined numerically. Furthermore, concrete recommendations on modeling and meshing shims are introduced. Finally experimental investigations are carried out to validate damping and stiffness characteristics using methods as described in SAE guideline J3001 (SAE-J3001, 2011).

30.2 Modeling of Shims

The dynamical characterization of shims includes modal parameters like natural frequencies, mode shapes and damping ratios. The focus is to map torsional and bending modes analytically. Therefore, mechanical models focusing on stiffness and damping behavior of shims are examined. Following most set-ups in literature and in order to prevent influence resulting from the support, which may affect natural frequencies and damping characteristics, a free-free support is utilized for modeling and experimental investigations (Ewins, 1984) in the following. These boundary conditions will of course change, if the resulting pad and shim model will be integrated into the model of the entire brake.

30.2.1 Continuous Mechanical Systems

For introduction classical models for torsional vibrations of bars as well as Euler-Bernoulli beams with a rectangular cross section are reconsidered in the following, where the corresponding formulas can be taken from textbooks, e.g. Hagedorn and DasGupta (2007). A free-free torsional bar is shown in Fig. 30.2, where G is the shear modulus, ϱ the mass density as well as I_{P} and I_{T} are the polar and torsional moments of inertia respectively. Describing the geometrical dimensions of the bar, the length l, the width b and height h are introduced. The following partial differential equation describes the torsional free vibrations $\vartheta(x,t)$ of the bar with uniform cross section by

$$\frac{\partial^2 \vartheta}{\partial t^2} - c^2 \frac{\partial^2 \vartheta}{\partial x^2} = 0, \tag{30.1}$$

where c is the wave propagation speed for the torsional vibrations with

$$c^2 = \frac{G\, I_{\mathrm{T}}}{\varrho\, I_{\mathrm{P}}}. \tag{30.2}$$

Taking the boundary conditions of the system $\vartheta'(0,t) = 0$ and $\vartheta'(l,t) = 0$ into account, the natural frequencies $f_{\mathrm{t}n}$ can be determined as

$$f_{\mathrm{t}n} = \frac{n\, c}{2\, l} \quad \forall n \in \mathbb{N}. \tag{30.3}$$

In general for isotropic materials, G can be expressed by the respective Young's modulus E and the Poisson's ratio ν

$$G = \frac{E}{2\,(1+\nu)}. \tag{30.4}$$

A rectangular Euler-Bernoulli beam executing lateral vibrations is sketched in Fig. 30.3. The corresponding partial differential equation describing lateral free vibrations $w(x,t)$ of an Euler-Bernoulli beam with uniform cross section is given by

$$\frac{\partial^2 w}{\partial t^2} + \frac{E I_{\mathrm{y}}}{\varrho A} \frac{\partial^4 w}{\partial x^4} = 0. \tag{30.5}$$

The equation includes the flexural rigidity with the geometrical moment of inertia I_{y} and the area A. With the free-free boundary conditions $w''(0,t) = 0$, $w''(l,t) = 0$, $w'''(0,t) = 0$ and $w'''(l,t) = 0$ the natural frequencies can be determined in the

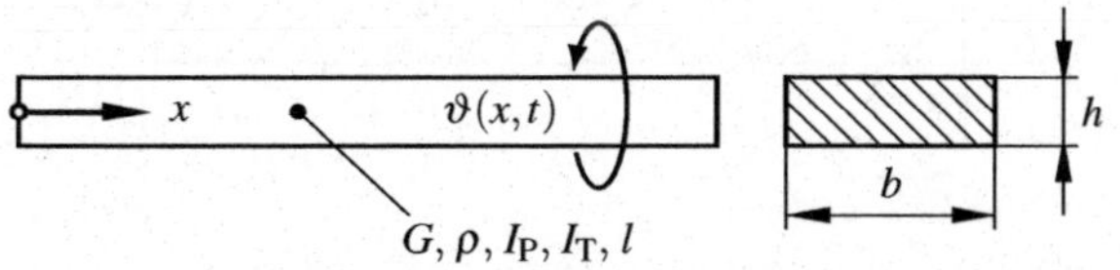

Fig. 30.2 Rectangular free-free bar

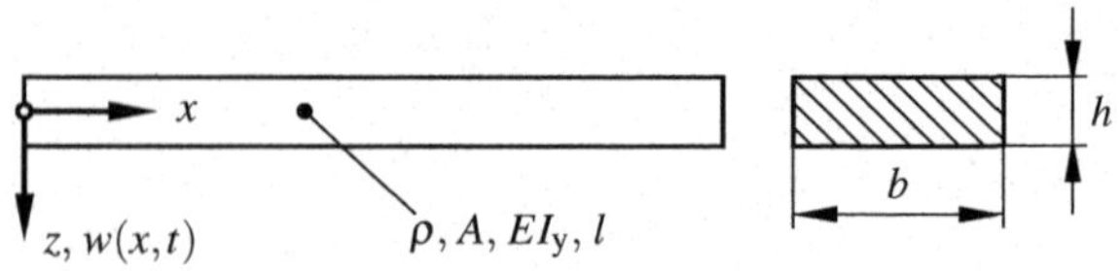

Fig. 30.3 Free-free Euler-Bernoulli beam

following form considering correction terms e_n Hagedorn and DasGupta (2007)

$$f_{bn} = \frac{1}{2\pi}\left(\frac{2n+1}{2}\pi + e_n\right)^2 \frac{1}{l^2}\sqrt{\frac{E I_y}{\varrho A}} \quad \forall n \in \mathbb{N} \tag{30.6}$$

with $e_1 = 0.01766$ and $e_2 = -0.00078$. Results determined using this approach are denoted as "analytical homogenized" (analytical hom) in Sect. 30.5.

30.2.2 Constrained Layer Damping Theory

In contrast to free layer damping, where the extension of the composite caused by length change due to bending is responsible for the damping behavior, constrained layer damping treatments, in the following denoted as CLD approach, are more complicated to describe. Shearing of the viscoelastic layer, which is the main mechanism of energy dissipation, is induced due to the deflection of both elastic layers. The investigated assembly is illustrated in Fig. 30.4. The following theory is based on the formulations of Rao Rao (1978). The damping mechanism of the elastomer layer is characterized by the complex dynamic shear modulus in Eq. (30.7), where η_2 is the core loss factor and G_2 the elastic shear modulus Leaderman (1949). Here and in the following all parameters marked with * denote complex numbers:

$$G_2^* = G_2(1 + \mathrm{i}\,\eta_2). \tag{30.7}$$

Classifying and comparing the computed results with data from literature, loss factors of metals lie between 10^{-4} and 10^{-3} approximately, whereas polymers can be expected to possess loss factors in the range of 10^{-1} to $2 \cdot 10^0$, see Oberst (1956)

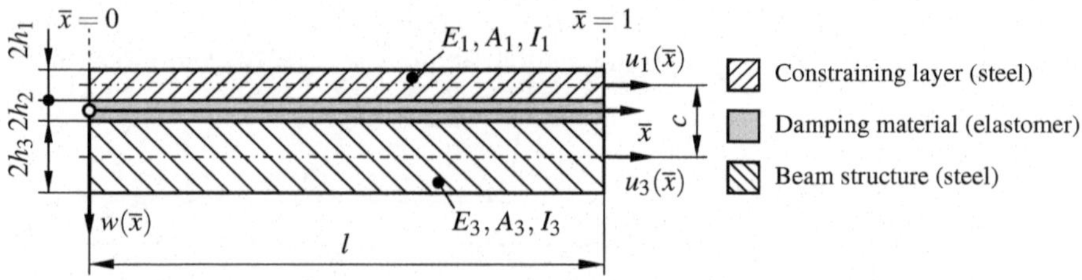

Fig. 30.4 Constrained layer damping with full coverage, according to Rao (1978)

and Beranek and Vér (1992); in some sources polymer loss factors are specified even up to 10^1 Ottl (1981). For describing flexural modes the main assumptions for the isotropic and homogeneous layers are according to Rao (1978):

- small beam deflections are considered which are determined using Euler-Bernoulli hypothesis
- the elastomer core layer is sheared which is the main energy dissipation mechanism
- longitudinal displacements of layers are continuous
- longitudinal and rotatory inertia effects are neglected

Ross, Kerwin and Ungar as well as other authors like Nokes and Nelson assumed, that mode shapes of the beam are unaffected by the damping treatment Kerwin Jr (1959) and Nokes and Nelson (1968). This fact was also confirmed experimentally in this contribution. Constrained layer damping theories consider the longitudinal displacements of the elastic layers u_1 and u_3 as well as the vertical displacement of the beam structure w. Characteristic sandwich equations can be formulated in terms of the transverse deflection $w(\overline{x}, \overline{t})$ only

$$-\frac{\partial^6 w}{\partial \overline{x}^6} + g^*(1+Y)\frac{\partial^4 w}{\partial \overline{x}^4} - \frac{\partial^4 w}{\partial \overline{t}^2 \partial \overline{x}^2} + g^* \frac{\partial^2 w}{\partial \overline{t}^2} = 0. \tag{30.8}$$

where $\overline{x}$ and $\overline{t}$ constitute normalized space and time coordinates as well as Y and g^* constitute geometric and shear parameters as

$$Y = \frac{(h_1 + h_3 + 2\,h_2)^2}{E_1\,I_1 + E_3\,I_3} \frac{E_1\,A_1\,E_3\,A_3}{E_1\,A_1 + E_3\,A_3}, \tag{30.9}$$

$$g^* = \frac{G_2^*\,A_2\,l^2}{4\,h_2^2} \frac{E_1\,A_1 + E_3\,A_3}{E_1\,A_1\,E_3\,A_3}. \tag{30.10}$$

Solving the sixth order partial differential equation the ansatz (30.11) can be used. A complex exponential ansatz for $w_m(\overline{x}, \overline{t})$ is assumed, where k_n^* are characteristic values, A_m the coefficients and Ω_m^* the complex frequency factors

$$w_m(\overline{x}, \overline{t}) = \sum_{n=1}^{6} A_m\, \mathrm{e}^{k_n^* \overline{x}}\, \mathrm{e}^{\Omega_m^* \overline{t}}. \tag{30.11}$$

Substituting (30.11) in (30.8) yields the characteristic equation

$$-{k_n^*}^6 + g^*(1+Y){k_n^*}^4 + {\Omega_m^*}^2({k_n^*}^2 - g^*) = 0. \tag{30.12}$$

Solving the polynomial (30.12) for k_n^*, three square-roots are obtained depending on the frequency factor Ω_m^*. This parameter includes the angular frequency Ω_m of the sandwich compound and the loss factor $\eta_{\text{struc},m}$ of the entire structure.

$$\Omega_m^* = \Omega_m \sqrt{1 + \mathrm{i}\,\eta_{\text{struc},\,m}} \tag{30.13}$$

The sixth order differential Eq. (30.8) for full coverage requires six boundary conditions, three for each end of the beam to determine the unknown coefficients A_1 to A_6. For a structure with unrestrained free ends there is no bending moment and shear force at the left end ($\overline{x} = 0$) and right end ($\overline{x} = 1$) of the beam. Additionally, the normal force is zero. The free-free unrestrained boundary conditions are given by

$$w_m^{\mathrm{IV}}(\overline{x}) - g^*(1+Y)w_m^{\mathrm{II}}(\overline{x}) - \Omega_m^{*2}\, w_m(\overline{x}) = 0, \tag{30.14}$$

$$w_m^{\mathrm{V}}(\overline{x}) - g^*(1+Y)w_m^{\mathrm{III}}(\overline{x}) - \Omega_m^{*2}\, w_m^{\mathrm{I}}(\overline{x}) = 0, \tag{30.15}$$

$$w_m^{\mathrm{IV}}(\overline{x}) - g^*\, Y\, w_m^{\mathrm{II}} - \Omega_m^{*2}\, w_m(\overline{x}) = 0. \tag{30.16}$$

These six equations can be transferred in a linear, homogeneous system of equations, where $\mathbf{M}_{\mathrm{A}}$ defines a square matrix and $\mathbf{a}$ is the vector of unknown coefficients A_1 to A_6

$$\mathbf{M}_{\mathrm{A}}\, \mathbf{a} = \mathbf{0}. \tag{30.17}$$

To obtain non-trivial solutions $\mathbf{a}$ the determinant of $\mathbf{M}_{\mathrm{A}}$ has to be zero. Analytical determined frequencies and loss factors are compared with experimental investigations to verify the prediction quality of this approach.

30.3 Experimental Investigations

Measuring modal parameters the test rig in Fig. 30.5 is used, considering standard methods as e.g. described in the SAE test procedure J3001 SAE-J3001 (2011). To prevent double hits during excitation, an automatic impulse hammer 1) is applied. The system response has been measured in point 2) by using a single point laser vibrometer detecting the velocity of the measurement objects in out of plane direction. To minimize the influence of the support, the test object is suspended (hung up) in a frame by strings. The investigations included shims bonded to rectangular steel

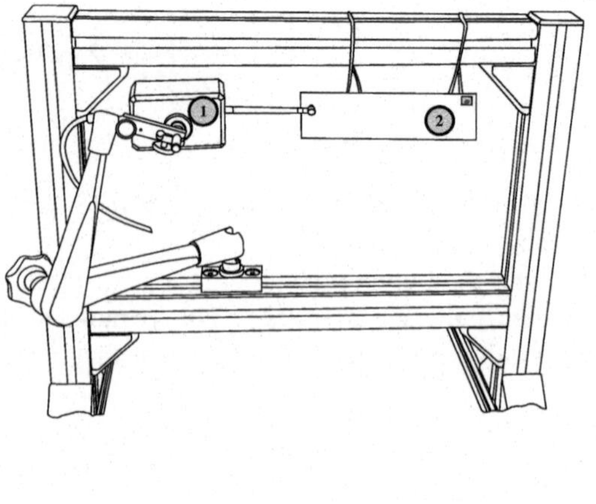

Fig. 30.5 Experimental set-up for shim investigations Schmid (2018b)

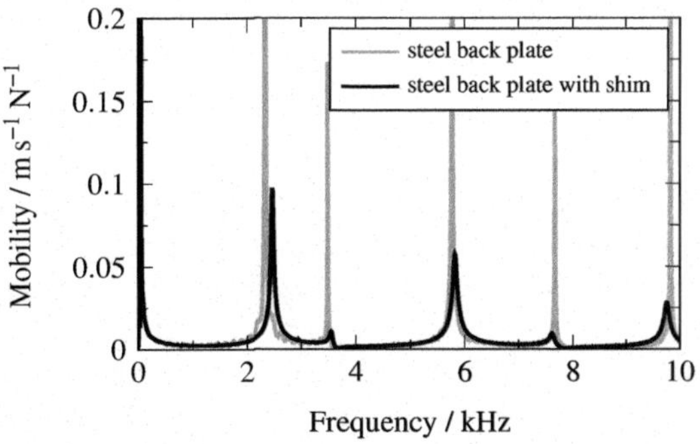

Fig. 30.6 Mobility of shims bonded to structure

plates with dimensions 180x50x5 mm^3 as well as brake pads with identical shims. Examining full coverage of the base structure shims are applied having equal length and width dimensions. A transfer function for one shim type bonded to brake pad backplates is illustrated in Fig. 30.6 in the range up to 10 kHz. These transfer functions are the basis for determining modal damping values for flexural and torsional mode shapes. The half-power bandwidth method has been applied for all damping ratios ϑ, considering a 3 dB decay logarithmically seen in equation

$$2\,\vartheta = \frac{1}{2}\left[\left(\frac{\Omega_{\mathrm{o}}}{\omega_{\mathrm{d}}} - \frac{\omega_{\mathrm{d}}}{\Omega_{\mathrm{o}}}\right) - \left(\frac{\Omega_{\mathrm{u}}}{\omega_{\mathrm{d}}} - \frac{\omega_{\mathrm{d}}}{\Omega_{\mathrm{u}}}\right)\right] \tag{30.18}$$

for nonsymmetrical transfer functions where ϑ is the modal damping ratio, ω_{d} the natural angular frequency and $\Omega_{\mathrm{o,\,u}}$ are angular frequencies determined from the transfer functions Beards (1983). The first four eigenmodes of a rectangular plate are presented in Fig. 30.7. Torsional and flexural mode shapes alternate for the first eight eigenfrequencies. Odd numbered natural frequencies represent bending modes, whereas even numbered modes correspond to torsion. Additional parameters playing an important role to obtain a more realistic mapping are the stress history and temperature impact on elastomers (Lazan, 1968), which we intend to address in future investigations. Also the rheological behavior of the viscoelastic layer may play an important role (Jones, 2001).

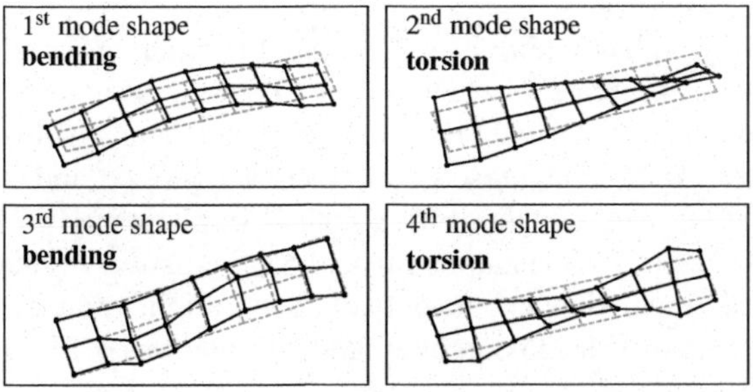

Fig. 30.7 Experimentally identified mode shapes of shims bonded to steel plates

30.4 Finite Element Approach

In the following multi- and single-layer shim structures are analyzed based on theories from literature applied to our shim problem. Corresponding FE results are denoted in Sect. 30.5 as "FE multilayer" and "FE hom" depending on the number of layers. The aim hereby is to map shims as homogenized entity to avoid modeling problems resulting from the layer thickness. As a simulation tool for modeling the shims and carrier structures, Abaqus CAE is used. All results have been produced using frequency and complex frequency steps of the implicit solver Abaqus/Standard. Specifically the Lanczos solver is used for this task Lanczos (1950).

30.4.1 Damping

Several aspects and problems of modeling damping in FE-models of brakes have been addressed in Gräbner et al (2015). Structural (30.19) as well as Rayleigh damping (30.20) are integrated for shim structures in Abaqus

$$\mathbf{M}\,\ddot{\mathbf{q}} + \mathbf{K}(1 + \mathrm{i}\beta_{\text{struc.}})\mathbf{q} = \mathbf{f}, \tag{30.19}$$

$$\mathbf{M}\,\ddot{\mathbf{q}} + (\alpha\,\mathbf{M} + \beta\,\mathbf{K})\dot{\mathbf{q}} + \mathbf{K}\,\mathbf{q} = \mathbf{f}. \tag{30.20}$$

These are linear systems of differential equations, where $\mathbf{q}$ is the displacement vector as a function of time, $\mathbf{M}$ the mass matrix, $\mathbf{K}$ the stiffness matrix and $\mathbf{f}$ the excitation vector. Instead of using a classical damping matrix for the energy dissipation, a complex stiffness matrix for structural damping is considered with the damping parameter $\beta_{\text{struc.}}$ of the material. For Rayleigh damping α and β are introduced to influence the damping behavior. This damping type is a mathematical construct weighting the impact of mass and stiffness matrices. The influence of α and β on the damping ratio ϑ can be calculated from Eq. (30.21) (Zienkiewicz, 1977), where ω_0 is the natural angular frequency of the system

$$\vartheta = \frac{\alpha + \beta\,\omega_0^2}{2\,\omega_0}. \tag{30.21}$$

Rayleigh damping is implemented for homogenized structures and the friction material separately, whereas layer bonding structures are described using structural damping for metal layers and viscoelastic layers. Figure 30.8 outlines a Rayleigh fit for experimental results which are shown as round marks. Therefore, a least squares fitting method (FindFit), which is implemented in Wolfram Mathematica is used. Using experimental modal analysis, statistical evaluations have shown, that samples of formal identical brake pads have a slight variance in natural frequencies and a serious variance in damping ratios. Deviations of almost 20% are identified regardless

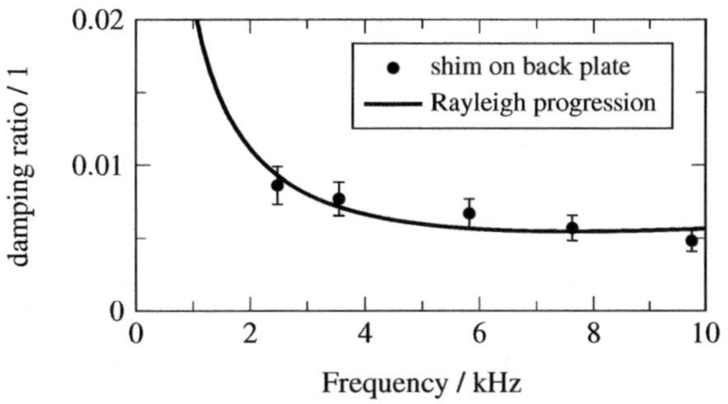

Fig. 30.8 Approximation of experimental modal damping ratios

of the mode shape. Hence, the implementation of the exact dissipation parameter is not expedient and justifies the Rayleigh approach.

30.4.2 Stiffness - Homogenization Theory

In the following the theory from Ross et al (1959) is used for the improved modeling of shims. The technique is based on summarizing elastic and viscoelastic layers up to one single layer with equivalent properties. This homogenization requires homogeneous single layers for creating a body with equal mechanical properties including density, thickness, stiffness and damping features, as shown in Fig. 30.9 schematically. Stiffness characteristics mainly come from the metal layers, in particular the base beam, whereas the structure loss factor depends on the shearing of the elastomer coating. This theory requires geometrical information about each layer as well as rheological specifications like material properties. Isotropic material behavior is required to all further computations. Therefore, the overall density and the replacement of Young's modulus of the compound is implemented for improved computations. Poisson ratio effects coming from the homogenization are neglectable for FE-simulations. The Poisson's ratio ν of the homogenization structure is assumed to be 0.3 corresponding to the elastic layers. In general the Poisson's ratio of elastomers is specified by 0.5 in literature. Avoiding numerical issues in FE calculations the Poisson's ratio of each viscoelastic layer is defined here as 0.49.

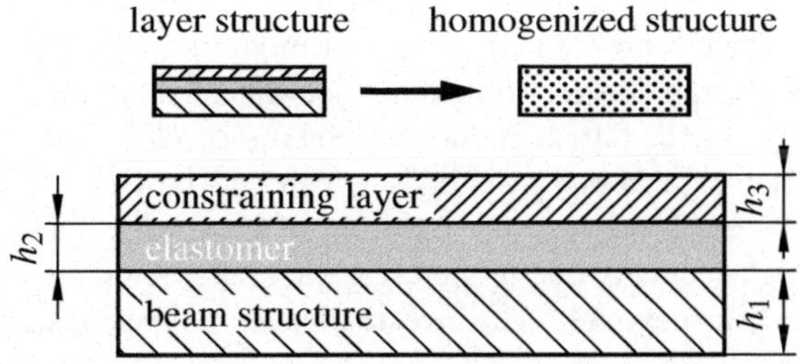

Fig. 30.9 Homogenization of a three layer compound

Shims, examined in more detail, consist of a stiff steel structure (constraining layer) and elastomer layers. Note that the homogenization includes the beam structure, the elastomer layer and the constraining layer. Adhesive layers and further top layers have to be modeled additionally. Equation (30.22) shows the flexural rigidity ratio $E\,I$ of the homogenized structure

$$\begin{aligned} E\,I = {} & E_1\frac{h_1^3}{12} + E_2\frac{h_2^3}{12} + E_3\frac{h_3^3}{12} - E_2\frac{h_2^2}{12}\left(\frac{h_{31}-D}{1+g}\right) + E_1\,h_1\,D^2 \\ & + E_2\,h_2\,(h_{21}-D)^2 + E_3\,h_3\,(h_{31}-D)^2 \\ & - \left(\frac{E_2\,h_2}{2}(h_{21}-D) + E_3\,h_3(h_{31}-D)\right)\left(\frac{h_{31}-D}{1+g}\right). \end{aligned} \tag{30.22}$$

The resulting composite with only one layer considers the thickness diversity of the steel and viscoelastic layer, where E is the Young's modulus, h the thickness of each layer shown in Fig. 30.9, I the corresponding geometrical moment of inertia, p the wave number (eigenvalue per length) and G_2 the shear modulus of the viscoelastic core as seen in Eq. (30.23) Nashif et al (1985).

$$D = \frac{E_2\,h_2\,(h_{21} - \frac{h_{31}}{2}) + g(E_2\,h_2\,h_{21} + E_3\,h_3\,h_{31})}{E_1\,h_1 + \frac{1}{2}E_2\,h_2 + g(E_1\,h_1 + E_2\,h_2 + E_3\,h_3)} \tag{30.23}$$

with the parameters:

$$h_{31} = \frac{1}{2}(h_1 + h_3) + h_2 \tag{30.24}$$

$$h_{21} = \frac{1}{2}(h_1 + h_2) \tag{30.25}$$

$$g = \frac{G_2}{E_3\,h_3\,h_2\,p^2}. \tag{30.26}$$

If there is no width difference between the single layers, the resulting Young's modulus only depends on thickness and elasticity ratios. The homogenization properties received are used for analytical and numerical models.

30.4.3 Modeling

Common problematic issues in modeling shims on brake pads, e. g. hourglass effects, may result from the selection of inappropriate element types as shown in Fig. 30.10. Often, reduced elements or elements without using hourglass control lead to zero energy modes which can be checked easily by inspecting the mode shapes. Moreover, the meshing of composite shim structures has to be done with high accuracy due to the layer thickness. A massive increase in degrees of freedom entails inevitably. The investigated shim type consists of one steel and two elastomer

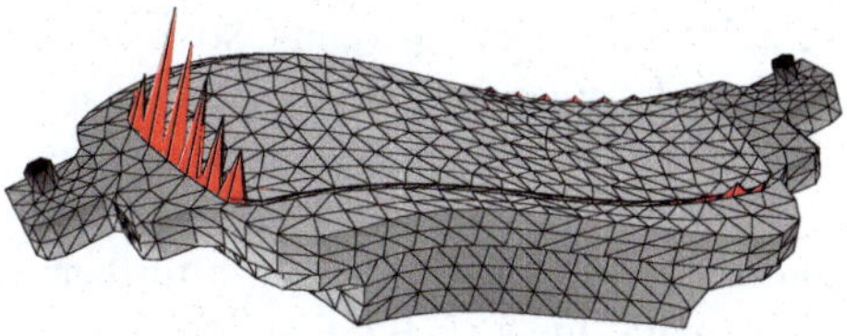

Fig. 30.10 Hourglass effect occuring at FE model with non-homogenized shim

layers one on each side, which is bonded on rectangular steel plates and backplates of brake pads. Figure 30.11 shows two test objects, the multilayer mesh compound and the homogenized structure. For layer bonding structures the components are as-

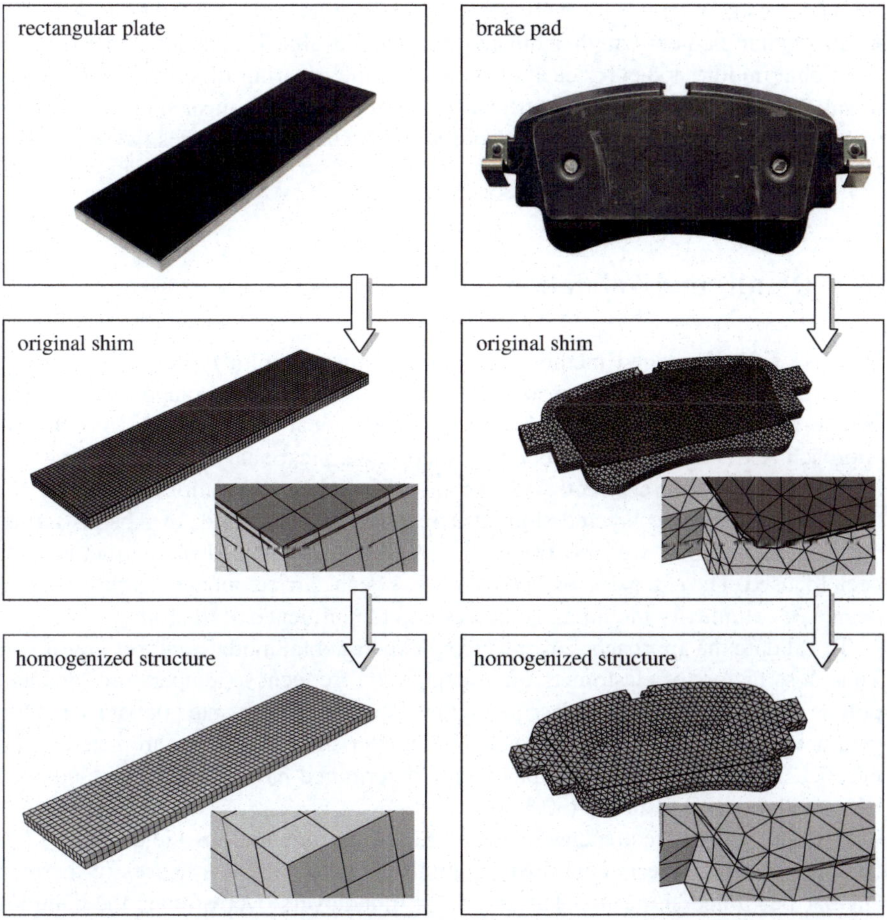

Fig. 30.11 FE modeling before and after having applied the homogenization theory to rectangular plates and brake pads

sembled using tie constraints. Solid as well as shell elements with eight nodes (S8R) are used for modeling steel plates and modified shim parts. Brake pad backplates are meshed exclusively with solid elements. When implementing the improved brake pad in a FE brake model, solid elements are recommended to be used due to several interactions of the brake pad in further steps. An examination on hexahedron elements with linear (C3D8) and quadratical (C3D20) approach has been carried out. To analyze the mesh influence a convergence analysis has been carried out with the original shim structure and after having applied the homogenization theory from Ross-Kerwin-Ungar. Thus, the element length is varied that leads to an aspect ratio of approximately 1:40 to 1:1 considering the thickness of the unmodified shim compound. The same element length has been applied for each layer. Quadratical elements show a very good convergence for computed natural frequencies even for a coarse mesh. Whereas the elements with linear approach need to be meshed with much smaller element length achieving the same results. This is due to the fact that additional middle nodes for each element map the shearing of structures more detailed. Note that higher mode shapes need to be computed with smaller aspect ratios to achieve a convergence behavior. Consequently for thin structures the quadratical approach is recommended to use.

30.5 Results and Validation

Results of the described methods for the investigated shim type in the range up to 5 kHz for a fully covered rectangular steel plate and in the range up to 10 kHz for a brake pad are shown in Table 30.1. An analytical constrained layer damping approach (CLD approach) for bending shapes, classical analytical calculations for homogenized structures (Sect. 30.2.1), Finite Element computations comparing the homogenized and the layered shim structure are demonstrated. In Abaqus implementing Rayleigh as well as structural damping in the investigated structures has been focused. The extension on brake pads has been carried out considering a higher degree of complexity including geometry and the influence of the lining.

To validate the approaches, results from experimental modal analyses are shown. Thus, loss factors of elastomers often depend on frequency (temperature) and amplitude Crandall (1970), all experimental investigations are carried out at room temperature (23^0). Natural frequencies and characteristic loss factors are listed in Table 30.1. The experimental loss factors are determined having regard to the power-bandwidth method from Eq. (30.18).

Overall, natural frequencies of Finite Element computations yield a very good compliance with experimental investigations. In particular loss factors of the rectangular beam and shim modeled as several monolayers overestimate the damping behavior from the second mode shape. Structural damping only showed good results for the backplate. In both homogenized single layer test structures the implemented Rayleigh-damping showed a very good agreement with experimental determined loss factors. Solid as well as shell elements deliver excellent results and map the

Table 30.1 Comparison of shim natural frequencies and loss factors for a free-free support

Object	Method	$f_1{}^a$ / Hz	η_1 -	$f_2{}^b$ / Hz	η_2 -	$f_3{}^a$ / Hz	η_3 -	$f_4{}^b$ / Hz	η_4 -	$f_5{}^a$ / Hz	η_5 -
steel plate	Experimental	819	0.008	1769	0.004	2244	0.004	3663	0.003	4381	0.001
steel plate	Experimental	872	0.014	1809	0.018	2341	0.019	3722	0.016	4472	0.022
with shim	FE multi layer*	878	0.014	1780	0.025	2353	0.025	3672	0.025	4494	0.027
	FE hom*	847	0.017	1788	0.014	2333	0.015	3714	0.019	4551	0.023
	FE hom**	847	0.017	1787	0.014	2332	0.015	3711	0.019	4548	0.022
	CLD approach	881	0.009	-	-	2367	0.019	-	-	4525	0.021
	Analytical hom	848	-	1763	-	2347	-	3526	-	4595	-
backplate	Experimental	2348	0.003	3483	0.002	5766	0.001	7664	0.001	9820	0.001
backplate	Experimental	2475	0.017	3548	0.015	5817	0.013	7622	0.011	9755	0.010
with shim	FE multi layer*	2447	0.018	3508	0.017	5810	0.016	7638	0.013	9726	0.012
	FE hom*	2416	0.019	3543	0.014	5838	0.011	7698	0.011	9780	0.011
brake pad	Experimental	4045	0.024	5334	0.029	7373	0.027	9900	0.024		
with shim	FE hom*	4362	0.023	5731	0.025	7409	0.024	10061	0.031		

^{a} bending mode, ^{b} torsional mode, * solid elements, ** shell elements

progression of the damping behavior correctly. Note that the first natural frequency in both cases is below the experimental determined frequency.

Regarding a complete brake pad the friction material behaves like a typical transversely isotropic material. Out-of-plane is the preferred direction of the lining characterized by less stiffness. The stiffness of the lining increases with piston pressure applied in normal direction. For Finite Element models engineering constants are used in Abaqus for describing the behavior for a certain pressure stage. Therefore experimental identified natural frequencies are lower than numerical computed ones listed here. Detailed information on this topic can be found e. g. in Hornig (2015). Beside this, the application of the homogenized shim structure enables to reduce the number of degrees of freedom. Figure 30.12 depicts the modeling advantage, whereby all components compared are meshed with the same element length. Two test objects, a rectangular plate with dimensions of 180x50x5 mm^3 fully covered with a shim and a backplate with the same shim type are listed. In particular quadratical solid elements (C3D20) and shell elements (S8R) are used for the considered structures. The number of elements needed for convergence is much less for homogenized models, which reduces the computation time drastically. The reduction of degrees of freedom is intended for large FE brake models often built with several million degrees of freedom. Achieving similar results using the homogenization theory, shims are highly recommended to be modeled as one layer over the thickness. Furthermore elements with full integration and second order approach are recommended preventing hourglass effects and shear locking as described in Flanagan and Belytschko (1981) and Bathe (1996). The analytical constrained layer damping approach provides a very good forecast quality for bending shapes. Solely the loss factor of the first bending mode deviates from the actual damping. Simple continuum mechanical approaches reflect only the stiffness characteristics of structures

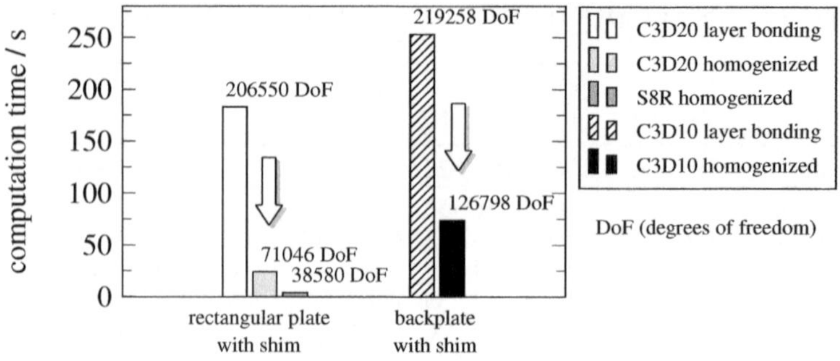

Fig. 30.12 Reduction of computation time due to improved single layer modeling

presented in Sect. 30.2.1. The results for torsional vibrations point out, that there has been simplified modeling for this application at increasing frequencies.

To sum up it can be said, that the results of the applied analytical and modified FE single layer approach conduct in very good compliance with the experimental investigations.

30.6 Conclusion and Outlook

The analytical constrained layer damping approach allows the prediction of natural frequencies and loss factors of bending mode shapes without considering relevant damping ratios carried out in prior investigations. Experimental evaluations of formal identical brake pads have shown, that there's a not neglectable deviation in the damping ability. Therefore the exact mapping of damping characteristics in Finite Element models is not feasible and favours the Rayleigh approach clearly. By modeling each layer separately the meshing becomes a challenging task due to the layer thickness. Therefore, the homogenization approach is the preferred modeling technique. A significant reduction of computation time for homogenized shim structures arose from a much lower number of degrees of freedom.

Primarily the advantage for future works is to reduce experimental investigations in a greater scope, improve the prediction quality of potential squeal frequencies and make the development of quiet brakes more efficient. The implementation of these homogenized shim structures in a FE complete model of the brake will be conducted in further steps. Furthemore the impact of varying temperature on shim loss factors has been depicted in prior examinations and can now be implemented in Finite Element calculations (Schmid et al, 2017).

Acknowledgements "This work is funded by the Deutsche Forschungsgemeinschaft (DFG, German Research Foundation) – WA 1427/27-1 within the PP 1897 "*Calm Smooth and Smart - Novel Approaches for Influencing Vibrations by Means of Deliberately Introduced Dissipation*". We would also like to thank Audi AG and Wolverine Advanced Materials for their support.

References

Bathe K (1996) Finite element procedures. Prentice-Hall Inc.

Beards C (1983) Structural vibration analysis. John Wiley and Sons New York

Beranek L, Vér I (1992) Noise and vibration control engineering: principles and applications. John Wiley & Sons, Inc.

Cantoni C, Cesarini R, Mastinu G, Rocca G, Sicigliano R (2009) Brake comfort - a review. Vehicle System Dynamics 47(8):901–947

Crandall S (1970) The role of damping in vibration theory. Journal of Sound and Vibration 11(1):3–18

DiTaranto R (1965) Theory of vibratory bending for elastic and viscoelastic layered finite-length beams. Journal of Applied Mechanics 32(4):881–886

Esgandari M, Olatunbosun O (2016) Computer aided engineering prediction of brake noise: modeling of brake shims. Journal of Vibration and Control 22(10):2347–2355

Ewins D (1984) Modal testing: theory and practice. Research studies press Letchworth

Festjens H, Chevallier G, Renaud F, Dion J, Lemaire R (2012) Effectiveness of multilayer viscoelastic insulators to prevent occurrences of brake squeal: A numerical study. Journal of Applied Acoustics 73(11):1121–1128

Flanagan D, Belytschko T (1981) A uniform strain hexahedron and quadrilateral with orthogonal hourglass control. International Journal for Numerical Methods in Engineering 17(5):679–706

Flint J (2002) Disc brake squeal. PhD thesis, University of Southern Denmark

Flint J, McDaniel J, Li X, Elvenkemper A, Wang A, Chen SE (2004) Measurement and simulation of the complex shear modulus of insulators. Tech. rep., SAE Technical Paper 2004-01-2799

Gräbner N, Tiedemann M, von Wagner U, Hoffmann N (2014) Nonlinearities in friction brake NVH - experimental and numerical studies (no. 2014-01-2511). SAE Technical Paper

Gräbner N, Gödecker H, von Wagner U (2015) On the influence of damping on brake vibrations. In: International Conference on Engineering Vibration

Gräbner N, Mehrmann V, Quraishi S, Schröder C, von Wagner U (2016) Numerical methods for parametric model reduction in the simulation of disk brake squeal. ZAMM-Journal of Applied Mathematics and Mechanics 96(12):1388–1405

Hagedorn P, DasGupta A (2007) Vibrations and waves in continuous mechanical systems. John Wiley & Sons

Hornig S (2015) Development of measurement and identification methods for automotive brake lining material properties in nvh relevant loading parameter range. PhD thesis, Technische Universität Berlin

Jones D (2001) Handbook of viscoelastic vibration damping. John Wiley & Sons

Kang J (2012) Finite element modelling for the investigation of in-plane modes and damping shims in disc brake squeal. Journal of Sound and Vibration 331(9):2190–2202

Kerwin Jr E (1959) Damping of flexural waves by a constrained viscoelastic layer. The Journal of the Acoustical Society of America 31(7):952–962

Kinkaid N, O'Reilly O, Papadopoulos P (2003) Automotive disc brake squeal. Journal of Sound and Vibration 267(1):105–166

Lall A, Asnani N, Nakra B (1988) Damping analysis of partially covered sandwich beams. Journal of Sound and Vibration 123(2):247–259

Lanczos C (1950) An iteration method for the solution of the eigenvalue problem of linear differential and integral operators. United States Governm. Press Office Los Angeles, CA

Lazan B (1968) Damping of materials and members in structural mechanics. Pergamon press Oxford

Leaderman H (1949) Proposed nomenclature for elastic and inelastic behavior of high polymers. Journal of Colloid Science 4(3):193–210

Marcelin JL, Shakhesi S, Pourroy F (1995) Optimal constrained layer damping of beams: experimental and numerical studies. Shock and Vibration 2(6):445–450

Markuš v (1974) Damping mechanism of beams partially covered by constrained viscoelastic layer. Acta Technica ČSAV 2:179–194

Mead D, Markuš v (1969) The forced vibration of a three-layer, damped sandwich beam with arbitrary boundary conditions. Journal of Sound and Vibration 10(2):163–175

Mead D, Markuš v (1970) Loss factors and resonant frequencies of encastré damped sandwich beams. Journal of Sound and Vibration 12(1):99–112

Nashif A, Jones D, Henderson J (1985) Vibration damping. John Wiley & Sons

Nokes D, Nelson F (1968) Constrained layer damping with partial coverage. Shock and Vibration Bulletin 38:5–10

Oberst H (1956) Werkstoffe mit extrem hoher innerer Dämpfung. Acta Acustica united with Acustica 6(1):144–153

Oberst H, Frankenfeld K (1952) Über die Dämpfung der Biegeschwingungen dünner Bleche durch fest haftende Beläge. Acta Acustica united with Acustica 2(4):181–194

Oberst H, Becker G, Frankenfeld K (1954) Über die Dämpfung der Biegeschwingungen dünner Bleche durch fest haftende Beläge II. Acta Acustica united with Acustica 4(1):433–444

Ottl D (1981) Schwingungen mechanischer Systeme mit Strukturdämpfung. VDI-Verlag

Rao D (1978) Frequency and loss factors of sandwich beams under various boundary conditions. Journal of Mechanical Engineering Science 20(5):271–282

Ross D, Ungar E, Kerwin Jr E (1959) Damping of plate flexural vibrations by means of viscoelastic laminae. Structural damping pp 49–97

SAE-J3001 (2011) Brake insulator damping measurement procedure. URL https://doi.org/10.4271/J3001_201102

Schmid D (2018a) Brake pad. "figshare.com/articles/Brake_pad/7046663", license CC BY 4.0, doi:110.6084/m9.figshare.7046663.v1

Schmid D (2018b) Shim test rig. "http://figshare.com/articles/Shim_test_rig/7046909", DOI "10.6084/m9.figshare.7046909.v1", license CC BY 4.0, doi:10.6084/m9.figshare.7046909.v1

Schmid D, Gräbner N, von Wagner U (2017) Experimental investigations of brake pad shim properties. PAMM 17(1):41–44

Sun C, Lu Y (1995) Vibration damping of structural elements, vol 1. Prentice Hall Englewood Cliffs, NJ

Sylwan O (1978) Calculation of partially covering damping layers of sandwich structures with some practical results. In: Proceedings from Inter-Noise-78, Institute of Noise Control Engineering, San Francisco, pp 219–224

Zienkiewicz O (1977) The finite element method, 3rd edn. McGraw-Hill London

Chapter 31
Teaching Mechanics: Inequalities in Statically Indeterminate Static Friction Problems

Patrick Schneider & Reinhold Kienzler

Abstract In this contribution,[1] we provide a mathematical treatment of the inequality systems arising from two toy examples of statically indeterminate, static friction problems with multiple contact planes. Both examples illustrate general phenomena that are not covered by introductory textbooks. The first one illustrates that a certain inequality might restrict the solution set for a certain choice of a parameter, while being not restrictive for another choice of the parameter. The example also illustrates why it is not sufficient to investigate only the limit case of impending motion $(H = \mu_0 N)$ for more complex problems.
The second example illustrates that the positiveness of contact forces can also be a limiting condition. It treats a problem where equilibrium might be lost by slipping or by losing contact, where it is a-priori unclear which condition is decisive.
Finally, we investigate the intriguing question how this knowledge can be exploited by a purely hypothetical evil professor—not to upset anyone, we give him a common German surname as a working name, say professor Müller—to set up test problems that are solved wrongly by the majority of students.

Keywords: Engineering didactics · Engineering mechanics · Statically indeterminate · Static friction · Contact · Inequality

---

Patrick Schneider
Institute for Lightweight Construction and Design (KLuB), Department of Mechanical Engineering, Technische Universität Darmstadt, Otto-Berndt-Straße 2, D-64287 Darmstadt, Germany,
e-mail: patrick.schneider@klub.tu-darmstadt.de

Reinhold Kienzler
Bremen Institute for Mechanical Engineering (bime), Department of Production Engineering, University of Bremen, Am Biologischen Garten 2, D-28359 Bremen, Germany,
e-mail: rkienzler@uni-bremen.de

[1] On the occasion of Wolfgang H. Müller's 60th birthday

B. E. Abali et al. (eds.), *New Achievements in Continuum Mechanics and Thermodynamics*, Advanced Structured Materials 108,
https://doi.org/10.1007/978-3-030-13307-8_31

31.1 Introduction

Although a variety of German textbooks treating the basic concepts of statics in Engineering Mechanics is available, we refer only to Gross et al (2006) in the following, which is used at several German universities. The arguments set forth, however, apply equally well to other textbooks written in either German or English language. Being quite compendious, the book teaches some concepts only at a few, easy examples. Static Friction is treated in a particular short section. A single example for a statically indeterminate static friction problem is provided (Example 9.2 b)), which is also the only example of a problem with more than one contact plane. Here only a graphical solution is provided with the comment that a mathematical solution is "not quite simple" because of the inequalities.

We do not question the merits or general quality of Gross et al (2006), we rather keep the style of this textbook in the following with the intention that our contribution might serve as an extension. Thus the aim of this paper is to provide two more toy examples for statically indeterminate static friction problems, both involving two contact planes, featuring a mathematical treatment of the inequality system, in order to enable students to deal with these kinds of problems in a general manner. Both toy examples are kept minimalistic enough to be easy to comprehend, while illustrating important phenomena that are not treated by Gross et al (2006), or any other basic course textbook known to the authors.

31.2 Aim of the First Example

The first given example basically extends example 9.1 from Gross et al (2006) to a second plane of contact, while keeping the kinematics as simple as in 9.1, by the use of a kinematic coupling.

The paragraph before example 9.1 in Gross et al (2006) closes with the remark that in the case of a system involving inequalities due to static friction ($|H| \leq \mu_0 N$) it is "often easier" to investigate only the equation system given by the limit case of impending motion ($H = \mu_0 N$). This advise can be misleading, since the information whether a bound acts as upper or lower bound is essential and lost following the advice. Furthermore, due to the nature of an inequality, a formal bound of the limit case equation system might not at all be restrictive to the solution set. Also it is unclear what happens if the bounds intersect, perhaps because of a singularity. All these phenomena happen in the example problem below, although the problem is easy to treat, if the inequalities are investigated in an adequate manner.

31.3 First Example

Let us consider the toy example of two bricks on top of each other (mass $m_1 > 0$ and $m_2 > 0$) on an inclined ramp (angle $0 < \alpha < \pi/2$) which are connected by a rope that runs over a pulley, as depicted in Fig. 31.1. The rope is assumed to be rigid and massless and the pulley is assumed to be frictionless. The coefficient of static friction between the ramp and the second brick, as well as between the bricks, shall be denoted by μ_0. The task is to determine whether or not the system is in static equilibrium.

Obviously the two bricks will not both slide upwards the ramp and cannot slide both downwards the ramp due to the rope. One of two possible motions is that the upper brick (I) slides downwards while the lower brick (II) slides upwards. To prevent this motion the static friction forces $H_1 > 0$ and $H_2 > 0$ act as depicted in the free-body diagram in Fig. 31.2, where $G_1 = m_1 g$ and $G_2 = m_2 g$.

The equilibrium of forces reads:

$$I: \nearrow: S + H_1 - G_1 \sin(\alpha) = 0, \tag{31.1}$$

$$\nwarrow: N_1 - G_1 \cos(\alpha) = 0, \tag{31.2}$$

$$II: \nearrow: S - H_1 - H_2 - G_2 \sin(\alpha) = 0, \tag{31.3}$$

$$\nwarrow: N_2 - N_1 - G_2 \cos(\alpha) = 0. \tag{31.4}$$

Solving equations (31.2) and (31.4) for the contact forces, we obtain

$$N_1 = G_1 \cos(\alpha), \tag{31.5}$$

$$N_2 = (G_1 + G_2) \cos(\alpha) \tag{31.6}$$

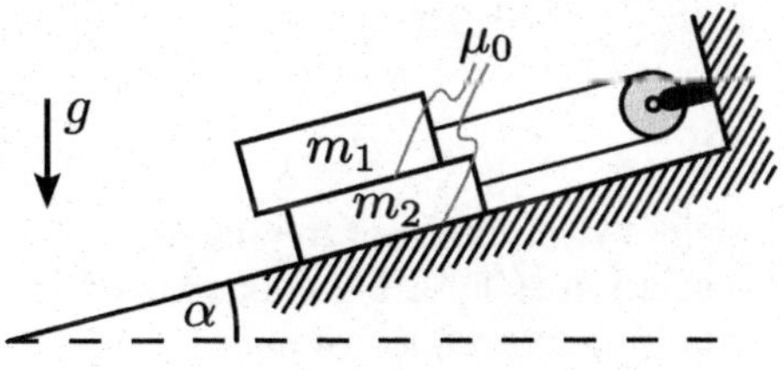

Fig. 31.1 Toy example with two planes of static friction.

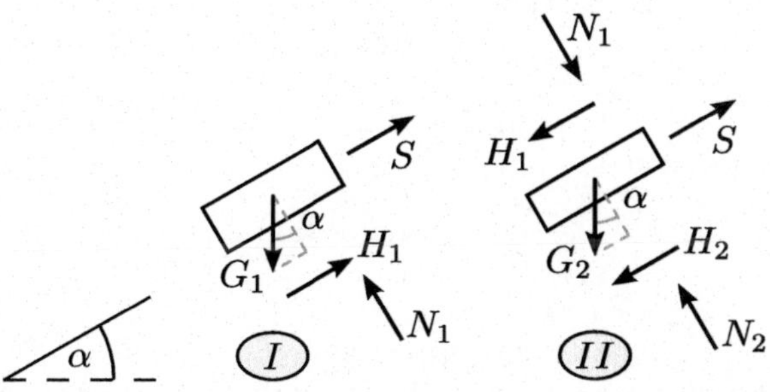

Fig. 31.2 Free-body diagram for prevented downwards motion of the upper brick (I) and upwards motion of the lower brick (II).

and eliminating the rope-normal force S from the equations (31.1) and (31.3) we get

$$2H_1 + H_2 = (G_1 - G_2)\sin(\alpha). \tag{31.7}$$

The inequalities of static friction (Coulomb's law)

$$|H_1| = H_1 \leq \mu_0 N_1,$$
$$|H_2| = H_2 \leq \mu_0 N_2,$$

have to be satisfied as well. Insertion of the inequalities into equation (31.7) and elimination of the contact forces by using equations (31.5) and (31.6) yields a single inequality that assures the prevention of the predicted motion

$$G_1\,(\tan(\alpha) - 3\mu_0) \leq G_2\,(\tan(\alpha) + \mu_0).$$

(Note that we divided the equation by $\cos(\alpha) > 0$.) While the right hand side is positive, the left hand side can be negative, if the round bracket is negative. In this case, the inequality is always satisfied. Only if the round bracket is positive, we obtain an upper bound for the mass ratio that has to be satisfied in order to prevent the predicted motion

$$\frac{G_1}{G_2} = \frac{m_1}{m_2} \leq \frac{\tan(\alpha) + \mu_0}{\tan(\alpha) - 3\mu_0}, \qquad \text{if } \tan(\alpha) > 3\mu_0. \tag{31.8}$$

The other possible motion is that brick (I) slides upwards while brick (II) slides downwards. To prevent this motion, the static friction forces have to act in the opposite direction, i.e. we have $H_1 < 0$ and $H_2 < 0$. In this case the inequalities of static friction are

$$-H_1 \leq \mu_0 N_1,$$
$$-H_2 \leq \mu_0 N_2.$$

Again, the insertion of the inequalities into equation (31.7) and elimination of the contact forces by using equations (31.5) and (31.6) yields a single inequality that assures the prevention of motion

$$G_1\,(\tan(\alpha) + 3\mu_0) \geq G_2\,(\tan(\alpha) - \mu_0).$$

It is always satisfied if $\tan(\alpha) \leq \mu_0$, otherwise we obtain a lower bound for the mass ratio

$$\frac{G_1}{G_2} = \frac{m_1}{m_2} \geq \frac{\tan(\alpha) - \mu_0}{\tan(\alpha) + 3\mu_0}, \qquad \text{if } \tan(\alpha) > \mu_0. \tag{31.9}$$

The mass ratio has to satisfy the bounds (31.9) and (31.8) simultaneously in order for the system to be in equilibrium. If one of the two bounds is violated the system will slide in the corresponding direction. The situation is depicted in Fig. 31.3 for different values of μ_0. In the leftmost graph the direction of movement is depicted that will take place, if the systems mass ratio is in the light or dark gray area. The area of equilibrium is hatched. Like indicated in the second graph, there is always a range ($0 < \alpha \leq \tan^{-1}(\mu_0)$) for the angle, where we have equilibrium independent of the mass ratio. This is known as the cone of friction. In the range $\tan^{-1}(\mu_0) < \alpha \leq \tan^{-1}(3\mu_0)$, we have only a lower bound - the upper brick has to be heavy enough to prevent downwards motion of the lower brick. For $\alpha > \tan^{-1}(3\mu_0)$ we have a range of admissible mass ratios that narrows towards the value 1 for $\alpha \to \pi/2$. The threshold $\alpha = \tan^{-1}(3\mu_0)$ is depicted by the vertical dashed line - it is the singular value of the upper bound. For $\alpha = \pi/2$ we can only have equilibrium for equal masses, independent of the coefficient of static friction, and on the other hand, we always have equilibrium for equal masses independent

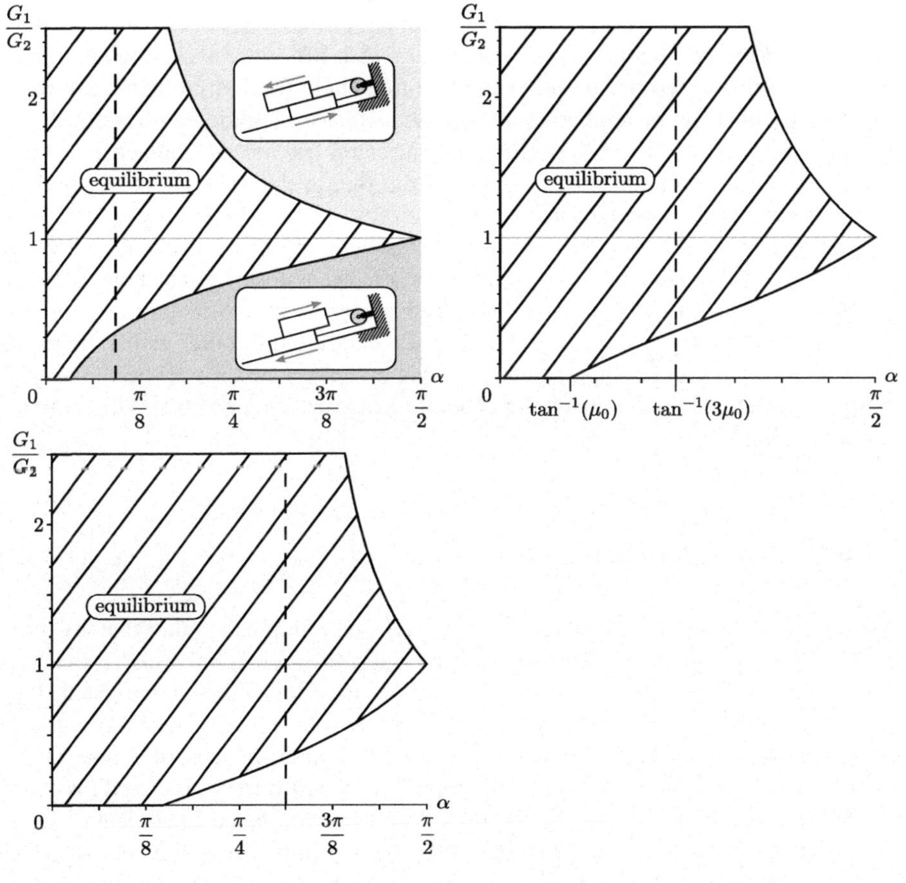

Fig. 31.3 The systems mass ratios for static equilibrium for $\mu_0 = \frac{1}{10}, \frac{3}{10}, \frac{5}{10}$ (left, right, bottom).

of the coefficient of static friction or the angle (horizontal gray line). Like we see by the three graphs, the corridor of equilibrium narrows if μ_0 gets smaller. For the limit case of frictionless materials equal masses are the only possibility to achieve equilibrium.

31.4 Aim of the Second Example

Like in Gross et al (2006), all basic course books teach students that contact forces N act normal to the contact plane, that they are headed towards the contact partners and have to be included accordingly in free-body diagrams. However, the condition that we have contact is also an inequality, i.e., $N \geq 0$, that needs to be satisfied.

In typical dry friction toy problems, it is mostly pretty obvious that no lift-off will occur. Anyway, it is a good advice to give to students to check at least a-posteriori, if the contact forces are positive once an example problem has been solved. If they are not (and the calculation is right), an unexpected lift of the system would occur and the solution is wrong. As an example, often static friction problems ask for a limit load F that can be applied to the system so that it is still in equilibrium. If for the limit load derived from the inequalities of static friction the contact forces are negative, the limit load has to be derived from the, in this case more demanding, condition of positive contact forces instead.

The objective of the second example is to provide the solution to a toy problem where it is a-priori unclear, if the equilibrium fails due to losing contact or sliding of the system. In a second step, we furthermore investigate under which parameter sets the lift-off condition gets decisive over the slip condition. In other words, how can an evil professor set up the parameters, so that an unexpected lift-off occurs and all of his students that do not check the contact forces for being positive will compute the wrong limit load.

31.5 Second Example

We consider the modified example of two bricks on top of each other that are connected by a rope. A positive force $F > 0$ attacks at the second brick under an angle $0 < \alpha < \pi/2$. The first brick will turn out to have a greater mass than the second one and is therefore depicted bigger in Fig 31.4, however we do not make any a-priori assumptions with respect to the bricks masses at this point, besides, of course, being positive ($m_1 > 0$, $m_2 > 0$). Again, the ropes are assumed to be rigid and massless and the pulleys are assumed to be frictionless. We have the same coefficient of static friction $\mu_0 > 0$ between the ground and the second brick, as well as between the bricks. First, we investigate once again the conditions under which the system is in a state of static equilibrium.

Fig. 31.4 Second toy example.

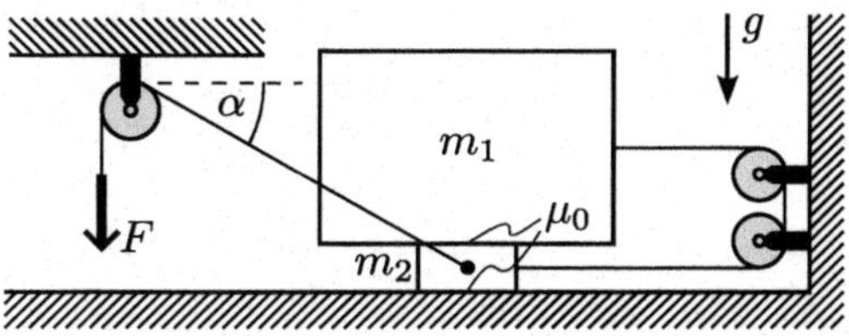

The only possible sliding motion of the system is given by the lower brick (II) sliding leftwards and the upper brick (I) sliding rightwards. In order to prevent the motion, the static friction forces $H_1 > 0$ and $H_2 > 0$ have to act as depicted in the free-body diagram in Fig. 31.5, where $G_1 = m_1 g$ and $G_2 = m_2 g$.

The equilibrium of forces reads:

$$I : \rightarrow: S - H_1 = 0, \tag{31.10}$$

$$\uparrow: N_1 - G_1 = 0, \tag{31.11}$$

$$II : \rightarrow: S + H_1 + H_2 - F\cos(\alpha) = 0, \tag{31.12}$$

$$\uparrow: N_2 - N_1 - G_2 + F\sin(\alpha) = 0. \tag{31.13}$$

From equation (31.11) we directly obtain a positive contact force

$$N_1 = G_1 > 0 \tag{31.14}$$

but the second contact force, derived from equation (31.13)

$$N_2 = G_1 + G_2 - F\sin(\alpha) \tag{31.15}$$

is only non-negative, if the vertical component of F is smaller or equal the combined weight of the bricks

$$G_1 + G_2 \geq F\sin(\alpha), \tag{31.16}$$

which is the first inequality that has to be fulfilled so that the system is in equilibrium. The second one is again obtained by eliminating S from the equations (31.10)

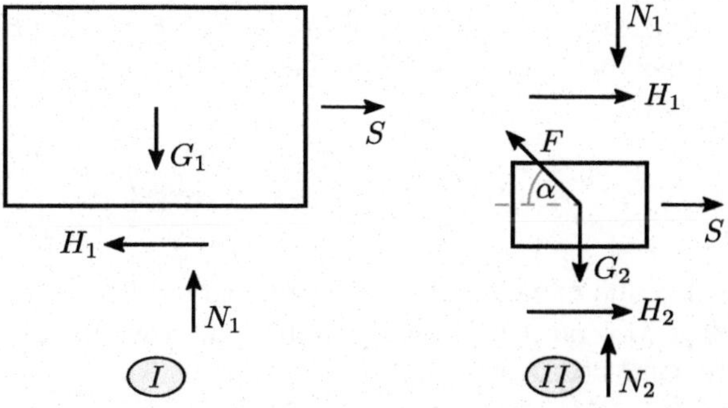

Fig. 31.5 Free-body diagram for prevented rightwards motion of the upper brick (I) and leftwards motion of the second, lower brick (II).

and (31.12)

$$2H_1 + H_2 = F\cos(\alpha)$$

and insertion of the inequalities of static friction

$$|H_1| = H_1 \leq \mu_0 N_1,$$
$$|H_2| = H_2 \leq \mu_0 N_2,$$

and subsequent insertion of the equations for the contact forces (31.14) and (31.15), eventually leading to the inequality

$$3G_1 + G_2 \geq F\left(\frac{1}{\mu_0}\cos(\alpha) + \sin(\alpha)\right). \tag{31.17}$$

If inequality (31.16) is harmed, the system will lift off, whereas, violation of inequality (31.17) results in the predicted sliding motion. More elegantly, equilibrium can be characterized by simultaneous fulfillment of the inequality system

$$\frac{G_1 + G_2}{F} \geq \sin(\alpha), \tag{31.18}$$
$$\frac{G_1}{F} \geq \frac{1}{2\mu_0}\cos(\alpha). \tag{31.19}$$

However, this representation does not answer the question when the inequality for contact becomes decisive over the inequality for prevented sliding - How has an evil professor to set up the parameters, so that not checking the positivity of the normal forces will lead to an erroneous result? In the evil case the negation of inequality (31.16) holds true simultaneously to the original inequality (31.17), so that we have to find a positive mass ration G_1/G_2 that fulfills

$$\frac{1}{3}\frac{F}{G_2}\left(\frac{1}{\mu_0}\cos(\alpha) + \sin(\alpha)\right) - \frac{1}{3} \leq \frac{G_1}{G_2} < \frac{F}{G_2}\sin(\alpha) - 1. \tag{31.20}$$

Actually, in many cases it is impossible to fulfill these inequalities. Note that upper and lower bound are linear functions in F/G_2 and that for $F/G_2 = 0$ the upper(!) bound starts at -1, and therefore lower than the lower bound, which starts at $-1/3$. So a necessary condition for lifting becoming decisive is that the inclination of the upper bound is greater than the inclination of the lower bound what happens only if

$$\alpha > \tan^{-1}\left(\frac{1}{2\mu_0}\right). \tag{31.21}$$

(Here the monotonicity of the tangent has been used). The situation is sketched in Fig. 31.6 on the left. Only in the light gray regime the inequality for contact can become decisive.

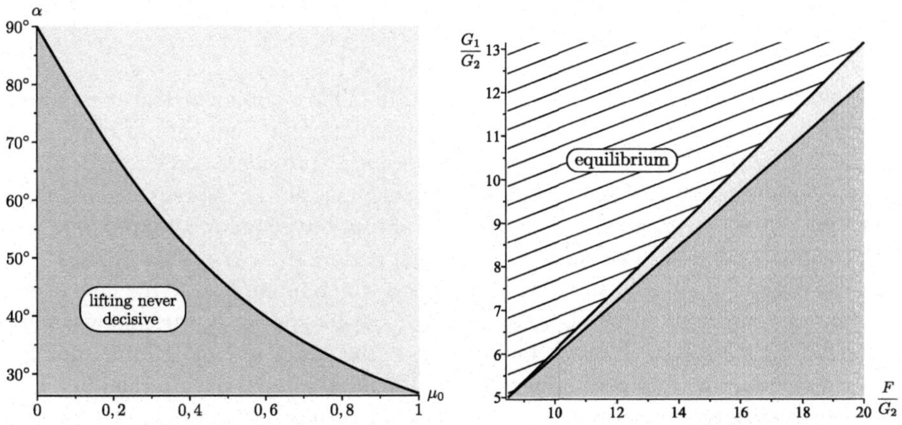

Fig. 31.6 On the left: Only in the light gray area lifting can be decisive due to (31.21). On the right: Bounds of (31.20) for $\alpha = \pi/4$ and $\mu_0 = 0,6$.

Having selected appropriate values of α and μ the parameter F/G_2 has to be selected greater or equal the intersection value of the bounds, which means

$$\frac{F}{G_2} \geq \frac{1}{\sin(\alpha) - \frac{1}{2\mu_0}\cos(\alpha)}. \tag{31.22}$$

The corresponding mass ratio at the intersection point is

$$\frac{G_1}{G_2} = \frac{1}{2\mu_0\tan(\alpha) - 1}, \tag{31.23}$$

which is always positive due to inequality (31.21). Therefore, we finally have a nonempty range for the selection of a positive G_1/G_2 by (31.20).

Note that the mass ratio (31.23) gets singular for the limit angle of inequality (31.21). This means that although it is possible to have decisiveness of the contact inequality in the high angle regime for low mass ratios (which is not surprising), it requires in general a high mass ratio near the boundary line of the left graph of figure 31.6, i.e., in the area where the lift-off is not necessarily expected.

As an example, let us pick a point not that close to the boundary line, say $\alpha = \pi/4$ and $\mu_0 = 0,6$. Then from equations (31.22) and (31.23), we find the intersection point to be ($F/G_2 = \sqrt{2}/12 \approx 8,49\,; G_1/G_2 = 5$). The bounds of (31.20) are depicted in figure 31.6 on the right. If we are above the upper bound of (31.20), we are in the equilibrium regime. In the light gray regime, we have loss of equilibrium due to a negative N_2 (lift-off), while the inequality for prevented sliding is still fulfilled. So the desired evil problem would be for example, to set the parameters to $\alpha = \pi/4$, $\mu_0 = 0,6$, $G_1/G_2 = 10$ and ask for the maximum F so that the system is still in static equilibrium.

31.6 Discussion

Usually, textbooks on statics, e.g., Gross et al (2006), comment that it is "often easier" to attack static friction problems by examining the limit case of impending motion $H = \mu_0 N$. The first example of this paper provides a case where the a-posteriori interpretation of the functional dependencies of the bounds derived from the limit equation system is at least difficult. From the equation system alone it is impossible. Note that, for example, in the first regime ($0 < \alpha \leq \tan^{-1}(\mu_0)$) both bounds for the positive mass ratio are negative, which might lead to the conclusion that no equilibrium can be achieved, however, the opposite is true, there are no bounds for the mass ratio at all. Going over to the easier to handle limit equation system of impending motion comes along with an actual loss of information! For a proper treatment it is necessary to investigate the actual system of inequalities like done in this paper.

The second example handles a problem where it is a-priori unclear if the equilibrium fails due to losing contact or sliding of the system. A case not discussed by the standard textbooks. We investigated under which conditions for the parameters the lift-off can actually happen and give a hint how this knowledge can be exploited by evil professors to set up test problems which will be solved wrongly by the majority of students.

Finally there is only one more thing left to say: "Happy birthday Professor Müller!"

References

Gross D, Hauger W, Schnell W, Schröder J, Wall W (2006) Technische Mechanik, Band 1: Statik, 9th edn. Springer-Verlag, Berlin, Heidelberg

Chapter 32
Initial Damage of Composite Materials

Vladimir S. Shorkin, Victoria Yu. Presnetsova, Vadim M. Presniakov, Sergey N. Romashin, Larisa Yu. Frolenkova & Svetlana I. Yakushina

Abstract In composite materials, damage occurs even without excessive loading, because of the discrepancy between material parameters of adhered materials. This damage (damage of nonconformity) is expressed as a violation of the continuity at the interface. A mathematical model of the interaction of inclusion particles and the composite matrix, based on taking into account nonlocal interactions of contacting materials particles is proposed. The quantifying parameter of damage is determined from the condition of stationarity of the energy of the elastic deformations of the discrepancy. The distribution of the nonconformity damage found in this way is proposed to be used to determine the initial distribution of the scattered damage of a homogeneous material that models composite material. Such scattered damage can be used to solve evolution equations in problems of fracture mechanics.

Keywords: Composite · Defects of discrepancy · Damage · Adhesion · Kinetic equation · Nonlocal multiparticle interaction · Second-order material

32.1 Introduction

Composite materials, especially nanostructured composites are widely used in electronics, medicine, various technical applications, for example in the aerospace craft (Valiev et al, 2007). When investigating the strength of these materials, models based on continuum mechanics of fracture are used. To evaluate the strength, an approach based on the application of damage parameter introduced by Kachanov (1974) and Rabotnov (1959) is used. This parameter is introduced explicitly into

Vladimir S. Shorkin · Victoria Yu. Presnetsova · Vadim M. Presniakov · Sergey N. Romashin · Larisa Yu. Frolenkova · Svetlana I. Yakushina
Orel State University named after. I.S. Turgenev, 29 Naugorskoe Shosse, 302020 Orel, Russia
e-mail: vshorkin@yandex.ru, alluvian@mail.ru, vadim1024@rambler.ru
sromashin@yandex.ru, Larafrolenkova@yandex.ru, jakushina@rambler.ru

B. E. Abali et al. (eds.), *New Achievements in Continuum Mechanics and Thermodynamics*, Advanced Structured Materials 108,
https://doi.org/10.1007/978-3-030-13307-8_32

the defining relationships for describing the distribution of damages in a homogeneous medium (the medium models a real heterogeneous composite (Volegov et al, 2016, 2017)). Currently, there are scalar, vector and tensor characteristics of damage (Aptukov, 2007; Astaf'ev VI, 2001; Lokoshchenko, 2012). The development of damageability is determined by evolutionary kinetic equations that reflect the irreversible nature of damage (Volegov et al, 2017; Lokoshchenko, 2012). To obtain a unique solution it is necessary to set the initial damage parameter value. This value corresponds to the absence of external influences on the studied body. Usually the damage at the initial time (initial damage) is considered to be absent. Although for the case when the damage characterizes the volume of formed pores, it is possible to distinguish the initial damage from zero (Kashtanov and Petrov, 2006).

Atomic lattices of inclusions and matrices are different and incoherent. When forming a composite they tend to coincide, continue each other and become coherent. In this case, stresses and deformations of the discrepancy develop. If coherence is violated, they relax by reducing the energy of the discrepancy. The areas of the contact surface of the inclusion particles and the matrix with broken coherence are considered to be the defects of discrepancy (Gutkin, 2012; Gutkin and Ovidko, 2009; Gutkin et al, 2007; Ovid'ko and Sheinerman, 2002).

In this paper, the areas of the contact surface of the inclusion particle and the matrix along which their lattices are coherent are called the areas of their adhesion. The areas of the contact surface, along which the coherence of the lattices is violated, are called damaged. The term "damage" in this paper means the absence of adhesion partly on the interface between inclusion and matrix. It is characterized by a scalar parameter β, which for brevity hereinafter is also called "damage." It quantifies relatively the missing bound on the interface. Using the parameter β defined along the boundary of one inclusion particle, we can determine the parameter ω that characterizes the scattered damage to the damage of a homogeneous medium that models a heterogeneous composite.

In both cases, the damage is considered to be initial if it is determined in the absence of external influences affecting the stress-strain state of the composite. A method for determining the initial value of the damage β is proposed. Along with the parameter $\beta \in [0,1]$, a parameter $\alpha = 1 - \beta \in [0,1]$ called "continuity" is used.

32.2 General Statements

A macroscopic analysis is used in this paper. Materials of inclusion particles and matrix are modeled as linear elastic materials. They are characterized by different sets of elastic constants. A macroscopic approach is the most common when describing defects of discrepancy. By minimizing the energy of a system of interacting bodies, this approach makes it possible to determine its critical parameters (Ovid'ko and Sheinerman, 2002). In our case, such parameters are the parameters β and ω.

Let the element of the composite $B_{(1)}$ be an inclusion particle, and the matrix $B_{(2)}$ be two homogeneous solid bodies, contacting along a surface $S_{(12)}$ that has a unit normal $\boldsymbol{n}_{(12)}$ directed from $B_{(1)}$ to $B_{(2)}$ and a normal $\boldsymbol{n}_{(21)} = -\boldsymbol{n}_{(12)}$ directed from $B_{(2)}$ to $B_{(1)}$.

The appearance of stresses, deformations, and defects of discrepancy—caused by adhesion between materials of bodies $B_{(1)}$ and $B_{(2)}$—is effected by the distortion of atomic lattices from their state far from the contact surface to the state at which they are coherent. In order to reflect this distortion in determining the parameter β, it is assumed that the bodies $B_{(1)}$ and $B_{(2)}$ were first obtained by a mental, and then by a real instantaneous release from the corresponding infinitely extended, homogeneous, isotropic elastic medium $\Omega_{(1)}$ and $\Omega_{(2)}$ at the same temperature. As a reference configuration of bodies $B_{(1)}$ and $B_{(2)}$, the configuration that they had in the composition $\Omega_{(1)}$ and $\Omega_{(2)}$ after the mental selection is accepted.

It is assumed that when a composite is formed, the deformations of the inclusion particles $B_{(1)}$ and the matrix $B_{(2)}$ are small enough that changes in the density of their materials can be neglected. At the moment considered, there are no external effects on the composite, the composite is in a state of equilibrium, the temperature is distributed evenly over it.

Let $\boldsymbol{u}_{(j)}$ be the displacement of body particles $B_{(j)}$ $(j = 1, 2)$ relatively to the reference state. It is believed that there are no temperature effects on the displacement fields and the density of the material. Temperature effects are generated only by the adhesion interaction of bodies $B_{(j)}$ along the surface $S_{(12)}$. It is assumed that the norms of the first and second displacement gradients satisfy the inequalities:

$$\left\|\nabla \boldsymbol{u}_{(j)}\right\| \ll 1 \quad h\left\|\nabla^2 \boldsymbol{u}_{(j)}\right\| \ll 1, \tag{32.1}$$

where h is the characteristic size of the composite-inclusion element.

It is also assumed that the stress-strain states of inconsistency that develop in the vicinity of different particles do not affect each other. For the adhesion of solid materials, it is characteristic that the atomic structure of one material continues the structure of another (Bakulin et al, 2011; Mamonova et al, 2016; Raynolds et al, 1996). It is obvious that the lattices of the contacting bodies are coherent.

The incoherence of atomic lattices in section $S_{1-\alpha} \equiv S_\beta \subset S_{(12)}$ corresponds to the absence of ideal adhesion at this site. Depending on the nature of defects of discrepancy along S_β , the type of conjugation conditions for the displacement and stress fields that develop in bodies $B_{(j)}$ in the vicinity $dV_{(j)}$ of the section S_β of their contact surface may be different.

It is assumed that on every elementary section $\mathrm{d}S_{(12)}$ of the contact surface $S_{(12)}$ there is an area of $\mathrm{d}S_{\alpha(12)}$ where there is adhesion and the area $\mathrm{d}S_{1-\alpha} \equiv \mathrm{d}S_\beta$ n which adhesion (coherence of the lattices) is disturbed Figure 32.1.

$$\mathrm{d}S_{(12)} = \mathrm{d}S_\alpha + \mathrm{d}S_\beta = \alpha \mathrm{d}S_{(12)} + \beta \mathrm{d}S_{(12)}. \tag{32.2}$$

On sections $\mathrm{d}S_\alpha$ and $\mathrm{d}S_\beta$ the conditions of interaction of bodies $B_{(1)}$ to $B_{(2)}$ are different. It is assumed that the displacements $\boldsymbol{u}_{\alpha(j)}$ and $\boldsymbol{u}_{\beta(j)}$ of material points on them are independent of each other. The first are defined only on $\mathrm{d}S_\alpha$, the second - only on $\mathrm{d}S_\beta$. Wherein:

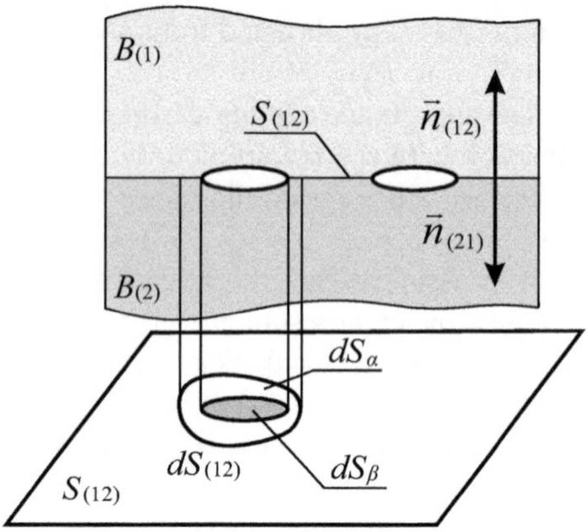

Fig. 32.1 Distribution of displacements on the elementary contact area.

$$\left(\boldsymbol{u}_{\alpha(j)}, \boldsymbol{u}_{\beta(j)}\right)\begin{cases}\neq 0, \boldsymbol{r} \in S_\alpha, S_\beta,\\ = 0, \boldsymbol{r} \in S_\beta, S_\alpha.\end{cases} \tag{32.3}$$

The area $\mathrm{d}S_{(12)}$ (if it is assumed to be continuous) is attributed to the average displacement $\boldsymbol{u}_{(j)}$ Figure 32.2

$$\boldsymbol{u}_{(j)} = \alpha\boldsymbol{u}_{\alpha(j)} + \beta\boldsymbol{u}_{\beta(j)} = (1-\beta)\,\boldsymbol{u}_{\alpha(j)} + \beta\boldsymbol{u}_{\beta(j)} \tag{32.4}$$

The transition from the volume values of the displacement of the material particles of each of the bodies $B_{(j)}$ to their surface values by $S_{(j)}$, (and therefore also on $S_{(12)} \subset S_{(j)}$) and vice versa is continuous and smooth. Therefore, it is assumed that the distributions of the resultant fields $\boldsymbol{u}_{(j)} = \boldsymbol{u}_{(j)}(\boldsymbol{r})$ over the areas $V_{(j)}$ occupied by the bodies $B_{(j)}$, have a structure described by the equality (32.4). Herewith, the field $\boldsymbol{u}_{\alpha(j)}$ for $\boldsymbol{r} \in V_{(j)}$ is generated by the surface field $\boldsymbol{u}_{\alpha(j)}$ for $\boldsymbol{r} \in dS_\alpha$. The field $\boldsymbol{u}_{\beta(j)}$ for $\boldsymbol{r} \in V_{(j)}$ is generated by the surface field $\boldsymbol{u}_{\beta(j)}$ for $\boldsymbol{r} \in dS_\beta$.

The conditions for the coherence of lattices, as well as its violations, are proposed to be described in the model of homogeneous isotropic linearly elastic materials of the second order that are in the state of adhesive contact. In this case, the energy of their elastic deformation depends not only on the first, but also on the second gradients of displacements. These quantities are considered to be generalized displacements, on which the generalized forces|two stress tensors $\boldsymbol{P}^{(1)}_{(j)}$ and $\boldsymbol{P}^{(2)}_{(j)}$,

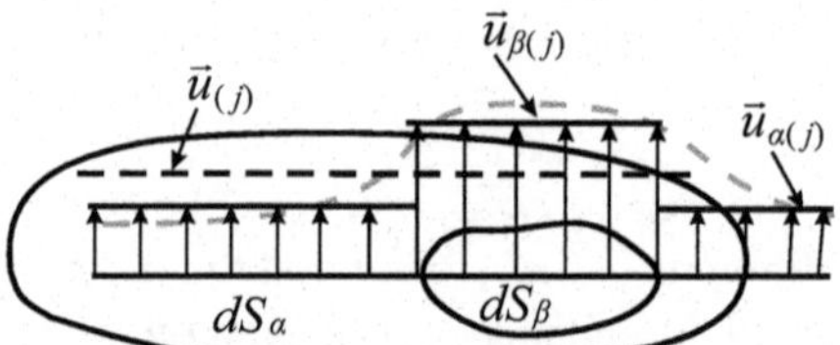

Fig. 32.2 The scheme of the damaged adhesion contact of the bodies B_1 and B_2.

respectively, the second and third ranks (Aifantis, 1992; Mindlin, 1965; Belov and Lurie, 2007b). It should be noted that a lot of papers have been devoted to constructing models in which the described surface effects are considered in continuous elastic medium, for example, Belov and Lurie (2009); Gurtin and Murdoch (1975); Povstenko (1993); Altenbach et al (2010, 2012); Eremeyev et al (2009); Belov and Lurie (2007a).

It is also assumed that the fields $\boldsymbol{u}_{(j)} = \boldsymbol{u}_{(j)}(\boldsymbol{r})$, $\boldsymbol{u}_{\alpha(j)} = \boldsymbol{u}_{\alpha(j)}(\boldsymbol{r})$ and $\boldsymbol{u}_{\beta(j)} = \boldsymbol{u}_{\beta(j)}(\boldsymbol{r})$ $(\boldsymbol{r} \in V_{(j)})$ are described by the same model of a linearly elastic isotropic medium. For the fields $\boldsymbol{u}_{(j)} = \boldsymbol{u}_{(j)}(\boldsymbol{r})$, the change in the potential energy of the body $B = B_{(1)} \bigcup B_{(2)}$ relative to the reference state is a functional that depends on distribution of the first $\nabla \boldsymbol{u}_{(j)}$ and the second $\nabla^2 \boldsymbol{u}_{(j)}$ displacement gradients in each of the parts $B_{(j)}$.

$$W_{(12)} = W_{(12)} \left(\nabla \boldsymbol{u}_{(1)}, \nabla \boldsymbol{u}_{(2)}, \nabla^2 \boldsymbol{u}_{(1)}, \nabla^2 \boldsymbol{u}_{(2)} \right). \tag{32.5}$$

For any fixed values of the parameter β, in the absence of the influence of external mechanical influences, the system $B = B_{(1)} \bigcup B_{(2)}$ and any part of it are in a state of equilibrium. In this case, the distributions of the arguments of the functional (32.5) in terms of their domains of definition correspond to its stationary value, when

$$\begin{aligned} \delta(W_{(12)}(\alpha)) &= \sum_{j=1}^{2} \int\limits_{V_{(j)}} \delta w_{(j)} \mathrm{d}V_{(j)} \\ &= \sum_{j=1}^{2} \int\limits_{V_{(j)}} \left[\boldsymbol{P}^{(1)}_{(j)} \cdot\cdot \delta(\nabla \boldsymbol{u}_{(j)})^T + \boldsymbol{P}^{(2)}_{(j)} \cdot\cdot\cdot \delta(\nabla^2 \boldsymbol{u}_{(j)})^T \right] = 0. \end{aligned} \tag{32.6}$$

Here $w_{(j)}$ is the volume density of the energy distribution of the elastic deformations caused in the bodies $B_{(j)}$ by the uneven distribution of the averaged displacements (32.4);

$$\boldsymbol{P}^{(n)}_{(j)} = \frac{\partial w_{(j)}}{\partial (\nabla^n \boldsymbol{u}_{(j)})^T} \; (n = 1, 2) \text{ - stress tensors.} \tag{32.7}$$

Integrating by parts, taking Ostrogradsky's theorem into account, on the basis of expression (32.6), we can obtain the differential equations of equilibrium for each of the bodies $B_{(j)}$, the boundary conditions for them on the free sections $S_{0(j)} = S_{(j)} \backslash S_{(12)}$ of their boundaries, and the variational condition of conjugation of the displacement and stress fields along the contact surface $S_{(12)}$. These expressions have the following form

$$\nabla \cdot (\boldsymbol{P}^{(1)}_{(j)} - \nabla \cdot \boldsymbol{P}^{(2)}_{(j)}) = 0 \qquad \boldsymbol{r} \in V_{(j)} \tag{32.8}$$

$$\boldsymbol{n}_{(j)} \cdot (\boldsymbol{P}^{(1)}_{(j)} - \nabla \cdot \boldsymbol{P}^{(2)}_{(j)}) - \nabla_S \cdot (\boldsymbol{n}_{(j)} \cdot \boldsymbol{P}^{(2)}_{(j)}) = 0 \qquad \boldsymbol{r} \in S_{0(j)} \tag{32.9}$$

$$(\boldsymbol{n}_{(j)} \boldsymbol{n}_{(j)}) \cdot\cdot \boldsymbol{P}^{(2)}_{(j)} \qquad \boldsymbol{r} \in S_{0(j)} \tag{32.10}$$

$$\int\limits_{S_{(12)}} \left\{ \left[\boldsymbol{n}_{(12)} \cdot \left(\boldsymbol{P}^{(1)}_{(1)} - \nabla \cdot \boldsymbol{P}^{(2)}_{(1)} \right) - \nabla_S \cdot \left(\boldsymbol{n}_{(12)} \cdot \boldsymbol{P}^{(2)}_{(1)} \right) \right] \cdot \delta \boldsymbol{u}_{(1)} \right.$$
$$\left. + \left[\boldsymbol{n}_{(21)} \cdot \left(\boldsymbol{P}^{(1)}_{(2)} - \nabla \cdot \boldsymbol{P}^{(2)}_{(2)} \right) - \nabla_S \cdot \left(\boldsymbol{n}_{(21)} \cdot \boldsymbol{P}^{(2)}_{(2)} \right) \right] \cdot \delta \boldsymbol{u}_{(2)} \right\} \mathrm{d}S_{(12)}$$
$$+ \int\limits_{S_{(12)}} \left\{ \left((\boldsymbol{n}_{(12)} \boldsymbol{n}_{(12)}) \cdot \cdot \boldsymbol{P}^{(2)}_{(1)} \right) \cdot \delta \left(\partial \boldsymbol{u}_{(1)} / \partial n_{(12)} \right) \right.$$
$$\left. + \left((\boldsymbol{n}_{(21)} \boldsymbol{n}_{(21)}) \cdot \cdot \boldsymbol{P}^{(2)}_{(2)} \right) \cdot \delta \left(\partial \boldsymbol{u}_{(2)} / \partial n_{(21)} \right) \right\} \mathrm{d}S_{(12)} = 0. \quad (32.11)$$

Here $\boldsymbol{n}$ is the unit vector of the outward normal to the section of the surface S defined by the corresponding lower index; $\partial \boldsymbol{u}/\partial n = \boldsymbol{n} \cdot (\nabla \boldsymbol{u})$ - the designation of the derivative in its direction; ∇_S - gradient along the surface S.

In the linear theory of elasticity of second-order materials, the tensors $\boldsymbol{P}^{(1)}_{(j)}$ and $\boldsymbol{P}^{(2)}_{(j)}$ are linear functions of the gradients $\nabla \boldsymbol{u}_{(j)}$ and $\nabla^2 \boldsymbol{u}_{(j)}$. Therefore, taking into account representation (32.4), we can obtain:

$$\boldsymbol{P}^{(n)}_{(j)} = \alpha \boldsymbol{P}^{(n)}_{\alpha(j)} + \beta \boldsymbol{P}^{(n)}_{\beta(j)} \ (n = 1, 2). \quad (32.12)$$

Substituting the representations (32.4) and (32.12) into the system (32.8) - (32.10), assuming that the parameter β is arbitrary, for tensors $\boldsymbol{P}^{(n)}_{\alpha(j)}$ and $\boldsymbol{P}^{(n)}_{\beta(j)}$ we can obtain two independent systems of equilibrium equations and boundary conditions that coincide with the expressions (32.8) - (32.10). Substituting (32.4) and (32.12) into expression (32.11), taking into account that $\alpha \mathrm{d}S_{(12)} = \mathrm{d}S_\alpha$ and $\beta \mathrm{d}S_{(12)} = \mathrm{d}S_\beta$ it is possible to obtain the integral variational condition of the displacement and stress fields' conjugation, satisfied by tensors $\boldsymbol{P}^{(n)}_{\alpha(j)}$, $\boldsymbol{P}^{(n)}_{\beta(j)}$ and vectors $\boldsymbol{u}_{\alpha(j)}$, $\boldsymbol{u}_{\beta(j)}$. On its basis, the conditions for conjugation of the fields of the listed objects on each of the sections $\alpha \mathrm{d}S_{(12)} = \mathrm{d}S_\alpha$ and $\beta \mathrm{d}S_{(12)} = \mathrm{d}S_\beta$ based on the following assumptions are constructed.

In areas $\mathrm{d}S_\alpha$ where adhesion is observed between the bodies $B_{(1)}$ and $B_{(2)}$, the coherence of their lattices is conserved not only in the equilibrium state of the system, but also in the variation of its stress-strain state. In this case

$$\delta \boldsymbol{u}_{\alpha(1)} = \delta \boldsymbol{u}_{\alpha(2)}, \qquad \delta \left(\frac{\partial \boldsymbol{u}_{\alpha(1)}}{\partial n_{(12)}} \right) = \delta \left(\frac{\partial \boldsymbol{u}_{\alpha(2)}}{\partial n_{(12)}} \right) \quad (32.13)$$

The displacements $\boldsymbol{u}_{\alpha(j)}$ and their gradients $\nabla \boldsymbol{u}_{\alpha(j)}$, defined on $S_\alpha \subset S_{(12)}$, taking into account (Frolenkova and Shorkin, 2013), satisfy the conditions:

$$\boldsymbol{u}_{\alpha(1)} - \boldsymbol{u}_{\alpha(2)} = 0, \qquad \nabla \boldsymbol{u}_{\alpha(1)} - \nabla \boldsymbol{u}_{\alpha(2)} = \boldsymbol{C}_{S(12)}. \quad (32.14)$$

The components of the vector $\boldsymbol{u}_{\alpha(12)}$ and the tensor $\boldsymbol{C}_{S(12)}$ are determined on the basis, for example, of the relations presented in Ovid'ko and Sheinerman (2002). These relations take into account the discrepancy between the parameters of atomic lattices of bodies $B_{(1)}$ and $B_{(2)}$.

In addition to conditions (32.13) and (32.14), the equalities in the areas of adhesive contact are:

$$\begin{aligned}\boldsymbol{n}_{(12)} \cdot \left(\boldsymbol{P}^{(1)}_{\alpha(1)} - \nabla \cdot \boldsymbol{P}^{(2)}_{\alpha(1)}\right) - \nabla_S \cdot \left(\boldsymbol{n}_{(12)} \cdot \boldsymbol{P}^{(2)}_{\alpha(1)}\right) \\ + \boldsymbol{n}_{(21)} \cdot \left(\boldsymbol{P}^{(1)}_{\alpha(2)} - \nabla \cdot \boldsymbol{P}^{(2)}_{\alpha(2)}\right) - \nabla_S \cdot \left(\boldsymbol{n}_{(21)} \cdot \boldsymbol{P}^{(2)}_{\alpha(2)}\right) = 0. \quad (32.15)\end{aligned}$$

$$\left(\boldsymbol{n}_{(12)}\boldsymbol{n}_{(12)}\right) \cdot \cdot \boldsymbol{P}^{(2)}_{\alpha(1)} + \left(\boldsymbol{n}_{(21)}\boldsymbol{n}_{(21)}\right) \cdot \cdot \boldsymbol{P}^{(2)}_{\alpha(2)} = 0. \quad (32.16)$$

In the areas of absence of adhesive contact (for $\boldsymbol{r} \in \mathrm{d}S_\beta$), it is assumed that there is no " disclosure " of the mismatch defect. Hence

$$\delta \boldsymbol{u}_{\beta(1)} = \delta \boldsymbol{u}_{\beta(2)}, \qquad \boldsymbol{u}_{\beta(j)} = \boldsymbol{u}_{\alpha(j)}. \quad (32.17)$$

It is assumed that the edges of the defect do not interact with each other. Since there are no other possible effects on them except for mutual ones, so on each of the edges for $\boldsymbol{r} \in dS_\beta$ the qualities are:

$$\boldsymbol{n}_{(12)} \cdot \left(\boldsymbol{P}^{(1)}_{\alpha(1)} - \nabla \cdot \boldsymbol{P}^{(2)}_{\alpha(1)}\right) - \nabla_S \cdot \left(\boldsymbol{n}_{(12)} \cdot \boldsymbol{P}^{(2)}_{\alpha(1)}\right) = 0, \quad (32.18)$$

$$\boldsymbol{n}_{(21)} \cdot \left(\boldsymbol{P}^{(1)}_{\alpha(2)} - \nabla \cdot \boldsymbol{P}^{(2)}_{\alpha(2)}\right) - \nabla_S \cdot \left(\boldsymbol{n}_{(21)} \cdot \boldsymbol{P}^{(2)}_{\alpha(2)}\right) = 0, \quad (32.19)$$

$$\left(\boldsymbol{n}_{(12)}\boldsymbol{n}_{(12)}\right) \cdot \cdot \boldsymbol{P}^{(2)}_{\alpha(1)} = 0, \quad (32.20)$$

$$\left(\boldsymbol{n}_{(21)}\boldsymbol{n}_{(21)}\right) \cdot \cdot \boldsymbol{P}^{(2)}_{\beta(2)} = 0. \quad (32.21)$$

Conditions (32.17) - (32.21) are sufficient for unambiguous definition $\boldsymbol{u}_{\beta(j)}$, including for $\boldsymbol{r} \in dS_\beta$. Therefore, it is assumed that

$$\delta\left(\frac{\partial \boldsymbol{u}_{\beta(1)}}{\partial n_{(12)}}\right) = 0, \qquad \delta\left(\frac{\partial \boldsymbol{u}_{\beta(2)}}{\partial n_{(12)}}\right) = 0. \quad (32.22)$$

Conditions (32.13) - (32.21) are sufficient to satisfy the conjugation condition (32.11). In this case, the problems of determining the fields $\boldsymbol{u}_{(j)}(\boldsymbol{r})$ and $\boldsymbol{u}_{\beta(j)}(\boldsymbol{r})$ are independent of each other. On the basis of their solution, we can obtain an expression for the discrepancy energy of the bodies $B_{(1)}$ and $B_{(2)}$.

$$W_{(12)}(\beta) = (1-\beta)^2 W_{\alpha\alpha} + \beta(1-\beta) W_{\alpha\beta} + \beta^2 W_{\beta\beta}. \quad (32.23)$$

Values W_{pm} $p, m = \alpha, \beta$ are the sums of the combined body $B = B_{(1)} \bigcup B_{(2)}$ works of internal generalized forces $\boldsymbol{P}^{(k)}_{(j)}$ $(k = 1, 2)$ on internal generalized displacements $\nabla^k \boldsymbol{u}_{(j)}$. They are defined by the equalities

$$W_{(pm)} = \sum_{j=1}^{2} \sum_{k=1}^{2} \int_0^{\nabla^k \boldsymbol{u}_{(j)}} \boldsymbol{P}^{(k)}_{p(j)} d(\nabla^k \boldsymbol{u}_{(j)}). \quad (32.24)$$

Assuming that the damage of the adhesive contact contributes to the achievement of the interaction energy of the inclusion particles and the composite matrix of the minimum value, we can obtain a formula by which we can calculate the value of the damage parameter

$$\beta = \frac{\partial W_{(12)}}{\partial \beta} = \left[1 + \frac{2W_{\beta\beta} - W_{\alpha\beta}}{2W_{\alpha\alpha} - W_{\alpha\beta}}\right]^{-1}. \tag{32.25}$$

It is proposed to use a second-order material model in the calculations, which is a consequence of the nonlocal model presented in Shorkin (2011); Shorkin et al (2011). The relationship between stresses and strains is given by

$$\boldsymbol{P}^{(k)}_{(j)} = \boldsymbol{P}^{0(k)}_{(j)} + \sum_{m=1}^{2} (\nabla^m \boldsymbol{u}_{(j)}) \overbrace{\cdots\cdot}^{n \text{ step}} \boldsymbol{C}^{m,k}_{j}. \tag{32.26}$$

Here $\boldsymbol{P}^{0(k)}_{(j)}$ are the tensors of the initial stresses caused by the properties of the material; $\boldsymbol{C}^{m,k}_{j}$ are the tensors of elastic constants caused by the properties of the material. These tensors are defined by equalities (Dolgov et al, 2015):

$$\boldsymbol{P}^{0(k)}_{(j)} = \int\limits_{V_{(j)}} \frac{1}{k!}\left(\frac{\mathrm{d}}{\mathrm{d}\boldsymbol{l}_{12}}\Phi^{(2)}_{(j)}\right)\boldsymbol{l}^k_{12}\mathrm{d}V_{2(j)} + \sum_{q=2}^{3}\int\limits_{V_{(j)}}\left[\int\limits_{V_{(j)}} \frac{1}{k!}\left(\frac{\mathrm{d}}{\mathrm{d}\boldsymbol{l}_{1q}}\Phi^{(3)}_{(j)}\right)\boldsymbol{l}^k_{1q}\mathrm{d}V_{2(j)}\right]\mathrm{d}V_{3(j)}. \tag{32.27}$$

$$\boldsymbol{C}^{(m,k)}_{(j)} = \int\limits_{V_{(j)}} \frac{1}{m!k!}\boldsymbol{l}^k_{12}\left(\frac{\mathrm{d}^2}{\mathrm{d}\boldsymbol{l}^2_{12}}\Phi^{(2)}_{(j)}\right)\boldsymbol{l}^m_{12}\mathrm{d}V_{2(j)} + \sum_{p,q=2}^{3}\int\limits_{V_{(j)}}\left[\int\limits_{V_{(j)}} \frac{1}{m!k!}\boldsymbol{l}^k_{1p}\left(\frac{\mathrm{d}^2}{\mathrm{d}\boldsymbol{l}_{1p}\mathrm{d}\boldsymbol{l}_{1q}}\Phi^{(3)}_{(j)}\right)\boldsymbol{l}^m_{1q}\mathrm{d}V_{2(j)}\right]\mathrm{d}V_{3(j)}. \tag{32.28}$$

In expressions (32.27), (32.28), the following terms:

$$\Phi^{(2)}_{(j)}(\boldsymbol{l}_{12})\mathrm{d}V_{1(j)}\mathrm{d}V_{2(j)} \quad \text{and} \quad \Phi^{(3)}_{(j)}(\boldsymbol{l}_{12},\boldsymbol{l}_{13})\mathrm{d}V_{1(j)}\mathrm{d}V_{2(j)}\mathrm{d}V_{3(j)}$$

are the potentials of pair and triple nonlocal interactions, respectively. The former evaluates the interaction of the particle $\mathrm{d}B_{2(j)}$ with $\mathrm{d}B_{1(j)}$; and the latter quantifies the interaction of particles $\mathrm{d}B_{2(j)}$ and $\mathrm{d}B_{3(j)}$ with $\mathrm{d}B_{1(j)}$. Both terms are proportional to their volumes and $\boldsymbol{l}_{1p}$ denotes the position vector of the particle position $\mathrm{d}B_{p(j)}$ relative to $\mathrm{d}B_{1(j)}$.

In order to illustrate the conclusions about the possibility of determining the damage of the discrepancy by the methods of mechanics of a deformable solid body, the value β is calculated for several pairs of inclusion materials and a matrix, which form a flat layered structure. It was assumed that the distance h between the contact surfaces of the material $B_{(1)}$ with the material $B_{(2)}$ and the radius of their curvature is much higher than the effective radius R of pair and ternary interparticle interactions. On the basis of this assumption, it was assumed that the value determined for two semi-infinite bodies contacting along the plane coincides with the same value determined for the real contact surfaces of real particles from materials $B_{(1)}$ and $B_{(2)}$.

A confirmation of the presence of dislocations of mismatch of crystal lattices, which in this work are considered to be damaged, is a picture Figure 32.3. One can cite a number of similar photos, for example, presented in Wang (2012).

In addition to the analysis of photos similar to those presented in Möbus and Trampert (2003), the analysis of the correspondence of the damage of the adhesive contact with the results of calculation by the molecular dynamics of the boundary structure Ni-Al, Cu-Au, Ni- Fe is proposed. These results are presented in Poletaev et al (2015).

The calculations were carried out based on the methods described in Romashin et al (2015); Frolenkova and Shorkin (2017); Romashin et al (2016). Their essence is as follows. It is assumed that the studied materials are homogeneous and isotropic. In this case, the potentials of pair and triple interactions of particles of contacting materials depend only on the distances between the particles. The type of these dependencies is introduced. For pair interactions, this is an analogue of the Morse potential. For triple interactions, the potential is proportional to the product of the potentials of pair interactions of two particles of a triple with the chosen as the main one. Presnetsova et al (2018) presents a rational for the stability of the system of elementary particles of a continuous medium, the interaction of which is described by such potentials.

The set of potentials of pair and triple interactions contains three unknown parameters. To determine them, the connections of the parameters of inter particle interactions with the parameters of lame obtained on the basis of (32.28) were used. In addition, the possibility of describing the nonlinearity of the dispersion law for acoustic plane waves in an elastic medium is used. The Lame parameters and the nonlinearity characteristics of the dispersion law are considered to be known (Kittel, 2004). The parameters of the inter particle potentials are found using the obtained bonds. The found parameters are used to calculate the characteristics of the elastic state of the studied materials additional to the Lame parameters.

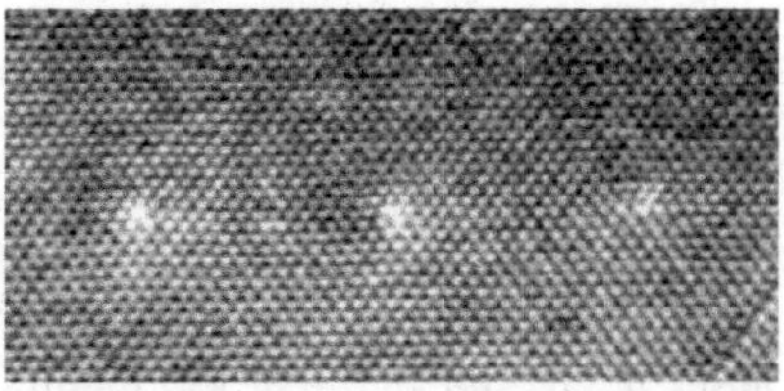

Fig. 32.3 Photo of mismatch dislocations at the interface of InSb (top) and GaAs (bottom) crystals in the plane (111) obtained by electron microscope.

Based on the data obtained, the characteristic of the rupture of the adhesive contact of the pairs structure Ni-Al, Cu-Au, Ni- Fe is calculated. To calculate the lame parameters, the following values of Young's modules and Poisson's ratios of the materials of these pairs were used (Grigoriev, 1991):

- For Ni: $E = 2.0 \cdot 10^{11}\mathrm{N/m}^2$, $\nu = 0.28$.
- For Al: $E = 0.7 \cdot 10^{11}\mathrm{N/m}^2$, $\nu = 0.35$.
- For Cu: $E = 1.3 \cdot 10^{11}\mathrm{N/m}^2$, $\nu = 0.34$.
- For Au: $E = 0.8 \cdot 10^{11}\mathrm{N/m}^2$, $\nu = 0.37$.
- For Fe: $E = 2.1 \cdot 10^{11}\mathrm{N/m}^2$, $\nu = 0.29$.

The difference of lattice parameters (Poletaev et al, 2015) (Ni-Al - 14.9%, Cu-Au - 12.8%, Ni-Fe - 3.2%) leads to the formation of mismatch dislocations at the interface: the higher this difference, the higher the dislocation density is obviously. As a result of calculation of damage of adhesive contact by methods of this work it is received for Ni-Al - $\beta = 0.17$, Cu-Au - $\beta = 0.09$, Ni-Fe - $\beta = 0.01$. The observed correlation of the discrepancy between the parameters of the grids of contacting materials and the damage of their adhesive contact testifies in favor of the fairness of the calculation methods of the latter.

If the value of the damage parameter β is high, it is proposed to check the existence of adhesion between the contacting materials on the basis of Presnetsova et al (2017). There are proposed criteria for the presence or absence of adhesion between materials for which the young modulus and Poisson's ratio are known. For example, the material talc is known for its property not to stick to solid materials. In Presnetsova et al (2017) this is confirmed quantitatively in the study of iron and talc adhesion, if for it (Vakhromeev et al, 1997; Erofeev et al, 2006): $E = 0.16 \cdot 10^{11}\mathrm{N/m}^2$, $\nu = 0$. It is established that the particles of talc when approaching the iron body will be repelled from it. In assessing the damage (assuming the presence of adhesion between iron and talc) can be obtained: $\beta = 0.87$. This corresponds to the lack of adhesion.

The possibility of finding the distribution of the damage mismatch $\beta = \beta(\boldsymbol{r})$ along the boundaries $S_{(12)}$ of the inclusion particles, which are distributed over the volume of the composite in the absence of external influences on it, allows us to introduce the initial distribution of the scattered damage $\omega = \omega(\boldsymbol{r})$. One of the options for determining the scattered damage is an option similar to that presented in Aptukov (2007).

$$\omega = \frac{1}{S} \int\limits_{\sum S_{(12)}} \beta \mathrm{d}S_{(12)}, \tag{32.29}$$

where S is the surface area of the representative volume of the composite.

Recently, there has been a growing interest in the study of materials with a nanocrystalline structure, since it was found that a decrease in the size of crystallites (or any other structural formations) below a certain threshold nanoscale value leads to a radical change in the physical properties of these materials. Equations (32.27) and (32.28) indicate that the method proposed for calculating the damage in this paper allows to describe this effect.

32.3 Conclusion

A mathematical model of the interaction of inclusion particles and the matrix of a composite material is proposed. It is based on the nonlocal model of an elastic medium developed by the authors, based on taking into account the paired and ternary interactions of its particles. The model makes it possible, on the basis of data on the elastic characteristics of the inclusion particles and matrix materials, to calculate the damage of their adhesion contact, which is caused by the mismatch of the parameters of their lattices. The damage area is considered to be the area of their contact surface along which the adhesion condition is violated - the preservation at its intersection not only the continuity but also the smoothness of the deformations of the material fibers. To build the model, as a consequence of the nonlocal model, the theory of elasticity of second-order materials is used. The possibility to find the distribution of damage mismatch along the boundaries of inclusion particles is proposed to be used to determine the initial distribution of the scattered damage of a homogeneous medium that models a heterogeneous composite. The assumption is made that taking into account the nonlocality of the interaction of particles of interacting materials will make it possible to describe the effects specific to nanocomposites.

Acknowledgements The work was performed within the framework of the basic part of the State task for 2017 - 2019, project code 1.5265.2017 / BCh.

References

Aifantis EC (1992) On the role of gradients in the localization of deformation and fracture. International Journal of Engineering Science 30(10):1279–1299

Altenbach H, Eremeyev VA, Lebedev LP (2010) On the existence of solution in the linear elasticity with surface stresses. ZAMM-Journal of Applied Mathematics and Mechanics/Zeitschrift für Angewandte Mathematik und Mechanik: Applied Mathematics and Mechanics 90(3):231–240

Altenbach H, Eremeyev VA, Morozov NF (2012) Surface viscoelasticity and effective properties of thin-walled structures at the nanoscale. International Journal of Engineering Science 59:83–89

Aptukov V (2007) Model' uprugo-povrezhdennoi ortotropnoi sredy [model of the elastic damaged ortotropny environment]. Vestnik Permskogo universiteta Seriia: Matematika Mekhanika Informatika 7(12):84–90

Astaf'ev VI SL Radaev YuN (2001) Nonlinear mechanics of destruction (in Russian). Izdatel'stvo "Samarskii universitet"

Bakulin A, Eremeev S, Tereshchenko O, Chulkov E, Kulkova S (2011) Adsorption of halogen atom (f, cl, i) on cation-rich gaas (001) surface. In: IOP Conference Series: Materials Science and Engineering, IOP Publishing, vol 23(1), p 012015

Belov P, Lurie S (2007a) Theory of ideal adhesion interactions. J Compos Mech Des 14:545–561

Belov P, Lurie S (2007b) To the general geometric theory of defect-containing media (in russian). Fiz Mezomekh 10(6):49–61

Belov P, Lurie S (2009) A continuum model of microheterogeneous media. Journal of Applied Mathematics and Mechanics 73(5):599–608

Dolgov N, Romashin S, Frolenkova LY, Shorkin V (2015) A model of contact of elastic bodies with account for their adhesion. Nanoscience and Technology: An International Journal 6(2):117–133

Eremeyev V, Altenbach H, Morozov N (2009) The influence of surface tension on the effective stiffness of nanosize plates. In: Doklady Physics, Springer, vol 54(2), pp 98–100

Erofeev LY, Vakhromeev G, Zinchenko V, Nomokonova G (2006) Physics of rock properties (in Russian)]. TPU Publ. House, Tomsk

Frolenkova LJ, Shorkin VS (2013) Method of calculating the surface and adhesion energies of elastic bodies. PNRPU Mechanics Bulletin 1:235–259

Frolenkova LY, Shorkin VS (2017) Surface energy and adhesion energy of elastic bodies. Mechanics of Solids 52(1):62–74

Grigoriev IS (1991) Physical quantities. Handbook. (in Russian)]. M.: Energoatomizdat

Gurtin ME, Murdoch AI (1975) A continuum theory of elastic material surfaces. Arch Rat Mech Anal 57:291–323

Gutkin MY (2012) Misfit stress relaxation in composite nanoparticles. International Journal of Engineering Science 61:59–74

Gutkin MY, Ovidko IA (2009) Defect structures on inner interfaces in nanocrystalline and polycrystalline films. Mater Phys Mech 6(2):108–148

Gutkin MY, Ovid'ko I, Skiba N (2007) Effect of inclusions on heterogeneous crack nucleation in nanocomposites. Physics of the Solid State 49(2):261

Kachanov LM (1974) Fundamentals of fracture mechanics (in Russian). Nauka, Moscow

Kashtanov A, Petrov YV (2006) Energy approach to determination of the instantaneous damage level. Technical physics 51(5):604–608

Kittel C (2004) Introduction to solid state physics. John Wiley and Sons Ltd, New York

Lokoshchenko A (2012) Application of kinetic theory to the analysis of high-temperature creep rupture of metals under complex stress. Journal of Applied Mechanics and Technical Physics 53(4):599–610

Mamonova MV, Prudnikov VV, Prudnikova IA (2016) Surface physics: theoretical models and experimental methods. CRC Press

Mindlin RD (1965) Second gradient of strain and surface-tension in linear elasticity. International Journal of Solids and Structures 1(4):417–438

Möbus G, Trampert A (2003) 3d display and analysis of strain fields at heterointerfaces. Microscopy and Microanalysis 9(S02):746–747

Ovid'ko I, Sheinerman A (2002) Dislocation dipoles in nanoscale films with compositional inhomogeneities. Philosophical Magazine A 82(16):3119–3127

Poletaev G, Sannikov A, Berdychenko A, Starostenkov M (2015) Molecular dynamics study of plastic deformation mechanisms near the interphase boundary in two-dimensional bimetallic systems. Materials Physics and Mechanics 22(1):15–19

Povstenko YZ (1993) Theoretical investigation of phenomena caused by heterogeneous surface tension in solids. Journal of the Mechanics and Physics of Solids 41(9):1499–1514

Presnetsova V, Romashin S, Frolenkova L, Shorkin V, Yakushina S (2017) The adhesive and antiadhesive non-local interaction of solids. In: IOP Conference Series: Materials Science and Engineering, IOP Publishing, vol 208(1), p 012034

Presnetsova VY, Romashin S, Frolenkova LY, Shorkin VS, Yakushina S (2018) A variant of describing adhesion interaction in the probe—sample system of an atomic-force microscope. Nanoscience and Technology: An International Journal 9(4)

Rabotnov YN (1959) On the long-term fracture mechanics (in russian). Issues of Strength of Materials and Structures, Izd-vo AN SSSR, Moscow pp 5–7

Raynolds JE, Smith JR, Zhao GL, Srolovitz DJ (1996) Adhesion in nial-cr from first principles. Physical Review B 53(20):13,883

Romashin S, Presnetsova VY, Frolenkova LY, Shorkin V (2015) A model of adhesive interaction of elastic bodies. In: Mechanics-Seventh Polyakhov's Reading, 2015 International Conference on, IEEE, pp 1–3

Romashin SN, Presnetsova VY, Frolenkova LY, Shorkin VS (2016) Method for calculating the characteristics of elastic state media with internal degrees of freedom. In: Generalized Continua as Models for Classical and Advanced Materials, Springer, pp 363–376

Shorkin V (2011) Nonlinear dispersion properties of high-frequency waves in the gradient theory of elasticity. Mechanics of solids 46(6):898–912

Shorkin VS, Frolenkova LY, Azarov AS (2011) The influence of the triple interaction of particles of the medium on the surface and adhesion properties of solid bodies. Materials Science 2:2–7

Vakhromeev GS, Yerofeev LY, Kanaikin VS, Nomokonova GG (1997) Petrophysics (in Russian)]. TPU Publ. House, Tomsk

Valiev RZ, Zhilyaev AP, Langdon TG (2007) Bulk nanostructured materials: fundamentals and applications. John Wiley & Sons

Volegov P, Gribov D, Trusov P (2016) Damage and fracture: Review of experimental studies. Physical Mesomechanics 19(3):319–331

Volegov P, Gribov D, Trusov P (2017) Damage and fracture: Classical continuum theories. Physical Mesomechanics 20(2):157–173

Wang Y (2012) Misfit dislocation and strain relaxation at large lattice mismatched iii-v semiconductor interfaces. PhD thesis, Université de Caen

Chapter 33
How the Properties of Pantographic Elementary Lattices Determine the Properties of Pantographic Metamaterials

Emilio Turco

Abstract In this paper we describe a three-scales homogenization process which we use to determine a macroscopic model for pantographic metamaterials. The smallest scale refers to the length at which the considered deformable mechanical system can be modeled as a Cauchy's continuum. Of course, at this scale, its geometry is rather complex. The meso-scale refers to a length at which the system can be modeled as a Hencky-type discrete system constituted by masses interconnected by extensional and rotational springs. At macro-scale the model to be used is a generalized plate whose deformation energy depends on geodesic curvature. While the direct identification from the smallest scale to the macro-scale seems rather difficult, the identification from smallest scale to meso-scale can be successfully obtained. The geometrical properties, along with Young and Poisson coefficients of the used isotropic material, at the smallest scale determine the extensional and rotational stiffnesses to be used at the meso-scale. On the other hand, the Piola-type identification process allows us to determine the stiffnesses of the macroscopic generalized plate model, *via* a simple asymptotic expansion. We have observed that this process is valid in both cases when the smallest scale is of the order of microns and when the smallest scale is of the order of tenth of millimeters. Some experimental and numerical results supporting this statement are exhibited.

Keywords: Mechanics of (meta)materials · Lagrangian models · Micro-meso and meso-macro parameters identification

---

Emilio Turco
Department of Architecture, Design and Urban planning (DADU), University of Sassari and M&MOCS International Research Center, Italy,
e-mail: emilio.turco@uniss.it

B. E. Abali et al. (eds.), *New Achievements in Continuum Mechanics and Thermodynamics*, Advanced Structured Materials 108,
https://doi.org/10.1007/978-3-030-13307-8_33

33.1 Introduction

In technical literature there are not clearly established homogenization methods (however, see the attempts Franciosi et al, 2018; Karamooz et al, 2014; Karamooz and Kadkhodaei, 2015; Saeb et al, 2016; Abali et al, 2016) leading to higher gradient continua starting from mechanical systems showing complex geometrical microstructures and highly inhomogeneous stiffnesses fields. Therefore, we have decided to use a three-scales homogenization process building the macroscopic continuum model for pantographic metamaterials in a three-steps process. The study of this class of mechanical systems has a great importance in design of metamaterials, see, *e.g.*, , the review (Barchiesi et al, 2018c). Some identification results are however available, see, for instance, Barchiesi and Placidi (2017); Placidi et al (2017b); Yang et al (2018); Placidi et al (2015, 2017a); Placidi and El Dhaba (2017); Placidi et al (2017c); Barchiesi et al (2018b); Lekszycki et al (2018).

The analysis becomes more difficult when the considered pantographic specimens undergo damage and plasticity phenomena: the identification of macro-mechanical properties seems beyond actual state-of-the-art. However, the results found in Placidi et al (2018a,b) supply an important guidance for this more general identification procedure, as they give a possible target macro-model.

At the smallest scale the considered deformable body is, at least from the theoretical point of view, mechanically homogeneous but its geometry is involved, showing, see Fig. 33.1, large gaps and narrow material connections (pivots) in which deformation energy may be concentrate. At this length scale the standard three-dimensional Cauchy continuum model can be usefully introduced. Unfortunately due to the complex geometry and the expected deformation patterns, to use this modeling one needs heavy numerical codes, involving, for simple specimens, even several millions of degrees of freedom. Therefore, also with a view towards technological applications and having in mind some interesting optimization problems, there is a need to find so-called reduced order models.

However, a direct deduction algorithm, even using some simple formal perturbation methods, has not been yet clearly and satisfactorily developed. Therefore, it has been proposed to introduce a meso-scale where the deformable body is modeled by

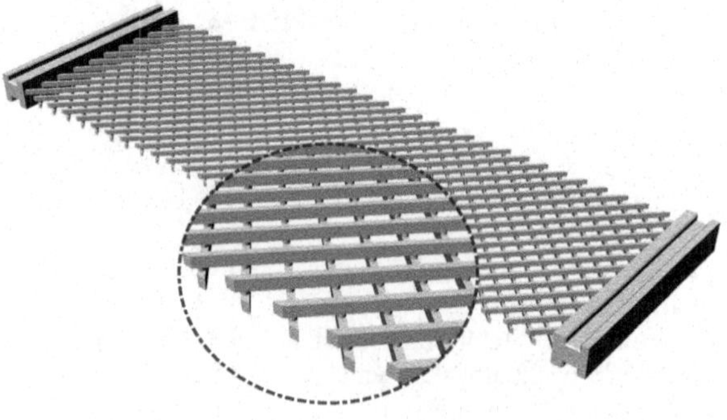

Fig. 33.1 Three-dimensional rendering of a pantographic lattice with a blow-up of a part.

means of a Hencky-type discrete Lagrangian system, see Turco et al (2016c) for the model and Turco et al (2016e,a,b) for the comparison, for some remarkable cases, between experiments and numerical simulations. This system is constituted by material points, in which one will concentrate masses when inertia has to be taken into account, interconnected by extensional and rotational springs, see Fig. 33.2. These springs will have deformation energies whose constitutive law can be linked with the geometrical and mechanical properties characterizing the system at the lowest scale. This identification is rather simple, at least from the theoretical point of view, and can be performed by using the methods already conjectured by Piola (dell'Isola et al, 2015, 2019) and developed by Hencky (1921) to obtain some estimates of the buckling load for *Elastica*.[1]

Finally, at the macro-scale the model which seems to us to be the most suitable is a generalized plate. In standard plate theories, the curvature is the only deformation measure due to the transverse displacement on the current configuration. Instead the true mechanical nature of pantographic sheets can be captured by a two-dimensional continuum only by introducing a deformation energy depending on so-called geodesic curvature (Steigmann and dell'Isola, 2015; Giorgio et al, 2016, 2015, 2017b, 2018) and on twisting of the material curves modeling the pantographic fibers.

Hencky-type discrete models, by using Piola's identification (see dell'Isola et al, 2016), based on the identification of finite differences with derivatives, immediately produces a macro-model which belongs to the just described class of generalized plates. Therefore even if the direct identification from the smallest scale to the macro-scale cannot be attained directly, the just described three steps process is rather successful, at least when limiting ourselves to numerical identifications. However the numerical tuning of constitutive parameters can be driven by some

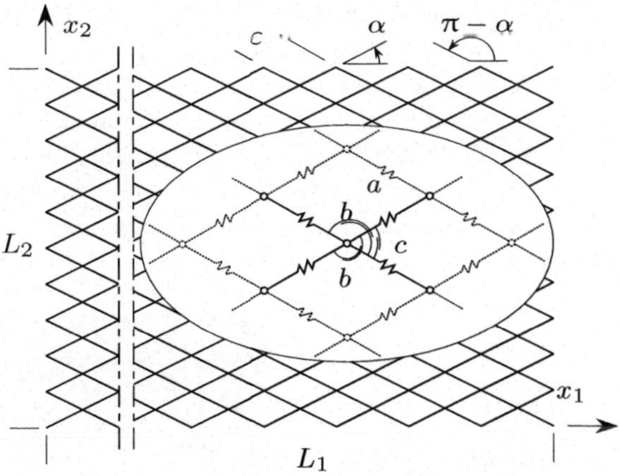

Fig. 33.2 Meso-mechanical model for the in-plane mechanical behavior of a pantographic lattice with non-orthogonal beams.

[1] The identification process is not so simple if we want to model the whole nonlinear mechanical behavior of the pantographic lattice by using, as we discuss later, only three parameters.

theoretical knowledge: indeed the main geometrical properties, together with the Young and Poisson coefficients of the Cauchy material used as a model at the smallest scale, can give an order of magnitude for the values of those extensional and rotational stiffnesses which are correct at the meso-scale.

The aforementioned general three-steps procedure is valid even if the lowest length scale has different values. Actually it has been experimentally proven that in both cases, when the smallest scale is of the order of microns and when the smallest scale is of the order of tenth of milli-meters, it can be applied successfully. Therefore the same numerical procedure for getting predictive models can be usefully applied in these two different cases simply varying a few constitutive parameters. We believe however that, when going down to lower length scales, some quantum effects may arise and in such cases the applicability of Cauchy models may be disputable.

In this work, after this brief introduction, we firstly describe in Sect. 33.2 the pantographic micro-structure starting from the presentation of the fundamental brick or unit. Successively, in Sect. 33.3 we propose a strategy to determine the triplets of stiffnesses which completely define the Lagrangian meso-mechanical model and we use, in Sect. 33.4, the obtained results to estimate the macro-mechanical parameters following the identification described in dell'Isola et al (2016). Some concluding remarks, in Sect. 33.5 close the paper along with some future challenges.

33.2 Description of Pantographic Units used to form Pantographic Micro-structures

The crucial feature of pantographic modules consists in the presence of pivots as interconnecting structural elements to link different, rectilinear or curvilinear, beams. Figure 33.3, for instance, is a shot of a pantographic lattice with orthogonal fibers inside a testing machine immediately before an elongation test. We remark that the obtained assembly of beams is, generally speaking, a lattice and not a truss. However, equivalent truss structures have been conceived having the same behavior as pantographic modules, see, *e.g.*, , Seppecher et al (2011); Turco et al (2017a); Khakalo and Niiranen (2018). The presence of pivots in a specifically designed geometrical pattern induce the existence of so-called *floppy modes*, *i.e.*, deformation modes whose associated deformation energy is vanishing or very small when compared

Fig. 33.3 Shot of a pantographic lattice with orthogonal fibers inside a testing machine immediately before an elongation test.

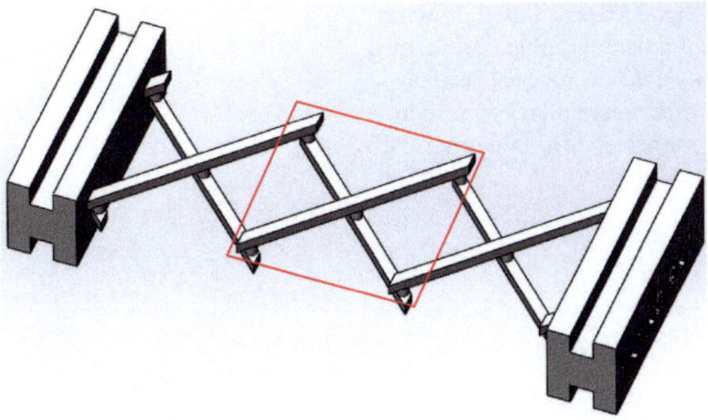

Fig. 33.4 Pantographic lattice formed by three units (the central unit is highlighted by means of a red box).

with the other deformation modes. The existence of floppy modes implies that at a macro-level the deformation energy exhibits some singularities and therefore some new mathematical tools were needed to start studying the well-posed problems for pantographic metamaterials, see, *e.g.*, , Boutin et al (2017).

Looking at Fig. 33.4 we distinguish three units each one formed by beams having rectangular cross-section and pivots, that is cylinders with circular cross-section. An exploded drawing of, for example, the central unit is reported in Fig. 33.5 highlighting the dimensions of the ℓ_b long beam with rectangular $b_b \times h_b$ cross-section and those of the circular cross-section cylinder (d_c and h_c are the diameter and the height, respectively).

Although in this work we refer to orthogonal families of rectilinear beam, there are examples, which show some interesting peculiarities, of non-orthogonal rectilinear beams, see Turco et al (2017b), and curvilinear beams, see Scerrato et al (2016); Giorgio et al (2016).

The composition of complex structures by using pantographic modules allows for the design of

1. pantographic beams: they can be obtained by a sequence of units which growths in one direction, see, *e.g.*, , Barchiesi et al (2018a); Birsan et al (2012);
2. pantographic sheets: they can be obtained by a sequence of units which growths in two directions, see, *e.g.*, , dell'Isola et al (2016); Eremeyev et al (2018); Khakalo et al (2018); Eremeyev and dell'Isola (2018) and the reviews Spagnuolo and Andreaus (2018); Laudato and Di Cosmo (2018); Golaszewski et al (2018);
3. pantographic blocks: they can be obtained by a sequence of lattices which growths in one direction, see, *e.g.*, , dell'Isola et al (2018).

33.3 How the unit Properties Determine the Meso-stiffnesses

Once the geometry is perfectly described we need the parameters which describe the mechanical behavior of the used material. Several specimens were built using 3D

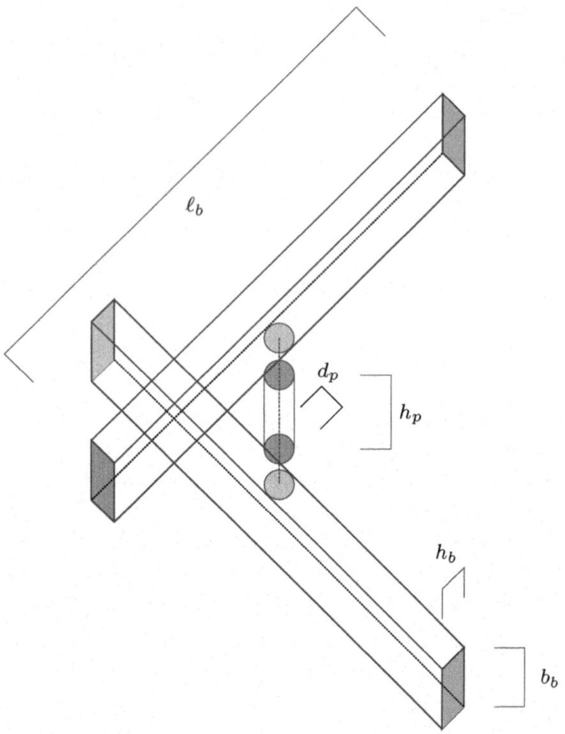

Fig. 33.5 Exploded drawing of a pantographic unit formed by two orthogonal beams with rectangular cross-section connected with a cylindrical pivot.

printing process both using as material the polyamide and aluminum or steel. Usual printing process are based on melting processes of pulverized material. This kind of process may alter the mechanical parameters of the used material. In addition, as we will discuss in the following, since it may introduce some hollows which surely could alter appreciably the mechanical parameters of the material. For the moment, we may start with the value of the Young modulus Y and the Poisson ratio ν given for the material used in the printing process. For example, the polyamide, following EN ISO 527, has $Y = 1700$ MPa and $\nu = 0.4$.

In Turco et al (2016c) a completely discrete Lagrangian model has been introduced to capture the in-plane mechanical response of pantographic lattice. It is based on three mechanical parameters for modeling the stretching, bending and shearing strain energy of the whole lattice[2]. The basic idea derives from the Hencky approach for modeling the bending strain energy. Roughly speaking, each beam is modeled by a series of rigid links and elastic joints (rotational springs of stiffness b). Increasing the number of the elastic joints, Hencky obtained estimates approaching to the bifurcation load of the *Elastica*. It is almost simple to prove, see Turco (2018); Turco et al (2018b), that using de Saint-Venant results for a beam under bending we might

[2] This idea has been enhanced, improving the bending energy by means of a h-refinement, in Turco et al (2018a,b).

estimate the parameter b form the geometry and the Young modulus. In practice, when the number of elastic joints is large enough[3], we have:

$$b = \frac{YJ}{\ell}, \quad \text{being} \quad J = \frac{b_b h_b^3}{12}. \tag{33.1}$$

In the same way we might estimate the spring stiffness a used to model the stretching strain energy. In formula:

$$a = \frac{YA}{\ell}, \quad \text{being} \quad A = b_b h_b. \tag{33.2}$$

The third parameter c is related to the torsional stiffness of the cylinder with circular cross-section or, briefly, of the pivot. Also in this case we can use the de Saint-Venant solution for the torsion to estimate c:

$$c = \frac{GI_p}{h_c}, \quad \text{being} \quad G = \frac{Y}{2(1+\nu)} \quad \text{and} \quad I_p = \frac{\pi d_p^4}{32}. \tag{33.3}$$

The aforementioned road to estimate the parameters is very simple, and therefore attractive, but it is based on hypotheses too far from the phenomenon which we want do describe. In particular:

1. de Saint-Venant results are accurate for beams and pivots long enough (respect to the dimensions of their cross-section); this is not verified for pantographic lattices where the ratios ℓ_b/h_b and h_p/d_p are approximatively, see again Fig. 33.5 for the meaning of the used symbols, 18.4 and 2.22, respectively;
2. the intrinsic porosity of the material obtained by 3D printing process reduces, often in a remarkable way, the Young modulus of the printed specimens; Fig. 33.6 shows a part of a beam using three different, and increasing, magnifications; it is almost clear the granular nature of the printed material, furthermore, micro X-ray computed tomography analysis show the presence of hollows of not negligible dimensions in the printed material;
3. the theoretical geometry is only an approximation of the true one, see again Fig. 33.6, however the presence of errors on the stiffness parameters gives stable results as is proved in Turco and Rizzi (2016);
4. in almost all the experiments pantographic lattices undergo large displacements, this is completely far from the de Saint-Venant results.

For all the aforementioned reasons the values suggested from the de Saint-Venant solutions can only be a starting point for an accurate enough estimation of the stiffness parameters of the discrete model. Refined values of these estimates can be derived comparing the results of numerical simulations[4] with those deriving from

[3] In Turco (2018) some quantitative results relative to the buckling load of an elastic beam are reported. Two and five rigid links estimate the buckling load with an error of 18.9% and 2.23%, respectively.

[4] Numerical simulations whose results are reported in Figs. 33.7 and 33.8 were performed by use an in-house made code which considers large displacements (but neglect material nonlinearities

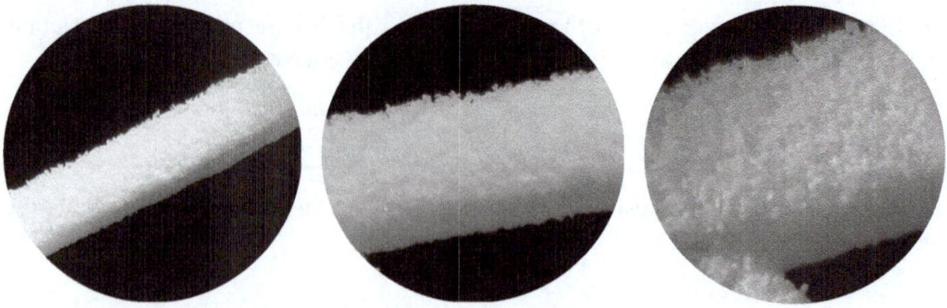

Fig. 33.6 Blow-up sequence of a pantographic lattice part built by a 3D printing process using polyamide powder; the sequence clearly shows the granular nature of the specimen.

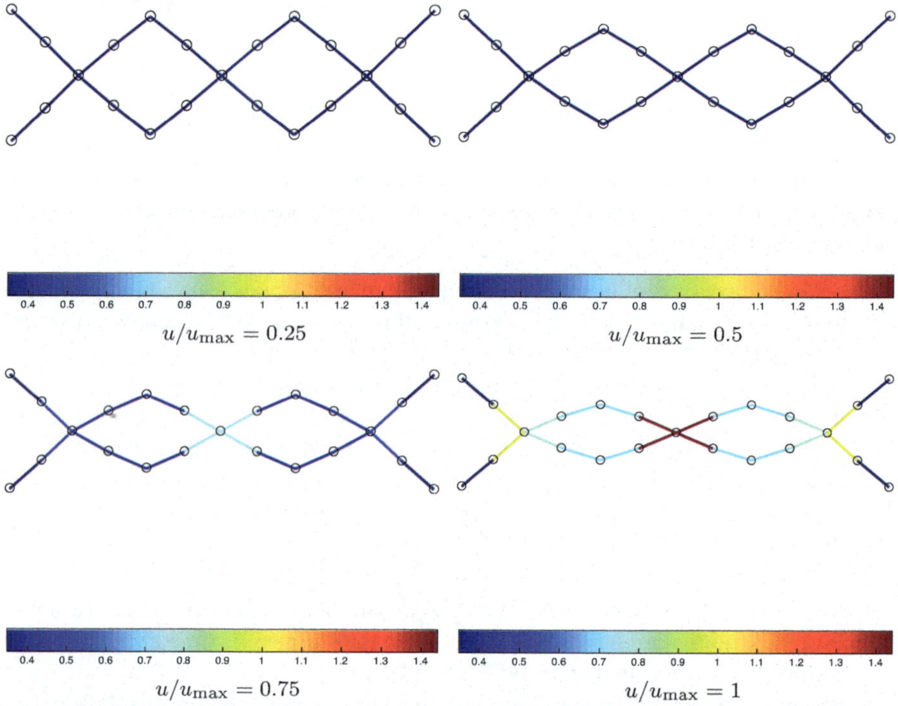

Fig. 33.7 Extensional test with controlled displacement u on the right side nodes: sequence of deformations for $u/u_{\max} = 0.25, 0.5, 0.75$ and 1 for a pantographic lattice formed by three units (colors show the strain energy level achieved on beams, small circles represent the nodes of the discrete model).

experiments following the methodologies reported in Turco (2017). In practice,

---

and viscous effects) modeling the pantographic lattices as a set of rigid links and elastic joints to approximate the bending and shearing strain energy. The bending contributions are improved by adding an intermediate node between two close pivots (pivots and additional nodes are drawn in Fig. 33.7 using small circles). In addition, the model considers also the stretching energy, see Turco by choosing the triplet (a, b, c) which fits the experimental results, see Fig. 33.8, *i.e.*, the load-displacement curve and the deformations at some prefixed points of the curve, see Fig. 33.7 which reports the displacements for $u/u_{\max} = 0.25, 0.5, 0.75$ and 1 along with the strain energy level achieved on beams by means of colors for the pantograph depicted in Fig. 33.4. We remark that the nonlinearity of the force-displacement curve is only due to the large displacements attained both in the experiment and in its numerical simulation. Even if the viscoelastic effect is, in general, present in the polyamide, here we have not considered this effect in the numerical simulations since the experimental loading rate was designed to remove viscous phenomena.

In other words minimizing the discrepancy between the experiment and its numerical simulation in a least square fashion:[5]

$$\arg\min_{(a,b,c)} \left(\|\mathbf{n}(a,b,c) - \mathbf{m}\|^2 \right) , \tag{33.4}$$

being $\mathbf{n}$ and $\mathbf{m}$ the vector collecting the numerical and measured, in some experiment, response, respectively. We remark that the vector $\mathbf{m}$, and consequently $\mathbf{n}$, can collect different kind of information such as the measured force and displacements in a prefixed number of points of the structure both corresponding to an assigned displacement (if we consider a load-displacement curve obtained by assigning the displacement).

In addition, since experimental measurements surely include errors, for example those deriving from the instrumental precision, the least square formulation (33.4) can be made able to filter these errors by an additional term which imposes, by means a Lagrangian multiplier λ, a desired condition, in formula:[6]

$$\arg\min_{(a,b,c)} \left(\|\mathbf{n}(a,b,c) - \mathbf{m}\|^2 + \lambda^2 \|\mathbf{c}\|^2 \right) , \tag{33.5}$$

having used the constrain condition $\mathbf{c}$. In several cases to filter the errors in $\mathbf{m}$ it is enough to choose the solution which has the minimum norm corresponding to $\mathbf{c} = [a\, b\, c]^T$, see again Turco (2017).

Particularly useful for a fixed pantographic lattice is the plot of the load-displacement, such that reported in Fig. 33.9, which immediately suggests the values of the triplet (a, b, c) which best fit the experimental load-displacement curve.

---

et al (2018b) for more details. Numerical simulations try to reproduce the experimental tests whose results are displacement-controlled and the whole test reproduce quasi-static results.

[5] This method is also known as Levenberg-Marquardt algorithm.

[6] This method is also known as damped Levenberg-Marquardt algorithm.

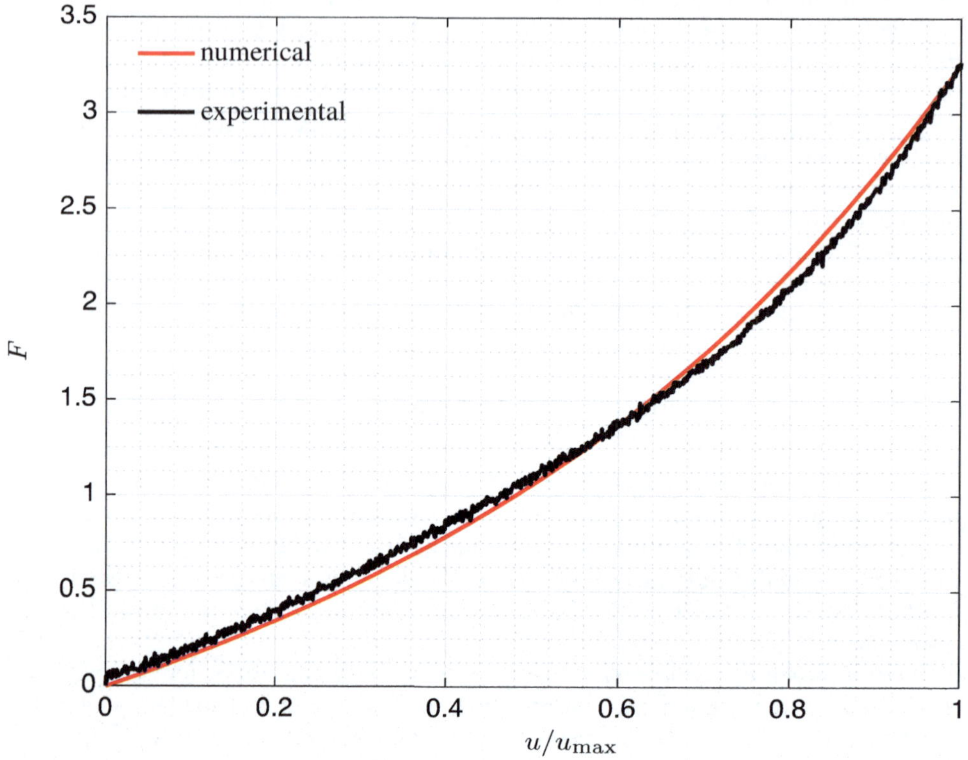

Fig. 33.8 Extensional test with controlled displacement u on the right side nodes: experimental, in black, and best fitting, in red, of the load-displacement curve for a pantographic lattice formed by three units.

33.4 Meso-macro Identification

The original work by Hencky deals with beams, see Hencky (1921). Hencky wanted to find a discrete Lagrangian system which approximates the Euler beam, when the number of used beams tends to infinity. The beams considered by Hencky are inextensible and the Euler bending stiffness is obtained by introducing suitable rotational springs. It is remarkable to note that already Piola (dell'Isola et al, 2014, 2019), although for different epistemological reasons, wanted to introduce a discrete Lagrangian system to approximate Euler continuum beam. Piola was motivated by the need of justifying physically the Euler's and Bernoulli's dependence of deformation energy on curvature. Therefore the correspondence between the Euler's continuum model and Hencky's discrete model has been established by Piola for justifying some modeling choices, while by Hencky for practical computing purposes.

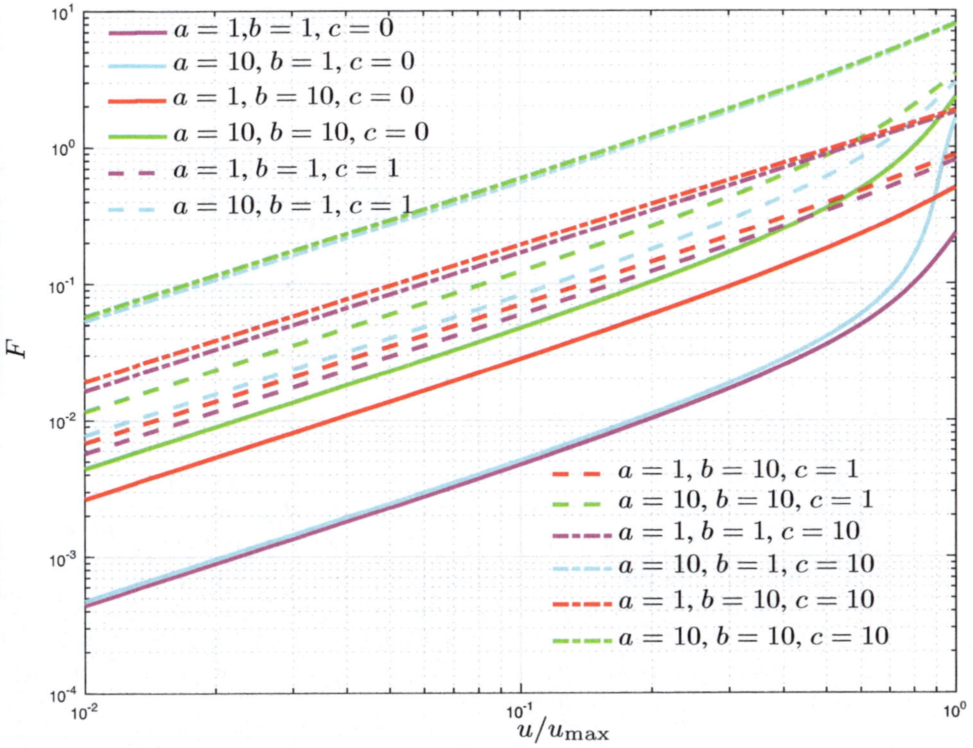

Fig. 33.9 Load-displacement curves useful for a quick estimate of the stiffness triplet (a, b, c).

The work of Piola and Hencky gave us the conceptual tool for finding, from the constitutive parameters of the meso-scale model the corresponding values of the constitutive parameters of the second gradient plate theory to be used at macro-level for pantographic sheets. The material coefficients for macro-generalised plate theory have to be determined in terms of the meso-springs stiffnesses. Here, it is crucial the procedure established in dell'Isola et al (2016). One has to remark that usually the identification procedures have been studied mainly in linear regimes (even if some remarkable results were obtained in Alibert and Della Corte (2015); Seppecher et al (2011); Braides et al (2018). The presented identification process allows for the treatment of problems in which large displacements occur.

The meso-stiffness parameters a, b and c identified by the process described in Sect. 33.3 can be used to identify the macro-stiffness parameters A, B and C to be used in an in-plane model such as well-described in dell'Isola et al (2016) or in an out-of-plane model such as presented in Giorgio et al (2018). Starting from Piola's ansatz, a straightforward procedure gives:

$$
\begin{aligned}
A &= a\,,\\
B &= b\,,\\
C &= \frac{c}{\varepsilon^2}\,,
\end{aligned}
\tag{33.6}
$$

being ε the distance between two successive elastic joints.[7] We remark that in dell'Isola et al (2016) the energy associated to the shearing part has a ρ exponent, *i.e.*, it is not quadratic in general, as that presented in Turco et al (2016c), therefore continuum and discrete models can be linked only in the case $\rho = 2$.

Equation (33.6) represents a formal statement which is assuring that the considered system, at macro-level, behaves as a second gradient material. It is a sufficient condition which, once verified at meso-level, assures that the macro-system must have a deformation energy depending on second gradient of displacement. In an intuitive way one usually says that: *to get a macro second gradient material one has to have high contrast in meso-stiffnesses*. In Equation (33.6) small letters characterize meso-stiffnesses of the Hencky-type springs, while the corresponding capital letters refer to macro elasticity coefficients, including second gradient ones. The characteristic length-scales at macro-level are given by ratio square roots $\sqrt{B/A}$ and $\sqrt{B/C}$. Therefore it is evident that such length-scales can be different from ε, when suitable choices for the values of meso-stiffnesses are made. In order to have large boundary layers where second gradients of displacement are concentrated, ratio square roots $\sqrt{b/a}$ and $\sqrt{b/c}$ must be very large. This statement substantiates the intuitive statement recalled few lines before.

33.5 Concluding Remarks and Future Challenges

The problem of designing novel metamaterials requires a modeling capability which includes the possibility to optimize structural parameters, see dell'Isola et al (2018). In the context of pantographic microstructures the optimization required involves the capacity to perform at high speed complex numerical calculations. Indeed the pantographic sheets are intended to supply a scientific concept suitable to obtain as technological output a metamaterial capable to undergo large deformations and displacements while remaining in elastic regimes.

Of course the existence of elongation floppy modes favors this kind of performances. Indeed the pantographic module has been conceived, see Alibert et al (2003) exactly to get an approximation of ideal pure second gradient materials exhibiting elongations with null or very small deformation energies. The initial motivation was purely scientific, however some demands from aeronautical engineering require the development of such kind of materials in the effort to build light composite structures for flying vehicles.

[7] The relation between macro- and micro-stiffnesses reported in Eq. (33.6) derive from the micro-macro identification based on the Piola's ansatz. A detailed explanation of the computations necessary to reach these results is reported in dell'Isola et al (2016).

The pure demand of existence of elongation floppy modes is not sufficient to calculate the most suitable microstructure for a given tailored purpose. Actually an optimization process is unavoidable. This process requires large computing burdens and, therefore, imposes the formulation of a modeling procedure which is the simplest possible. While it is undoubted that the modeling at the lowest scale is both more detailed and more complex if one wants to obtain predictive specifications, the macro continuum model, although of scientific interest, do requires a formulation of a discretized version to be implemented into a numerical code. It is therefore our opinion that the most suitable modeling procedure is what we have called meso-model, as we believe to have proven in the argumentations presented in this paper.

Finally, we list some open problems which we will tackle in the next future:

1. even if the proposed method seems to produce reliable results, we are strongly interested to any strategies able to correlate experimental measurements to the identification of stiffnesses, *e.g.*, , that reported in Placidi et al (2015);
2. it will be interesting to consider problems where the inertia forces are non negligible and therefore the hypothesis of quasi-static application of external loads or given displacements is not close enough to well-describe the underline problem; in these cases, following the guidelines reported in Giorgio et al (2017a); Engelbrecht and Berezovski (2015); Javili et al (2015), might be useful to verify if the proposed parameter identification furnishes results equivalent in accuracy to those obtainable in quasi-static loading, see, *e.g.*, , di Cosmo et al (2018); Abd-alla et al (2017); Berezovski et al (2016);
3. in this work we only considered the simplest law for describing the strain energy; however, the proposed strategy is open to consider more sophisticated models as in Braides et al (2018); Atai and Steigmann (1997); Challamel et al (2014); Placidi and Barchiesi (2018); Turco et al (2016d);
4. since the meso-mechanical model able to represent the behavior of pantographic structures can be considered as a Representative Elementary Volume (REV) for a continuum models, see, for example, the attempt Andreaus et al (2018), and that this has to be surely discretized, modern interpolation laws such as based on B-splines and NURBS, see Piegl and Tiller (1997); Cottrell et al (2009); Greco et al (2017); Greco and Cuomo (2013); Cuomo et al (2014); Balobanov and Niiranen (2018); Cazzani et al (2016), look like interesting;
5. discrete models are the fundamental brick to construct continuum models, see, *e.g.*, , Shirani et al (2018), able to treat plane or three-dimensional problems, see Turco (2018); Eugster et al (2014), or to consider problems involving foams using the suggestions reported in De Masi et al (2008, 2009); Grimmett (2016);
6. finally, it appears interesting and fruitful to consider as an alternative approach the peridynamic formulation, see *e.g.*, Diyaroglu et al (2015); Meo et al (2016).

References

Abali BE, Wu CC, Müller WH (2016) An energy-based method to determine material constants in nonlinear rheology with applications. Continuum Mechanics and Thermodynamics 28(5):1221–1246

Abd-alla AN, Alshaikh F, Del Vescovo D, Spagnuolo M (2017) Plane waves and eigenfrequency study in a transversely isotropic magneto-thermoelastic medium under the effect of a constant angular velocity. Journal of Thermal Stresses 40(9):1079–1092

Alibert JJ, Della Corte A (2015) Second-gradient continua as homogenized limit of pantographic microstructured plates: a rigorous proof. Zeitschrift für Angewandte Mathematik und Physik 66(5):2855–2870

Alibert JJ, Seppecher P, dell'Isola F (2003) Truss modular beams with deformation energy depending on higher displacement gradients. Mathematics and Mechanics of Solids 8(1):51–73

Andreaus U, Spagnuolo M, Lekszycki T, Eugster SR (2018) A Ritz approach for the static analysis of planar pantographic structures modeled with nonlinear euler–bernoulli beams. Continuum Mechanics and Thermodynamics pp 1–21

Atai A, Steigmann DJ (1997) On the nonlinear mechanics of discrete networks. Archive of Applied Mechanics 67(5):303–319

Balobanov V, Niiranen J (2018) Locking-free variational formulations and isogeometric analysis for the Timoshenko beam models of strain gradient and classical elasticity. Computer Methods in Applied Mechanics and Engineering 339:137–159

Barchiesi E, Placidi L (2017) A review on models for the 3d statics and 2d dynamics of pantographic fabrics. In: Sumbatyan MA (ed) Wave Dynamics and Composite Mechanics for Microstructured Materials and Metamaterials, Advanced Structured Materials, vol 59, Springer, Singapore, pp 239–258

Barchiesi E, dell'Isola F, Laudato M, Placidi L, Seppecher P (2018a) A 1D continuum model for beams with pantographic microstructure: asymptotic micro-macro identification and numerical results. In: dell'Isola F, Eremeyev V, Porubov A (eds) Advances in Mechanics of Microstructured Media and Structures. Advanced Structured Materials, Springer

Barchiesi E, Ganzosch G, Liebold C, Placidi L, Grygoruk R, Müller WH (2018b) Out-of-plane buckling of pantographic fabrics in displacement-controlled shear tests: experimental results and model validation. Continuum Mechanics and Thermodynamics

Barchiesi E, Spagnuolo M, Placidi L (2018c) Mechanical metamaterials: a state of the art. Mathematics and Mechanics of Solids 24(1):212–234

Berezovski A, Giorgio I, Della Corte A (2016) Interfaces in micromorphic materials: wave transmission and reflection with numerical simulations. Mathematics and Mechanics of Solids 21(1):37–51

Birsan M, Altenbach H, Sadowski T, Eremeyev VA, Pietras D (2012) Deformation analysis of functionally graded beams by the direct approach. Composites Part B: Engineering 43(3):1315–1328

Boutin C, dell'Isola F, Giorgio I, Placidi L (2017) Linear pantographic sheets. Asymptotic micro-macro models identification. Mathematics and Mechanics of Complex Systems 5(2):127–162

Braides A, Causin A, Solci M (2018) A homogenization result for interacting elastic and brittle media. Proceedings of the Royal Society of London A: Mathematical, Physical and Engineering Sciences 474(2218)

Cazzani A, Malagù M, Turco E (2016) Isogeometric analysis of plane curved beams. Mathematics and Mechanics of Solids 21(5):562–577

Challamel N, Lerbet J, Wang CM (2014) On buckling of granular columns with shear interaction: Discrete versus nonlocal approaches. Journal of Applied Physics 115(23):234,902

Cottrell JA, Hughes TJR, Bazilevs Y (2009) Isogeometric Analysis: Toward Integration of CAD and FEA. Wiley

Cuomo M, Contraffatto L, Greco L (2014) A variational model based on isogeometric interpolation for the analysis of cracked bodies. International Journal of Engineering Science 80:173–188

De Masi A, Merola I, Presutti E, Vignaud Y (2008) Potts models in the continuum uniqueness and exponential decay in the restricted ensembles. Journal of Statistical Physics 133(2):281–345

De Masi A, Merola I, Presutti E, Vignaud Y (2009) Coexistence of ordered and disordered phases in Potts models in the continuum. Journal of Statistical Physics 134(2):243–306

dell'Isola F, Maier G, Perego U, Andreaus U, Esposito R, Forest S (2014) The complete works of Gabrio Piola: Volume I - Commented English Translation. Springer International Publishing

dell'Isola F, Andreaus U, Placidi L (2015) At the origins and in the vanguard of peridynamics, non-local and higher-gradient continuum mechanics: An underestimated and still topical contribution of Gabrio Piola. Mathematics and Mechanics of Solids 20(8):887–928

dell'Isola F, Giorgio I, Pawlikowski M, Rizzi NL (2016) Large deformations of planar extensible beams and pantographic lattices: Heuristic homogenisation, experimental and numerical examples of equilibrium. Proceedings of the Royal Society of London A: Mathematical, Physical and Engineering Sciences 472(2185):1–23

dell'Isola F, Seppecher P, Alibert JJ, Lekszycki T, Grygoruk R, Pawlikowski M, Steigmann DJ, Giorgio I, Andreaus U, Turco E, Gołaszewski M, Rizzi N, Boutin C, Eremeyev VA, Misra A, Placidi L, Barchiesi E, Greco L, Cuomo M, Cazzani A, Della Corte A, Battista A, Scerrato D, Zurba Eremeeva I, Rahali Y, Ganghoffer JF, Muller W, Ganzosch G, Spagnuolo M, Pfaff A, Barcz K, Hoschke K, Neggers J, Hild F (2018) Pantographic metamaterials: an example of mathematically driven design and of its technological challenges. Continuum Mechanics and Thermodynamics 10.1007/s00161-018-0689-8

dell'Isola F, Maier G, Perego U, Andreaus U, Esposito R, Forest S (2019) The complete works of Gabrio Piola: Volume II - Commented English Translation. Springer International Publishing

di Cosmo F, Laudato M, Spagnuolo M (2018) Acoustic metamaterials based on local resonances: Homogenization, optimization and applications. In: Altenbach H, Pouget J, Rousseau M, Collet B, Michelitsch T (eds) Generalized Models and Non-classical Approaches in Complex Materials 1, Advanced Structured Materials, vol 89, Springer International Publishing, Cham, pp 247–274

Diyaroglu C, Oterkus E, Oterkus S, Madenci E (2015) Peridynamics for bending of beams and plates with transverse shear deformation. International Journal of Solids and Structures 69:152–168

Engelbrecht J, Berezovski A (2015) Reflections on mathematical models of deformation waves in elastic microstructured solids. Mathematics and Mechanics of Complex Systems 3(1):43–82

Eremeyev V, dell'Isola F (2018) A note on reduced strain gradient elasticity. In: Altenbach H, Pouget J, Rousseau M, Collet B, Michelitsch T (eds) Generalized Models and Non-classical Approaches in Complex Materials 1, Advanced Structured Materiasl, vol 89, Springer International Publishing, Cham, pp 301–310

Eremeyev VA, dell'Isola F, Boutin C, Steigmann D (2018) Linear pantographic sheets: Existence and uniqueness of weak solutions. Journal of Elasticity 132(2):175–196

Eugster SR, Hesch C, Betsch P, Glocker C (2014) Director-based beam finite elements relying on the geometrically exact beam theory formulated in skew coordinates. International Journal for Numerical Methods in Engineering 97(2):111–129

Franciosi P, Spagnuolo M, Salman OU (2018) Mean Green operators of deformable fiber networks embedded in a compliant matrix and property estimates. Continuum Mechanics and Thermodynamics pp 1–32

Giorgio I, Grygoruk R, dell'Isola F, Steigmann DJ (2015) Pattern formation in the three-dimensional deformations of fibered sheets. Mechanics Research Communications 69:164–171

Giorgio I, Della Corte A, dell'Isola F, Steigmann DJ (2016) Buckling modes in pantographic lattices. Comptes Rendus - Mécanique 344(7):487–501

Giorgio I, Della Corte A, dell'Isola F (2017a) Dynamics of 1D nonlinear pantographic continua. Nonlinear Dynamics 88(1):21–31

Giorgio I, Rizzi NL, Turco E (2017b) Continuum modelling of pantographic sheets for out-of-plane bifurcation and vibrational analysis. Proceedings of the Royal Society A: Mathematical, Physical and Engineering Sciences 473(20170636):1–21

Giorgio I, Harrison P, dell'Isola F, Alsayednoor J, Turco E (2018) Wrinkling in engineering fabrics: a comparison between two different comprehensive modelling approaches. Proceedings of the Royal Society A: Mathematical, Physical and Engineering Sciences 474(20180063):1–20

Golaszewski R Mand Grygoruk, Giorgio I, Laudato M, di Cosmo F (2018) Metamaterials with relative displacements in their microstructure: technological challenges in 3D printing, experiments and numerical predictions. Continuum Mechanics and Thermodynamics

Greco L, Cuomo M (2013) B-Spline interpolation of Kirchhoff–Love space rods. Computer Methods in Applied Mechanics and Engineering 256:251–269

Greco L, Cuomo M, Contraffatto L, Gazzo S (2017) An efficient blended mixed B-spline formulation for removing membrane locking in plane curved Kirchhoff rods. Computer Methods in Applied Mechanics and Engineering 324:476–511

Grimmett GR (2016) Correlation inequalities for the Potts model. Mathematics and Mechanics of Complex Systems 4(3-4):327–334

Hencky H (1921) Über die angenäherte lösung von stabilitätsproblemen im raum mittels der elastischen gelenkkette. PhD thesis, Engelmann

Javili A, Dortdivanlioglu B, Kuhl E, Linder C (2015) Computational aspects of growth-induced instabilities through eigenvalue analysis. Computational Mechanics 56(3):405–420

Karamooz MR, Kadkhodaei M (2015) A computationally efficient modeling approach for predicting mechanical behavior of cellular lattice structures. Journal of Materials Engineering and Performance 24(1):245–252

Karamooz MR, Kadkhodaei M, Badrossamay M, Rezaei R (2014) Numerical investigation on mechanical properties of cellular lattice structures fabricated by fused deposition modeling. International Journal of Mechanical Sciences 88:154–161

Khakalo S, Niiranen J (2018) Form II of Mindlin's second strain gradient theory of elasticity with a simplification: for materials and structures from nano- to macro-scales. European Journal of Mechanics A/Solids (to appear)

Khakalo S, Balobanov V, Niiranen J (2018) Modelling size-dependent bending, buckling and vibrations of 2D triangular lattices by strain gradient elasticity models: applications to sandwich beams and auxetics. Journal of Engineering Science 127:33–52

Laudato M, Di Cosmo F (2018) Euromech 579 Arpino 3-8 April 2017: Generalized and microstructured continua: new ideas in modeling and/or applications to structures with (nearly) inextensible fibers - a review of presentations and discussions. Continuum Mechanics and Thermodynamics pp 1–15

Lekszycki T, Di Cosmo F, Laudato M, Vardar O (2018) Application of energy measures in detection of local deviations in mechanical properties of structural elements. Continuum Mechanics and Thermodynamics

Meo DD, Diyaroglu C, Zhu N, Oterkus E, Siddiq MA (2016) Modelling of stress-corrosion cracking by using peridynamics. International Journal of Hydrogen Energy 41(15):6593–6609

Piegl L, Tiller W (1997) The NURBS Book, 2nd edn. Springer-Verlag Berlin Heidelberg

Placidi L, Barchiesi E (2018) Energy approach to brittle fracture in strain-gradient modelling. Proceedings of the Royal Society A: Mathematical, Physical and Engineering Sciences 474(20170878):1–19

Placidi L, El Dhaba AR (2017) Semi-inverse method à la Saint-Venant for two-dimensional linear isotropic homogeneous second-gradient elasticity. Mathematics and Mechanics of Solids 22(5):919–937

Placidi L, Andreaus U, Della Corte A, Lekszycki T (2015) Gedanken experiments for the determination of two-dimensional linear second gradient elasticity coefficients. Zeitschrift für Angewandte Mathematik und Physik (ZAMP) 66(6):3699–3725

Placidi L, Andreaus U, Giorgio I (2017a) Identification of two-dimensional pantographic structure via a linear D4 orthotropic second gradient elastic model. Journal of Engineering Mathematics 103(1):1–21

Placidi L, Barchiesi E, Battista A (2017b) An inverse method to get further analytical solutions for a class of metamaterials aimed to validate numerical integrations. In: dell'Isola F, Sofonea M,

Steigmann D (eds) Mathematical Modelling in Solid Mechanics, Springer Singapore, Singapore, pp 193–210

Placidi L, Barchiesi E, Della Corte A (2017c) Identification of two-dimensional pantographic structures with a linear D4 orthotropic second gradient elastic model accounting for external bulk double forces. In: dell'Isola F, Sofonea M, Steigmann D (eds) Mathematical Modelling in Solid Mechanics, Springer Singapore, Singapore, pp 211–232

Placidi L, Barchiesi E, Misra A (2018a) A strain gradient variational approach to damage: a comparison with damage gradient models and numerical results. Mathematics and Mechanics of Complex Systems 6(2):77–100

Placidi L, Misra A, Barchiesi E (2018b) Two-dimensional strain gradient damage modeling: a variational approach. Zeitschrift für angewandte Mathematik und Physik 69(3):56

Saeb S, Steinmann P, Javili A (2016) Aspects of computational homogenization at finite deformations: A unifying review from Reuss' to Voigt's bound. Applied Mechanics Reviews 68:050,801

Scerrato D, Giorgio I, Rizzi NL (2016) Three-dimensional instabilities of pantographic sheets with parabolic lattices: numerical investigations. ZAMP - Journal of Applied Mathematics and Physics 67(53):1–19

Seppecher P, Alibert JJ, dell'Isola F (2011) Linear elastic trusses leading to continua with exotic mechanical interactions. In: Journal of Physics: Conference Series, IOP Publishing, vol 319(1), p 012018

Shirani M, Luo C, Steigmann DJ (2018) Cosserat elasticity of lattice shells with kinematically independent flexure and twist. Continuum Mechanics and Thermodynamics pp 1–11

Spagnuolo M, Andreaus U (2018) A targeted review on large deformations of planar elastic beams: extensibility, distributed loads, buckling and post-buckling. Mathematics and Mechanics of Solids p 108128651773700

Steigmann DJ, dell'Isola F (2015) Mechanical response of fabric sheets to three-dimensional bending, twisting, and stretching. Acta Mechanica Sinica 31(3):373–382

Turco E (2017) Tools for the numerical solution of inverse problems in structural mechanics: review and research perspectives. European Journal of Environmental and Civil Engineering 21(5):509–554

Turco E (2018) Discrete is it enough? the revival of Piola–Hencky keynotes to analyze three-dimensional elastica. Continuum Mechanics and Thermodynamics 30(5):1039–1057

Turco E, Rizzi NL (2016) Pantographic structures presenting statistically distributed defects: numerical investigations of the effects on deformation fields. Mechanics Research Communications 77:65–69

Turco E, Barcz K, Pawlikowski M, Rizzi NL (2016a) Non-standard coupled extensional and bending bias tests for planar pantographic lattices. Part I: numerical simulations. Zeitschrift für Angewandte Mathematik und Physik 67(122):1–16

Turco E, Barcz K, Rizzi NL (2016b) Non-standard coupled extensional and bending bias tests for planar pantographic lattices. Part II: comparison with experimental evidence. Zeitschrift für Angewandte Mathematik und Physik 67(123):1–16

Turco E, dell'Isola F, Cazzani A, Rizzi NL (2016c) Hencky-type discrete model for pantographic structures: numerical comparison with second gradient continuum models. Zeitschrift für Angewandte Mathematik und Physik 67(4):1–28

Turco E, dell'Isola F, Rizzi NL, Grygoruk R, Müller WH, Liebold C (2016d) Fiber rupture in sheared planar pantographic sheets: numerical and experimental evidence. Mechanics Research Communications 76:86–90

Turco E, Golaszewski M, Cazzani A, Rizzi NL (2016e) Large deformations induced in planar pantographic sheets by loads applied on fibers: experimental validation of a discrete Lagrangian model. Mechanics Research Communications 76:51–56

Turco E, Giorgio I, Misra A, dell'Isola F (2017a) King post truss as a motif for internal structure of (meta)material with controlled properties. Royal Society Open Science 4(171153)

Turco E, Golaszewski M, Giorgio I, D'Annibale F (2017b) Pantographic lattices with non-orthogonal fibres: experiments and their numerical simulations. Composites Part B: Engineering 118:1–14

Turco E, Misra A, Pawlikowski M, dell'Isola F, Hild F (2018a) Enhanced Piola–Hencky discrete models for pantographic sheets with pivots without deformation energy: numerics and experiments. International Journal of Solids and Structures 147:94–109

Turco E, Misra A, Sarikaya R, Lekszycki T (2018b) Quantitative analysis of deformation mechanisms in pantographic substructures: experiments and modeling. Continuum Mechanics and Thermodynamics pp 1–15

Yang H, Ganzosch G, Giorgio I, Abali BE (2018) Material characterization and computations of a polymeric metamaterial with a pantographic substructure. Zeitschrift für angewandte Mathematik und Physik 69(4):105

Chapter 34
Metallic Interconnection Technologies for High Power Vertical Cavity Surface Emitting Lasers Modules

Constanze Weber, Lena Goullon, Matthias Hutter & Martin Schneider-Ramelow

Abstract Highly reliable power VCSEL (Vertical Cavity Surface Emitting Lasers) array systems with an optimized optical output require a plan parallel assembly for a homogeneous radiation and an advanced packaging design to ensure good heat dissipation and an overall reliable performance. The aim of this study is to evaluate if metallic interconnection technologies like soldering and silver sintering can meet these requirements. Therefore, GaAs dies with VCSEL arrays of more than 2000 single lasers were mounted on substrates by soldering using AuSn20 and SnAg3 solder as well as by applying pressure assisted silver sintering. The samples were analyzed using ultrasonic microscopy (C-SAM), X-ray microscopy and 3D laser profilometry. Cross-sections of selected samples were made and analyzed using light- and scanning electron microscopy (SEM). Soldered and silver sintered samples were subjected to thermal cycling between -55°C and +125°C to validate the reliability of the metallic interconnects. Furthermore, it was tested, if it is possible to assemble a DCB onto a micro channel water cooler made of copper by pressure assisted silver sintering in order to enable an advanced heat transfer of the high power VCSEL module.

Keywords: Soldering · Silver sintering · Optoelectronic · Reliability

Constanze Weber · Lena Goullon · Matthias Hutter
Fraunhofer-Institut für Zuverlässigkeit und Mikrointegration IZM, Gustav-Meyer-Allee 25, 13355 Berlin, Germany,
e-mail: Constanze.Weber@izm.fraunhofer.de, Lena.Goullon@izm.fraunhofer.de, Matthias.Hutter@izm.fraunhofer.de

Martin Schneider-Ramelow
Technische Universität Berlin, Fachgebiet Werkstoffe der Hetero-Systemintegration, Gustav-Meyer-Allee 25 13355 Berlin, Germany,
e-mail: Martin.Schneider-Ramelow@izm.fraunhofer.de

B. E. Abali et al. (eds.), *New Achievements in Continuum Mechanics and Thermodynamics*, Advanced Structured Materials 108,
https://doi.org/10.1007/978-3-030-13307-8_34

34.1 Introduction

VCSELs are highly reliable and have a high potential for advanced and robust applications. They have been used for low power applications like laser mice or data sensing for more than 10 years already. By upscaling a single VCSEL to an array of more than 2000 VCSELs high power modules, like the one shown in Fig. 34.1, can be realized. Using these kind of arrays the application field of VCSELs can be en-

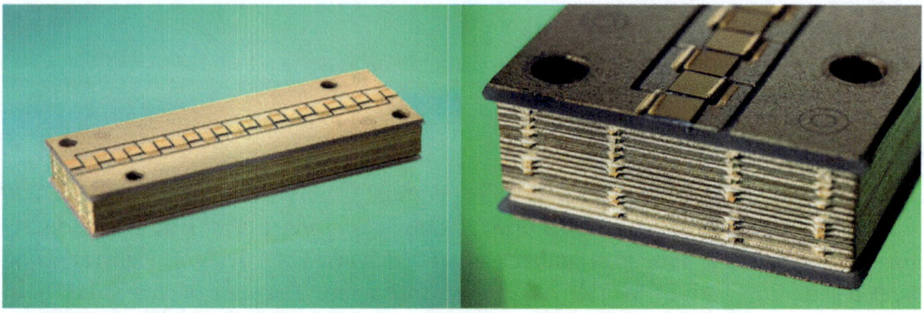

Fig. 34.1 High power VCSEL array module.

larged for example to build high power VCSEL pump modules for surface cleaning or infrared radiant heating (Moench et al, 2014).

The VCSEL module shown is built up with a DCB (direct copper bonded) assembled with 23 GaAs dies on top (1st level) and mounted on a copper micro channel water cooler (2nd level). The gap between the dies is 200 μm. Each GaAs die has a VCSEL array of more than 2000 single lasers as shown in Fig. 34.2.

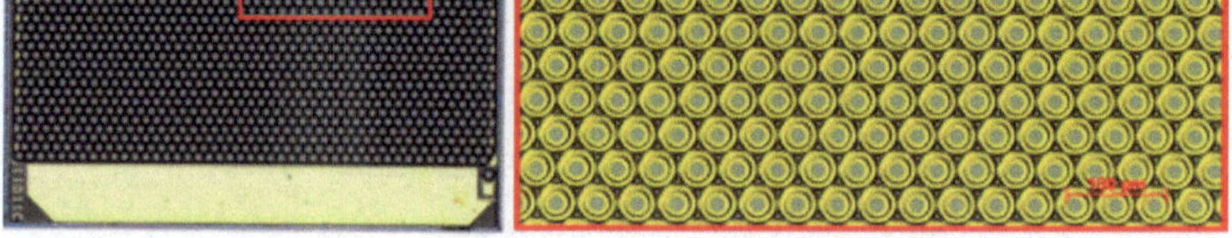

Fig. 34.2 GaAs die (4 mm^2) with VCSEL array of more than 2000 VSCELs.

The heat produced by such high integrated VCSEL arrays needs to be effectively transferred to a heat spreader and further to a cooling system. Therefore, high power VCSEL array systems require mounting technologies, which are reliable and enable an optimized heat transfer at 1st as well as at 2nd level.

Metallic interconnection technologies like soldering using AuSn20 and SnAg3 or silver sintering can meet these requirements (Hutter, 2009; Weber et al, 2014; Hutter et al, 2016).

However, depending on the interconnection technology chosen individual challenges have to be considered in order to obtain a reliable interconnect of high quality. In addition, direct VCSEL attach requires a very plan parallel assembly of the die in respect to the surface of the substrate. While for silver sintered joints this goal is rather easily achievable conducting soldering requires the use of very thin solder lines or the usage of spacers so that tilting cannot happen to a large scale. On the other hand, while for soldering the gaps between the dies on 1st level (down to 200 μm) are not worth mentioning for silver sintering the gap size is a challenging factor. Processes need to be adjusted in order to ensure that no shortcut will by generated during the application of the Ag sintering material or during the application of the dies. All these issues are demanding tasks to improve the reliability of the three different 1st level interconnects.

In addition, applying silver sintering as 2nd level interconnection technology to improve the thermal heat transfer of the VCSEL module, is challenging as well, since current silver sintering pastes and processes are recommended for a maximum area of about 100 mm^2 only. Therefore, processes have to be developed and adjusted to ensure high quality large area interconnects between the ceramic-based substrate and the micro channel water cooler of the VCSEL module.

The aim of this work is to increase the lifetime of a high power VCSEL pump module and assure a very high reliability by optimizing the thermal properties and quality of the 1st and 2nd level interconnects.

Therefore, high power VCSEL pump modules were built by applying metallic joining technologies like soldering using eutectic AuSn20 and SnAg3 solder at 1st level and applying silver sintering at 1st and 2nd level.

The joining processes were adjusted in order to optimize the quality of the interconnects considering the individual challenges described beforehand. In order to validate the quality of the interconnects ultrasonic microscopy (C-SAM), X-ray microscopy and 3D laser profilometry were conducted. Furthermore, cross-sections were prepared and analysed using light- and scanning electron microscopy (SEM). Furthermore, the reliability of soldered and silver sintered 1st level interconnects was tested by thermal cycling in a range of -55°C to +125°C.

34.2 1st Level Interconnection—Mounting GaAs Dies with VCSEL Array onto Ceramic-based Substrate

34.2.1 Application of Metallic Interconnection Technologies

For the 1st level die attach of the GaAs dies three interconnection technologies have been used, namely soldering using eutectic AuSn20, soldering using SnAg3 preforms with Ni spacers and silver sintering using preforms.

Soldering with eutectic **AuSn20 solder** means that the interconnects consist of the intermetallic phases Au_5Sn (ζ') and AuSn (δ) as shown in the binary phase diagram in Fig. 34.3. Such interconnects exhibit a very high creep resistance (Hutter, 2009; Goullon et al, 2015).

For the die attach of the GaAs dies with VCSEL array using AuSn20 the solder was provided by Au and Sn multilayers on the substrate side, which were deposited by thin film technologies. Besides that other approaches like electroplating AuSn layers from the die side or by using preforms are often used. However, both variants are not feasible for the application of this study. The GaAs dies with VCSEL array were provided as single dies and the deposition of the galvanic layers would have to be done on wafer-level. Using AuSn20 preforms a rather thick solder line is generated. This leads to two main issues. On the one hand, a thick intermetallic bond line can lead to a critical stress state. On the other hand, the thicker the bond line the more likely tilting can happen on a large scale during soldering. In order to join the GaAs dies to a substrate serving as a heat spreader, the dies were placed on the

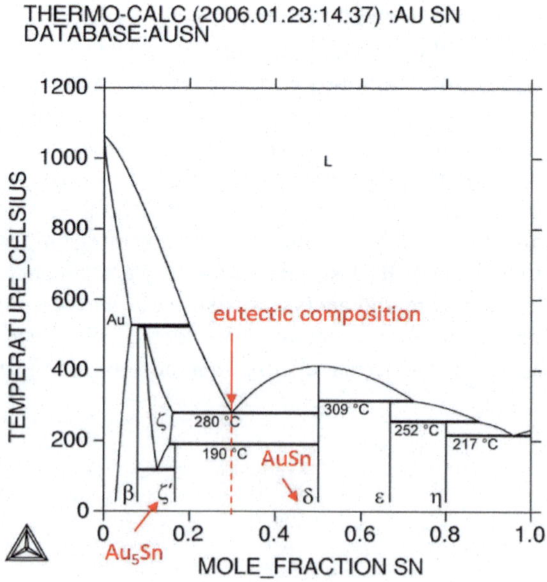

Fig. 34.3 Binary phase diagram of Au-Sn with Au5Sn and AuSn phases highlighted, adapted from TCS (1999).

Au-rich solder depots of the substrate using a fine placer. The soldering was carried out in a reflow batch oven. For reduction of oxides gaseous flux was used.

As shown in Fig. 34.4, the soldering process was adjusted optimal to generate an almost void free solder joint. The Sn was transformed totally into Au_5Sn and AuSn during the soldering. The final bond line thickness is about 5 μm. A tilt of the die of less than 1 μm could be generated.

Sn-based solders are used most commonly for die attach. The solder is provided either as paste or as preform. 50 μm-100 μm are typical values for the bond line thickness. Therefore, the tilt of the die can be double- or even three-digit. In order to ensure a homogeneous emission pattern of the VCSEL array a plan parallel assembly of the GaAs is an important requirement. In order to reduce the process related tilt of the die during soldering with the SnAg3 solder, preforms with additional Ni spacers were used. Adding spacers can prevent the occurrence of a major tilt, especially if the diameter of the Ni particles is of the same size as the targeted solder bond line (see Fig. 34.5). For the die attach of the GaAs dies SnAg3 preforms with a

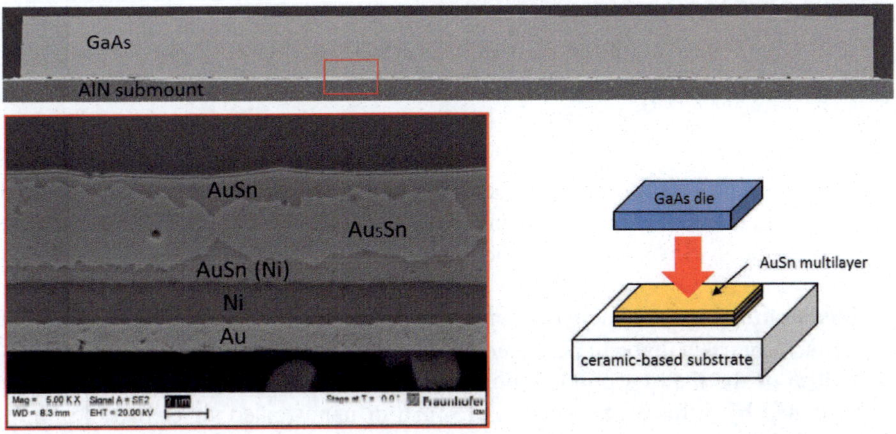

Fig. 34.4 Cross-section of GaAs die mounted on ceramic-based substrate by soldering with AuSn20 provided by Au and Sn thin film multilayers on the substrate side.

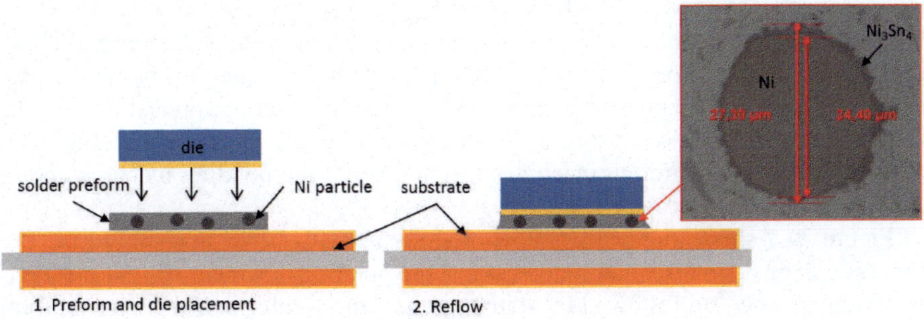

Fig. 34.5 Schematic sketch of process flow using SnAg3 solder preform with Ni spacers.

thickness of 30 μm and Ni spacers with a diameter of 30 μm were used. The soldering was carried out in a reflow batch oven. For the reduction of oxides gaseous flux was used. In Fig. 34.6, light microscopic images of a cross-section as well as a 3D profilometry picture of the assembly are shown. Based on the results the soldering

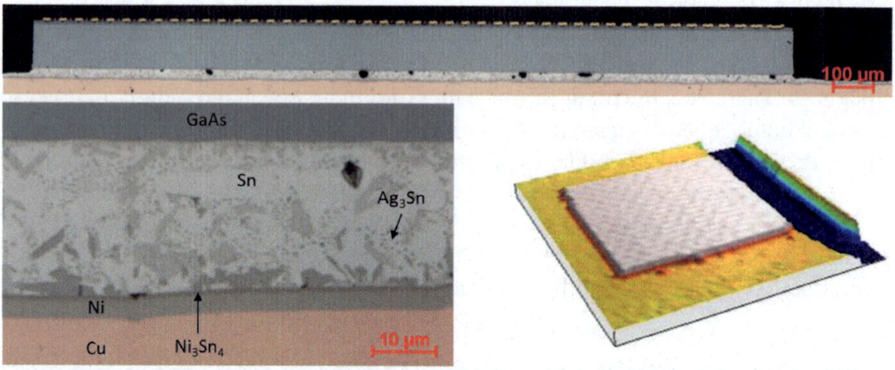

Fig. 34.6 Cross-section of GaAs die mounted on ceramic-based substrate by soldering using a SnAg3 preform with Ni spacers (top and bottom left) and 3D laser profilometry measurement of the assembly (bottom right).

process was adjusted optimal to generate an almost void free solder joint having a bond line thickness of about 25 μm and a tilt of about 9 μm. Compared to the soldering process using SnAg3 solder without spacers the tilt of the die could be reduced by a factor of 10.

Silver sintering is an interconnection technology, which provides monometallic silver interconnects that can be generated by applying joining temperatures of less than 40% of the melting point of pure silver ($T_m = 961°C$). The interconnection is generated by diffusion between micro- and or nano-scaled silver particles only. Therefore, no melting occurs during the entire joining process.

The high melting point, the increased high thermal and electrical conductivity as well as the resulting high creep resistance of such interconnects are the key advantages compared to common solder joints.

Just like solders silver sinter materials are available as paste or as preform. However, silver sinter preforms are less common and relatively rare on the market. By using preforms the main advantage is that the silver material application and the sintering step can be done using only one machine, while for silver sintering with pastes three to four different machines (printer, oven, die bonding machine and/or sinter press) are needed.

Figure 34.7 shows the process flow of sintering with silver sinter preform. The process can be done fully automatic using a die bonding machine. At first the die is picked and heated up followed by stamping the sinter preform. The sinter preform remains on the backside of the die and is then used to make the die attach by applying temperatures above 200°C and pressure of at least 10 MPa for a few seconds or

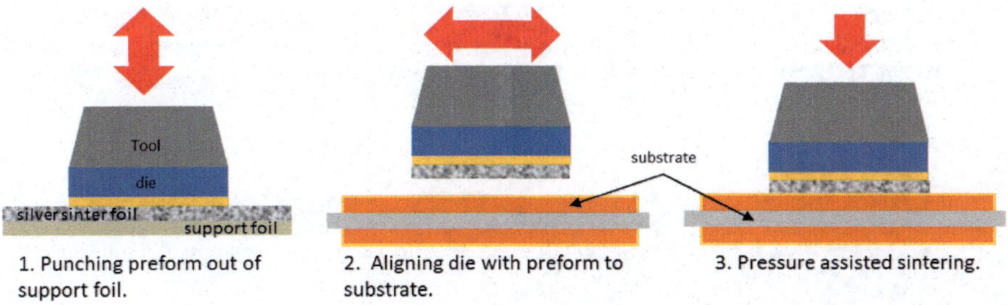

Fig. 34.7 Schematic sketch of the process flow of silver sintering using Ag sinter preform.

even minutes. A more important aspect, which lead to choose silver sinter preforms instead of a silver sinter paste, was due to the design of the high power VCSEL array module. The gap between the dies is 200 μm. Applying silver sinter paste without causing shortcuts would have been difficult or even impossible.

As shown in Fig. 34.8, using a silver sinter preform based process the dies can be placed and sintered with high accuracy and without causing any shortcuts. In order to generate a high quality joint; but still ensure that the VCSELs will not be damaged during the pressure assisted sintering, the sintering parameter were adjusted. By optimizing the main process parameters (temperature and time) the pressure applied could be reduced down to 5 MPa.

As shown in Fig. 34.9 using the adjusted parameter set a very good adhesion between the silver sinter layer and the GaAs die as well as between the sintered silver and the substrate can be generated. As visible in the light microscopic image (Fig. 34.9, bottom left) the porosity and the bond line thickness can mainly be influenced by the surface roughness of the substrate. The average porosity of the preform in the initial state is about 35%. After the sintering step the remaining porosity av-

Fig. 34.8 GaAs dies with VCSEL array mounted on DCB by pressure assisted silver sintering using preforms.

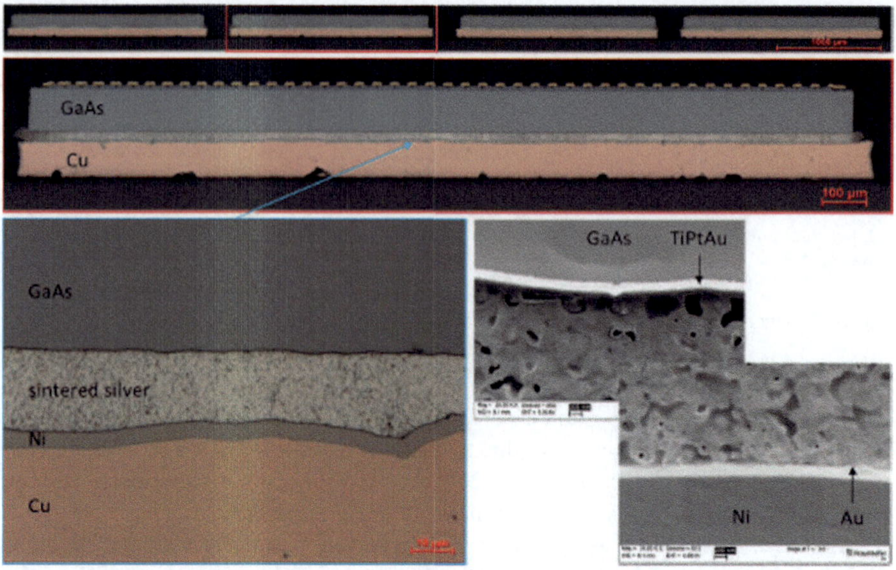

Fig. 34.9 Cross-section of GaAs dies mounted on DCB by pressure assisted silver sintering using preforms.

erages 18%. The final bond line thickness averages 25 µm. A tilt of the die of less than 1 µm could be generated.

34.2.2 Reliability Testing of Soldered and Silver Sintered 1st Level Interconnects

In order to evaluate the reliability of the three different joining technologies (soldering with AuSn20, soldering with SnAg3 and pressure assisted silver sintering) for each variant three samples with a single GaAs die with VCSEL array was mounted onto a DCB (see Fig. 34.10). The samples were subjected to thermal cycling be-

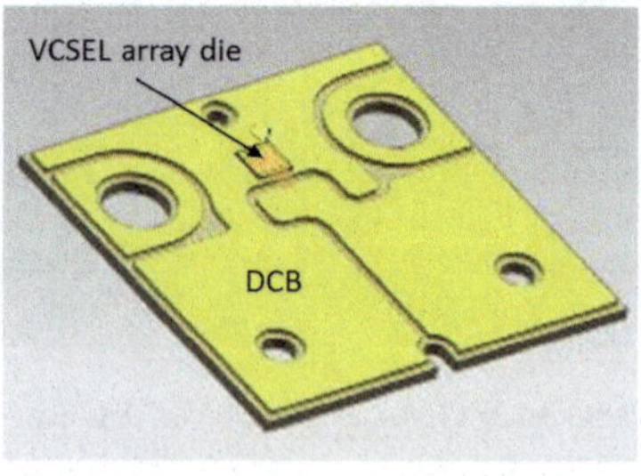

Fig. 34.10 Sketch of the test samples used for reliability testing design made by project partner Philips.

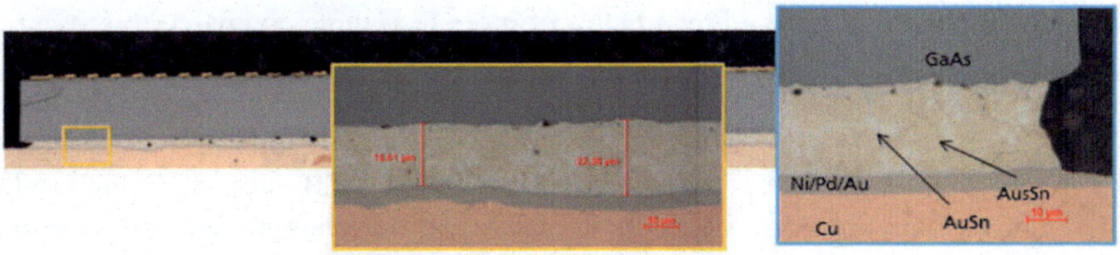

Fig. 34.11 Cross-section of GaAs die mounted on DCB by soldering with AuSn20 after 500 thermal cycles.

tween -55°C and +125°C in a three chamber oven. Initially, after 100, 500 and finally after 1000 thermal cycles the samples were analyzed using C-SAM. Based on the non-destructive analysis method no failures could be detected up to and including 500 thermal cycles.

In Fig. 34.11, a cross-section image of an AuSn solder joint after 500 thermal cycles is shown. Based on the light microscopic analyses no solder fatigue or fracture of the GaAs die is present. All samples were tested non-destructively using C-SAM and X-ray microscopy. C-SAM is used in order to detect defects like delaminations, weak adhesions or cracks in particular. X-ray microscopy is used to detect three-dimensional defects like voids and cracks.

Even after 1000 thermal cycles neither the die nor any of the die attaches tested were damaged. Representatively, Fig. 34.12 shows C-SAM images of the die attach generated by soldering with AuSn20 at the initial state and after 1000 thermal cycles. Based on the results of the C-SAM analyses after 1000 thermal cycles no failures could be detected. The light grey area reflects the bonding area having a good adhesion. The dark grey spots are voids in the joint as the comparison with the X-ray image (Fig. 34.12, top left) shows. Although no damages occurred at the die attaches tested, as shown in Fig. 34.12 (bottom right) the end of life of the modules tested was due to failures in the ceramic-based substrates. The DCB is the bottleneck of the system, since a delamination between copper and ceramic occurred.

34.3 2nd Level Interconnection—Mounting DCB onto a Micro Channel Water Cooler by Pressure Assisted Silver Sintering

The main challenge for the 2nd level interconnect using silver sintering technology was to generate a high quality joint having a homogeneous densification of the silver particles over a bonding area of more than 800 mm^2. Compared to soldering silver sintering is a rather new interconnection technology, which is mainly used for 1st level interconnects having bonding areas of 100 mm^2 at the maximum. In addition, to the significant enlargement of the bonding area the pressure applied during the sintering step needed to be adjusted in order to ensure that the micro channel cooler will not be damaged during the pressure assisted sintering. Therefore, the pressure applied was reduced by a factor of more than three compared to pressures of 30-40 MPa, which are typically applied for 1st level interconnects. However, the pressure needed to generate a dense and reliable 2nd level interconnect is still too high to use common pick and place machines to apply a sintering process using silver sinter preforms. Therefore, in order to mount the DCB of the high power VCSEL module onto the micro channel water cooler a sintering process using a silver sinter paste was the method of choice. The paste was deposited by stencil printing using a 150 µm thick stencil. After the deposition of the paste onto the backside of the DCB, the substrate with the printed depot was pre-dried in an oven to combust the main organics of the paste. Then the DCB was placed onto the cooler as shown in Fig. 34.13. The joining of the DCB to the cooler was done by pressure assisted sintering using a hydraulic press. Cross-sectioning and SEM was carried out to evaluate the quality of the sintered joint. The results show, that the sintered

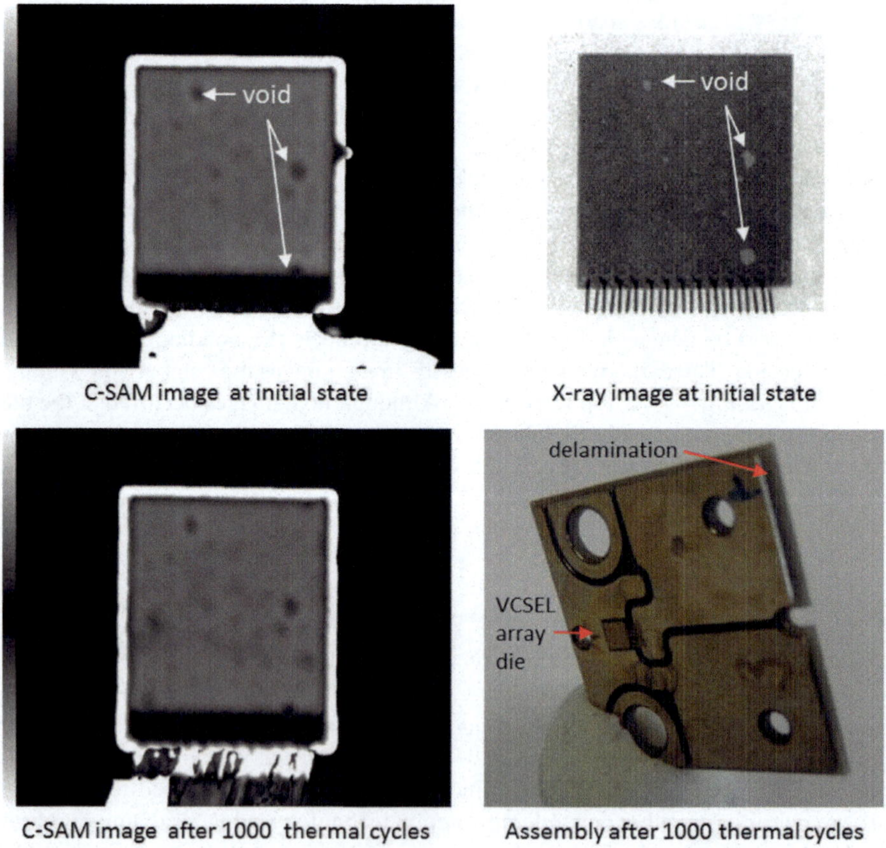

Fig. 34.12 GaAs die with VCSEL array mounted on DCB by soldering with AuSn20 at the initial state (C-SAM and X-ray image) and after 1000 thermal cycles (C-SAM image and photograph).

silver interconnect has a homogeneous sinter line thickness of about 70 μm over the whole area. Furthermore, a very good adhesion to the joining partners was generated, as shown in Fig. 34.14. The remaining porosity is less than 20%. Studies have shown, that comparable silver sintered probes in fact have a lower thermal conductivity than pure silver. However, by replacing Sn-based solder with sintered silver the significant increase of the thermal conductivity is still beneficial regarding the transfer of the heat produced by the high integrated VCSEL arrays (Ras, 2015).

34.4 Conclusions

The aim of this work was to increase the lifetime of high power VCSEL pump modules and assure a very high reliability by optimizing the thermal properties and quality of the 1st and 2nd level interconnects. Based on the results solder (AuSn20 and SnAg3) as well as silver sintered joints have been proven viable in order to act

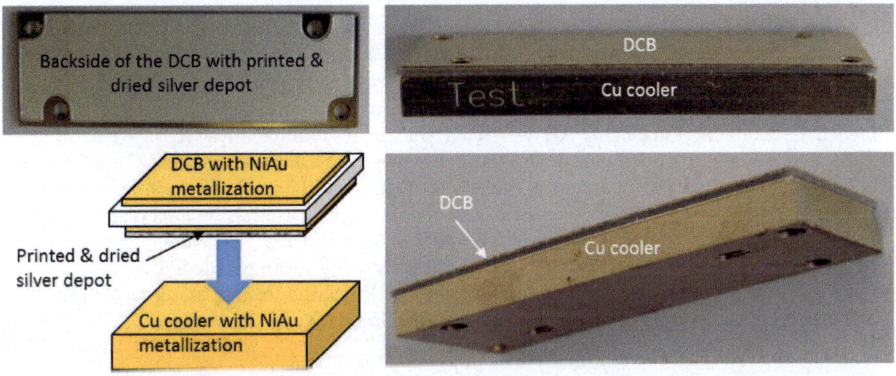

Fig. 34.13 Photographs of DCB backside after drying the printed silver sinter paste and after mounting the DCB onto the micro channel cooler by pressure assisted silver sintering.

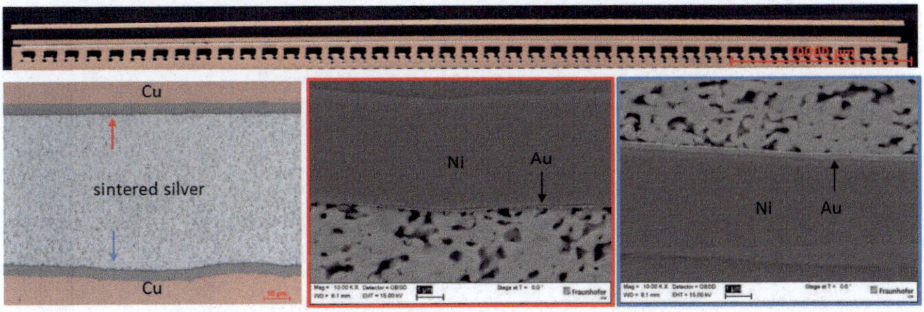

Fig. 34.14 Cross-section of DCB mounted on micro channel cooler by pressure assisted silver sintering.

as reliable 1st level interconnects. Even after 1000 thermal cycles between -55°C and +125°C neither the die nor any of the die attaches tested were damaged. The end of life of the modules tested was due to failures in the ceramic-based substrates. In addition, it was shown, that pressure assisted silver sintering can be applied as 2nd level interconnection technology in order to further improve the heat transfer of the high power VCSEL pump module.

Acknowledgements The authors would like to thank the German Federal Ministry of Education and Research (BMBF) for funding under contract FKZ 13N12470. Furthermore, the authors would like to thank the partners of the project "VORTEIL" for the close cooperation.

References

Goullon L, Jordan R, Braun T, Bauer J, Becker KF, Hutter M, Schneider-Ramelow M, Lang KD (2015) Large area led package. Photonics West, San Francisco, USA

Hutter M (2009) Verbindungstechnik höchster Zuverlässigkeit für optoelektronische Komponenten. Fraunhofer-Verlag

Hutter M, Weber C, Ehrhardt C, Lang KD (2016) Comparison of different technologies for the die attach of power semiconductor devices conducting active power cycling. In: CIPS 2016; 9th International Conference on Integrated Power Electronics Systems; Proceedings of, VDE, pp 1–7

Moench H, Kolb JS, Engelhardt AP, Gerlach P, Jaeger R, Pollmann-Retsch J, Weichmann U, Witzigmann B (2014) Optimized VCSELs for high-power arrays. In: Vertical-Cavity Surface-Emitting Lasers XVIII, International Society for Optics and Photonics, vol 9001

Ras MA (2015) Thermal characterization of tim and die attach materials by tima platform. ECPE Workshop Advances in Thermal Materials and Systems of Electronics, Nuremberg, Germany

TCS (1999) TCS Alloys Mobility Database, v2.0 (provided by Thermo-Calc Software)

Weber C, Hutter M, Ehrhardt C, Oppermann H (2014) Reliability testing of Ag sinter joints. In: Nuremberg, Germany: ECWC13–13th Electronic Circuits World Convention

Chapter 35
Coupled Thermal and Electrochemical Diffusion in Solid State Battery Systems

Marek Werner & Kerstin Weinberg

Abstract In the light of today's extensive research on rechargeable batteries a electro-chemically diffusion model for a temperature sensitive multi-phase solid is presented. The derivation of the model is based on the framework of the Thermodynamics of Irreversible Processes with the assumption of a local equilibrium. The physical effects which are accounted for are: the flux of ions including chemical reactions, the heat flux, the electrical current, and their coupling resulting, e.g. in phase decomposition, thermal diffusion, and thermoelectric effects. Two numerical examples illustrate typical applications and demonstrate the versatility of the coupled multi-physics model.

Keywords: Phase-field · Non-equilibrium thermodynamics · Thermal and electrochemical diffusion

35.1 Introduction

Given the rise of portable electronics, mobile devices and especially electric cars, the demand for secondary batteries is expected to grow rapidly in the following years (Aifantis et al, 2010). Currently, the most successful secondary batteries are lithium based. In the automotive industry alone estimates indicate that the market for lithium batteries will grow from its current size of \$10 billion in 2015 to \$50 billion by 2020 (Berger, 2012). This extraordinary potential explains the strong interest in the optimization of high energy-density batteries in these day.

A typical lithium-ion battery consists of a positive lithium metal electrode, a negative graphite carbon electrode and a polymeric electrolyte in between, see Fig. 35.1.

Marek Werner · Kerstin Weinberg
Universität Siegen, Lehrstuhl für Festkörpermechanik, Department Maschinenbau, Paul-Bonatz-Str. 9-11, 57076 Siegen,
e-mail: marek.werner@uni-siegen.de, kerstin.weinberg@uni-siegen.de

B. E. Abali et al. (eds.), *New Achievements in Continuum Mechanics and Thermodynamics*, Advanced Structured Materials 108,
https://doi.org/10.1007/978-3-030-13307-8_35

During charging, the elementary chemical reaction at the carbon anode is

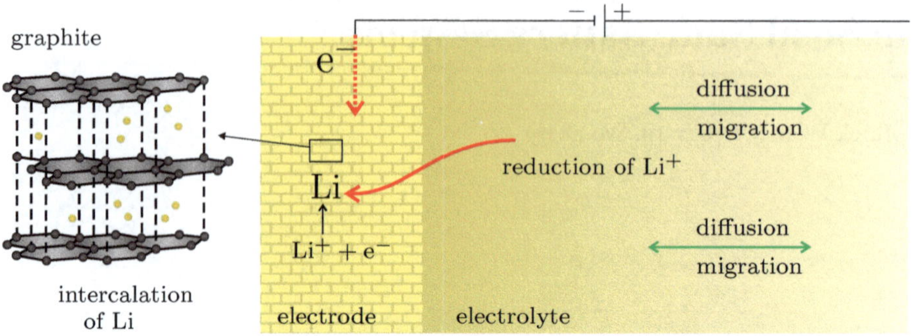

Fig. 35.1 Charging process at the lithium anode

$$Li^+ + e^- + C_6 \rightleftharpoons LiC_6 \,.$$

In this treatise the process of charging and discharging in a lithium-ion battery electrode will be modeled by recourse to a classical Boltzmann continuum approach (Truesdell and Toupin, 1960). The physical mechanisms which are assumed to contribute are: the particle exchange between the anode and the electrolyte, the diffusion in the anode, the thermal conduction, and heat flux, and the energetic forces associated with the elastic and electrostatic fields of the system.

The formal derivation of our electrochemo-thermo-mechanical model is based on concepts from the Thermodynamics of Irreversible Processes (TIP) in Eckart (1940) and on the Non-Equilibrium Thermodynamics (NET) in the spirit of De Groot and Mazur (1962). In this sense we assume a local equilibrium within the level of a small representative volume element. This implies that the globally defined balance relations are valid for the material point at $\boldsymbol{x} \in \Omega$ and in its infinitesimal neighborhood within the domain, $\Omega \subset \mathbb{R}^3$.

The remaining of the paper is organized as follows: In Sect. 35.2 we derive the electrochemical-thermo-mechanical model of lithium transport through the solid in full detail. An extended diffusion model involving electrical, thermal, mechanical, and chemical diffusion contributions is formulated and specified to in Sect. 35.3 where also some remarks on the solution procedure are made. In Sect. 35.4 two numerical examples demonstrate the capability and versatility of the presented approach. The paper concludes with a short summary.

35.2 Electrochemo-thermo-mechanical Diffusion

The constitutive equations, which characterize the material properties of continuous media, must conform with the basic laws of thermodynamics namely the conservation of energy and the entropy principle. To ensure this, we start with basic balance equations and introduce the total mass density ρ of our n-component system and the normalized concentration c_k of the kth component by

$$\rho = \sum_{k=1}^{n} \rho_k \quad \text{and} \quad c_k = \frac{\rho_k}{\rho}. \tag{35.1}$$

The conservation of mass can be expressed as

$$\rho \frac{\mathrm{d}c_k}{\mathrm{d}t} = -\operatorname{div} \boldsymbol{J}_k + \sum_{j=1}^{r} \upsilon_{kj} \mathfrak{J}_j, \quad k \in \{1, 2, \ldots, n\}, \tag{35.2}$$

where $\boldsymbol{J}_k$ represents the diffusive mass flow density. The parameter $\mathfrak{J}_j$ is called the reaction rate of the jth of r chemical reactions and υ_{kj} denotes a parameter which is proportional to the stoichiometric coefficient that weights the contribution of component k, i.e. formulation (35.2) incorporates mass production. Since mass is conserved in each separate chemical reaction, it holds

$$\sum_{k=1}^{n} \upsilon_{kj} = 0, \quad j \in \{1, 2, \ldots, r\}. \tag{35.3}$$

For later reference we introduce the chemical affinity A_j of the jth reaction

$$A_j = \sum_{k=1}^{n} \upsilon_{kj} \mu_k, \quad j \in \{1, 2, \ldots, r\} \tag{35.4}$$

where μ_k is the chemical potential per unit mass of component k.

The electric field $\boldsymbol{e}\,(\boldsymbol{x}, t)$, which is assumed to be given by the gradient of an electric potential, $\boldsymbol{e} = -\boldsymbol{\nabla} \phi_e$, is proportional to the electric displacement field (induction) $\boldsymbol{d}$ through the vacuum permittivity. A common constitutive relation assumes a corrected permittivity ε to account for the specific dielectric material which results in

$$\boldsymbol{d} = \varepsilon\, \boldsymbol{e}\,. \tag{35.5}$$

For a local electric charge density ρz, with z being the charge per unit mass, the equilibrium equations of electrostatics (Gauss's flux theorem and the Maxwell–Faraday equation) must hold,

$$\operatorname{div} \boldsymbol{d} = \rho z \quad \text{and} \quad \boldsymbol{\nabla} \times \boldsymbol{e} = \boldsymbol{0}\,. \tag{35.6}$$

In a mixture of n components the total current density $\boldsymbol{I}$ is composed of a convective component $\rho z \boldsymbol{v}$ with barycentric velocity $\boldsymbol{v}$ and a conductive current density $\boldsymbol{i}$,

$$\boldsymbol{I} = \rho z \boldsymbol{v} + \boldsymbol{i} = \rho z \boldsymbol{v} + \sum_{k=1}^{n} z_k \boldsymbol{J}_k \,. \tag{35.7}$$

Comparing the total charge density $\rho z = \sum_{k=1}^{n} \rho_k z_k$ with (35.1) gives the charge, $z = \sum_{k=1}^{n} c_k z_k$. With (35.2) the law of conservation of charge follows,

$$\rho \frac{\mathrm{d}z}{\mathrm{d}t} = -\operatorname{div} \boldsymbol{i} \,. \tag{35.8}$$

The linear momentum has now, aside of the convective flow and the mechanical stress tensor $\boldsymbol{\sigma}$, also an electric contribution which is known as Maxwell stress tensor $\boldsymbol{\sigma}^M = \varepsilon \boldsymbol{e} \otimes \boldsymbol{e} - \frac{1}{2}\varepsilon \boldsymbol{e}^2 \mathbf{I}$, cf. Landau et al (2013). The symbol $\mathbf{I}$ denotes the identity tensor. We state the change of momentum density

$$\frac{\mathrm{d}\rho \boldsymbol{v}}{\mathrm{d}t} = -\operatorname{div}\left(\rho \boldsymbol{v} \otimes \boldsymbol{v} + \boldsymbol{\sigma} - \boldsymbol{\sigma}^M\right) \,. \tag{35.9}$$

The equation of motion which has now an additional term following from the current density (35.7).

$$\rho \frac{\mathrm{d}\boldsymbol{v}}{\mathrm{d}t} = -\operatorname{div} \boldsymbol{\sigma} + \sum_{k=1}^{n} \rho_k \boldsymbol{F}_k + \sum_{k=1}^{n} \rho_k z_k \boldsymbol{e} \tag{35.10}$$

Here the $\boldsymbol{F}_k$ are conservative external forces (dead loads) and $z_k \boldsymbol{e}$ are the non-conservative Lorentz forces acting on a component k, both per unit mass. The $\boldsymbol{F}_k$ are derived from a stationary potential ψ_k with $\boldsymbol{F}_k = -\boldsymbol{\nabla} \psi_k$.

35.2.1 First Law of Thermodynamics

Multiplying Eq. (35.10) with the barycentric velocity field $\boldsymbol{v}$ and making use of the tensor differential relation $\operatorname{div}(\boldsymbol{\sigma} \cdot \boldsymbol{v}) = \operatorname{div} \boldsymbol{\sigma} \cdot \boldsymbol{v} + \boldsymbol{\sigma} : \boldsymbol{\nabla} \boldsymbol{v}$ gives the balance equation for the kinetic energy density

$$\rho \frac{\mathrm{d}\frac{1}{2}\boldsymbol{v}^2}{\mathrm{d}t} = -\operatorname{div}(\boldsymbol{\sigma} \cdot \boldsymbol{v}) + \boldsymbol{\sigma} : \boldsymbol{\nabla} \boldsymbol{v} + \sum_{k=1}^{n} \rho_k \boldsymbol{F}_k \cdot \boldsymbol{v} + \rho z \boldsymbol{v} \cdot \boldsymbol{e} \tag{35.11}$$

The first two terms on the right hand side comprise the density of mechanical work flow, in particular, $\boldsymbol{\sigma} : \boldsymbol{\nabla} \boldsymbol{v}$ is the power density of deformation. The third term summarizes the conservative forces which result from the potential energy per unit volume $\rho\psi$. The rate of change of $\rho\psi = \sum_{k=1}^{n} \rho_k \psi_k$ leads to a flux term $\sum_k \psi_k \boldsymbol{J}_k$ and a remaining source term $-\sum_k \boldsymbol{J}_k \cdot \boldsymbol{F}_k$, where $\boldsymbol{F}_k$ is the dead load acting on component k, and $\boldsymbol{J}_k$ is the corresponding mass flux. A detailed derivation of these terms can be found in Anders and Weinberg (2012). Additionally, the Poynting theorem of energy conservation for electromagnetic fields is known, $\frac{1}{2}\partial \boldsymbol{e}^2/\partial t = \boldsymbol{I} \cdot \boldsymbol{e}$, with current density (35.7). With all this at hand we may restate (35.11) as an equation for the rate of change of the kinetic, the potential, and the electric energy

$$\frac{\partial\left(\rho(\frac{1}{2}\boldsymbol{v}^2+\psi)+\boldsymbol{e}^2\right)}{\partial t} = -\operatorname{div}\left(\rho\left(\frac{1}{2}\boldsymbol{v}^2+\psi\right)\boldsymbol{v}+\boldsymbol{\sigma}\cdot\boldsymbol{v}+\sum_k \psi_k \boldsymbol{J}_k\right) + \boldsymbol{\sigma} : \nabla\boldsymbol{v} - \sum_{k=1}^{n} \boldsymbol{J}_k \cdot \boldsymbol{F}_k - \boldsymbol{i}\cdot\boldsymbol{e}\,. \tag{35.12}$$

From this formulation it can be seen, that the sum of kinetic, potential, and electric energy is not a conserved quantity: a source term appears at the right-hand side. The total energy per unit mass e within an arbitrary control volume Ω, however, needs to be conserved and according to the first law of thermodynamics it can only change due to an energy flux rate $\boldsymbol{J}_e$ through the boundary $\Gamma = \partial\Omega$,

$$\frac{\mathrm{d}}{\mathrm{d}t}\int_\Omega \rho e\,\mathrm{d}\Omega = \int_\Omega \frac{\partial \rho e}{\partial t}\,\mathrm{d}\Omega = -\int_{\partial\Omega} \boldsymbol{J}_e \cdot \boldsymbol{n}\,\mathrm{d}\Gamma \tag{35.13}$$

where $\boldsymbol{n}$ denotes the unit outward normal on $\partial\Omega$. An application of the Gauss–Ostrogradsky theorem provides the local form of energy conservation,

$$\frac{\partial \rho e}{\partial t} = -\operatorname{div}\boldsymbol{J}_e \tag{35.14}$$

where $\boldsymbol{J}_e$ is a total energy flux rate density. It has to include a convective term $\rho e\boldsymbol{v}$, an energy flux rate density $\boldsymbol{\sigma}\cdot\boldsymbol{v}$ due to mechanical work performed on the system, a potential energy flux rate density if applicable, and a heat flow density $\boldsymbol{J}_\theta$.

$$\boldsymbol{J}_e = \rho e\boldsymbol{v} + \boldsymbol{\sigma}\cdot\boldsymbol{v} + \sum_{k=1}^{n} \psi_k \boldsymbol{J}_k + \boldsymbol{J}_\theta \tag{35.15}$$

To specify the internal energy per unit mass u we can as well attribute the difference between e and $\rho(\frac{1}{2}\boldsymbol{v}^2+\psi)+\boldsymbol{e}^2$ in (35.12),

$$\rho u = \rho e - \left(\rho(\frac{1}{2}\boldsymbol{v}^2+\psi)+\boldsymbol{e}^2\right)\,. \tag{35.16}$$

Then, subtracting (35.12) from (35.14), we obtain the balance equation for the internal energy

$$\frac{\partial \rho u}{\partial t} = -\operatorname{div}\left(\rho u\boldsymbol{v}+\boldsymbol{J}_\theta\right) - \boldsymbol{\sigma} : \nabla\boldsymbol{v} + \sum_{k=1}^{n} \boldsymbol{J}_k\cdot\boldsymbol{F}_k + \boldsymbol{i}\cdot\boldsymbol{e} \tag{35.17}$$

This equation shows that the internal energy is not conserved in this constellation as well. Again there is a source term, which is equal but of opposite sign to the source term in the balance equation (35.12) of the sum of kinetic, potential and electric energy. Therefore such a formulation of the balance equation for the internal energy inherently guarantees the conservation of total energy.

In our notation it is convenient to write Eq. (35.17) in an alternative form. For this purpose we split the total pressure tensor $\boldsymbol{\sigma}$ into a hydrostatic part $p\mathbf{I} = \mathrm{tr}\,(\boldsymbol{\sigma})\,/3\mathbf{I}$ and a remaining deviatoric part $\boldsymbol{S}$:

$$\boldsymbol{\sigma} = p\mathbf{I} + \boldsymbol{S} \tag{35.18}$$

With relations (35.18) Eq. (35.17) becomes[1]

$$\rho\frac{\mathrm{d}u}{\mathrm{d}t} = -\,\mathrm{div}\,\boldsymbol{J}_\theta - p\,\mathrm{div}\,\boldsymbol{v} - \boldsymbol{S} : \nabla\boldsymbol{v} + \sum_{k=1}^{n}\boldsymbol{J}_k\cdot\boldsymbol{F}_k + \boldsymbol{i}\cdot\boldsymbol{e}\,. \tag{35.20}$$

Here we have used that $\mathbf{I} : \nabla\boldsymbol{v} = \mathrm{div}\,\boldsymbol{v}$. Now we can employ another version of the mass continuity equation in terms of the specific volume $\nu \equiv \rho^{-1}$

$$\rho\frac{\mathrm{d}\nu}{\mathrm{d}t} = \mathrm{div}\,\boldsymbol{v} \tag{35.21}$$

to establish

$$\frac{\mathrm{d}u}{\mathrm{d}t} = -\nu\,\mathrm{div}\,\boldsymbol{J}_\theta - p\frac{\mathrm{d}\nu}{\mathrm{d}t} - \nu\boldsymbol{S} : \nabla\boldsymbol{v} + \nu\sum_{k=1}^{n}\boldsymbol{J}_k\cdot\boldsymbol{F}_k + \nu\boldsymbol{i}\cdot\boldsymbol{e}\,. \tag{35.22}$$

The last terms of Eq. (35.20) and (35.22) represent the amount of electric energy transformed into internal energy per unit volume and per unit time. The acting Lorentz forces are non-conservative forces.

35.2.2 *Second Law of Thermodynamics*

The entropy per unit mass s is a function of numerous variables which are necessary to define the macroscopic state of the system entirely. The total differential of s in state of equilibrium is given by the common Gibbs relation

[1] By means of the rate form of the continuity mass equation $\mathrm{d}\rho/\mathrm{d}t = -\rho\,\mathrm{div}\,\boldsymbol{v}$ we get for any scalar field α the identity

$$\rho\frac{\mathrm{d}\alpha}{\mathrm{d}t} = \frac{\partial\,(\rho\alpha)}{\partial t} + \mathrm{div}\,(\rho\alpha\boldsymbol{v})\,. \tag{35.19}$$

$$T\mathrm{d}s = \mathrm{d}u + p\mathrm{d}\nu - \sum_{k=1}^{n} \mu_k \mathrm{d}c_k, \tag{35.23}$$

where p is the ambient pressure and T the absolute temperature. Following the local equilibrium hypothesis of the classical TIP, the local version of Eq. (35.23) becomes

$$T\frac{\mathrm{d}s}{\mathrm{d}t} = \frac{\mathrm{d}u}{\mathrm{d}t} + p\frac{\mathrm{d}\nu}{\mathrm{d}t} - \sum_{k=1}^{n} \mu_k \frac{\mathrm{d}c_k}{\mathrm{d}t}. \tag{35.24}$$

We intend to find an explicit form of this relation and to bring it into the typical structure of a balance equation

$$\rho\frac{\mathrm{d}s}{\mathrm{d}t} = -\operatorname{div} \boldsymbol{J}_s + \pi_s, \tag{35.25}$$

where $\boldsymbol{J}_s$ is a general entropy flux rate and π_s is an entropy source strength. According to the second law of thermodynamics the entropy source π_s vanishes for reversible (equilibrium) thermodynamic processes and it holds $\pi_s > 0$ for irreversible processes. Consequently, it must hold $\pi_s \geq 0$ for a general thermodynamic process.

To specify Eq. (35.24) we insert the expressions for the internal energy (35.22) and the mass balance (35.2). For brevity we drop the dead loads in the following, $\boldsymbol{F}_k = 0$, and obtain with the chemical affinity (35.4) the expression

$$\rho T\frac{\mathrm{d}s}{\mathrm{d}t} = -\operatorname{div} \boldsymbol{J}_\theta - \boldsymbol{S} : \boldsymbol{\nabla v} + \sum_{k=1}^{n} \mu_k \operatorname{div} \boldsymbol{J}_k + \boldsymbol{i}\cdot\boldsymbol{e} - \sum_{j=1}^{r} A_j \mathfrak{J}_j\,. \tag{35.26}$$

By means of simple differential relations we obtain the entropy balance equation in the required form (35.25),

$$\begin{aligned}\rho\frac{\mathrm{d}s}{\mathrm{d}t} = &-\operatorname{div}\left(\frac{1}{T}\left(\boldsymbol{J}_\theta - \sum_{k=1}^{n}\mu_k\boldsymbol{J}_k\right)\right) - \frac{1}{T^2}\boldsymbol{J}_\theta\boldsymbol{\nabla}T\\ &-\frac{1}{T}\sum_{k=1}^{n}\boldsymbol{J}_k\cdot\left(T\boldsymbol{\nabla}\left(\frac{\mu_k}{T}\right) - z_k\boldsymbol{e}\right) - \frac{1}{T}\boldsymbol{S}:\boldsymbol{\nabla v} - \frac{1}{T}\sum_{j=1}^{r}A_j\mathfrak{J}_j.\end{aligned} \tag{35.27}$$

From comparison with (35.25) it is possible to identify the entropy flux rate as

$$\boldsymbol{J}_s = \frac{1}{T}\left(\boldsymbol{J}_\theta - \sum_{k=1}^{n}\mu_k\boldsymbol{J}_k\right) \tag{35.28}$$

and the entropy source term as

$$\pi_s = -\frac{1}{T^2}\boldsymbol{J}_\theta \boldsymbol{\nabla} T - \frac{1}{T}\sum_{k=1}^{n} \boldsymbol{J}_k \cdot T\boldsymbol{\nabla}\left(\frac{\mu_k}{T}\right)$$
$$+ \frac{1}{T}\sum_{k=1}^{n} z_k \boldsymbol{J}_k \cdot \boldsymbol{e} - \frac{1}{T}\boldsymbol{S} : \boldsymbol{\nabla}\boldsymbol{v} - \frac{1}{T}\sum_{j=1}^{r} A_j \mathfrak{J}_j \geq 0. \qquad (35.29)$$

At the first glance the separation of relation (35.27) into an entropy flux rate $\boldsymbol{J}_s$ and an entropy source π_s is arbitrary, however, the quantities have to satisfy a number of requirements which validate this specific separation. For example, the entropy source expression has to vanish in the thermodynamic equilibrium and it has to be invariant under coordinate transformations and rigid movements which both is the case here. The entropy source (35.29) has five contributions which all are related to a flow quantity. The first term of (35.29) arises from heat conduction and is connected to heat flow $\boldsymbol{J}_\theta$, the second is a weighted mass diffusion flow $\boldsymbol{J}_k$, the third term stems from the mass diffusion of electric charges of the n components, the fourth one comprised the inelastic mechanical work, and the last term is a sum of chemical rates $\mathfrak{J}_j$ multiplied by their affinities A_j.

Expression (35.28) shows that the entropy flow rate consists of a heat flow rate and a current due to diffusion. In order to separate all thermodynamic forces related to the temperature gradient we will take the second term of (35.29), apply the chain rule, and modify flow rate $\boldsymbol{J}_\theta$ to a generalized heat flow rate $\boldsymbol{J}'_\theta$,

$$\boldsymbol{J}'_\theta = \boldsymbol{J}_\theta - \sum_{k=1}^{n} h_k \boldsymbol{J}_k \qquad (35.30)$$

where $h_k := \mu_k - T\partial\mu_k/\partial T$ is the partial specific enthalpy of component k. This heat flow can now be used to restate the entropy flux as

$$\boldsymbol{J}_s = \frac{1}{T}\boldsymbol{J}'_\theta + \sum_{k=1}^{n} s_k \boldsymbol{J}^r_k \qquad (35.31)$$

where $s_k = -(\mu_k - h_k)/T$ is the partial specific entropy of component k. In this way the entropy flux comprises the heat flow, the flow of enthalpy and, additionally to (35.28), a transport of partial entropies s_k relative to the barycentric velocity $\boldsymbol{v}$ with $\boldsymbol{J}^r_k = \rho_k(\boldsymbol{v}_k - \boldsymbol{v})$.

We consider a solid-state system at rest and write $\boldsymbol{J}^r_k = \boldsymbol{J}_k = \rho_k \boldsymbol{v}_k$. For example, we obtain for a metal with a flux of electrons e^- in a stationary lattice,

$$\boldsymbol{J}_s = \frac{1}{T}\boldsymbol{J}'_\theta + s_{e^-}\boldsymbol{J}_{e^-}, \qquad (35.32)$$

and with current $\boldsymbol{I} = z_{e^-}\boldsymbol{J}_{e^-}$ this gives for the entropy source

$$\pi_s = -\frac{1}{T}\boldsymbol{J}_s \boldsymbol{\nabla} T - \frac{1}{T}\boldsymbol{I}\left(\boldsymbol{\nabla}\frac{\mu_e}{z_e} - \boldsymbol{e}\right) \qquad (35.33)$$

The entropy source (35.29) and (35.33) seem to indicate that the thermodynamic forces conjugate to the mass diffusion fluxes $\boldsymbol{J}_k$ do not only consist of the chemical potential but instead an additional term of electrostatic origin appears. This contribution can also be attributed to the chemical potential of component k by recourse to the definition $\boldsymbol{e} = -\boldsymbol{\nabla}\phi_e$ and

$$\mu_k^* = \mu_k + z_k\phi_e \,. \tag{35.34}$$

In this way the electrochemical potential consists of the ordinary microscopic chemical potential and an additional long-range, macroscopic electric potential. The internal energy may also be redefined as $u^* = u + z\phi_e$, which would give an additional term in Eq. (35.24), i.e.

$$T\frac{\mathrm{d}s}{\mathrm{d}t} = \frac{\mathrm{d}u^*}{\mathrm{d}t} + p\frac{\mathrm{d}\nu}{\mathrm{d}t} - \sum_{k=1}^{n} \mu_k^* \frac{\mathrm{d}c_k}{\mathrm{d}t} - z\frac{\mathrm{d}\phi_e}{\mathrm{d}t} \,. \tag{35.35}$$

In the following we will leave the general scenario in order to reduce the general form of the entropy source (35.29) to a multicomponent isotropic mixture in an isobaric, mechanically equilibrated systems without external forces and chemical reactions. Because the incorporation of mechanical effects follows the well known strategies of continuum mechanics and the effect of chemical reactions has been studied in detail in Weinberg et al (2018), see also Goddard (2015). We focus here on the reduced entropy source in a non-moving reference system,

$$\pi_s = -\frac{1}{T}\boldsymbol{J}_s \boldsymbol{\nabla} T - \frac{1}{T}\sum_{k=1}^{n} \boldsymbol{J}_k \cdot (\boldsymbol{\nabla}\mu_k - z_k \boldsymbol{e}) \,, \tag{35.36}$$

and we treat mass flux $\boldsymbol{J}_k$ and current $\boldsymbol{I}_k = z_k \boldsymbol{J}_k$ separately to illuminate the electrochemical coupling effects in a solid mixture.

35.3 Constitutive Relations

In general TIP applications the terms in the entropy source are classified into *thermodynamic fluxes* and quantities which multiply the fluxes are called *thermodynamic forces* or *affinities*. In this sense we regard the entropy source as a linear combination of thermodynamic fluxes $\boldsymbol{J}_\bullet$ multiplied by their corresponding affinities $\mathfrak{X}_\bullet$:

$$\pi_s = \boldsymbol{J}_s \mathfrak{X}_s + \sum_{k=1}^{n-1} \boldsymbol{J}_k \mathfrak{X}_k \tag{35.37}$$

The assumption of a homogeneous linear dependency between the fluxes and affinities with material parameters L_{ij} (Onsager coefficients),

$$\boldsymbol{J}_\bullet = \boldsymbol{L}_{\bullet\bullet}\boldsymbol{\mathfrak{X}}_\bullet + \sum_{j=1}^{n-1} \boldsymbol{L}_{\bullet j}\boldsymbol{\mathfrak{X}}_j\,, \tag{35.38}$$

leads to a quadratic expression for the entropy source strength. Since we study here isotropic systems, the Onsager coefficients are scalars; they would be tensor valued quantities for anisotropic crystalline systems.

If we specifically consider an electroneutral system of n charged or uncharged components, of which the entropy source has been derived above, we may write this expression as

$$\pi_s = -\frac{1}{T}\boldsymbol{J}_s\boldsymbol{\nabla}T - \frac{1}{T}\sum_{k=1}^{n}\boldsymbol{J}_k\cdot\boldsymbol{\nabla}\mu_k + \frac{1}{T}\boldsymbol{I}\cdot\boldsymbol{e} \tag{35.39}$$

with $\boldsymbol{I} = \sum_{k=1}^{n}\boldsymbol{I}_k$ being the total current as a sum of partial currents $\boldsymbol{I}_k = z_k\boldsymbol{J}_k$ of component k. Expressed as a closed system at rest, we may formulate the phenomenological equations for the fluxes as

$$\boldsymbol{J}_s = -L_{ss}\boldsymbol{\nabla}T - \sum_{j=1}^{n-2} L_{sj}\boldsymbol{\nabla}\hat{\mu}_j + L_{se}\boldsymbol{e}\,, \tag{35.40}$$

$$\boldsymbol{J}_k = -L_{ks}\boldsymbol{\nabla}T - \sum_{j=1}^{n-2} L_{kj}\boldsymbol{\nabla}\hat{\mu}_j + L_{ke}\boldsymbol{e}\,, \quad k\in\{1,2,\dots,n-2\}, \tag{35.41}$$

$$\boldsymbol{I} = -L_{es}\boldsymbol{\nabla}T - \sum_{j=1}^{n-2} L_{ej}\boldsymbol{\nabla}\hat{\mu}_j + L_{ee}\boldsymbol{e}\,. \tag{35.42}$$

Here the two depended mass fluxes are already eliminated by the definition of the current and with the condition of mass conservation, $\sum_{k=1}^{n-1}\boldsymbol{J}_k = 0$. Correspondingly, $\hat{\mu}_k$ denotes the relative chemical potentials. In this way we obtain the constitutive equations for a system coupling thermal diffusion, mass diffusion, and electric current. The specific n-component material enters these equations through the definition of the chemical potentials of the components and via the material parameter $L_{ss}, L_{sj}, \dots L_{ee}$. For these coefficients the Onsager reciprocal relations mandate symmetry,

$$L_{js} = L_{sj}, \quad L_{kj} = L_{jk}, \quad L_{ej} = L_{je}, \qquad j\in\{1,2,\dots,n-2\}\,, \tag{35.43}$$

and, due to the positive definiteness of the entropy source, it also holds

$$L_{ss}\geq 0, \quad L_{kk}\geq 0, \quad L_{ee}\geq 0, \quad L_{ii}L_{jj}\geq\frac{1}{4}\left(L_{ij}+L_{ji}\right)^2. \tag{35.44}$$

The material coefficients of the equations (35.40 -35.42) are related to the different effects of heat conduction, diffusion, and electric conductivity. They are combinations and modifications of coefficients which are known from simple empirical relations. For example, L_{ss} is the heat conductivity of the system per temperature λ/T (*Fourier's law*); L_{sj} is related to the *Dufour effect*, where a heat flow is caused by concentration gradients; and L_{se} is the *Peltier* coefficient π/T. The coupling coefficients L_{ks} characterize the phenomenon of thermal diffusion (*Ludwig–Soret effect*), where a mass diffusion current is caused by temperature gradients; L_{kj} are mass diffusion coefficients and the electric transport coefficients L_{ke} is given by the transference number of the component t_k divided by z_k. The thermopower parameter L_{es} is the negative Seebeck coefficient S which quantifies the thermoelectric effect, i.e. the amount of direct conversion of a temperature differences to electric voltage (*Seebeck effect*); L_{ej} is the partial diffusion potential of component j and L_{ee} directly refers to the electric resistivity of the system (*Ohm's law*).

For practical applications these relations are commonly simplified and may result in the empirical equations of thermal diffusion and thermoelectricity. The main advantage of the outlined strict derivation in the sense of TIP is a clear interaction of the coupling effects. For all the coupling coefficients the Onsager relations (35.43-35.44) hold which implies, e.g. $S = -\pi/T$. However, this does not say anything about the magnitude of the coupling terms. For example thermophoresis can clearly be observed in polymers, where different particle types move differently under the force of the temperature gradient, whereas the cross contribution from the Dufour phenomenon on the temperatures field is hardly measurable. The same holds for the electric field, where a temperature gradient induces an electric Seebeck current $\boldsymbol{I} \propto -S\boldsymbol{\nabla}T$ whereas the inverse Peltier effect $\boldsymbol{J}_\theta = \pi/T\boldsymbol{e}$ might be negligible.

What remains is to define the chemical potential which, in the simplest case, is determined by the concentration gradient between the n components (Natsiavas et al, 2016). Here, in the battery anode, the material mixture tends to form phases of different composition, e.g. LiC_6 and Li_2C_2. We choose a variational approach to determine μ_k. Then the free energy of a mixture is of the general form

$$\int_\Omega \rho\Psi \,\mathrm{d}\Omega = \int_\Omega \rho \left[\Psi^{\mathrm{chm}} + \Psi^{\mathrm{sur}}\right] \mathrm{d}\Omega$$

where Ψ^{chm} denotes the chemical free energy and Ψ^{sur} is the contribution of the interphase surface energy, all per unit mass of the bulk. Additional energy contributions resulting from elasticity and/or viscous flow may be added, cf. Anders and Weinberg (2011); Anders et al (2012); Anders and Weinberg (2018). Specifically we consider the energy of a binary mixture where it simply holds $c_2 = 1 - c_1 =: c$. For the chemical energy we make use of a Porter's-type model which is known from Flory–Huggins thermodynamics of mixing.

$$\begin{aligned}\Psi^{\mathrm{chm}}(c,T) &= \sum_{i=1}^{n=2}\Big\{g_i c_i + \theta RT c_i \ln(c_i) + \frac{1}{2}\sum_{j\neq i}^{n=2}\chi_{ij}RTc_ic_j\Big\} \\ &= g_1(1-c) + g_2 c + \theta RT\left[c\ln(c) + (1-c)\ln(1-c)\right] \\ &\quad + \chi RTc(1-c)\end{aligned} \tag{35.45}$$

The terms $g_1(1-c)$ and $g_2 c$ denote the enthalpic contribution to the Gibbs free energy density per unit mass of the individual components; θ is a temperature dependent material parameter, R is the universal gas constant, and the interaction parameter $\chi = (\chi_{12} + \chi_{21})/2$ characterizes the chemical interaction between the constituents of the mixture. In particular, for $\chi < 2\theta$ the energy is convex which corresponds to a homogeneous solution. A separation of phases will be observed only for $\chi > 2\theta$, when the chemical energy (35.45) turns into a double-well function, i.e., it has two relative minima and a concave region in between. The co-existence of two phases causes surface energy term,

$$\Psi^{\text{sur}}(\nabla c, T) = \sum_{i=1}^{n=2} \frac{\kappa_i(T)}{2} \|\nabla c_i\|^2 = \frac{\kappa(T)}{2} \|\nabla c\|^2 \tag{35.46}$$

where $\kappa = \kappa_1 + \kappa_2$ is related to specific surface energy of the interfacial regions between the domains of each phase. The variational derivative of the free energy density with respect to c_k, gives the chemical potential μ_k per unit mass,

$$\mu_k = \partial_{c_k} \Psi^{\text{chm}} - \nabla \cdot [\partial_{\nabla c_k} (\Psi^{\text{sur}})] \tag{35.47}$$

which can be specified for the binary mixture to

$$\mu = \mu_2 = \partial_c \Psi^{\text{chm}} - \nabla \cdot [\partial_{\nabla c} (\Psi^{\text{sur}})] = \partial_c \Psi^{\text{chm}} - \kappa \Delta c - \frac{\partial \kappa}{\partial T} \nabla T \nabla c \, . \tag{35.48}$$

Without additional electric contributions the uncoupled model for mass diffusion ($\boldsymbol{J}_s = 0$, $\boldsymbol{I} = 0$ in Eqs. 35.40-35.42) corresponds to the classical Cahn–Hilliard model in Cahn (1965). Please note that the derivative of the interfacial energy density introduces a biharmonic operator which requires a special treatment for numerical solution.

35.4 Numerical Examples

In the following, we present two- and three-dimensional computational studies of binary and ternary systems in order to illustrate the application of the derived general model. Because we tackle diffusion problems, i.e. large deformations are not involved, the method of finite differences is employed to provide a fast solution of the systems; for a discussion of different solution algorithms we refer to Anders et al (2012). In the first example the influence of the temperature on the concentration field will be demonstrated and in the second example the system will be exposed to an isothermal environment where the concentration fields are subjected to reactive interactions.

35.4.1 Thermal Diffusion

For the formulation of the kinetics and thus the temporal evolution of the normalized concentration field we need to shift all remaining fields and potentials to a dimensionless description. Therefore, the spatial field $\boldsymbol{x}$ is scaled by a characteristic length ℓ and rewritten as $\boldsymbol{x} = \ell\tilde{\boldsymbol{x}}$. The time t is treated in the same way and reformulated as $t = \tau\,\tilde{t}$, the temperature field as $T = T_0\vartheta$, and the chemical potential (35.48) as $\mu = (RT_0/M_m)\tilde{\mu}$, where M_m is the molar mass. With this substitutions, it follows for the notation of a dimensionless temporal derivative that $\overset{\circ}{\bullet} = \tau\dot{\bullet}$ and for the dimensionless spatial derivative that $\tilde{\nabla}\bullet = \ell\,\nabla\bullet$. By inserting (35.2) into (35.42) and exploiting Fourier's law for heat conduction both, the chemical field and the temperature field follow an extended Cahn–Hilliard equation in our modeling. For parameter identification see Schuß et al (2018).

$$\overset{\circ}{c} = M_1\tilde{\nabla}\cdot[c(1-c)(\tilde{\nabla}\tilde{\mu})_T - M_{21}\tilde{\nabla}\vartheta] \tag{35.49}$$

$$\overset{\circ}{\vartheta} = M_3\tilde{\nabla}\cdot[c(1-c)(\tilde{\nabla}\tilde{\mu})_T - M_{43}\tilde{\nabla}\vartheta] \tag{35.50}$$

The gradient of the chemical potential is obtained at constant temperature (index T). M_1 is a normalized diffusivity D, M_{21} is a weighted Soret coefficient S_T, M_3 is a normalized Dufour coefficient M_d, M_{43} is a normalized fraction of thermal conductivity k to the Dufour coefficient and c_p is the isobaric molar specific heat capacity. The binary density field is given by $\rho(c) = \rho_1/f(c)$ where $f(c) = 1 + c(r-1)$ and $r = \rho_1/\rho_2$ is the ratio of pure densities, cf. Anders et al (2012). We use the numerical values which are listed in Table 35.1 for a simulation on a unit cube with $N = 128$ equidistant grid points in each dimensional direction and a normalized time increment $\Delta\tilde{t} = 5\cdot 10^{-3}$.

The initial concentration is set to $c = 0.8$ within the domains

Table 35.1 Normalized simulation parameters

M_1	M_{21}	M_3	M_{43}	$\tilde{\theta}$	$\tilde{\chi}$	$\tilde{\kappa}$
$\frac{\tau D}{\ell^2}$	$S_T T_0$	$-\frac{\tau}{\ell^2}\frac{RM_d}{c_p\rho_1}$	$-\frac{kT_0}{M_d}$	θM_m	$2.5\tilde{\theta}$	$\frac{\kappa\tau}{D\ell^2}$
10^{-3}	$5\cdot 10^{-2}$	$-5\cdot 10^{-5}$	$-2\cdot 10^{-1}$	1	2.5	$6\cdot 10^{-5}$

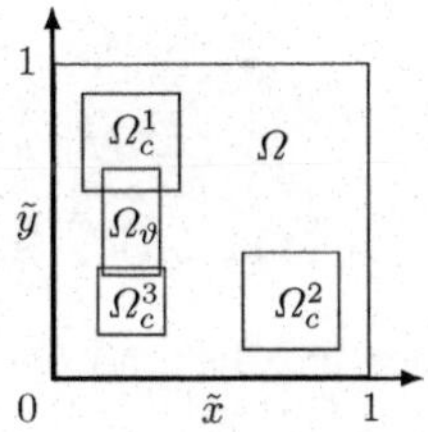

Fig. 35.2 Visualization of $\Omega \supset (\Omega_c^1 \cup \Omega_c^2 \cup \Omega_c^3 \cup \Omega_\vartheta)$

$$\Omega_c^1 = \{(\tilde{x}, \tilde{y}) \in \mathbb{R}^2 \mid 0.1 \leq \tilde{x} \leq 0.4, 0.6 \leq \tilde{y} \leq 0.9\} \tag{35.51}$$

$$\Omega_c^2 = \{(\tilde{x}, \tilde{y}) \in \mathbb{R}^2 \mid 0.6 \leq \tilde{x} \leq 0.9, 0.1 \leq \tilde{y} \leq 0.4\} \tag{35.52}$$

$$\Omega_c^3 = \{(\tilde{x}, \tilde{y}) \in \mathbb{R}^2 \mid 0.15 \leq \tilde{x} \leq 0.3, 0.15 \leq \tilde{y} \leq 0.3\} \tag{35.53}$$

and to $c = 0.2$ on $\Omega \backslash (\Omega_c^1 \cup \Omega_c^2 \cup \Omega_c^3)$. We perform two simulations.

In the first simulation we choose an isothermal setting; it is displayed in the upper panel in Fig. 35.3. It is known from the standard Cahn–Hilliard model that the energy minimization leads to round shaped islands and according to the Ostwald ripening both larger phases grow on the expense of the smaller one.

In a second simulation we choose the same setup for the concentration field and allow locally for an increasing temperature field on Ω_ϑ which is located between one of the larger phases and the smaller phase. The applied temperature bridge enables a preferential flux direction and as a consequence a predominantly growth of the larger connected phase; it is displayed in the lower panel in Fig. 35.2. The mentioned heat generation source is placed within the squared region

$$\Omega_\vartheta = \{(\tilde{x}, \tilde{y}) \in \mathbb{R}^2 \mid 1/6 \leq \tilde{x} \leq 1/3, 1/3 \leq \tilde{y} \leq 2/3\} \tag{35.54}$$

and raises within the first 3000 normalized time increments ($\tilde{t} \leq 15$) such that the initially normalized global unit temperature ϑ is increased by 10^{-4} per time step and in total by 30% on Ω_ϑ. Note that this procedure leads to an increasing average temperature on Ω. Thus the temperature is not a conserved quantity. Moreover, due to the missing biharmonic operator in (35.50) no temperature islands are formed, although on the first glance the governing equations (35.49) and (35.50) look similarly.

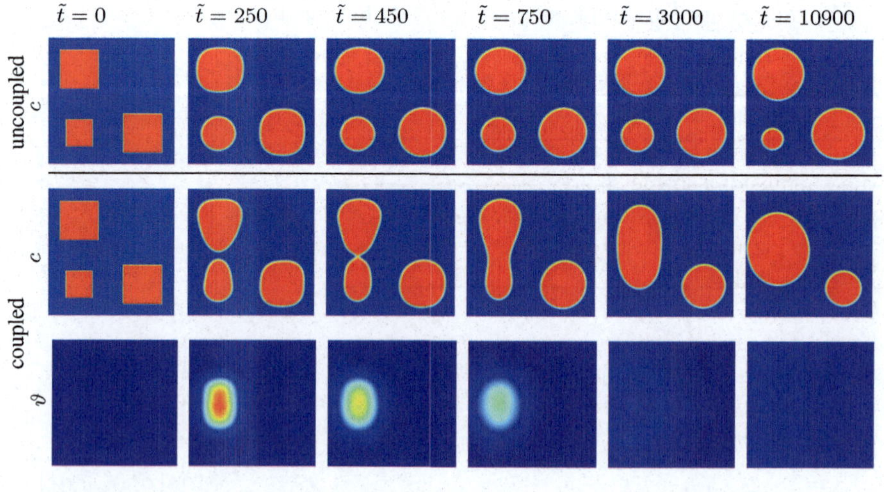

Fig. 35.3 Influence of locally increased temperature on chemical decomposition. Color coding: $c \in [0, 1]$ and $\theta \in [1, 1.3]$ where higher values are shown in red and lower values in blue.

35.4.2 Multifield Chemical Reactions

In this example we study a ternary mixture with concentrations $c_1 = c_A$, $c_2 = c_B$, $c_3 = c_C$, in an isothermal environment. The constituents are subjected to a two-way bimolecular reaction (self-reaction).

$$n_A A + n_B B \rightleftharpoons n_C C \tag{35.55}$$

Both, the forward reaction $n_A A + n_B B \rightarrow n_C C$ and the backward reaction $n_C C \rightarrow n_A A + n_B B$, have rate coefficients k^+ and k^-, respectively. The corresponding reaction rate $\mathfrak{z}_1 = k^+ c_1^{n_A} c_2^{n_B} - k^- c_3^{n_C}$ is considered in Eq. (35.2). Similar characteristic quantities as in the first example are used and, furthermore, the hard constraint of mass conservation $c_1 + c_2 + c_3 = 1$ is employed. Then each of the two remaining fields follows an extended Cahn–Hilliard equation.

$$\overset{\circ}{c}_1 = \tilde{\nabla} \cdot (\boldsymbol{M} \nabla \mu_1) + v_{11} \left\{ k^+ c_1^{n_A} c_2^{n_B} - k^- [1 - (c_1 + c_2)]^{n_C} \right\} \tag{35.56}$$

$$\overset{\circ}{c}_2 = \tilde{\nabla} \cdot (\boldsymbol{M} \nabla \mu_2) + v_{21} \left\{ k^+ c_1^{n_A} c_2^{n_B} - k^- [1 - (c_1 + c_2)]^{n_C} \right\} \tag{35.57}$$

Both chemical potentials are calculated by Eq. (35.48) and we assume $n_A = n_B = n_C = 1$ and $k^+ = k^- = k$.

For the computation we use a uniform mesh consisting of 128 grid points in each direction. The uniformly perturbed initial concentration values are $c_1^j = 0.22 \pm 0.01X^j$ and $c_2^j = 0.21 \pm 0.01X^j$ on Ω, where $X^j \sim \mathcal{U}(0,1), \forall j \in \{1, \ldots, 128^3\}$. Periodic boundary conditions apply. The normalized time increment is set to $\Delta\tilde{t} = 10^{-2}$. The mobility is $M = 1$, the gradient energy coefficients are $\kappa_1 = \kappa_2 = \kappa_3 = 128^{-3}, \theta = 0.35$, and the interaction parameters are $\chi_{ij} = 1.25$ for all $1 \leq i, j \leq 3$.

We compare two different scenarios with each other, see Fig 35.4. In the first case with zero reaction rate ($k = 0$) we observe c_3 to be the dominant phase which forms a foam like matrix structure. The smaller phases tend to reduce their surface energy by the formation of nuggets. One should keep in mind, that the shown snapshot is an arbitrarily chosen point in time and not a final stable state of decomposition in this dynamic process. Of course, the system continues to decompose until all subdomains of a species merge to a single one. The displayed arrangements of each phase give some insight into the complexity of the system.

A different situation is observed for the second case with a bimolecular reaction. Please note that reactants are counted as negative numbers by convention, here we have $kv_{11} = kv_{21} = -600/128^3$. In this reactive case, phase c_3 is quenched in between c_1 and c_2 which can hardly be seen in the surface plot. The single phase arrangements illustrate the pattern. The fact that the C-rich phase arranges at the interface between specimen A and B follows from reaction (35.55), which states that specimens of type A and of type B are needed to produce the specimen of type C.

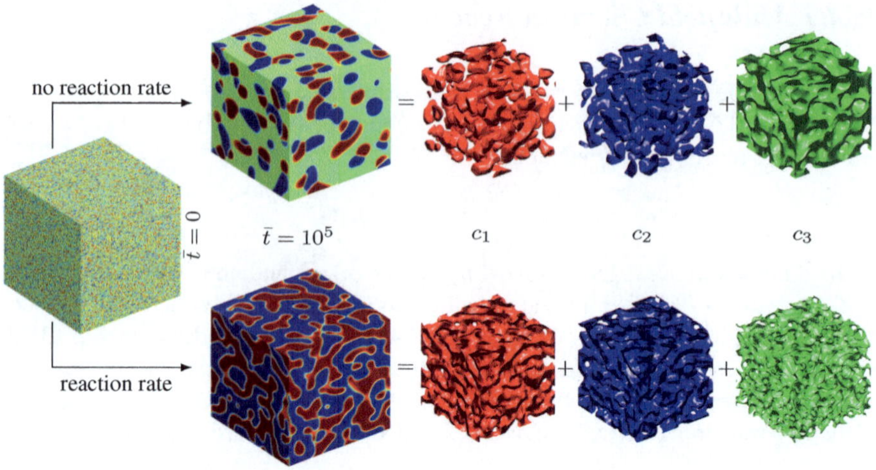

Fig. 35.4 From left to right: initial configuration, snapshot at the early stage (after $\bar{t} = 10,000$) of resulting morphologies for both, a ternary non-reactive system (top line) and a reactive system (bottom line), and extraction of its single phases c_1, c_2, and c_3.

35.5 Conclusion

In this work an extended diffusion model involving electrical, thermal, mechanical, and chemical contributions is formulated and applied to two illustrative examples. The derivation of the coupled multi-field model is based on the Thermodynamic of Irreversible Processes (TIP), cf. De Groot and Mazur (1962).

Finally, we want to mention that the appropriate derivation of the entropy-flux $\boldsymbol{j}^s$ is subject to many discussions, see Cimmelli et al (2014); Muschik and Ehrentraut (1996) among others. At one hand, the local equilibrium hypothesis is the cornerstone of the classical strategy, i.e. thermodynamic state variables in non-equilibrium states are considered to be the same as in equilibrium. It is still under debate that other variables, not found at equilibrium, might be able to influence non-equilibrium processes. On the other hand, the standard procedure, which implies that the globally defined balance relations are valid for a material point at $\boldsymbol{x}$ and its infinitesimal neighborhood, has the severe drawback, that weakly nonlocal continuum theories which take the gradients of the unknown variables as independent state variables, result to be not compatible with second law, (Ván, 2003). This applies, e.g., for higher gradient elasticity, phase-field fracture approaches or similar models. For our model of diffusion in solid state battery systems, however, we have no reason to assume additional state variables and thus the derived TIP based model is comprehensive.

Acknowledgements The authors gratefully acknowledge the support of the Deutsche Forschungsgemeinschaft (DFG) under the project WE-2525/8-1.

References

Aifantis K, Hackney S, Kumar R (2010) High Energy Density Lithium Batteries: Materials, Engineering, Applications. Wiley-VCH

Anders D, Weinberg K (2011) A variational approach to the decomposition of unstable viscous fluids and its consistent numerical approximation. ZAMM - Journal of Applied Mathematics and Mechanics / Zeitschrift für Angewandte Mathematik und Mechanik 91(8):609–629

Anders D, Weinberg K (2012) Thermophoresis in binary blends. Mechanics of Materials 47:33 – 50

Anders D, Weinberg K (2018) A Thermodynamically Consistent Approach to Phase-Separating Viscous Fluids. Journal of Non Equilibrium Thermodynamics 43:185–191

Anders D, Hesch C, Weinberg K (2012) Computational modeling of phase separation and coarsening in solder alloys. International Journal of Solids and Structures 49(13):1557–1572, DOI 10.1016/j.ijsolstr.2012.03.018

Berger R (2012) Technology & market drivers for stationary and automotive battery systems. Tech. rep., Strategy Consultants, URL http://www.rechargebatteries.org/wp-content/uploads/2013/04/Batteries-2012-Roland-Berger-Report1.pdf

Cahn JW (1965) Phase separation by spinodal decomposition in isotropic systems. The Journal of Chemical Physics 42(1):93–99

Cimmelli VA, Jou D, Ruggeri T, Ván P (2014) Entropy principle and recent results in non-equilibrium theories. Entropy 16(3):1756–1807

De Groot SR, Mazur P (1962) Non-equilibrium thermodynamics. North-Holland, Amsterdam

Eckart C (1940) The thermodynamics of irreversible processes. ii. fluid mixtures. Phys Rev 58:269–275

Goddard JD (2015) Dissipation potentials for reaction-diffusion systems. Industrial & Engineering Chemistry Research 54(16):4078–4083

Landau LD, Lifshitz EM, Pitaevskii LP (2013) Electrodynamics of Continuous Media. Course of Theoretical Physics, Elsevier Science

Muschik W, Ehrentraut H (1996) An amendment to the second law. Extracta Mathematicae 11(1):163–171

Natsiavas PP, Weinberg K, Rosato D, Ortiz M (2016) Effect of prestress on the stability of electrode–electrolyte interfaces during charging in lithium batteries. Journal of the Mechanics and Physics of Solids 95:92 – 111

Schuß S, Weinberg K, Hesch C (2018) Thermomigration in snpb solders: Material model. Mechanics of Materials 121:31 – 49

Truesdell C, Toupin R (1960) The classical field theories. In: Flügge S (ed) Principles of Classical Mechanics and Field Theory / Prinzipien der Klassischen Mechanik und Feldtheorie, Springer, Berlin, Heidelberg, pp 226–858

Ván P (2003) Weakly nonlocal irreversible thermodynamics. Annalen der Physik 12(3):146–173

Weinberg K, Werner M, Anders D (2018) A chemo-mechanical model of diffusion in reactive systems. Entropy 20(2)

Chapter 36
Nonclassical Bending Behavior of Thin Strips of Photochromic Liquid Crystal Elastomers Under Light Illuminations

Yang Zhang & Yongzhong Huo

Abstract Photochromic liquid crystal elastomers (LCEs) bend when irradiated by light of suitable wavelength. However, due to the rotation of the liquid crystal director, rather large shear strains are inevitably produced and some basic assumptions of the classical simple beam theory of Euler-Bernoulli fail to be satisfied. In this work, we use the first-order shear deformation beam theory of Timoshenko to model the unusual quasi-soft bending behavior of soft LCEs under light illuminations. The results show that in addition to the large shear strain, the effect of initial effective length ratio makes a great difference to the deflections due to the rotation of director. This represents the first direct verification that Euler-Bernoulli beam theory fails to deal with such nonclassical bending of soft LCEs, while Timoshenko beam model can work sufficiently well, which also gives a possible method to measure the effective opto bending moment experimentally.

Keywords: Liquid crystal elastomer · Photochromic materials · Beam theory

36.1 Introduction

Photochromic LCEs doped with rod-like groups, such as azobenzenes, which undergo trans-cis isomerization on absorption of UV photons, are found to contract when irradiated at suitable wavelengths since the local order is disrupted by the kinked dopant groups (Warner and Terentjev, 2007; Finkelmann et al, 2001). Since light is absorbed by the material (Corbett and Warner, 2006, 2008), the reduction in intensity through the thickness of a cantilever gives a gradient of response and then

Yang Zhang · Yongzhong Huo
Institute of Mechanics and Computational Engineering, Department of Aeronautics and Astronautics, Fudan University, Shanghai 200433, China,
e-mail: yzhuo@fudan.edu.cn, 14110290008@fudan.edu.cn

B. E. Abali et al. (eds.), *New Achievements in Continuum Mechanics and Thermodynamics*, Advanced Structured Materials 108,
https://doi.org/10.1007/978-3-030-13307-8_36

non-uniform strains lead to the bending of a cantilever of the active material as in Ikeda et al (2003); Camacholopez et al (2004); Yu et al (2003); Jin et al (2006).

The light induced bending behavior of LCEs with a beam model based on simple bending assumptions has been studied by several authors (Jin et al, 2006; Warner and Mahadevan, 2004; van Oosten et al, 2007; Jin et al, 2010a,b, 2011; Dunn, 2007; Dunn and Maute, 2009; He, 2007; Modes et al, 2011; Warner et al, 2010; Warner and Corbett, 2010; Zeng et al, 2010). Besides, due to the unusual soft or semi-soft behavior of LCEs, the rotation of the LC director can have a strong effect on the mechanical response of the materials as shown in Warner and Terentjev (2007); Jin et al (2010b, 2011). Lin et al (2012) proposed the constitutive equation including the effect of the photo isomerization, and found that the opto-mechanical behaviors are also affected by the soft behavior . Large shear strains occur in the quasi-soft bending due to the anisotropy and its very special mechanical properties (soft elasticity) of LCEs. However, on this occasion, straight lines normal to the mid-plane of LCE beams before deformation won't remain normal to the mid-plane after deformation, which finally leads the classical Euler-Bernoulli beam assumption to fail as discussed in Lin et al (2012).

The first-order shear deformation beam theory of Timoshenko allows for the effect of transverse shear deformation which is neglected in the Euler-Bernoulli beam theory. In the first-order shear deformation theory, the transverse shear strain is assumed to be constant with respect to the thickness coordinate, so shear correction factors are introduced to correct for the discrepancy between the actual transverse shear force distributions and those computed using the relations of the TBT in Timoshenko (1921); Reddy et al (1997).

In this paper, the Timoshenko beam model for quasi-soft bending of photochromic LCEs under light illuminations is presented. Based on the assumption of the form of the displacement and the stress field, the governing equations and the general solutions of rotations and deflections of beams are obtained. The finite element results are compare with the theoretical results of TBT model for various external loads. A numerical method is used to evaluate shear correction factor introduced in TBT.

36.2 TBT Model for Optical-mechanical Bending of Beam Shaped Specimens

36.2.1 Optical-mechanical Constitutive Relations

As shown in Fig. 36.1, we consider a uniform LCE beam with length L, thickness h and width w. The director of the sample is parallel with x direction, i.e. $\mathbf{n}_0 = (1, 0, 0)^{\mathrm{T}}$ and it is illuminated upward by unpolarized light along the y direction from the bottom. Here a linearized opto-mechanical constitutive relation of soft LCEs for infinitesimal deformations obtained by one of the authors, see Lin et al (2012), is applied to study the light induced bending behavior of photochromic LCEs. As $\mathbf{n}_0 = (1, 0, 0)^{\mathrm{T}}$, the components of Cauchy stress take the form

$$\begin{aligned}
\sigma_{xx} &= \frac{E}{1+\nu}\left(\varepsilon_{xx} - \varepsilon^r_{xx}\right) - p, \\
\sigma_{yy} &= \frac{E'}{1+\nu'}\left[\left(\varepsilon_{yy} - \varepsilon^r_{yy}\right) + \frac{\nu - \nu'}{(1+\nu)(1-\nu')}\left(\varepsilon_{xx} - \varepsilon^r_{xx}\right)\right] - p, \\
\sigma_{zz} &= \frac{E'}{1+\nu'}\left[\left(\varepsilon_{zz} - \varepsilon^r_{zz}\right) + \frac{\nu - \nu'}{(1+\nu)(1-\nu')}\left(\varepsilon_{xx} - \varepsilon^r_{xx}\right)\right] - p, \\
\sigma_{xy} &= 2G'\left(\varepsilon_{xy} - \frac{r_0 - 1}{r_0 + 1}\omega_{xy}\right), \\
\sigma_{xz} &= 2G'\left(\varepsilon_{xz} - \frac{r_0 - 1}{r_0 + 1}\omega_{xz}\right), \\
\sigma_{yz} &= 2G\varepsilon_{yz},
\end{aligned} \tag{36.1}$$

where ε_{ij} are the Cauchy strains, ω_{ij} are the antisymmetric parts of the displacement gradient, ε^r_{ij} are the light-induced strains and p is the lagrangian multiplier introduced due to the incompressibility. The elastic constants and the light-induced strains are given by

$$\begin{gathered}
\beta = \frac{r}{r_0},\ E = \mu\beta^{\frac{1}{3}}\frac{2+\beta}{\beta},\ E' = 2\mu\beta^{\frac{1}{3}}\frac{2+\beta}{1+\beta},\ \nu' = \frac{1}{1+\beta},\ \nu = \frac{1}{2}, \\
G = \frac{E'}{2(1+\nu')},\ G' = \frac{1}{2}\frac{r_0+1}{r_0-1}\left(\frac{1}{\beta} - 1\right), \\
\varepsilon^r_{xx} = -\frac{1-\beta}{2+\beta},\ \varepsilon^r_{yy} = \varepsilon^r_{zz} = -\nu\varepsilon^r_{xx},
\end{gathered} \tag{36.2}$$

where μ is the effective shear moduli, r and r_0 denote the anisotropy of the shape distribution of nematic network in the current configuration and in the reference configuration, respectively. Under light illuminations, the anisotropy denoted by r

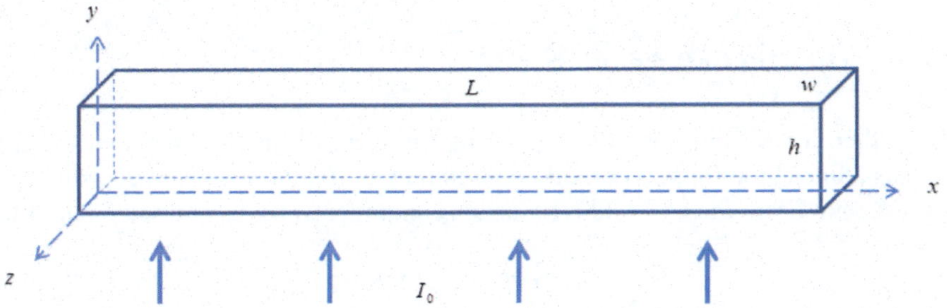

Fig. 36.1 The schematic of the beam shaped specimen under upward unpolarized light.

decreases due to the photo isomerization and decays with the penetration depth of light. Here we assume that the change is very small, that is, $(r_0 - r)/r_0 \ll 1$. Thus β is less than but approximately equal to 1. The parameters used for calculations in the following part are the same as those in Lin et al (2012). Note that for this unusual constitutive relation in Eq. (36.1), the first terms of three normal stresses represent the deviatoric part of stress tensor and p is the spherical part, which can be determined by using the incompressibility constraint. More details on the derivation of Eq. (36.1) has been given in Lin et al (2012).

If there is no light illumination, which means $r = r_0$, the light induced strain ε^r_{xx} is zero and the soft material behaves as an isotropic and incompressible Hookean material, except for its vanished in-plane shear moduli G', which is often referred as the soft behavior as predicted by the neo-classical elastic energy. However, the material behavior is rather different under light illumination. Due to $\beta \neq 1$, two elastic moduli and two Poisson's ratio arise in constitutive relation of Eq. (36.1), and single domain LCEs become transverse isotropy in the plane perpendicular to the director. In addition, the light induced change of the effective length ratio $r_0 - r$ will produce nonzero shear moduli G' and light induced strain ε_r.

Besides, the Young's moduli and light induced strain depend on the effective length ratio , which are affected by the light illumination conditions (incident light intensity i_0 and light decay distance d). Since light is absorbed by the material, the reduction in light intensity through the thickness of a cantilever gives a gradient of response and non-uniform strains lead to the bending of a cantilever of the active material. Therefore, the LCE material under light illumination becomes a functional gradient material. Moreover, the light induced decrease of the effective length ratio r implies the light induced anisotropy.

As observed in experiments and discussed in several theoretical works, the elastic moduli of single domain LCEs are anisotropic and depend strongly on the temperature. However, it is necessary to take into consideration that the stress induces a biaxiality of the liquid crystal molecules in order to obtain this anisotropy. In the present paper, the biaxiality is neglected for simplicity. Thus, the elastic moduli are taken as isotropic under mechanical loading and the anisotropy is induced by the light illumination.

36.2.2 Timoshenko Beam Model

Beam theories are developed by assuming the form of the displacement or stress field as a linear combination of unknown functions and the thickness coordinate. In Timoshenko beam theory (TBT), for stress components, we have basic assumption

$$\sigma_{yy} = \sigma_{zz} = 0. \tag{36.3}$$

Combining (36.3) and the incompressibility $tr(\varepsilon) = 0$, we get

$$p = \frac{E'}{2\left(1+\nu'\right)}\varepsilon_{yy} = \frac{E'}{2\left(1+\nu'\right)}\varepsilon_{zz} = -\frac{E'}{\left(1+\nu'\right)}\varepsilon_{xx}. \tag{36.4}$$

Hence, we can obtain the following constitutive equations for the bending of LCE beams

$$\sigma_{xx} = E\left(\varepsilon_{xx} - \varepsilon^{r}_{xx}\right),\ \sigma_{xy} = 2G'\left(\varepsilon_{xy} - \frac{r_0 - 1}{r_0 + 1}\omega_{xy}\right). \tag{36.5}$$

The Timoshenko beam theory (TBT) is based on the in-plane displacement field at $z = 0$

$$u\left(x, y\right) = u_0\left(x\right) + \left(y - \bar{y}\right)\phi\left(x\right),\ v\left(x, y\right) = v_0\left(x\right) + \tilde{v}\left(y\right), \tag{36.6}$$

where $u_0\left(x\right)$ and $v_0\left(x\right)$ is the displacements of the point $\left(x, \bar{y}\right)$ on plane $z = 0$, $\phi\left(x\right)$ denotes the rotation of straight lines normal to the mid-plane about z axes and $\tilde{v}\left(y\right)$ denotes the difference of displacements between the two points $\left(x, y\right)$ and $\left(x, \bar{y}\right)$. In view of the displacement field given in Eq. (36.6), the in-plane strains and rotation components are given by

$$\begin{aligned}
\varepsilon_{xx} &= \frac{\partial u}{\partial x} = \frac{\partial u_0}{\partial x} + \phi'\left(x\right)\left(y - \bar{y}\right) = \varepsilon^{0}_{xx} + \phi'\left(x\right)\left(y - \bar{y}\right),\\
\varepsilon_{yy} &= \frac{\partial v}{\partial y} = \frac{\partial \tilde{v}}{\partial y},\\
\varepsilon_{xy} &= \frac{1}{2}\left(\frac{\partial u}{\partial y} + \frac{\partial v}{\partial x}\right) = \frac{1}{2}\left(\phi\left(x\right) + v_0'\left(x\right)\right),\\
\omega_{xy} &= \frac{1}{2}\left(\frac{\partial u}{\partial y} - \frac{\partial v}{\partial x}\right) = \frac{1}{2}\left(\phi\left(x\right) - v_0'\left(x\right)\right).
\end{aligned} \tag{36.7}$$

Insert expressions of Eq. (36.7) into constitutive equations Eq. (36.5) and we can express the bending moment M_{xx} and shear force Q_x in terms of $v_0\left(x\right)$ and $\phi\left(x\right)$

$$\begin{aligned}
M_{xx} &= \iint_A \sigma_{xx}\left(y - \bar{y}\right)\mathrm{d}A = \iint_A E\left(\varepsilon^{0}_{xx} + \phi'\left(x\right)\left(y - \bar{y}\right) - \varepsilon^{r}_{xx}\right)\left(y - \bar{y}\right)\mathrm{d}A\\
&= D\phi'\left(x\right) + M_{\text{eff}},\\
Q_x &= \iint_A \sigma_{xy}\mathrm{d}A = \iint_A \frac{2G_1}{1 + r_0}\left(\phi + r_0 v_0'\right)\mathrm{d}A = K_{\mathrm{s}}A_{xy}\left(\phi + r_0 v_0'\right),
\end{aligned} \tag{36.8}$$

where

$$\begin{aligned}
D &= \iint_A E(y - \bar{y})^2\mathrm{d}A,\\
M_{\text{eff}} &= -\iint_A E\varepsilon^{r}_{xx}\left(y - \bar{y}\right)\mathrm{d}A,\\
A_{xy} &= \iint_A \frac{2G'}{1 + r_0}\mathrm{d}A.
\end{aligned} \tag{36.9}$$

and K_s is the shear correction factor that has been introduced to compensate for the error caused by assuming a constant shear stress distribution through the beam depth. The shear correction factor depends not only on the material and geometric parameters but also on the loading and boundary conditions. Here,

$$\bar{y} = \frac{\iint_A Ey\mathrm{d}A}{\iint_A E\mathrm{d}A} \approx \frac{h}{2}. \tag{36.10}$$

From Eq. (36.8), we can write the relations of generalized displacement field and externally applied loads as

$$\phi'(x) = \frac{M_{xx} - M_{\mathrm{eff}}}{D}, \; v_0'(x) = -\frac{1}{r_0}\phi(x) + \frac{Q_x}{r_0 K_\mathrm{s} A_{xy}}. \tag{36.11}$$

Then institute Eq. (36.11) into the balance equations of moments and forces

$$\frac{\mathrm{d}M_{xx}}{\mathrm{d}x} = Q_x, \; -\frac{\mathrm{d}Q_x}{\mathrm{d}x} = q, \tag{36.12}$$

we can get the governing equation of deflections

$$v_0''(x) = \frac{M_{\mathrm{eff}} - M_{xx}}{r_0 D} + \frac{q}{K_\mathrm{s} A_{xy} r_0}. \tag{36.13}$$

Note that without considering the last term in Eq. (36.13), the solutions of Eq. (36.13) are reduced to Euler-Bernoulli beam theory when $r_0 = 1$. In other words, there exist large discrepancies between the two theories EBT and TBT in the bending of soft LCE beams.

Furthermore, the solutions for the Timoshenko beam under the light actuations and external distributing loads q may be readily obtained by integrating the fourth-order differential equation and using two boundary conditions from at each end of the beam to evaluate the integration constants. By integrating Eqs. (36.12) and (36.11) with respect to x field, we can express general solutions of the bending moments, shear forces, rotations and deflections of beams as

$$\begin{aligned}
M_{xx} &= M_{xx}^L - Q_x^L(L - x) - \int_x^L \int_\xi^L q\mathrm{d}\eta\mathrm{d}\xi, \\
Q_x &= Q_x^L + \int_x^L q\mathrm{d}\xi, \\
\phi(x) &= \phi^0 + \int_0^x \frac{1}{D}(M_{xx} - M_{\mathrm{eff}})\mathrm{d}\xi, \\
v_0(x) &= v_0^0 - \frac{\phi^0}{r_0}x + \int_0^x \frac{1}{r_0}\int_0^\xi \frac{1}{D}(M_{\mathrm{eff}} - M_{xx})\mathrm{d}\eta\mathrm{d}\xi + \int_0^x \frac{Q_x}{r_0 K_\mathrm{s} A_{xy}}\mathrm{d}\xi.
\end{aligned} \tag{36.14}$$

where v_0^0, ϕ^0, M_{xx}^L, Q_x^L are constants of integration. These constants are to be determined by using the boundary conditions of the particular beam.

For free (F), simply supported (S) and clamped (C) ends, boundary conditions are given by

$$F: Q_x = M_{xx} = 0,\ S: v_0 = M_{xx} = 0,\ C: \phi = v_0 = 0. \tag{36.15}$$

36.3 Examples of Cantilever Beams and Numerical Results

The most important class of problems involves cantilever beams, which are usually tested experimentally in mechanics. Here, we take cantilever beams for examples and use three simple cases to demonstrate our theoretical model. The first case with $q = 0$ shows only the effect of light illuminations, and the other two with a point load $q = f\delta(x - L)$ and uniformly distributed loads $q = q_0$ represent the coupled effect of optical and mechanical loads.

In all the three cases, the cantilever beams is clamped at $x = 0$ and is free at $x = L$. Thus according to Eqs. (36.15), the boundary conditions are set by $v_0^0 = \phi^0 = 0$ and $M_{xx}^L = Q_x^L = 0$. Substitue the conditions into Eqs. (36.14) and we obtain the solutions to cantilever beams as

$$\begin{aligned}\phi(x) &= \int_0^x \frac{1}{D}(M_{xx} - M_{\text{eff}})\mathrm{d}\xi,\\ v_0(x) &= \int_0^x \frac{1}{r_0}\int_0^\xi \frac{1}{D}(M_{\text{eff}} - M_{xx})\mathrm{d}\eta\mathrm{d}\xi + \int_0^x \frac{Q_x}{r_0 K_s A_{xy}}\mathrm{d}\xi.\end{aligned} \tag{36.16}$$

For brevity, we assume that beams have the uniform anisotropy at the initial state, i.e. r_0 is a constant, and we only consider uniform cantilever beams under uniform light illuminations, so D and M_{eff} are independent of x coordinate with the form $D = D_0$ and $M_{\text{eff}} = M_{\text{eff}}^0$. Besides, notice that the shear correction factor should be taken into account if nonzero shear forces are present.

Besides, the finite element method proposed by Lin et al (2012) is used to model the deformation of the specimen under light illuminations and finite element results are used to compare with the theoretical results of TBT model. To investigate whether straight lines normal to the mid-plane of LCE beams will remain straight after deformation or not, the displacements in the axial direction vs. y coordinates are plotted in Fig. 36.2. It shows the displacement u changes linearly with the y coordinate and the plane cross-section assumption is still valid.

36.3.1 First Case: no Load

In this case, beams are only driven by optical loads, i.e. $q = Q_x = M_{xx} = 0$ and thus solutions read

$$\phi(x) = -\frac{M_{\text{eff}}^0}{D_0}x,\ v_0 = \frac{M_{\text{eff}}^0}{2r_0 D_0}x^2. \tag{36.17}$$

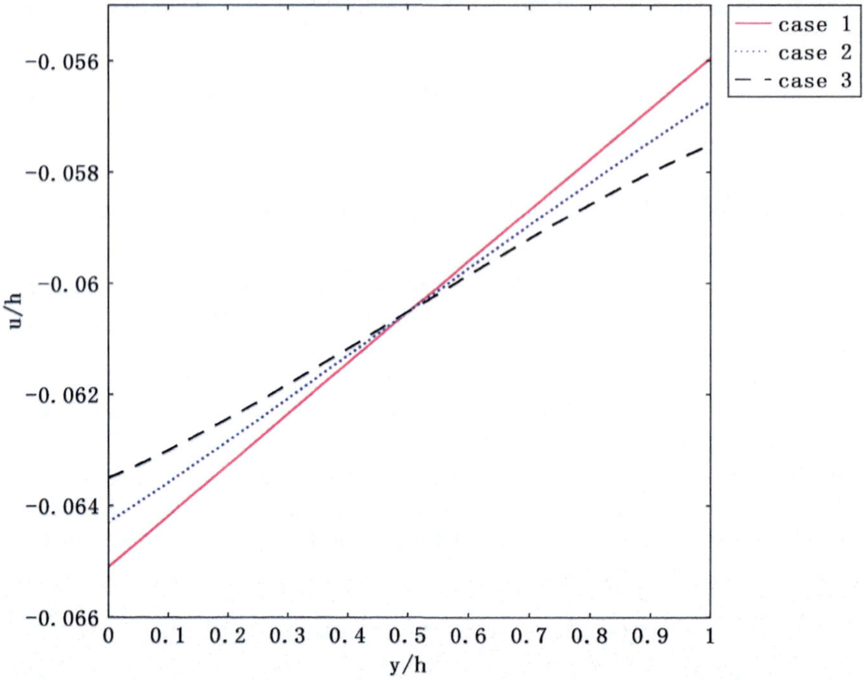

Fig. 36.2 Displacements $u$of cross section along y axis for three cases. $(r_0 = 3,\ i_0 = 2,\ d/h = 1)$

Obviously, the maximum deflection occurs at $x = L$

$$v_0^{\max} = \frac{M_{\mathrm{eff}}^0}{2r_0 D_0} L^2. \tag{36.18}$$

In an experiment, we can obtain the effective optical bending moment through measuring the maximum deflection of cantilever beams.

The following expression can well describe the relation of the solution of EBT and the solution of TBT in this case

$$v_0^L = \frac{1}{r_0} v_0^E, \tag{36.19}$$

where the superscript "T" and the superscript "E" respectively denotes the quantity in TBT and the quantity in EBT. It's obvious that the solutions of TBT is reduced to Euler-Bernoulli beam solutions if $r_0 = 1$. For anisotropy LCEs, it holds $r_0 > 1$, which implies that the effect of r_0 finally leads the classical EBT to fail. Figures 36.3 and 36.4 indicate that theoretical results of TBT model agree well with the finite element results for different initial anisotropy r_0 and different dimensionless incident light intensites i_0.

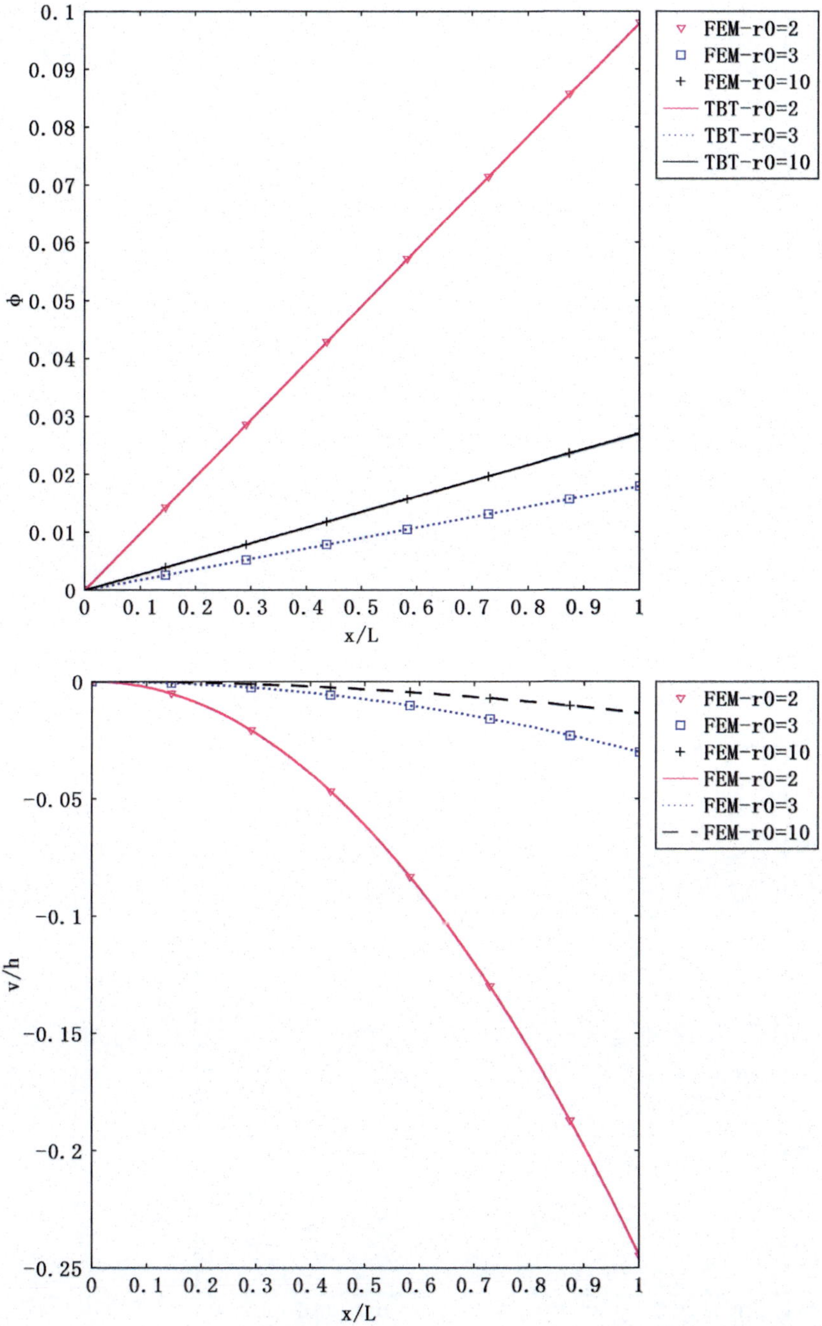

Fig. 36.3 (a) Rotation curves and (b) deflection curves for different r_0. ($i_0 = 2,\ d/h = 1$)

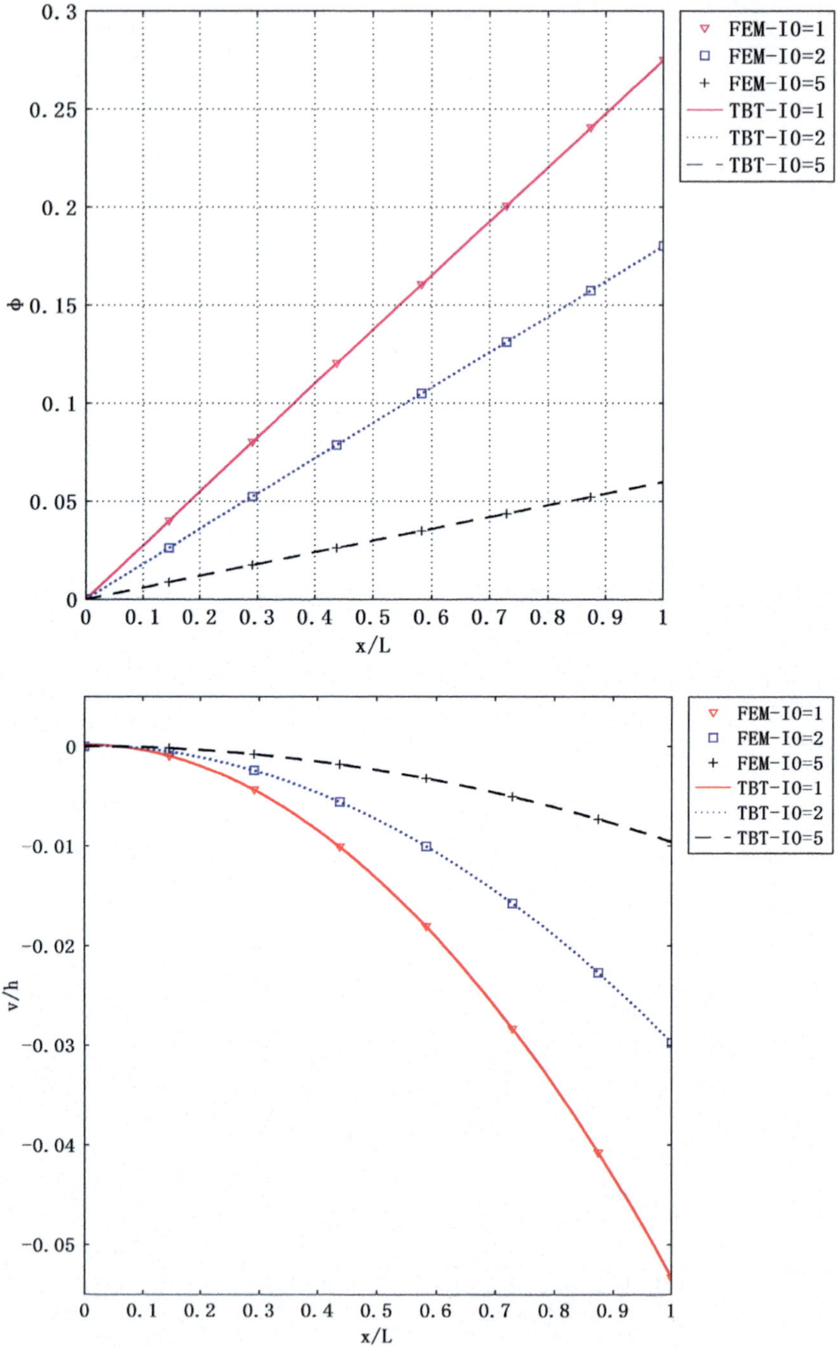

Fig. 36.4 Rotation curves and deflection curves for different i_0. ($i_0 = 3,\ d/h = 1$)

36.3.2 Second Case: a Point Load

In this case, we consider a cantilever beam with a concentrated load f applied at the free end. The solutions are given by

$$\begin{aligned}\phi(x) &= -\frac{M_{\text{eff}}^{0}}{D_0}x - \frac{f}{D_0}x\left(L - \frac{x}{2}\right),\\ v_0(x) &= \frac{M_{\text{eff}}^{0}}{2r_0 D_0}x^2 + \frac{f}{2r_0 D_0}x^2\left(L - \frac{x}{3}\right) + \frac{f}{r_0 K_s A_{xy}}x.\end{aligned} \tag{36.20}$$

Notice that shear correction factor K_s is introduced in the expression of deflections due to nonzero shear forces, which however, does not arise in the expression of rotations. Figure 36.5 indicates that the solutions of rotations in TBT model agree well with the finite element results in this case.

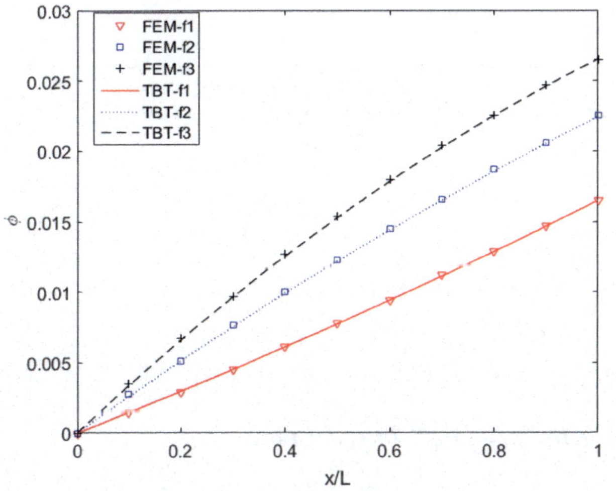

Fig. 36.5 Rotation curves for different concentrated loads. ($r_0 = 3,\ i_0 = 2,\ d/h = 1$)

Here, a numerical method is used to evaluate the shear correction factor. For beams with different length, we can obtain the forces f, which satisfy $v_0(L) = 0$. From the expression of Eq. (36.20), we have the following expression if $v_0(L) = 0$ holds

$$\underbrace{-\frac{M_{\text{eff}}^{0}}{2fL}}_{Y} = \frac{1}{3} + \frac{1}{K_s}\underbrace{\left(\frac{D_0}{A_{xy}h^2}\right)\left(\frac{h}{L}\right)^2}_{X}. \tag{36.21}$$

So in FEM, we can find out a unique force f that can make the free end satisfy $v_0(L) = 0$ and the obtained points for different length of beams (X, Y) are plotted in Figure 36.6. It shows that the shear correction factor is independent of length of

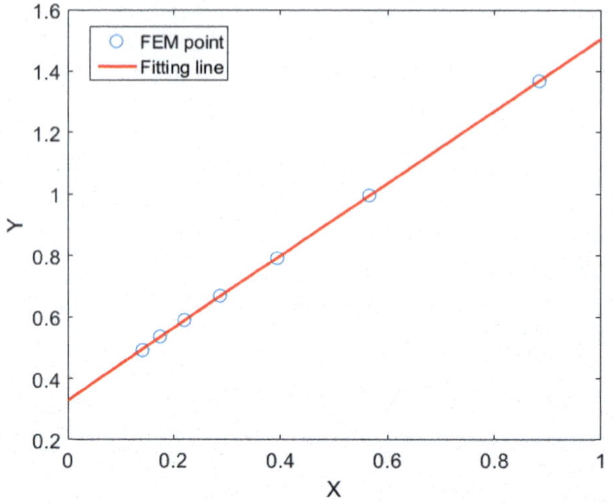

Fig. 36.6 Points of FE resutls are fitted by a line using the linear least square method. $(r_0 = 3,\ i_0 = 2,\ d/h = 1)$

beams and loads. And the points are fitted with the line $y = 0.328 + 1.177x$ by linear least square method, which implies that $K_s = 1.177^{-1} = 0.850$. Figure 36.7 indicates that the theoretical results of deflections fit well with the finite element results in this case when $K_s = 0.850$.

36.3.3 Third Case: Uniformly Distributed Load

The solutions of cantilever beams with uniformly distributed load $q = q_0$ are given by

$$\begin{aligned} \phi(x) &= -\frac{M^0_{\text{eff}}}{D_0}x + \frac{q_0}{2D_0}\left(-L^2x + Lx^2 - \frac{x^3}{3}\right), \\ v_0(x) &= \frac{M^0_{\text{eff}}}{2r_0D_0}x^2 + \frac{q_0}{2r_0D_0}x^2\left(\frac{1}{2}L^2 - \frac{1}{3}Lx + \frac{1}{12}x^2\right) \\ &+ \frac{q_0}{r_0K_sA_{xy}}x\left(L - \frac{x}{2}\right). \end{aligned} \tag{36.22}$$

Figure 36.8 shows that numerical comparisons of both rotations and deflections between the theory and finite element results show good agreement when the shear correction factor $K_s = 0.850$ has been taken into consideration.

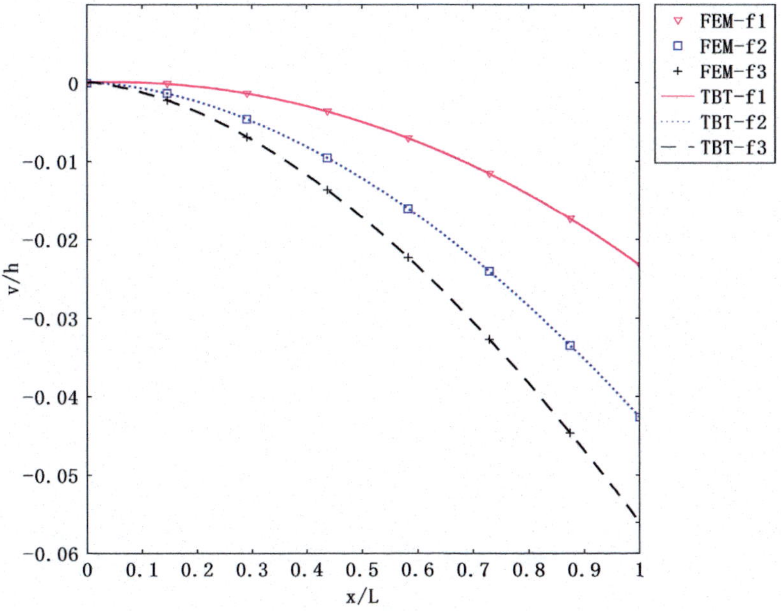

Fig. 36.7 Deflection curves for different concentrated loads. ($r_0 = 3,\ i_0 = 2,\ d/h = 1$)

36.4 Discussion About Shear Correction Factor

One of the main difficulties in using Timoshenko beam theory is the proper selection of the shear correction factor, since in TBT the shear correction factor is introduced to allow for the fact that the shear stress is not uniform over the cross section. In history, many authors have published definitions of the shear correction factor and have proposed various methods to calculate it. Most of these approaches fall into one of two categories. The first approach is to use the shear correction factor to match the frequencies of vibration of various beam constructions with exact solutions to the theory of elasticity. The second approach is to use the shear correction factor to account for the difference between the average shear or shear strain and the actual shear or shear strain using exact solutions to the theory of elasticity. At the present stage of theories and experiments, Timoshenko's expression in Timoshenko (1921) and Cowper's one in Cowper (1966) will be the most probable ones. Although not explicitly written in Timoshenko (1921), the shear correction factor obtained in the first manner for a rectangular beam is

$$K_\mathrm{s} = \frac{(5 + 5\nu)}{(6 + 5\nu)}, \tag{36.23}$$

where ν is the Poisson's ratio.

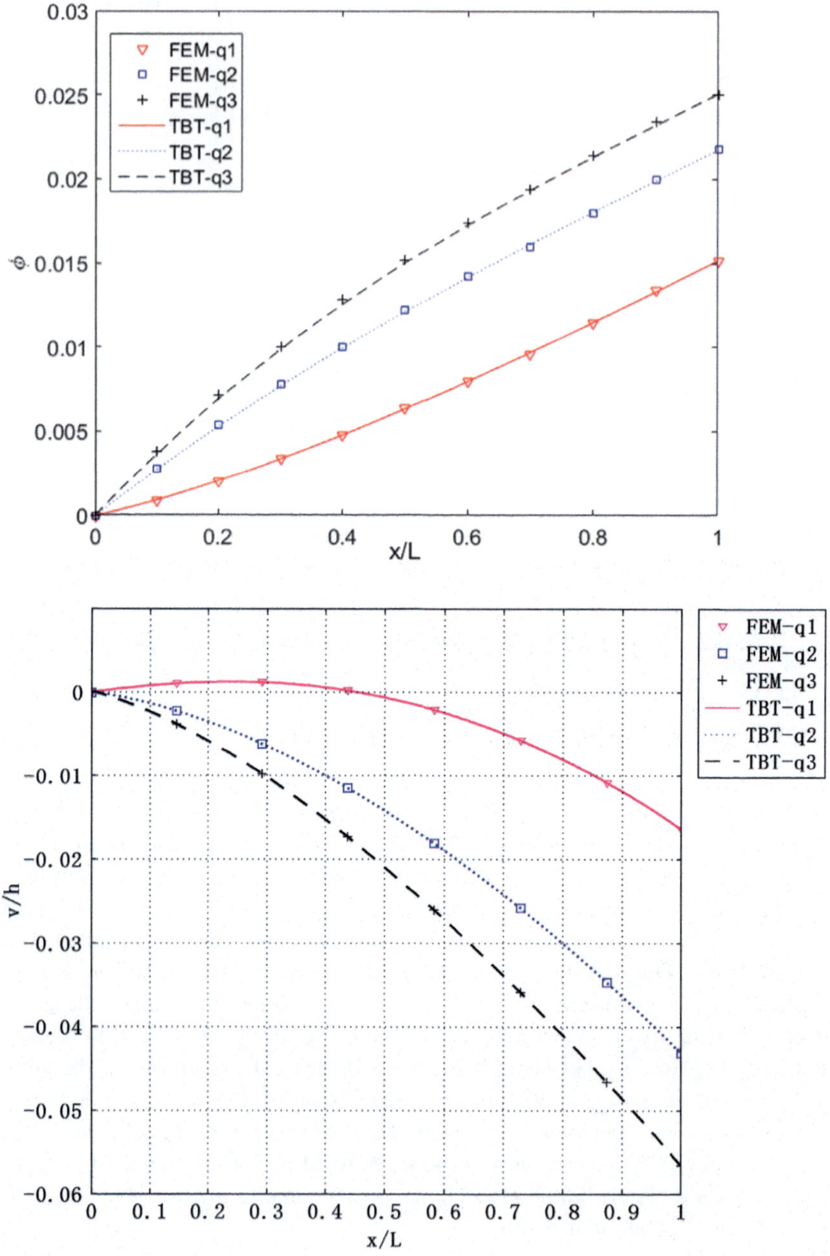

Fig. 36.8 (a) Rotation and (b) deflection curves for various uniformly distributed loads. ($r_0 = 3,\ i_0 = 2,\ d/h = 1$)

Cowper (1966) calculated the shear correction factor using an approach from the second category described above. For a rectangular isotropic homogeneous beam, Cowper found the following shear correction factor:

$$K_s = \frac{10\,(1+\nu)}{12+11\nu}. \tag{36.24}$$

With regard to our incompressible materials of LCEs, the Poisson's ratio is close to 0.5, as indicated in Eq. (36.2). Hence, according to the Timoshenko's expression and Cowper's, the shear correction factor for the rectangle is respectively 0.882 and 0.857. Our numerical results indicate that of the shear correction factor is about 0.850 very close to Cowper's formula Eq. (36.24).

36.5 Conclusions

Photochromic LCE is a currently developed smart material, which can contract and bend under suitable light illuminations. However, due to the unusual soft or semi-soft behavior of LCEs, the rotation of the LC director can have strong effect on the mechanical response of the materials. Large shear strains occur in the quasi-soft bending due to its very special mechanical properties (soft elasticity) of LCEs, which finally lead the classical Euler-Bernoulli beam assumption to a failure even for slender strips.

In this paper, the first-order shear deformation beam theory model of Timoshenko for quasi-soft bending of photochromic LCEs under light illuminations has been presented, which allows for the effect of transverse shear deformation. General solutions of the bending moments, shear forces, rotation and deflections of beams subjected to optical and mechanical loads are given, and the solutions show much difference between EBT and TBT. The effect of r_0 arises due to the free rotation of director of LCEs.

In TBT, the shear correction factor has to been taken into consideration due to the assumption of a constant shear stress distribution through the beam depth. The shear correction factor evaluated in the numerical method is 0.850, which shows good agreement with the value predicted by Cowper's formula. Numerical results indicate that TBT model we presented fits very well with finite element results for different geometric parameters and different loading and boundary conditions of beams.

Acknowledgements We gratefully acknowledge the support of this research by the National Natural Science Foundation of China (11461161008, 11772094) and Joint Funds of National Natural Science Foundation of China (No. Y81GLW1101).

References

Camacholopez M, Finkelmann H, Palffymuhoray P, Shelley M (2004) Fast liquid-crystal elastomer swims into the dark. Nature Materials 3(5):307–10

Corbett D, Warner M (2006) Nonlinear photoresponse of disordered elastomers. Physical Review Letters 96(23):237,802

Corbett D, Warner M (2008) Polarization dependence of optically driven polydomain elastomer mechanics. Physical Review E Statistical Nonlinear & Soft Matter Physics 78(6 Pt 1):061,701

Cowper GR (1966) The shear coefficients in Timoshenko's beam theory. Journal of Applied Mechanics 33(2):335–340

Dunn ML (2007) Photomechanics of mono- and polydomain liquid crystal elastomer films. Journal of Applied Physics 102(1):307

Dunn ML, Maute K (2009) Photomechanics of blanket and patterned liquid crystal elastomer films. Mechanics of Materials 41(10):1083–1089

Finkelmann H, Nishikawa E, Pereira GG, Warner M (2001) A new opto-mechanical effect in solids. Physical Review Letters 87(1):015,501

He LH (2007) Surface deformation of nematic elastomers under striped illumination. Physical Review E Statistical Nonlinear & Soft Matter Physics 75(1):041,702

Ikeda T, Nakano M, Yu Y, Tsutsumi O, Kanazawa A (2003) Anisotropic bending and unbending behavior of azobenzene liquid-crystalline gels by light exposure. Advanced Materials 15(3):201–205

Jin L, Xin J, Huo Y (2006) Light-induced nonhomogeneity and gradient bending in photochromic liquid crystal elastomers. Science in China 49(5):553–563

Jin L, Yan Y, Huo Y (2010a) A gradient model of light-induced bending in photochromic liquid crystal elastomer and its nonlinear behaviors. International Journal of Non-Linear Mechanics 45(4):370–381

Jin L, Zeng Z, Huo Y (2010b) Thermomechanical modeling of the thermo-order–mechanical coupling behaviors in liquid crystal elastomers. Journal of the Mechanics & Physics of Solids 58(11):1907–1927

Jin L, Lin Y, Huo Y (2011) A large deflection light-induced bending model for liquid crystal elastomers under uniform or non-uniform illumination. International Journal of Solids & Structures 48(22):3232–3242

Lin Y, Jin L, Huo Y (2012) Quasi-soft opto-mechanical behavior of photochromic liquid crystal elastomer: Linearized stress–strain relations and finite element simulations. International Journal of Solids & Structures 49(18):2668–2680

Modes CD, Bhattacharya K, Warner M (2011) Gaussian curvature from flat elastica sheets. Proceedings Mathematical Physical & Engineering Sciences 467(2128):1121–1140

van Oosten CL, Harris KD, Bastiaansen CW, Broer DJ (2007) Glassy photomechanical liquid-crystal network actuators for microscale devices. Eur Phys J E Soft Matter 23(3):329–336

Reddy JN, Wang CM, Lee KH (1997) Relationships between bending solutions of classical and shear deformation beam theories. International Journal of Solids & Structures 34(26):3373–3384

Timoshenko SP (1921) On the correction for shear of the differential equation for transverse vibrations of prismatic bars. Philosophical Magazine 41(245):744–46

Warner M, Corbett D (2010) Suppression of curvature in nematic elastica. Proceedings of the Royal Society A 466(2122):3561–3578

Warner M, Mahadevan L (2004) Photoinduced deformations of beams, plates, and films. Physical Review Letters 92(13):134,302

Warner M, Terentjev EM (2007) Liquid Crystal Elastomers. Oxford University Press,

Warner M, Modes CD, Corbett D (2010) Curvature in nematic elastica responding to light and heat. Proceedings Mathematical Physical & Engineering Sciences 466(2122):2975–2989

Yu Y, Nakano M, Ikeda T (2003) Directed bending of a polymer film by light. Nature 425(6954):145

Zeng Z, Jin L, Huo Y (2010) Strongly anisotropic elastic moduli of nematic elastomers: Analytical expressions and nonlinear temperature dependence. Eur Phys J E Soft Matter 32(1):71–79

Chapter 37
A Simple Qualitative Model for the Pressure-induced Expansion and Wall-stress Response of Fluid-filled Biological Channels

Tarek I. Zohdi

Abstract This work investigates the effects of a pressure increase in deformable fluid-filled biochannels, such as arteries and veins. Simple qualitative expressions are developed relating pressure-induced changes to the biochannel expansion, volumetric flow rate, and biochannel wall stress. Such relations are necessary for a rapid analysis in potential applications such as post-traumatic stress, hemorrhagic strokes, atherosclerotic plaque buildup, etc. The relations are based on the development of functions that correct classical pressurized thin-tube expressions for hoop stress for finite deformations.

Keywords: Pressure increase · Biochannels · Fluid flow

37.1 Introduction

This work studies the pressure-induced expansion and stress increase in deformable fluid-filled biochannels, such as arteries, vein, etc. This is motivated by interest in hypertension, hemorrhagic strokes and recently wide-spread interest in the effects of body-blows to pressure-induced biochannel rupture, arising from contact sports, such as boxing, football, ice-hockey, etc. Simple expressions are developed relating the pressure-induced changes to the biochannel expansion, volumetric flow rate and biochannel wall stress. Intended applications include post-traumatic stress, hemorrhagic strokes, atherosclerotic plaque buildup. The expressions developed allow for rapid analysis of such systems, circumventing the use of computationally-intensive numerical methods for detailed studies. The long-term objective is to couple such models to kinematic systems developed in Zohdi (2017) to simulate a wide range

---

Tarek I. Zohdi
Department of Mechanical Engineering, 6195 Etcheverry Hall, University of California, Berkeley, CA, 94720-1740, USA
e-mail: zohdi@berkeley.edu

B. E. Abali et al. (eds.), *New Achievements in Continuum Mechanics and Thermodynamics*, Advanced Structured Materials 108,
https://doi.org/10.1007/978-3-030-13307-8_37

of induced forces involving fist-to-head and fist-to-chest force calculations in order to determine the connections to possible channel expansion and wall-stress, leading to arterial rupture [1]. However, in certain circumstances, the fluid-induced shear stress may decrease, which increases the tendency of atherosclerotic plaque buildup (Zohdi, 2005, 2004, 2014). These scenarios are discussed further in the paper.

37.2 Classical Pressure-flow Relations

We consider a relatively simple model problem comprised of a biochannel which is filled with a fluid (such as blood, Fig. 37.1). By following Coleman et al (2012, Sect. 13.i), we consider a steady helical flow; by taking an annular element and summing the pressure and shear forces in the axial x-direction, we obtain

$$-\frac{\partial P}{\partial x} + \frac{1}{r}\frac{\partial (r\tau)}{\partial r} = 0 \Rightarrow \frac{1}{r}\frac{\partial (r\tau)}{\partial r} = \frac{\partial P}{\partial x}, \tag{37.1}$$

where P is the pressure and τ is the shear stress (in physical coordinates). Under the assumption that the pressure gradient is constant along the radius, integrating yields

$$\tau = \frac{r}{2}\left(\frac{\partial P}{\partial x}\right) + \frac{C_1}{r} = \mu\frac{\partial v}{\partial r}, \tag{37.2}$$

where v denotes the velocity along the axial direction and μ is the (shear) viscosity of the filling fluid. Integrating again yields

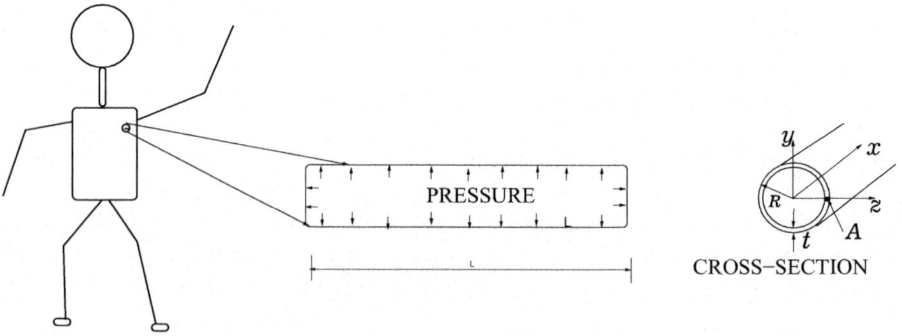

Fig. 37.1 Nomenclature for a simplified flow and stress analysis.

[1] This approach employs a combined kinematic and energy analysis, by drawing on methods used in the robotics literature (for example, see Hunt, 1978; Hartenberg and Denavit, 1964; Howell, 2001; McCarthy and Soh, 2010; McCarthy, 1990; Reuleaux, 1876; Sandor and Erdman, 1984; Slocum, 1992; Suh and Radcliffe, 1978; Uicker Jr et al, 2003).

$$v(r) = \frac{1}{\mu}\left(\frac{r^2}{4}\left(\frac{\partial P}{\partial x}\right) + C_1 \ln r\right) + C_2. \quad (37.3)$$

$v(r = 0)$ must be finite, thus $C_1 = 0$, and $v(r = R) = 0$ yields

$$v(r) = -\frac{R^2}{4\mu}\left(\frac{\partial P}{\partial x}\right)\left(1 - \left(\frac{r}{R}\right)^2\right). \quad (37.4)$$

The stress becomes

$$\tau(r) = \mu\frac{\partial v(r)}{\partial r} = \frac{r}{2}\frac{\partial P}{\partial x}. \quad (37.5)$$

The stress at the wall becomes

$$\tau_w = -\tau(r = R) = -\frac{R}{2}\frac{\partial P}{\partial x}. \quad (37.6)$$

An important observation is that if the radius of the channel grows, and the pressure gradient remains constant or grows, then the shear induced wall stress decreases. However, the flow rate can also be computed to reveal

$$\begin{aligned} Q &= \int_A v\, dA = -\int_A \frac{R^2}{4\mu}\left(\frac{\partial P}{\partial x}\right)\left(1 - \left(\frac{r}{R}\right)^2\right) r dr d\theta = \\ &= -\frac{2\pi R^2}{4\mu}\left(\frac{\partial P}{\partial x}\right)\left(\frac{r^2}{2} - \frac{r^4}{4R^2}\right)\Bigg|_{r=0}^{r=R} = -\frac{1}{\mu}\left(\frac{\partial P}{\partial x}\right)\frac{\pi R^4}{8}, \end{aligned} \quad (37.7)$$

thus indicating that decreasing R decreases the flow rate, if the pressure gradient does not increase appropriately. The implications of this are discussed further in the paper.

37.3 Simple Approximations of Radial Deformation

We now consider the radial deformation of the biochannel as a function of the pressure in the fluid (Fig. 37.1). We make the simplifying assumption that it is a thin-walled circular tube which expands self-similarly (uniformly) to a larger circular tube. At any point along the tube, the radial expansion is simplified by postulating it to be a linear function of the length-averaged mean pressure differential, $\Delta P^m = P^m - P^m_o$ with the nominal pressure P^m_o, of the form:

$$\frac{R}{R_o} = 1 + \mathcal{K}_w(P^m - P^m_o) = 1 + \mathcal{K}_w \Delta P^m, \quad (37.8)$$

where R is the deformed radius, R_o is the nominal (at $\Delta P^m = 0$) radius and $\mathcal{K}_w$ is a constant that represents the compliance of radial expansion. In order to determine the constant, consider a *thin-walled* cylindrical tube of mean radius R_o and thickness t_o is pressurized internally with ΔP^m. We also make the classical assumption that the tube is (eventually) closed at both ends. To calibrate/approximate the wall compliance constant, we can resort to its infinitesimal deformation response and we modify *the classical thin-walled tube approximations*, as explained next.

37.3.1 Estimate of Wall Stresses

We consider a tube with deformed radius R, thickness t and length L and initial radius R_o, thickness t_o and length L_o. For the thin-walled tube approximations, the stress components at point A in the wall (Fig. 37.1, far from the edges) of the tube, as a function of the applied pressure arise from the hoop (circumferential) stresses and the longitudinal stresses, leading to

$$[\boldsymbol{\sigma}] = \begin{bmatrix} \sigma_{xx} & \sigma_{xy} & \sigma_{xz} \\ \sigma_{yx} & \sigma_{yy} & \sigma_{yz} \\ \sigma_{zx} & \sigma_{zy} & \sigma_{zz} \end{bmatrix} = \begin{bmatrix} \Delta P^m R/2t & 0 & 0 \\ 0 & \Delta P^m R/t & 0 \\ 0 & 0 & 0 \end{bmatrix}. \tag{37.9}$$

37.3.2 Determination of the Compliance Constant

In order to calibrate the constant $\mathcal{K}_w$, we first assume a self-similar infinitesimal deformation, ignoring end-effects, with stresses given by

$$[\boldsymbol{\sigma}] = \begin{bmatrix} \sigma_{xx} & \sigma_{xy} & \sigma_{xz} \\ \sigma_{yx} & \sigma_{yy} & \sigma_{yz} \\ \sigma_{zx} & \sigma_{zy} & \sigma_{zz} \end{bmatrix} = \begin{bmatrix} \Delta P^m R_o/2t_o & 0 & 0 \\ 0 & \Delta P^m R_o/t_o & 0 \\ 0 & 0 & 0 \end{bmatrix}, \tag{37.10}$$

and linear elasticity, isotropic and homogeneous with Young's modulus E and Poisson ratio ν. The strains in the tube at point A can be computed to be, using Hooke's law:

$$[\epsilon] = \begin{bmatrix} \epsilon_{xx} & \epsilon_{xy} & \epsilon_{xz} \\ \epsilon_{yx} & \epsilon_{yy} & \epsilon_{yz} \\ \epsilon_{zx} & \epsilon_{zy} & \epsilon_{zz} \end{bmatrix} = \begin{bmatrix} \frac{\Delta P^m}{E}\frac{R_o}{t_o}(\frac{1}{2}-\nu) & 0 & 0 \\ 0 & \frac{\Delta P^m}{E}\frac{R_o}{t_o}(1-\frac{\nu}{2}) & 0 \\ 0 & 0 & -\frac{\Delta P^m}{E}\frac{R_o}{t_o}\frac{3\nu}{2} \end{bmatrix} \tag{37.11}$$

The change in the tube radius

$$\Delta R/R_o = \frac{R-R_o}{R_o} \approx \epsilon_{yy},$$

by relating the perimeters:

$$2\pi R - 2\pi R_o \approx 2\pi R_o \epsilon_{yy} \Rightarrow \frac{R - R_o}{R_o} \approx \epsilon_{yy}. \tag{37.12}$$

Thus, one may immediately write

$$\frac{R - R_o}{R_o} = \frac{\Delta P^m}{E}\frac{R_o}{t_o}(1 - \frac{\nu}{2}) \Rightarrow R = R_o(1 + \frac{\Delta P^m}{E}\frac{R_o}{t_o}(1 - \frac{\nu}{2})) \tag{37.13}$$

Thus, an estimate of the compliance to radial expansion is

$$\mathcal{K}_w = \frac{R_o}{t_o E}(1 - \frac{\nu}{2}) \tag{37.14}$$

We assume that the wall compliance remains constant over the ΔP^m regimes of interest. At point A (the problem is radially symmetric), as a function of ΔP^m, the change in thickness is $\Delta t/t_o \approx \epsilon_{zz}$, which leads to

$$\frac{t - t_o}{t_o} = -\frac{\Delta P^m}{E}\frac{3\nu R_o}{2t_o} \Rightarrow t = t_o(1 - \frac{\Delta P^m}{E}\frac{3\nu R_o}{2t_o}). \tag{37.15}$$

37.3.3 Stress Correction Factors

For the finite deformation case, we approximate the stresses by

$$\sigma_{xx} = \frac{\Delta P^m R}{2t} \approx \frac{\Delta P^m R_o}{2t_o} \underbrace{\left[\frac{\left(1 + \frac{\Delta P^m}{E}\frac{R_o}{t_o}(1 - \frac{\nu}{2})\right)}{(1 - \frac{\Delta P^m}{E}\frac{3\nu R_o}{2l_o})}\right]}_{\text{correction factor} \stackrel{\text{def}}{=} \phi} = \frac{\Delta P^m R_o}{2t_o}\phi \tag{37.16}$$

and

$$\sigma_{yy} = \frac{\Delta P^m R}{t} \approx \frac{\Delta P^m R_o}{t_o}\left[\frac{(1 + \frac{\Delta P^m}{E}\frac{R_o}{t_o}(1 - \frac{\nu}{2}))}{(1 - \frac{\Delta P^m}{E}\frac{3\nu R_o}{2t_o})}\right] = \frac{\Delta P^m R_o}{t_o}\phi. \tag{37.17}$$

37.3.4 Corrected Material Failure Criteria

There are obviously many possible models for material failure. The most appropriate for a tubelike failure (longitudinal rupture) would likely be a hoop-stress failure criteria based on

$$\sigma_{yy} = \frac{\Delta P^m R}{t} \leq \sigma_H^* \tag{37.18}$$

so that

$$\frac{\Delta P^m R_o}{t_o} \leq \frac{\sigma_H^*}{\phi}, \tag{37.19}$$

where the correction factor ϕ by Eq. (37.16) is a function of ΔP^m. In order to isolate ΔP^m, we write inequality (37.19) by setting

$$A(\Delta P^m)^2 + B\Delta P^m + C = 0, \tag{37.20}$$

where

- $A = 1$,
- $B = \frac{t_o}{R_o c_2}\left(\frac{R_o}{t_o} + c_1\sigma_H^*\right)$, where $c_1 = \frac{R_o 3\nu}{2Et_o}$ and $c_2 = \frac{R_o}{Et_o}\left(1 - \frac{\nu}{2}\right)$ and
- $C = -\frac{\sigma_H^* t_o}{c_2 R_o}$.

Consequently, we have

$$\Delta P^m \leq \frac{1}{2A}\left(-B \pm \sqrt{B^2 - 4AC}\right) \tag{37.21}$$

which on taking the positive root leads to

$$\frac{R_o \Delta P^m}{t_o} \leq \frac{E\gamma}{2-\nu} \tag{37.22}$$

where γ is given by

$$\begin{aligned}\gamma &\stackrel{\text{def}}{=} -\left(1 + \frac{3\nu\sigma_H^*}{2E}\right) + \sqrt{\left(\left(1 + \frac{3\nu\sigma_H^*}{2E}\right)^2 + \frac{\sigma_H^* 2(2-\nu)}{E}\right)} = \\ &= -\left(1 + \frac{3\nu\sigma_H^*}{2E}\right) + \sqrt{1 + \frac{\sigma_H^*\nu}{E} + \frac{4\sigma_H^*}{E} + \left(\frac{3\nu\sigma_H^*}{2E}\right)^2}.\end{aligned} \tag{37.23}$$

We may then write

$$\frac{\Delta P^m R_o}{t_o} \leq \frac{E\gamma}{2-\nu} = \frac{E}{2-\nu}\left(-\left(1+\frac{3\nu\sigma_H^*}{2E}\right) + \right.$$
$$\left. + \sqrt{1+\frac{\sigma_H^*\nu}{E}+\frac{4\sigma_H^*}{E}+\left(\frac{3\nu\sigma_H^*}{2E}\right)^2}\right) \stackrel{\text{def}}{=} \sigma_H^{*,\text{corr}}, \tag{37.24}$$

which is a "corrected" failure criteria. We have a number of observations:

- **Observation #1:** In special cases, such as $\nu = 0$ (no transverse contraction),

$$\gamma = -1 + \sqrt{\left(1+\frac{4\sigma_H^*}{E}\right)}, \tag{37.25}$$

thus

$$\frac{\Delta P^m R_o}{t_o} \leq \frac{E}{2}\gamma = \frac{E}{2}\left(\sqrt{1+\frac{4\sigma_H^*}{E}} - 1\right). \tag{37.26}$$

One can linearize γ around $\sigma_H^* = 0$, yielding

$$\gamma = -1 + \sqrt{\left(1+\frac{4\sigma_H^*}{E}\right)} \approx \frac{2}{E}\sigma_H^*, \tag{37.27}$$

thus recovering

$$\frac{\Delta P^m R_o}{t_o} \leq \sigma_H^*, \tag{37.28}$$

for small values of σ_H^*.

- **Observation #2:** The change in the domain length given by $\Delta L/L_o \approx \epsilon_{xx}$ tends to zero as the material becomes volume preserving, $\nu \to 1/2$, thus $L = L_0$. In this isochoric or incompressible case[2] of $\nu = \frac{1}{2}$ (incompressible)

$$\frac{\Delta P^m R_o}{t_o} \leq \frac{2E}{3}\left(-\left(1+\frac{3}{4}\frac{\sigma_H^*}{E}\right) + \sqrt{1+\frac{9\sigma_H^*}{2E}+\left(\frac{3\sigma_H^*}{4E}\right)^2}\right). \tag{37.29}$$

- **Observation #3:** Although for soft tissue, a criterion based on von Mises equivalent stress would not be most appropriate, an estimate for the maximum allowable pressure, based on the von Mises (distortion energy) criterion is

$$\begin{aligned} 3||\boldsymbol{\sigma}'||^2 &= (\sigma_{xx}-\sigma_{yy})^2 + (\sigma_{xx}-\sigma_{zz})^2 + (\sigma_{yy}-\sigma_{zz})^2 + 6(\sigma_{xy}^2+\sigma_{xz}^2+\sigma_{yz}^2) \\ &= (\Delta P^m R/2t)^2 + (\Delta P^m R/2t)^2 + (\Delta P^m R/t)^2 + 6\tau_w^2 \\ &\leq 2\sigma_o^2, \end{aligned} \tag{37.30}$$

[2] Of course, an incompressible soft matter would be modeled by a hyperelastic material model stemming from energy description, herein we explain the physical significance by observing a volume preserving deformation.

where, σ_o is a material constant (failure stress) determined from a standard uniaxial failure tension test. There are, of course, numerous other criteria for failure.

37.4 Subsequent Flow Changes

Due the change in the radius, the fluid flow changes according to

$$\sigma_{xz} = \tau_w = -\frac{R}{2}\frac{\partial P}{\partial x} = -\frac{R_o}{2}\underbrace{\left(1 + \frac{\Delta P^m}{E}\frac{R_o}{t_o}\left(1 - \frac{\nu}{2}\right)\right)}_{\lambda}\frac{\partial P}{\partial x}, \tag{37.31}$$

where λ can be interpreted as a fluid-flow correction factor.

37.5 Closing Remarks

This work developed simple expressions between pressure change and mechanical response of the soft tissue filled with a fluid. The main results of the paper were, under some simplifying assumptions (self-similar expansion) at finite deformations:

- An expression relating the change in pressure
  - to the expansion of the biochannel radius,
  - to the reduction of the biochannel wall thickness,
  - to the wall stress of the biochannel,
- A flow correction relation for a biochannel with changing radius.

These relations are based on the development of functions that correct classical pressurized thin-tube expressions (ϕ) for hoop stress for finite deformations. Possible applications are to stroke and post-traumatic stress and, in particular, hemorrhagic strokes and alimentary rupture. The expressions developed allow for rapid analysis of such systems, reserving the direct use of computationally-intensive numerical methods for detailed studies as for example in Abali (2017). In closing, we make a few more observations with respect to flow changes and fluid-induced shear stresses, which were alluded to earlier in the paper. We note that $v(r)$ is a maximum where

$$\frac{\partial v}{\partial r} = 0 = \frac{r}{2\mu}\frac{\partial P}{\partial x}, \tag{37.32}$$

which is at $r = 0$. Thus,

$$v_{\max} = v(r=0) = -\frac{R^2}{4\mu}\left(\frac{\partial P}{\partial x}\right) \Rightarrow v(r) = v_{\max}\left(1 - \left(\frac{r}{R}\right)^2\right) \tag{37.33}$$

Relating this to the flow rate yields:

$$Q = \int_A v\,dA = \frac{\pi v_{\max} R^2}{2} \Rightarrow v_{\max} = \frac{2Q}{\pi R^2}, \tag{37.34}$$

and we obtain

$$v(r) = \frac{2Q}{\pi R^2}\left(1 - \left(\frac{r}{R}\right)^2\right) \tag{37.35}$$

The stress becomes

$$\tau(r) = \mu \frac{\partial v(r)}{\partial r} = -\frac{4\mu Q r}{\pi R^4}. \tag{37.36}$$

The stress at the wall becomes

$$\tau_w = -\tau(r = R) = \frac{2\mu v_{\max}}{R} = \frac{4\mu Q}{\pi R^3}. \tag{37.37}$$

Explicitly, the shear stress becomes:

$$\begin{aligned}\sigma_{xz} = \tau_w &= -\frac{R}{2}\frac{\partial P}{\partial x} = -\frac{R_o}{2}(1 + \frac{\Delta P^m}{E}\frac{R_o}{t_o}(1 - \frac{\nu}{2}))\frac{\partial P}{\partial x} = \frac{4\mu Q}{\pi R^3} \\ &= \frac{4\mu Q}{\pi\left(R_o(1 + \frac{\Delta P^m}{E}\frac{R_o}{t_o}(1 - \frac{\nu}{2}))\right)^3}.\end{aligned} \tag{37.38}$$

Thus, unless Q increases appropriately, the fluid-induced shear stress at the wall will decrease. For example, consider an increase in volumetric flow rate due to the change in lumen (cavity of the artery) diameter of the following form

$$Q(\Delta P) = \pi R^2 v^m, \tag{37.39}$$

where $R = R(\Delta P)$ and v^m (the mean velocity) is constant, which implies from Equation 37.7 that

$$v^m = -\frac{1}{\mu}\frac{\partial P}{\partial x}\frac{R^2}{8}, \tag{37.40}$$

which leads to

$$\tau_w = \frac{4\mu Q}{\pi R^3} = \frac{4\mu \pi R^2 v^m}{\pi R^3} = \frac{4\mu v^m}{R}. \tag{37.41}$$

Thus, the wall shear stress will decrease. Low wall shear stress is associated with the growth of plaque buildup (Zohdi, 2005, 2004, 2014; Zohdi et al, 2004), due to the accumulation of material in diseased arteries. This is often the initial stage of arterial occlusive growth processes (Ambrosi et al, 2011; Göktepe et al, 2010; Menzel and Kuhl, 2012; Kuhl et al, 2007; Zöllner et al, 2012). For surveys of plaque-related work, see Chyu and Shah (2001); Davies et al (1993); Corti et al (2002); Kaazempur-Mofrad et al (2005, 2004, 2003); Libby (2001); Libby et al (2001, 2002); Libby and Aikawa (2002); Loree et al (1992); Richardson et al (1989); Shah (1997); van der Wal and Becker (1999). Thus, in addition to coronary diseases, the accumulation of material subsequently reduces the cross-sectional area of the biochannel, which can lead to dementia-like symptoms, potentially due to the build up of calcium and fatty deposits on biochannel walls (Wenk et al, 2010; Klepach et al, 2012; Lee et al, 2013; Weinberg et al, 2009). This is under current investigation by the author.

References

Abali BE (2017) Computational Reality: Solving nonlinear and coupled problems in continuum mechanics. Springer Nature

Ambrosi D, Ateshian G, Arruda E, Cowin S, Dumais J, Goriely A, Holzapfel GA, Humphrey JD, Kemkemer R, Kuhl E, et al (2011) Perspectives on biological growth and remodeling. Journal of the Mechanics and Physics of Solids 59(4):863–883

Chyu KY, Shah PK (2001) The role of inflammation in plaque disruption and thrombosis. Reviews in cardiovascular medicine 2(2):82–91

Coleman BD, Markovitz H, Noll W (2012) Viscometric flows of non-Newtonian fluids: theory and experiment, vol 5. Springer Science & Business Media

Corti R, Badimon L, Fuster V, Badimon J (2002) Assessing and modifying the vulnerable atherosclerotic plaque, chapter endothelium, flow, and artherothrombosis. American Heart Association

Davies MJ, Richardson PD, Woolf N, Katz DR, Mann J (1993) Risk of thrombosis in human atherosclerotic plaques: role of extracellular lipid, macrophage, and smooth muscle cell content. Heart 69(5):377–381

Göktepe S, Abilez OJ, Parker KK, Kuhl E (2010) A multiscale model for eccentric and concentric cardiac growth through sarcomerogenesis. Journal of theoretical biology 265(3):433–442

Hartenberg R, Denavit J (1964) Kinematic Synthesis of Linkages, McGraw-Hill Book Company. McGraw-Hill, New York

Howell LL (2001) Compliant mechanisms. John Wiley & Sons

Hunt KH (1978) Kinematic geometry of mechanisms, vol 7. Oxford University Press, USA

Kaazempur-Mofrad M, Younis H, Patel S, Isasi A, Chung C, Chan R, Hinton D, Lee R, Kamm R (2003) Cyclic strain in human carotid bifurcation and its potential correlation to atherogenesis: Idealized and anatomically-realistic models. Journal of Engineering Mathematics 47(3-4):299–314

Kaazempur-Mofrad M, Isasi A, Younis H, Chan R, Hinton D, Sukhova G, LaMuraglia G, Lee R, Kamm R (2004) Characterization of the atherosclerotic carotid bifurcation using mri, finite element modeling, and histology. Annals of biomedical engineering 32(7):932–946

Kaazempur-Mofrad M, Wada S, Myers J, Ethier C (2005) Blood flow and mass transfer in arteries with axisymmetric and asymmetric stenoses. Int J Heat Mass Transfer 48:4510–4517

Klepach D, Lee LC, Wenk JF, Ratcliffe MB, Zohdi TI, Navia JL, Kassab GS, Kuhl E, Guccione JM (2012) Growth and remodeling of the left ventricle: a case study of myocardial infarction and surgical ventricular restoration. Mechanics research communications 42:134–141

Kuhl E, Maas R, Himpel G, Menzel A (2007) Computational modeling of arterial wall growth. Biomechanics and modeling in mechanobiology 6(5):321–331

Lee LC, Wenk JF, Zhong L, Klepach D, Zhang Z, Ge L, Ratcliffe MB, Zohdi TI, Hsu E, Navia JL, et al (2013) Analysis of patient-specific surgical ventricular restoration: importance of an ellipsoidal left ventricular geometry for diastolic and systolic function. Journal of applied physiology 115(1):136–144

Libby P (2001) Current concepts of the pathogenesis of the acute coronary syndromes. Circulation 104(3):365–372

Libby P, Aikawa M (2002) Stabilization of atherosclerotic plaques: new mechanisms and clinical targets. Nature medicine 8(11):1257–1262

Libby P, Ridker PM, Maseri A (2002) Inflammation and atherosclerosis. Circulation 105(9):1135–1143

Libby P, et al (2001) The vascular biology of atherosclerosis (ed. E. Braunwald and D. P. Zipes and P. Libby), vol Chap. 30. Saunders, Philadelphia, Pennsylvania

Loree HM, Kamm RD, Stringfellow RG, Lee RT (1992) Effects of fibrous cap thickness on peak circumferential stress in model atherosclerotic vessels. Circulation research 71(4):850–858

McCarthy JM (1990) Introduction to theoretical kinematics. MIT press

McCarthy JM, Soh GS (2010) Geometric design of linkages, vol 11. Springer Science & Business Media

Menzel A, Kuhl E (2012) Frontiers in growth and remodeling. Mechanics research communications 42:1–14

Reuleaux F (1876) Book Review: The Kinematics of Machinery (trans. and annotated by A. B. W. Kennedy). reprinted by Dover, New York (1963)

Richardson PD, Davies M, Born G (1989) Influence of plaque configuration and stress distribution on fissuring of coronary atherosclerotic plaques. The Lancet 334(8669):941–944

Sandor GN, Erdman AG (1984) Mechanism design: analysis and synthesis, vol 1. Prentice-Hall New Delhi

Shah PK (1997) Plaque disruption and coronary thrombosis: new insight into pathogenesis and prevention. Clinical cardiology 20(11 Suppl 2):II–38

Slocum AH (1992) Precision machine design. Society of Manufacturing Engineers

Suh C, Radcliffe C (1978) Kinematics and mechanisms design. john willey & sons. New York

Uicker Jr J, Pennock G, Shigley J (2003) Theory of mechanisms and machines

van der Wal AC, Becker AE (1999) Atherosclerotic plaque rupture–pathologic basis of plaque stability and instability. Cardiovascular research 41(2):334–344

Weinberg EJ, Schoen FJ, Mofrad MR (2009) A computational model of aging and calcification in the aortic heart valve. PLoS One 4(6):e5960

Wenk JF, Papadopoulos P, Zohdi TI (2010) Numerical modeling of stress in stenotic arteries with microcalcifications: a micromechanical approximation. Journal of biomechanical engineering 132(9):091,011

Zohdi T (2004) A computational framework for agglomeration in thermochemically reacting granular flows. Proceedings of the Royal Society of London A: Mathematical, Physical and Engineering Sciences 460(2052):3421–3445

Zohdi T (2005) A simple model for shear stress mediated lumen reduction in blood vessels. Biomechanics and modeling in mechanobiology 4(1):57–61

Zohdi T (2014) Mechanically driven accumulation of microscale material at coupled solid–fluid interfaces in biological channels. Journal of The Royal Society Interface 11(91):20130,922

Zohdi T (2017) On the biomechanical analysis of the calories expended in a straight boxing jab. Journal of The Royal Society Interface 14(129):20170,153

Zohdi T, Holzapfel G, Berger S (2004) A phenomenological model for atherosclerotic plaque growth and rupture. Journal of theoretical biology 227(3):437–443

Zöllner AM, Tepole AB, Kuhl E (2012) On the biomechanics and mechanobiology of growing skin. Journal of theoretical biology 297:166–175

Printed by Printforce, the Netherlands